GEOLOGICAL SOCIETY SPECIAL PUBLICATION NO 21

North Atlantic Palaeoceanography

EDITED BY

C.P. SUMMERHAYES
Stratigraphy Branch, BP Research Centre,
Sunbury-on-Thames

AND

N.J. SHACKLETON
Department of Quaternary Research,
Godwin Laboratory,
Cambridge University

1986

Published for The Geological Society by
Blackwell Scientific Publications
OXFORD LONDON EDINBURGH
BOSTON PALO ALTO MELBOURNE

Published by

Blackwell Scientific Publications
Osney Mead, Oxford, OX2 0EL
8 John Street, London, WC1N 2ES
23 Ainslie Place, Edinburgh, EH3 6AJ
52 Beacon Street, Boston,
Massachusetts 02108, USA
667 Lytton Avenue, Palo Alto,
California 94301, USA
107 Barry Street, Carlton,
Victoria 3053, Australia

First published 1986

Printed in Great Britain at
The Alden Press, Oxford

DISTRIBUTORS

USA and Canada
Blackwell Scientific Publications Inc
PO Box 50009, Palo Alto
California 94303

Australia
Blackwell Scientific Publications
(Australia) Pty Ltd
107 Barry Street,
Carlton, Victoria 3053

British Library Cataloguing in Publication Data

North Atlantic palaeoceanography.
1. Geology—North Atlantic Ocean
I. Summerhayes, C.P.
II. Shackleton, N.J.
III. Geological Society of London
551.46′08′0931 QE350.22.N65

ISBN 0-632-01516-0

Contents

Circulation, unconformities and sedimentation

Neogene deep and surface water palaeoceanography

Mesozoic palaeoceanography and black shales

Preface

Through the activities of the Deep Sea Drilling Project (DSDP) since 1969, palaeoceanography—the study of the effects of oceanic and atmospheric circulation on oceanic sedimentation—has recently become an active quasi-independent area of the earth sciences. DSDP having drawn to a close at the end of 1983 with a series of sampling cruises around the North Atlantic, we thought the time ripe for a review of North Atlantic Palaeoceanography. This review, on behalf of the Geological Society's Marine Studies Group, duly took place at a two-day meeting at the Society's premises in November 1984. Most of the significant papers presented at that meeting are contained in this book.

Organising the meeting was very satisfying, because we encountered such a strong interest in taking part. Marine geologists, biostratigraphers, sedimentologists, and geochemists all offered manuscripts, and it is their contributions that dictate the division of the volume into its three discrete sections; one on circulation, unconformities, and sedimentation; one on Neogene deep and surface water palaeoceanography; and one on Mesozoic palaeoceanography and black shales.

We were delighted at the range of nations represented by the contributors at the meeting (10 countries in all). But perhaps even more important was the range of institutions represented: university geology departments, oceanographic institutions, and industrial groups were all actively interested. As the UK does not include any academic oceanographic institution of the type that is found in the USA, we were very pleased to see the wide degree of interest, largely based on past participation in deep sea drilling cruises from geology or earth science departments in British Universities. This broad participation highlighted the fact that although palaeoceanography is currently a focus for the attention of scientists from many different geological disciplines, it is ultimately a part of geology, and contributes a great deal to the solution of geological problems relating to the Mesozoic rocks that are commonly examined by the occupants of land-locked geology departments. We believe that this volume will be of interest not only to those for whom a drilling ship is a routine sampling tool, but also to those who have no intention of sampling with anything except a hammer.

Despite the absence of contributors from the petroleum industry (excluding those from the Institut Française du Petrole), industry's strong interest in the results of deep ocean drilling was obvious from the attendance of many petroleum geologists at the meeting. Several of the contributions in this volume, especially those in Mesozoic black shales, the potential source rocks for petroleum, will be of direct interest to petroleum geologists assessing deepwater hydrocarbon potential around the UK and elsewhere in the North Atlantic.

There is no doubt that deep ocean drilling has been a cheap and beneficial source of useful information for petroleum exploration offshore UK and elsewhere. The DSDP has made a major and lasting contribution to petroleum geology by providing, for instance, major revisions in biotratigraphy, improvements in the geological time scale, documenting the nature and extent of potential petroleum source rocks in deep water, evaluating the hydrocarbon potential of parts of the continental margin unsampled by industry, providing data for calibrating industry's long seismic lines, and improving our understanding of the evolution of continental margin basins. In effect, DSDP holes are like free COST wells for industry. Industry has shown its interest in the DSDP by regularly providing shipboard geochemists, biostratigraphers, and sedimentologists, and by doing shorebased laboratory work for publication in DSDP reports. This public interest is the thin end of the wedge: most of industry's extensive use of DSDP data is confidential.

This brings us back to the central theme of the meeting. All but a few of the findings reported did depend on the drilling ship *Glomar Challenger* having been available as a sampling tool. The International Phase of Ocean Drilling of the DSDP, during which these results were obtained, was a marvellous example of the value that derives from a well-planned international project. It opened new avenues world-wide in the pursuit of knowledge. The follow-up project that has just begun, using a new and significantly better drilling ship, promises another decade of exciting geology, full of implications for the academic and industrial communities. As proofs were being corrected it was announced that Britain has decided to participate in the new project, the Ocean Drilling Programme. This will give British scientists the chance to take part in some of the most exciting research that is taking place in the earth sciences today.

C.P. Summerhayes, Sunbury-on-Thames
N.J. Shackleton, Cambridge
July 1985

Acknowledgements

First, our thanks go to the Secretary, staff, and housekeepers of the Geological Society, whose help in organising and running the Conference in November 1984 made the meeting on which this book is based a great success. No less important were our financial backers, for, without their support in providing aid for air fares for overseas contributors, that meeting and this book would not have had the truly international flavour that both required. For their generosity we thank (in alphabetical order) ARCO, BP, BRITOIL, CHEVRON, ELF, LASMO, and SHELL, as well as the Geological Society and the Royal Society. Our speakers and poster presenters excelled themselves at the meeting, and the contents of this volume are a fair reflection of the high standards of November's presentations. Not everyone was able to contribute a paper to this proceedings volume; some who were willing were unable to keep to the tight schedules that we imposed deliberately to ensure publication as soon as possible after the meeting. While we regret the lack of complete representation of the presentations from the meeting, the papers contained herein are a good cross section of the topics and themes discussed. Our reviewers deserve considerable praise—not only Professor Brian Funnell who, for consistency, bravely reviewed every paper, but also a large anonymous team of capable volunteers and draftee experts who reviewed one or two papers each. Finally, we must thank our authors, for sticking at it unflinchingly, while the editors cracked the whip from the sidelines, and for providing us with such a tasty mixture of ingredients.

Circulation, unconformities and sedimentation

Late Mesozoic and Cenozoic sediment flux to the central North Atlantic Ocean

Jörn Thiede & Werner U. Ehrmann

SUMMARY: A history of Mesozoic and Cenozoic palaeoenvironments of the North Atlantic Ocean has been developed based on a detailed analysis of the temporal and spatial distribution of major pelagic sediment facies, of hiatuses, of bulk sediment accumulation rates, and of concentrations and fluxes of the main deep-sea sediment components. The depositional history of the North Atlantic can be subdivided into three major phases: (a) Late Jurassic and Early Cretaceous phase: clastic terrigenous and biogenic pelagic sediment components accumulated rapidly under highly productive surface water masses over the entire ocean basin; (b) Late Cretaceous to Early Miocene phase: relatively little terrigenous and pelagic biogenic sediment reached the North Atlantic Ocean floor, intensive hiatus formation occurred at variable rates, and wide stretches of the deep-ocean floor were covered by slowly accumulating terrigenous muds; (c) Middle Miocene to Recent phase: accumulation rates of biogenic and terrigenous deep-sea sediment components increased dramatically up to Quaternary times, rates of hiatus formation and the intensity of deep-water circulation inferred from them seem to have decreased. However, accumulation rate patterns of calcareous pelagic sediment components suggest that large scale reworking and displacement of deep-sea sediments occurred at a variable rate over wide areas of the North Atlantic during this period.

Over 150 drill holes have partly or completely penetrated the deep-sea sediment layer in the central and North Atlantic; they enable us to describe temporal and spatial patterns of sediment fluxes to this ocean basin during the past 150 million years. As one of a series of syntheses of the history of the main sub-basins of the World Ocean we have now analysed information available from the North Atlantic (see van Andel *et al.* 1977, for a comparable study of the South Atlantic). The bulk of the information obtained from the North Atlantic study has been published by Ehrmann & Thiede (1985); and only a few essential aspects are highlighted in this paper.

In this study the authors point out temporal relationships between hiatus formation and sediment fluxes to the North Atlantic Ocean during the past 150 Ma. Data hitherto available seemed to suggest a strong correlation between the fluxes of individual North Atlantic deep-sea sediment components, and a negative correlation between the intensity of hiatus formation and sediment accumulation rates. Here these relationships are traced in more detail than previously, because they seem to document some important properties of the North Atlantic deep-sea depositional environment.

Ways of quantifying sediment fluxes

The authors have used methods developed by van Andel *et al.* (1975) to estimate temporal distributions of hiatuses and calculate bulk sediment and component accumulation rates for all central and North Atlantic deep-sea drill sites (Figs. 1–3). The presence of hiatuses has been assumed where bulk sediment accumulation rates fell below 100 g $cm^{-2}Ma^{-1}$, or where stratigraphies suggested the absence of sediments representing time spans longer than 2 Ma. Identified hiatuses have not been included in the calculation of sedimentation rates. The resultant data have been used to reconstruct sediment fluxes to the North Atlantic deep-sea floor, displaying them as time series or synoptic time slices both for individual sub-basins and for the entire ocean. The palaeogeographic and palaeobathymetric movements of the individual data points were backtracked by using the methods of Sclater *et al.* (1977) and Berger (1972). It is clear from Fig. 1 and Fig. 2 that both the temporal and spatial coverage of the North Atlantic's depositional history is inhomogeneous, and that the early part of this history is poorly documented. However, despite these deficiencies these data provide a basis for mapping sediment fluxes on palinspastic maps such as Fig. 4.

For each of the individual drill sites it has been possible to calculate the vertical movements (almost always subsidence) of the sediment surface with time. This has enabled the authors to compile information on the nature of the sediments on age versus palaeodepth diagrams for the central and North Atlantic Ocean and its major sub-basins (Fig. 5). These diagrams comprise data for all drill sites available from the region

From SUMMERHAYES, C.P. & SHACKLETON, N.J. (eds), 1986, *North Atlantic Palaeoceanography*, Geological Society Special Publication No. 21, pp. 3–15.

FIG. 1. Distribution of Deep-Sea Drilling Project drill sites analysed for this sudy.

under study, and provide the best regional coverage presently attainable. The bulk sediment accumulation rate data thus compiled have also been used to establish time series of average values of sediment fluxes for 1 Ma time increments, and of hiatus frequencies in the North Atlantic Ocean.

The information required for this study has been extracted from the Initial Reports of the Deep-Sea Drilling Project (DSDP), and from the shipboard site reports. The quality of the data in the DSDP Initial Reports is highly variable. Also, drill site density varies with distance from the adjacent continents, and is particularly low in the

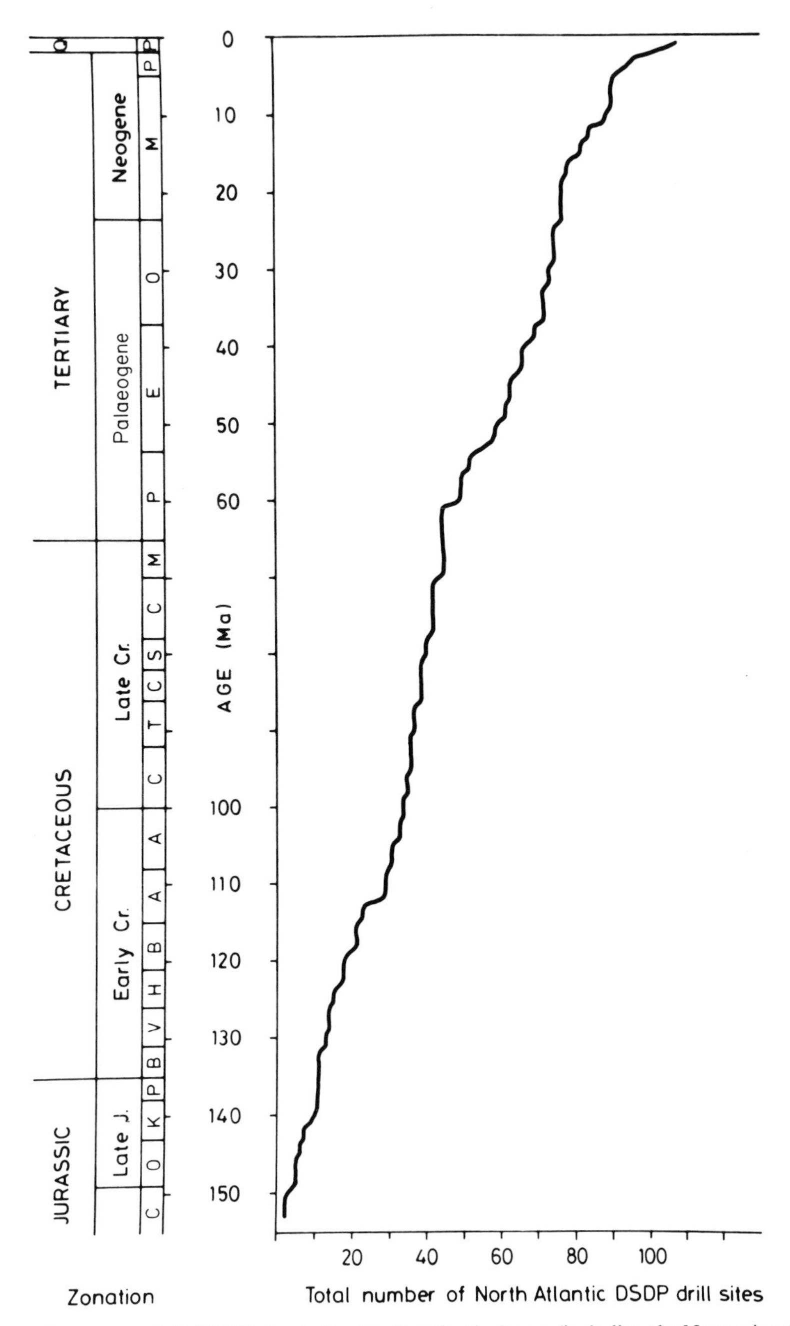

FIG. 2. Temporal coverage of all DSDP sites in the North Atlantic Ocean (including the Norwegian-Greenland Sea). The absolute time-scale (in Ma) is based on available biostratigraphic zonations (van Hinte 1976, 1978; van Couvering & Berggren 1977; Berggren *et al.* 1978); absolute ages of stratigraphic boundaries as used in this study are indicated in this figure.

central parts of the east and west North Atlantic. Most drill sites were chosen to address a specific problem or feature; only a few were selected to obtain a record of normal oceanic basement and its sediment cover. Therefore, many of the records may be documenting atypical depositional environments.

The task of standardizing and in part revising

and correcting the North Atlantic data base, which had been accumulated during 15 years of deep-sea drilling, was difficult and time consuming. It was necessary: (1) to check the stratigraphic data and their validity; (2) to generate a lithologic data base which allowed comparison of sediment data from sites drilled during the early days of deep-sea drilling with those obtained from the latest North Atlantic deep-sea drill sites; (3) to collect the physical property data needed for the calculation of bulk sediment accumulation rates and of individual sediment component accumulation rates; and (4) to assess the importance and length of hiatuses.

Temporal and spatial variability of sediment fluxes

Figures 3–6 show that the sediment flux to the North Atlantic deep-sea floor has been highly variable in space and time. As an example of the variability at one location data from Site 369 are presented (Fig. 3). Sediment flux at this and other sites has been sporadic, and the continuity of sedimentation has been interrupted by numerous, sometimes long hiatuses. Intervals inbetween hiatuses have maximal bulk accumulation rates. A sporadic, discontinuous influx of sediments seems to have been the rule, rather than the exception in the North Atlantic during the Late Mesozoic and Cenozoic. Similar observations have been made about other oceans (Moore *et al.* 1978), but the hiatus records of the different ocean basins have yet to be compared.

Figure 4 (a and b) presents data from two Palaeogene time slices plotted onto the corresponding palinspastic maps to show the spatial variability of bulk accumulation rates. These examples illustrate that sediment fluxes have been low for most parts of the deep central and North Atlantic, values usually being 500 g $cm^{-2}Ma^{-1}$. Only a few areas, mostly close to the continental margins (proximal), received sediment at rates which exceeded 3000–5000 g $cm^{-2}Ma^{-1}$, and which are up to an order of magnitude higher than in the central ocean basin (distal).

The use of sediment traps to measure bulk sediment accumulation rates in modern oceans (Honjo 1978, 1982) has revealed rates many

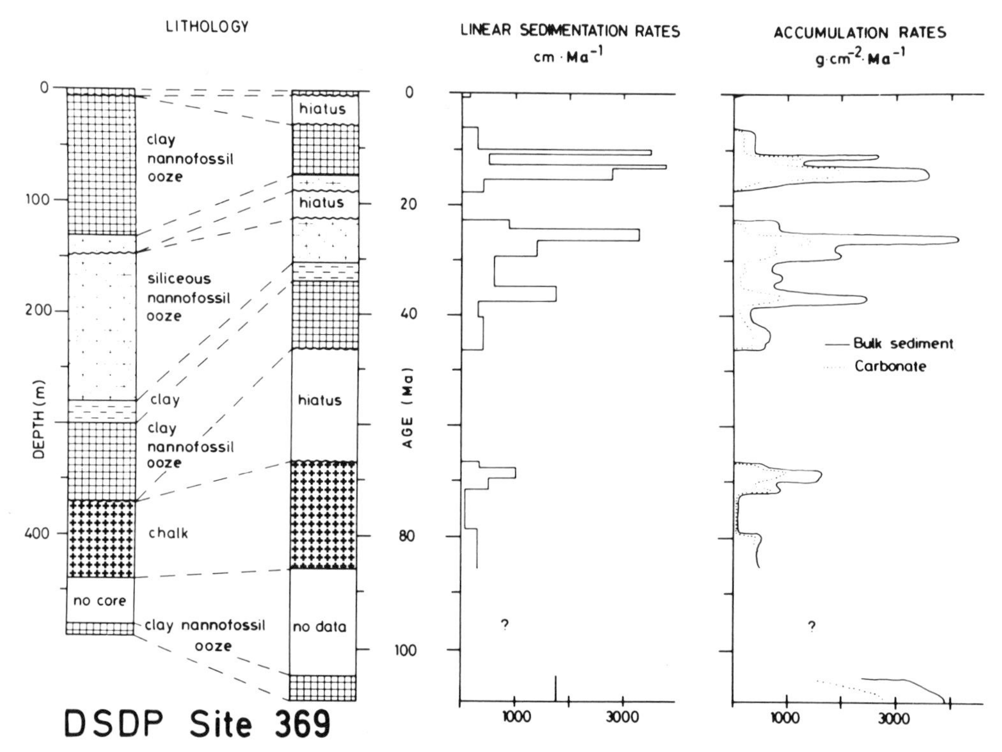

FIG. 3. Graphical example of the authors' data processing: the left column provides the lithostratigraphic information (simplified), as found in the Initial Reports of the Deep-Sea Drilling Project; in the second column lithology is plotted versus age. By correcting linear sedimentation rates for compaction the accumulation rates for every million year time interval have been calculated.

orders of magnitude higher than those determined for North Atlantic deep-sea sediments. Thus, it seems likely that other processes than the original vertical sediment flux are documented in the bulk sediment accumulation rates which have been reconstructed for North Atlantic DSDP sites.

The regional differentiation described above (Fig. 4) cannot be observed in the Jurassic to mid-Cretaceous time slices while the North Atlantic was part of the Tethyan ocean regime (Bernoulli 1984). At this time the (poorly documented) depositional environment seems to have been quite uniform throughout the entire basin. Since mid-Cretaceous times high sediment fluxes have been restricted to isolated centres like those shown in Fig. 4, although the importance and the position of these centres have often changed (Ehrmann & Thiede, 1985).

To assess the average sediment flux to the North Atlantic during the past 150 Ma the accumulation rate data have been plotted onto

FIG. 4. Accumulation rates of bulk sediment in the North Atlantic in Palaeogene times (g $cm^{-2}Ma^{-1}$, averaged). The data have been plotted onto palinspastic maps taken from Thiede (1979). DSDP Sites: ● = data; ○ = hiatus; × = no data. (a) 50–46 Ma; (b) 30–26 Ma.

Fig. 4 (b).

palaeodepth versus age diagrams (Fig. 5). To make these diagrams the North Atlantic was divided into eastern and western parts along the mid-Atlantic ridge, excluding Norwegian-Greenland Sea sites. This approach oversimplifies the regional variability by implying that the sediment flux of both basins was uniformly distributed throughout the basin. Thus Fig. 5 (a and b) reveals only major temporal and spatial distribution patterns, and obscures much of the small scale variability.

Both sub-basins of the North Atlantic are

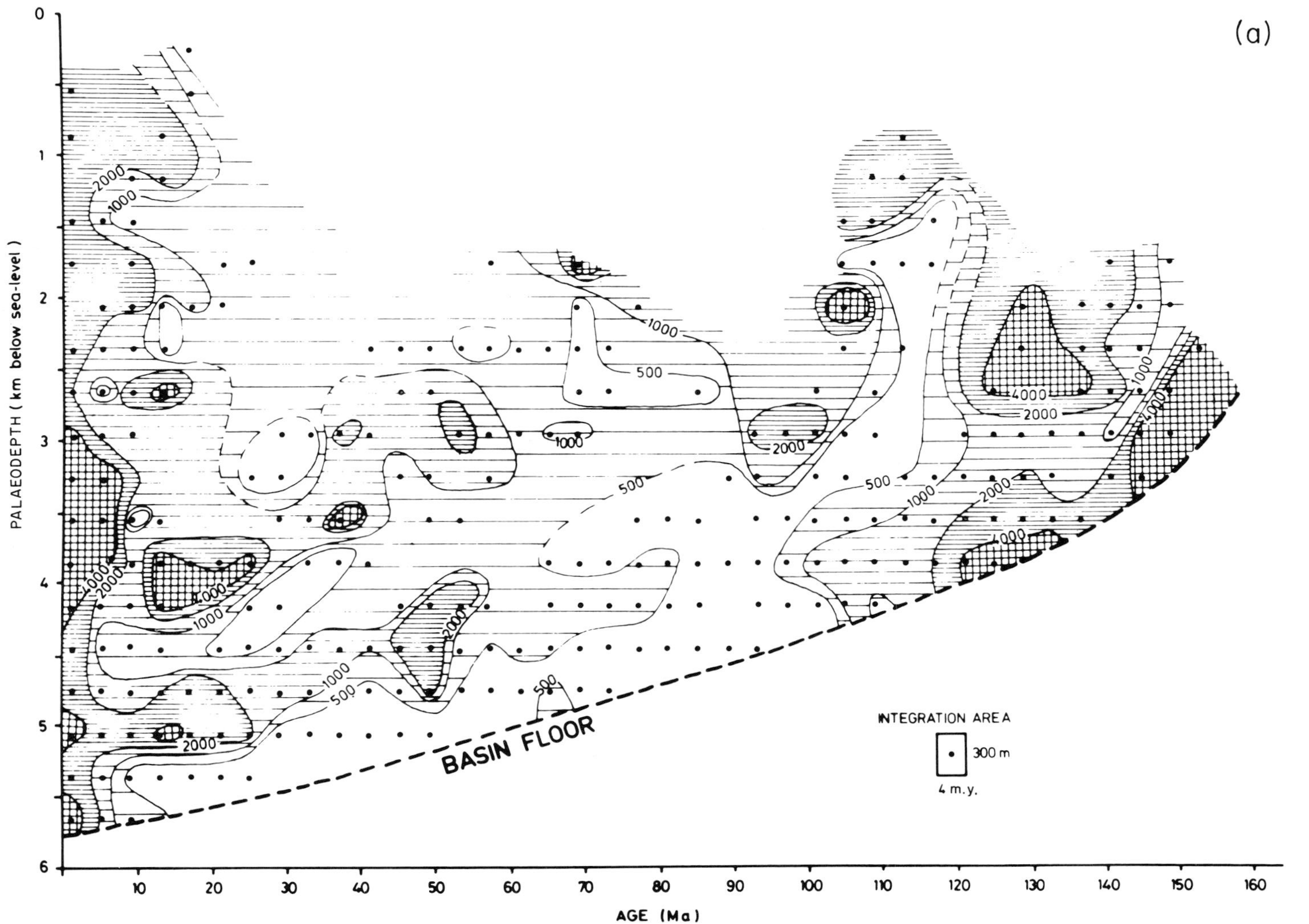

FIG. 5. Accumulation rates (g cm^{-2}Ma^{-1}) of bulk sediment plotted versus age and palaeodepth for (a) the western, and (b) the eastern basins of the North Atlantic Ocean.

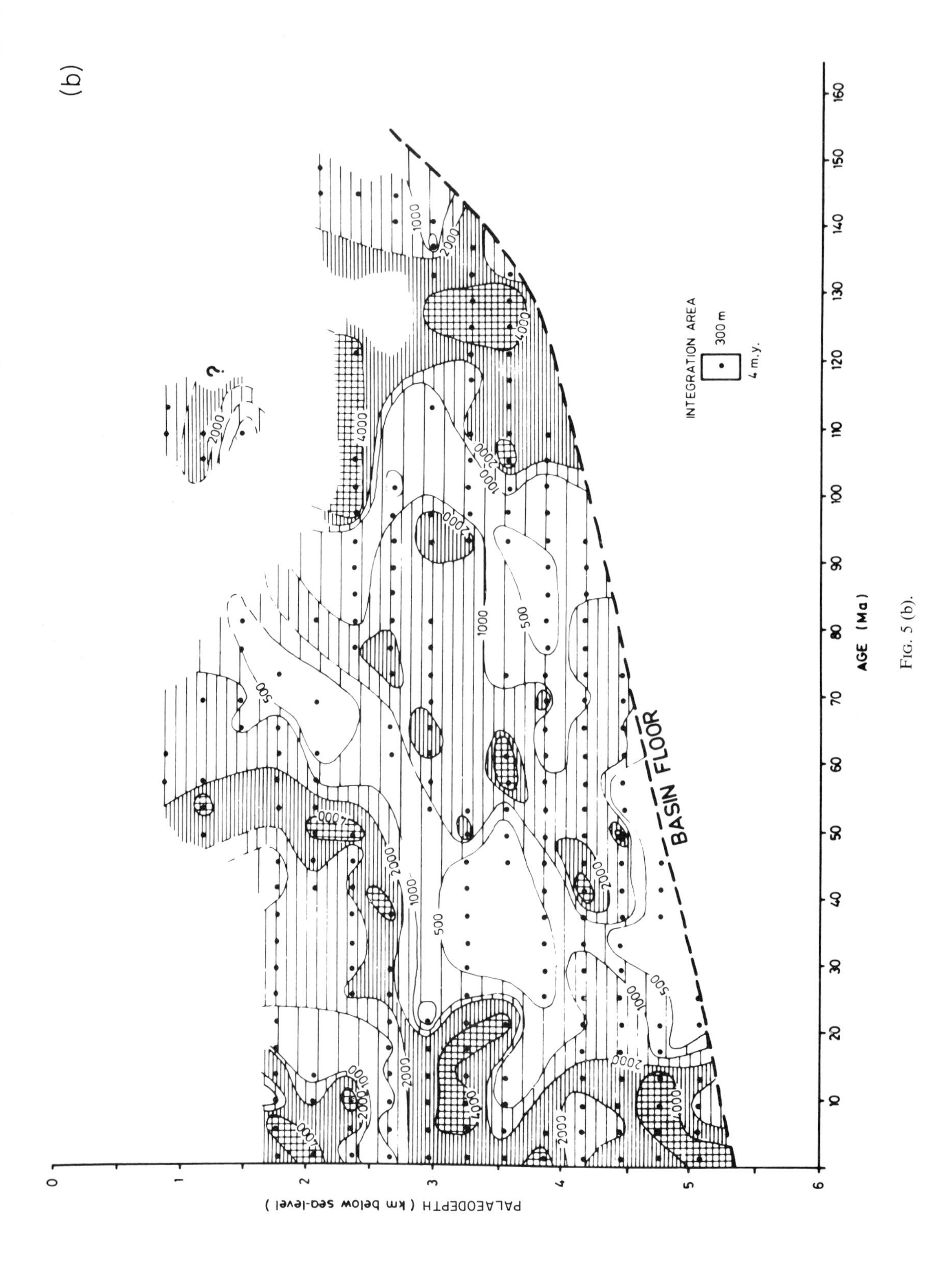
(b)
PALAEODEPTH (km below sea-level)
0
1
2
3
4
5
6
AGE (Ma)
10
20
30
40
50
60
70
80
90
100
110
120
130
140
150
160
BASIN FLOOR
INTEGRATION AREA
300 m
4 m.y.
500
1000
2000
4000
?

FIG. 5 (b).

characterized by high bulk sediment accumulation rates during their early history, ending approximately 100–110 Ma ago and apparently extending throughout the entire water column (Fig. 5). This early phase is succeeded by a long interval of variable, but generally low, bulk sediment accumulation rates, which also seem to suggest a distinct vertical zonation in the sediment flux. At about 10–20 Ma ago bulk sediment accumulation rates gradually rose again in both basins to similar or even higher values than those reached in the early North Atlantic phase. This last phase seems to have started somewhat earlier in the east than in the west North Atlantic basins.

The palaeodepth distribution of the bulk sediment accumulation rates also reveals a distinct pattern, although the North Atlantic DSDP drill sites only permitted the description of palaeodepth intervals in water depths >1500 m. The early high bulk accumulation rate phase can be traced across the entire water column, with no suggestions of any depth stratification. Even though the ensuing phases of low and high bulk sediment accumulation rates are highly variable in detail, between the east and west North Atlantic sub-basins, as well as within the same basin, they seem to suggest some stratification and henceforth some type of vertical zonation of the rate of preservation of the sedimentary record reflecting the action of different water masses. The authors believe that this difference is highly significant and that it represents a signal of some basic characteristic of the oceanic water column. The authors also note that the early 'non-depth stratified' phase of high bulk sediment accumulation rates coincided with the repeated develop-

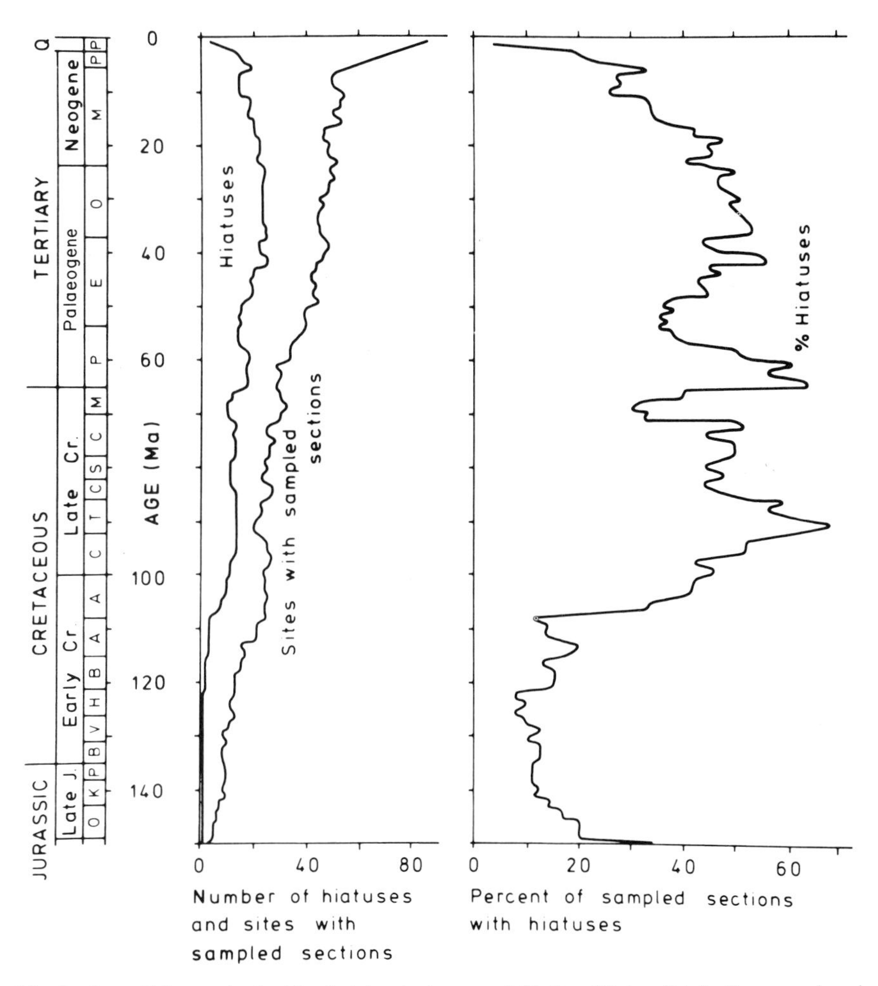

FIG. 6. Distribution of hiatuses in the North Atlantic deep-sea drill sites. Hiatus distributions are given in absolute and relative figures.

ment of oxygen deficient depositional environments (de Graciansky *et al.* 1984; Arthur & Dean, in press).

On the relationship between hiatuses and bulk sediment fluxes

To investigate further the temporal variability of sediment fluxes the hiatus frequencies (Fig. 6) and average bulk sediment accumulation rates (Fig. 7) versus time only have been re-plotted. Although these data were generated independently, both data sets seem to support a threefold subdivision in the depositional history of the North Atlantic.

Hiatuses are rare in sediments older than 100–110 Ma in both the eastern and western sub-basins of the North Atlantic (Fig. 6). The time span from 100 Ma to about 20 Ma is generally characterized by high, but variable hiatus frequencies. Pronounced maxima occur close to 90 Ma, 65 Ma, and 40 Ma. After 40 Ma, and clearly after 20 Ma, hiatus frequencies decreased to their modern minimum. The maxima are separated from each other by equally pronounced minima at 70 Ma and 50–60 Ma. The authors interpret these data as showing that the erosion of deep-sea sediments has fluctuated extensively through

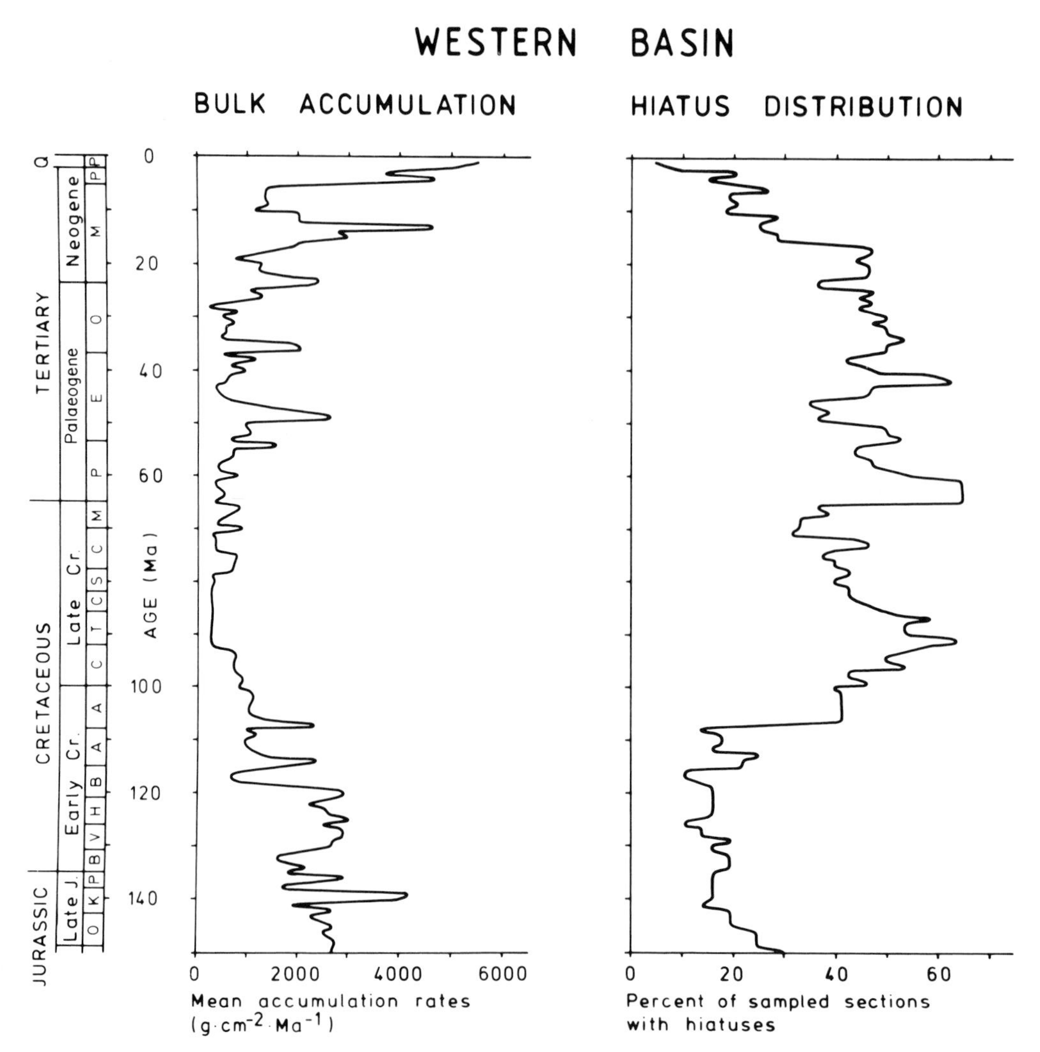

FIG. 7. Average bulk sediment accumulation rates and hiatus distributions versus time in the western sub-basin of the North Atlantic.

time. This has already been pointed out by Moore *et al.* (1978) for the Cenozoic pelagic deposits of most major ocean basins. For the North Atlantic it is now possible to use hiatus distributions to precisely pinpoint intervals of strengthening and weakening of deep-sea erosion. As outlined by Ehrmann & Thiede (1985), approximately 30–50% of the time which might be represented by North Atlantic deep-sea deposits cannot be documented properly because of the development of hiatuses. At present it is difficult to know where the sediment representing the hiatus intervals is situated.

Evaluating the importance of hiatuses is complicated by the fact that bulk sediment fluxes were high during times of low hiatus frequencies, and generally low during times of high hiatus frequencies. In Fig. 7 and Fig. 8 data have been plotted representing average values for 1 Ma time increments for bulk sediment accumulation rates, and hiatuses versus time. Previously it has been pointed out that hiatuses identified and represented in the hiatus plots, have been excluded when calculating bulk sediment accumulation rates. Despite these precautions an inverse relationship between sediment flux and hiatus formation can still be seen.

The above coincidence has been investigated by calculating correlation coefficients between bulk sediment accumulation rates and hiatus frequencies (Tables 1 and 2). The authors discovered that (a) phases of high correlation between these two parameters alternated with phases of low or no correlation; (b) correlations

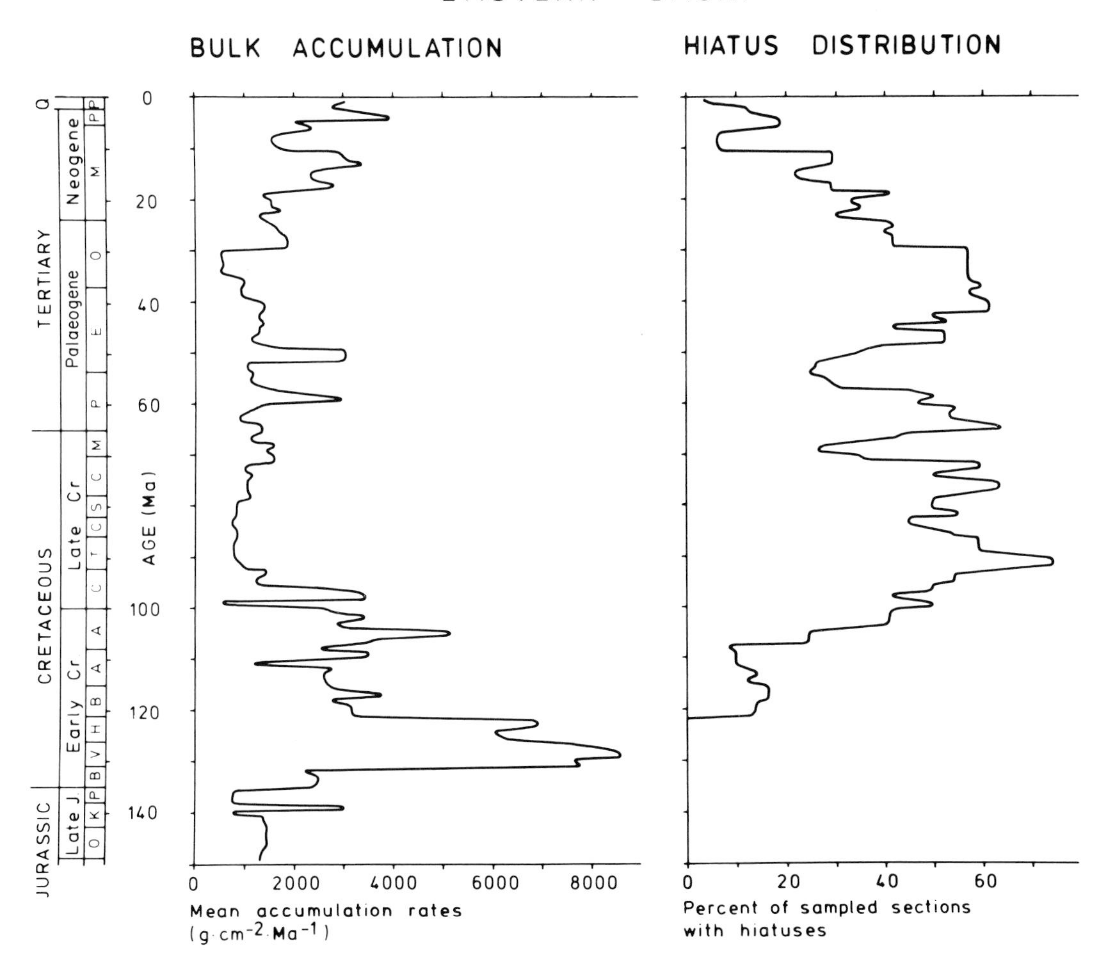

FIG. 8. Average bulk sediment accumulation rates and hiatus distribution versus time in the eastern sub-basin of the North Atlantic.

TABLE 1. *Correlation of bulk sediment accumulation rates (BAR) and hiatuses (correlation coefficients)*

		Correlation coefficients	
Time span	(Ma)	Western basin	Eastern basin
Total record		−0.69 (1–150 Ma)	−0.62 (1–120 Ma)
Tertiary	2–65	−0.73	−0.57
Neogene	2–24	−0.54	−0.14
Palaeogene	24–65	−0.53	−0.41
Late Cretaceous	65–100	−0.34	−0.30
Early Cretaceous	100–135	−0.43	0.17
Jurassic	135–150	0.06	—
Miocene	5–24	0.25	0.05
Oligocene	24–37	0.39	−0.92
Eocene	37–54	−0.56	−0.38
Palaeocene	54–65	0.50	0.08
Maastrichtian-Coniacian	65–86	−0.40	−0.52
Turonian-Albian	86–108	−0.77	−0.74
Aptian-Barremian	108–121	0.33	0.37
Hauterivian-Berriasian	121–135	−0.58	—

TABLE 2. *Correlation between the eastern and western sub-basins of the North Atlantic Ocean in respect to accumulation of bulk sediment and hiatus occurrence*

		Correlation coefficients	
Time span	(Ma)	Bulk sediment	Hiatuses
Total record		0.53 (1–139 Ma)	0.82 (1–121 Ma)
Tertiary	2–65	0.64	0.76
Neogene	2–24	0.73	0.79
Palaeogene	24–65	0.13	0.35
Late Cretaceous	65–100	0.53	0.75
Early Cretaceous	100–135	0.63	0.85
Jurassic	135–150	—	—
Miocene	5–24	0.59	0.78
Oligocene	24–37	0.18	0.72
Eocene	37–54	0.28	0.05
Palaeocene	54–65	−0.21	0.87
Maastrichtian-Coniacian	65–86	0.37	0.64
Turonian-Albian	86–108	0.69	0.90
Aptian-Barremian	108–121	0.19	−0.32
Hauterivian-Berriasian	121–135	0.65	—

between these parameters were variable, but similar in both sub-basins of the North Atlantic during the time spans prior to 65 Ma, and (c) correlations between these parameters were variable, but quite different from each other in the east and west Atlantic sub-basins over the past 65 Ma. The authors interpret these observations to suggest a relationship between the depositional processes controlling the bulk sediment flux to the ocean floor, and hiatus formation.

The authors have also attempted to relate their time series of hiatus frequencies and bulk sediment accumulation rates to evidence presented by Vail & Hardenbol (1979) and Vail *et al.* (1977) for relative eustatic sea-level changes during the Late Mesozoic and Cenozoic. The authors have been unable to detect any direct and easily recognizable relationship between eustatic sea-level changes, hiatus frequencies and bulk sediment accumulation rates, in contrast with previous authors (e.g. Worsley & Davies 1979).

Results and conclusions

(1) The available data allow us to subdivide the sedimentary history of the North Atlantic Ocean into three major phases, with high sediment fluxes in Late Jurassic-Early Cretaceous and late Cenozoic times, but low sediment fluxes in between.

(2) Interruptions of the sediment flux (=hiatuses) are of little importance during times of high sediment flux. They have made it difficult to document the North Atlantic's history over wide regions and for long time spans during the Late Cretaceous and the main part of the Tertiary.

(3) Regional distributions of bulk sediment accumulation rates have been highly variable. Centres of sediment fluxes have usually been located close to the continental margins of NW Africa, NW Europe, Greenland and NE America. Their location and intensity changed rapidly and frequently.

(4) An inverse correlation between sediment flux and hiatus frequencies suggests that bulk sediment accumulation rates are an expression of the rates and amount of preservation of the original sediment flux to the sea floor rather than the original sediment flux itself.

(5) The depositional environments of the east and west North Atlantic basins were very similar during the time span from 150 to 100 Ma. These basins became differentiated during the Late Cretaceous and Cenozoic.

(6) During the early high bulk sediment accumulation rate phase of the North Atlantic (150–100 Ma) the authors did not find in the sediment flux data any indications of stratified water columns. However, the sediment flux and hiatus data suggest that the deep North Atlantic was well stratified since that time.

ACKNOWLEDGMENTS: The research on which this paper is based was supported by the German Research Foundation (DFG). The data were compiled from the Initial Reports of the Deep-Sea Drilling Project.

References

ARTHUR, M.A. & DEAN, W.E., in press. Cretaceous paleoceanography. *In*: TUCHOLKE, B. & VOGT, P. (eds), *DNAG Western North Atlantic Synthesis*. Geol. Soc. Am.

BERGER, W.H. 1972. Deep sea carbonates: dissolution facies and age depth constancy. *Nature* **236,** 392–5.

BERGGREN, W.A., MCKENNA, M., HARDENBOL, J. & OBRADOVICH, J. 1978. Revised Paleogene polarity time scale. *J. Geol.* **86,** 67–81.

BERNOULLI, D. 1984. The early history of the Atlantic-Tethyan system. *Ann. Geophys.* **2** (2), 133–6.

EHRMANN, W.U. & THIEDE, J. 1985. History of Mesozoic and Cenozoic sediment fluxes to the North Atlantic Ocean. *Contr. Sediment.* **15,** 1–109.

GRACIANSKY, P.C. DE, DEROO, G., HERBIN, J.P., MONTADERT, L., MÜLLER, C., SCHAAF, A. & SIGAL, J. 1984. Ocean-wide stagnation episode in the Late Cretaceous. *Nature* **308,** 346–9.

HONJO, S. 1978. Sedimentation of materials in the Sargasso Sea at a 5,367 m deep station. *J. mar. Res.* **36** (3), 469–92.

——1982. Seasonality and interaction of biogenic and lithogenic particulate flux at the Panama Basin. *Science* **218,** 883–4.

MOORE, T.C., VAN ANDEL, TJ. H., SANCETTA, C. & PISIAS, N. 1978. Cenozoic hiatuses in pelagic sediments. *Micropaleontology* **24** (2), 113–38.

SCLATER, J.G., HELLINGER, S. & TAPSCOTT, C. 1977. The paleobathymetry of the Atlantic Ocean from the Jurassic to the Present. *J. Geol.* **85,** 509–52.

THIEDE, J. 1979: History of the North Atlantic Ocean: evolution of an asymmetric zonal paleo-environment in a latitudinal basin. *In*: TALWANI, M., HAY, W.W. & RYAN, W.B.F. (eds), *Deep Drilling Results in the Atlantic Ocean: Continental Margins and Paleoenvironment*. Maurice Ewing Series 3. Am. Geophys. Union. 275–96.

VAIL, P.R., MITCHUM, R.M. & THOMPSON, S. 1977. Global cycles of relative changes of sea level. *Am. Ass. Petrol. Geol. Mem.* **26,** 63–97.

——& HARDENBOL, J. 1979. Sea level changes during the Tertiary. *Oceanus* **22,** 71–9.

VAN ANDEL, TJ. H., HEATH, G.R. & MOORE, T.C. 1975. Cenozoic history and paleoceanography of the central equatorial Pacific. *Geol. Soc. Am. Mem.* **143,** 134 pp.

——, THIEDE, J., SCLATER, J.G. & HAY, W.W. 1977. Depositional history of the South Atlantic Ocean during the last 125 million years. *J. Geol.* **85,** 651–98.

VAN COUVERING, J.A. & BERGGREN, W.A. 1977. Biostratigraphical basis of the Neogene time scale. *In*: KAUFFMAN, E.G. & HAZEL, J.E. (eds), *Concepts and Methods of Biostratigraphy*. Dowden, Hutchinson & Ross, Stroudsburg, Pa. 283–305.

VAN HINTE, J.E. 1976. A Cretaceous time scale. *Am. Ass. Petrol. Geol. Bull.* **60,** 498–516.

—— 1978. A Jurassic time scale. *Am. Assoc. Petrol. Geol. Stud. Geol.* **6,** 289–97.

WORSLEY, T.R. & DAVIES, T.A. 1979. Sea-level fluctuations and deep-sea sedimentation rates. *Science* **203,** 455–6.

JÖRN THIEDE & WERNER U. EHRMANN, Geologisch-Paläontologisches Institut und Museum, Christian-Albrechts-Universität, Olshausenstrasse 40, D-2300 Kiel, Federal Republic of Germany.

Evidence for changes in Mesozoic and Cenozoic oceanic circulation on the south-western continental margin of Ireland: DSDP/IPOD Leg 80

P.C. de Graciansky, C. Wylie Poag, E.A. Hailwood, R.W.O'B.Knox, D.G. Masson, L. Montadert, C. Ravenne, C. Müller, J.C. Sibuet, J. Sigal, S.W. Snyder, D.W. Waples & R. Cunningham

SUMMARY: A transect of four coreholes, drilled by the *Glomar Challenger* across the Irish continental margin at the Goban Spur, evidences a dynamic palaeoceanographic regime during the late Mesozoic and Cenozoic. Shallow marine waters invaded the rift-stage grabens of the Goban Spur in the early Barremian. Thereafter, the margin subsided rapidly, producing a pelagic depositional regime by late Barremian time. Deep marine conditions were maintained as sea-floor spreading began in the early Albian, and chiefly pelagic deposition continued to the present.

Among a series of significant post-rift oceanographic changes, one of the most notable is the familiar fluctuation of oxic and anoxic sea-floor environments during the Cenomanian and Turonian. Another marked change took place during the late Palaeocene, when cooler, oxygen-rich, northern bottom waters reached the Goban Spur as a consequence of rifting and sea-floor spreading between Greenland, Rockall Plateau, and Norway. Later during the Cenozoic, the initial production of Antarctic bottom water, several accelerations of polar ice-cap growth, and fluctuating eustatic sea-level produced a variety of circulatory shifts on the Goban Spur. A particularly significant sedimentological consequence of these interacting processes was the widespread creation of numerous erosional and non-depositional unconformities.

Four holes were drilled during Leg 80 of the International Phase of Ocean Drilling (IPOD) along a transect across the Goban Spur, on the south-western continental margin of Ireland. The Goban Spur, a prominent physiographic and structural feature of the Celtic Margin forms the southern boundary of the Porcupine Seabight (Fig. 1). Locating the transect on this sediment-starved margin has afforded an exceptional opportunity to: (a) study the onset of oceanic circulation during the earliest stages of marine deposition; (b) distinguish local effects from ocean-wide influences on sedimentation; and (c) follow the migration of different water masses across the margin during the long phase of sea-floor spreading.

Initiation of a pelagic regime during the Barremian

Site 549 of Leg 80 was drilled on top of the Pendragon Escarpment, at 2525 m water-depth on the middle continental slope (Fig. 2). Here, approximately 300 m of Barremian deposits lie directly on the subaerially eroded surface of the Hercynian basement. The authors infer from seismic profiles (Figs 3 and 4) that no sediment much older than Barremian was deposited in the deepest parts of the adjacent grabens. The sequence drilled at Site 549 reflects a progressive deepening from hyposaline to open-marine, outer sublittoral palaeoenvironments (Snyder *et al.* 1985; Rat *et al.* 1985; Fig. 5). This deepening took place during a relatively short period comprising the lower two-thirds of the Barremian (uppermost Barremian not represented).

During this time, sedimentation was clearly controlled by local sources and physiography, i.e. (1) terrigenous and bioclastic sedimentary influxes and (2) the presence of a complex network of islands and partly closed shallow basins. The physiography was created by the tilting of basement blocks along listric normal faults, a consequence of the rifting process (Montadert *et al.* 1979). By the end of the Barremian, many subaerially eroded surfaces had been submerged beneath pelagic environments. Subsequently, sedimentary starvation of the young margin, combined with accelerated subsidence that accompanied sea-floor spreading, minimized local depositional control and promoted greater influence of basin-wide events.

From Summerhayes, C.P. & Shackleton, N.J. (eds), 1986, *North Atlantic Palaeoceanography*, Geological Society Special Publication No. 21, pp. 17–33.

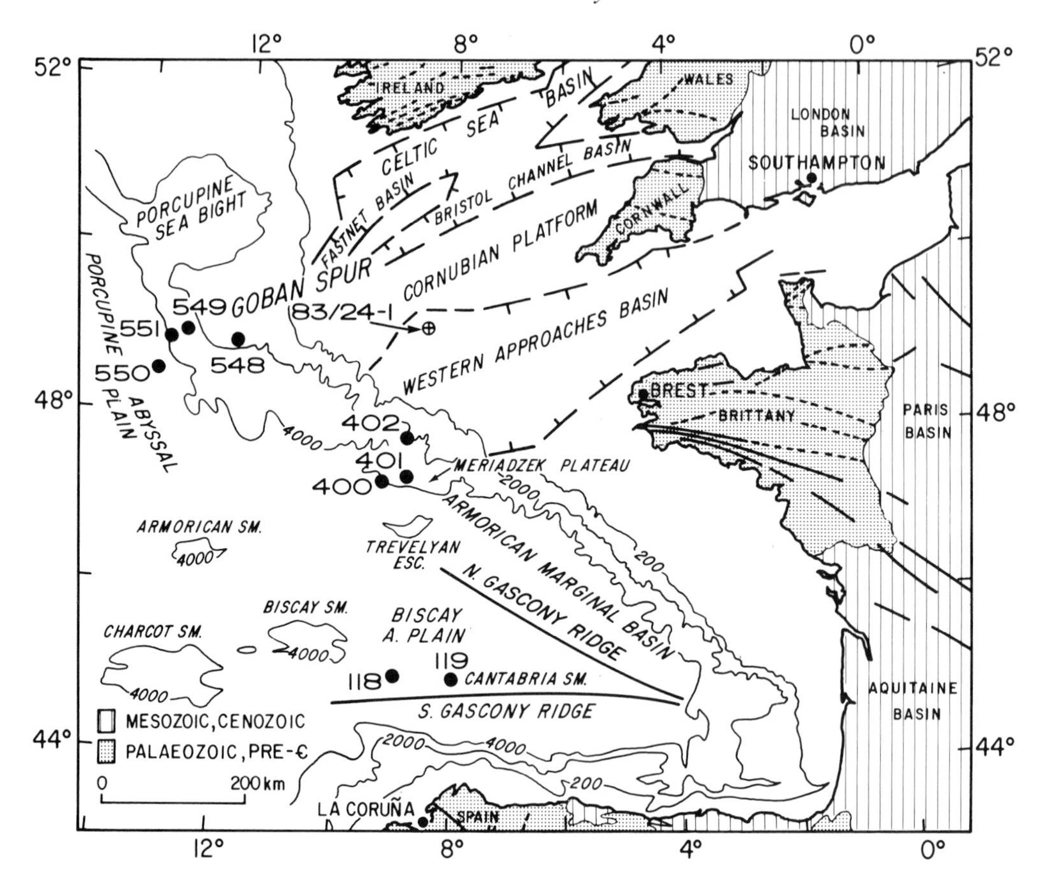

FIG. 1. Location of selected DSDP boreholes in the NE Atlantic.

Cyclic stagnation during late Cretaceous times

Cyclic deposits of black shale produced by periodic anoxic bottom conditions during the Cenomanian were found at Site 550 at the foot of the Goban Spur (Figs 1 and 2). The anoxia was probably produced by local environmental changes. In contrast, at Sites 549 and 551, located farther shoreward on the slope (Fig. 1), a different type of organic enrichment recorded an ocean-wide stagnation event of late Cenomanian–early Turonian age.

Cenomanian cyclic anoxia on the Porcupine Abyssal Plain

Alternating white chalky limestones and dark organic limestones of middle and late Cenomanian age were drilled at Site 550 (Figs 1 and 2). The white limestones contain little organic carbon, but marine organic matter (as much as 2% total organic carbon [TOC]) is present in the darker lithologies (Waples & Cunningham 1985). This relatively high TOC documents episodic anoxia on the sea-floor.

The rugged morphology of the sea-floor, characterized by basins and ridges inherited from the rifting period, probably trapped anoxic waters in the deeper basins. Anoxia could have been induced also, in part, by more widespread events, as suggested by the distribution or rhythmical anoxic sediments throughout the North Atlantic during mid-Cretaceous time (Graciansky, Brosse *et al.* 1984). Of particular interest on Goban Spur is the fact that anoxic environments (Fig. 6(a)) at depositional depths of 2000 to 2500 m (Site 550) were contemporaneous with well-aerated environments at 1000 m water depth (Site 549) and shallower (outer Cornubian Platform; Evans *et al.* 1981; Fig. 1). The upper part of the water column, therefore, was continuously well-oxygenated, while below 1800–2000 m, periodic stagnation took place. The rhythmic stagnation could

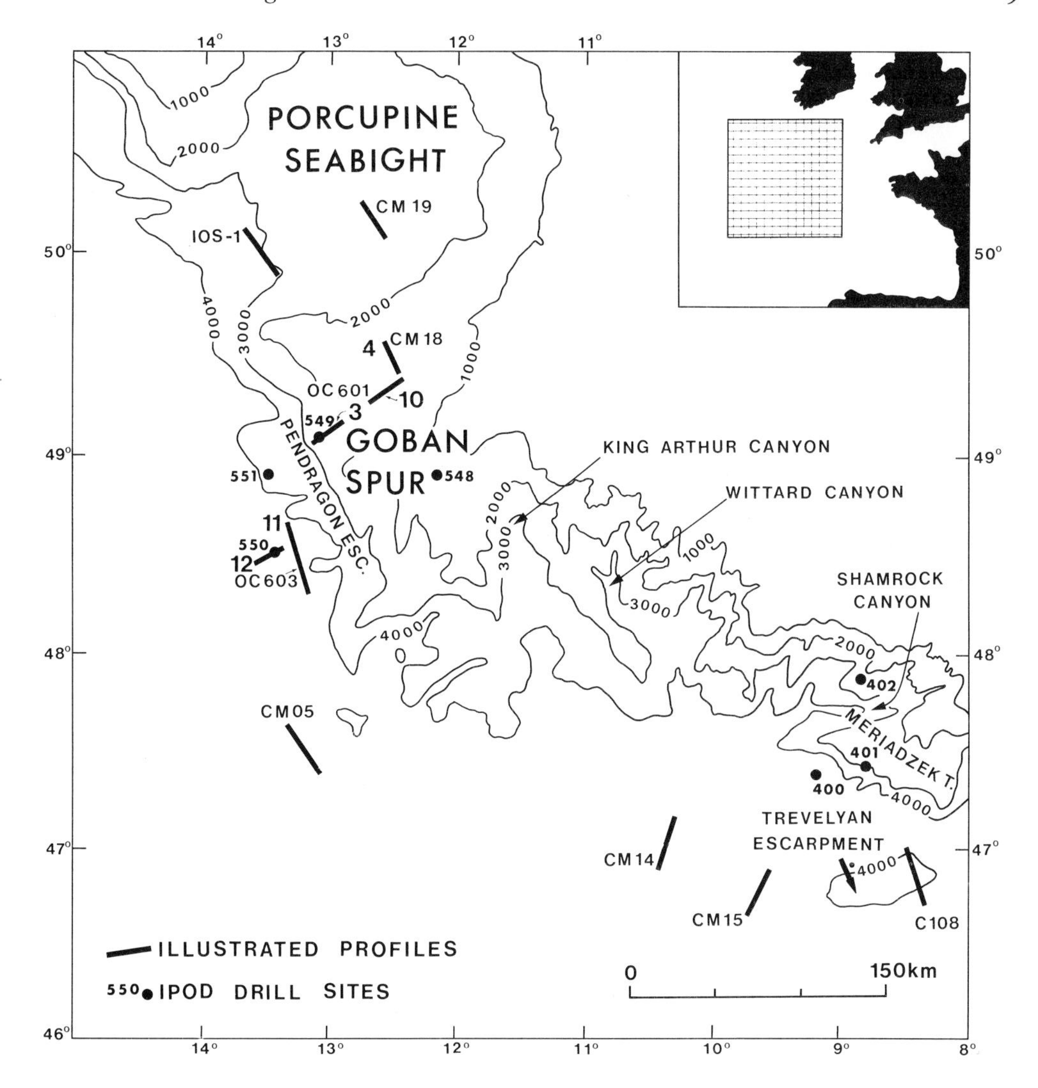

FIG. 2. Locations of selected seismic reflection profiles (bathymetric chart from Roberts *et al.* 1981, with depths in metres). Profiles IOS 1, CM 19, CM 14, CM 15, OC 108 are illustrated by Masson & Parson (1983). They all document folds disconformably overlain by late Palaeocene–early Eocene brownish siliceous chalks.

have been a consequence of periodic stratification and turnover of water in the newly formed ocean (Graciansky & Gillot 1985). For example, the accumulation of the Cenomanian 'Upper Greensands' on land, and in the Celtic Sea and Fastnet Basins (Robinson *et al.* 1981) suggests that an especially active meteoric run-off from the adjacent land-mass could have provided enough fresh water to induce stratification (Rossignol-Strick *et al.* 1982; Pratt 1984). On an even broader scale, the periodicity of bedding in the middle-to-late Cenomanian interval at Site 550 approximates cyclical perturbations of the Earth's orbit (Schwarzacher & Fischer 1982).

The Cenomanian–Turonian anoxic event

At sites of intermediate water depth on the Goban Spur (549, 551; Fig. 6(b)), a thin (7 m-thick) black shale is enclosed by white chalks of late Cenomanian to early Turonian age. This black shale has a maximum TOC of 10%, which is partly of marine origin (Waples & Cunningham 1985). The main biogenic component is siliceous

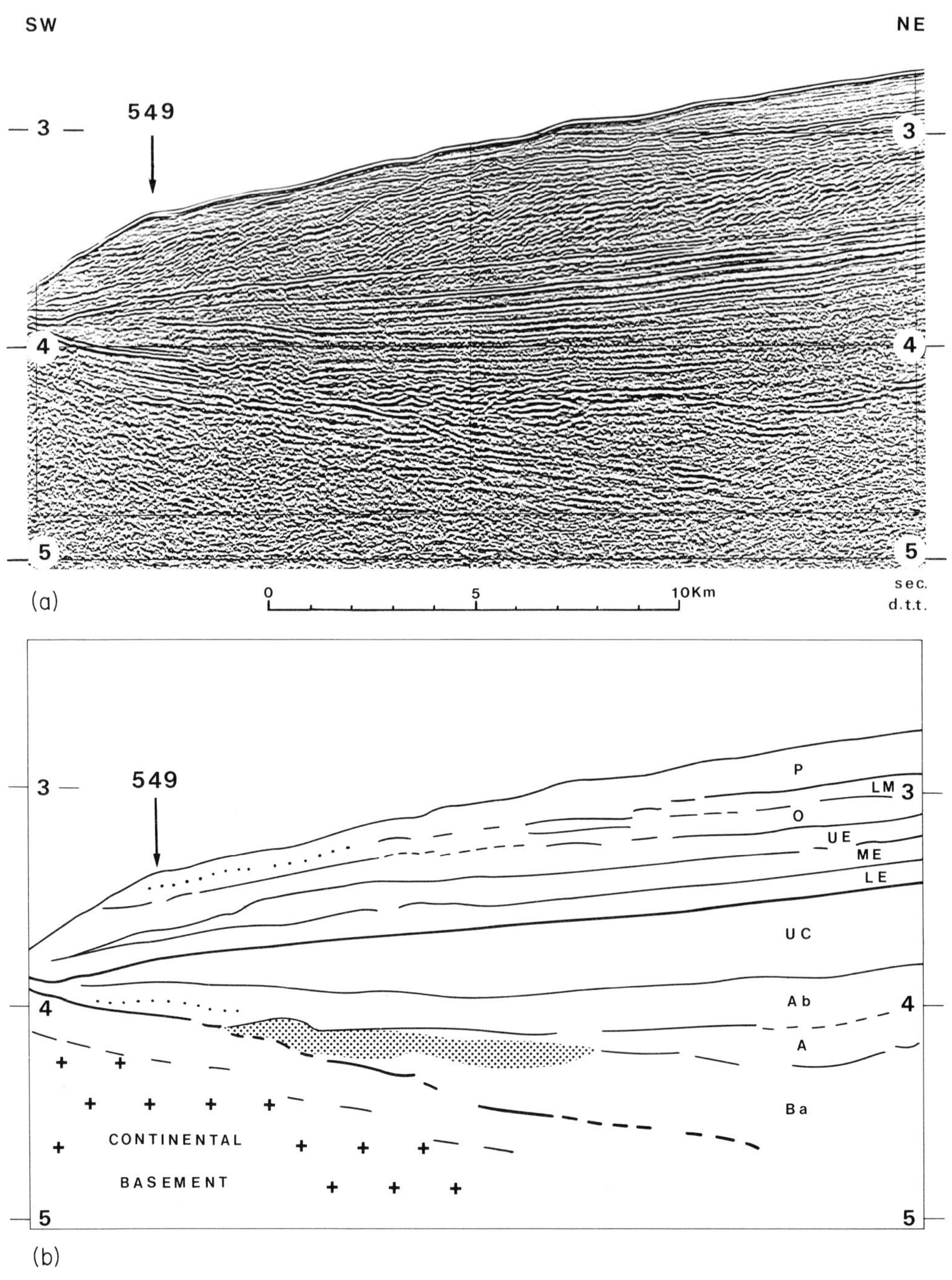

FIG. 3. Migrated seismic reflection profile through site 549 (a). (b) is an interpretation of (a). Ba: Barremian; A: Aptian(?); Ab: Albian; UC: Upper Cretaceous; LE: lower Eocene; ME: middle Eocene; UE: upper Eocene; O: Oligocene; LM: lower Miocene; P: Pliocene to Recent. Unpublished high resolution profile IFP OC 601. Location 3 on Fig. 2. Note the change in depositional style between Barremian-Aptian(?) (synrift) and overlying (post-rift) sequence. Barremian and Albian have been documented by drilling at Site 549. The presence of Aptian layers in the adjacent half graben is inferred from this record. A mass flow or a broad channel-fill is indicated in the synrift sequence by the dense pattern of dots. The synrift deposits have a wedge-shape which pinches out on the elevated side of the titled block.

A clear acoustic reflector separates the Late Cretaceous chalks from the late Palaeocene–early Eocene siliceous chalks (Ba). Discontinuous reflector corresponds to the middle/late Miocene unconformity. Oblique reflectors in the upper Eocene sequence are truncated by the lower surface of the Oligocene chalks.

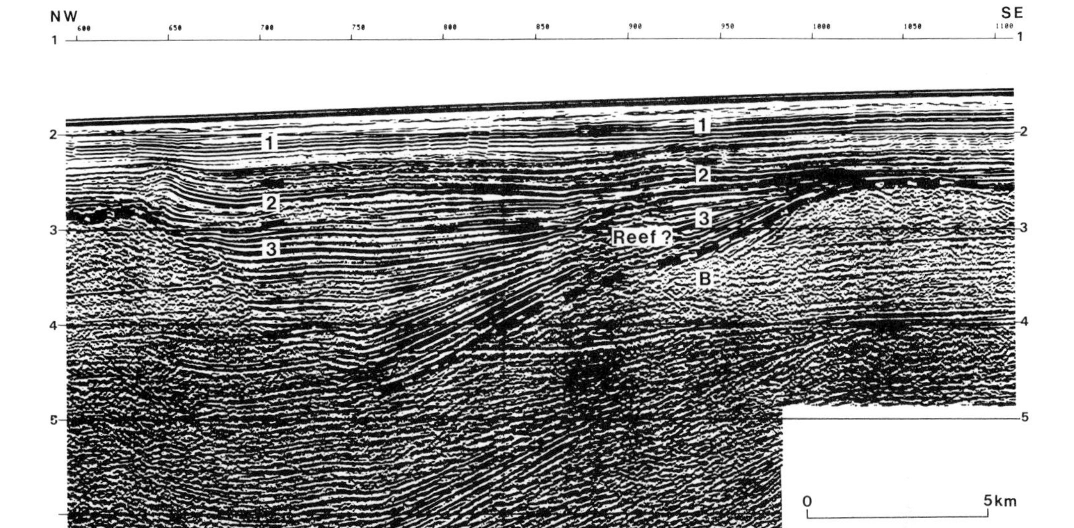

FIG. 4. Migrated seismic reflection profile across a half-graben at mid-depth on the Goban Spur. Sequence 1: middle Eocene to Recent chalks; Sequence 2: Albian shales, Upper Cretaceous to lower Eocene chalks; Sequence 3: sandstones, siltstones and limestones of Barremian and Aptian(?) age in the synrift sequence. Profile CM 18. Location 4 on Fig. 2. See also Masson *et al.* (1985). The divergent reflector pattern within the half-graben is typical of synsedimentary tilting of a basement block. Note a reefal(?) body towards the elevated tip of the tilted block. The one kilometre-wide fold within seismic sequence 2 near the north-west edge of the profile is related to late Paleocene–early Eocene reactivation of the fault which bounds the half-graben.

(e.g. the carbonate content is nearly zero at Site 551, where the reconstructed depositional depth was ~1300 m).

The distribution of sediments across the Goban Spur clearly documents the vertical disposition of water masses during the Cenomanian–Turonian anoxic event. At the shallowest drill site (548), there is no Mesozoic sediment older than Campanian. But shoreward, on the northern side of the Cornubian Platform (Fig. 1) white, glauconitic chalk was deposited during the Cenomanian and Turonian, and no stratigraphic gap or lithologic change was recorded in the section (Zephyr Well 83/24-1; Evans *et al.* 1981). However, at the deepest site (550) of Leg 80, a hiatus spanning the Turonian and Coniacian interval, separates Cenomanian silty limestones from dark, smectitic, abyssal mudstone of Santonian age.

The total Cenomanian–Turonian depositional record across the Goban Spur suggests that a thick oxygen-minimum zone was located at mid-depth (Sites 549, 551), below a well-oxygenated, highly productive surface layer (high productivity indicated by biogenic silica in the anoxic sediments; Fig. 6(b)). Site 550 may have lain below the oxygen-minimum zone. The postulated oxygenation of bottom waters at Site 550 could have been the result of very slow sediment accumulation caused by the combined effects of carbonate dissolution and low detrital influx (Waples 1985).

The layer of Cenomanian–Turonian black shale beneath the Goban Spur is the northernmost evidence in the Atlantic of an apparently ocean-wide episode of anoxia (Fig. 7). Although the origin of the black shales is poorly understood, conditions prevailing in the oceans must have caused general stagnation below water depths of a few hundred metres, accompanied by a temporary rise of the CCD and general sediment-starvation. In continental-margin or epicontinental palaeoenvironments in North America, Africa and Europe, the Cenomanian–Turonian boundary is often marked by either a stratigraphic gap or a condensed stratigraphic section. These features are correlative with a postulated early Turonian high stand of sea-level (Vail & Hardenbol 1979), which some authors estimate to have been as much as 450 m higher than present sea-level (Hancock & Kauffman 1979). Further implications of this peculiar layer, which may have been unique in the Mesozoic–Cenozoic stratigraphic record, have been discussed by Graciansky, Deroo *et al.* (1984).

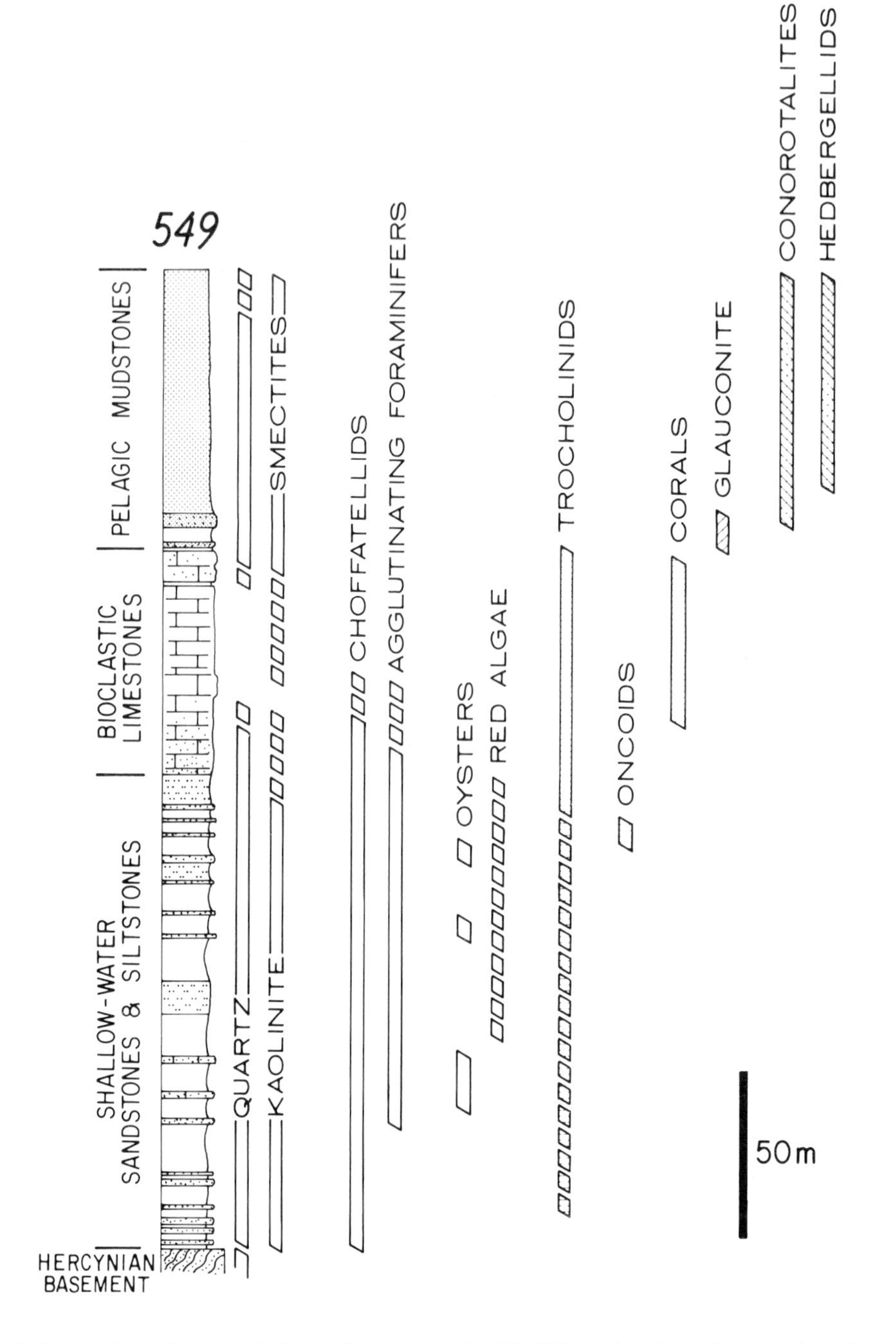

FIG. 5. Facies column for the early Barremian strata at site 549. This section shows the gradation of palaeoenvironments from very shallow water to open-sea pelagic environments that took place during the early synrift phase at Goban Spur.

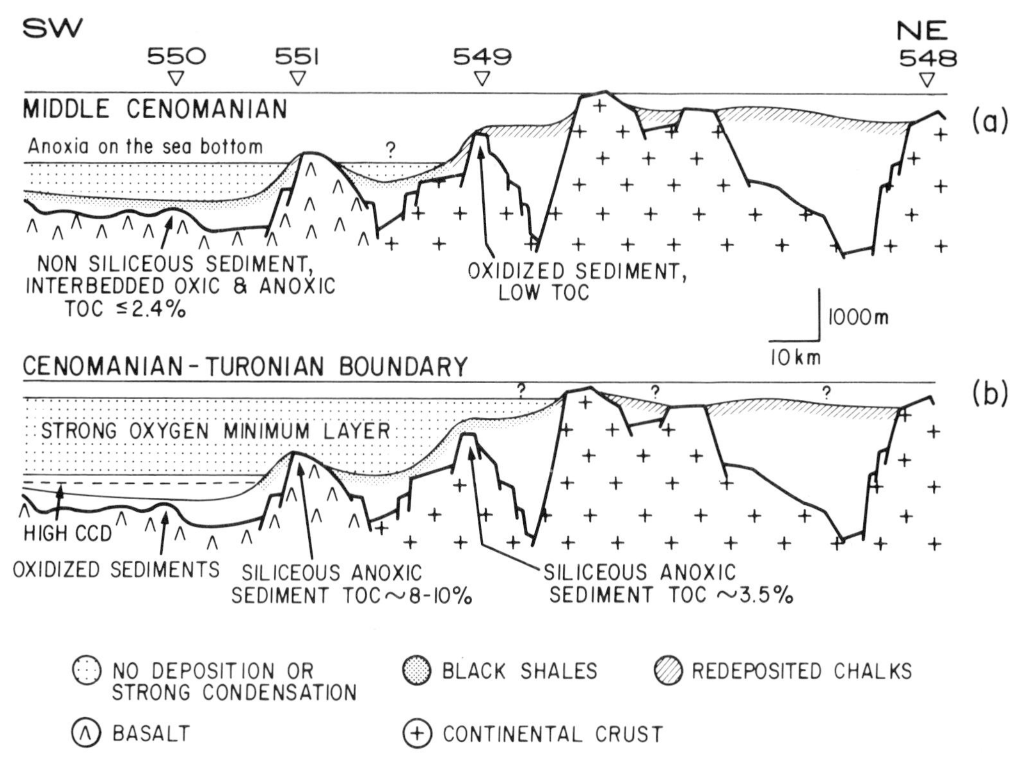

FIG. 6. Anoxia in oceanic waters of the Goban Spur during Cenomanian–Turonian time (from Gillot 1983; Waples 1985). (a) Middle Cenomanian anoxic waters on the deep sea bottom rhythmically displaced by oxic waters. (b) Expanded oxygen minimum layer located at mid-depth during the latest Cenomanian to early Turonian.

Tectonic, magmatic, and oceanographic events in the late Palaeocene–early Eocene

As much as 80 m of brown siliceous chalks was deposited during the late Palaeocene–early Eocene on the Goban Spur (Figs 8 and 9). These dark chalks differ from underlying and overlying white chalks by their higher contents of biosilica, detrital quartz, and terrigenous clay minerals (kaolinite, illite) as well as by the presence of volcanic debris and trace quantities of manganese, iron and other metals (Karpoff *et al.* 1985). Similar late Palaeocene–early Eocene strata have been cored at sites widely scattered throughout the North Atlantic (Site 119, Laughton *et al.* 1972; Site 398, Sibuet *et al.* 1979; Site 400, Montadert *et al.* 1979; Zephyr well 83/24-1, Evans *et al.* 1981), which suggests that a basin-wide event initiated this specific sedimentation. The precise dating of the onset of this sequence was accomplished at Site 550, where the lower part of the upper Palaeocene (nannofossil Zone NP 4) was identifed at its base. Elsewhere on the Goban Spur transect, the brown siliceous chalks are underlain by an unconformity (Fig. 8).

The influx of terrigenous minerals on the Goban Spur cannot be correlated directly with the southward progradation of a late Palaeocene (nannofossil zone NP 9) deltaic platform in the adjacent Porcupine Basin (Roberts *et al.* 1981), nor with late Palaeocene–early Eocene fluviatile deposits disconformably overlying marine sediments in the Celtic Sea Basin (Naylor & Shannon 1982). All are related to broad-scale palaeogeographic changes in the north-east Atlantic, during which the Irish landmass was uplifted (George 1967). Offshore, mild tectonism took place in conjunction with the Pyrenean orogeny (Masson & Parson 1983; Sibuet *et al.* 1985). A number of east–west trending compressive structures and strike-slip faults trending north-west or north-east have been documented beneath the Goban Spur and vicinity. This tectonism created extensive angular unconformities (Fig. 10), folds (see profile CM 14 of Montadert *et al.* 1979, p. 1056),

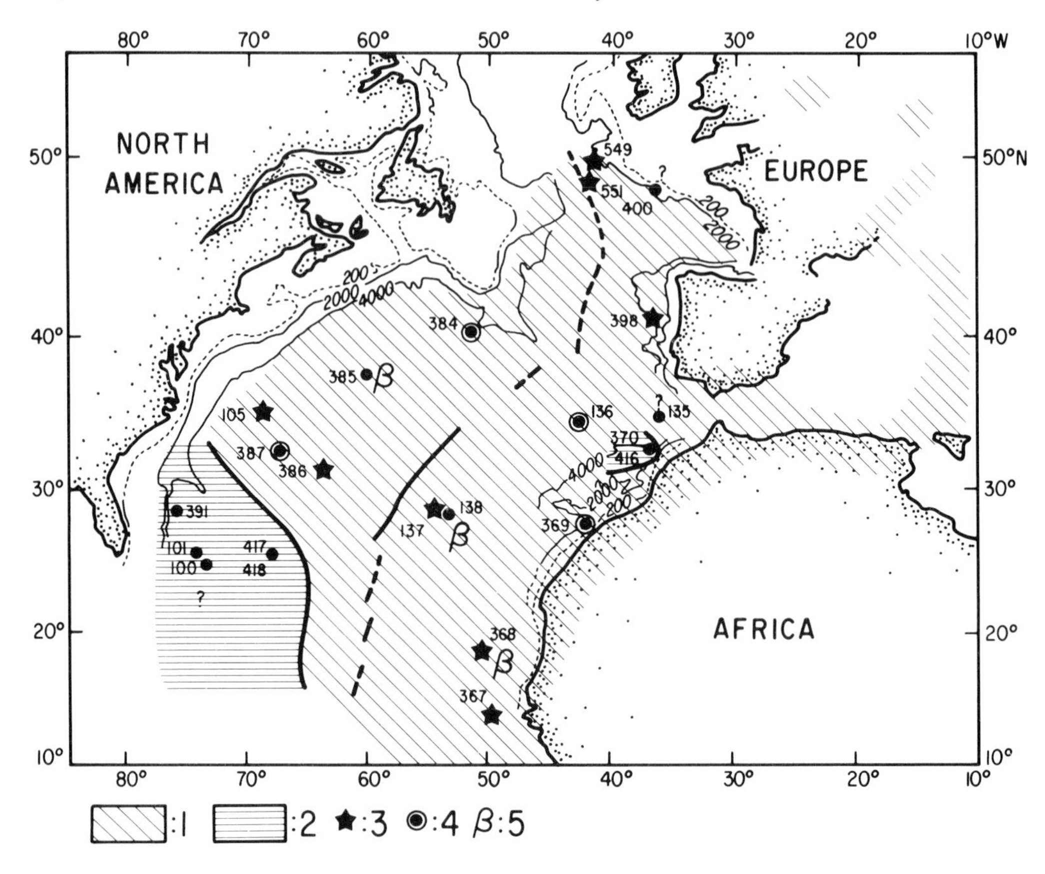

FIG. 7. Geographical extent of the Cenomanian–Turonian anoxic layer in the North Atlantic and western Europe (from Graciansky, Brosse *et al.* 1984). 1. Presence of the anoxic layer. 2. Extent of Upper Cretaceous deposits removed by subsequent erosion. 3. Presence of the anoxic layer at DSDP corehole. 4. Hiatus dated as Cenomanian–Turonian at DSDP corehole. 5. Corehole bottoming in oceanic basalts.

reverse faults, and related physiographic features on the sea-floor (Fig. 11). A particularly notable feature is the Pastouret Ridge, which runs seaward from the foot of the Pendragon Escarpment between Sites 550 and 551 (Sibuet *et al.* 1984; Sibuet *et al.* 1985; and unpublished single-trace seismic reflection profiles of the R.V. *Jean Charcot*; cruises NORESTLANTE I, October, 1983, and NORESTLANTE II, March, 1984; Fig. 12). Formation of these features is tentatively dated as late Palaeocene to middle Eocene.

Ashes and bentonite layers are present in the uppermost Palaeocene (nannofossil Zone NP 9) and the lower Eocene (Zone NP 10); the latter can be correlated with similar volcanic deposits known in Holland, northern Germany, Denmark, and south-eastern England (Knox and Morton 1983; Knox 1985), in the North Sea (Jacqué & Thouvenin 1975), and in the Meriadzek area (Fig. 1; Site 401; Montadert *et al.* 1979). On the western side of Rockall Plateau, the first accretion of oceanic crust is believed to be recorded by eruptive volcanic strata that accumulated during deposition of nannofossil Zone NP 10. This regional volcanic event is also recorded in the North Sea, where rifting commenced at the boundary between nannofossil zones NP 5 and NP 6 (Roberts *et al.* 1981).

The separation of the Rockall region from Greenland probably was accompanied by the southward incursion of cold northern waters. This separation is documented by palaeomagnetic, palaeontological and geochemical characteristics of late Palaeocene–early Eocene sediments on the Goban Spur. The alignment of the long axes of magnetic grains in most sediments at Sites 548 and 549 is dominantly north-east–south-west, indicating downslope current flow (Hailwood & Folami 1985). In the late Palaeocene at Site 548, however, the magnetic fabric changes to

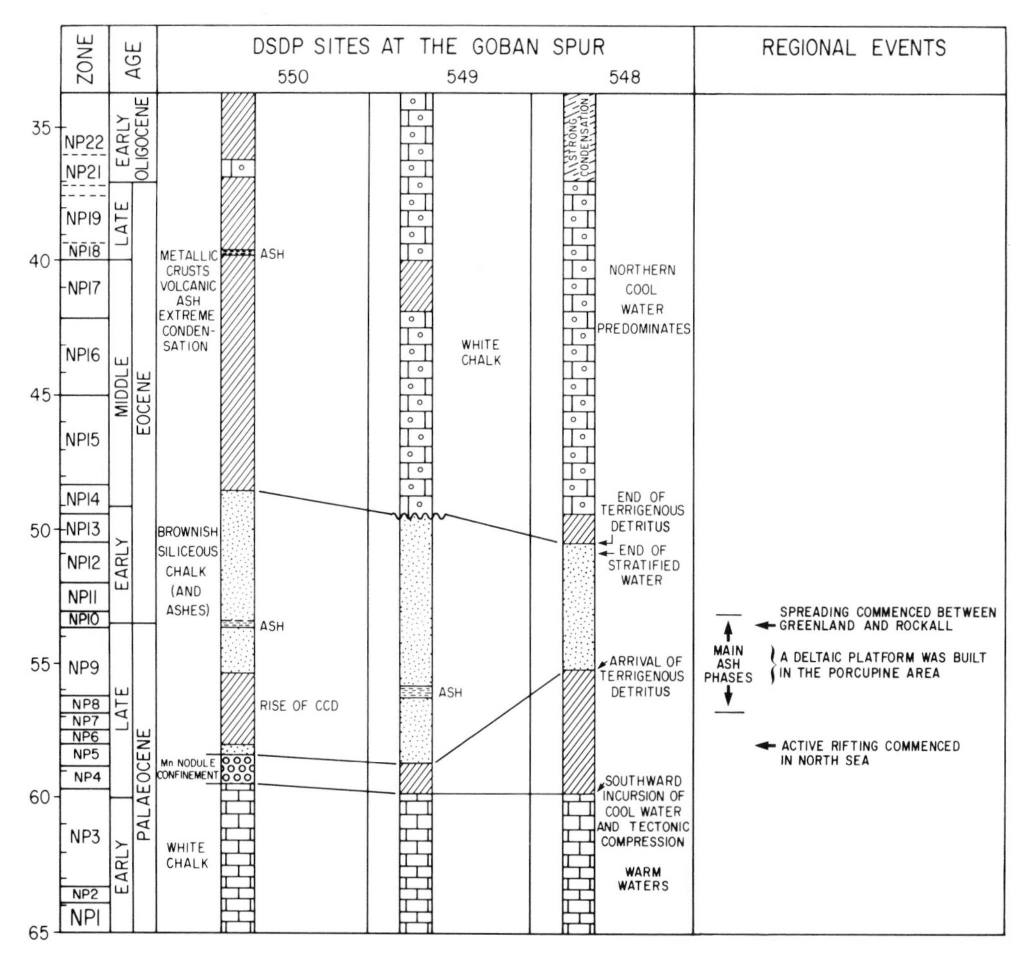

FIG. 8. Relationships of early Tertiary oceanographic events to sedimentary facies beneath the south-western Irish margin.

a trend parallel to the slope contours (NW–SE; Fig. 13), suggesting the possible influence of contour-currents brought about by introduction of corrosive northern bottom-waters into the Goban Spur region. Such corrosive bottom water could account for the marked decrease in calcareous fossils and the coincident increase in siliceous fossils (sponge spicules, diatoms, silicoflagellates) observed at Site 550 (Snyder *et al.* 1985; Müller 1985; Karpof *et al.* 1985). The presence of cool surface waters at Site 549 is indicated by well-preserved, low-diversity associations of early Palaeocene planktonic foraminifers and nannofossils (Snyder *et al.* 1985).

Detailed biostratigraphic and sedimentological studies further document the complex changes in circulation that took place during the late Palaeocene–early Eocene (Figs 8 and 9). At Site 550, late Palaeocene sedimentation began with a 60 cm thick agglomerate of buck-shot-size manganese nodules (Karpoff *et al.* 1985). The nodules originated as fecal pellets, enriched in organic matter. Subsequent to their accumulation, they were replaced by well-crystallized manganese, exceptionally rich in barium sulphate. Smectite and iron-oxides developed within the clayey matrix, surrounding the pellets.

A slightly anoxic environment conducive to the preservation of organic matter within the pellets could have been brought about by temporary stratification resulting from the emplacement of cold northern bottom water (Haq 1981) beneath warm surface waters. Warm surficial waters have been inferred from foraminiferal associations in late Palaeocene turbiditic sediments (nannofossil Zone NP 9) at Site 548 (Poag *et al.* 1985). Later diagenetic alteration and metallogenesis could have taken place within the sediment, in the

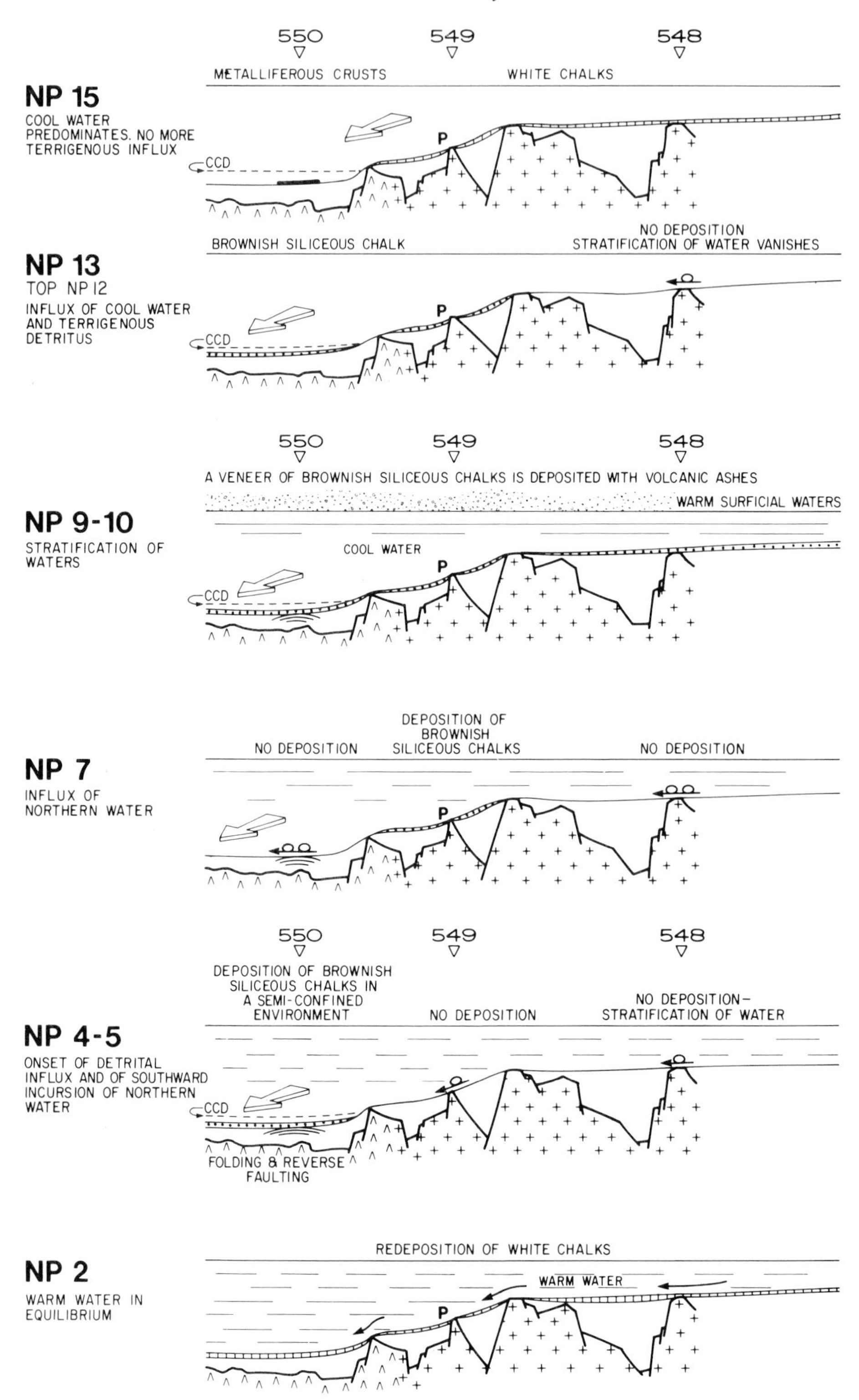
550
549
548
METALLIFEROUS CRUSTS
WHITE CHALKS
NP 15
COOL WATER PREDOMINATES. NO MORE TERRIGENOUS INFLUX
P
CCD
BROWNISH SILICEOUS CHALK
NO DEPOSITION
STRATIFICATION OF WATER VANISHES
NP 13
TOP NP 12
INFLUX OF COOL WATER AND TERRIGENOUS DETRITUS
550
549
548
A VENEER OF BROWNISH SILICEOUS CHALKS IS DEPOSITED WITH VOLCANIC ASHES
WARM SURFICIAL WATERS
NP 9-10
STRATIFICATION OF WATERS
COOL WATER
DEPOSITION OF BROWNISH SILICEOUS CHALKS
NO DEPOSITION
NO DEPOSITION
NP 7
INFLUX OF NORTHERN WATER
550
549
548
DEPOSITION OF BROWNISH SILICEOUS CHALKS IN A SEMI-CONFINED ENVIRONMENT
NO DEPOSITION
NO DEPOSITION–STRATIFICATION OF WATER
NP 4-5
ONSET OF DETRITAL INFLUX AND OF SOUTHWARD INCURSION OF NORTHERN WATER
FOLDING & REVERSE FAULTING
REDEPOSITION OF WHITE CHALKS
NP 2
WARM WATER IN EQUILIBRIUM
WARM WATER

presence of corrosive interstitial water. Cooling of surficial waters at Site 548 took place just prior to a period of sea-floor erosion in the late early Eocene (Poag *et al.* 1985). During this first incursion of northern bottom waters into the Goban Spur region, sea-floor erosion took place in several other parts of the North Atlantic (Fig. 7).

The oceanographic changes of the late Palaeocene–early Eocene were contemporaneous with a number of tectonic and climatic changes (Fig. 8), including: (1) uplift of the Irish land mass; (2) reactivation of old Caledonian fractures; (3) compressive tectonism on the Bay of Biscay margin (Montadert *et al.* 1979) the Goban Spur (Sibuet *et al.* 1985) and the Porcupine Seabight (Masson & Parson 1983); (4) active rifting in the North Sea and Rockall area; (5) initiation of spreading between Norway and Greenland; and (6) explosive volcanism in north-western Europe.

Oceanographic and climatic changes in the middle to late Miocene

At all sites of the Goban Spur transect, a 5–6 m.y. hiatus has been recognized, spanning the late part of the middle Miocene and early part of the late Miocene (nannofossil zones NN7–NN10 are missing).

Modification of sea-floor environmental conditions across this unconformity at Site 548 have been documented from micropalaeontological data (Poag *et al.* 1985). Sediments below the unconformity were deposited in a well-oxygenated, relatively cool and stable, epibathyal environment, which had been maintained since the onset of the Miocene. Above the unconformity, sediments were deposited also at epibathyal depths, but within an oxygen-minimum zone, characterized by abundant and diverse bolivinids, and low overall benthic diversity (Poag & Low 1985). It is presumed that lowered sea-level during the hiatus (Vail *et al.* 1977) depressed the cool and stable water mass, replacing it with a warmer, oxygen-poor water mass derived from the Mediterranean Basin (similar to the modern Mediterranean outflow water). Later, the Miocene benthic foraminiferal diversity increased and the bolivinids decreased as sea-level rose and oxygen-enriched waters returned.

Extensive oceanographic changes took place in the North Atlantic in the middle to late Miocene, as documented by an episode of deep-sea erosion (Ruddiman 1972; Miller & Tucholke 1983; see Figs 10, 11 and 15), by a reorganization of deep-sea benthic foraminifer communities (Schnitker 1979), and by a marked eustatic sea-level drop (Vail *et al.* 1977). These oceanographic events cannot be correlated with regional tectonic events. They can be related, however, to climatic cooling, locally represented by an upsection decrease in smectite and quartz enrichment at Site 548 (recording a climatic change on the adjacent land-mass; Cheunaux *et al.* 1985). This cooling was associated with an expansion of the Antarctic ice cap (Savin *et al.* 1975; Berger *et al.* 1981). The induced thermal gradient may have caused vigorous bottom circulation responsible for the sea-floor erosion.

Conclusions

The oceanographic evolution of the south-western continental margin of Ireland and the adjacent Porcupine Abyssal Plain was marked initially (early to middle Barremian) by a rapid transition from shallow hypohaline to open-marine, outer sublittoral environments. This transition occurred during the rifting that preceded opening of the North Atlantic between Ireland and the Grand Banks. The authors' data indicate that the transition from shallow to deep pelagic circulation took probably no longer than 3–4 m.y.; distensional rifting was not preceded by uplift or regional doming, as it was in the Rockall area.

Oceanic changes during the post-rift phase, beginning in the early Albian, were probably

FIG. 9. (opposite) Changes in late Palaeocene–early Eocene circulation across the Goban Spur. This model of palaeoceanic circulation shows the progressive invasion of cool northern water through the deeper parts of the young ocean during the late Palaeocene and early Eocene. Simultaneously, compressive deformation in the underlying basement and uplift of the adjacent landmass induced an influx of detrital minerals, which ended in the early Eocene–Oligocene interval, cool water predominated and white chalk containing a characteristic foraminiferal and nannofossil association was laid down on the continental slope. On the abyssal plain, several superimposed metalliferous crusts, rich in Mn, Ba, Ni, Co, and Cu, were deposited together with volcanic ashes and zeolites. Water mass stratification is documented by the preservation of warm planktonic foramin feral species reworked from shallower depths; this lasted until the end of the early Eocene (Zone NP 12).

Environments were favourable to the preservation of organic compounds in the late Palaeocene at Site 550. The progressive strengthening of cool, deep circulation by overflow of Norwegian Sea bottom waters, caused condensed sedimentation in oxidizing conditions and encrustation of metallic oxides at deepest sites.

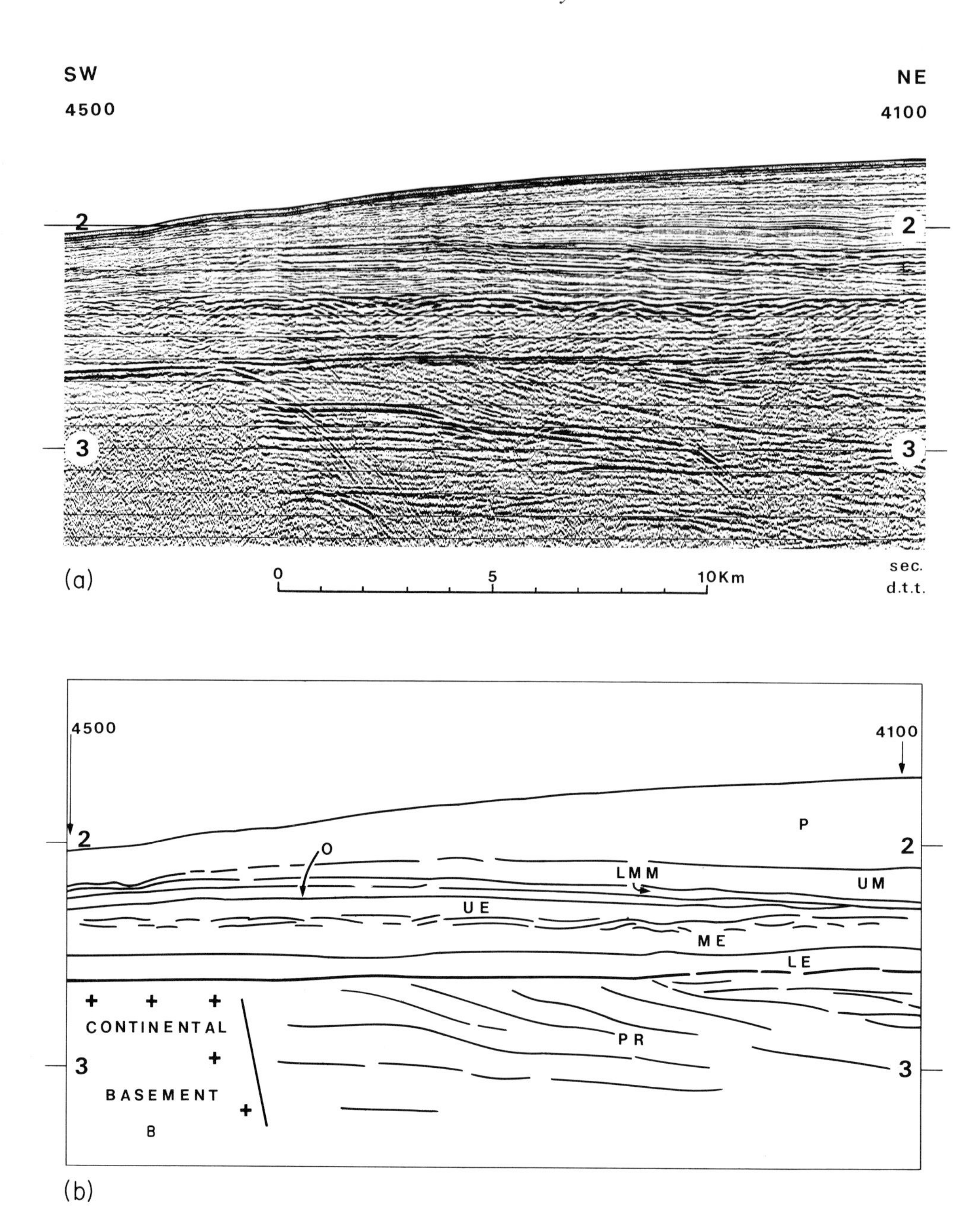

FIG. 10. Migrated seismic reflection profile at mid-slope on the Goban Spur (a). (b) is an interpretation of (a). B: Hercynian basement; RP: (?) prerift sequence; LE: lower Eocene; ME: middle Eocene; UE: upper Eocene; O: Oligocene; LMM: lower-middle Miocene; UM: upper Miocene; P: Pliocene to Recent. Unpublished high resolution profile IFP OC 601 located at 10 on Fig. 2. Note the broad extent of the angular unconformity at the base of the lower Eocene sequence, and the distinct acoustic reflector between lower middle Miocene and upper Miocene.

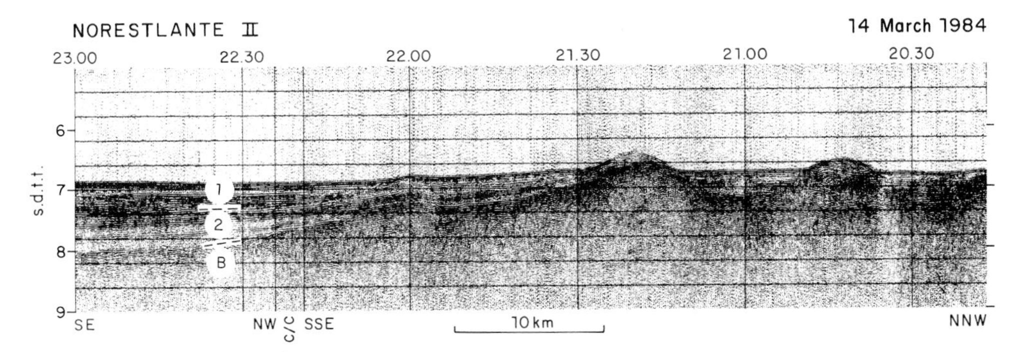

FIG. 11. Single-channel seismic reflection profile in the vicinity of the ocean/continent boundary south-west of the Goban Spur. The west–south-west–east–north-east trending line of hills developed at the foot of the Goban Spur during the Pyrenean orogenic phase dated as late Palaeocene–Eocene. They have not been buried by subsequent deposition. Unpublished profile from a recent cruise of the R.V. *Jean Charcot*, located at 11 on Fig. 2. Acoustic unit 1 is late Oligocene to Recent; unit 2 is late Albian to Eocene; B is oceanic basement.

largely the result of interaction between southern and northern water masses. During middle Cretaceous time in the north-eastern Atlantic, there was no long period of restriction or anoxia such as that documented on the African margin or in the South Atlantic. During the middle and late Cenomanian, however, partial anoxia developed at depths greater than 1800–2000 m (Site 550; reconstructed palaeodepth) on the Irish margin. This anoxia was cyclically interrupted by low-energy bottom currents bringing oxic waters into a generally quiet environment. Later, the short-lived, but ocean-wide Cenomanian–Turonian anoxic event (a few hundred-thousand years' duration) was characterized by a temporary oxygen minimum layer at mid-depths. For the remainder of the Cretaceous, oceanic circulation was strong enough to allow continuous oxygenation of the water column.

During the Cenozoic, cool, oxic, northern bottom waters repeatedly invaded the Goban Spur region, commencing in the late Palaeocene. Concurrently, north–south compressional tectonism was taking place in the North Sea, Rockall Plateau and Norwegian Sea areas, accompanied by explosive volcanism.

The Eocene/Oligocene transition coincided with the onset of polar bottom water production, and surface waters were dramatically cooled near the Oligocene/Miocene transition. Between the middle and late Miocene, a global climatic and oceanographic disturbance, related to rapid growth of the Antarctic ice cap and lowered sea-level, caused temporary stagnation at shallow depths on the Goban Spur (shallower than ~1200 m).

The major unconformities documented at the Goban Spur can be correlated with regional tectonic movements and with global climatic and oceanographic events, such as shifts in sea-level, in deep circulation patterns, or in the location of the CCD. All these mechanisms appear to have been linked together in a delicately balanced relationship. Any general explanation of the phenomena that has been documented or inferred, must take all these causal mechanisms into account.

ACKNOWLEDGEMENTS: The authors are grateful to P. Loubere, J.M. Mazzullo, K. Otsuka, L.A. Reynolds, H.A. Townsend, S.P. Vaos, B. Ancel, J.M. Auzende, M. Bourbon, M.H. Caralp, J. Duprat, S.L. Folami, J.P. Foucher, E. Gillot, P. Guennoc, P.M. Hunter, M. Labracherie, D. Low, B. Mathis, K.G. Miller, L. Pastouret, C. Pujol, J.F. Saliege, and C. Vergnaud Grazzini, who analysed the interpreted various data sets from coring, dredging, and geophysical surveying of the Goban Spur and vicinity. John S. Schlee and Page C. Valentine kindly reviewed the manuscript.

These results were made possible largely through support of the Deep Sea Drilling Project, Scripps Institution of Oceanography and the United States National Science Foundation.

References

BERGER, W.H., VINCENT, R.E. & THIERSTEIN, H.R. 1981. The deep sea record: major steps in Cenozoic ocean evolution. *In*: Warm, J.E., Douglas, R.G. & Winterer, E.L. (eds), *The Deep Sea Drilling Project: A Decade of Progress. Soc. Econ. Paleontol. Min. Spec. Publ.*, 32. 489–504.

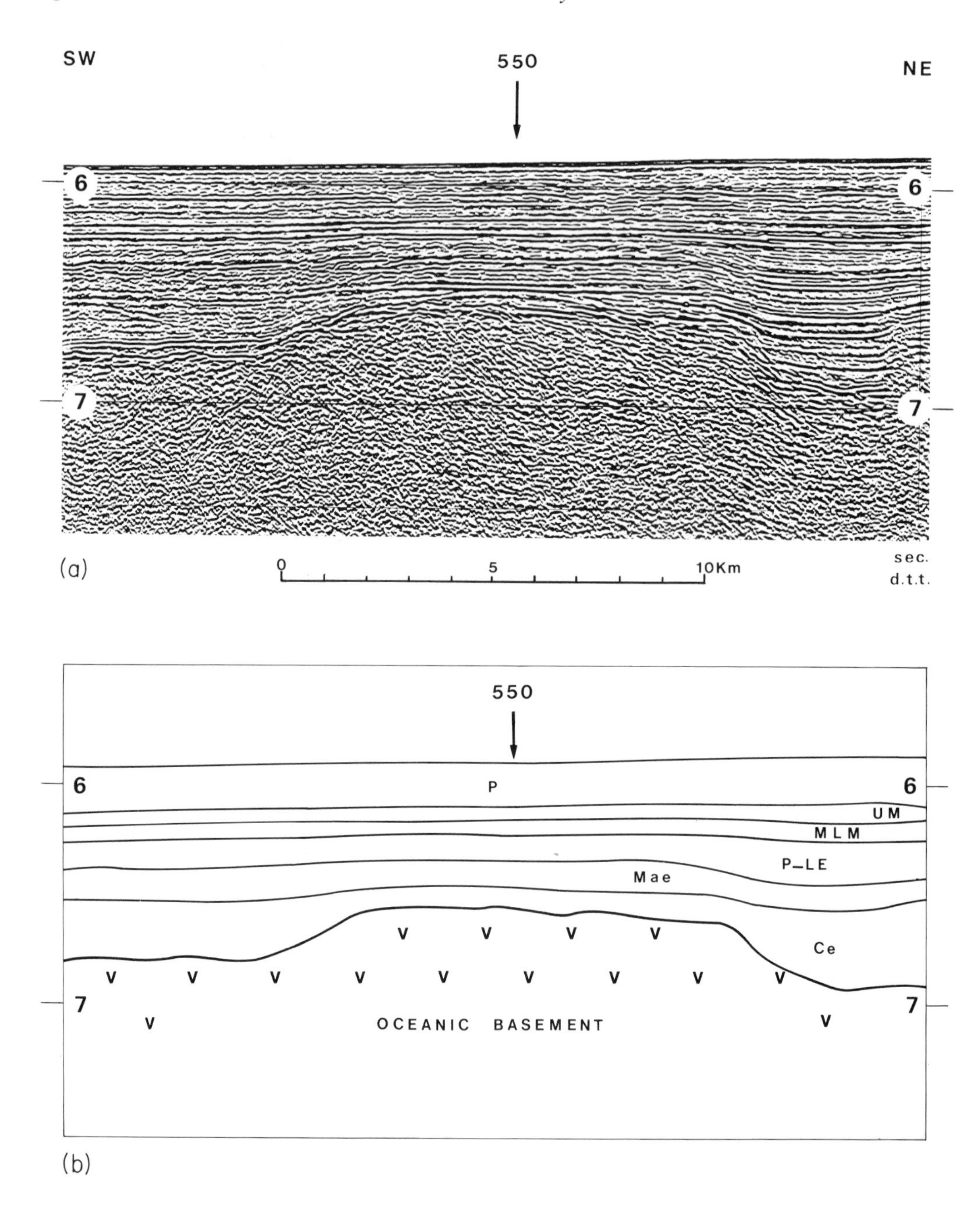

FIG. 12. Migrated seismic reflection profile through Site 550 on the Porcupine Abyssal Plain (a). (b) is an interpretation of (a). Ce: Cenomanian carbonaceous mudstone; Mae: Maestrichtian turbiditic sequence; P–LE: Palaeocene–lower Eocene; MLM: middle and lower Miocene; UM: upper Miocene chalks; P: Pliocene to Recent. Unpublished high-resolution profile IFP OC 601, located on Fig. 2. Strong reflectors record the main unconformities at Site 550: (1) unconformity between Cenomanian carbonaceous siltstones and uppermost Cretaceous chalks, (2) late Palaeocene unconformity, (3) unconformity between lower Eocene siliceous chalks and Miocene chalks, (4) intra-Miocene unconformity, and (5) uppermost Miocene unconformity.

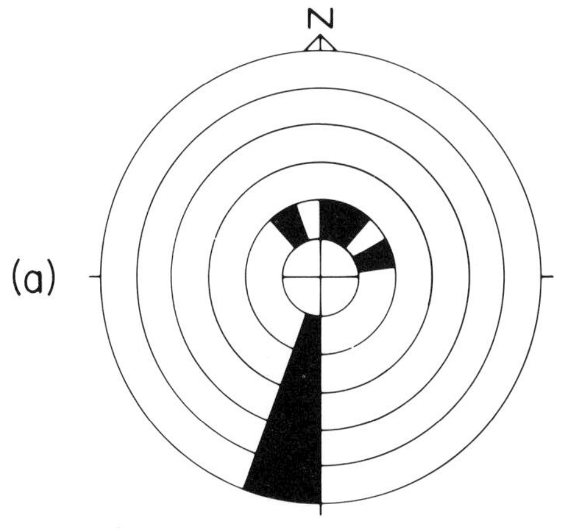

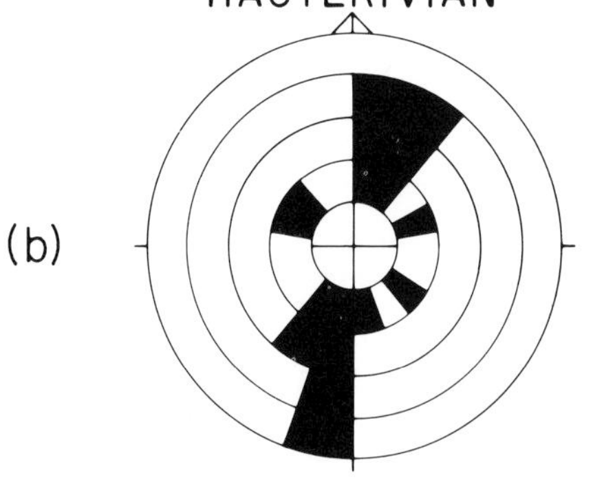

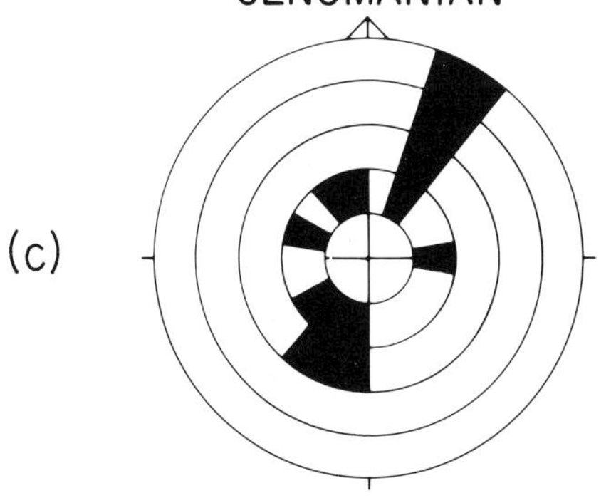

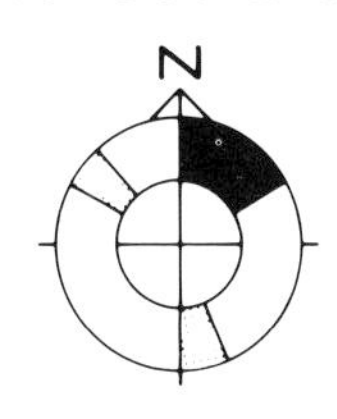

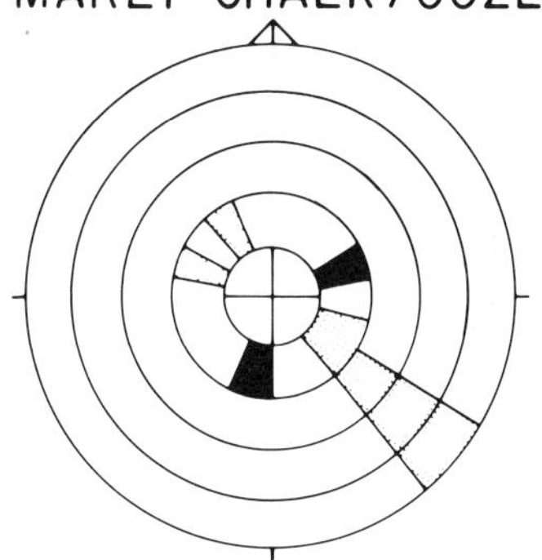

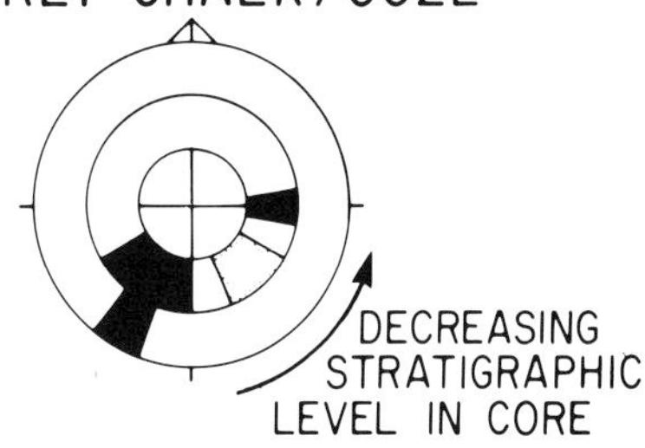

FIG. 13. Rose diagrams showing azimuthal distribution of grain long-axes and their variation with time (from Hailwood & Folami 1985), inferred from directions of maximum magnetic susceptibility.

CHENNAUX, G., ESQUEVIN, J., JOURDAN, A., LATOUCHE, C. & MAILLET, N. 1985. X-ray mineralogy and mineral geochemistry of Cenozoic strata (Leg 80) and petrographic study of associated pebbles. *In*: GRACIANSKY, P.C. DE, POAG, C.W. *et al.*, Init. Repts. DSDP **80.** US Govt. Print. Off., Washington, DC. 1019–46.

EVANS, C.D.R., LOTT, E.K. & WARRINGTON, G. (compilers) 1981. The Zephyr (1977) wells, South-western Approaches and western English Channel. *Inst. Geol. Sci., Rept.* **81/1,** 44 p.

GEORGE, T.N. 1967. Landform and structure in Ulster. *Scottish J. Geol.* **3,** 413–48.

GILLOT, E. 1983. *La Marge Celtique au Cretacé d'aprés la Campagne 80 du DSDP–IPOD (Atlantique NE).* These 3eme cycle, Fac. Sc. Univ. Dijon, France. 198 p.

GRACIANSKY, P.C. DE & GILLOT, E. 1985. Sedimentological study of mid-Cretaceous carbonaceous limestones at Sites 549 and 550, northeast Atlantic. *In*: Graciansky, P.C. de, Poag, C.W. *et al., Init. Repts* DSDP **80.** US Govt. Print. Off. Washington, DC. 885–97.

GRACIANSKY, P.C. DE, BROSSE, E., DEROO, G., HERBIN, J.P., MONTADERT, L., MÜLLER, C., SIGAL, J. & SCHAAF, A. 1985. Organic rich sediment and palaeoenvironmental reconstructions of the Cretaceous North Atlantic. *In*: BROOKS, J. & FLEET, A. (eds), *Marine Petroleum Source Rocks*. Spec. Publ. Geol. Soc. London, Blackwell Scientific Publication, Oxford.

GRACIANSKY, P.C. DE, DEROO, G., HERBIN, J.P., MONTADERT, L., MÜLLER, C., SCHAFF, A. & SIGAL, J. 1984. Ocean-wide stagnation episode in the Late Cretaceous. *Nature* **308,** 346–9.

HAILWOOD, E.A. & FOLAMI, S.L. 1985. Magnetic fabric of Quaternary, Tertiary, and Cretaceous sediments from Goban Spur, Leg 80: implications for sediment transport processes. *In*: GRACIANSKY, P.C. DE, POAG, C.W. *et al.*, Init. Repts DSDP **80,** US Govt. Print. Off., Washington, DC. 415–21.

HANCOCK, J.M. & KAUFFMAN, E.G. 1979. The great transgressions of the Late Cretaceous. *J. Geol. Soc., London*, **136,** 175–86.

HAQ, B.U. 1981. Paleogene palaeoceanography: early Cenozoic oceans revisited. *Proceed. 26th Internat. Geol. Congr., Oceanol. Acta. Suppl.* **4,** 33–44.

JAQUÉ, M. & THOUVENIN, J. 1975. Lower Tertiary tuffs and volcanic activity in the North Sea. *In*: Woodland, A.W. (ed.), *Petroleum and the Continental Shelf of Northwest Europe*, Volume 1, *Geology*. Applied Science Publishers, London. 445–65.

KARPOFF, A.M., BOURBON, M., ANCEL, B. & GRACIANSKY, P.C. DE 1985. Diagenetic polymetallic crusts at Sites 550 and 548 of Leg 80, northeastern Atlantic Ocean. *In*: GRACIANSKY, P.C. DE & POAG, C.W., *Init. Repts. DSDP* **80.** US Govt. Print. Off., Washington, DC. 823–44.

KNOX, R.W.O'B & MORTON, A.C. 1983. Stratigraphical distribution of the early Paleogene pyroclastic deposits in the North Sea Basin. *Proc. Yorkshire Geol. Soc.* **44,** 355–63.

KNOX, R.W.O'B. 1985. Stratigraphic significance of volcanic ash in Paleocene and Eocene sediments at Sites 549 and 550. *In*: GRACIANSKY, P.C. DE, POAG, C.W. *et al., Init. Repts DSDP* **80.** US Govt. Print. Off., Washington, DC. 845–50.

LAUGHTON, A.S., BERGGREN, W.A., *et al.* 1972. *Init. Repts. DSDP* **12.** US Govt. Print. Off., Washington DC.

MASSON, D.G., MONTADERT, L. & SCRUTTON, R.A. 1985. Regional geology of the Goban Spur continental margin. *In*: GRACIANSKY, P.C. DE, POAG, C.W. *et al., Init. Repts. DSDP* **80.** US Govt. Print. Off., Washington. 1141–53.

MASSON, D.G. & PARSON, L.M. (1983. Eocene deformation on the continental margin SW of the British Isles. *J. Geol. Soc. London*, **140,** 913–20.

MILLER, K.G. & TUCHOLKE, B.E. 1983. Development of Cenozoic abyssal circulation south of Greenland-Scotland Ridge. *In*: BOTT, M., SAXON, S., TALWANI, M. & THIEDE, J. (eds), *Structure and Development of the Greenland-Scotland Ridge*. Plenum Press, New York. 549–89.

MONTADERT, L., DE CHARPAL, O., ROBERTS, D.G., GUENNOC, P. & SIBUET, J.C. 1979. Northeast Atlantic continental margins: rifting and subsidence processes. *In*: TALWANI, M., HAY, W.W. & RYAN, W.B.F. (eds), *Deep Drilling Results in the Atlantic Ocean: Continental Margins and Palaeoenviroments.* American Geophysical Union, Washington, DC. 154–86.

MONTADERT, L., ROBERTS, D.G. *et al.* 1979. *Init. Repts. DSDP* **48.** US Govt. Print. Off. Washington, DC.

MÜLLER, C. 1985. Biostratigraphic and palaeoenvironmental interpretation of the Goban Spur region based on a study of calcareous nannoplankton. *In*: GRACIANSKY, P.C. DE, POAG, C.W. *et al., Init. Repts. DSDP* **80,** US Govt. Print. Off., Washington, DC. 573–99.

NAYLOR, D. & SHANNON, P.H. 1982. *The Geology of Offshore Ireland and West Britain.* Graham & Trotman, London. 161 p.

POAG, C.W. & LOW, D. 1985. Environmental trends among Neogene benthic foraminifers at Deep Sea Drilling Project Site 548, Irish continental margin. *In*: GRACIANSKY, P.C. DE, POAG, C.W. *et al., Init. Repts. DSDP* **80.** US Govt. Print. Off., Washington, DC. 489–503.

POAG, C.W., REYNOLDS, L.A., MAZZULLO, J.M. & KEIGWIN, L.D., JR. 1985. Foraminiferal, lithic, and isotopic changes across four major unconformities at Deep Sea Drilling Project Site 548, Goban Spur. *In*: GRACIANSKY, P.C. DE, POAG, C.W. *et al., Init. Repts. DSDP* **80.** US Govt. Print. Off., Washington, DC. 539–55.

PRATT, L.M. 1984. Influence of paleoenvironmental factors on preservation of organic matter in middle Cretaceous Greenhorn Formation, Pueblo, Colorado. *Am. Ass. Petrol. Geol. Bull.* **68,** 1146–59.

RAT, P., GILLOT, E., MAGNIEZ, F. & PASCAL, A. 1985. Paleoenvironmental study of Barremian-Albian sediments at Deep Sea Drilling Project Site 549 in the eastern North Atlantic. *In*: GRACIANSKY, P.C. DE, POAG, C.W. *et al., Init. Repts. DSDP* **80,** US Govt. Print. Off., Washington, DC. 905–25.

ROBERTS, D.G., MASSON, D.G., MONTADERT, L. & DE CHARPAL, O. 1981. Continental margin from the Porcupine Seabight to the Armorican marginal basin. *Proc. Conf. Petrol. Cont. Shelf Northwest Europe*, 455–73.

ROBINSON, K.W., SHANNON, P.M. & YOUNG, D.G.D. 1981. The Fastnet Basin: an integrated analysis. *Proc. Conf. Petrol. Geol. Cont. Shelf Northwestern Europe*, 444–54.

ROSSIGNOL-STRICK, M., NESTEROFF, W., OLIVE, P. & VERGNAUD-GRAZZINI, C. 1982. After the deluge: Mediterranean stagnation and sapropel formation. *Nature* **295,** 105–10.

RUDDIMAN, W.F. 1972. Sediment redistribution on the Reykjanes Ridge, seismic evidence. *Geol. Soc. Am. Bull.* **83,** 2039–69.

SAVIN, S.M., DOUGLAS, R.G. & STEHLI, F.G. 1975. Tertiary marine paleotemperatures. *Geol. Soc. Am. Bull.* **86,** no. 11, p. 1499–1510.

SCHNITKER, D. 1979. Cenozoic deep water benthic foraminifers, Bay of Biscay. *In*: MONTADERT, L., ROBERTS, D.G. *et al.*, *Init. Repts. DSDP* **48.** US Govt. Print. Off., Washington, DC. 377–89.

SCHWARZACHER, W. & FISCHER, A.G. 1982. Limestone-shale bedding and perturbations of the Earth's orbit. *In*: EINSELE, G. & SEILACHER, A. (eds), *Cyclic and Event Stratification*. Springer Verlag, Berlin. 72–95.

SIBUET, J.C., RYAN, W.B.F. *et al.* 1979. *Init. Repts. DSDP* **47B.** US Govt. Print. Off., Washington, DC. 787 p.

SIBUET, J.C., MATHIS, B., PASTOURET, L., AUZENDE, J.M., FOUCHER, J.P., HUNTER, P.M., GUENNOC, P., GRACIANSKY, P.C. DE, MONTADERT, L. & MASSON, D.G. 1985. Morphology and basement structures of the Goban Spur continental margin (northeastern Atlantic) and the role of the Pyrenean orogeny. *In*: GRACIANSKY, P.C. DE, POAG, C.W. *et al.*, *Init. Repts. DSDP* **80.** US Govt. Print. Off., Washington, DC. 1153–65.

SIBUET, J.C., MATHIS, B. & HUNTER, P. 1984. La ride Pastouret (plaine abyssal de Porcupine): une structure Eocene. *CR Acad. Sci Paris*, **299,** II, 20, 1391–6.

SNYDER, S.W., MÜLLER, C., SIGAL, J., TOWNSEND, H. & POAG, C.W. 1985. Biostratigraphic, paleoenvironmental, and paleomagnetic synthesis of the Goban Spur region, Deep Sea Drilling Project, Leg 80. *In*: GRACIANSKY, P.C. DE, POAG, C.W. *et al.*, *Init. Repts. DSDP* **80.** US Govt. Print. Off., Washington, DC. 1169–86.

VAIL, P.R. & HARDENBOL, J. 1979. Sea-level changes during the Tertiary. *Oceanus*, **22,** 71–9.

VAIL, P.R., MITCHUM, R.M., JR. & THOMPSON, S. III. 1977. Seismic stratigraphy and global changes of sea level, Part 4: global cycles of relative changes of sea level. *In*: PAYTON, C.E. (ed.), *Seismic Stratigraphy—Applications to Hydrocarbon Exploration*. Am. Ass. Petrol. Geol. Mem. 26. 83–97.

WAPLES, D.W. 1985. A reappraisal of anoxia and richness of organic material, with emphasis on the Cretaceous North Atlantic. *In*: GRACIANSKY, P.C. DE & POAG, C.W. *et al.*, *Init. Repts. DSDP* **80.** US Govt. Print. Off., Washington, DC. 999–1016.

WAPLES, D.W. & CUNNINGHAM, R. 1985. Leg 80 shipboard organic geochemistry. *In*: GRACIANSKY, P.C. DE, POAG, C.W. *et al.*, *Init. Repts. DSDP* **80.** Govt. Print. Off., Washington, DC. 949–68.

P.C. DE GRACIANSKY, Ecole Nationale Superieure des Mines, Paris, France
C. WYLIE POAG, US Geological Survey, Woods Hole, Massachusetts, USA
E.A. HAILWOOD, Institute of Oceanography, University of Southampton
R.W. O'B. KNOX, British Geological Survey, Nottingham
D.G. MASSON, Institute of Oceanographic Sciences, Wormley, Godalming, Surrey GUB SUB
L. MONTADERT & C. RAVENNE, Institut Francais du Pétrole, Rueil- Malmaison, France
C. MULLER, Geologisch-Paläontologisches Intitute, Johann-Wolfgang-Goethe Universität, Frankfurt/Main, Federal Republic of Germany
J.C. SIBUET, Centre Océanologique de Bretagne, Brest, France
J. SIGAL, Vincennes, France
S.W. SNYDER, Department of Geology, East Carolina University, Greenville, North Carolina, USA
D.W. WAPLES, Resource Reconnaisance Incorporated, Dallas, Texas, USA
R. CUNNINGHAM, Exxon Production Research Company, Houston, Texas, USA

Mesozoic–Cenozoic clastic depositional environments revealed by DSDP Leg 93 drilling on the continental rise off the eastern United States

Sherwood W. Wise, Jr., Jan E. Van Hinte, Gregory S. Mountain, Brian N.M. Biart, J. Mitchener Covington, Warren S. Drugg, Dean A. Dunn, John Farre, Daniel Habib, Malcolm B. Hart, Janet A. Haggerty, Mark W. Johns, Thomas H. Lang, Philip A. Meyers, Kenneth G. Miller, Michel R. Moullade, Jay P. Muza, James G. Ogg, Makoto Okamura, Massimo Sarti & Ulrich von Rad

SUMMARY: Prior to Deep Sea Drilling Project (DSDP) Leg 93 (1983), drill data along the continental rise of the Atlantic margin of the United States were quite limited compared to those of the adjacent continental shelf and the deeper, more seaward expanses of the North American Basin. Interpretations of the geologic history and of processes that controlled sedimentation along the rise were strongly dependent on studies of seismic reflection profiles. DSDP Leg 93 drilled deep holes on both the lower and upper rise, allowing correlation with commercial wells on land and offshore (as well as with subsequent DSDP Leg 95 holes along the 'New Jersey Transect') and providing the first down dip suite of drill holes across a passive continental margin from the coastal plain to the abyssal plain.

Site 603 on the lower rise 270 miles east of Cape Hatteras was cored nearly continuously over 1585 m to Berriasian pelagic limestones. It intersected an extensive Lower Cretaceous deep-sea fan complex which provides new information on the petroleum potential of the rise. Hauterivian to early Aptian in age, this 208 m interval of interbedded limestones, sand and blackshale turbidites calls into question the existence of any post Valanginian reefs along the Baltimore Canyon Trough. Less extensive terrigenous turbidites were encountered as far up in the section as the Cretaceous-Tertiary (K–T) boundary. The K–T boundary is marked by a current-laminated sand rich in dark, 1 mm diameter spherules which may denote an extraterrestrial impact event.

DSDP Sites 604 and 605 on the upper rise, the first along the 'New Jersey Transect', are located some 100 miles south-east of Atlantic City, New Jersey. Hole 605, drilled 816.7 m down to mid-Maestrichtian limestones, penetrated a near complete Cretaceous–Tertiary boundary section, above which 20 m of lower Palaeocene are separated by a disconformity from an expanded 175 m Palaeocene sequence. Terrigenous silts and glauconite at the K–T boundary and immediately above suggest either significant sea-level change, increased current erosion along the adjacent shelf and slope, increased terrigenous input caused by decreased vegetation, a high energy event (tsunami?), or some combination of these possible factors.

Site 604, 3 miles seaward of Site 605, was terminated by unstable hole conditions at 294.5 m within a unit of Miocene glauconitic sands and debris flows. Emplaced largely during the Tortonian (8.2–10.0 Ma; Vail cycle TM3.1), these upper Miocene sediments contain shelf-derived gravels, exotic blocks of Eocene chalk (up to 50 cm across) eroded from the adjacent slope, and clasts of middle and upper Miocene carbonates or silts derived from canyon walls or shallow water strata upslope. Study of closely spaced, high resolution seismic profiles suggests that large-scale regional erosion (canyon cutting), which is related to the debris flows, began during the late middle Miocene.

On the lower rise, turbiditic silts and clays began to accumulate rapidly during the middle Miocene. Under the influence of a strengthening Western Boundary Undercurrent, these were deposited as muddy contourites in antidune-like sediment waves which, at site 603, grew rapidly with no appreciable break in sedimentation until at least early Pleistocene times to form the present Lower Continental Rise Hills of the Hatteras Outer Ridge (HOR). The somewhat elevated edge of the Lower Continental Rise Terrace formed as a natural levee behind which the coarser portions of the terrigenous turbidites were ponded to form the terrace. No coarse clastics that bypassed the pond were deposited with the clays of the HOR at this locality.

Throughout the study, seismic sequence boundaries of the upper and lower continental rise were calibrated and correlated with continental margin unconformities as well as with deep sea reflection horizons.

From Summerhayes, C.P. & Shackleton, N.J. (eds), 1986, *North Atlantic Palaeoceanography*, Geological Society Special Publication No. 21, pp. 35–66.

In their comprehensive review and synthesis of the geologic history of the US Atlantic continental margin, Mountain & Tucholke (1985) interpret sedimentation in terms of the interplay of three main processes: (1) pelagic sedimentation, (2) down slope (or cross slope) detrital sedimentation and erosion, and (3) along slope (or 'contour') detrital sedimentation and erosion. As outlined in their Fig. 8.1 (cf. our Fig. 20) and reviewed briefly here, there is a hierarchy of controls over these processes. In shallow marine environments, changes in sea-level, riverine input, reefs, faulting, slumping, and local diapirism are among the dominant controls, whereas along the lower continental rise these factors are often overshadowed by abyssal currents and vertical fluctuations in the calcite compensation depth (CCD). The balance among these various controls shifts when traversing from shallow-marine to deep-marine environments. Along the continental slope and rise, this balance has also shifted through time during the evolution of the North American Basin. Specifically, those controlling factors which produced turbidity currents and regulated pelagic sedimentation were most influential on Mesozoic and Palaeogene sedimentation patterns, whereas after the Eocene, those factors which generated contour-following bottom currents have played an equally important, if not dominant, role.

In general, sedimentation along the continental rise has been stongly affected by both shallow and deep marine processes, thus its history is relatively complex. Until Deep Sea Drilling Project Leg 93, however, drill data along the rise of the US Atlantic margin were limited compared to those of the continental shelf and the deeper, more seaward expanses of the North American Basin, particularly for pre-Neogene sediments. Past interpretations, therefore, have been strongly dependent on studies of seismic reflection profiles, such as those presented by Schlee *et al.* (1976), Tucholke & Mountain (1979), Klitgord & Grow (1980), Hutchinson *et al.* (1983), Ewing & Rabinowitz (1984), and Mountain & Tucholke (1985). In order to narrow this data gap, DSDP Leg 93 initiated a series of deep holes on both the lower and upper continental rise which was continued by DSDP Leg 95. The Leg 93 results, which are being made available as expeditiously as possible (see Van Hinte *et al.* 1983; Leg 93 Scientific Party/Leg 94 Scientific party, 1984; von Rad *et al.* 1984; Van Hinte *et al.* in 1985a, b), have expanded our understanding of the geologic history and the processes which have shaped the North Atlantic continental margin.

The purpose of this paper, which was written midway through the investigations leading to the publication of the Deep Sea Drilling Project (DSDP) Leg 93 *Initial Reports* volume (Van Hinte, Wise *et al.*, in prep.), is to trace the history of clastic sedimentation beneath the continental rise as revealed by our drill cores. This we do in the context of the backdrop provided by the Mountain & Tucholke (1985) regional synthesis as well as in terms of the principal controls on sedimentation and erosion which they discuss. Because the authors' results at this writing are preliminary, some changes in the interpretations presented here, particularly in the recognition of hiatuses and in the ages assigned to some sedimentary units, should be expected by the time the DSDP Leg 93 *Initial Reports* are published.

In this discussion, the long (1576 m) section cored beneath the lower continental rise at Site 603 (Fig. 1) is described first, concentrating primarily on the rather persistent Cretaceous turbidites recovered there. Next the Cretaceous–Tertiary boundary and Palaeogene sections recovered on the upper rise at Site 605 will be discussed, before moving on to the upper Neogene debris flows encountered at nearby Site 604. Last, the authors will return to Site 603 to describe the Neogene sediment drift that underlies the Lower Continental Rise Hills. This narrative is based on the authors' unpublished *Preliminary Report* written aboard ship and updated at the postcruise meeting of May, 1984. Parts of that *Preliminary Report* appear elsewhere (see references above) and are repeated here with only slight modifications in wording in order to maintain better scientific consistency among the various progress reports. Additional information contributed by some shore-based investigators is also included, particularly that pertaining to the debris flows at Site 604 and the K–T boundary at Sites 603 and 605.

Site 603, lower continental rise

Site 603 consisted of three holes drilled in 4634–4639 m of water on the lower continental rise 270 miles east of Cape Hatteras, North Carolina (Fig. 1; Table 1). The principal scientific objectives were:

(1) to sample and identify the several prominent 'seismic reflectors' in the Mesozoic and Cenozoic sedimentary section (Beta, Km, A*, Au and X);
(2) to sample the upper sedimentary section by HPC (hydraulic piston corer) to understand active current-controlled depositional processes which predominated in the area during the late Cenozoic;
(3) to sample the oldest sediments deposited

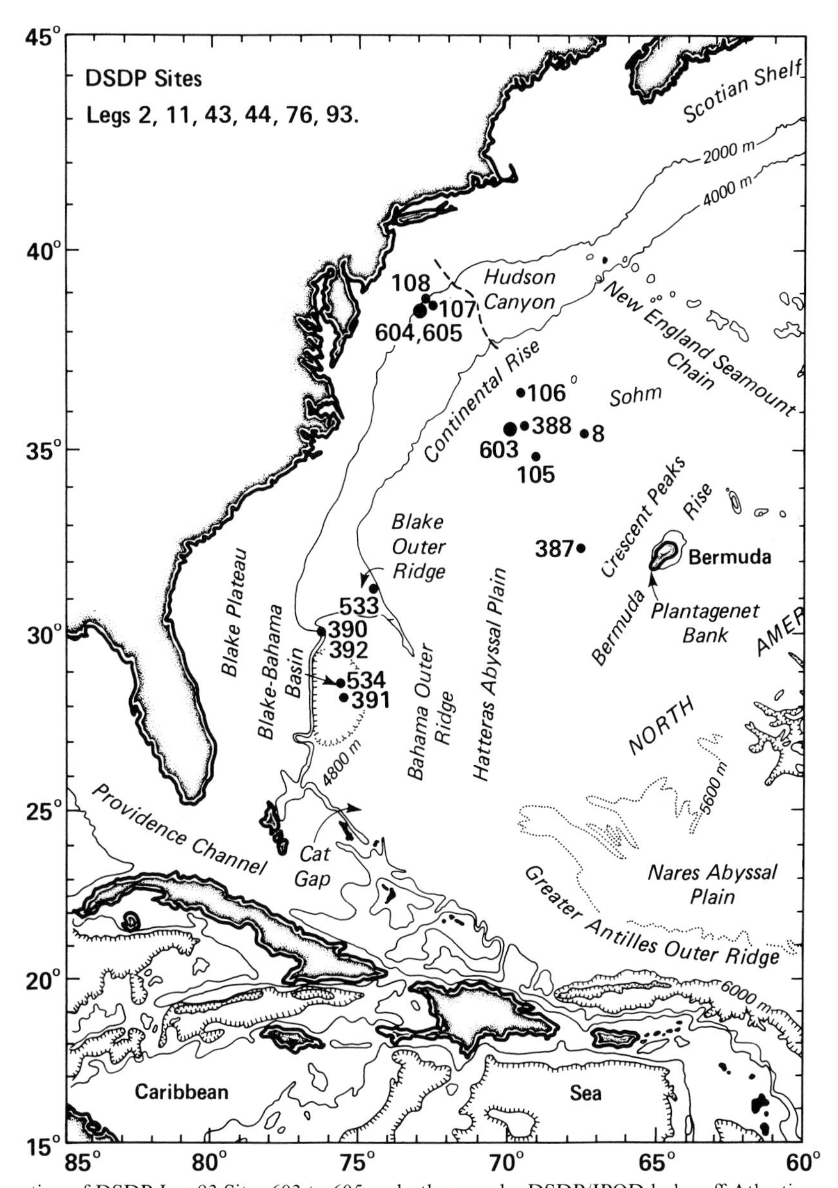

FIG. 1. Location of DSDP Leg 93 Sites 603 to 605 and other nearby DSDP/IPOD holes off Atlantic margin of the United States.

on oceanic crust at a location landward of nearby DSDP Site 105 where Upper Jurassic sediments had been cored over a prominent basement high; and

(4) to sample basement.

Stratigraphy

The seismic sequences and lithologic units penetrated at Site 603 are depicted in Figs 2 and 3. Figure 4 provides a detailed view of the major lithologic components in the Cretaceous section. Despite facies differences due to the influx of terrigenous sediments from the nearby continent, general correlations can be made between the local lithologic units and the oceanic formations proposed for the North American Basin by Jansa *et al.* (1979) (see Fig. 3). Units IB to IVC are essentially devoid of calcareous material. In descending order, these units and their correlations are:

Unit I: 960 m of lower Pleistocene–lower/

TABLE I. *Site 603 coring summary*

Hole	Dates (1983)	Latitude	Longitude	Water depth*	Penetration	No of. cores	Metres cored	Metres recovered	Percent of recovery
603	5–11 May	35°29.66′N	70°01.70′W	4634.00 m	832.6 m	41 (13)	393.0	226.38	58
603A	11–12 May	35°29.69′N	70°01.69′W	4633.00 m	0.0 m	0	0.0	0.00	0
603B	12–31 May	35°29.71′N	70°01.71′W	4642.50 m	1585.2 m	75 (8)	683.6	484.44	71
603C	31 May–6 June	35°29.78′N	70°01.86′W	4639.50 m	366.0 m	40	366.0	314.44	86
						156 (21)	1442.6	1025.26	215

* Water depth from sea-level, as measured by drill string length.
Note: the numbers inside parentheses are additional cores obtained from washed intervals.

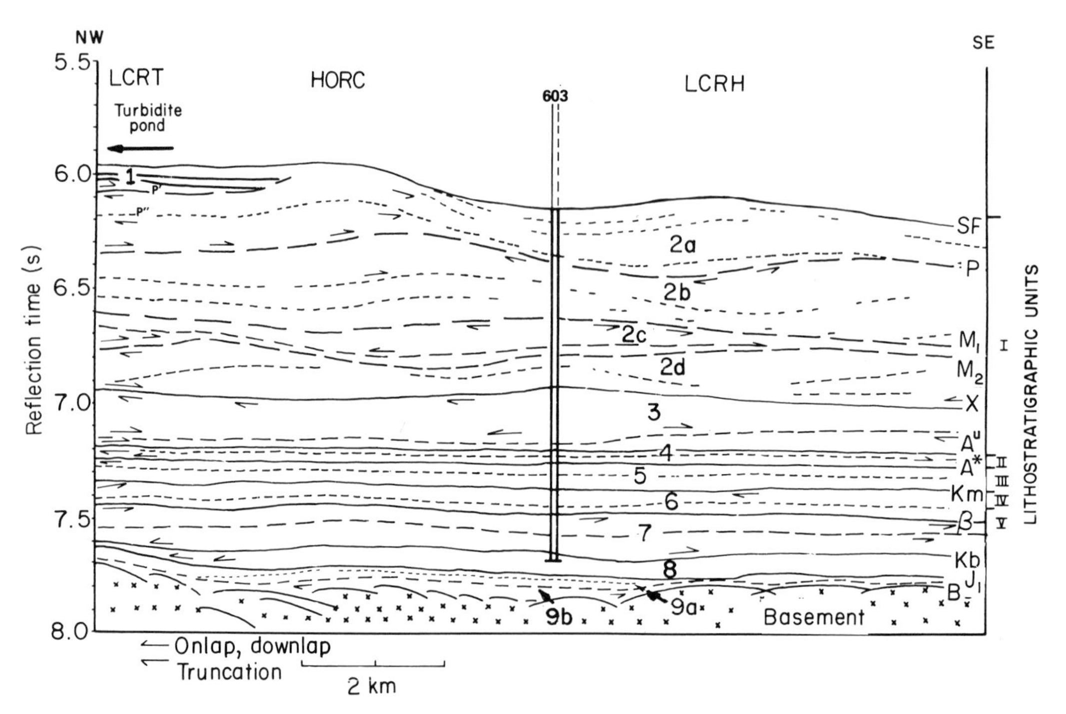

FIG. 2. Line drawing after multichannel seismic reflection profile Conrad 2101, Line 77 (after Van Hinte *et al.* 1985a). Units I to V are local lithostratigraphic units. Units 1 to 9 are local seismostratigraphic units; Units 1 (turbidite pond to the west) and 9 (the Jurassic) were not drilled during Leg 93. Reflection horizons (sequence boundaries) are shown on the right. LCRT = Lower Continental Rise Terrace. HORC = Hatteras Outer Ridge Crest. LCRH = Lower Continental Rise Hills.

middle Miocene hemipelagic clay and claystone (Blake Ridge Formation).

Unit I is subdivided on the basis of lithology and colour into four subunits. The dark green homogeneous muds of subunit IA contain nannofossils and planktonic foraminifers deposited above the calcite compensation depth. Greenish-grey muds of subunits IB and IC are largely devoid of calcareous microfossils except for the more dissolution resistant discoasters. Subunit IC contains radiolarians in some numbers at selected intervals as well as the first unambiguous evidence of sporadic silt and sand turbidites below 720 m (Core 603–44–2, 64–65 cm). In contrast to the green clays above, subunit ID is composed of yellowish-brown to brown clays barren of microfossils except for fish teeth which one of the authors (MBH) has found to be at least middle

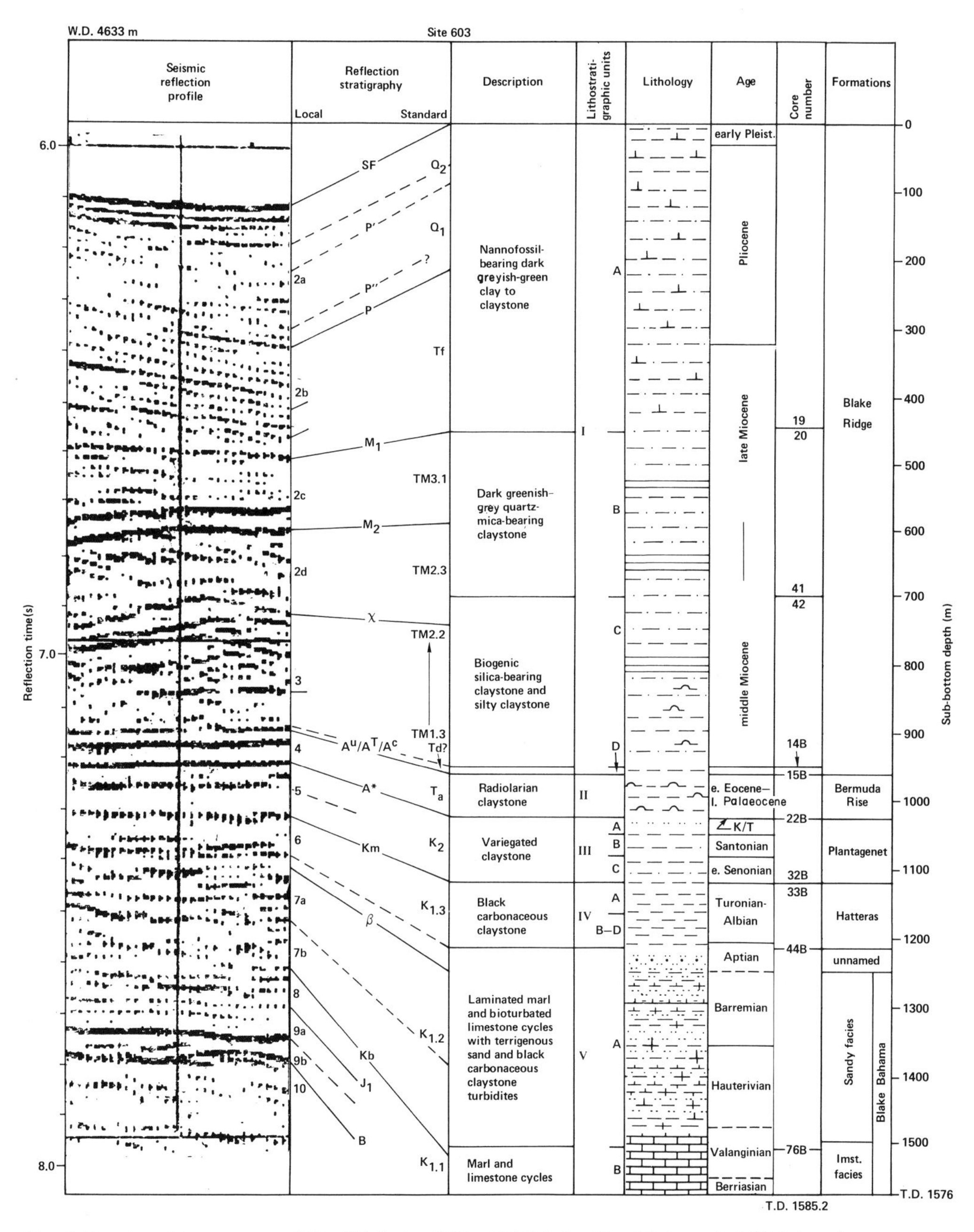

FIG. 3. Stratigraphic summary of Site 603 (holes 603 to 603C) (after Van Hinte *et al.* 1985a). Local lithostratigraphic and seismostratigraphic units numbered as in Fig. 2 above; standard seismic sequence notation after Vail *et al.* (1980).

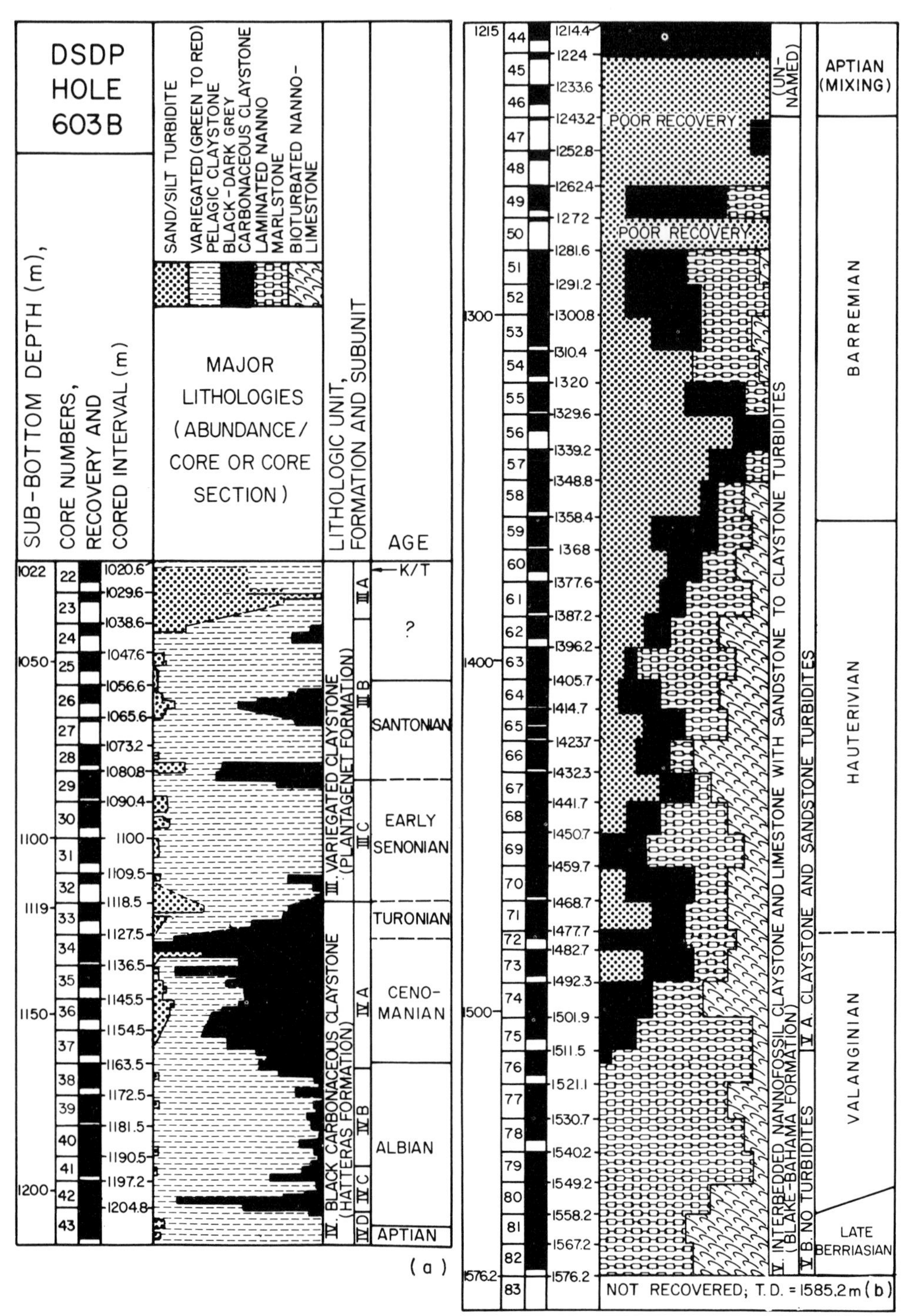

FIG. 4. Lithologies, lithologic units, oceanic formations, and ages for the Cretaceous section cored in Hole 603B. (a) Cores 22 to 43 (lithologic units III and IV). (b) Cores 44 to 83 (lithologic unit V).

Miocene in age in Hole 603B, and early Miocene in age in Hole 603E which was drilled by DSDP Leg 95 (Leg 95 Scientific party, 1984).

Unit II: 63 m of variegated lower Eocene and upper Palaeocene radiolarian claystone (Bermuda Rise Formation).

This unit is extremely colourful at the top, with hues ranging from pale green, greyish-green, reddish-brown, yellow-brown, to pale orange. Radiolarian content varies from 50% to less than 10%. Incipient diagenesis has converted some of the biogenic opal to opal-CT, but not in quantities exceeding 50% of the rock, therefore the sediment is not porcelainitic 'chert' in the strict sense. A lithium spike noted in interstitial water samples may be due to the dissolution of biogenic opal.

Unit III: 97 m of mostly Senonian (and lowest Palaeocene?) variegated claystone (Plantagenet Formation).

This unit is subdivided into three subunits based on the presence or absence of dark grey carbonaceous claystone. These organic-rich clays first appeared in subunit IIIB (Santonian) and were subsequently encountered in each of the underlying lithologic units (units IV and V) as well. Carbonaceous claystones are not present in sediments from subunit IIIA or IIIC. Nearly all cores of unit III contain 1–10% terrigenous silt and sand. Considered turbiditic in origin, these are also found in all underlying lithologic units. This comparatively landward locality, however, is the only DSDP site in which terrigenous clastics have been reported from the Plantagenet Formation.

Near the top of Unit III at sample interval 603B–22–3, 73 cm (1023.3 m) is a 4.5 cm thick current bedded sandy layer (Fig. 5(a)) which contains numerous black spherules each about 1 mm in diameter. One of the authors (JEVH) immediately suspected aboard ship that this spectacular bed marks the Cretaceous–Tertiary boundary impact event (Alvarez *et al.* 1980; Smit & Hertogen 1980).

Unit IV: 106 m of mid-Cretaceous (Aptian–Turonian) black carbonaceous claystones (Hatteras Formation).

This unit is subdivided on the occurrence or relative abundance of pelagic reddish-brown claystones which are present in subunits IVB and IVD. It contains the highest concentration of 'Black shales' (carbonaceous claystones) in the section. Wood fragments (charcoal) are common. Organic carbon contents range up to 13.6%. The organic geochemistry of cores from this and the other Cretaceous units is discussed in detail by Degens *et al.*, Meyers, and Rullkotter (this volume). Sharp basal contacts and graded silts at the bases of the black claystones suggest that they were emplaced as mud turbidites. Occasional well-defined silt turbidites (Bouma Tc) are also present (Fig. 5(b)).

Unit V: 261 m of upper Berriasian–Aptian interbedded nannofossil clays and limestones with sandstone to claystone turbidites (Blake-Bahama Formation plus an unnamed sand unit at the top of the sequence).

This unit is distinguished from Unit IV by the presence of abundant calcareous nannofossils. It is subdivided into two subunits based on the presence or absence of turbidites. Subunit VA includes 30 m of unconsolidated coarse sand at the top and is further characterized by abundant claystone, siltstone, and sandstone turbidites and debris flows (Fig. 5(c)). These are of two main types: siltstone-sandstone and organic matter-rich claystone. The presence of intermediate types and complete sandstone to claystone graded sequences indicates that the two textural types are related. Subunit VB is composed exclusively of *in situ* pelagic carbonates which consist of rhythmic alternations of laminated and bioturbated nannofossil chalks and limestones. These lithologies also occur in unit VA (Fig. 4(b)).

Cretaceous depositional history

The oldest sediments (Berriasian) recovered at Site 603 are characterized by carbonate cycles consisting of laminated marls that alternate with well bioturbated homogeneous chalks and limestones. Entirely pelagic, these represent alternating periods of oxic and anoxic bottom conditions, respectively, perhaps caused by periodic changes in the intensity of ocean circulation, upwelling and oxygen replenishment in the somewhat narrow, restricted incipient North Atlantic ocean basin.

It may be possible that the cycles reflect some type of oscillations of the northern edge of the equatorial belt of high productivity. Such oscillations could be caused by variations in the earth's orbital parameters. Work is currently underway by one of us (JGO) to determine if there is a periodicity among the limestone couplets which can be related to climate through such mechanisms as the Milankovitch cycle.

The pelagic limestone sequence is first interrupted in the upper Valanginian by organic-rich claystone turbidites, the first indication of down slope depositional processes in the section drilled. The turbidites contain exceptionally well preserved calcareous nannofossils which were probably derived from the slope or upper rise where deposition could have been taking place within a broad oxygen minimum zone. Whole ammonites

FIG. 5. Cretaceous cores from DSDP Hole 603B. (a) Black 1 mm diameter spherules in a 4.5 cm thick, current bedded sandy layer near the top of the Plantagenet Formation (unit III, 603B-22-3, 65–74.4 cm, 1023.3 m subbottom). This bed may record the Cretaceous–Tertiary boundary impact event. (b) Turbidite (Bouma interval Tc) from unit IVA (Hatteras Formation, Core 603B-36-3, Cenomanian) consists of laminated and current-ripple cross-laminated, medium grey siltsone bound within laminated black claystone (laminae expanded due to reaction of clay minerals with fresh water used to clean split core surface). (c) Core 603B-58 from lithostratigraphic unit VA (Blake Bahama Formation, Barremian). Grey laminated limestone ('L') and white bioturbated limestone cycles ('B') are interrupted by black carbonaceous claystone turbidites ('C') and terrigenous sandstone turbidites ('S'). 'W' = wood fragments (charcoal). Sandstone beds in Sections 58-2 and 58-4 exceed 90 cm in thickness. Whitish bases of some sandstone beds (example, Section 1, 65 cm) denote presence of calcite cement; otherwise, these sandstones are quite friable.

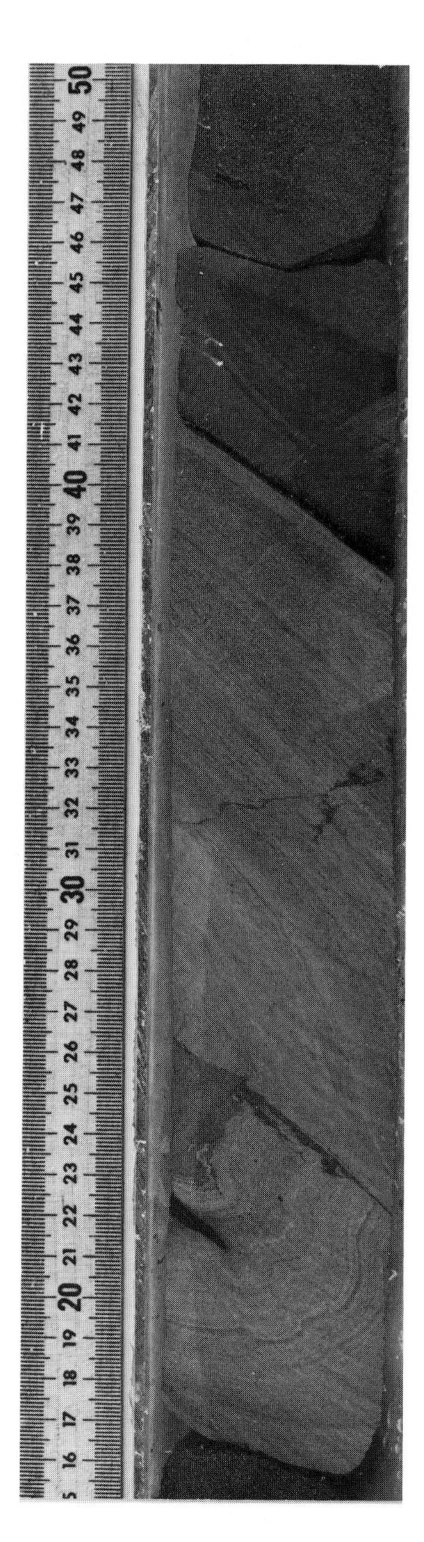

and aptychi from unsplit cores provide additional biostratigraphic control for this part of the section. One specimen collected aboard ship and originally thought to be an aptychus has since been found to be the lower jaw of a marine vertebrate (P. Hoedemaker, pers. comm., 1984), the first jaw recovered by deep sea drilling.

Silt and sandstone turbidites first appear in the upper Hauterivian, reach a peak in the Barremian and, following what may have been an erosive, massive sand depositional event in the Aptian, continue intermittently into the Senonian of unit III (Fig. 4). Over a 218 m interval (unit VA), they comprise about 47% of the recovered section. Largely unconsolidated coarse terrigenous sands with shallow water shell fragments (disturbed by drilling) constitute nearly all of the sediment recovered from Cores 603B–45 and –46. Below Core 48, individual turbidites range up to 2 m in thickness and are dominated by subangular quartz with abundant feldspar, mica, heavy minerals, opaques, wood fragments (locally up to 20%), glauconite and shallow-water bioclastics. Poorly cemented except at the base of some beds (example, Fig. 5(c), whitish sandstone at 603B–58–1, 65 cm), the sands can usually be split under the pressure of one's fingernail. Collectively, these turbidites exhibit the entire range of Bouma structures; however, the basal sequences (Ta to Tc) are dominant and, in general, the turbidites are disorganized and characterized by incomplete Bouma sequences. Associated with these are intraclast-rich debris flows and, more rarely, plastically deformed blocks of fine-grained sediments up to 30 cm across (Fig. 6). The latter could denote slumped overbank deposits which would indicate the presence of well defined fan channels (similar features have recently been cored in the middle/lower Mississippi fan [Leg 96 Scientific Party, 1984]). A detailed analysis of these turbidites is given by Sarti *et al.* (1984). Apparently Hole 603B intersected a complex of one or more deep-sea fans which perhaps form part of an apron of clastic-rich sediments of this age along the lower continental margin of eastern North America.

Fan deposition culminated with the emplacement of the loose sands recovered in Cores 45 and 46. These are the coarsest in the section, and their emplacement must have been a near instanta-

FIG. 6. Slump fold in laminated nannofossil claystone, black claystone, and sandstone of unit VA (Blake Bahama Formation, Core 603B-58-1, 16–49 cm, Barremian). This feature may be associated with slumped overbank deposits similar to those cored recently on the Mississippi fan (Leg 96 Scientific Party 1984).

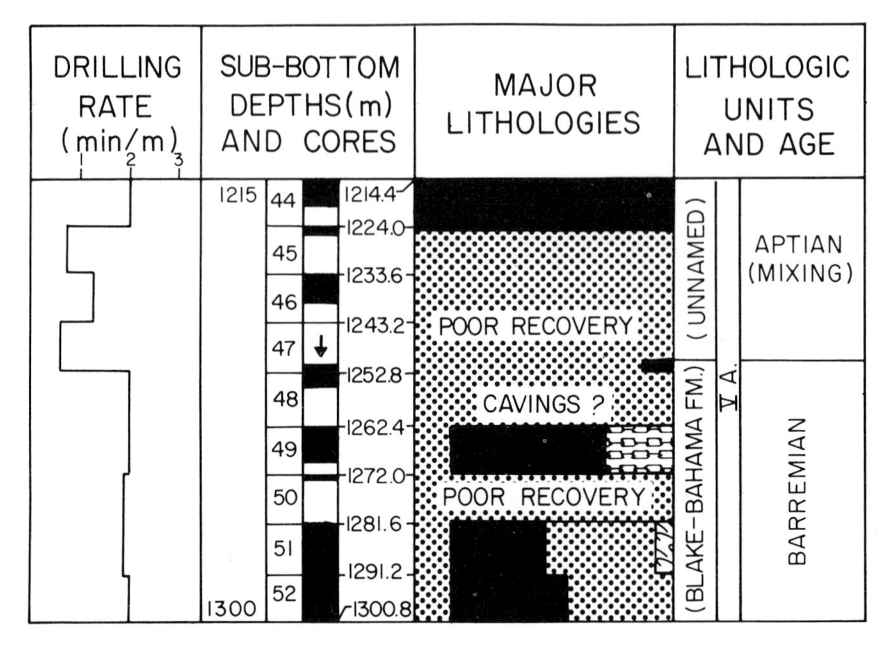

FIG. 7. Reinterpretation of the upper portion of Fig. 4 (b) to show probable extent of the massive, unconsolidated 30 m Aptian sand unit near the top of Lithologic unit VA. The recovered portion of Core 603B-47 has been moved to the bottom of the cored interval (arrow) rather than to the top as in Fig. 4(b), thereby allowing better correlation of the unconsolidated sand unit with the drilling rate log (left).

neous geologic event. If, as in Fig. 7, we consider the short sediment section of consolidated sandstones recovered in Core 47 to have been drilled at the *bottom* of the 9.5 m cored interval (not as represented arbitrarily by DSDP convention at the top of the interval as in Fig. 4(b)), then the total interval of loose sand would equal some 30 m. As shown in Fig. 7, the drilling rate log supports this supposition (about 1 min/m for Cores 45 through 47, vs 2 or more min/m for the more consolidated section above and below). It follows that the soft coarse sand recovered in Core 48 and at the top of Core 49 could be cavings.

Palaeontologic dating of unit VA is fraught with difficulty, particularly when using calcareous nannofossils, due firstly to turbidites displacing older microfossils into younger units, and secondly to contamination of the section through caving associated with the drilling process. These problems are especially evident in the upper portion of unit VA. The dinoflagellate stratigraphy of Habib and Drugg is used to draw the Barremian–Aptian boundary between Cores 46 and 47. Cores 44 to 46 contain numerous Barremian coccoliths such as *Nannoconus colomii* (de Lapparent) that are here considered to have been reworked upwards by turbidites. The prominent upper Aptian coccolith index species *Lithastrinus floralis* Stradner is not present, and the bulk of the nannoflora appears to be latest Barremian to earliest Aptian in age. A poorly preserved form that superficially resembles the upper Aptian *Rhagodiscus angustus* Stradner is present in these cores, but is also present sporadically down to Core 57. This form is apparently either a mimic or an ancestral form of *P. angustus.*

Based on the present somewhat ambiguous evidence, several possible interpretations for the massive unconsolidated sands at the top of Unit VA can be considered. One possibility is that they may have been emplaced as the result of the mid-late Aptian eustatic sea-level drop of Vail *et al.* (Vail *et al.* 1977; Vail & Mitchum 1979; Vail *et al.* 1980). These sands would then represent what Vail *et al.* (1984) described as a Type I unconformity event. A Type I event is caused by a rapid fall of sea-level, is characterized by both subaerial and submarine erosion, and results in the construction of a deep sea fan. The sands at Site 603 do contain displaced inner neritic foraminifers and shell debris. The lack of a definable upper Aptian nannofossil assemblage in these sands or in the silty sediment above them (Core 44) might arise if the sands represent slightly older material eroded and/or slumped from the shelf and slope in sufficient quantity to dilute below the level of detection any slowly accumulating upper Aptian pelagic material. Another possibility could be that Vail's Type I unconformity event occurred during the early Aptian (at least prior to the evolution of *Lithastrinus floralis*) rather than during the late Aptian as his charts suggest. Alternatively, the phenomenon being examined may be a 'nonsynchronous' relative sea-level change of the type discussed by Parkinson & Summerhayes (1985) which does not require a short-period fluctuation of global sea-level. Such a sea-level change may be recorded at somewhat different times in basins around the world depending on the subsidence history of each basin.

Another possible explanation that might explain in part the turbidite sequence at Site 603 and the massive sands that cap it is tectonic activity in the source area, particularly that associated with thermal doming. Thermal doming might have accompanied the emplacement of the Great Stone Dome (Fig. 8), a mafic plutonic body which intruded the Baltimore Canyon Trough during the Early Cretaceous (Libby-French 1984). Grow (1980) noted that its emplacement created a local unconformity of approximately Aptian age. The interaction of such tectonic activity with sea-level changes as discussed above could produce a complex pattern of turbidite deposition, perhaps not unlike that seen at Site 603 (P. R. Vail, pers. comm. 1985).

At the end of Unit V deposition, the CCD shoaled rapidly leading to the abrupt loss of calcareous nannofossils. This was the natural result of the sharp sea-level rise predicted by the Vail *et al.* coastal onlap curve for the late Aptian. This sea-level rise shifted the locus of carbonate deposition from the deep ocean basins to the newly reflooded continental margins. It also served to trap coarse clastics high on the continental shelves, thereby shutting off, in essence, the supply of terrigenous sands to the North American Basin and causing the abrupt cessation of deep sea fan deposition at the study site. This in turn set the stage for the accumulation of black shales, which followed during the exceptionally high sea-level stands of the mid-Cretaceous.

During the Cenomanian–Turonian, brief fleeting blooms of calcareous nannofossils represented only by the genus *Nannoconus* (Core 603B–35) suggest highly restricted surface water environments at the peak of 'black shale' (carbonaceous claystone) deposition (subunit IVA). The absence of benthic foraminifers in the black claystone turbidites suggests a greatly expanded oxygen minimum zone on the shelf and slope. Reddish lithologies and the absence of any organic matter at some intervals indicate that the deep basin environment itself was, at least at times (subunits IVB and IVC), strongly oxidizing.

Dark carbonaceous turbidites continue into the overlying Senonian of Unit III. This lithology accumulated over a longer period of time (Hauterivian to Santonian or later [Core 603B–24]) at this relatively landward site.

Despite the occasional influx of sand, silt and mud turbidites, clastic starved and carbonate free sedimentation during the Late Cretaceous was only 5 m/Ma. In this 'red clay' sequence, there may well be hiatuses which could not be detected due to the lack of calcareous microfossils. For instance, the strking Cretaceous–Tertiary (K/T) boundary sand with its small spherules (Fig. 5(a)) represents a high energy event and appears to have an eroded base. The same holds true for the basal sand of unit III. This period of sediment starvation was fostered by high sea-levels that continued to favour deposition on the continental shelves at the expense of the abyssal plains.

The boundary between units III and II is marked by an unconformity (Horizon A*) which was followed by the accumulation of the upper Palaeocene–lower Eocene biosiliceous claystones of unit II. Erosion along the A^u disconformity

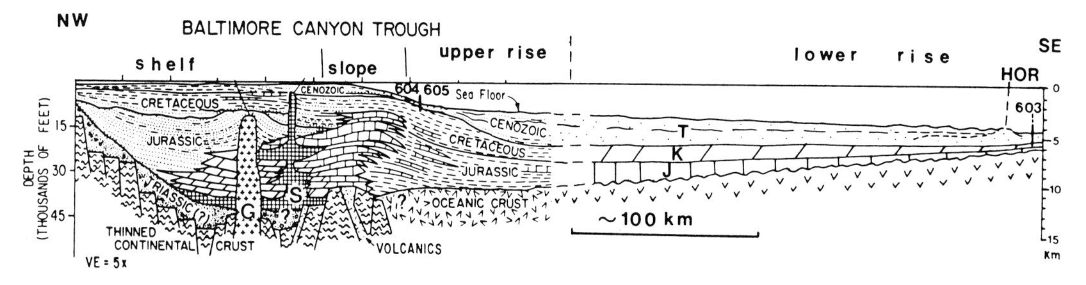

FIG. 8. Schematic section through the New Jersey continental margin and DSDP Sites 604/605 extended to Site 603 on the lower continental rise (from von Rad *et al.* 1984, as modified from Jansa & Wiedmann 1982). G = Great Stone Dome, S = salt diapir, HOR = Hatteras Outer Ridge (Continental Rise Hills), J = Jurassic, K = Cretaceous, T = Tertiary.

following unit II deposition accounts for the sharp contact between unit II and unit I. Middle to upper Eocene units were eroded at this site, and the entire Eocene is absent just to the south and east of this locality (Jansa *et al.* 1979, Fig. 15).

Significance of the 'Cape Hatteras' deep sea fan complex and other clastic deposits in the Cretaceous section at Site 603

The discovery of a major Lower Cretaceous passive margin deep-sea fan complex which consists of up to 47% sand over a 218 m interval and is located this far seaward (320 km) of the Cretaceous shelf edge was quite unexpected. The possibility that coarse clastics may have bypassed the Lower Cretaceous shelf break to accumulate at the foot of the slope had only been suggested speculatively in recent assessments of the stratigraphy and petroleum potential of the eastern North American continental margin (Jansa & McQueen 1978; Mattick *et al.* 1978). Tucholke & Mountain (1979) had observed that the distinct, high-amplitude Reflector Beta seen in profiles at great distance from the North American margin deteriorated landward of the lower rise; they concluded that terrestrial contributions during the Hauterivian–Barremian were the cause. No such clastic sediments had been found, however, in rocks of this age cored at DSDP Site 105 just 60 miles south-east of Site 603, and the possibility of coarse clastics in any quantity reaching these distant localities was not generally anticipated. The only firm indication that such a deposit may have existed is given by the presence of coeval coarse turbidites at DSDP Sites 391 and 534 to the south (Blake-Bahama Basin; Sheridan *et al.* 1983).

The deposition of deep-sea siliciclastics at Sites 603, 391 and 534 coincides with a major progradation of terrestrial deltas on the adjacent continental shelf (Libby-French 1984). A similar phenomenon has been recorded in the southern Wessex Basin of England (Wealden Beds), and deep-sea fans of this age have been cored off north-west Africa at DSDP Sites 370 and 416 (Lancelot *et al.* 1980). The authors' estimate of the extent of these deposits is given in Fig. 9. Hallam (1984; this volume) reviews the occurrence of this 'Wealden-type facies' throughout the Tethyan-mid-eastern realm. He attributes the initiation of this widespread siliciclastic deposi-

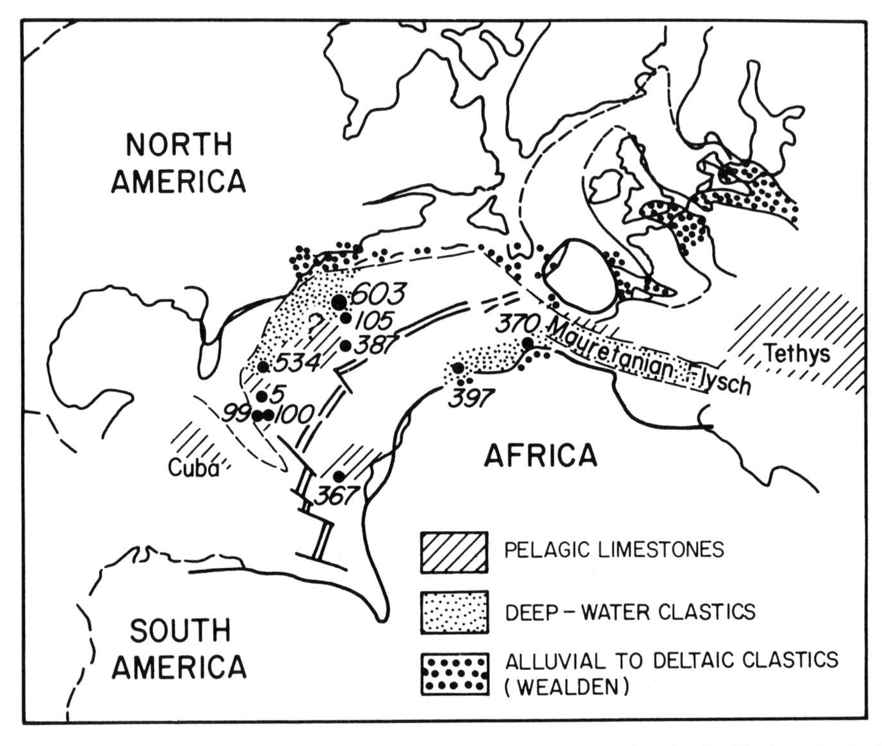

FIG. 9. Early Cretaceous palaeogeography of the North Atlantic region showing the distribution of pelagic limestone, deep water clastic, and alluvial to deltaic (Wealden) facies. Generalized from Von Rad & Arthur (1979) and Robertson & Bernoulli (1982) and supplemented using DSDP Leg 76 and 93 data.

tion in the circum-North Atlantic to climatic change from arid conditions prevalent during the Late Jurassic to more humid environments during the post-Berriasian Cretaceous. As stated previously, the authors attribute the termination of this 'Wealden facies' deposition in their study area to the late Aptian sea-level rise. This is not, however, to rule out the possibility of drainage diversion in the watershed.

Until recently, it was also believed by many that an extensive reef/carbonate bank complex existed along the outer continental shelf from Mexico to Nova Scotia during the Early Cretaceous. This reef/carbonate bank system, plus the presumed rise in eustatic sea-level (Sliter 1977; Vail *et al.* 1980), were thought to have effectively confined most terrigenous clastics to the inner continental shelf (particularly in depocentres such as the Baltimore Canyon Trough, Fig. 8). Presumably, little clastic material would have bypassed the reefs to the deep-sea environment. Schlee *et al.* (1976), however, noted the absence of evidence for reefs on their multichannel line 3 across the southern Baltimore Canyon Trough, and Mattick *et al.* (1978) commented that carbonate deposition there might be incompatible with the prograding deltas in that region. More recently, Poag (1982) and Libby-French (1984) have suggested that deltaic clastics probably did over-step the shelf edge during the Hauterivian–Barremian in both the Georges Bank and Baltimore Canyon areas, thereby shedding detrital sediments down the slope beyond. For the seismic interval between J_1 and Beta, Uchupi *et al.* (1984a, b) map these sands as extending about one-half the distance from the Mesozoic shelf break to Site 603. For the interval immediately above Beta (which would include the massive sands in our cores 603–44 to 48), they map no sands beyond the Cretaceous shelf break.

The drilling results clearly show that large amounts of terrigenous Hauterivian–Aptian sands did bypass the outer shelf to spread over the the abyssal plain hundreds of kilometres beyond. They were interbedded with organic-rich claystones in deep-sea fan complexes along the continental rise and abyssal plain, later to be covered by younger deposits consisting predominantly of clays. Among other things, this demonstrates that many of the ingredients considered favourable for petroleum accumulation are present in this passive margin, deep-sea environment. Although no mature hydrocarbons were found at the relatively shallow burial depths at Site 603, these might be present under deeper burial conditions as would exist at more landward localities such as along the nearby Carolina Trough (see Fig. 13 of Hutchinson *et al.* 1983).

The implications of the above are multifaceted and are the subject of the authors' continuing research. At this point, however, it appears that:

(1) Large deep-sea fan complexes are present along the older passive margin of eastern North America.
(2) The continental rise environments of passive margins should not be neglected as potential petroleum provinces.
(3) A progressive but slow worldwide sea-level rise from the Hauterivian to the mid-Aptian did not prevent the outpouring of terrestrial clastics that occurred along the shelves and rises of the Noth American Atlantic margin. The presumption by Vail *et al.* (1980) of an Aptian drop and subsequent rise in sea-level may be supported and recorded in a massive sand deposit which is followed by the cessation of sand turbidite deposition and a sharp rise in the CCD at Site 603. Tectonic activity in the source area may also have played a role in the sand deposition. Interestingly, the Horizon Beta can be traced to the base of this massive sand unit. Whether the exact timing of the Aptian sea-level drop (upper Aptian according to Vail *et al.* 1980) needs to be recalibrated to any extent based on our data remains to be seen.
(4) The extent of Lower Cretaceous reef development should be reassessed, particularly considering the turbid water conditions which must have prevailed on the shelf during the period in question. Such conditions should have inhibited reef development. Furthermore, no fragments or traces of framework building organisms (such as rudistids or coralline algae) were found in any of the turbiditic materials swept down from the shelf to the drill site. The authors strongly question (Van Hinte *et al.* 1983), therefore, whether any significant reef development occurred along the mid-eastern seaboard of the United States after the Valanginian.
(5) The absence of sand turbidites at DSDP site 105 is probably due to its location on a basement high. The density currents would have flowed around such a topographic high.
(6) Lateral and vertical sedimentary facies changes are far more rapid and complex beneath the continental rise than in the more seaward environments traditionally explored by deep-sea drilling. More deep penetration holes will be necessary to understand this important continental margin province.

DSDP sites 604 and 605 and subsequent investigations of the upper continental rise and slope along the 'New Jersey Transect'

Due in part to the recent advent of commercial petroleum exploration to the area, the geologic framework and depositional history of the outer continental shelf and slope along the Baltimore Canyon Trough are relatively well known and have been summarized by Schlee *et al.* (1976), Schlee (1981), and Libby-French (1984). In addition, investigators such as Schlee *et al.* (1979), Robb *et al.* (1981b, 1983), McGregor *et al.* (1982), Stubblefield *et al.* (1982), Prior *et al.* (1984) Robb (1984) and Stanley *et al.* (1984) have undertaken a variety of geological, geophysical, or manned submersible surveys to explore the continental slope and canyon systems. In contrast, data on the continental rise, particularly in intercanyon areas, are more limited, and early attempts at deep sea drilling there recovered only a few shallow cores (Hollister *et al.* 1972).

To help tie the Cretaceous–Pleistocene sequence at DSDP Site 603 on the lower continental rise (Fig. 1) into the onshore and offshore geology of the Baltimore Canyon Trough area, the authors began a series of shallow holes on the upper continental rise some 100 miles (161 km) south-east of Atlantic City, New Jersey. This 'New Jersey Transect', completed on Leg 95 (Leg 95 Scientific Party 1984), consists of sites along or near USGS multichannel line 25, which trends south-eastward across the continental slope and rise. These sites can be joined with existing wells along the shelf and in the deep sea to form the first comprehensive down dip transect of a passive continental margin from the coastal plain to the abyssal plain. The chief scientific objectives of the overall study, as formulated by the IPOD Planning Committee and the main proponent of the transect, Dr. C. Wylie Poag, are:

(1) to document the presence of unconformities, and to determine their nature, age, correlation with seismic discontinuities, and relationships to sea-level fluctuations;
(2) to examine the relationship between the faunal and lithic compositions and the variable character of seismic sequences seen on line 25;
(3) to describe and correlate litho- and biofacies sequentially along the transect, and
(4) by virtue of the above, to document the upbuilding, outbuilding, and subsidence of this passive margin.

DSDP Site 604 was located in 2364 m of water on the uppermost continental rise, 97 miles (156.5 km) south-east of Atlantic City, New Jersey (Fig. 1, Table 2) and 231 miles (372 km) from Site 603 (Fig. 10). Site 605 was spudded in 2197 m of water, washed to 154 m subbottom, and continuously cored to 816.7 m (Figs 11 and 12; Table 2).

Stratigraphy

The lithologic units penetrated at Site 604 between 0 and 295 m are shown in Fig. 11. From top to bottom they are:

Unit I: 84 m of Holocene (?) to Pleistocene grey to dark greenish-grey clays and silts in alternating sequences.

This unit contains reworked Eocene and Neogene microfossils, has some slumped or redeposited Eocene biosiliceous chalk in the upper 35 m (subunit IA) and has internal slump structures in the lower part (subunit IB). Displaced shallow water benthic foraminifers are particularly common in beds characterized by cold water planktonics (glacial stages).

Unit II: 155 m of lower Pleistocene to upper Miocene greenish-grey clay with variable amounts of glauconitic shelf sand turbidites and biogenic silica. This unit is divided into four

TABLE 2. *Site 604 and 605 coring summaries*

Hole	Dates (1983)	Latitude	Longitude	Water depth*	Penetration	No of cores	Metres cored	Metres recovered	Percent of recovery
604	8–10 June	38°42.79′N	72°32.95′W	2364.00 m	294.5 m	31	294.5	117.58	40
604A	10 June	38°43.08′N	72°33.64′W	2340.00 m	384.4 m	4 (1)	34.8	2.17	6
605	11–17 June	38°44.53′N	72°36.55′W	2197.00 m	186.7 m	70 (1)	662.4	532.08	80
						105 (2)	991.7	651.83	126

* Water depth from sea-level, as measured by drill string length.
Note: the numbers inside parentheses are additional cores obtained from washed intervals.

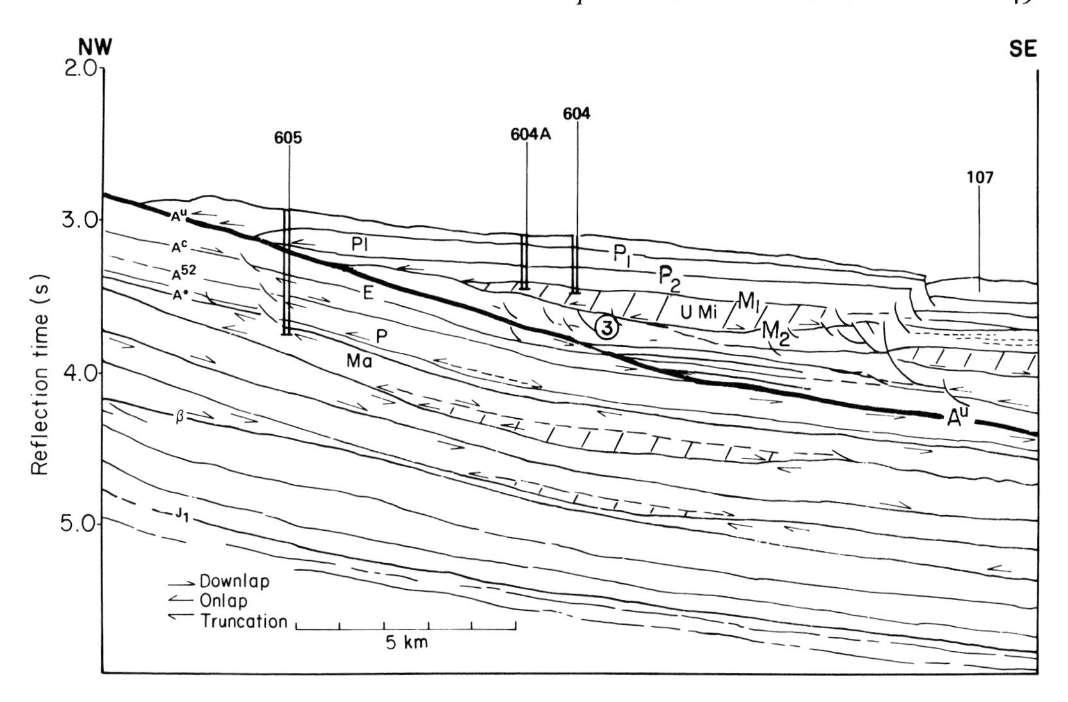

FIG. 10. Line drawing after multichannel seismic reflection profile US Geological Survey Line 25 (after Van Hinte *et al.* 1985b). Leg 93 holes and nearby DSDP Site 107 are shown. Pl = Pleistocene, UMi = upper Miocene, E = Eocene, P = Palaeocene, Ma = Maestrichtian. Reflection horizons (sequence boundaries) indicated to the left and right. Circled '3' = local seismostratigraphic unit.

subunits based on the presence or absence of sand and biogenic silica.

Unit III: 56 m of upper Miocene glauconitic, biosiliceous silty claystone, sand and conglomerate.

In this unit, the sands are inferred from the drilling characteristics. The conglomerates contain rounded quartz pebbles up to 1 cm long, clasts of claystone, chalk, and limestone up to 10 cm in diameter, and rare shell fragments. Elsewhere in the unit are found exotic blocks of nannofossil chalk up to 50 cm thick. These and the conglomerates appear to represent debris flows, whereas some of the sands could be turbiditic. Further details of these debris flows are given below.

Lithologic units penetrated at Site 605 between 0 and 816.7 m are depicted in Figs 11 and 12. In descending order they are:

Unit IA. 198 m of Pleistocene grey silt-rich clay.

Unit IB. Pleistocene green biosiliceous and calcareous clay. This 30 cm unit is separated by a disconformity from the underlying unit and contains some reworked siliceous microfossils from that unit.

Unit II. 153 m of lower middle Eocene biosiliceous nannofossil chalk rich in foraminifers, radiolarians and diatoms.

Unit III. 214 m of lower Eocene greenish-grey nannofossil limestone with varying amounts of foraminifers and calcified radiolarians.

Unit IV. 176 m of upper Palaeocene dark greenish-grey clayey nannofossil-foraminiferal limestone (marl).

Unit V: 77 m of lower Palaeocene to middle Maestrichtian olive grey, clayey nannofossil-foraminiferal limestone.

All of the Palaeogene to Maestrichtian units are strongly bioturbated. Preservation of trace fossils is excellent and several examples of whole *Zoophycos* burrows are noted. The preservation of siliceous and calcareous microfossils is generally good in unit II (except for foraminifers), poor in unit III, and much improved in units IV and V. Recovery was excellent throughout most of the section, particularly in the long (200 m) Palaeocene interval.

Maestrichtian–Eocene depositional history, Site 605

The oldest sediment cored at Sites 604 and 605 was middle to upper Maestrichtian clayey lime-

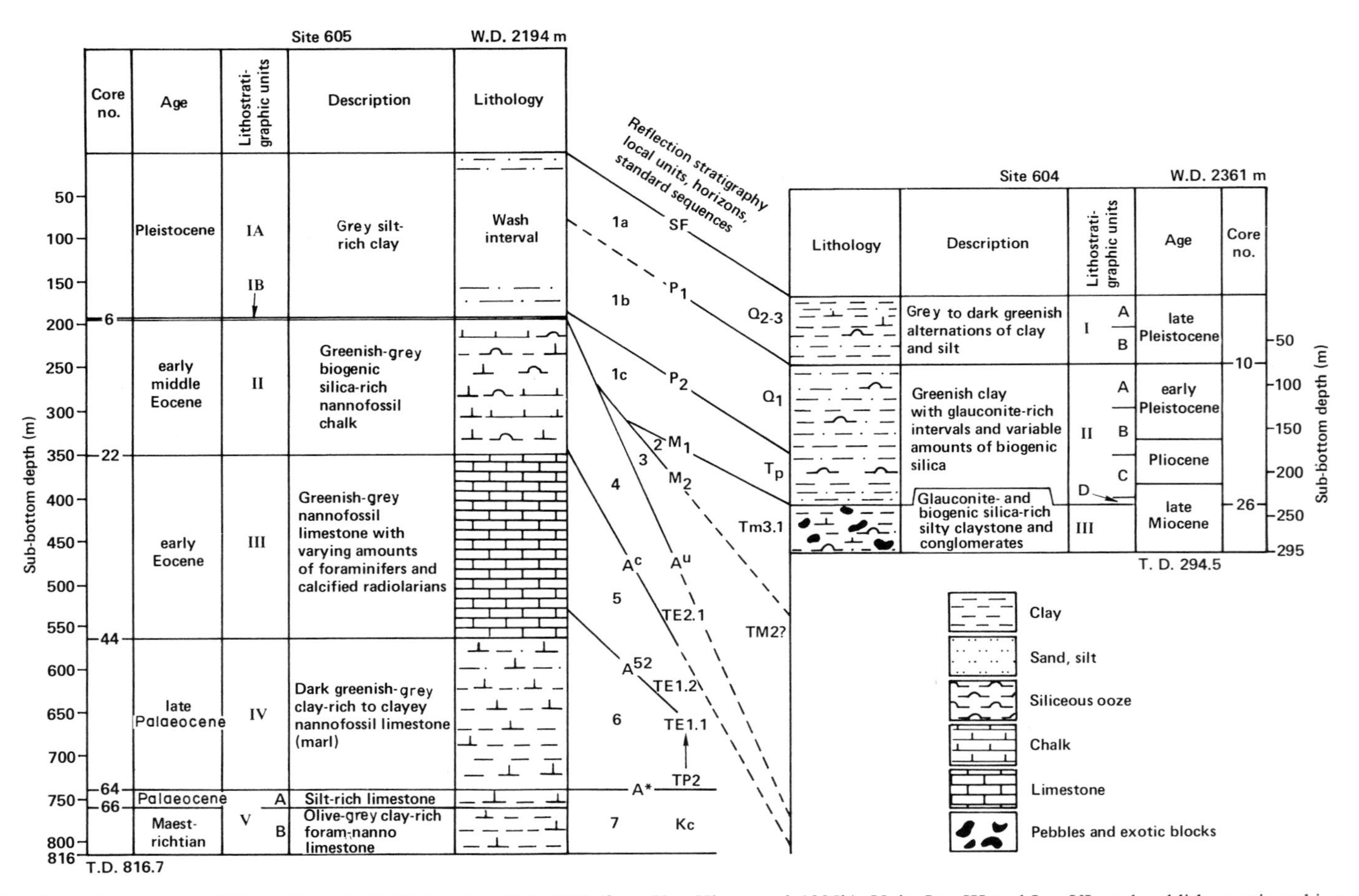

FIG. 11. Stratigraphic summary of Sites 604 and 605 (Holes 604, 604A, 605) (from Van Hinte *et al.* 1985b). Units I to III and I to VI are local lithostratigraphic units. Standard seismic sequence notation after Vail *et al.* (1980).

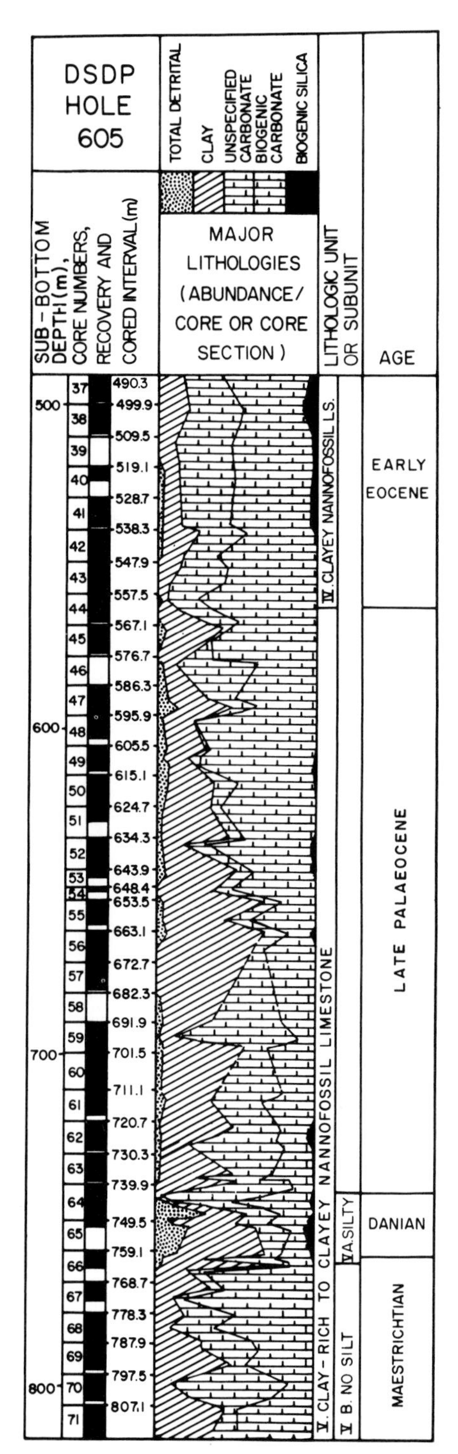

FIG. 12. Lithologies, lithologic units and ages for the Maestrichtian to lower Eocene section cored in Hole 605 (Cores 37 to 71, lithologic units IV and V). Note that the high detrital content (terrigenous micaceous- rich silts) of subunit VA begins near the Cretaceous–Tertiary boundary.

stone (Hole 605, subunit VB). The terrigenous silt content of this material is quite low, averaging about 3%, whereas the carbonate content is high, usually greater than 60%. It contains only trace amounts of glauconite. Going upsection it was evident that a reasonably complete, but only partially recovered (due to fracturing of the core), Cretaceous/Tertiary boundary sequence in an expanded section had been penetrated. At the contact, foraminiferal *Guembelitria cretacea* zone (P_0) and the coccolith *Zygodiscus sigmoides* Zone (CP1) overlies the foraminiferal *Abathomphalus mayaroensis* and coccolith *Nephrolithus frequens* Zones, while other lower Danian zones (Zones Pla and Plb) have not been found in this fragmented portion of the core. At that point there is no visual lithologic difference between the Maestrichtian and the lower Palaeocene section which, on the other hand, is disconformably overlain by an unusually thick upper Palaeocene section.

Unlike all but the uppermost (45 cm) Maestrichtian limestone below them, the earliest Danian sediments (subunit VA) show an anomalously high percentage of terrigenous silt (15–20%; see Fig. 12) and glauconite (2–4%). Assuming that these micaceous silts and glauconites are land- and shelf-derived, the authors can at this point only speculate on their possible significance. Hypotheses regarding their origin which are to be tested during shorebased laboratory and regional studies include:

(1) A terminal Cretaceous extraterrestrial impact (Alvarez *et al.* 1980; Smit & Hertogen 1980) which may have induced exceptional current-wave (tsunami?) activity, an explanation which would also account for the deep water, current-bedded K–T sand with spherules at Site 603 (Fig. 5(a)). The effects of such an impact would have decimated plant life, thereby increasing hinterland erosion and terrigenous riverine output, hence the high detrital content of subunit VA.
(2) Strong current erosion along the outer continental shelf which would also account in part for a well documented upper Maestrichtian–Eocene hiatus there (see Olsson 1980, Fig. 2).
(3) Coincidental outbuilding of some local delta (R. K. Olsson, oral comm., 1984; R. E. Sheridan, pers. comm., 1984).

In general, the Maestrichtian–Eocene sequence (lithologic units II to V) is heavily bioturbated, indicating a steady input of nutrients (either via surface productivity or terrestrial influx) into the sediments. Productivity in the overlying waters increased dramatically beginning with unit II deposition during the early middle Eocene, as shown by the sharp increase in biosiliceous material (diatoms, radiolarians, silicoflagellates). Although this contrast may have been further accentuated by silica diagenesis, this is not likely to be a significant factor. The clay content drops markedly going into unit II, yet the biogenic opal is quite well preserved. A direct relationship between preservation and higher clay content might be expected if productivity were not the overriding factor. The cause for the enhanced productivity at the beginning of the middle Eocene may well have been upwelling, perhaps associated with the incursion of a proto-Labrador current into this area.

Because of drilling problems at Site 604 which terminated both holes drilled there within the upper Miocene debris flow sequence, DSDP Leg 93 obtained no direct information on the Oligocene–middle Miocene interval. Reworked upper Eocene and middle Miocene microfossils in the upper Miocene sediments of Site 604 indicate that those units are present in the area. No reworked Oligocene material was identified, and at Site 605 these sediments are missing. There, Pleistocene sediments are separated from the middle Eocene by a pronounced disconformity which can be traced as the coalesced A^u seismic reflection horizon.

Miocene erosion and debris flows

Debris flows cored at Site 604

In this section further details on the debris flows cored at Site 604 are provided. The best recovery was in Core 604–26 which penetrated the top of the debris flows beginning at Section 2, 46 cm (arrows, Fig. 13(a); white arrowhead, Fig. 13(b)). The sandy greenish-grey clay immediately above that contact contains coccoliths which belong to the upper Miocene *Discoaster quinqueramus* Zone (CN9) of Okada & Bukry (1980). The interval from the contact down through Core 27 (Fig. 13(b)) belongs to the *Discoaster neohamatus* Zone (CN8, Tortonian), as do the lowest sands recovered in Core 3 at Hole 604A. For the remainder of Hole 604, recovery was meagre with no recovery in core barrel 604-29 and less than a metre of siltstones and claystones in Cores 604–28 and 30. These latter belong to the coccolith *Discoaster hamatus* Zone (CN7). The debris flows sampled at Site 604 before drilling was terminated, therefore, fall within nannofossil Zones CN7 and CN8 which Miller *et al.* (1985) and Berggren *et al.* (1985) now correlate with the lower to mid Tortonian (upper Miocene) and assign an age of about 8.2–10.0 Ma. By that correlation, these debris flows would have been emplaced during cycle TM3.1 of Vail *et al.* (1977), which represents a significant sea-level drop leading up to the Messinian low stand.

The portion of Core 604-26 below the top of the debris flows (Fig. 13(a)) contains a wide assortment of pebbles, cobbles, and shell debris in a matrix sometimes quite rich in siliceous microfossils (Fig. 13(b)). A number of pebbles and a portion of a bivalve shell are shown in Fig. 14(a). The bivalve appears to be a clam displaced from shallow water. Pebbles 'b' and 'c' are mafic in composition as is pebble 'e' which is a peridotite composed of coarse-grained olivine. Pebble 'f' is a Palaeozoic chert fragment 3 cm in length which contains recognizable portions of trilobite exoskeletons and what appears to be a brachiopod (Fig. 14(b)). This highly angular pebble (which was sectioned on one side by the saw used to split the core) was probably derived from lower Palaeozoic strata of the Appalachians (C. Schreiber, pers. comm. 1984), whereas the peridotite would be from the structural core of the same mountain belt.

Cobbles in Cores 26, 27, and 30 represent a variety of ages ranging from early Eocene to middle and late Miocene. Those from Core intervals 604-26-3, 52–54 cm (Fig. 13(b)) and 604-27-1, 23–25 cm (Fig. 14(d)) are middle Miocene or younger, both having been derived from shallower water. For instance, the first sample mentioned, seen in thin section in Fig. 14(c), contains shallow water benthic foraminifers characteristic of outer neritic environments (cibicidids, hanzawaids, etc.). Particularly interesting, however, is the fact that the highly angular clastic grains (predominantly quartz and glauconite) are supported by a fine-grained carbonate matrix. This would indicate an original aragonite mud environment in the near shore; however, no such environment is known from the Miocene of the New Jersey coastal plain. This is probably due to the fact that most of the middle to upper Miocene section is missing in the coastal plain record due to a hiatus. Missing, therefore, are those strata which would have represented the peak of the middle Miocene transgression (according to the Vail *et al.* 1980, coastal onlap curve) when Miocene climates should have been most equitable, the most likely time to expect aragonite mud production this far north (R. K. Olsson, pers. comm. 1984). Thus, the Miocene cobbles in the debris flow at Site 604 may be the only specimens available which represent that shallow marine environment.

The displaced blocks of Eocene carbonate are similar to lithologies drilled at Sites 108 and 605, therefore their sources were probably much closer to Site 604 than those of the Miocene cobbles.

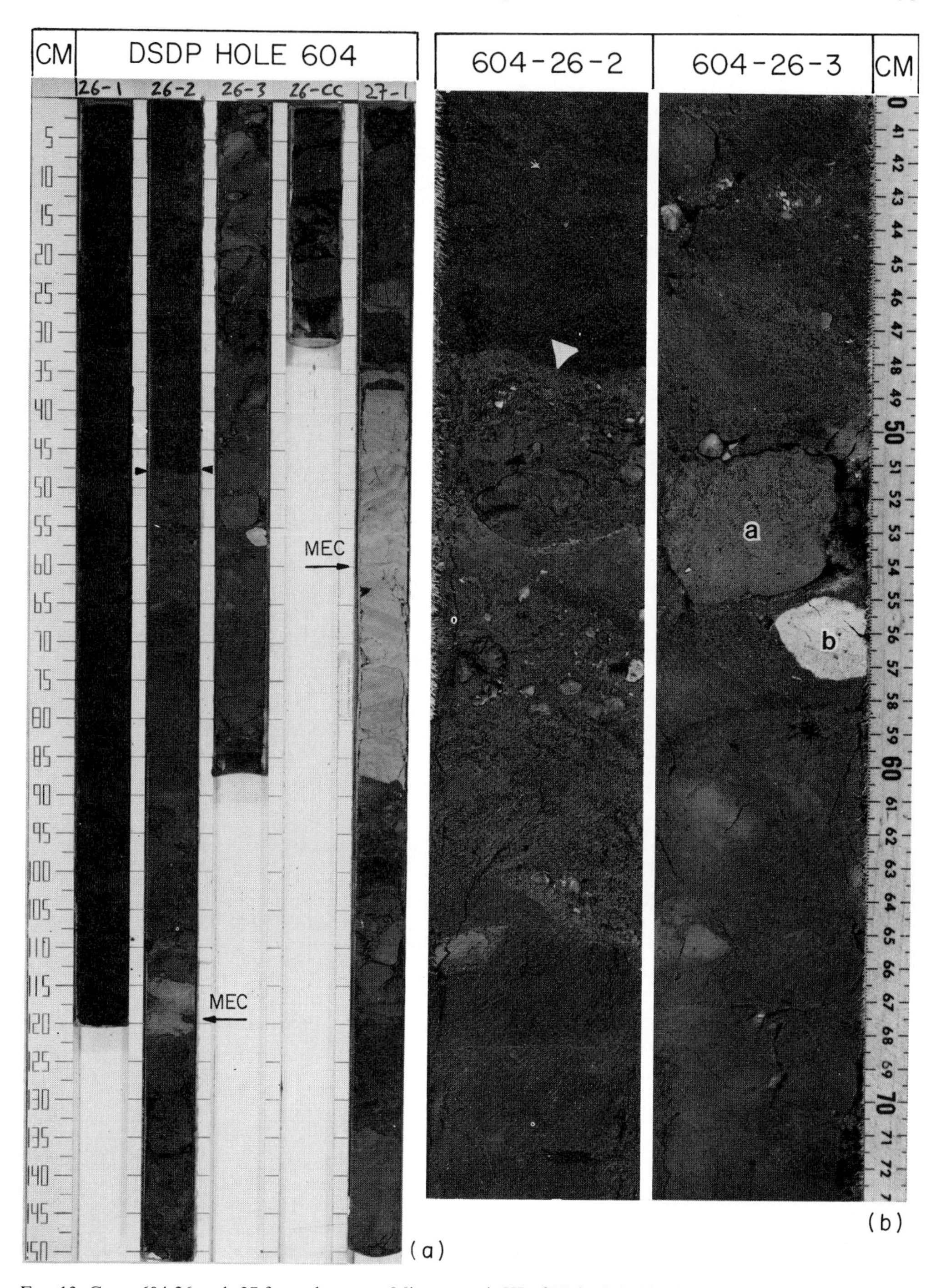

FIG. 13. Cores 604-26 and -27 from the upper Miocene unit III of Hole 604. (a) The top of the youngest debris flow is indicated by the arrowheads at 604-26-2, 46 cm. MEC = middle Eocene chalk. (b) Details of upper Miocene debris flows in Core 604-26, Sections 2 and 3. The top of the youngest debris flow is marked by white arrowhead in Section 2. Note numerous pebbles in both sections and exotic cobble sized clasts in Section 3. Clast 'a' at 51 to 55 cm in Section 2 is middle Miocene or younger in age; clast 'b' is Eocene chalk.

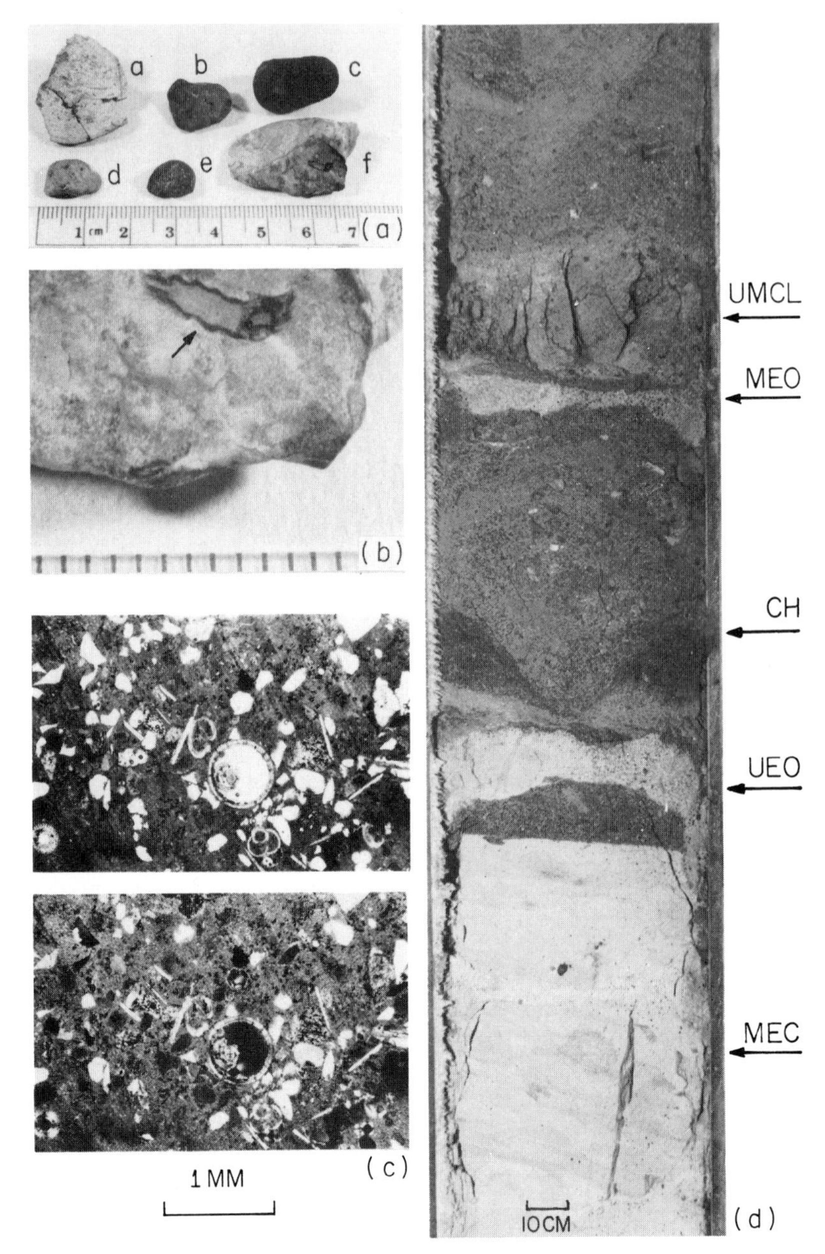

FIG. 14. Further details of debris flows in Cores 604-26 and -27. (a) Shell and pebbles from upper Miocene debris flows in Core 604-26; a = clam shell (604-26-3, 16 cm); b and c = mafic pebbles (604-26-2, 109–110 cm); d = quartz pebble; e = coarse-grained peridotite from 604-26-3, 18 cm; f = angular pebble of lower Palaeozoic chert containing trilobite exoskeletons and a brachiopod. (b) Detail of brachiopod (arrow) in Palaeozoic chert pebble (Fig. 14(a) above). (c) Thin section of middle (or upper) Miocene cobble ('a') shown in Fig. 13(b); 604-26-3, 52–54 cm. Note foraminiferal tests and highly angular clastic grains (quartz, glauconite, feldspars and mica) supported by a fine-grained carbonate matrix. Original matrix thought to have been aragonite mud; (top) transmitted light; (bottom) cross-polarized light. (d) Detail of the interval 604-27-1, 17–46 cm from Fig. 13 above showing an erosional cut-and-fill channel (CH) that marks the contact between two debris flows, and juxtaposed sediments of various lithologies and ages overlying the 50 cm thick slab of middle Eocene chalk. MEC = middle Eocene chalk; UEC = upper Eocene nannofossil ooze; MEO = middle Eocene nannofossil ooze; the dark lithologies are upper Miocene clastics (coccolith Zone CN8), including one large (2.5 cm thick) clast (UMCL) which contains relatively shallow water (upper bathyal) benthic foraminifers.

Today in the study area, a broad belt of middle to upper Eocene strata crops out along the lower slope (Robb *et al.* 1981) from which blocks or slabs of the material can be shed onto the rise below. The most spectacular example of such a slab in the Miocene section is the 50 cm block of middle Eocene limestone recovered in Core Section 27-1 (Fig. 13(a)), the internal coherence of which was disturbed by drilling. A detailed view of the core immediately above this block reveals the complex nature of the debris flows (Fig. 14(d)). Laminae containing sediments or microfossils of quite different ages, compositions, and consistencies are closely juxtaposed.

Perhaps the most important feature shown in Fig. 14(d) is the cut-and-fill channel indicated by arrow 'CH'. This erosional surface marks the termination of one debris flow and the beginning of the next, a clear indication that more than one debris flow event is represented in lithologic unit III. Despite the minimum recovery in this unit, at least three separate debris flows could be distinguished on a tentative basis by the Leg 93 shipboard sedimentologists. This delineation of separate and distinguishable debris flows at Site 604 indicates that the debris flows within lithologic unit III (seismic reflection unit 2) have a complex history, that sample coverage through the interval drilled is incomplete, and that only the uppermost portion of the unit was penetrated. From the evidence at Site 604, it is not possible to say when the channel (or canyon) that contains the debris flow fill was cut, or when the first debris flows were deposited in that channel. To answer these questions, other seismic and drill hole data in the region must be considered.

DSDP Sites 612 and 613

Approximately two months after the termination of DSDP Leg 93, DSDP Leg 95 returned to the New Jersey transect and drilled two additional holes (Leg 95 Scientific Party 1984; Thorne & Watts 1984; Poag 1985). Hole 613 was located 3.8 miles (6.1 km) NNE of Site 604 and about 3.5 miles (5.6 km) NNE of USGS multichannel line 25. There the troublesome upper Miocene debris flow unit appeared on seismic profiles to be only about 10 m thick where it crossed the top of an interchannel ridge. This proved to be the case and a 675 m hole penetrated to the lower Eocene while avoiding the drilling problems which had plagued operations at Site 604. The other Leg 95 hole in the area, Site 612, was sited directly on USGS line 25 at a point on the mid continental slope (water depth = 1404 m) some 8.9. miles (14.1 km) WNW of Site 605.

In order to tie the various Leg 93 and 95 sites together, the USGS line 25 plus additional crossing MCS lines 34 and 35 (Grow *et al.* 1980) have been studied. In addition, two of the authors (GSM and JF) have examined a tight grid of single channel airgun seismic profiles collected by the United States Geological Survey (Robb 1980). The close line spacing (1 km between dip lines, 2–3 km between strike lines) and the relatively high resolution of the data have enabled complex relationships to be recognized in the Neogene and Quaternary stratigraphy of the continental slope not detectable on available MCS profiles.

The seismic profiles show deep erosional surfaces cut into the upper slope which exhibit considerable along-strike variability. For example, a seismic unconformity can be recognized on these high resolution profiles beneath DSDP Site 612 that, when traced 1 km south-west to dip line 68 (Fig. 15), rises to reveal 0.1 seconds of section not sampled at Site 612. This intervening section of lower Oligocene–upper Miocene age (?) is shown in white in Fig. 15; stippled patterns highlight the units dated at Site 612. Numerous unconformities can be recognized in the seismically chaotic interval above this major hiatus.

Glauconitic, micaceous muds of nannofossil Zone CN7 were cored above this unconformity at Site 612 (Poag 1985). Radiolaria are sparse but include taxa normally found in contemporaneous neritic environments of the coastal plain (Palmer 1983). Hence, the seismic character, lithology and microfauna all suggest cross-shelf and downslope transport of sediment following a major erosional event that occurred between the base of the Oligocene and some point in the late Miocene. The records obtained at DSDP Sites 604, 605, and 613 show that the upper rise was the major depocentre for these sediments. No profile has yet been found to suggest that sediment considerably older than that at Site 612 rests on the unconformity along the upper slope. Until further studies in progress are completed, the authors are satisfied that the erosional event was initiated in the late middle Miocene.

Seismic observations—rise

Mountain & Tucholke (1985) have argued that bottom current erosion near the Eocene–Oligocene boundary cut into Eocene sediments along the rise, and may have contributed to slope failure due to undercutting. They postulated that a second erosional event in the late middle Miocene (foraminiferal Zone N14) cut into the base of the slope, and in places removed all sediment that had accumulated since the Oligocene. The first event is marked by reflector A^u, the second by reflector

Merlin (Fig. 16). In the vicinity of DSDP Site 613, Merlin cut down to A^u, and consequently upper Miocene sediments rest unconformably on the middle Eocene (Poag 1985). (Note that in Fig. 16, GSM and JF also show a coalesced Merlin/A^u in the vicinity of DSDP Site 604, an interpretation which differs somewhat from that of JEVH in Fig. 10.)

As explained previously, the first attempt to drill through these coalesced unconformities was thwarted by the coarse pebbly mudstone debris flows which halted drilling at 294 m subbottom at DSDP Site 604. These strata correspond to the high amplitude reflectors at 3.50 seconds on Line 35 (Fig. 16). Site 605 (Fig. 15) was accordingly located 6 km landward, beyond the updip pinch-out of these deposits against the erosional top of the Eocene section. Similarly, the anticipated thick debris flows were avoided on Leg 95 by locating Site 613 near a pinchout, but this time along strike from 604 where reflector Merlin is a local high (Fig. 16). As expected, only 10 m of debris flow deposits were found immediately above the Eocene section, and were penetrated successfully (Poag 1985).

The debris flow deposits as sampled at Sites 604 and 613 are of early late Miocene age (Tortonian; nannofossil Zones CN7 to CN8). The thicker section at 604 reveals a wider variety of clasts within the mudstone matrix (Figs 13 and 14). Evidently, deep erosion of the continent occurred in the late Miocene and, as material bypassed the slope, sediment was removed from canyon walls and from the Eocene outcrop belt along the lower slope, all of this being rapidly deposited on the upper rise.

The authors consider it likely that the first of these debris flows occurred in the late middle Miocene, coincident with the time of upper slope erosion argued earlier. This could have been confirmed if Site 604 had been successfully drilled into the axis of the broad trough above reflector Merlin. The thickest unsampled interval between the highly reflective debris flows and reflector Merlin occurs at shotpoint (SP) 750 on line 35. At this location the interval is between 3.60 and 3.88 seconds reflection time, and is approximately 240 m thick. For all of this to have accumulated between the base of N14 (the assumed age of Merlin) and the top of CN8 (3.1 Ma: Berggren *et al.* 1985), average sedimentation was at least 72.5 m/Ma.

If one accepts this average accumulation rate as reasonable, thereby adopting N14 as a likely age for the formation of reflector Merlin, it is also likely that Merlin represents the down-slope continuation of late middle Miocene slope canyons out onto the rise. This is in conflict with the initial interpretation cited above (Mountain & Tucholke 1985) that Merlin represents an along-slope current and erosional event. The size and distribution of broad undulations of Merlin are similar to those of modern canyons in the same region. For example, examine the trough marked by Merlin at SP 750 between 3.75 and 3.88 seconds on Line 35 (Fig. 16). It is roughly 6 km wide and 115 m deep, similar to Lindenkohl Canyon, which crosses the same profile at SP 870. It remains for future seismic analysis to determine whether or not Merlin corresponds to a surface of buried canyons.

Conclusions

Significant erosion of the continental margin off New Jersey occurred sometime between 11.3 and 8.2 Ma (late middle to late Miocene times). Deep erosion of the craton is evidenced by metamorphic rock fragments carried out to the continental edge, where deep canyons were cut into the slope and probably a short distance out onto the rise. As seen at Site 604, chaotic debris flows containing these and other rock fragments cut away from the slope and canyon walls continued to maintain irregular and complicated patterns of accumulation through at least mid-Tortonian times (nannofossil Zone CN8) at this site. Subsequently, glauconite-rich sands along with occasional influxes of reworked Eocene calcareous materials continued to accumulate in varying quantities from the latest Miocene (nannofossil Zone CN9) through the Pleistocene.

The emplacement of the debris flows sampled at Site 604 centred about a time (10.0–8.2 Ma) when the effects of the late Miocene glaciations on the Antarctic continent were beginning to be felt

FIG. 15. (Opposite) Original profile (top) and interpreted line drawing (bottom) of single channel USGS line 68, a dip line across the slope and uppermost rise off New Jersey. The five drill sites on the interpreted section are located 1 to 13 km off line 68 and are projected onto the profile. The drill consequently encountered slightly different sections than this one due to along-strike variability (compare especially DSDP Sites 612 and 613 in this profile with those in Fig. 16). On the uppermost rise, latest middle Miocene and younger strata rest directly on pre-early Oligocene strata, the unconformable contact being represented by the merged reflectors Merlin and A^u. Coarse-grained debris flows of late Miocene age (the age of the oldest canyon fill on the upper slope) onlap this Merlin/A^u surface. Note also the crossing of USGS line 35.

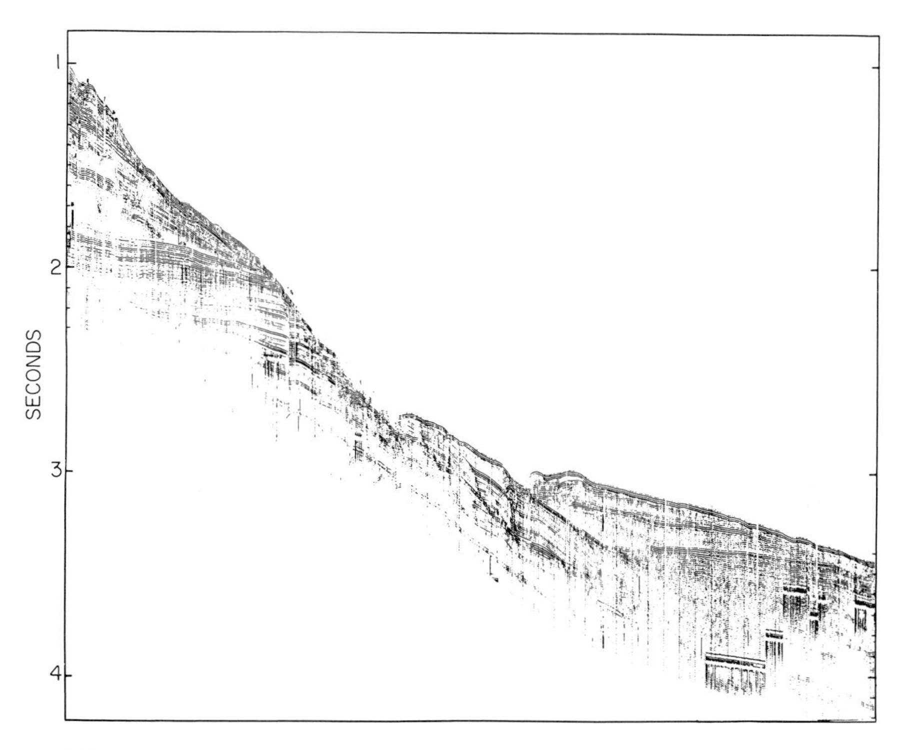
1
2
3
4
SECONDS

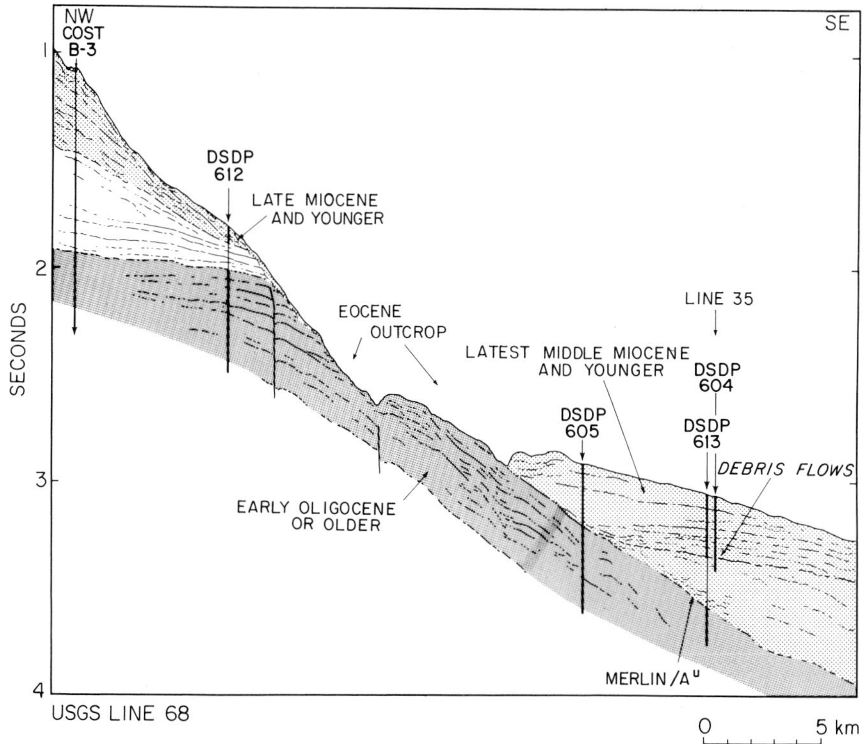
NW
COST
B-3
SE
DSDP
612
LATE MIOCENE
AND YOUNGER
EOCENE
OUTCROP
LINE 35
LATEST MIDDLE MIOCENE
AND YOUNGER
DSDP
604
DSDP
605
DSDP
613
DEBRIS FLOWS
EARLY OLIGOCENE
OR OLDER
MERLIN/A^u
1
2
3
4
SECONDS
USGS LINE 68
0
5 km

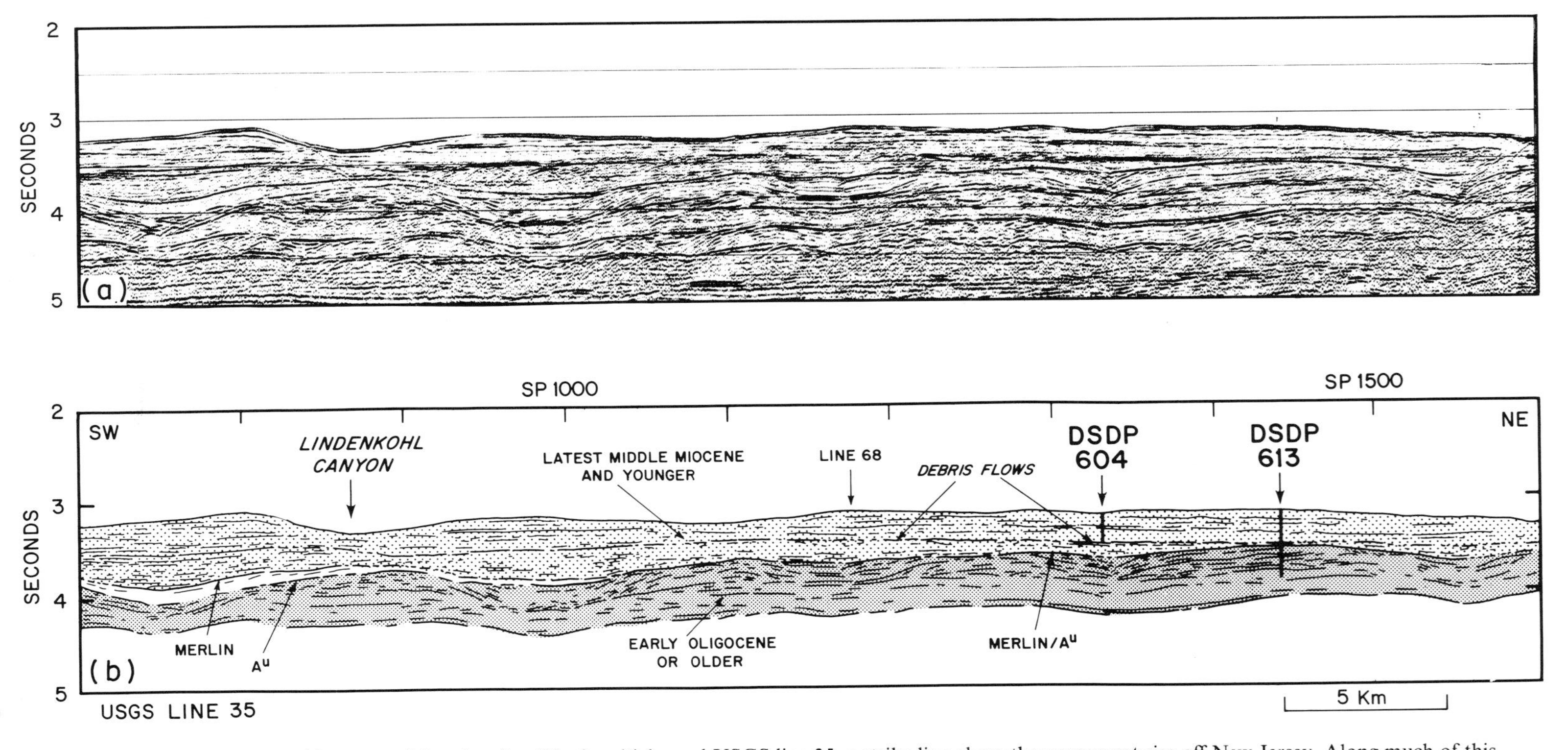

FIG. 16. Original profile (a) and interpreted line drawing (b) of multichannel USGS line 35, a strike line along the uppermost rise off New Jersey. Along much of this profile the uppermost middle Miocene and younger section rests unconformably on strata that are early Oligocene or older in age, separated by the merged reflectors Merlin and A^u. South-west of SP 1050 these reflectors diverge, revealing an upper Oligocene(?)–lower Miocene(?) section not found at DSDP Site 613. Note the highly reflective debris flows at the bottom of Hole 604, and their pinchout against a local high of reflector Merlin at Site 613. Note also the crossing of USGS line 68 (Fig. 15).

in the more northerly reaches of the Southern Ocean. For the area north-east of the Falkland Plateau (DSDP Site 513), Ciesielski & Weaver (1983) document a major abyssal erosional event which they believe occurred in the early late Miocene (prior to 9.5 or 10.0 Ma). This is followed by the initiation of ice rafting at 8.7 Ma which Ciesielski & Weaver (1983) consider indicative of the establishment of large Antarctic ice shelves, which in turn probably led for the first time to the formation of a grounded West Antarctic ice sheet.

The Miocene glacial events on the Antarctic continent discussed above apparently induced significant eustatic sea-level fluctuations. Upper Miocene sediments are practically absent in offshore boreholes along the east coast of the United States and have not been identified in land outcrops of the coastal Plain. During this time of lowered sea-levels, the shoreline migrated toward the shelf break where rivers spilled their loads, and submarine canyons eroded headward to receive them. The Tortonian regression represented by Vail cycle TM3.1 seems to have been particularly sharp and marks the first time since the early middle Miocene that sea-level stood below present day levels (Vail *et al.* 1979). The debris flows and turbidites drilled at Site 604 resulted from this period of deposition in an unstable shelf edge environment. Diatomaceous sediments associated with these deposits indicate further upwelling in this area, perhaps again related to incursions of the Labrador Current system. Glauconitic sands, shallow water foraminifers and shell fragments, and occasional displaced Eocene material in lithostratigraphic units I and II reflect further unstable sea-level and current conditions into the Pliocene and Pleistocene, particularly since the beginnings of Northern Hemisphere glaciation.

Before leaving this discussion of the New Jersey transect, it should be pointed out that evidence for continued instability and downslope transport of sediments is quite evident in this area. In particular, the authors wish to call attention to the highly contorted and convolute patterns in the Pleistocene sediments of subunit IB at Site 604. These are interpreted as internal slump structures, features which indicate unstable slope conditions. Slumping within Quaternary deposits has been widely observed on the continental slope in the detailed seismic surveys by Robb *et al.* (1981b), and these downslope, gravity driven processes are thought to have continued in this area from the Holocene to the present (Stanley *et al.* 1984).

Having described Neogene clastic sedimentation on the slope and upper rise off New Jersey, the coeval sedimentation on the lower rise at Site 603 will now be examined. A physiographic sketch relating the Leg 93 drill sites and others in the two study areas is given in Fig. 17.

Neogene clastic sedimentation at site 603 and the evolution of the Hatteras Outer Ridge

Earlier the discussion of Palaeogene sedimentation at Site 603 was ended with the observation that strong erosion on a regional scale along the A^u disconformity had followed the deposition of the Eocene biosiliceous units (see Jansa *et al.* 1979, Fig. 15). This marked a radical departure from the pelagic depositional regime which had dominated sedimentation on both the slope and rise of the study area during the Eocene. For the first time, there is evidence of strong abyssal currents operating along the contour of the western boundary of the North American Basin.

The first sediments deposited above the A^u disconformity at Site 603 are the lower Miocene, light yellowish-brown to brown hemipelagic clays of unit ID (Fig. 3). Sedimentation rates were sufficiently low to allow oxidation of these clays. Biogenic productivity was also low in that only fish teeth accumulated at these depths. Abyssal currents had obviously waned, thereby allowing sediments to accumulate again; climate had also ameliorated somewhat.

Downslope processes, however, soon began to dominate sedimentation at Site 603 during the middle Miocene (nannofossil Zone CN5 of Okada & Bukry 1980; 10.9–14.4 Ma according to Berggren *et al.* 1985), just as they did at about this time on the upper rise and slope. Sedimentation rates increased markedly with the deposition of dark hemipelagic clays and occasional sand or silt turbidites (unit IC, Fig. 3) which were fed by the outflow of terrigenous material (including wood debris, mica, etc.) from the shelf. Turbidity currents that cut into Eocene strata of the slope or upper rise were able on occasion to erode and transport delicate Eocene coccoliths without apparent damage (Wash Core Sample 603B-6M, CC [831.0–850.2 m]). Generally flat-lying to mounded at first, the sediments of subunit IC began to be caught and shaped into a large-scale (50 × 500 km) elongate drift deposit by the Western Boundary Undercurrent (WBUC) by the time subunit IB was deposited. This signalled a change in Atlantic bottom circulation and marked the birth of the feature now called the Hatteras Outer Ridge (HOR). As the material was built up above the level of the CCD (which was falling at an

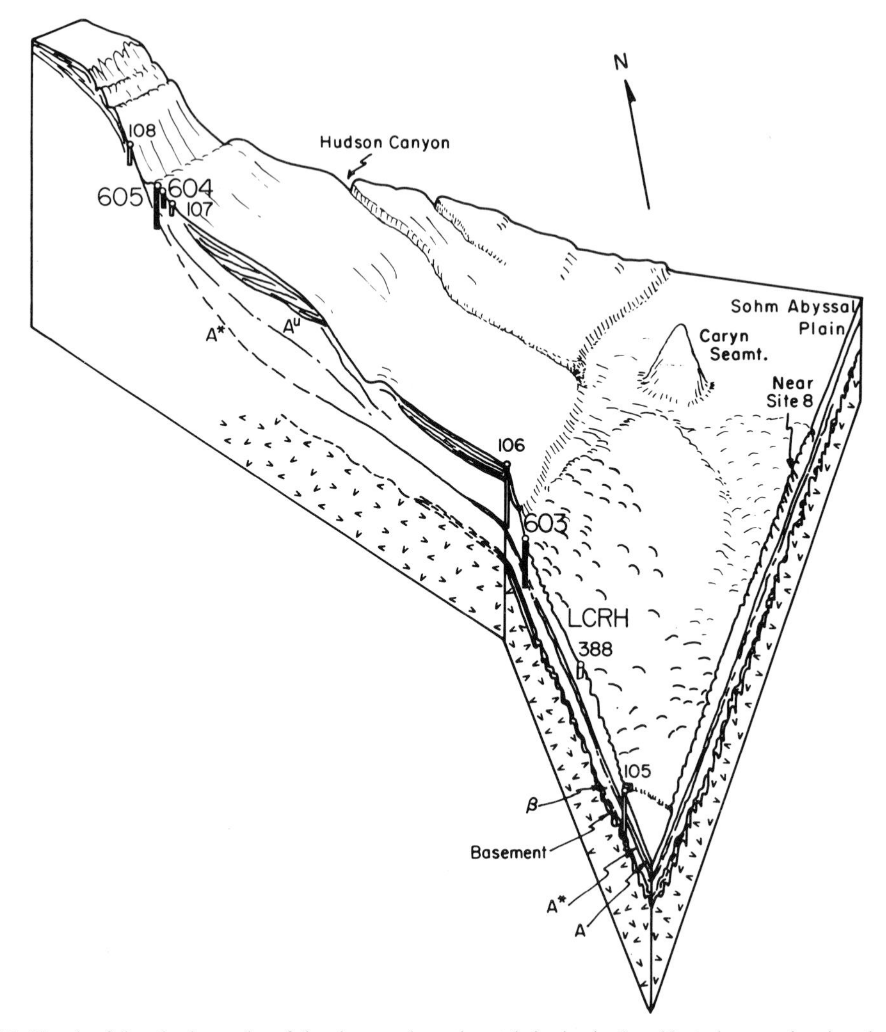

FIG. 17. Sketch of the physiography of the slope and continental rise in the Leg 93 study area showing the relationship between Sites 604 and 605 on the upper rise and Site 603 on the lower rise. Other drill sites and principal seismic reflection horizons also shown (after Ewing & Hollister 1972, Fig. 11).

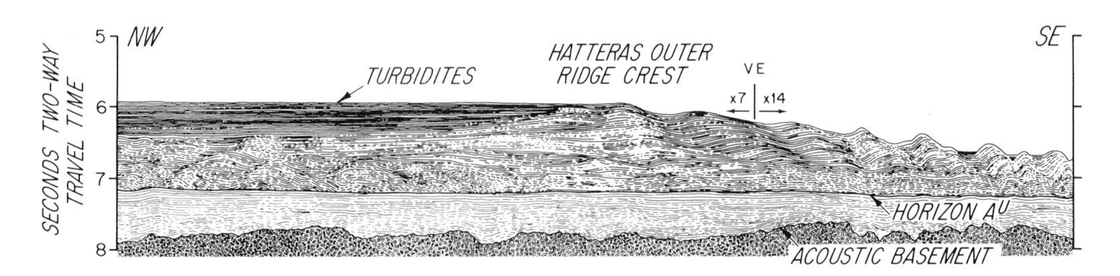

FIG. 18. Tracing of single-channel record from Conrad 21 multichannel seismic line across the Hatteras Outer Ridge crest which separates the tubidite pond to the north-west from the Lower Continental Rise Hills to the south-east (from Tucholke & Laine 1982, Fig. 4).

increased rate by the end of middle Miocene times), appreciable numbers of calcareous microfossils began to accumulate, the first since the site passed beneath the CCD following the Barremian.

As suggested by Tucholke & Laine (1982), a relatively constant NE–SW bottom current and a W.–E. turbidite input from the margin led to the formation of antidune-like sediment waves which grew to form the present Lower Continental Rise Hills (Figs 2 and 17). The crest of the HOR grew apace where the turbidity currents were intercepted by the WBUC, which formed a natural levee behind which coarse terrigenous turbidites have been ponded, particularly during glacial times (DSDP Site 106; Ewing *et al.* 1972). Remarkably, no coarse clastics bypassing the pond have been deposited with the clays of the HOR at the study locality, the crest of which, unlike today, was situated topographically above the sediment pond. As a result, no sandy contourites are present in our section and sediments of the

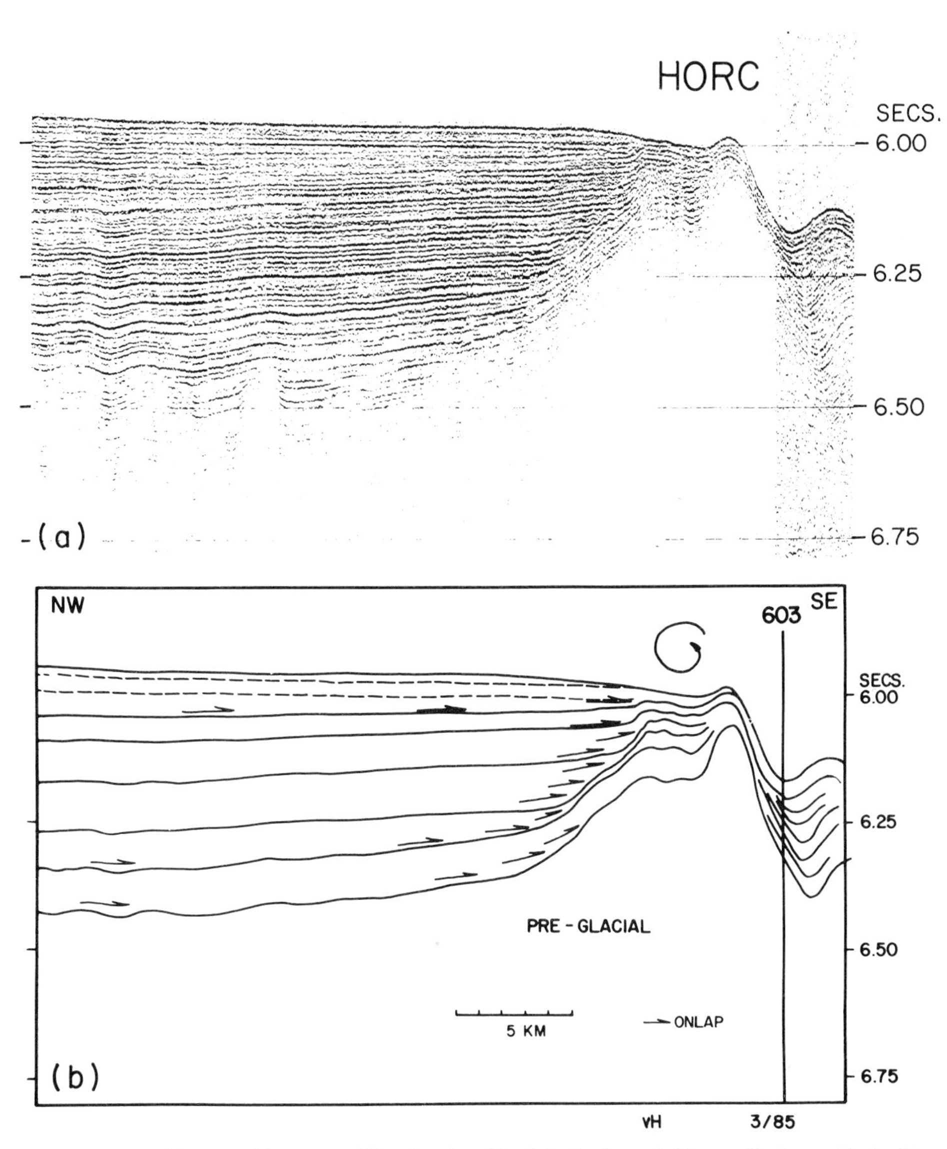

FIG. 19. Original profile (a) and interpreted line drawing (b) of single channel *Glomar Challenger* Cruise 93 seismic reflection profile across DSDP Site 603, the crest of the Hatteras Outer Ridge (HORC), and the turbidite pond to the west.

HOR are monotonously uniform, consisting of a homogeneous mixture of clay and silt (muddy contourites of Stow & Lovell 1979). Our seismic analysis (Fig. 19) suggests that the two grew contemporaneously at least into the early Pleistocene (*Helicosphaera sellii* Zone of Gartner 1977 [1.45–1.37 Ma according to Backman & Shackleton 1983]), and our core data show no appreciable break in sedimentation at Site 603 until that time. The ridge, however, lost much of its prominence as it grew slower than the massive glacial turbidite accumulations. Today at our location, the ridge crest is actually slightly below the main terrace and would not stand out if it were not for a moat behind it. It is expected that the main current runs in the valley where it has prevented deposition since early Pleistocene time (there are no signs of erosion) and that a side eddy has created the moat behind the ridge (Fig. 19).

Conclusion

This progress report on Mesozoic-Cenozoic clastic depositional environments revealed by DSDP Leg 93 drilling off the Atlantic margin of the US is concluded by presenting a modification made by one of the authors (JEVH) of Mountain & Tuckolke's (1985) Fig. 8.1. As seen in Fig. 20, the relationship between eustatic sea-level and detrital influx and ocean circulation has been emphasized. This is a reflection of the belief by some of

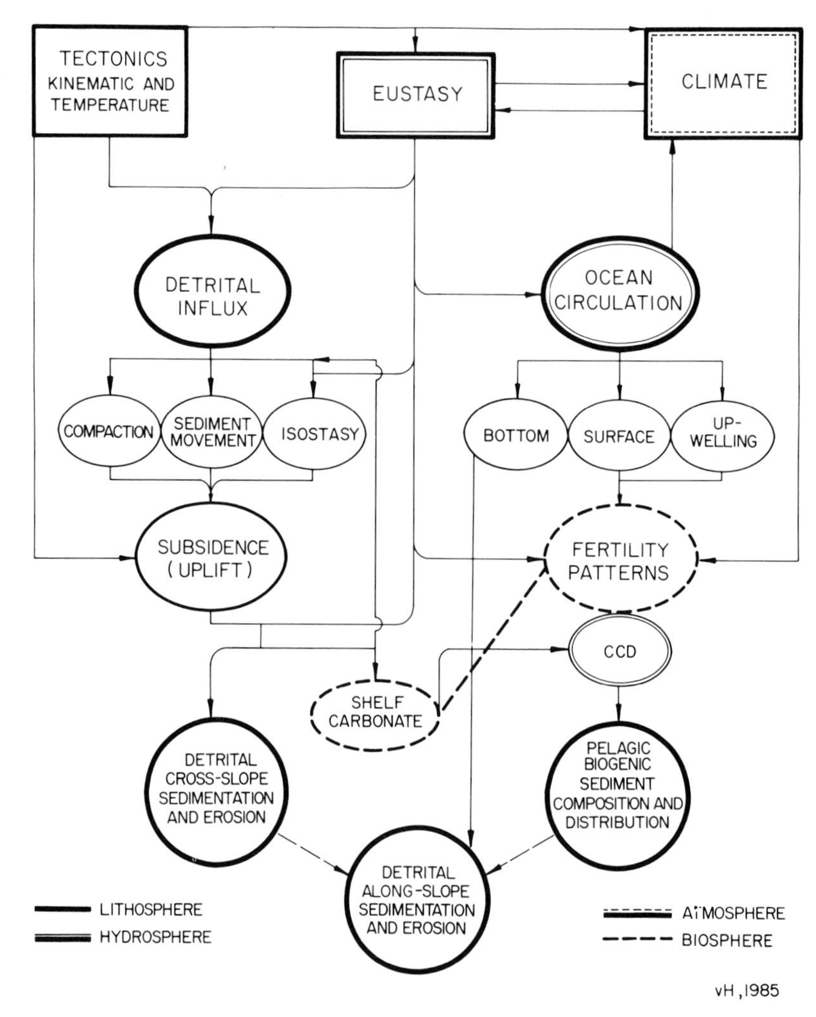

FIG. 20. Summary of the hierarchy and interaction of the principal controls on sedimentation and erosion along the US Atlantic slope and rise (modified from Mountain & Tucholke 1985).

the authors that effects of processes which produced the coastal onlap curve of Vail and his colleagues can be detected in the deep sea, particularly during the Cenozoic as suggested by Vail *et al.* (1980) and more recently by Poag and others (see summary by Kerr 1984). This is not to ignore contrary opinions and other modes of seismic analysis and modelling which are being applied to the study area (Watts 1982; Thorne & Watts 1984; Watts & Thorne 1984; see also discussion by Miller *et al.* 1985, which seeks to reconcile some of the divergent viewpoints). Although our Leg 93 seismic evidence is not discussed extensively in the present paper, it is described in more detail by Van Hinte *et al.* (1985a, b).

In Fig. 20 feedback loops have been added which emphasize the relationships between changes in climate, eustatic sea-level and ocean circulation. The diagram further stresses the crucial role of the biosphere in sedimentation processes. Note, for instance, that eustasy influences the oceanic fertility pattern not only indirectly by its influence on ocean circulation, but also directly. In particular, during times of falling sea-level, the ocean receives an increased amount of nutrients and a growing rain of plankton skeletons adds to the deep sea sediment, thereby suppressing the CCD. The reverse process takes place in the deep sea during times of rising sea-level when nutrients and carbonates are trapped on the widening shelves. This effect is particularly noticeable during extreme high stands and rapid sea level rises and can be seen at Site 603 in the starved Senonian record of unit III and in the Aptian–Albian (unit IV) and early–middle Miocene (unit ID) respectively.

ACKNOWLEDGMENTS: The authors are grateful to the Deep Sea Drilling Project for the opportunity to participate on Legs 93 and 95, to Captain Sidney Schuman and the support staffs of the *Glomar Challenger* for their untiring efforts on our behalf, and to the Marine Studies Committee of the Geological Society of London for the invitation to present and publish the results in this symposium. Discussions with Charlotte B. Schreiber and Peter R. Vail have been most helpful, and critical reviews by Colin Summerhayes and Brian Funnell measurably improved the manuscript. Kim Riddle, Rosemary Raymond, Jim Breza and the DSDP production staff assisted generously in the preparation of the manuscript.

References

ALVAREZ, L.W., ALVAREZ, W., ASARO, F. & MICHEL, H.V. 1980. Extraterrestrial cause for the Cretaceous-Tertiary extinction. *Science* **208,** 1095–108.

BACKMAN, J. & SHACKLETON, N.J. 1983. Quantitative biochronology of Pliocene and early Pleistocene calcareous nannofossils from the Atlantic, Indian, and Pacific Oceans. *Mar. Micropaleont.* **8,** 141–70.

BERGGREN, W.A., KENT, D. & VAN COUVERING, J.A. 1985. The Neogene: Part 2. Neogene geochronology and chronostratigraphy. *In*: SNELLING, N.J. (ed.), *The Chronology of the Geological Record. Mem. geol. Soc. Lond.* **10,** 211–60. Published by Blackwell Scientific Publications, Oxford.

CIESIELSKI, P.F. & WEAVER, F.M. 1983. Neogene and Quaternary paleoenvironment history of Deep Sea Drilling Project Leg 71 sediments, southwest Atlantic Ocean. *In*: LUDWIG, W.J., KRASHENINNIKOV, V.A. *et al., Init. Repts. DSDP*, **71,** US Govt. Print. Off., Washington, DC 461–77.

EWING, J.I. & HOLLISTER, C.D. 1972. Regional aspects of deep sea drilling in the western North Atlantic. *In*: HOLLISTER, C.D., EWING, J.I. *et al., Init. Repts. DSDP*, **11.** US Govt. Print. Off., Washington, DC. 951–73.

—— & RABINOWITZ, P.D. (eds) 1984. *Eastern North American Continental Margin and Adjacent Ocean Floor, 34° to 41°N and 68° to 78°W.* Ocean Margin Drilling Program, Regional Atlas Series, 4. Marine Science International, Woods Hole, Massachusetts.

GARTNER, S. 1977. Calcareous nannofossil biostratigraphy and revised zonation of the Pleistocene. *Mar. Micropaleont.* **2,** 1–25.

GROW, J.A. 1980. Deep structure and evolution of the Baltimore Canyon Trough in the vicinity of the COST B-3 well. *Circ. US geol. Surv.* **833,** 117–25.

——, SCHLEE, J.S. & DILLON, W.P. 1980. MCS reflection profiles collected along the US Continental Margin 1978. *Open File-Rept. US geol. Surv.* **80–834.**

HALLAM, A. 1984. Continental humid and arid zones during the Jurassic and Cretaceous. *Palaeogeogr., Palaeoclimatol., Palaeoecol.* **47,** 195–223.

HOLLISTER, C.D., EWING, J.I. *et al.* 1972. *Init. Repts. DSDP* **11,** US Govt. Print. Off., Washington, DC. 1077 p.

HUTCHINSON, D.R., GROW, J., KLITGORD, K. & SWIFT, B.A. 1983. Deep structure and evolution of the Carolina Trough. *In*: WATKINS, J.S. & DRAKE, C.L. (eds), *Studies in Continental Margin Geology. Mem. Am. Ass. Petrol. Geol.* **34,** 129–52.

JANSA, L.F., ENOS, P., TUCHOLKE, B., GRADSTEIN, F. & SHERIDAN, R.E. 1979. Mesozoic-Cenozoic sedimentary formations of the North American Basin; western North Atlantic. *In*: TALWANI, M., HAY, W. & RYAN, W.B.F. (eds), *Deep Drilling Results in the Atlantic Ocean: Continental Margins and Paleoenvironment. Maurice Ewing Series 3.* Am. Geophys. Union., Washington DC. 1–57.

—— & MacQueen, R.W. 1978. Stratigraphy and hydrocarbon potential of the central North American Basin. *Geosci. Can.* **5,** 173–83.

—— & Wiedmann, J. 1982. Mesozoic-Cenozoic development of the Eastern North American and Northwest African Continental Margins: A comparison. *In*: Von Rad, U., Hinz, K., Sarnthein, M. & Seibold, E. (eds), *Geology of the Northwest African Continental Margin*. Springer-Verlag, Berlin. 215–69.

Kerr, A.R. 1984. Vail's sea-level curves aren't going away. *Science* **226,** 677–8.

Klitgord, K.D. & Grow, J.A. 1980. Jurassic seismic stratigraphy and basement structure of the western Atlantic magnetic quiet zone. *Bull. Am. Ass. Petrol. Geol.* **64,** 1658–80.

Lancelot, Y., Winterer, E.L. *et al.* 1980. *Init. Repts DSDP* **50.** US Govt Print. Off., Washington, DC. 868 p.

Leg 93 Scientific Party/Leg 94 Scientific Party 1984. Deep Sea Drilling Project drills margin and studies paleoclimate. *Geotimes* **29,** No. 4, 16–18.

Leg 95 Scientific Party 1984. Deep Sea Drilling Project adds data on the Atlantic margin. *Geotimes* **29,** No. 5, 14–16.

Leg 96 Scientific Party 1984. Challenger drills Mississippi Fan. *Geotimes* **29,** No. 5, 15–17.

Libby-French, J. 1984. Stratigraphic framework and petroleum potential of northeastern Baltimore Canyon Trough, mid-Atlantic outer continental shelf. *Bull. Am. Ass. Petrol. Geol.* **68,** 50–73.

Mattick, R.E., Girard, O.J. Jr., Scholle, P. & Grow, J.A. 1978. Petroleum potential of United States Atlantic slope, rise and abyssal plain. *Bull. Am. Ass. Petrol. Geol.* **62,** 592–608.

McGregor, B., Stubblefield, W., Ryan, W. & Twichell, D.C. 1982. Wilmington Submarine Canyon: A marine fluvial-like system. *Geology* **10,** 27–30.

Miller, K.G., Mountain, G. & Tucholke, B.E. 1985. Oligocene glacio-eustasy and erosion on the margins of the North Atlantic. *Geology* **13,** 10–13.

Mountain, G.S. & Tucholke, B.E. 1985. Mesozoic and Cenozoic stratigraphy of the United States Atlantic continental slope and rise. *In*: Poag, C.W. (ed.), *Geologic Evolution of the United States Atlantic Margin*. Van Nostrand-Reinhold, Inc., Stroudsburg, Pa., 293–341.

Okada, H. & Bukry, D. 1980. Supplementary modification and introduction of code numbers to the 'low-latitude coccolith biostratigraphic zonation' (Bukry 1973, 1975). *Mar. Micropaleont.* **5,** 321–5.

Olsson, R.K. 1980. The New Jersey coastal plain and its relationship with the Baltimore Canyon Trough. *In*: Manspeizer, E. (ed.), *Field Studies New Jersey Geology and Guide to Field Trips of 52nd Annual Meeting New York State Geological Association*. Rutgers University Newark, New Jersey. 116–29.

Palmer, A.A. 1983. Biostratigraphic and paleoenvironmental results from Neogene radiolarians, U.S. mid-Atlantic coastal plain. *Bull. Am. Ass. Petrol. Geol.* **67,** 528–9.

Parkinson, N. & Summerhayes, C.P. 1985. Synchronous global sequence boundaries. *Bull. Am. Ass. Petrol. Geol.* **69,** 685–7.

Poag, C.W. 1982. Stratigraphic reference section for Georges Bank Basin depositional model for New England passive margin. *Bull. Am. Ass. Petrol. Geol.* **66,** 1021–41.

—— 1985. Cenozoic and Upper Cretaceous sedimentary facies and depositional systems of the New Jersey slope and rise. *In*: Poag, C.W. (ed.), *Geologic Evolution of the United States Atlantic Margin*. Van Nostrand-Reinhold, Inc., Stroudsburg, Pa., 343–365.

Prior, D.B., Coleman, J.M. & Doyle, E.H. 1984. Antiquity of the continental slope along the middle-Atlantic margin of the United States. *Science* **223,** 926–8.

Robb, J.M. 1980. High-resolution seismic reflection profiles collected by the R/V James M. Gills, Cruise GS 7903–4 in the Baltimore Canyon Outer Continental Shelf area, offshore New Jersey. *Open-File Rept. US geol. Surv.* **80–934.** 3 pp.

—— 1984. Spring sapping on the lower continental slope, offshore New Jersey. *Geology* **12,** 278–82.

——, Hampson, J.C., Kirby, J.R., & Twichell, D.C. 1981a. Geology and potential hazards of the Continental Slope between Lindenkohl and South Toms Canyons offshore mid-Atlantic United States. *Open File-Rept. US geol. Surv.* **81–600.**

——, Hampson, J.J. Jr. & Twichell, D.C. 1981b. Geomorphology and sediment stability of a segment of the United States continental slope off New Jersey. *Science* **211,** 935–7.

——, Kirby, J., Hampson, J.J. Jr., Gibson, P. & Hecker, B. 1983. Furrowed outcrops of Eocene chalk on the lower continental slope offshore New Jersey. *Geology* **11,** 182–6.

Robertson, A.H.F. & Bernoulli, D. 1982. Stratigraphy, facies and significance of late Mesozoic and early Tertiary sedimentary rocks of Fuerteventura (Canary Islands) and Maio (Cape Verde Islands). *In*: Von Rad, U., Hinz, K., Sarnthein, M. & Seibold, E. (eds), *Geology of the Northwest African Continental Margin*. Springer-Verlag, Berlin. 498–525.

Sarti, M. *et al.* 1984. Early Cretaceous turbidite sedimentation at DSDP Site 603, off Cape Hatteras (USA) (DSDP/IPOD Leg 93). *Abstracts for North Atlantic Palaeoceanography, Geological Society of London*. London, November, 1984.

Schlee, J.S. 1981. Seismic stratigraphy of Baltimore Canyon Trough. *Bull. Am. Ass. Petrol. Geol.* **65,** 26–53.

——, Behrendt, J., Grow, J., Robb, J. & Mattick, R.E. 1976. Regional geologic framework off northeastern United States. *Bull. Am. Ass. Petrol. Geol.* **60,** 926–51.

——, Dillon, W. & Grow, J.A. 1979. Structure of the continental slope off the eastern United States. *Spec. Publ. Soc. econ. Paleont. Miner.* **27,** 95–117.

Sheridan, R.E., Gradstein, F.M. *et al.* 1983. *Initial Repts DSDP* **76.** US Govt. Print. Off., Washington. DC. 943 p.

Sliter, W.V. 1977. Cretaceous foraminifers from the southwestern Atlantic Ocean, Leg 36, Deep Sea Drilling Project. *In*: Barker, P., Dalziel, I.W.D. *et al.*, *Init. Rept. DSDP*, **36.** US Govt. Print. Off., Washington, DC. 519–74.

SMIT, J. & HERTOGEN, J. 1980. An extraterrestrial event at the Cretaceous-Tertiary boundary. *Nature* **285,** 198–200.
STANLEY, D.J., NELSEN, T. & STUCKENRATH, R. 1984. Recent sedimentation on the New Jersey slope and rise. *Science* **226,** 125–33.
STOW, D.A.V. & LOVELL, J.P.B. 1979. Contourites: their recognition in modern and ancient sediments. *Earth Sci. Rev.* **14,** 251–91.
STUBBLEFIELD, W.L., MCGREGOR, B., FORDE, E., LAMBERT, D. & MERRILL, G.F. 1982. Reconnaissance in DSRV Alvin of a 'fluvial-like' meander system in Wilmington Canyon and slump features in South Wilmington Canyon. *Geology* **10,** 31–6.
THORNE, J. & WATTS, A.B. 1984. Seismic reflectors and unconformities at passive continental margins. *Nature* **311,** 365–8.
TUCHOLKE, B.E. & LAINE, E.P. 1982. Neogene and Quaternary development of the lower continental rise off the current United States east coast. *In*: WATKINS, J.S. & DRAKE, C., (eds.), *Studies in Continental Margin Geology. Mem. Am. Ass. Petrol. Geol.* **34,** 295–305.
—— & MOUNTAIN, G.S. 1979. Seismic stratigraphy, lithostratigraphy and paleosedimentation patterns in the North American basin. *In*: TALWANI, M., HAY, W. & RYAN, W.B.F. (eds), *Deep Drilling Results in the Atlantic Ocean: Continental Margins and Paleoenvironment. Maurice Ewing Series 3.* Am. Geophys. Union, Washington, DC. 58–86.
UCHUPI, E., SANCETTA, C., EUSDEN, J.J. JR., BOLMER, S.J. JR., MCCONNELL, R. & LAMBIASE, J.J. 1984a. Lithofacies of top of J/J1 to Beta sequence. *In*: EWING, J.I. & RABINOWITZ, P.D. (eds), *Ocean Margin Drilling Program, Regional Atlas Series, 4.* Marine Science International, Woods Hole, Massachusetts.
—— 1984b. Lithofacies of lower part of Beta to A* sequence. *In*: EWING, J.I. & RABINOWITZ, P.D. (eds), *Ocean Margin Drilling Program, Regional Atlas Series, 4.* Marine Science International, Woods Hole, Massachusetts.
VAIL, P.R., HARDENBOL, J. & TODD, R.G. 1984. Jurassic unconformities, chronostratigraphy, and sea-level changes from seismic stratigraphy and biostratigraphy. *In*: SCHLEE, J.S. (ed.), *Interregional Unconformities and Hydrocarbon Accumulation. Mem. Am. Ass. Petrol. Geol.* **36,** 129–44.
—— & MITCHUM, R.M. JR. 1979. Global cycles of relative changes of sea level from seismic stratigraphy. *In*: WATKINS, J.S., MONTADERT, L. & DICKERSON, P.W. (eds), *Geophysical and Geological Investigations of Continental Margins. Mem. Am. Ass. Petrol. Geol.* **29,** 469–72.
———, SHIPLEY, T. & BUFFLER, R.T. 1980. Unconformities of the North Atlantic. *Phil. Trans. R. Soc. Lond.* **294,** 137–55.
—— *et al.* 1977. Seismic stratigraphy and global changes of sea level. *In*: PAYTON, C.E. (ed.), *Seismic Stratigraphy-Applications to Hydrocarbon Exploration. Mem. Am. Ass. Petrol. Geol.* **26,** 49–212.
VAN HINTE, J. *et al.* 1983. The continental rise off North America. *Nature* **305,** 386.
——— 1985a. DSDP Site 603: First deep penetration (> 1000 m) of the continental rise along the passive margin of eastern North America. *Geology* **13,** 392–396.
——— 1985b. Deep sea drilling on the upper continental rise off New Jersey: DSDP Sites 604 and 605. *Geology* **13,** 397–400.
VON RAD, U. & ARTHUR, M.A. 1979. Geodynamic, sedimentary and volcanic evolution of the Cape Bojador continental margin (NW Africa). *In*: TALWANI, M., HAY, W. & RYAN, W.B.F. (eds). *Deep Drilling Results in the Atlantic Ocean: Continental Margins and Paleoenvironments. Maurice Ewing Series 3 Am. Geophys. Union,* Washington, DC. 187–203.
—— *et al.* 1984. Kretazisch-Kanozische Stratigraphie und Palaeoenvironment-Entwicklungam kontinentalfuss von dem ostlichen Nordamerika erste Ergebnisse von DSDP Leg 93. *Geologisches Jahrbuch* **75,** 237–259.
WATTS, A.B. 1982. Tectonic subsidence, flexure and global changes of sea level. *Nature* **297,** 469–74.
—— & THORNE, J. 1984. Tectonics, global changes in sea level and their relationship to stratigraphical sequences at the US Atlantic continental margin. *Mar. Perol. Geol.* **1,** 319–39.

SHERWOOD W. WISE, JR., Department of Geology, Florida State University, Tallahassee, Florida 32306, USA.
JAN E. VAN HINTE, Vrije Universiteit, Amsterdam, The Netherlands.
GREGORY S. MOUNTAIN, Lamont-Doherty Geological Observatory, Palisades, New York 10964, USA.
BRIAN N.M. BIART, Open University, Milton Keynes MK7 6AA.
J. MITCHENER COVINGTON, Florida State University, Tallahassee, Florida 32306, USA.
WARREN S. DRUGG, Chevron Oil Field Research Company, La Habra, California 90631, USA.
DEAN A. DUNN, University of Southern Mississippi, Hattiesburg, Mississippi 39401, USA.
JOHN FARRE, Lamont-Doherty Geological Observatory, Palisades, New York 10964, USA.
DANIEL HABIB, Queens College of the City University of New York, Flushing, New York 11367, USA.
JANET A. HAGGERTY, University of Tulsa, Tulsa, Oklahoma 74104, USA.
MARK W. JOHNS, Texas A & M University, College Station, Texas 77843, USA.
THOMAS H. LANG, Florida State University, Tallahassee, Florida 32306, USA.
PHILIP A. MEYERS, University of Michigan, Ann Arbor, Michigan 48109, USA.

KENNETH G. MILLER, Lamont-Doherty Geological Observatory, Palisades, New York 10964, USA.
MICHEL R. MOULLADE, Universite de Nice, 06034 Nice Cedex, France.
JAY P. MUZA, Florida State University, Tallahassee, Florida 32306, USA.
JAMES G. OGG, University of California, San Diego, California 92093, USA.
MAKOTO OKAMURA, Kochi University, Kochi City, Japan.
MASSIMO SARTI, Universita di Ferrara, 44100 Ferrara, Italy.
ULRICH VON RAD, Bundesanstalt fur Geowissenschaften und Rohstoffe, Hannover 51, Federal Republic of Germany.

Late Cretaceous anoxic events, sea-level changes and the evolution of the planktonic foraminifera

Malcolm B. Hart & Kim C. Ball

SUMMARY: This paper investigates the relationship, if any exists, between the evolution of the planktonic foraminifera and the presence of anoxic water masses in the water column. The relationship with changes in sea-level over the expanded Cretaceous shelves is also discussed. New data from DSDP Legs 80 and 95, coupled with that from southern England, north-east England and the North Sea Basin will be used to assess the evolution of the microfauna.

Living planktonic foraminifera are loosely stratified in the upper levels of the water column into three broad faunas (Bé 1977); 'shallow-water' (mature tests at <50 m), 'intermediate-water' (mature tests between 50 and 100 m) and 'deep-water' (mature tests below 100 m). All species migrate diurnally within their preferred depth range as a response to light and feeding cycles. All juveniles are usually found in surface waters shallower than 50 m. As they mature they migrate downwards to their preferred water depth. This depth is controlled by a wide range of physical and biological parameters, which may have been different in the geological past.

Oceanic sediment samples from above the lysocline tend to (or should) contain representatives of the shallow, intermediate and deep-water faunas. Such samples give information about latitudinal variations in the fauna, but provide no data about the depths at which particular species lived. However, in shelf seas we would only expect to find those species associated with the 'shallow-water' fauna (or perhaps the 'intermediate-water' fauna at a time of eustatic highstand), providing that depth was the only environmental restriction.

Recently Hart & Bailey (1979), Hart (1980a) and Caron (1983a) have suggested: (1) that the Cretaceous fauna (and probably that of the Cenozoic also) was similarly stratified in the water column, and (2) that these ancient homeomorphs of the modern fauna would probably have occurred in the same depth order in that water column, though not necessarily at the same absolute depths, as today's fauna.

Cretaceous planktonic foraminifera

Between 1976 and 1983 the 'European Working Group on Planktonic Foraminifera' attempted to stabilize some aspects of the taxonomy and stratigraphic distribution of the mid-Late Cretaceous fauna (Robaszynski & Caron 1979; Robaszynski *et al.* 1984). This work provides a set of reliable and consistent data on those genera and species considered by the working group. The work led to a zonation (Fig. 1) that is usable over large areas of the globe. The distribution of species within each zone shows several features of note:

(1) an abrupt increase in diversification of the fauna in the mid-Cenomanian. This change is synchronous with the mid-Cenomanian 'non-sequence' of Carter & Hart (1977) and coincides with a distinct eustatic event (Hart 1980b);

(2) an abrupt fall in diversity near the Cenomanian–Turonian boundary, which is followed very closely by a sharp recovery. The level at which the diversity falls is widely reported (Hancock & Kauffman 1979) as a sea-level maximum, and is different, therefore, to (1) above;

(3) a fall in diversity in the Coniacian–Santonian with only a slow recovery during the Campanian. The Late Campanian sea-level maximum (Hancock 1976; Hancock & Kauffman 1979) does not equate with the highest diversity levels.

(4) While there is only a slight drop in diversity in the Late Maastrichtian, the rate of appearance of new taxa is declining markedly, well in advance of the 'terminal Cretaceous event'.

The most distinctive, and short-lived, event is that associated with the Cenomanian–Turonian boundary; it will be used as a test case.

The Cenomanian–Turonian boundary event

In this paper the authors accept for this boundary the stratigraphic level defined by Rawson *et al.* (1978). Just below this stratigraphic level in the UK is the *A. plenus* Marl, and its correlative, the Black Band. This stratum is recognized in North

From Summerhayes, C.P. & Shackleton, N.J. (eds), 1986, *North Atlantic Palaeoceanography*, Geological Society Special Publication No. 21, pp. 67–78.

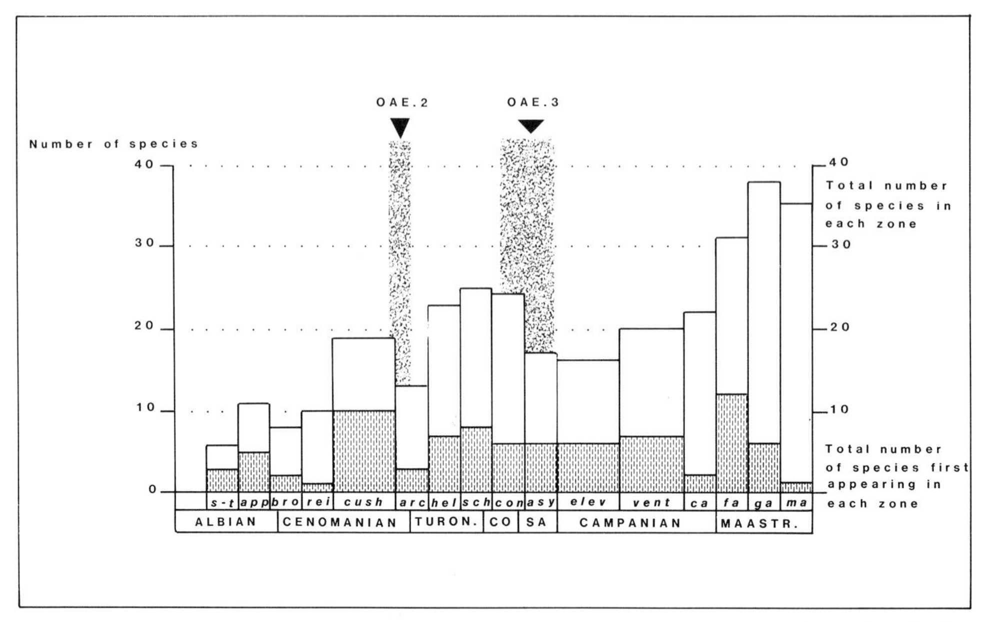

FIG. 1. The distribution of species within the zones of the mid-Late Cretaceous. The graphs are based only on the genera and species considered by the European Working Group on Planktonic Foraminifera (Robaszynski & Caron 1979; Robaszynski *et al.* 1984). The values plotted for each zone are a compromise, as within each zone certain taxa both appear and disappear. The zones are: *s-t, Rotalipora subticinensis* (Gandolfi); *app. R. appenninica* (Renz); *bro, R. brotzeni* (Sigal); *rei, R. reicheli* (Mornod); *cush, R. cushmani* (Morrow); *arc, Whiteinella archaeocretacea* Pessagno; *hel, Praeglobotruncana helvetica* (Bolli); *sch, Marginotruncana schneegansi* (Sigal); *con, Dicarinella concavata* (Brotzen); *asy, D. asymetrica* (Sigal); *elev, Globotruncanita elevata* (Brotzen); *vent, Globotruncana ventricosa* White; *ca, Globotruncanita calcarata* (Cushman); *fa, Globotruncana falsostuarti* Sigal; *ga, Gansserina gansseri* (Bolli); *ma, Abathomphalus mayaroensis* (Bolli).

Sea exploration wells by a distinctive, positive, kick on gamma-ray logs, and has been identified also at DSDP Sites 549 and 550 (Fig. 2). Using the gamma-ray logs and microfaunal data, successions on the Goban Spur (DSDP Leg 80) can be correlated with south-west England and the North Sea Basin (Fig. 3). This 'event' is the sedimentological manifestation of what has been termed 'Oceanic Anoxic Event 2' (Schlanger & Jenkyns 1976; Jenkyns 1980; Graciansky *et al.* 1982; Müller *et al.* 1983).

The Black Band represents anoxic depositional conditions, with a much-reduced benthonic fauna of rather simple, arenaceous foraminifera (Hart & Bigg 1981). The abundance of *Chondrites*-type bioturbation in the *A. plenus* Marl, and in equivalent horizons in northern Germany has also been taken to be a sign of anoxia (Bromley & Ekdale 1984).

The abundant, well-preserved, Late Cenomanian fauna (with *Rotalipora cushmani* (Morrow) and *R. greenhornensis* (Morrow)) disappears immediately below a dark mudstone at Site 551 (Fig. 4). The mudstone itself contains only a poor, etched planktonic fauna together with rare fish scales and quite abundant (though badly preserved) radiolarians. Over most of southern England the *A. plenus* Marl contains a diagnostic microfauna (Jefferies 1962; Carter & Hart 1977). *R. greenhornensis* always becomes extinct between Beds 1 and 2, while *R. cushmani* goes out abruptly between Beds 3 and 4. This can be seen particularly well at Shillingstone (Dorset), which is located (Carter & Hart 1977) only a few miles from the site of the Winterbourne Kingston borehole (the log of which is shown in Fig. 3).

The appearance of OAE.2 within the shelf succession of north-west Europe indicates that a well-developed oxygen minimum zone has moved upwards relative to the sea floor. This could be the result of:

(1) an increase in the width of the oxygen minimum zone, which brought it up to the appropriate depth, or
(2) an eustatic rise in sea-level, which would bring the oxygen minimum zone upwards by a similar amount—while still remaining at the same position in the water column.

The evidence of Hancock (1976), Hancock & Kauffman (1979) and Carter & Hart (1977) has

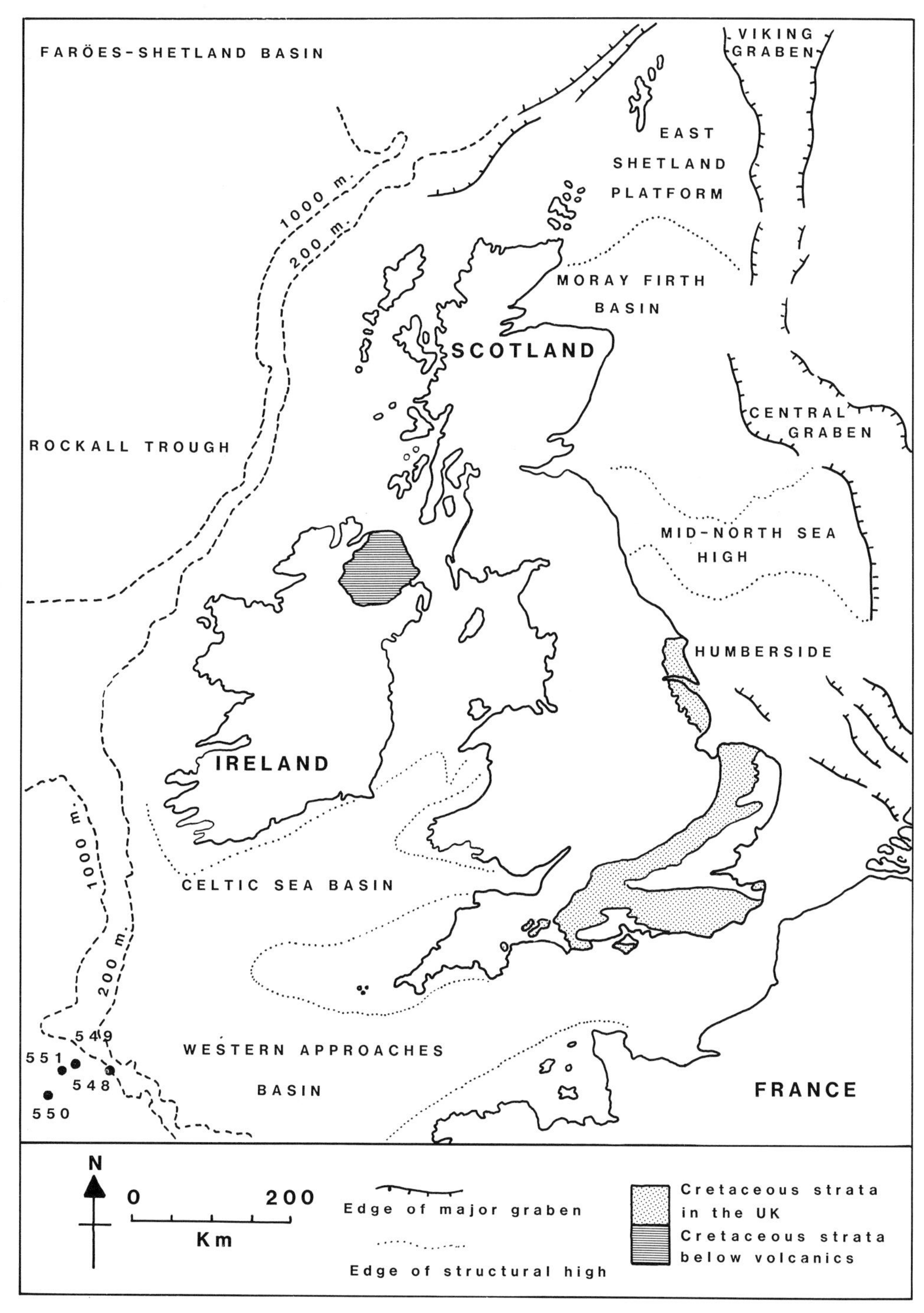

FIG. 2. The north-west European continental shelf. The outcrop of Cretaceous strata in the UK is shown, together with the principal structural trends. The sites of DSDP Leg 80 boreholes (548–551) are also indicated on the Goban Spur.

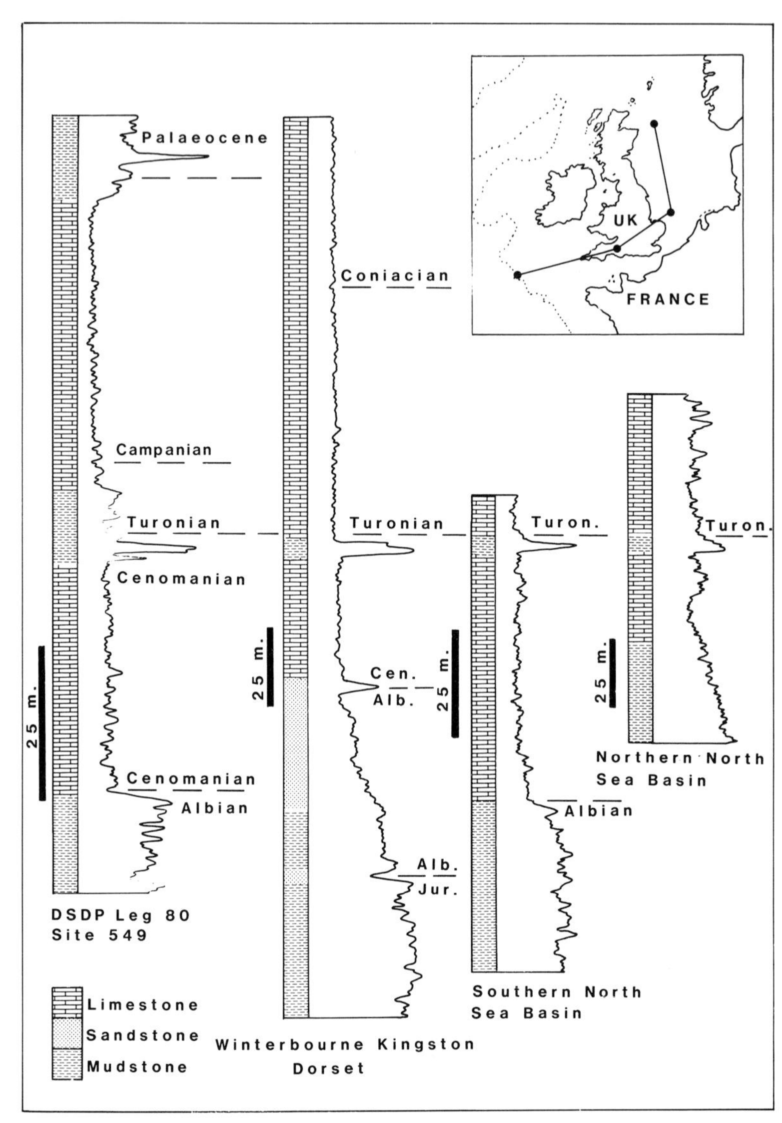

FIG. 3. Gamma-ray log correlation of four sites across the north-west European continental shelf. The data from the Northern North Sea Basin is taken from Burnhill & Ramsay (1981), while that from the southern North Sea Basin was kindly provided by Shell UK Exploration & Production Ltd. The log from the Winterbourne Kingston borehole is from Rhys *et al.* (1981).

shown that an eustatic rise did occur at this level. This can be confirmed by a study of detrital minerals in the carbonate sediments of south-west England (Hancock 1969). The synchronous disappearance of *R. cushmani* and *R. greenhornensis* at Site 551 is therefore preservational, as both species are still extant in the water column above the oxygen-depleted layer. Eventually the oxygen minimum zone must have expanded, and as a result the life cycle of *R. greenhornensis*, then that of *R. cushmani*, became affected. The latter must be prior to the appearance of the very black mudstone and associated marls of the Humberside succession (Hart & Bigg 1981). At the same level there are major changes in the remainder of the planktonic fauna; *Hedbergella* (with a more extraumbilical aperture) is largely replaced by the newly evolved *Whiteinella* (with a more typically extraumbilical-umbilical aperture), and the new genera *Marginotruncana* and *Dicarinella* expanded considerably in numbers.

Immediately above the black mudstone (of

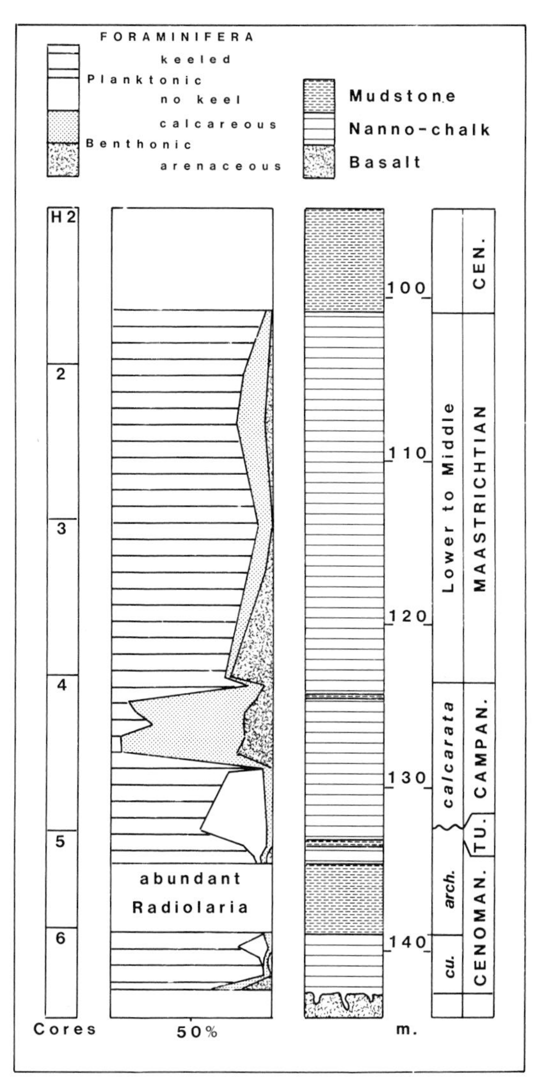

FIG. 4. The mid-Late Cretaceous succession at DSDP Site 551 on the Goban Spur. The Late Cenomanian mudstone with abundant radiolaria is OAE 2. The foraminiferal data is based on total counts of the 250–500 μm size fraction.

Late Cenomanian age) at Site 551 is a beautifully preserved planktonic fauna belonging to the *Whiteinella archaeocretacea* Pessagno Zone, which contains forms transitional between highly rugose whiteinellids and *'Praeglobotruncana' praehelvetica* (Trujillo). True *P. helvetica* (Bolli) is not recorded at Site 551 as the *W. archaeocretacea* Zone is cut out by a major hiatus that is overlain by sediments of Late Campanian age. In the more complete successions of southern England *P. helvetica* appears shortly after *P. praehelvetica* (Hart 1982).

Patterns of evolution

The authors consider it probable that the anoxic event near the Cenomanian–Turonian boundary had a major effect on the planktonic Foraminifera (Hart 1980a; Wonders 1980; Caron 1983b). In their evolutionary model for the mid-Cretaceous, Hart & Bailey (1979) argued that the *Rotalipora* fauna of the Late Cenomanian was adapted to adult maturity at a particular level in the water column, and suggested that the migration of an oxygen minimum zone upwards in the water column effectively removed this preferred depth habitat. Some members of the shallow-water *Hedbergella* fauna underwent morphological change (related largely to the position of the aperture) and are thereby classified as *Whiteinella*. Changes in the *Praeglobotruncana* fauna are less obvious, and this genus may have been more tolerant of the environmental stress caused by a period of oxygen depletion.

With the dominant element of the deeper-water fauna removed, new taxa were rapidly developed, particularly within the genera *Dicarinella* and *Marginotruncana*. Just as the *Dicarinella* lineage began to diversify, part of the *Whiteinella* population began to develop a faint keel, together with a plano-convex morphology, which by analogy with the modern fauna, suggests evolution to a deeper water habitat for the adult form. The authors suggest that the anoxic event was followed by rapid re-colonization of the deeper-water environments once normal oceanographic conditions were re-established within the range of living planktonic foraminifera. The evolution of *Whiteinella* to *P. helvetica* is shown in Fig. 5. This plano-convex morphotype, with a single keel (or a closely-spaced double keel) and ventrally inflated 'globigerine' chambers, can be seen to a greater or lesser degree in *Dicarinella primitiva* (Dalbiez), *D. concavata* (Brotzen), *D. asymetrica* (Sigal), *Globotruncanita elevata* (Brotzen), *G. calcarata* and *Gansserina gansseri* (Bolli). Figure 5 shows that in most of these cases there is no *direct* genetic connection; the plano-convex morphotype develops from either biconvex or whiteinellid-type forms, presumably by repeated colonization of lower levels of the water column. The evolution of the *Gansserina* group from an *Archaeoglobigerina* stock in the mid-Maastrichtian provides a repeat of the *P. helvetica* trend. Some confirmation of the deeper-water habitat of the plano-convex morphotype comes from the palaeotemperature work of Douglas & Savin (1978) and Bé (1977). Douglas & Savin place *G. elevata* firmly within the deep-water fauna and Boersma & Shackleton (1981) also record low temperatures for plano-convex morphotypes (*G. calcarata* 18°C, *G. subspinosa* 18°C, *G. elevata* 18°C, *G. gansseri* 14°C).

There are, however, problems arising from some palaeotemperature data. The biological data (Bé 1977) indicates that non-keeled, thin-shelled, 'globigerine' species normally reside in the shallow-water fauna, while heavily calcified, keeled species dominate the deeper-water faunas. Palaeotemperature data (Boersma & Shackleton 1981; Boersma & Premoli Silva 1983) indicate the

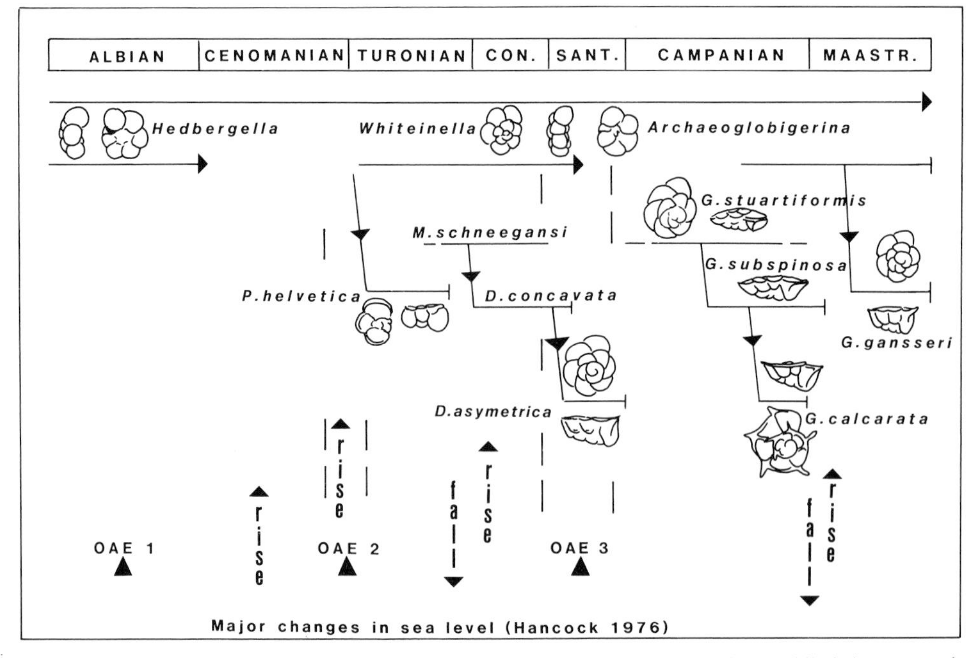

FIG. 5. An evolutionary model for some of the plano-convex planktonic foraminifera. While it is suggested that each trend was the result of evolution to a deeper-water habitat, there is no suggestion that all the species shown are in the correct relative positions. This is purely a diagrammatic presentation.

reverse; keeled species in the Palaeocene give warmer values than non-keeled species, while in the Cretaceous, hedbergellid groups record the lowest temperatures. This latter example is a major concern as hedbergellid taxa are often the only planktonic species found in what are accepted shallow water sediments (based on sedimentological criteria). This apparent conflict must be resolved with some urgency. It must also be remembered that as planktonic foraminifera migrate through various levels in the water column during growth they will inevitably secrete chambers at different temperature levels. An isotopic temperature determination will be based on several specimens, not the individual chambers.

Ocean–shelf comparisons

Drilling at Sites 548–551 on the Goban Spur allows a comparison of faunas from the North Sea Basin, the UK and the North Atlantic Basin, testing our ideas concerning the relationship between morphotype and water depth. In samples of Late Campanian age stenohaline faunas are found across the whole region. At that time palaeotemperatures were almost uniform (Urey *et al* 1951; Lowenstam & Epstein 1954) across the area. The European shelf sea fauna differs markedly from those found in the area of the Goban Spur (Fig. 6), that contain species such as *G. calcarata, G. subspinosa, G. stuartiformis* (Dalbiez) and *Rosita patelliformis* (Gandolfi). None of these taxa have been seen by the authors in samples of this age from southern England. In the North Sea Basin they are also absent, although the fauna (with a dominance of *Globotruncana arca* (Cushman) and *Rosita fornicata* (Plummer)) contains more planktonic Foraminifera than those from on-shore southern England.

G. subspinosa and *G. calcarata* are end members of an evolutionary lineage (Fig. 5), with the latter species being distinctly plano-convex. The authors would argue that this, again, indicates evolution to a deeper-water morphotype, and in this example the palaeotemperature data (Boersma & Shackleton 1981) would appear to be in agreement.

R. patelliformis is a derivative of the *R. fornicata* lineage (Fig. 7) and is in turn the ancestral form of *R. contusa* (Cushman). *R. contusa* is massive, heavily calcified, and often highly irregular in shape; a fact which led Troelson (1955) to suggest that some individuals had settled on the sea floor while still alive. In the Late Maastrichtian chalks of north-west Europe *R. contusa* is

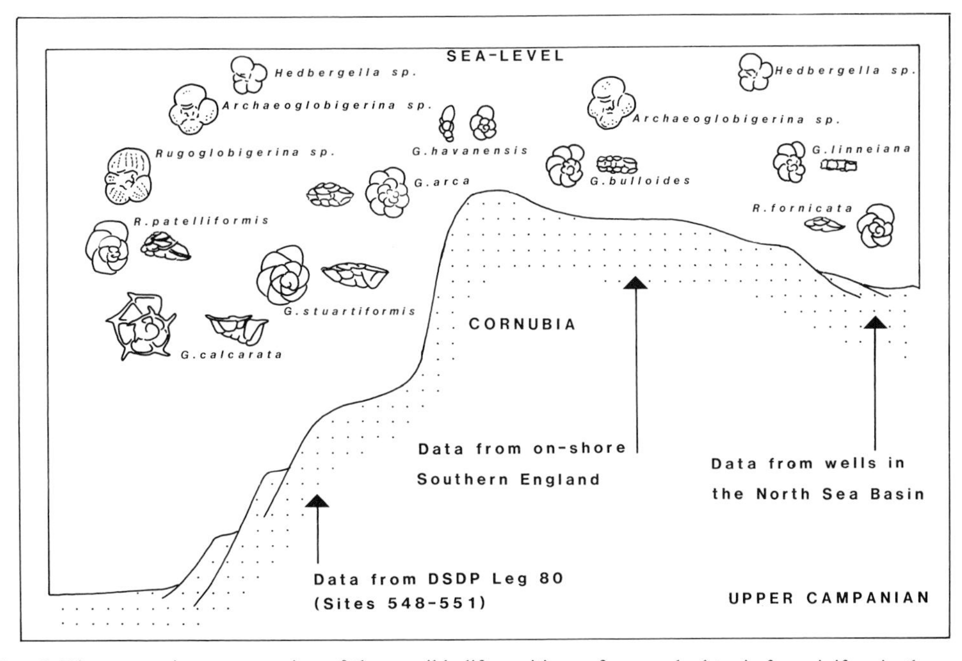

FIG. 6. Diagrammatic representation of the possible life positions of some planktonic foraminifera in the Late Campanian. This is based on the interpretation of the faunas found in samples from the areas indicated.

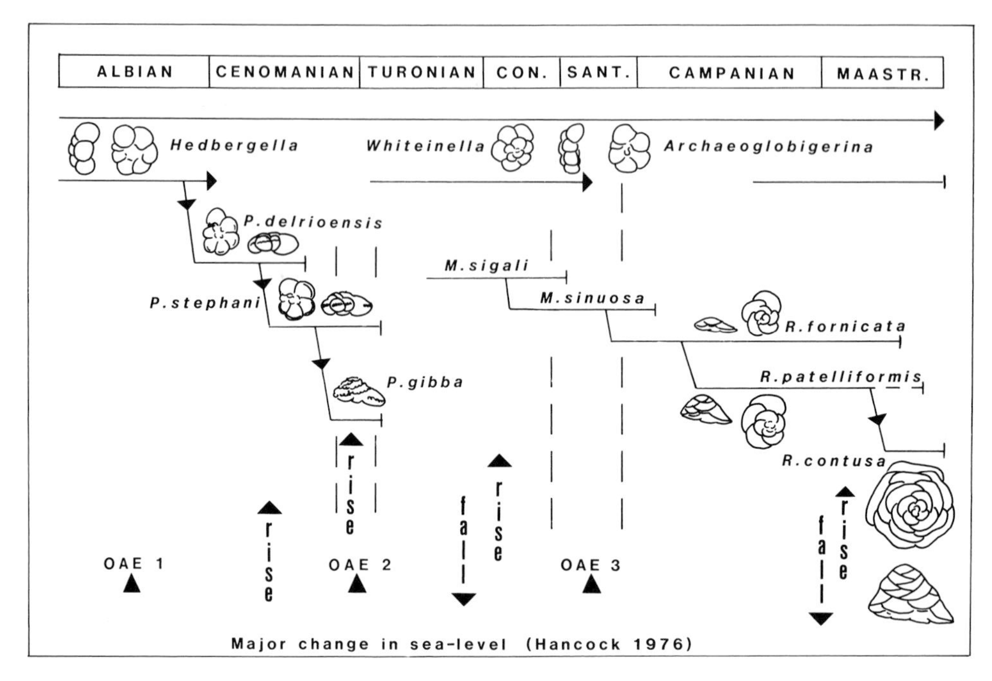

FIG. 7. An evolutionary model from some of the high-spired planktonic foraminifera. The diagram is drawn on the same assumptions as indicated in Fig. 5, and there is no suggestion that *R. contusa* ultimately ended up in deeper water than *P. gibba*, although on purely biological criteria this would appear to have been the case.

most abundant in the deeper parts of the North Sea Basin, and this deeper-water life habitat would seem (biologically) appropriate for such a heavily calcified taxon. This conclusion does, however, conflict with the isotope data of Boersma & Shackleton (1981). The morphological trends seen in the *fornicata–patelliformis–contusa* lineage is a Campanian to Maastrichtian re-run of the *Praeglobotruncana delrioensis* (Plummer)–*P. stephani* (Gandolfi)–*P. gibba* (Klaus) lineage seen in the Cenomanian and Turonian (Klaus 1960a, b).

Evidence of Campanian anoxia

The zone characterized by *G. calcarata* in the Late Campanian records a marked drop in the rate of appearance of new taxa. This level is widely accepted as the maximum transgressive pulse of the Late Cretaceous (Hancock 1976; Hancock & Kauffman 1979). Nyong & Olsson (1984) have shown how Santonian–Campanian transgressive/regressive events can be recognized in the sediments of the Baltimore Canyon Trough (north-west Atlantic Ocean). They also indicate that an oxygen minimum zone was developed in the Late Santonian and Early Campanian. Site 612 (DSDP Leg 95), located in slightly deeper water than the COST wells used by Nyong & Olsson, records an unusual Late Campanian fauna dominated by arenaceous benthonic foraminifera (Fig. 8). A similar, though less dramatic, Late Campanian event is also recorded at Site 551 on the Goban Spur (Fig. 4). This evidence from the Campanian, coupled with that of Nyong & Olsson (1984), Arthur & Schlanger (1979) and Jenkyns (1980) confirm that varying levels of anoxia in the water column were typical of the Late Cretaceous as a whole, and that periodic migration or expansion of the oxygen minimum zone may have interfered with the oceanic plankton on several occasions. The recently published views of Caron (1983b), which reflect the work of the 'European Working Group on Planktonic Foraminifera', have been modified slightly in Fig. 9 to draw attention to the authors' model. Some genera have been omitted for the sake of clarity, but inserted are the generally accepted levels of oxygen minimum zone expansion.

Patterns of evolution

It is clear from Fig. 9 that there are two important levels of taxonomic change, both of which appear

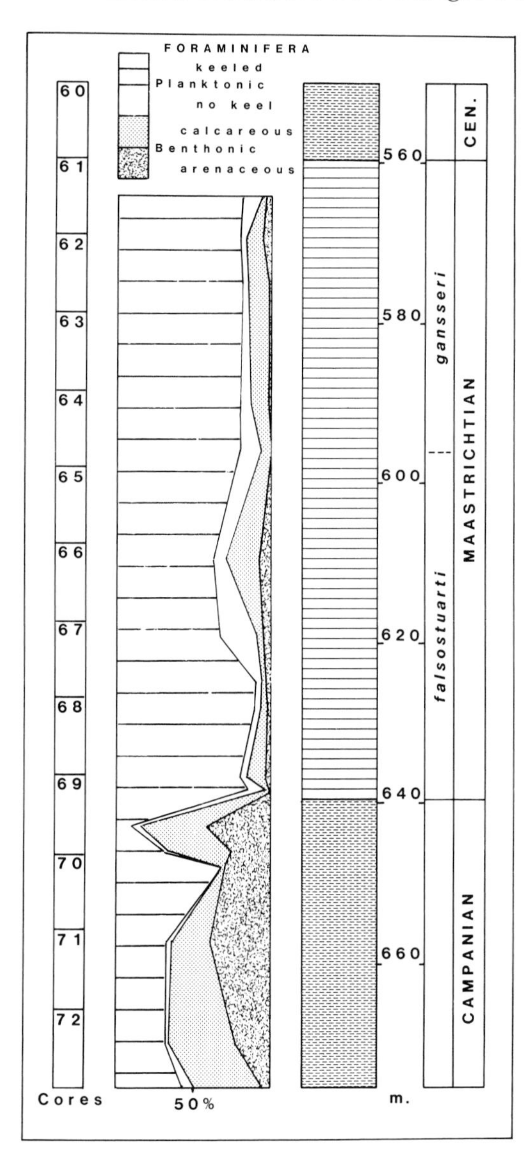

FIG. 8. The Late Cretaceous succession at DSDP Site 612 in the Baltimore Canyon Trough (NW Atlantic Ocean). The dark mudstone of Late Campanian age displays an anomalous foraminiferal fauna for such an environment, and possibly represents some depletion of oxygen in the water column and on the sea floor. The foraminiferal data is based on total counts of the 250–500 μm size fraction. The dark colour in the Late Campanian mudstone comes from the abundant dinoflagellate debris (Dr B. Tocher, pers. comm.).

to coincide with anoxic events. Between these there is progressive speciation, but no major structural changes in the fauna. Characters developed at OAE.2 (the extraumbilical-umbilical aperture typical of the *Whiteinella* and *Marginotruncana* lineages) are progressively passed on to all later taxa, until OAE.3 when the next major change occurs. At that level the true umbilical aperture appears; a character that is passed on to all later taxa. It is significant that this character appears simultaneously in four parallel lineages. In a recently published, theoretical model, Brasier (1982) has related foraminiferal architecture to 'communication distances' within the test. The structural changes in the form of the aperture outlined in this paper could be interpreted as a means of improving protoplasmic communications within the test. That these changes are closely associated with anoxic events must be significant.

Conclusions

The authors believe that there is strong evidence to support their claim that the movement of the oxygen minimum zone in the water column has affected the evolution of the planktonic foraminifera. The times of maximum impact—OAE.2 and OAE.3—saw major morphological changes in the fauna. Both these events are also closely related to sea-level changes (Hart & Bailey 1979; Hart 1980b; Jenkyns 1980; Wonders 1980) although direct proof is lacking. The authors' work on the Cretaceous faunas of north-west Europe certainly points to a water depth control on the *distribution* of the various taxa, and the overwhelming coincidence of *evolutionary change* with sedimentologically proven variations in sea-level must surely indicate a causal relationship.

ACKNOWLEDGEMENTS: The authors acknowledge the assistance they have received from the Natural Environment Research Council and Plymouth Polytechnic. Both authors wish to thank their palaeontological colleagues for fruitful discussion of the planktonic foraminifera.

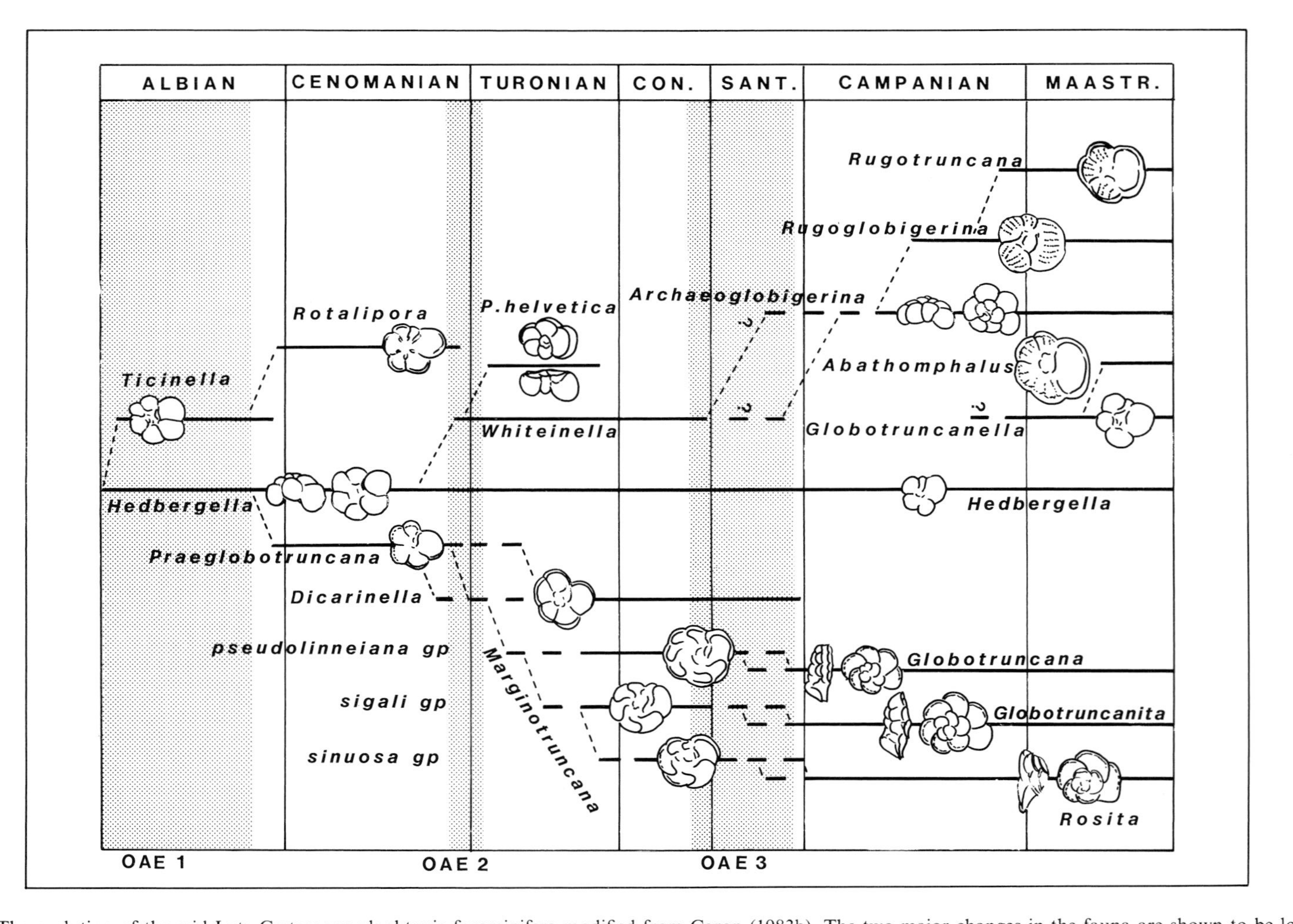

FIG. 9. The evolution of the mid-Late Cretaceous planktonic foraminifera modified from Caron (1983b). The two major changes in the fauna are shown to be located within the major anoxic events. While the Cenomanian–Turonian event is placed with some precision, the dating of the Coniacian–Santonian event is still imprecisely known. As indicated in the text the modification, in parallel, of four lineages in OAE 3 is quite striking.

References

ARTHUR, M.A. & SCHLANGER, S.O. 1979. Cretaceous 'oceanic anoxic events' as causal factors in development of reef-reservoired giant oil fields. *Am. Ass. Petrol. Geol. Bull.* **63,** 870–85.

BÉ, A.W.H. 1977. An ecological, zoogeographic and taxonomic review of recent planktonic foraminifera. *In*: RAMSAY, A.T.S. (ed.), *Oceanic Micropalaeontology*, Vol. 1. Academic Press, London. 1–100.

BIRKELUND, T., HANCOCK, J.M., HART, M.B., RAWSON, P.F., REMANE, J., ROBASZYNSKI, F., SCHMID, F. & SURLYJ, F. 1984. Cretaceous Stage boundaries. *Bull. Geol. Soc., Denmark* **33,** 3–20.

BOERSMA, A. & PREMOLI SILVA, I. 1983. Paleocene planktonic foraminiferal biogeography and the paleoceanography of the Atlantic Ocean. *Micropaleontology* **29**(4), 355–81.

BOERSMA, A. & SHACKLETON, N.J. (1981. Oxygen and carbon isotope variations and planktonic foraminiferal depth habitats, Late Cretaceous to Paleocene, Central Pacific, DSDP Sites 463 and 465, Leg 62. *In*: *Init. Repts DSDP*, **62.** US Govt. Print Off., Washington, DC. 513–26.

BRASIER, M.D. 1982. Architecture and evolution of the foraminiferid test—a theoretical approach. *In*: BANNER, F.T. & LORD, A.R. (eds), *Aspects of Micropalaeontology*. George Allen & Unwin, London. 1–41.

BROMLEY, R.G. & EKDALE, A.A. 1984. *Chondrites:* a trace fossil indicator of anoxia in sediments. *Science* **244,** 872.

BURNHILL, T.J. & RAMSAY, W.V. 1981. Mid-Cretaceous palaeontology and stratigraphy, Central North Sea. *In*: ILLING, L.V. & HOBSON, G.D. (eds), *Petroleum Geology of the Continental Shelf of North-West Europe*. Heyden, London. 245–54.

CARON, M. 1983a. La spéciation chez les Foraminifères planctiques: une réponse adaptée aux constraintes de l'environment. *Zitteliana* **10,** 671–6.

CARON, M. 1983b. Taxonomie et phylogenie de la famille des Globotruncanidae. *Zitteliana* **10,** 677–81.

CARTER, D.J. & HART, M.B. 1977. Aspects of mid-Cretaceous stratigraphical micropalaeontology. *Bull. Br. Mus. nat. Hist (Geol.)* **29,** 1–135.

DOUGLAS, R.G. & SAVIN, S.M. 1978. Oxygen isotopic evidence for the depth stratification of Tertiary and Cretaceous planktic Foraminifera. *Mar. Micropaleont.* **3,** 175–96.

GRACIANSKY, P.C. DE, BROSSE, E., DEROO, G. HERBIN, J.-P., MONTADERT, L., MÜLLER, C., SIGAL, J. & SCHAAF, A. 1982. Les formations d'âge Crétacé de l'Atlantique Nord et leur matière organique: paléogéographie et milieux de dépôt. *Rev. Inst. fr. Petrole* **37,** 275–336.

HANCOCK, J.M. 1969. The transgression of the Cretaceous sea in the south-west of England. *Proc. Ussher Soc.* **2,** 61–83.

HANCOCK, J.M. 1976. The petrology of the Chalk. *Proc. Geol. Ass.* **86,** 499–535.

HANCOCK, J.M. & KAUFFMAN, E.G. 1979. The great transgressions of the Late Cretaceous. *J. geol. Soc., Lond.* **136,** 175–86.

HART, M.B. 1980a. A water depth model for the evolution of the planktonic Foraminiferida. *Nature* **286,** 252–4.

HART, M.B. 1980b. The recognition of mid-Cretaceous sea-level changes by means of Foraminifera. *Cret. Res.* **1,** 289–97.

HART, M.B. 1982. Turonian foraminiferal biostratigraphy of Southern England. *In:* Colloque sur le Turonien, *Ent. Mus.*, Paris, 203–7.

HART, M.B. & BAILEY, H.W. 1979. The distribution of planktonic Foraminiferida in the mid-Cretaceous of N.W. Europe. *Aspekte der Kreide Europas, IUGS Ser. A* **6,** 527–42.

HART, M.B. & BIGG, P.J. 1981. Anoxic events in the Late Cretaceous Chalk seas of North-West Europe. *In*: NEALE, J.W. & BRASIER, M.D. (eds), *Microfossils of Recent and Fossil Shelf Seas*. Horwood, Chichester. 177–85.

JEFFERIES, R.P.S. 1962. The palaeoecology of the *Actinocamax plenus* Subzone (Lowest Turonian) in the Anglo-Paris Basin. *Palaeontology* **4,** 609–47.

JEFFERIES, R.P.S. 1963. The stratigraphy of the *Actinocamax plenus* Subzone (Turonian) in the Anglo-Paris Basin. *Proc. Geol. Ass.* **74,** 1–33.

JENKYNS, H.C. 1980. Cretaceous anoxic events: from continents to oceans. *J. geol. Soc., Lond.* **137,** 171–86.

KLAUS, J. 1960a. Le 'Complexe schisteux intermédiare dans le synclinical de la Gruyère (Préalpes médianes). Stratigraphie et micropaléontologie, avec l'étude spéciale des Globotruncanides de l'Albien, du Cénomanien et du Turonien. *Eclog. geol. Helv.* **52,** 753–851.

KLAUS, J. 1960b. Etude biométrique et statistique de quelques éspeces de Globotruncanides: 1. Les éspeces du genre *Praeglobotruncana* dans le Cenomanien de la Breggia. *Eclog. geol. Helv.* **53,** 285–308.

LOWENSTAM, H.A. & EPSTEIN, S. 1954. Palaeotemperatures of the post-Aptian Cretaceous as determined by the oxygen isotope method. *J. Geol.* **62,** 207–48.

MÜLLER, C., SCHAAF, A. & SIGAL, J. 1983. Biochronostratigraphie des formations d'âge Crétacé dans les forages du DSDP dans l'océan Atlantique Nord. *Rev. Inst. fr. Pètrole* (part 1), **38,** 683–708; (part 2), **39,** 3–23.

NYONG, E.E. & OLSSON, R.K. 1984. A paleoslope model of Campanian to Lower Maestrichtian Foraminifera in the North American Basin and adjacent continental margin. *Mar. Micropaleont.* **8,** 437–77.

RAWSON, P.F., CURRY, D., DILLEY, F.C., HANCOCK, J.M., KENNEDY, W.J., NEALE, J.W., WOOD, C.J. & WORSSAM, B.C. 1978. A correlation of Cretaceous rocks in the British Isles. *Geol. Soc. Lond., Spec. Rept.* **9,** 70 pp.

RHYS, G.H., LOTT, G.K. & CALVER, M.A. (eds) 1981. The Winterbourne Kingston borehole, Dorset, England. *Rep. Inst. Geol. Sci.* **81/3,** 196 pp.

ROBASZYNSKI, F. & CARON, M. 1979. Atlas of Mid-Cretaceous planktonic foraminifera (Boreal Sea and Tethys). *Cah. Micropal. CNRS*, part 1, 1–185; part 2, 1–181. (In English and French.)

ROBASZYNSKI, F., CARON, M., GONZALEZ DONOSO,

J.M. & Wonders, A.A.H. (eds) 1984. Atlas of Late Cretaceous Globotruncanids. *Rev. Micropaléont.* **26,** 145–305.

Schlanger, S.O. & Jenkyns, H.C. 1976. Cretaceous anoxic events: causes and consequences. *Geol. Mijnb.* **55,** 179–84.

Troelsen, J.C. 1955. *Globotruncana contusa* in the White Chalk of Denmark. *Micropaleont.* **1,** 76–82.

Urey, H.C., Epstein, S., Lowenstam, H.A. & McKinney, C.R. 1951. Measurement of palaeotemperatures and temperatures of the Upper Cretaceous of England, Denmark, and the south-eastern United States. *Bull. geol. Soc. Am.* **62,** 399–416.

Wonders, A.A.H. 1980. Middle and Late Cretaceous planktonic foraminifera of the Western Mediterranean area. *Utrecht Micropaleont. Bull.* **24,** 1–158.

Malcolm B. Hart, Department of Geological Sciences, Plymouth Polytechnic, Drake Circus, Plymouth PL4 8AA, Devon.

Kim C. Ball, Paleoservices Ltd., Unit 15, Paramount Industrial Estate, Sandown Road, Watford WD2 4UB, Herts.

Unconformities in the Cenozoic of the North-East Atlantic

I. Pearson & D. Graham Jenkins

Unconformities in the North-East Atlantic and Norwegian Greenland Sea have been examined from 60 DSDP boreholes, and correlated to a single biostratigraphic time-scale of the Cenozoic. The unconformities can be divided into two types: (a) widespread events with equivalent episodes outside this study area, and (b) locally restricted events.

Unconformities equivalent to those recognized by Vail & Hardenbol (1979) and Keller & Barron (1983) are present in the late Pliocene, Miocene, mid-Oligocene and late to mid-Eocene. Hiatuses corresponding to other unconformities noted by Vail & Hardenbol are difficult to recognize, being encompassed within large local unconformities.

Hiatuses have been formed either by (a) a change in bottom water activity or (b) tectonic events, diapirism and slope instability. Recognition of these types through a series of filters allows greater accuracy in defining circulatory events.

Since the recognition of an early Eocene to Cretaceous unconformity during DSDP Leg 1 (Ewing & Worzel 1969), the identification of hiatuses in the deep sea sedimentary record has assumed increasing importance for the explanation of changes in bottom water circulation and oceanographic events.

Several authors have produced papers dealing with the Cenozoic record of hiatuses (Rona 1973; Berggren & Hollister 1974; Vail *et al.* 1977, 1980; Moore *et al.* 1978; Hardenbol *et al.* 1981; Tucholke 1981; Keller & Barron 1983; Miller & Tucholke 1983) and there are numerous references to unconformities in Initial Reports of the Deep Sea Drilling Project (e.g. Poag *et al.*, in press).

Correlation of hiatuses between different DSDP Legs is complicated by the constant revision of the magnetic time-scale, and associated revisions of the ages of biostratigraphic zones.

The present work reviews the ages of unconformities in the NE Atlantic during the Cenozoic (Fig. 1), recalibrating them to the most recent time-scale of Berggren *et al.* (1984a, b). Due to the importance of hiatuses in the reconstruction of oceanographic events and palaeocirculation, this paper attempts to recognize and gradually exclude, through a series of filters, those hiatuses caused by non-circulation events, such as tectonics and slope instability.

Unconformities

Primary data have been extracted from (1) the site, biostratigraphy and synthesis chapters of DSDP volumes 12, 13, 14, 38, 47, 48, 49 and 50, (2) from the initial core descriptions of Legs 80 and 81, (3) from data provided by shipboard workers, Legs 81 and 94, and (4) from other published literature (Perch-Nielson 1971; Backman 1978, 1979; Berggren 1978; Berggren & Aubert 1976; Berggren & Schnitker 1983; Miller & Tucholke 1983).

An unconformity is defined as being a gap in the sedimentary record. As stated by Vail *et al.* (1980) 'all rocks below the unconformity are older than the rocks above it'. The age of the sediments immediately below an unconformity is controlled by the amount of erosion into previously deposited sediments. The time interval during which the unconformity was formed is determined at a locality where the hiatus time gap is at a minimum. This constraint may not represent the actual period of unconformity production, due to the inability to define the interaction between non-deposition, solution and erosion at a particular place. Hence an unconformity can only be reliably dated by its upper boundary, the age at which sediment deposition recommenced.

Several mechanisms have been proposed to account for deep-sea unconformities. These mechanisms produce two classes of unconformity. Members of the first class are caused by changes in circulation, which include changes in the relative intensity of bottom water circulation giving rise to non-deposition and/or erosion; other oceanographic changes, including variations in the CCD, bring about solution of previously deposited sediments. In this study no distinction has been made between these mechanisms.

The second class of unconformity, whose differentiation has sometimes been overlooked in circulation studies, is caused by tectonic events, diapirism or slope instability.

Methods

To determine which hiatuses for each NE Atlan-

From Summerhayes, C.P. & Shackleton, N.J. (eds), 1986, *North Atlantic Palaeoceanography*, Geological Society Special Publication No. 21, pp. 79–86.

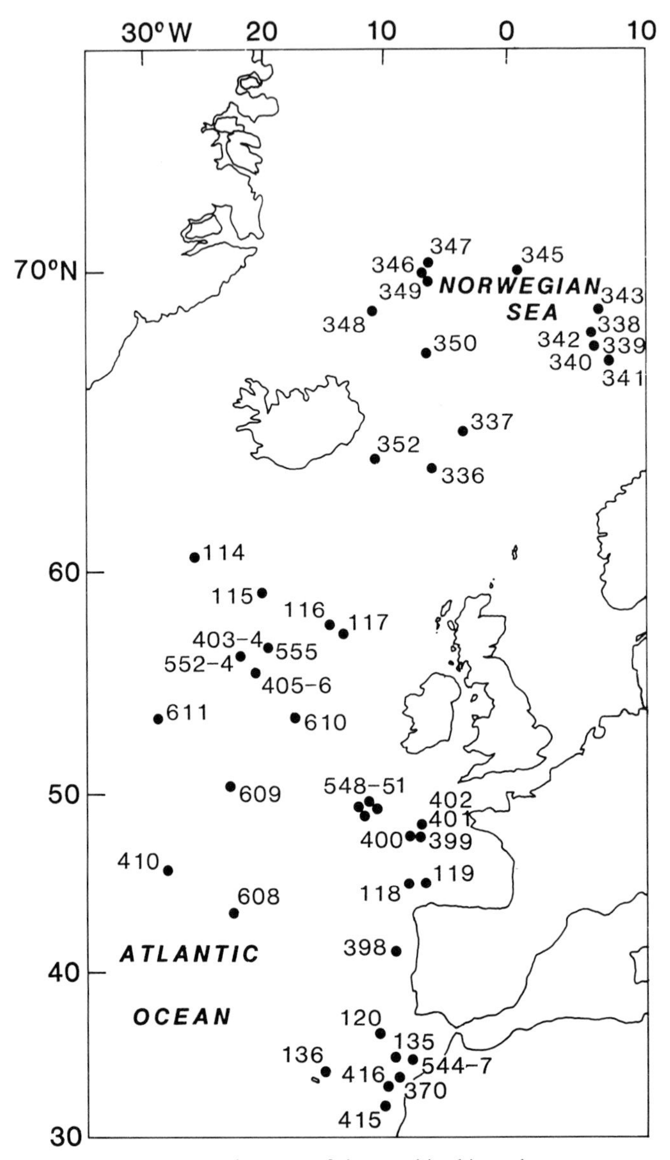

FIG. 1. Location map of sites used in this study.

tic site were probably produced by circulation effects, a series of screening filters was employed to remove successively various types of non-circulatory hiatuses.

To consider the correlation of hiatuses between sites it is first necessary to have confidence in the dating of the hiatus intervals at each individual site. The methods used to recalibrate the data from each site, and the criteria used are discussed further in Pearson *et al.* (in press). This recalibration acted as the first filter, removing those sites with a poor coring record, poor biostratigraphic control or a complete sedimentary record (Table 1).

Filter 2 removed those hiatuses assigned to be of tectonic origin by the DSDP Initial Report volumes (Table 2). A further study of the relevant seismic reflection profile for each site acted as filter 3, and removed probably tectonically formed hiatuses showing evidence of an angular unconformity between overlying and underlying sediments (Table 3). Finally, filter 4 involved the examination of the probable palaeogeographic setting of each hiatus in order to identify areas

TABLE 1. *Filter 1: sites removed due to poor coring, poor biostratigraphic control or having no hiatuses*

Site	Filter control
114	Poor coring record
115	No hiatuses
120	Poor coring in Tertiary sediments
341	Large slump section
344	Poor biostratigraphy. No hiatuses.
347	Poor coring record
348	No hiatuses
350	Poor coring record. Poor biostratigraphy.
399	No hiatuses
410	No hiatuses
411	Mid Ocean Ridge Crest
412	Pleistocene only
413	Pleistocene only
414	Pleistocene only
558	No hiatuses
606	No hiatuses
607	No hiatuses
609	No hiatuses
610	No hiatuses
611	No hiatuses
551	Poor coring record in Tertiary
339	Poor biostratigraphy

TABLE 2. *Filter 2: hiatuses removed where DSDP Initial Report volumes state they are of tectonic origin*

Site	Hiatuses removed
349	Upper Oligocene to Middle Miocene
549	Middle Eocene and older
550	Palaeocene
608	All
118	All
119	All
546	All
547	All
340	All
346	Early Oligocene

where slope instability was likely to have been the causal factor (Table 4).

NE Atlantic unconformities

The area under investigation has been divided into three regions.

(a) Norwegian-Greenland Sea and Iceland-Faroe Ridge.

TABLE 3. *Filter 3. hiatuses removed due to evidence of an angular unconformity on seismic reflection profiles*

Site	Hiatuses removed
398	Middle Eocene
415	Early Eocene to Middle Eocene
416	Early Palaeocene
544	All
545	All
548	All

TABLE 4. *Filter 4: hiatuses removed where sediment instability on slopes was likely to have been the causal factor in their formation*

Site	Hiatuses removed
337	All
401	Early Pliocene to Upper Oligocene
549	Recent to Early Oligocene
550	Recent to Palaeocene
402	All

(b) Rockall-Goban Spur and Biscay.

(c) Portuguese and NW African margin and the Kings Trough.

When examining the diagrams produced by the removal of all non-circulatory hiatuses (Figs 2, 3 and 4), the immediate impression is of non-uniformity of hiatus upper boundaries. This is the result of a large number of very long hiatus intervals, some of which were probably caused by a series of erosional events. It is therefore necessary to make the assumption that the hiatus upper boundary was present at a site with a long hiatus when the presence of the equivalent boundary is detectable in adjacent sites. For example the mid-Oligocene hiatus in the Rockall area at 30 Ma can be seen in Sites 403, 552 and 406 and its presence is assumed in holes 553, 404 and 555 where a younger hiatus has eroded to below the 30 Ma level (Fig. 3).

Norwegian-Greenland Sea and Iceland-Faroes Ridge (Fig. 2)

After the removal of Sites 337 and 340 and the mid-Miocene and early Oligocene hiatuses from Sites 349 and 346 by the filtering process (Tables 2 and 4) only one laterally extensive hiatus can be attributed to circulation effects. This commences

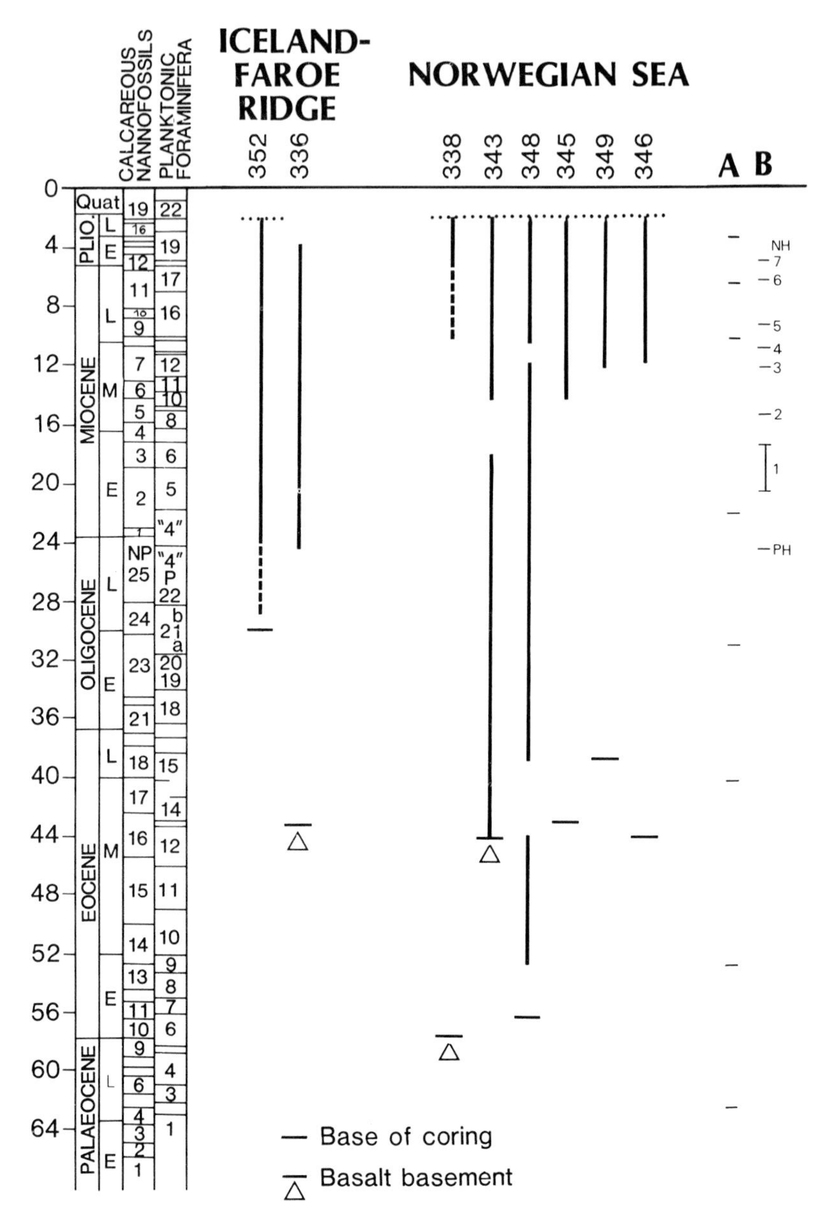

FIG. 2. Hiatuses (heavy vertical lines) present in the Norwegian–Greenland Sea and Iceland–Faroe Ridge area that are probably due to circulatory effects. Light horizontal lines = probable correlated events. A = major unconformities after Vail & Hardenbol (1979). B = Miocene unconformities after Keller & Barron (1983).

in the late Pliocene at *c*.2.5 Ma in all the Norwegian Sea sites (Fig. 2). Downcutting of this hiatus varies between sites but the hiatus is probably constrained between 2.5 Ma and 6 Ma at Site 338. In addition three localized hiatuses occur in Site 343 at 12 Ma and 44 Ma and at Site 342 at 18 Ma. The two youngest hiatuses may have been obliterated by erosion in 342, 345 and in 348.

The late Pliocene hiatus at site 352 on the Iceland-Faroes Ridge can probably be correlated with late Pliocene hiatus sedimentation recommencing at 2 Ma in the Norwegian Sea. At site 336, sedimentation recommenced earlier, at 4

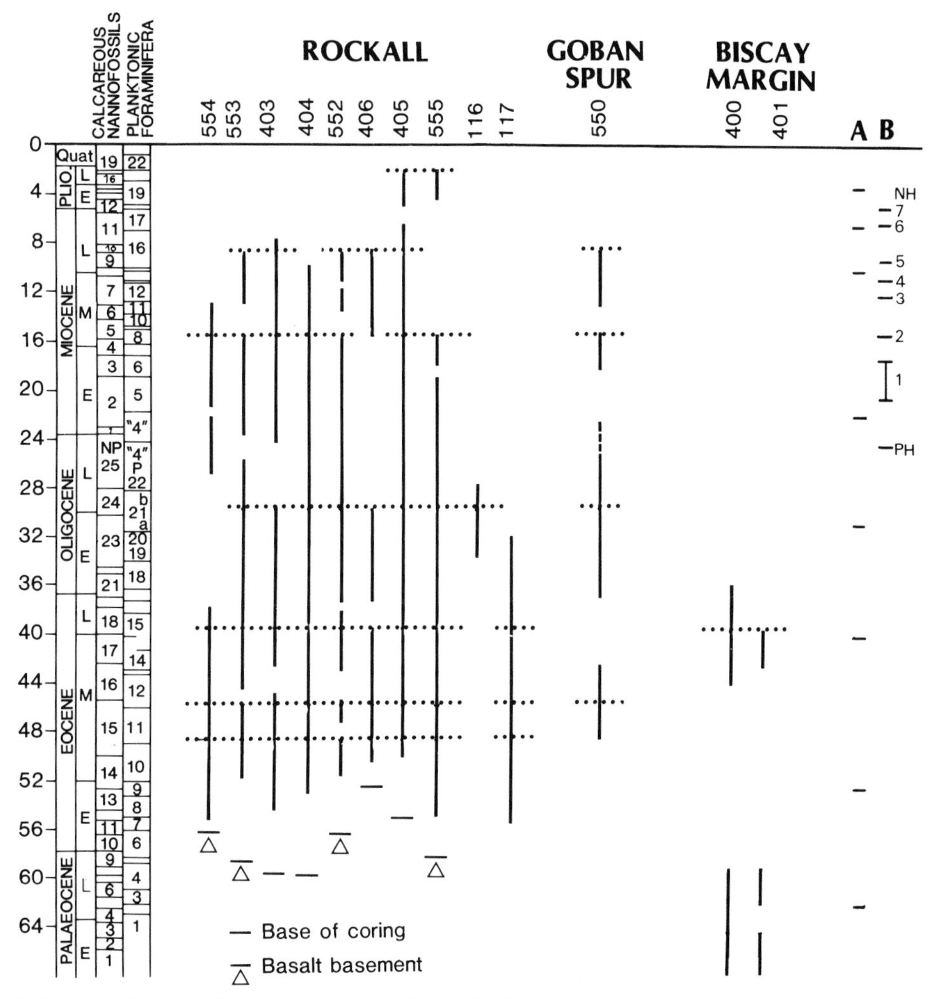

FIG. 3. Hiatuses (heavy vertical lines) present in the Rockall, Goban Spur and Biscay Margin areas that are probably due to circulatory effects. See Fig. 2 for key.

Ma, probably reflecting localized variability in the current speed of NADW flowing over the ridge. It is therefore suggested that the erosional capability was greatest near to Iceland, at least during the period 4 Ma to 2 Ma.

Rockall-Goban Spur and Biscay (Fig. 3)

The filtering process removed Sites 116 and 117 (Table 2) and Pliocene to Oligocene hiatuses at Sites 549 and 401. In addition, early Eocene to Palaeocene hiatuses were removed at Sites 549 and 550. There is a good correlation of upper hiatus boundary ages between the Rockall area and the Goban Spur from the middle Miocene (Fig. 3), suggesting a uniformity of palaeoceanographic events affecting these two areas.

Sites 405 and 555 have a hiatus in the late Pliocene at 2–2.5 Ma which correlates with hiatuses in the Norwegian Sea area. The position of these holes between the Rockall and Hatton banks suggest that this trough was one of the main passages for the southward flow of NADW during this period. The late to middle Miocene has several concordant upper hiatus boundaries. Three distinct surfaces occur between the period 8–12 Ma, although local conditions appear to have influenced their distribution. These surfaces correlate well with previously detected hiatuses NH5 and NH4 (Keller & Barron 1983), and one at 9.8 Ma (Vail & Hardenbol 1979). A major correlatable surface found in all sites in Rockall and the Goban Spur except at Site 406, occurs at 15 Ma, and is equivalent to Keller & Barron's NH2.

The middle Oligocene hiatus at 30 Ma detailed

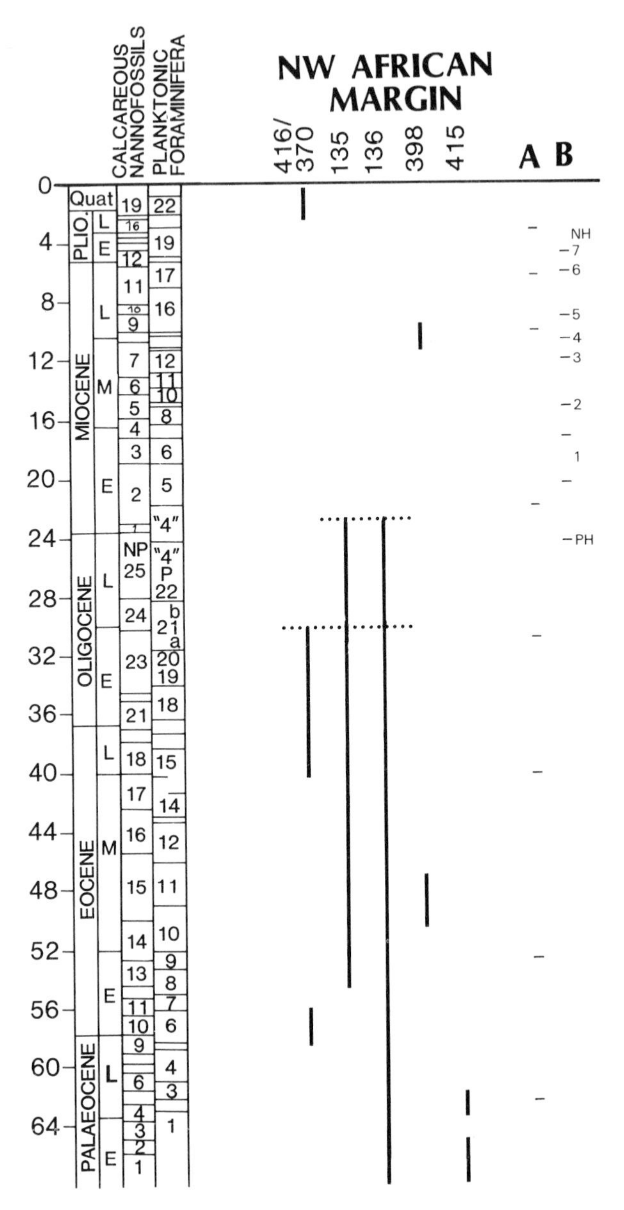

FIG. 4. Hiatuses (heavy vertical lines) present in the NW African Margin area that are probably due to circulation effects. See Fig. 2 for key.

by Poag *et al.* (in press) can be traced through all the Rockall and Goban Spur sites except Site 554. It correlates to Vail's mid-Oligocene sea-level fall, which is though to be of glacial origin (Keller & Barron 1983).

Rockall sites show three correlatable surfaces during the Eocene that are only partially detectable outside this area. The well documented late Eocene hiatus, *c*.40 Ma, can be traced in every Rockall site and those on the Biscay Margin, but is conspicuously absent from the Goban Spur. Two middle Eocene hiatus intervals are also seen in the Rockall area at 45 Ma and 49 Ma. These three Eocene surfaces can be detected on seismic records (Masson & Kidd, in press). The late Eocene hiatus has been ascribed to a circulation event on a global scale (Vail & Hardenbol 1979), although in this study area the hiatus is of dual origin, being a circulation effect in the Rockall area but caused by tectonic activity associated with the Pyrennean event in the south (Pearson *et al.*, in press).

Sites from the Goban Spur and Biscay Margin show two small hiatus producing events in the Palaeocene at 60 Ma and 65 Ma.

Portuguese and NW African Margins and Kings Trough (Fig. 4)

The late Eocene to late Oligocene hiatus at site 608 was removed by filter 2, being tectonic in origin. This also applied to a long hiatus commencing in the early Miocene at Site 415 and two smaller hiatuses in the middle Eocene of Site 398 and the Palaeocene of Site 416/370.

A middle Miocene hiatus at site 398 correlates with Keller & Barron's NH5 and equivalent hiatuses in the Rockall and Goban Spur area (Fig. 4).

A hiatus at 23 Ma at Sites 135, 136, 544, 545 and 547 is equivalent to PH (Keller & Barron 1983) and Vail's 22.5 Ma hiatus (Vail & Hardenbol 1979), and can possibly also be recognized in Sites 550 and 118. This appears to be a major event on the NW African margin. The middle Oligocene hiatus recognized in Rockall and the Goban Spur is also present at Site 416/370 and is probably also present at those sites with an early Miocene hiatus.

Conclusions

Inter-regional unconformities are rare. Three possible exceptions are (1) the late Pliocene unconformity at 2.5 Ma, in the Norwegian Sea, Rockall and Goban Spur areas, (2) an early Miocene unconformity at 23 Ma in areas south of the Goban Spur, and (3) a middle Oligocene unconformity at 30 Ma in areas from Rockall southwards. However, these events are not represented at every site in the regions where they occur. This, together with the difficulty of detecting Vail's unconformities associated with sea-level changes, suggests that local circulation effects (cf. Hollister & McCave 1984) have a greater effect on Cenozoic stratigraphic sequences recovered from cores in the NE Atlantic than do supposedly 'global' eustatic events.

The major difficulty in accurately correlating unconformities lies in detecting the presence of individual unconformities within long hiatuses. Future work should follow the example of Vail *et al.* (1980), and concentrate on the rigorous investigation of seismic profiles in the areas around each drill site to determine the lateral stratigraphic relationships of beds that coalesce onto each unconformity surface.

ACKNOWLEDGEMENTS: The authors thank Dr R.B. Kidd for performing the filter analysis of data and for helpful suggestions. Also many thanks go to Professor B.M. Funnell and Dr C. Summerhayes for useful comments and suggestions during the review process. Thanks also to Carol Whale for typing the manuscript and to John Taylor and Helen Boxall for the cartography. This work was performed while D.G. Jenkins was in receipt of an NERC research grant.

References

BACKMAN, J. 1978. Miocene-Pliocene nannofossils and sedimentation rates in the Hatton-Rockall Basin. N.E. Atlantic Ocean. *Stock. Contrib. Geol.* **XXXVI** (1), 1–91.

—— 1979. Pliocene biostratigraphy of DSDP sites 111 and 116 from the North Atlantic Ocean and the age of northern hemisphere glaciation. *Stock. Contrib. Geol.* **XXXII** (3), 115–37.

BERGGREN, W.A. 1978. Recent advances in Cenozoic planktonic foraminiferal biostratigraphy, biochronology and biogeography, Atlantic Ocean. *Micropaleontology* **24,** 337–70.

—— & HOLLISTER, C.D. 1974. Paleogeography, paleobiology and the history of circulation in the Atlantic Ocean. *Soc. econ. Paleont. Miner. Spec. Publ.* **20,** 126–86.

—— & AUBERT, J. 1976. Late Paleogene (Late Eocene and Oligocene) benthonic foraminiferal biostratigraphy and paleobathymetry of Rockall Bank and Hatton-Rockall Bank. *Micropaleontology* **22,** 307–26.

—— & SCHNITKER, D. 1983. Cenozoic marine environments in the North Atlantic and Norwegian-Greenland Sea. *In*: BOTT, M.H.P., SAXOV, S., TALWANI, M. & THIEDE, J. (eds), *Structure and Development of the Greenland-Scotland Ridge*. NATO Conference Series. Plenum Press, New York.

——, KENT, D.V. & FLYNN, J. 1984a. Jurassic to Paleogene: Part 2. Paleogene geochronology and chronostratigraphy. *In*: Snelling, N.J. (ed.), *The Chronology of the Geological Record. Mem. Geol. Soc. Lond.* **10,** 141–95 Blackwell Scientific Publications, Oxford.

——, KENT, D.V. & VAN COUVERING, J.A. 1984b. The Neogene: Part 2. Neogene geochronology and chronostratigraphy. *In*: Snelling, N.J. (ed.), *The Chronology of the Geological Record. Mem. Geol. Soc. Lond.* **10,** 211–60. Blackwell Scientific Publications, Oxford.

EWING, M. & WORZEL, J.L. 1969. *Init. Rept. DSDP*, **1.** US Govt. Print. Off., Washington, DC.

HARDENBOL, J., VAIL, P.R. & FERRER, J. 1981. Interpreting paleoenvironments, subsidence history, and sea level changes of passive margins from seismic and biostratigraphy. *Oceanol. Acta, Colloq*, C3, Geology of Continental Margins, 33–44.

HOLLISTER, C.D. & MCCAVE, I.N. 1984. Sedimentation under deep-sea storms. *Nature* **309,** 220–5.

KELLER, G. & BARRON, J.A. 1983. Paleoceanographic implications of Miocene deep-sea hiatuses. *Geol. Soc. Am. Bull.* **84,** 590–613.

MASSON, D.G. & KIDD, R.B., in press. Revised Tertiary seismic stratigraphy of the southern Rockall Trough. *Init. Repts. DSDP.* **81.** US Govt. Print. Off., Washington, D.C.

MILLER, K.G. & TUCHOLKE, B.E. 1983. Development of Cenozoic abyssal circulation south of the Greenland Scotland Ridge. *In:* BOTT, M.H.P., SAXOV, S., TALWANI, M. & THIEDE, J. (eds), *Structure and Development of the Greenland-Scotland Ridge.* NATO Conference Series. Plenum Press, New York.

MOORE, T.C., VAN ANDEL, T.J.H., SANCETTA, C. & PISIAS, N.G. 1978. Cenozoic hiatuses in pelagic sediments. *Micropaleontology* **24,** 113–38.

PEARSON, I., KIDD, R.B., BIART, B. & JENKINS, D.G., in press. Cenozoic unconformities of the North-East Atlantic: formational processes with reference to DSDP/IPOD Leg 94. *Init. Repts DSDP*, **94.**

PERCH-NIELSEN, K. 1971. Einige neve Coccolithen aus dem Paleozan der Bucht von Biskaya. *Bull. Geol. Soc. Denmark* **20,** 347–61.

POAG, C.W., REYNOLDS, L.A., MAZZULO, J.M. & KEIGWIN, L.D. jr., in press. Foraminiferal, Lithic and Isotopic changes across four major unconformities at DSDP–IPOD site 548, Goban Spur. *Init. Repts DSDP* **80.** US Govt. Print. Off., Washington, DC.

RONA, P.A. 1973. Worldwide unconformities in marine sediments related to eustatic changes of sea-level. *Nature Phys. Sci.* **244,** 25–6.

TUCHOLKE, B.E. 1981. Geological significance of seismic reflection in the deep western North Atlantic. *In:* Warme, J.E., Douglas, R.G. & Winterer, E.L. (eds), *The Deep Sea Drilling Project: A Decade of Progress. Soc. econ. Paleo. Miner. Spec. Publ.* **32,** 23–38.

VAIL, P.R. & MITCHUM, R.M. jr. *et al.* 1977. Seismic stratigraphy and global changes of sea level. *In:* Payton, C.E. (ed), *Seismic Stratigraphy—Applications to Hydrocarbon Exploration. Am. Ass. Petrol. Geol. Mem.* **26,** 49–205.

—— & HARDENBOL, J. 1979. Sea-level changes during the Tertiary. *Oceanus* **22,** 71–9.

——, MITCHUM, R.M. jr, SHIPLEY, T.H. & BUFFLER, R.T. 1980. Unconformities of the North Atlantic. *Trans. R. Soc. Lond.* **A 294,** 137–55.

I. PEARSON & D. GRAHAM JENKINS, Department of Earth Sciences, The Open University, Walton Hall, Milton Keynes MK7 6AA.

Sedimentation on mid-ocean sediment drifts

Robert B. Kidd & Philip R. Hill

SUMMARY: The paper reports results from DSDP Leg 94, where holes were drilled in the Feni and Gardar sediment drifts, with particular emphasis placed on studying sediment waves that ornament the surfaces of the drifts. The Feni Drift is a 1500–1700 m thick accumulation of sediment constructed on an early Eocene to late Oligocene seismic reflector. The Gardar Drift is 1300–1600 m thick, and is probably built directly on late Eocene oceanic basement. At Site 610 on Feni Drift, the deepest hole penetrated to upper Miocene sediments at 720 m, and at Gardar Drift, to upper Miocene sediment at 500 m. Pre-site surveys showed that the sediment wave fields on the drift surfaces were more irregular than Gloria sidescan records suggest, consisting of interfingered *en echelon* mounds. The waves did not show any well-defined sense of migration. The sediments recovered at Sites 610 and 611 show that the drifts consist of pelagic ooze and calcareous mud, with very few primary sedimentary structures preserved. No significant differences were observed between lithologies on sediment wave crests and those in trough locations. On the other hand, systematic differences in sedimentation rates between crest and trough holes at Gardar Drift support the hypothesis that the waves migrated during the Pliocene but were evenly draped during the Pleistocene.

The effects of the deep-ocean circulation on the distribution and deposition of hemipelagic and pelagic sediments are recognized most clearly in the development of major North Atlantic sediment drifts through the Cenozoic (Fig. 1; Hollister *et al.* 1978). Sediment drifts are anomalous sediment accumulations, frequently over 1 km thick, that are recognized as positive accumulations on seismic reflection profiles. They may be of three kinds (Lonsdale 1982): projections from continental rises (e.g. the Bahama Outer Ridge); isolated ridges detached from a continental margin (e.g. the Greater Antilles Outer Ridges); and mid-ocean sediment ridges banked up against pre-existing topography (e.g. the Bermuda Ridge). Smaller-scale sediment drifts are widespread on many continental margins, where they are built around upstanding topography that interrupts along-slope flow of bottom water (Heezen & Hollister 1971).

Many sediment drifts are characterized by surface sediment waves, which are generally tens of metres in amplitude and up to a few kilometres in wavelength (Flood 1978). Sediment wave fields are also found on continental rises and other parts of the continental margins (Kidd & Roberts 1982). Understanding the development of the sediment wave fields and their relationship (if any) to the present-day circulation is a precursor to understanding the history of sedimentation of the major drifts (Roberts & Kidd 1979).

During Deep Sea Drilling Project Leg 94 in the North Atlantic, we had a unique opportunity to examine the record of sedimentation on the sediment wave fields of Feni and Gardar drifts over much longer time-scales than had hitherto been possible. Hydraulic piston-core holes were located at Sites 610 and 611, so that wave crest and trough sequences could be examined.

The purpose of this paper is to review the nature and origin of sediment drifts and their associated sediment wave fields in the light of preliminary data from Leg 94. More detailed results from Leg 94 will be published in the *Initial Reports of the Deep Sea Drilling Project* (Kidd *et al.*, in prep.).

Modern circulation

The near-bottom circulation pattern shown in Fig. 1 is a much generalized overview of net flow in the north-east Atlantic. The detailed flows are much more complex, but relatively few long-term current meter data are available to interpret the details of bottom-water flow. The overflow of Norwegian Sea Water at several locations between Iceland and Scotland, into the North Atlantic basin, plays a major role in the overall Atlantic circulation. In the Rockall Trough–Feni Ridge area, the Norwegian Sea Overflow Water (NSOW) is largely diluted by North Atlantic Deep Water (NADW) and Antarctic Bottom Water (AABW) (McCave *et al.* 1980; Dickson & Kidd, in prep.). These waters also flow along the flanks of the Iceland Basin and Reykjanes Ridge but there the influence of NSOW increases, so that at the Gardar Ridge the influence of NADW and AABW is much reduced.

From SUMMERHAYES, C.P. & SHACKLETON, N.J. (eds), 1986, *North Atlantic Palaeoceanography*, Geological Society Special Publication No. 21, pp. 87–102.

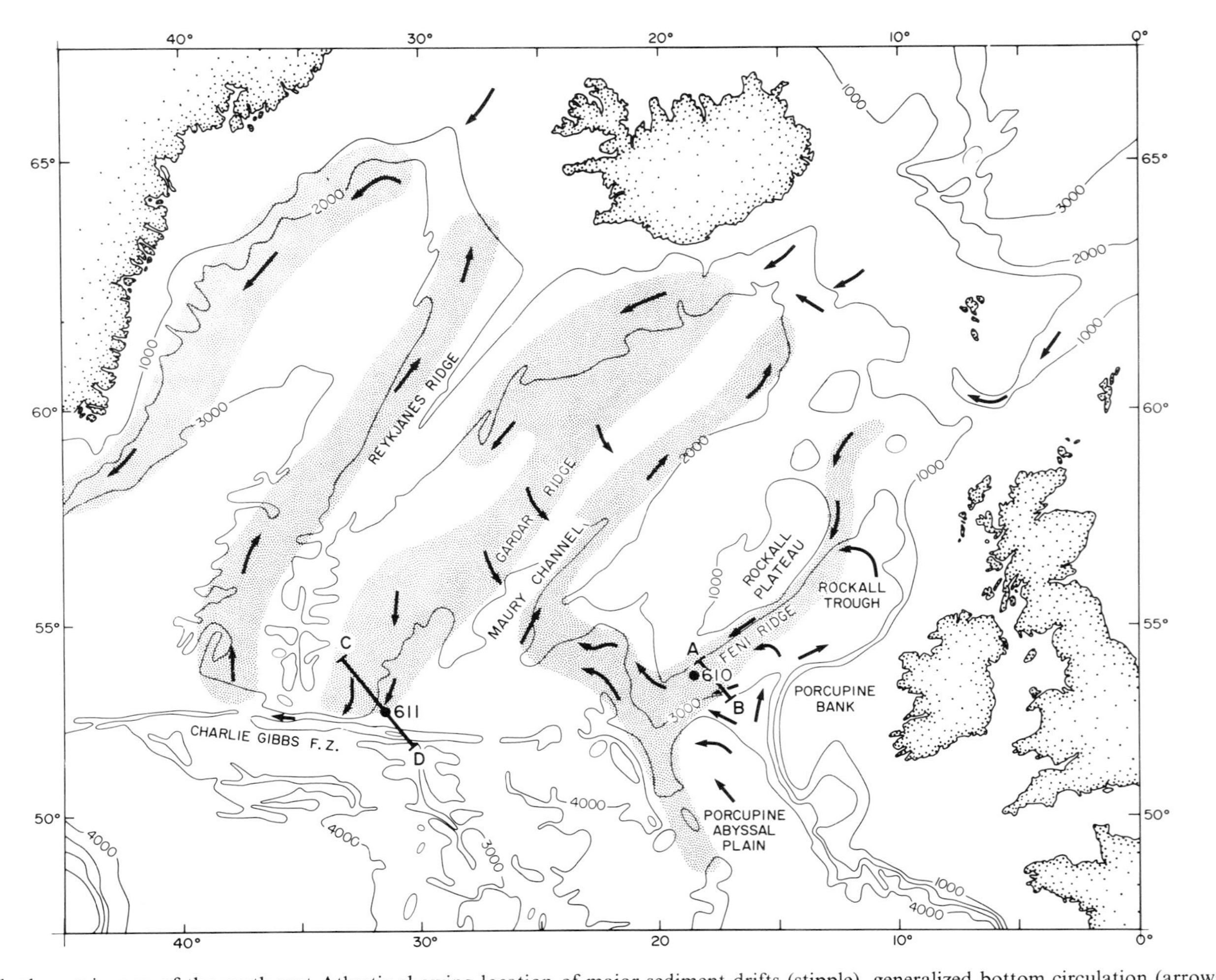

Fig. 1. Outline bathymetric map of the north-east Atlantic showing location of major sediment drifts (stipple), generalized bottom circulation (arrows), and location of DSDP Sites 610 and 611. The seismic sections illustrated in Figs 2 and 3 are shown as A–B and C–D.

Morphology and stratigraphy of the drifts

Feni Ridge (Figs 1 and 2) was first identified as a sediment accumulation against the south-eastern flank of Rockall Plateau by Jones *et al.* (1970). The ridge extends 520 km from around 55.5°N., 14.5°W., where it first becomes 'detached' from the slopes of Rockall Plateau, to around 52.5°N., 20.5°W., where the crest intersects the Isengard Ridge (Laughton *et al.* 1982), which is also presumed to be a sediment drift. The Feni Ridge crest is sinuous. In northern parts, the crestal depths are less than 2300 m, while the ridge flanks extend towards Rockall Trough, down to 2900 m in the north and to 4000 m in the south.

Sediment thicknesses to basement reach over 3.5 seconds (two-way travel time) under the ridge crest on multichannel seismic profiles (Masson & Kidd, in prep.). Feni Drift as a positive sediment accumulation is recognized only in the upper 1.5–1.7 seconds (1500–1700 m) of the records (Fig. 2) and is built on a regional reflector referred to as the 'brown reflector' by Masson & Kidd (in prep.). It is considered by them to be early Eocene to late Oligocene in age.

The ridge crest bifurcates on profiles to the south-west of Site 610, with a short branch between 53.5° and 54°N. extending back to the slopes of Rockall Plateau. A more northern crestal bifurcation may be represented by a mid-trough 'high' recognized between 54° and 55°N. This feature appears to be a later development than the main ridge since it is characterized on seismic profiles as a positive accumulation above a reflector at approximately 0.7 second depth (the yellow reflector of Masson & Kidd, in prep.).

The Gardar Drift was recognized by Jones *et al.* (1970) just east of the Reykjanes Ridge flank. The sediment drift extends over 1000 km (Figs 1 and 3) from 61.5°N., where it begins as a prolongation of the Icelandic insular margin, to around 52.5°N., where it abuts the topography of the Charlie-Gibbs Fracture Zone. Site 611 is located at the extreme southern end in a lower flank setting.

The Gardar Ridge crest displays a sinuous trace that over much of its length is shallower than 3000 m. The eastern flanks extend down to the Maury Channel, a mid-ocean turbidity current channel that follows the approximate axis of the Iceland Basin and extends from Iceland to the ridge flank topography west of the Porcupine Abyssal Plain (Cherkis *et al.* 1973).

Unlike Feni Ridge, the surface of Gardar Drift is intercepted, especially in its northern parts, by a number of abyssal hills and seamounts, presumably ridge-flank basement topography that has still not been blanketed by sediment deposition (Johnson *et al.* 1971). Moating occurs around these features (Ruddiman 1972), attesting to the effects of differential deposition by the bottom water circulation. The influence of the abyssal hills on Gardar Drift accumulation makes it difficult to ascertain whether any bifurcation of the ridge crest occurs, as on the Feni Drift.

Gardar Ridge sediment thicknesses vary from 1.3–1.6 seconds to acoustic basement (1300–1600 m thick). Because of the lack of good airgun or multi-channel seismic profiles, it is not clear whether the ridge is a positive accumulation on any particular regional reflector. Johnson *et al.* (1971) noted that a reflector traced at 0.6 seconds sub-bottom was sub-parallel to the ridge surface crestal elevation. It remains possible that one of the reflectors that is not easily resolved near basement is the true base of the drift, but it is just as likely that the drifted sediment may make up the entire column above basement.

The generalized stratigraphies of the two Sites 610 and 611 are summarized in Fig. 4. The deepest hole at Site 610 on the Feni Drift penetrated to upper Miocene sediments at 720 m below sea-bed. The yellow, purple and green reflectors that Masson & Kidd (in prep.) identified in multi-channel seismic profiles can be correlated with abrupt changes in sonic velocity measured on the cores recovered (Masson & Kidd, in prep.). The yellow and purple reflectors are dated as upper and middle Miocene, respectively, while the green reflector, which was originally interpreted as the regional north-east Atlantic R4 reflector, probably of Eocene–Oligocene age (Roberts 1975), is now dated as uppermost lower Miocene. Thus, the deeper, brown reflector, upon which the drift has accumulated as a positive feature, may instead be the Eocene–Oligocene R4 and the upper Feni Ridge sequence is much younger than previously thought.

Such detailed control on the seismic stratigraphy is not available at Gardar. The projected magnetic age of the crust at an estimated depth of 1300–1600 m below sea-bed is late Eocene. The holes penetrated upper Miocene sediments at 500 m below sea-bed. Extrapolating sedimentation rates, the authors suspect that Neogene drifted sediment makes up almost the entire column above basement.

Sediment wave fields

Roberts & Kidd (1979) used long-range sidescan sonar (GLORIA II) to map sediment wave fields that occur along the eastern flanks of Feni Ridge (Fig. 5). Individual longitudinal wave crests could

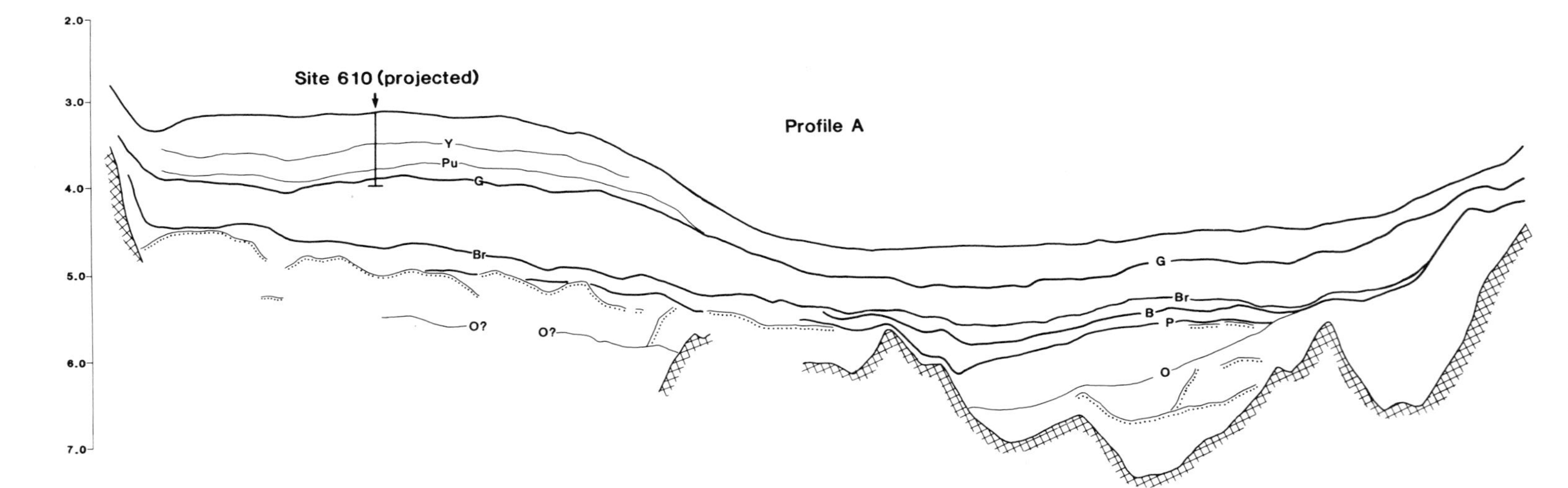

FIG. 2. Interpretative line drawing of a multi-channel seismic line across Southern Rockall Trough and the Feni Drift, Profile A from Masson & Kidd, in prep. (See Fig. 1 for location of the line A–B.) Their colour-coded reflectors are Y = yellow; Pu = purple; G = green; Br = brown; B = blue; P = pink; O = orange. Stippled discontinuous reflectors are presumed lavas. Basement is cross-hatched. Scale shows seconds two-way.

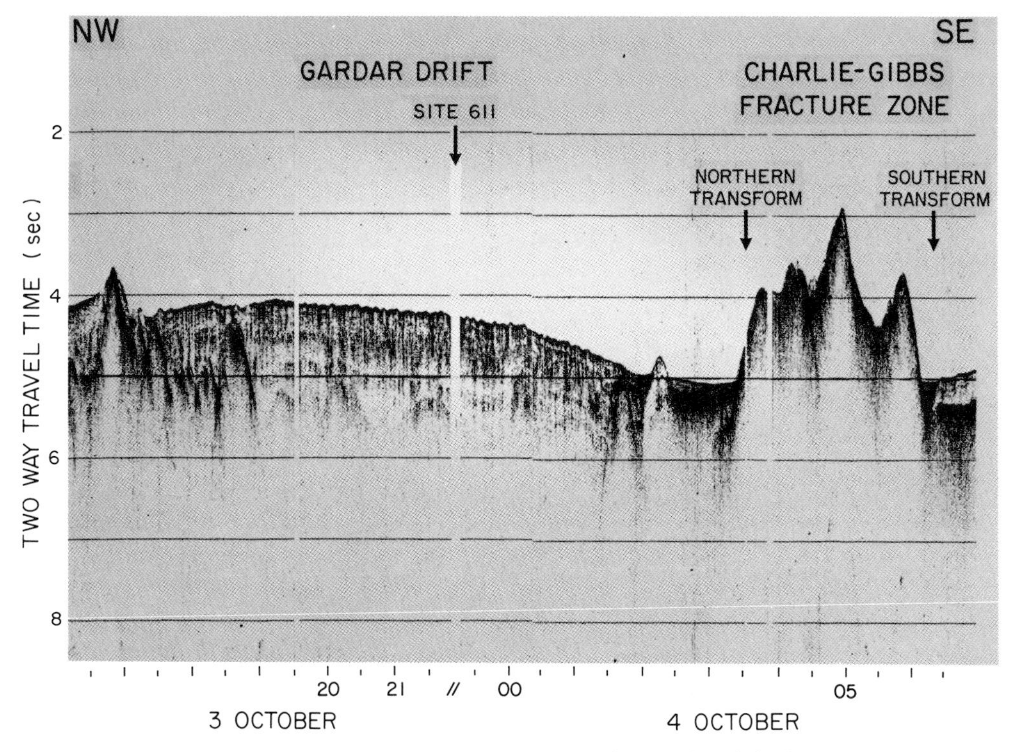

FIG. 3. Airgun seismic profile across the Gardar Drift. See Fig. 1 for location of the line (C–D).

be followed for up to 26 km, and wave amplitudes ranged from 25–50 m and wavelengths from 1–4 km. Along most of this survey of the ridge flank, the wave fields were well developed. Roberts (1975), however, on the basis of echo-sounder profiles, noted that the waves are sparse where the ridge is poorly developed and its east flank slopes gently, whereas large waves are found where the ridge is well developed and the east flank slopes steeply troughwards. In the Site 610 area near the ridge axis, GLORIA sonographs displayed only subtle lineations that are interpreted as longitudinal wave crests. Despite this, *Glomar Challenger* profiles show sediment waves with amplitudes of about 20 m. The single 3.5-kHz profile taken along the GLORIA survey track indicated that some of the waves were migrating upslope (Roberts & Kidd 1979), as had been observed on other sediment drifts (Flood 1978).

During the detailed pre-site survey aboard *Glomar Challenger* (Fig. 6), a number of crossings were made over the sediment wave on which the beacon had been dropped (Fig. 7), and it became clear that there had been a number of misconceptions about this wave field. The varying trends mapped from the GLORIA sonographs can generally be followed across the area shown in Fig. 6; however, detailed mapping shows that the 'axes' mapped from the sonographs are not all of individual waves that can be traced track-to-track in the survey area. The detailed bathymetric map of the area around Site 610 (Fig. 8) shows the authors' interpretation of the waves as irregular interfingered features arranged in a generally *en echelon* pattern. The subtle lineations, interpreted as continuous axes on the sonographs, must represent an over-emphasis of lineations that are parallel to track, one possible artifact of the GLORIA sonar system (Belderson *et al.* 1972). At Site 610 the sediment waves trend almost east–west.

The sediment waves at Feni do not show a well pronounced sense of migration in the 3.5-kHz survey crossings (Fig. 7). Attempts to use wave asymmetry and sequence thickening to determine apparent migration directions failed to give any clear migration direction for the waves, despite several crossings on different headings (Fig. 6). The authors conclude that there is no direct evidence for recent wave migration at the Feni site.

Very few GLORIA II sidescan profiles have

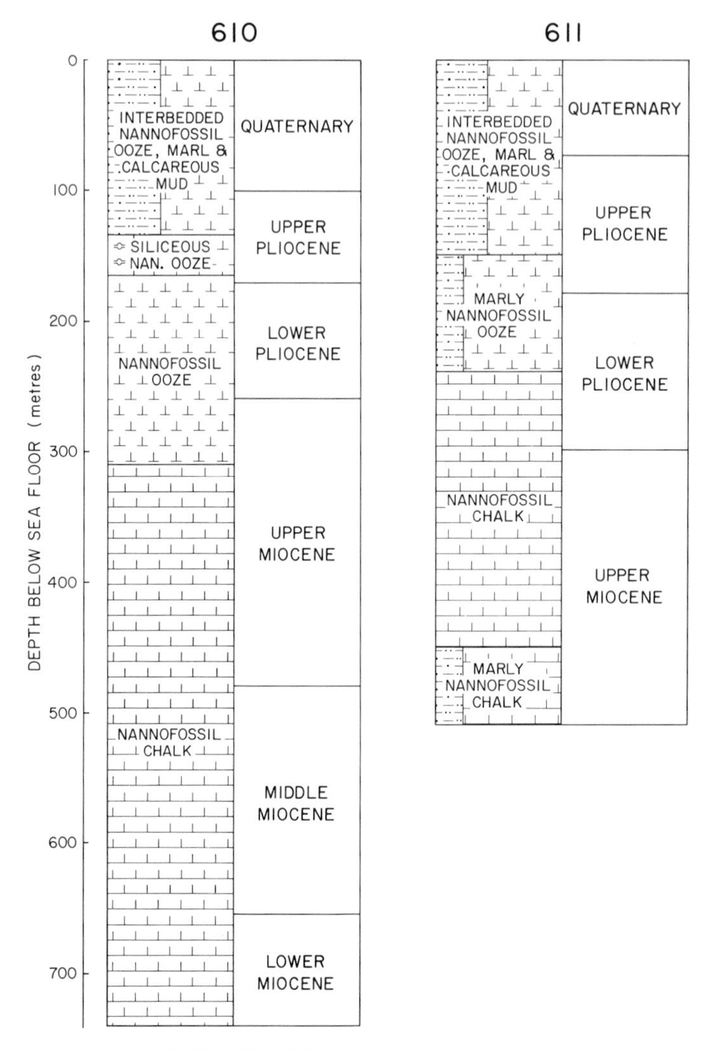

FIG. 4. Stratigraphic summaries of DSDP Sites 610 and 611.

been collected over the Gardar Ridge. Some very short traverses are available from the Reykjanes Ridge and Charlie-Gibbs Fracture Zone. These records displayed lineations on the drift surface that are again interpreted as sediment wave axes. Jacobs (in prep.) shows that wave crests north of Charlie-Gibbs Fracture Zone can be followed on the sonographs for over 6 km in the vicinity of Site 611 and they represent apparent wave amplitudes of around 10 m and wavelengths of about 1.5 km.

The survey lines around Site 611 are shown in Fig. 9, and a 3.5-kHz profile across the hole locations in Fig. 10. The waves are very irregular in section, and again do not show any obvious sense of migration. On airgun profiles, the largest waves are located over basement highs, which suggests that some waves originated from irregularities in basement topography.

The wave crests have been correlated across several airgun and 3.5-kHz profiles near to the site (Fig. 9). Although these correlations can be made and the wave axes appear linear, results from the Feni Drift suggest that the waves may not be as continuous as indicated in Fig. 9. Indeed, when attempting to offset from a wave crest to a trough location at 611, the irregularity of the selected wave again became very apparent (Kidd *et al.*, in prep.).

As a comparison the authors plotted histograms of apparent wave amplitude and length for both sites (Fig. 11). The data used represent

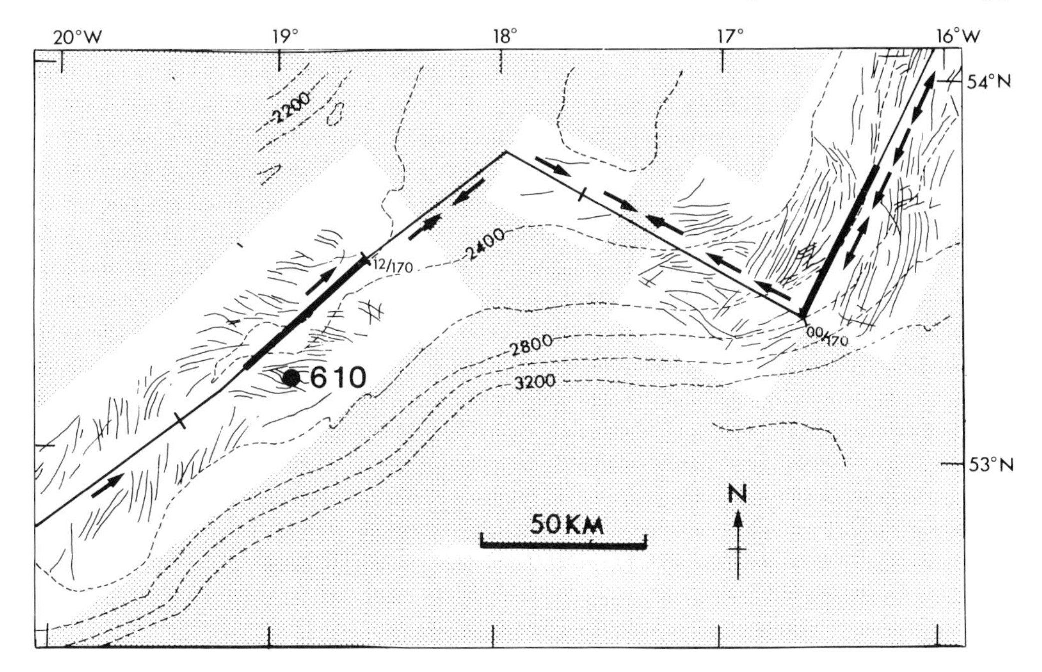

FIG. 5. Interpretative sketch-map (from Roberts & Kidd 1979) of sediment wave axes on the Feni Drift, in the vicinity of DSDP Site 610, from GLORIA long-range sidescan sonar coverage (unstippled areas). Arrows represent their interpretations of wave migration from an along-track 3.5-kHz profile.

apparent values, being oblique to crest line crossings. Frequently, values from more than one crossing of the same wave are plotted. Over one hundred crossings of wave axes were made during the survey of an area covering 4630 km^2 (Fig. 9). The information is of value in comparing these waves with data available from other wave fields (Flood & Hollister 1974; Gardner & Kidd 1983). Sediment waves near the crest of Feni Ridge have an average apparent amplitude of around 23 m and an average apparent wavelength of 2.2 km. The wave upon which the study's deep hole was drilled was approximately 28 m high and had a 1.4 km wavelength. The Gardar waves have very similar proportions, although they have more limited range of wavelengths. The wave drilled at Gardar was approximately 35 m high and again 1.4 km in wavelength.

Drillsite descriptions

At both Sites 610 and 611, holes were drilled on both the crest and trough of a single sediment wave, with the objective of comparing sediment lithologies, sedimentary structures, and sedimentation rates at the two locations. The holes were planned so that the sediment wave sampled on the Feni Drift was located close to the crest of the Feni Ridge, while at the Gardar Drift a wave farther down the flank of the ridge was sampled.

At Feni, a wave whose crest could be traced laterally for several kilometres was selected for drilling (Figs 6 and 8). The crest site was located in 2417 m of water and four holes (610, 610A, 610B, and 610C) were drilled, recovering a continuous section to the late Pliocene at approximately 190 m. The two trough holes (610D and 610E) were drilled approximately 600 m WNW of the crest site in 2445 m of water, giving a vertical offset of 28 m. Only the late Quaternary and late Miocene intervals were sampled at the trough site.

On the Gardar Drift, the crest site was located in 3203 m of water, while the trough site was offset to the south-west by approximately 1 km in 3227 m of water, giving a vertical offset of 24 m. The drilling strategy was slightly altered at Site 611. It was decided that a detailed sampling of both crest and trough holes was necessary to verify the hypothesis that sedimentation rate curves at the two locations were significantly different, thus indicating the possible migration of the sediment wave. Therefore, both crest (611, 611A, 611D, and 611E) and trough (611B and 611C) holes were sampled continuously to the early Pliocene. At the crest the drilling terminated at 244 m in the early Pliocene, whereas in the

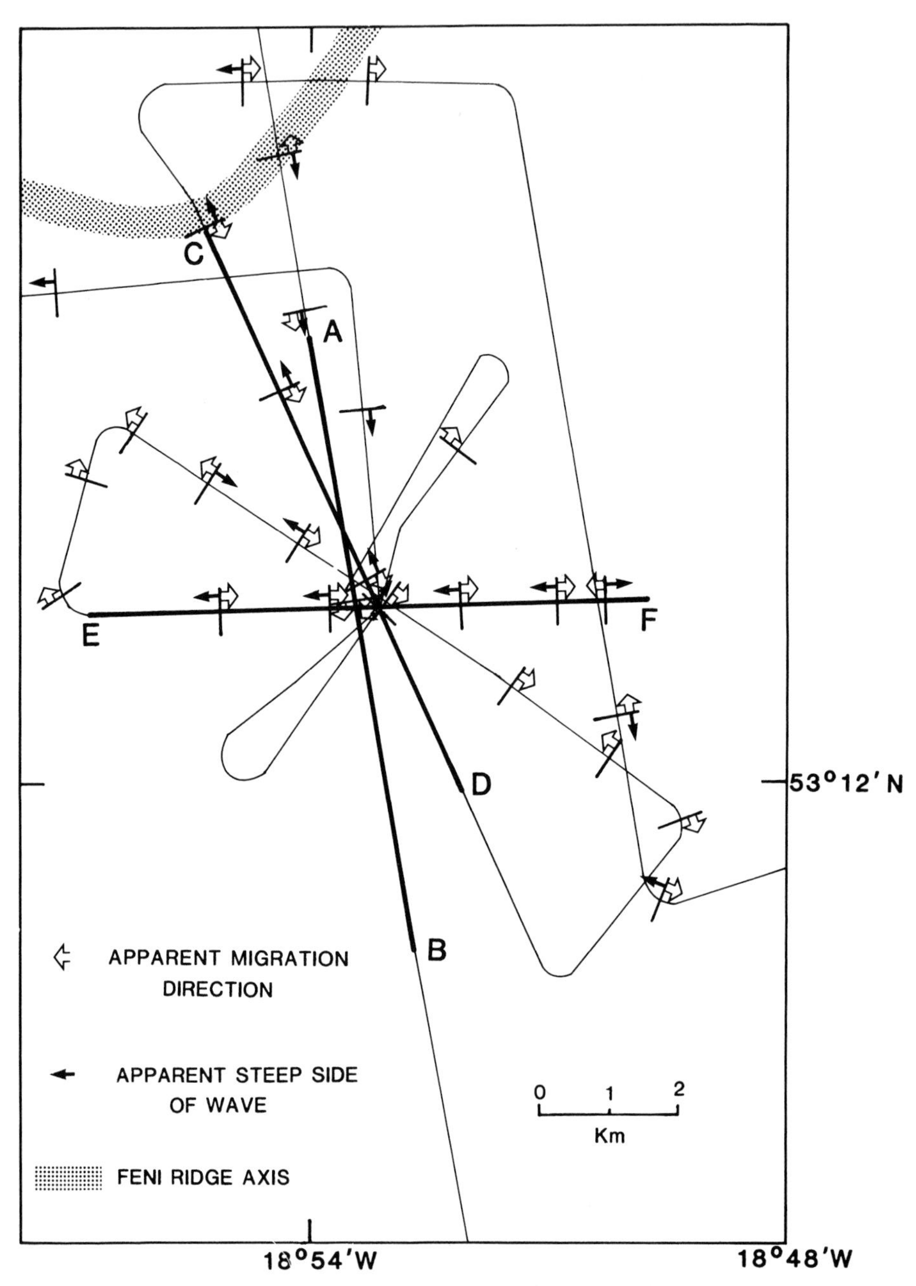

FIG. 6. *Glomar Challenger* 3.5-kHz and airgun seismic survey of area around DSDP Site 610. Broad arrows indicate apparent migration directions of sediment wave. Profiles A–B, C–D, and E–F are illustrated in Fig. 7.

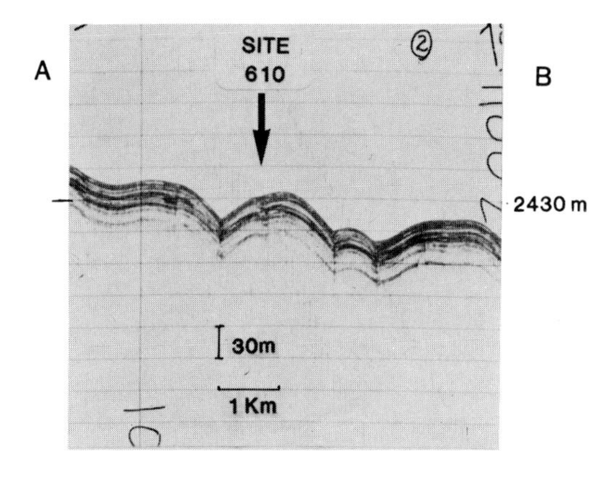

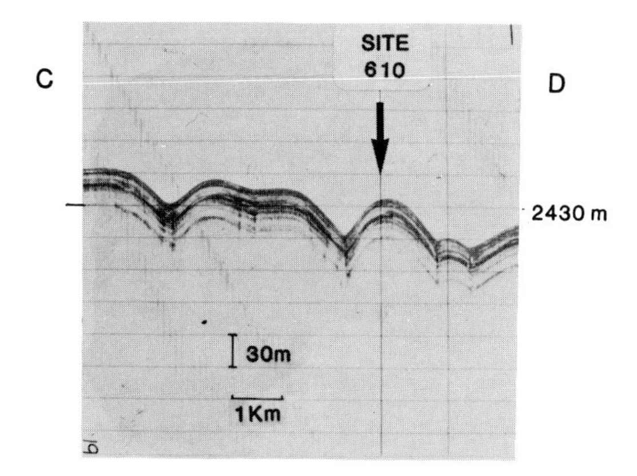

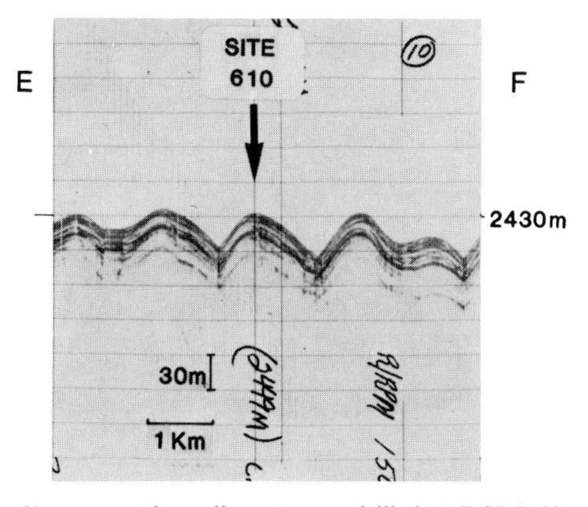

FIG. 7. 3.5 kHz sounder profiles across the sediment wave drilled at DSDP Site 610.

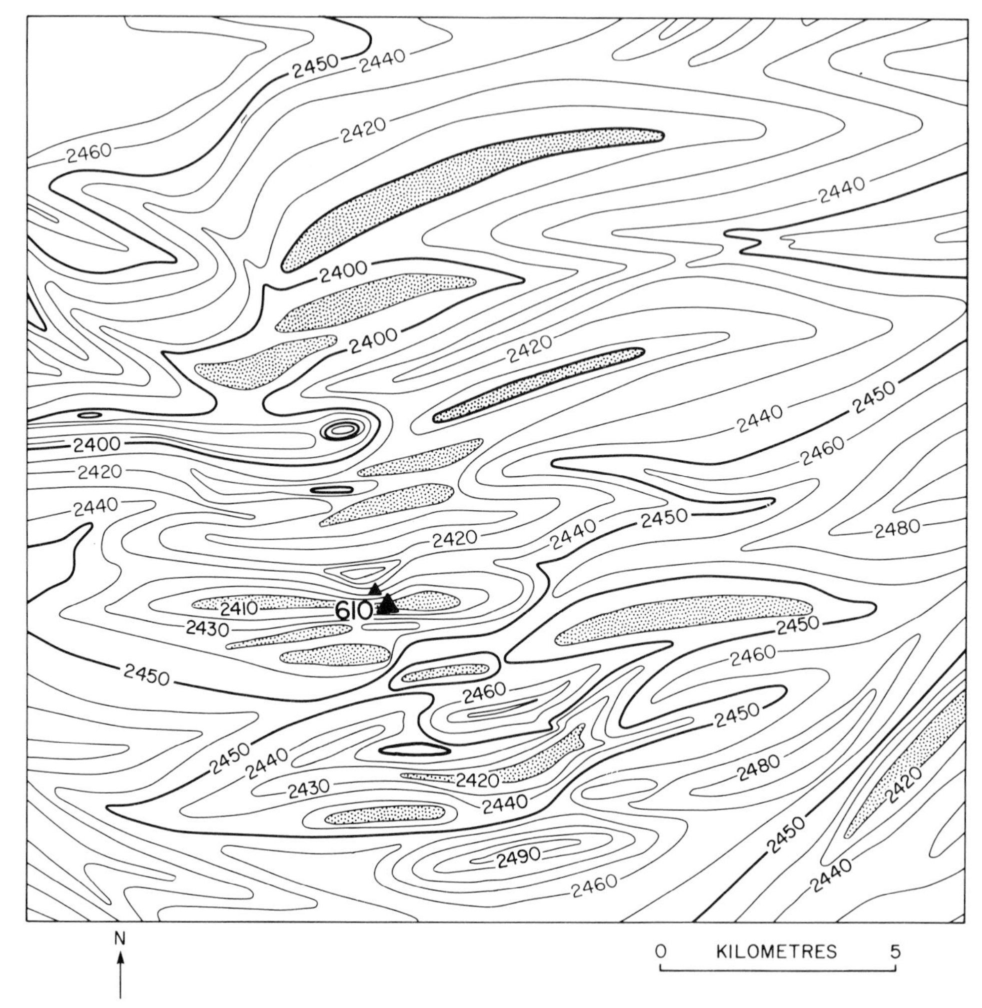

FIG. 8. Detailed bathymetric map of area in the immediate vicinity of DSDP Site 610. Major sediment wave crests are stippled.

trough discontinuous sampling was continued to the middle Miocene at 512 m.

Lithologies

A detailed description of the sedimentary characteristics of the drift sites is given in Hill (in prep.). The sequences at the two sites are essentially similar, consisting of interbedded nannofossil ooze and calcareous mud throughout the late Pliocene and Quaternary sections, and of nannofossil ooze alone in the Miocene to late Pliocene. The Gardar site is characterized by abundant volcanic clasts and grains in the coarse fraction of the calcareous muds, and a relatively high organic silica component in the nannofossil ooze.

The sequence is extensively bioturbated, with burrows, including *Zoophycos* and *Planolites*, most visible at the main lithological boundaries. There are few primary sedimentary structures apart from some occurrences of vague, horizontal lamination (Hill, in prep.). In some calcareous mud beds, erratic gravel clasts are concentrated in discrete layers, possibly as lag deposits.

Bulk grain size changes significantly over transitions from calcareous mud to nannofossil ooze (Fig. 12). These changes are not in phase with compositional changes, and so cannot be related simply to changes in sediment supply caused by glacial/interglacial climate changes. The compositional changes could have been caused by dissolution of foraminiferal tests, by changes in foraminiferal productivity, or by sorting by cur-

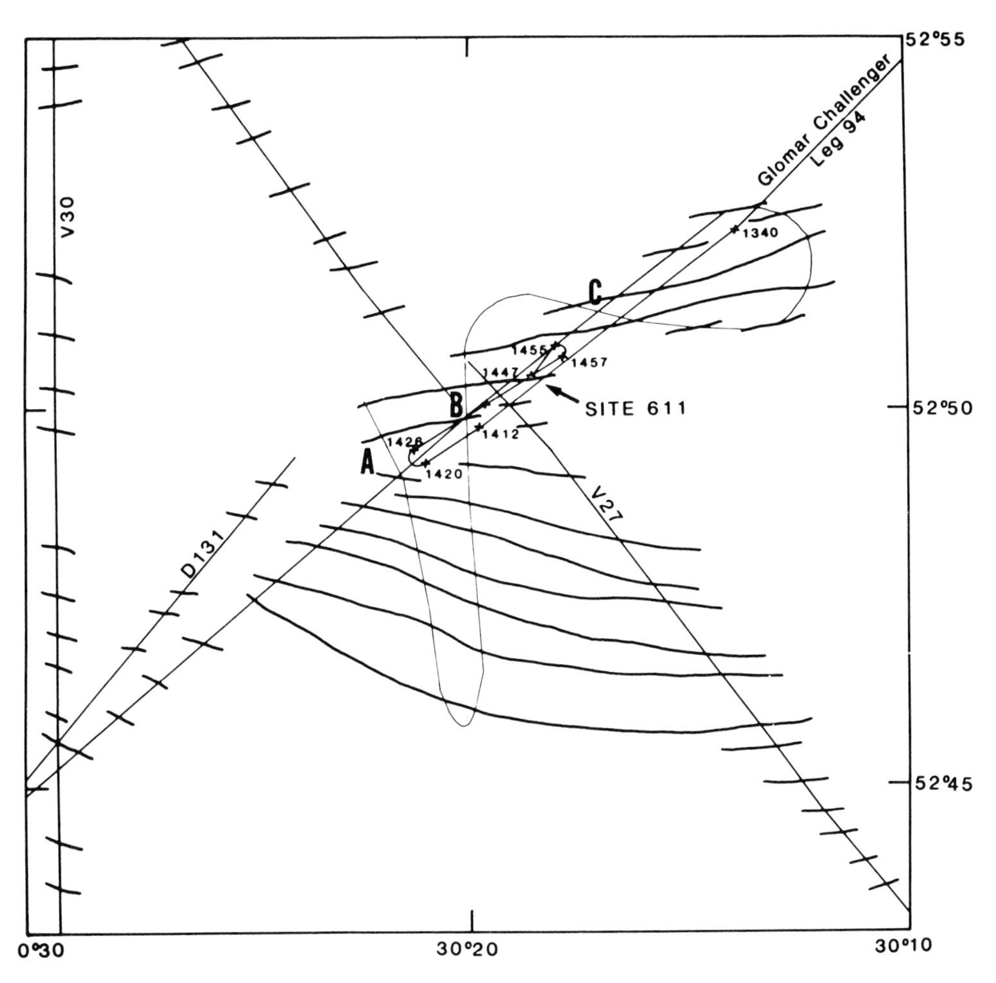

Axes of Sediment Waves from High Resolution (3.5Khz) Profiles

Scale 1:100 000

FIG. 9. 3.5 kHz and airgun seismic survey of area around DSDP Site 611. Attempts to correlate wave crests between profiles indicated by solid lines.

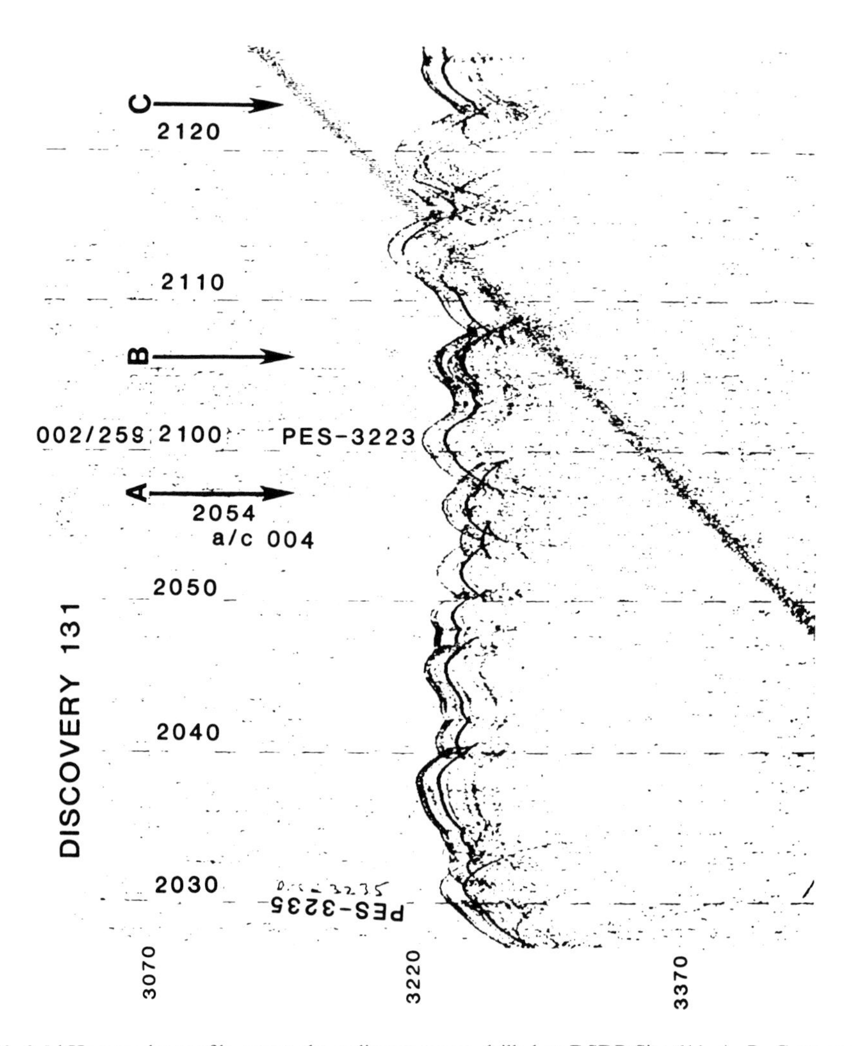

FIG. 10. 3.5 kHz sounder profile across the sediment waves drilled at DSDP Site 611. A, B, C represent turns in the *RRS Discovery* track.

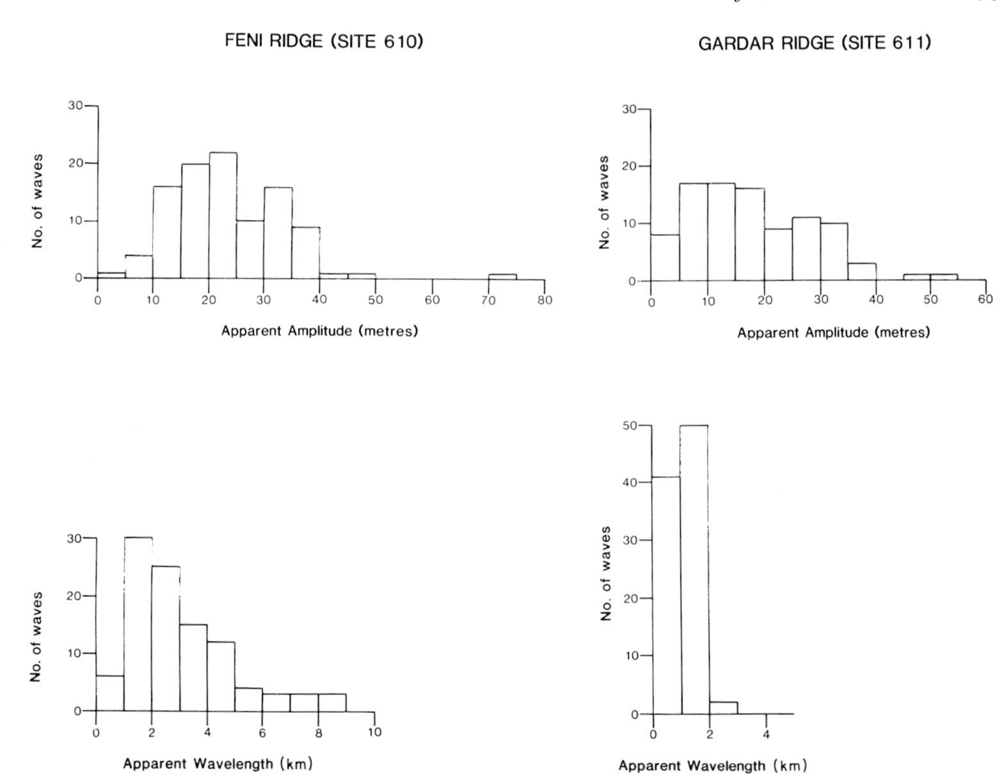

FIG. 11. Histograms of apparent amplitude and wavelength frequencies near DSDP Sites 610 and 611.

rents. Current sorting cannot explain the complete absence of foraminiferal tests at certain levels in the sequence, especially when laminations indicative of dynamic sorting are absent.

The correlation of individual beds from the crest of a sediment wave to the adjacent trough is relatively simple in the Plio-Pleistocene glacial sequences. These correlations show that at Site 611, short-term sedimentation rates have fluctuated during this time interval. Fluctuations between crest and trough holes are essentially not different from those between closely spaced holes on the crest at the same sites. Many such fluctuations are caused by coring disturbance (Ruddiman *et al.*, in prep.). However, as detailed lithological correlations can be made, the authors infer that sedimentation was continuous through the Plio-Pleistocene.

These results do not provide any evidence for current-controlled deposition on the sediment waves and drifts. A more detailed analysis of grain size and composition, linked to an isotopic analysis of glacial–interglacial fluctuations, might allow identification of separate dissolution, productivity and current-related effects.

Sedimentation rates

Compared to the average rate of Neogene and Late Quaternary pelagic sedimentation in the North Atlantic (2–40 m/Ma; Davies *et al.* 1977), rates of sedimentation on Feni and Gardar Drifts were relatively high (between 50 and 60 m/Ma). They are not as high as the 70 m/Ma rate measured at Leg 94 Site 609. However, the sequence at Site 609 is ponded between bedrock highs, and probably includes redeposited sediment from the highs. The sedimentation rates at both Sites 610 and 611 fluctuate significantly through time (Fig. 13), and reach 98 m/Ma on the Gardar Drift.

Drilling at the Gardar Drift enabled the authors to test the idea that the migration of sediment waves should be reflected in systematic differences in sedimentation rate between the crest and an adjacent trough of the wave. Excellent control on the sedimentation rates was provided by a combination of high resolution magnetostratigraphy and biostratigraphy using nannofossils, foraminifers, and diatoms (Kidd *et al.*, in prep.).

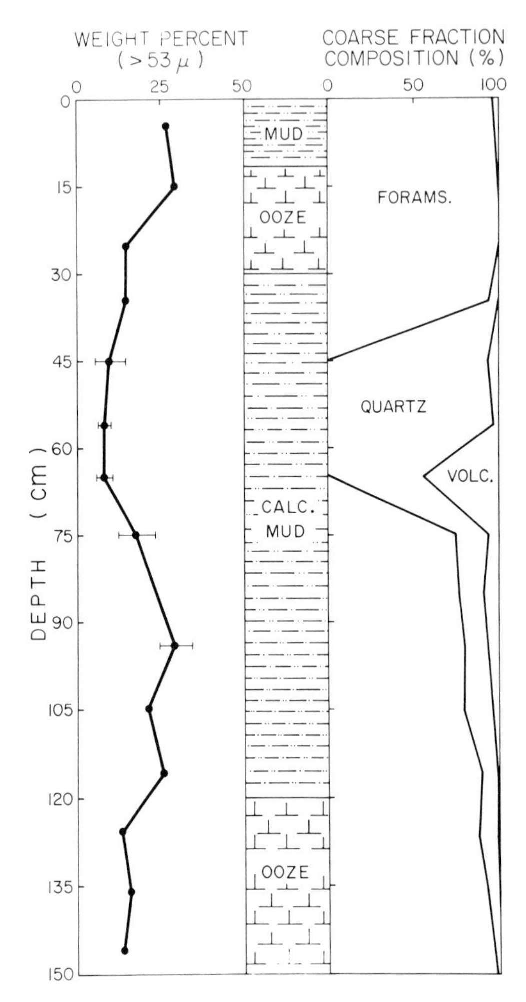

FIG. 12. Detailed grain size and composition changes through an ooze-mud-ooze sequence at DSDP Site 611.

Errors on the sedimentation rate curves are generally less than a few metres.

The curves show that, during the Pleistocene, sedimentation rates were approximately equal at the crest (611, 611D) and trough (611C) of the wave (Fig. 13(a)). However, during the late Pliocene, both locations experienced higher sedimentation rates, and the rate of accretion at the crest was significantly higher than in the trough. The two curves cross at approximately 3.5 Ma. Unfortunately, the crest holes were terminated at this level, 230 m below sea-bed, so that early Pliocene and Miocene rates cannot be compared. The rate at the trough site increases very slightly over this interval.

Figure 13(b) shows an interpretation of these curves based on sediment wave migration. The depths below sea-level of the two locations are plotted for several time planes. A sediment wave, based on present-day morphology, is fitted to the two points for each time plane. The results suggest that sediment waves migrated during the late Pliocene. At the Plio-Pleistocene boundary it is inferred that a major change in conditions made migration halt and led to deposition of a uniform drape of sediment during the Pleistocene. The authors cannot determine the mechanisms responsible for this change in sedimentation. The switch may have been brought about by a change in palaeocirculation, or by the interference of two sets of sediment waves. In support of a migration model, we can say that: (a) the migration in the Pliocene involves an increase in sedimentation rate at the present-day crest site, without any corresponding decrease at the trough site; (b) the scale of migration with respect to wavelength and amplitude of the waves is comparable to that of sediment waves where migration can be identified from seismic profiles; and (c) the interpretation of Pleistocene draping is supported by seismic profiles of the sediment waves at Gardar, where no migration directions can be determined.

Summary and conclusions

The results of the preliminary Leg 94 studies allow a number of conclusions to be drawn about the Feni and Gardar Drifts.

The base of the Feni Drift is younger and deeper than previously estimated. Accumulation of the drift seems to have been initiated in late Eocene or early Oligocene times. The sediment wave fields at both sites are even more complex and irregular than previously supposed from GLORIA records, although they maintain an overall sub-parallel orientation to linear axes observed on the sidescan records. There is little acoustic evidence for migration of the waves in either area, although variations in sediment accumulation rates between wave crest and trough suggest that migration may have taken place during the Pliocene and halted in the Pleistocene.

The lithologies observed in boreholes are fundamentally pelagic throughout in both Feni and Gardar Drift sequences. Crest to trough variations in texture and fabric are slight. There is no primary sedimentary structural evidence for current-controlled deposition, nor any grain-size variation that can be unambiguously related to current strength.

Many of the result of this study are negative and contradict some previously held ideas about

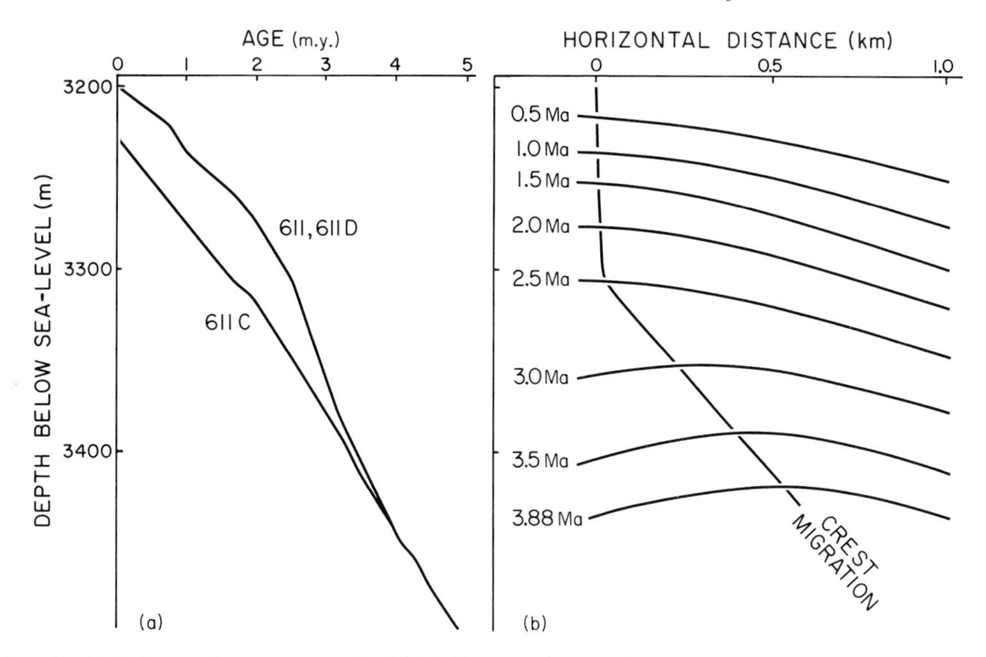

FIG. 13. (a) Sedimentation rate curves for DSDP Site 611. The trough site, 611C, is offset on the vertical axis by 32 metres, equivalent to the amplitude of the sediment wave. (b) Interpretation of the sedimentation rate curves, indicating possible Pliocene migration and Quaternary drape.

sediment drifts and sediment waves. Both Feni and Gardar Drifts have remarkably continuous sequences with very few, if any, discernible hiatuses through the Plio-Pleistocene. Both sites have provided excellent material for high-resolution studies of palaeoceanography.

References

BELDERSON, R.H., KENYON, N.H., STRIDE, A.H. & STUBBS, A.R. 1972. *Sonographs of the Seafloor: A Picture Atlas*. Elsevier, Amsterdam. 185 pp.

CHERKIS, N.Z., FLEMMING, H.S. & FEDEN, R.H. 1973. Morphology and structure of Maury Channel, Northeast Atlantic Ocean. *Geol. Soc. Am. Bull.* **84,** 1601–6.

DAVIES, T.A., HAY, W.W., SOUTHAM,. J.R. & WORSLEY, T.R. 1977. Estimates of Cenozoic oceanic sedimentation rates. *Science* **197,** 53–5.

DICKSON, R.R. & KIDD, R.B., in prep. Deep Circulation in the Southern Rockall Trough—the oceanographic setting of DSDP Site 610. *In*: KIDD, R.B., RUDDIMAN, W.F., THOMAS, E. *et al.*, *Initial Reports of the Deep Sea Drilling Project* **94.** US Govt. Print. Off., Washington, DC.

FLOOD, R.D. 1978. *Studies of Deep-sea Sedimentary Microtopography in the North Atlantic Ocean*. Ph.D. dissertation, Woods Hole, Massachusetts. MIT-WHOI Joint Program in Oceanography, WHOI Technical report, WHOI-78-64.

—— & HOLLISTER, C.D. 1974. Current-controlled topography on the continental margin off the eastern United States. *In:* BURK, C.A. & DRAKE, C.A. (eds), *The Geology of Continental Margins*. Springer-Verlag, New York. 197–205.

GARDNER, J.V. & KIDD, R.B. 1983. Sedimentary processes on the Iberian continental margin viewed by long-ranged sidescan sonar. Part 1, Gulf of Cadiz. *Oceanol. Acta* **6,** 245–54.

HEEZEN, B.C. & HOLLISTER, C.D. 1971. *The Face of the Deep*. Oxford University Press, New York. 659 pp.

HILL, P.R., in prep. Sedimentary characteristics of drift sequences from Sites 610 and 611, DSDP Leg 94. *In:* KIDD, R.B., RUDDIMAN, W.F., THOMAS, E. *et al.*, *Init. Repts of DSDP* **94.** US Govt. Print. Off., Washington, DC.

HOLLISTER, C.D., FLOOD, R.D., & MCCAVE, I.N. 1978. Plastering and decorating in the North Atlantic. *Oceanus* **21,** 5–13.

JACOBS, C., in prep. Site surveys for DSDP Sites 608, 610, and 611, eastern North Atlantic Ocean. *In:* KIDD, R.B., RUDDIMAN, W.F., THOMAS, E. *et al.*, *Init. Repts DSDP* **94.** US Govt. Print. Off., Washington, DC.

JOHNSON, G.L., VOGT, P.R. & SCHNEIDER, E.D. 1971. Morphology of the Northeastern Atlantic and Labrador Sea. *Dtsch. Hydrogr. Z.* **24,** 49–73.

JONES, E.J.W., EWING, M.J.I. & EITTREIM, S.L. 1970. Influences of Norwegian Sea overflow water on sedimentation in the northern North Atlantic and Labrador Sea. *J. Geophys. Res.* **75,** 1655–80.

KIDD, R.B. & ROBERTS, D.G. 1982. Long-range sidescan sonar studies of large-scale sedimentary features in the North Atlantic. *Bull. Inst. Geol. Bassin d'Aquitaine, Bordeaux* **31,** 11–29.

—— RUDDIMAN, W.F., THOMAS, E. *et al.*, in prep. *Init. Repts DSDP* **94.** Govt. Print. Off., Washington, DC.

LAUGHTON, A.S., ROBERTS, D.G. & HUNTER, P.M. 1982. *Bathymetry of the Northeast Atlantic.* Map Sheet 2. Institute of Oceanographic Sciences, Wormley, UK.

LONSDALE, P. 1982. Sediment drifts of the Northeast Atlantic and their relationship to the observed abyssal currents. *Bull. Inst. Geol. Bassin d'Aquitaine, Bordeaux* **31,** 141–9.

MASSON, D.G. & KIDD, R.B., in prep. Tertiary seismic stratigraphy of the Southern Rockall Trough. *In:* KIDD, R.B., RUDDIMAN, W.F., THOMAS, E. *et al.*, *Init. Repts DSDP* **94.** US Govt. Print. Off., Washington, DC.

MCCAVE, I.N., LONSDALE, P.F., HOLLISTER, C.D. & GARDNER, W.D. 1980. Sediment transport over the Hatton and Gardar contourite drifts. *J. sedim. Petrol.* **50,** 1049–62.

ROBERTS, D.G. 1975. Marine geology of the Rockall Plateau and Trough. *Phil. Trans. R. Soc. London*, Ser. A, **278,** 447–509.

ROBERTS, D.G. & KIDD, R.B. 1979. Abyssal sediment-wave fields on Feni Ridge, Rockall Trough: long range sonar studies. *Mar. Geol.* **33,** 175–91.

RUDDIMAN, W.F. 1972. Sediment distribution of the Reykjanes Ridge: Seismic evidence. *Bull. Geol. Soc. Am.* **83,** 2039–62.

——, CAMERON, D. & CLEMENT, B., in prep. Sediment disturbances and correlations of HPC offset holes: Leg 94. *In:* KIDD, R.B., RUDDIMAN, W.F., THOMAS, E. *et al.*, *Init. Repts DSDP* **94.** US Govt. Print. Off., Washington, DC.

ROBERT B. KIDD, Institute of Oceanographic Sciences, Wormley, Godalming, Surrey GUB 5UB. Present address: Ocean Drilling Program, Texas A & M University College Station, Texas 77843-3469, USA.

PHILIP R. HILL, Geological Survey of Canada, Atlantic Geoscience Centre, PO Box 1006, Dartmouth, Nova Scotia, Canada B2Y 4A5.

Late Neogene submarine erosion events along the north-east Atlantic continental margin

R. Stein, M. Sarnthein & J. Suendermann

SUMMARY: During the last 5 Ma, bottom currents have formed two major hiatuses along the continental margin off north-west Africa. In addition, slumping may have caused hiatuses at Site 397. The hiatuses extend from Gibraltar to the equator between 2800 and 4000 m water depth. At the hiatuses there has been up to 30 m erosion, or reduced sedimentation, at DSDP Sites 544B, 397, 141, and 366. Bio- and magneto-stratigraphy and oxygen-isotope curves relate the hiatuses to cooling events of the North Atlantic Deep Water at about 2.75 and 1 Ma.

A dynamic ocean model of deep-water circulation provides insights into how climate induced changes in the deep-water current regime.

Today, the north-east Atlantic continental margin is generally regarded as a region of sluggish deep-water circulation (Jacobi & Hayes 1982; Mantyla & Reid 1983). The occurrence of two spectacular stratigraphic gaps in the last 5 Ma at several DSDP Sites between Gibraltar and the equator (Figs 1 and 2, Tables 1 and 2) suggests that circulation was not always sluggish there. Some geologists have related these hiatuses in water depths of 2800 m to 4000 m (Table 1) to submarine erosion by strong bottom-water currents (Arthur *et al.* 1979; Cita & Ryan 1979; Stein 1984; Stein & Sarnthein 1984a). However, calcium-carbonate dissolution, turbidity currents, and slumping could have formed the hiatuses too (Moore *et al.* 1978; Table 3).

In this paper, a detailed study made to find out how the two hiatuses formed is discussed. Based on a detailed investigation of DSDP Sites 141, 366, 397, and 544B (Stein 1984, 1985; Stein & Sarnthein 1984a; 1984b), the stratigraphic ranges of the hiatuses are defined, and some sediment features that may record the mode of hiatus formation are listed. In addition, a dynamic five-layer ocean model is applied, for the first time (Suendermann 1985), in order to obtain insights into climatically induced changes in the palaeo-deep water current regime that may have caused the stratigraphic gaps.

Chronostratigraphic definition of hiatuses

The identification of the two hiatuses and their age range are based on a precise time-scale. The time-scale combined (1) a set of first (FAD) and last (LAD) appearance datums of both planktonic and coccolith species with (2) magnetic stratigraphy and (3) oxygen-isotope records. The authors have used the bio- and magneto-stratigraphic data of the Shipboard Scientific Parties of Legs 14, 41, 47A, and 79 (1972, 1978, 1979, and 1984), supplemented with biostratigraphic data derived by E. Martini, U. Pflaumann, and C. Samtleben (all pers. comm. 1983) (for details of age assignment, see Stein 1984).

The four DSDP sites all have hiatuses or distinctly reduced sedimentation rates near the Early–Late Pliocene boundary, and during the Early Quaternary (Table 2). The stratigraphic evidence is summarized in Fig. 2.

The Pliocene hiatus is from 2.1 Ma long (Site 544B) to 0.3 Ma long (Site 397). At Site 366, it is identified by the LAD of *D. tamalis*, *G. altispira*, and *G. margaritae* in Core 6–1 (Cepek *et al.* 1978; Pflaumann & Krasheninnikov 1978), but occurs between two cores and is thus somewhat questionable. At Site 141, i.e. in water depths below 4100 m, there is a distinct reduction in the sedimentation rate, rather than a hiatus (Table 2, Fig. 2). Although the age range of the Pliocene hiatus differs markedly at Sites 544B, 397, and 366, its upper limit appears almost contemporaneous at all Sites, near 2.5 Ma (Figs 2 and 3). The thickness of the lost sediment sequences at the different sites has been estimated from the sedimentation rates above and below the hiatus (Table 2). Despite the different age range of the Pliocene hiatus at Sites 397 and 544B, the thickness of the eroded sediment sequence at both Sites is surprisingly similar (about 33 m; Table 2). In general, the lost sediment sequence decreases from north (33 m) to south (11 m).

The Quaternary hiatus is from 0.63 m.y. long (Site 397) to 0.17 m.y. long ('Meteor'—Core 13519/Site 366) (Table 2). It was impossible to say whether Site 141 was an area of erosion or of reduced sedimentation, because the uppermost 5 m near to the surface was not cored (Fig. 2(d)). The lost sediment sequences vary from 75 m (Site

From SUMMERHAYES, C.P. & SHACKLETON, N.J. (eds), 1986, *North Atlantic Palaeoceanography*, Geological Society Special Publication No. 21, pp. 103–118.

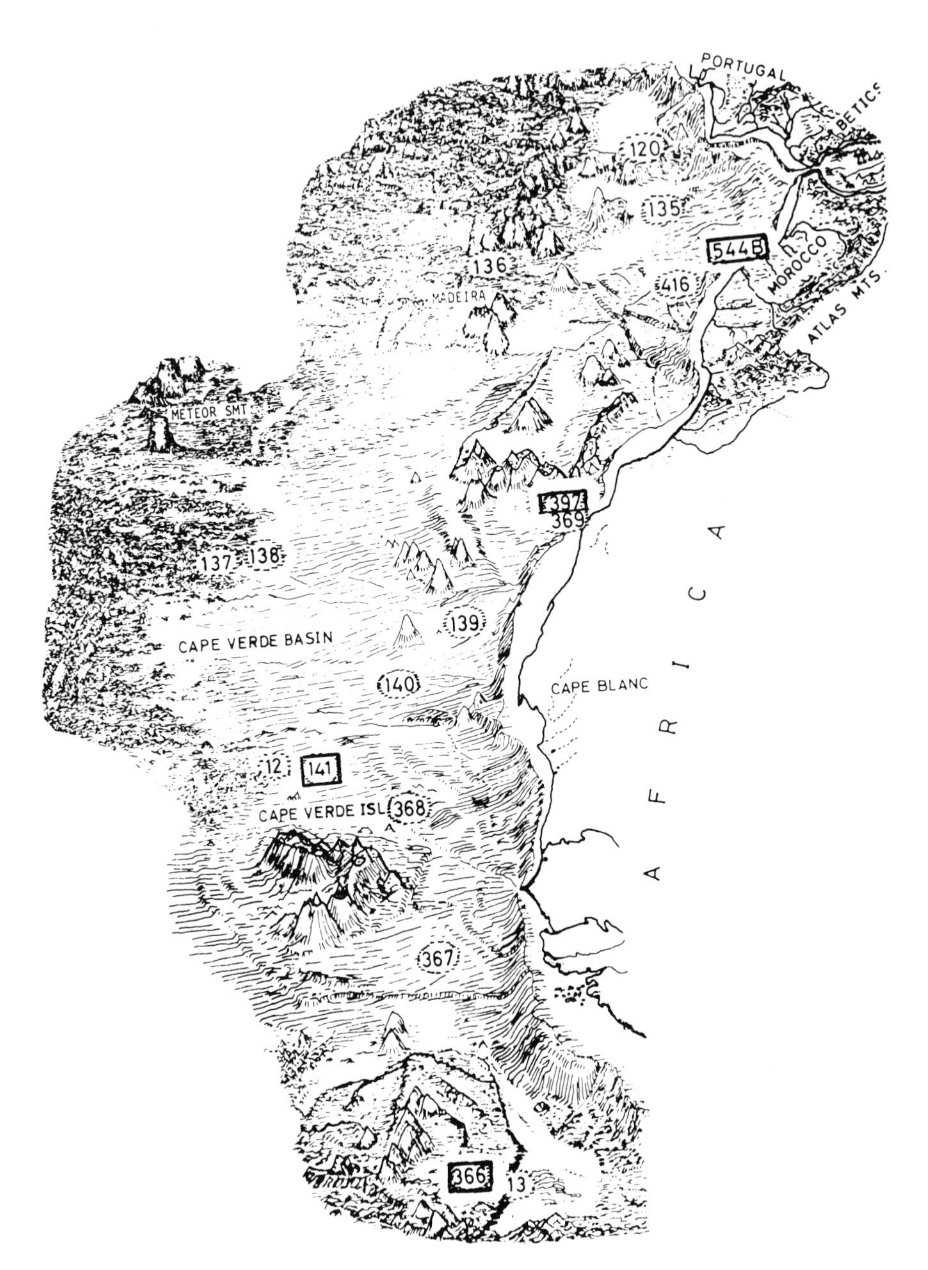

FIG. 1. Positions of DSDP Sites 141, 366, 397, and 544B in the north-east Atlantic Ocean. Base map redrawn from Berger & von Rad (1972) and Sarnthein *et al.* (1982).

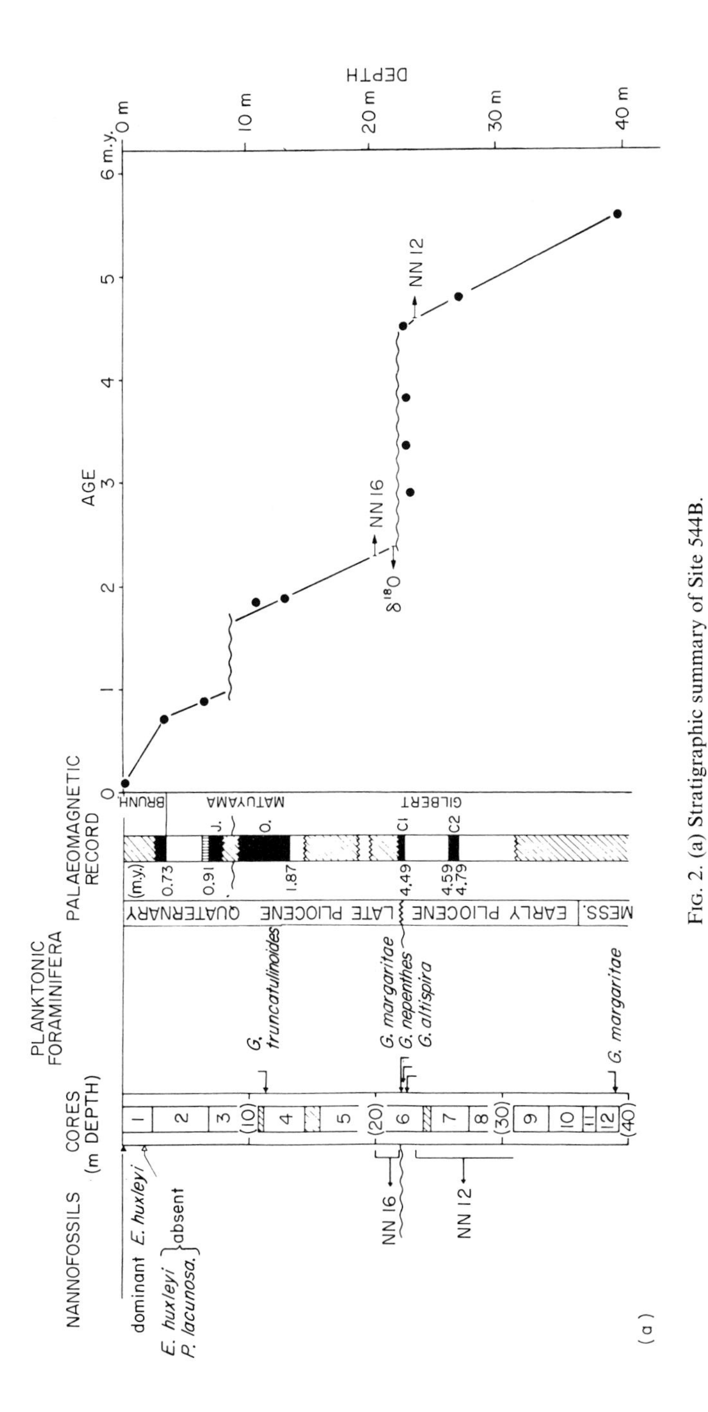

Fig. 2. (a) Stratigraphic summary of Site 544B.

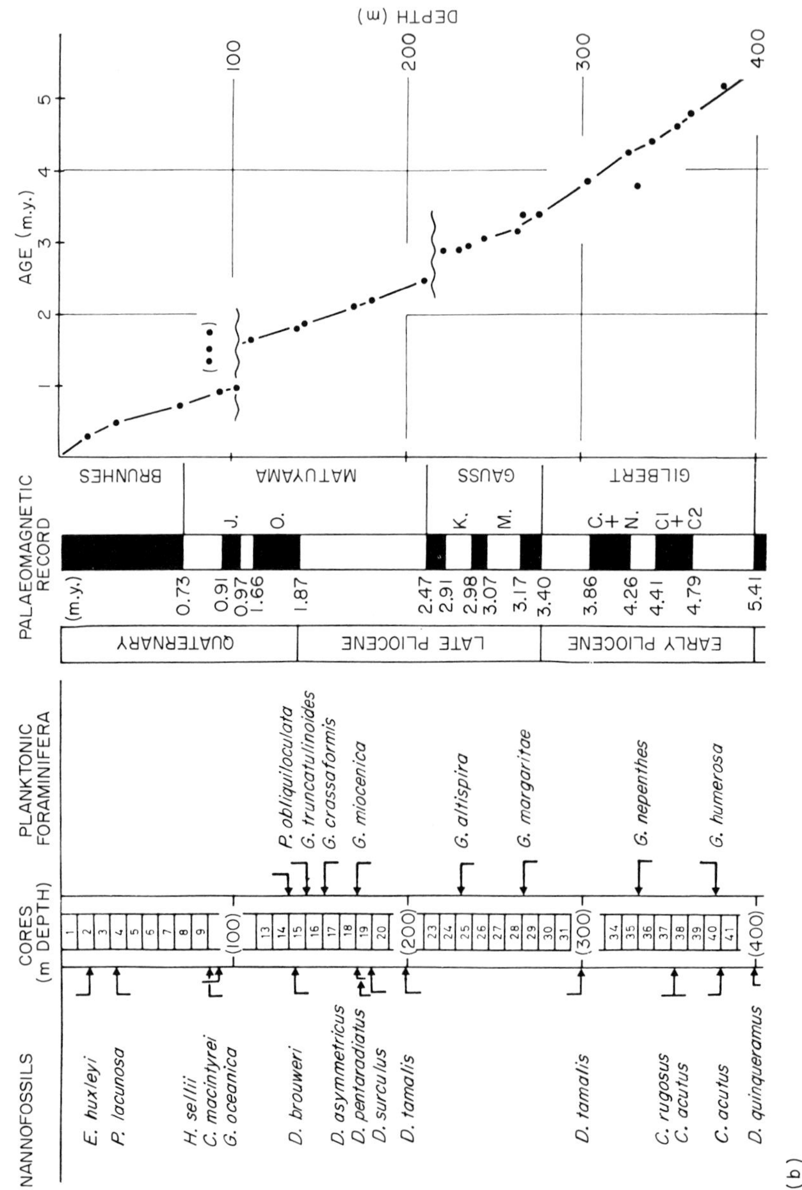

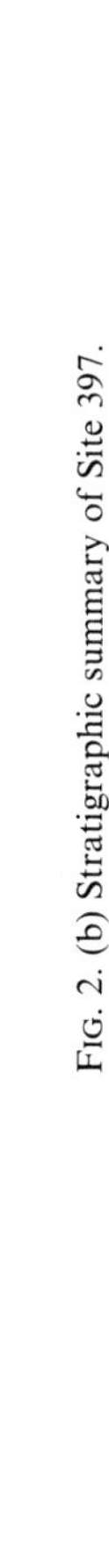
FIG. 2. (b) Stratigraphic summary of Site 397.

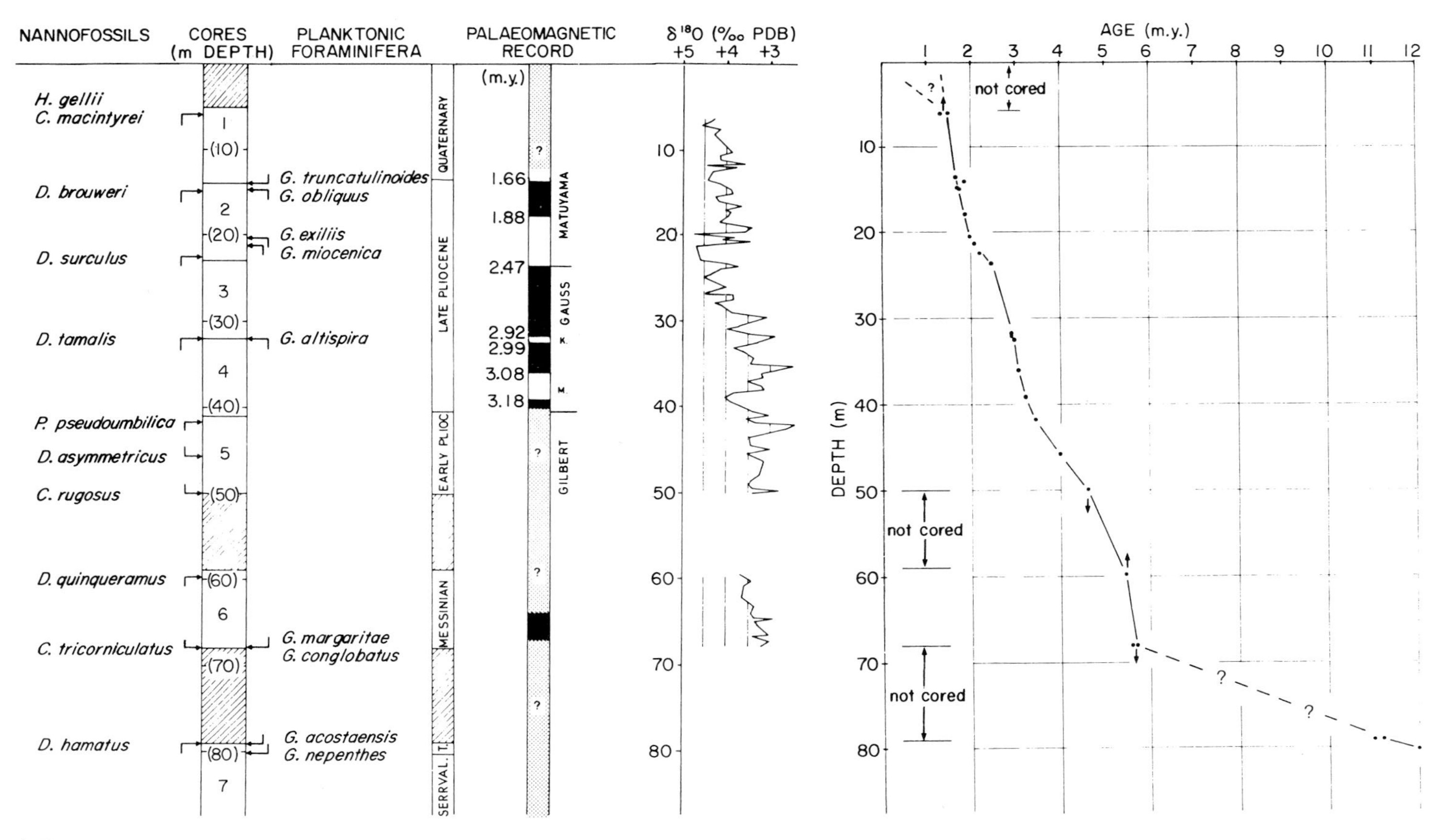

FIG. 2. (c) Stratigraphic summary of Site 141.

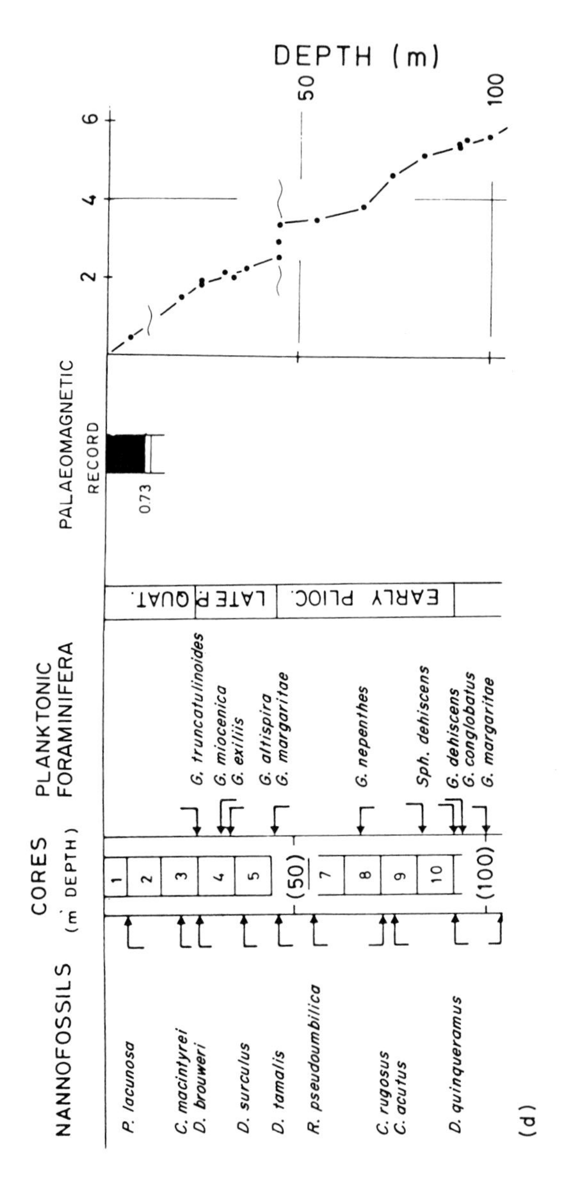

FIG. 2. (d) Stratigraphic summary of Site 366. Summaries of (a), (b), (c), (d) according to Shipboard Scientific Parties 1972, 1978, 1979, and 1984; details of age assignment in Stein 1984.

TABLE 1. *Positions and water depths of DSDP Sites 141, 366, 397, and 544B*

Site	Position	Water depth
544B	33° 46.13′ N./9° 24.29′ W. (seaward margin of a structural high at the rise of the Mazagan Plateau)	3607m
397	26° 50.7′ N./15° 10.8′ W. (uppermost continental rise south of the Canary Islands)	2900 m
141	19° 25.16′ N./23° 59.91′ W. (on top of a diapiric structure north of the Cape Verde Islands)	4148 m
366	5° 40.7′ N./19° 51.1′ W. (top plateau of the Sierra Leone Rise)	2853 m

397) to 2 m in thickness ('Meteor'—Core 13519/Site 366).

Variations of sediment composition and texture along with hiatuses

Both the Pliocene and Quaternary hiatuses occur at times of distinct shifts in the benthos oxygen-isotope composition towards heavier values (Fig. 3), i.e. they occur during periods of onset or expansion of major polar ice sheets and/or major cooling events of the North Atlantic Deep Water (NADW) (e.g. Shackleton & Cita 1979). The Pliocene hiatus at most Sites ends some 100 000 years prior to the first extreme glacial event at about 2.4 Ma (Shackleton *et al.* 1984).

At the northernmost Site, 544B, both stratigraphic gaps are associated with erosional sediment structures in the core (Stein & Sarnthein 1984b). They coincide with a distinct coarsening of both the bulk sediment (i.e. mainly planktonic foraminifera), and the terrigenous sediment fractions (Stein 1984; Stein & Sarnthein 1984b). The Pliocene hiatus is further characterized by a marked increase in chlorite, and a decrease of $CaCO_3$ (Stein & Sarnthein 1984b).

At Site 397, both hiatuses give rise to distinct seismic reflectors (R-1 and R-2; Shipboard Scientific Party 1979). The Pliocene hiatus lies within a markedly disturbed core (23) and is paralleled by a change in lithology from marly nannofossil ooze to siliceous foraminiferal nannofossil ooze (Shipboard Scientific Party 1979). Above the hiatus, Diester-Haass (1979) has described three thin layers of coarse-grained terrigenous and shallow-water sand (Cores 23–3, 22–5, and 21–6), possibly turbidites. Further evidence of reworking comes from the occurrence of *R. pseudoumbilica*, with a LAD of 3.56 Ma (Backman & Shackleton 1983) in Core 23–4 (Cepek & Wind 1979). This core is dated as less than 2.5 Ma (Fig. 2(c)). The increased flux of biogenic opal and the distinct lightening of $\delta^{13}C$ values at Site 397 during the Late Pliocene (Stein 1985) suggest increased palaeoproductivity and flux of organic matter during that time. The Quaternary hiatus at this site is also associated with a 10 cm thick, sandy-silty, quartz-rich sediment layer (Core 11–5; Shipboard Scientific Party 1979), caused by either turbidity flow, or contour current activity.

Despite a decrease in the mean $CaCO_3$ content

TABLE 2. *Position and age range of stratigraphic gaps at DSDP Sites 141, 366, 397, and 544B. Lost sediment sequences are calculated using minimum/maximum sedimentation rates of the preceding and subsequent time intervals (Stein 1984). Data of the Quaternary hiatus of Site 366 from 'Meteor' core 13519, a gravity core raised from the same position at Site 366 (Sarnthein et al. 1984)*

Pliocene Hiatus	544B	397	141	366
Core	6	23	no hiatus, but distinctly reduced sedimentation rates between 2.5 and 2.2 Ma	5/6
Sub-bottom depth	22.1 m	215 m		44 m
Age range	4.5–2.43 Ma	2.85–2.55 Ma		3.4–2.6 Ma (?)
Lost sediment sequence	33 m /34 m	26 m /44 m		14 m /8 m
Quaternary Hiatus				
Core	3	11	hiatus or distinctly reduced sedimentation rates near 1.4 Ma	2
Sub-bottom depth	8.7 m	103 m		10.5 m
Age range	1.6–1.05 Ma	1.6–0.97 Ma		0.95–0.78 Ma
Lost sediment sequence	9 m /9 m	74 m /139 m		2 m /3 m

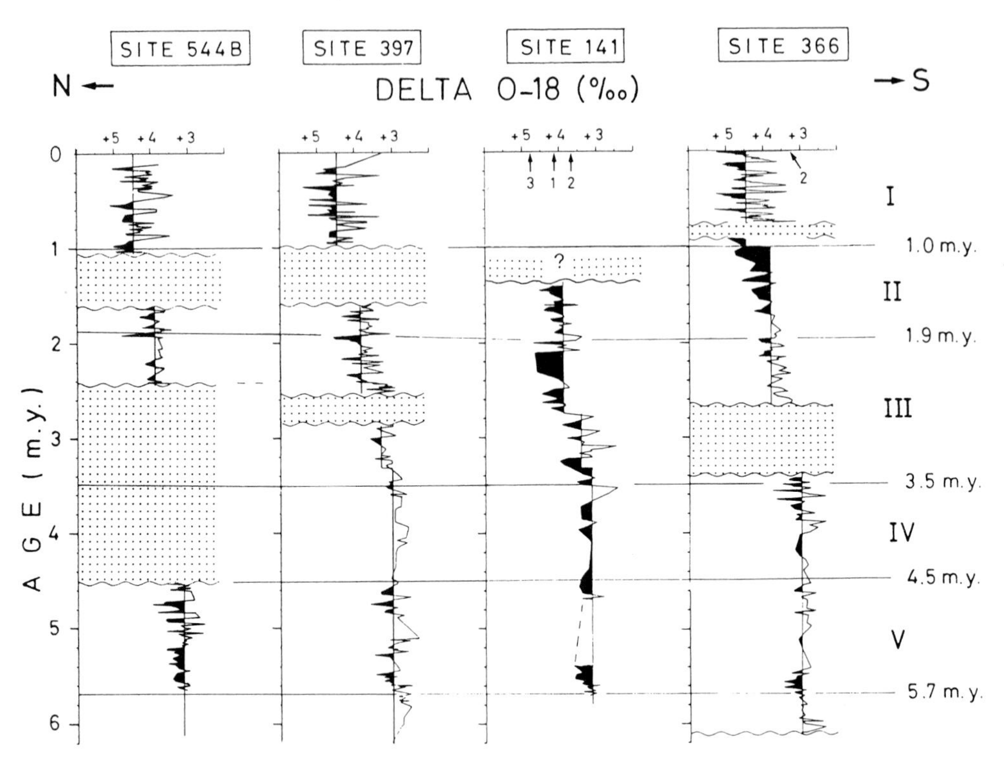

FIG. 3. Benthos oxygen-isotope curves (values in ‰ to PDB). All data are adjusted to *U. peregrina*, following Shackleton & Cita (1979). Data of Site 397 from Shackleton & Cita (1979), and of Site 366 from Blanc & Duplessy (1982), both supplemented by own data (Stein 1984). Records of Sites 141 and 544B from own data (Stein 1984; Stein & Sarnthein 1984b). Data from the topmost 10 m section of Site 366 derived from 'Meteor' core 13519 (Sarnthein *et al.* 1984). Small arrows indicate δ ^{18}O values of (1) the Present, (2) the last climatic optimum at 6000 years. BP, and (3) the last glacial maximum at 18000 years BP. Roman numerals indicate long-term climatic intervals (Stein 1984; Stein & Sarnthein 1984a). Dotted intervals mark hiatuses.

above the Pliocene hiatus at Site 366, no other sediment variables changed across the hiatus at this Site (Stein 1984; Stein & Sarnthein 1984a).

A dynamic mode of Atlantic deep-water circulation

A numerical model of the Atlantic Ocean has been used for simulating the palaeo-deep water circulation (Suendermann 1985). It is based on the three-dimensional non-linear adiabatic Boussinesq-equations. The circulation is driven by the windstress, by the heat flux from the atmosphere to the ocean, and by the Antarctic Circumpolar Current. It is assumed that there is no exchange of water with the Northern Polar Sea or with the Mediterranean. The input data from the geological record are palaeo-wind speeds and sea-surface temperatures (SST) near the equator and at high latitudes (Blanc & Duplessy 1982; Stein 1984). The authors assume that the principal structures of SST and the wind field were like those of today.

This model was not developed for the Neogene, so its use for interpreting submarine erosion is doubtful. Furthermore the grid spacing of 5° in the horizontal dimensions of the model, and its division of the ocean into five vertical layers, may be too coarse to represent local current fields off north-west Africa. Nevertheless, the degree of resolution appears adequate for the authors' purposes.

The results of the dynamic ocean model, palaeo-temperatures, and speeds and directions of the palaeo-deep water circulation are presented in Fig. 4 for three different time slices related to the Pliocene hiatus:

(1) 3.5 Ma (first onset of Early–Late Pliocene climatic deterioration)
(2) 2.75 Ma (time of distinct expansion of major Arctic ice sheets and/or major cooling of deep water)

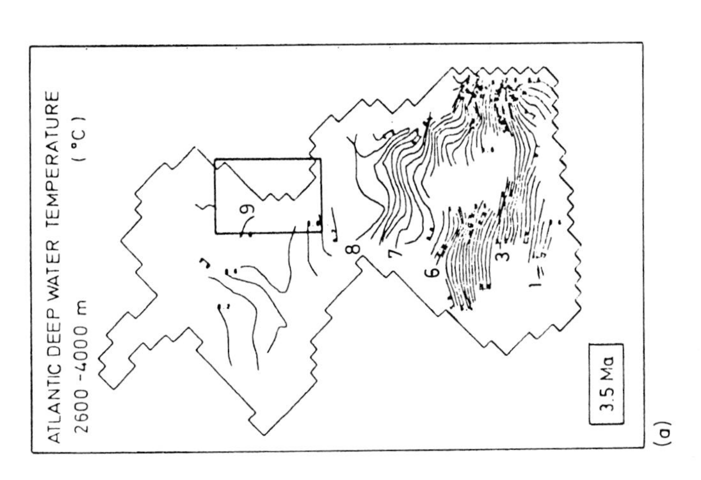

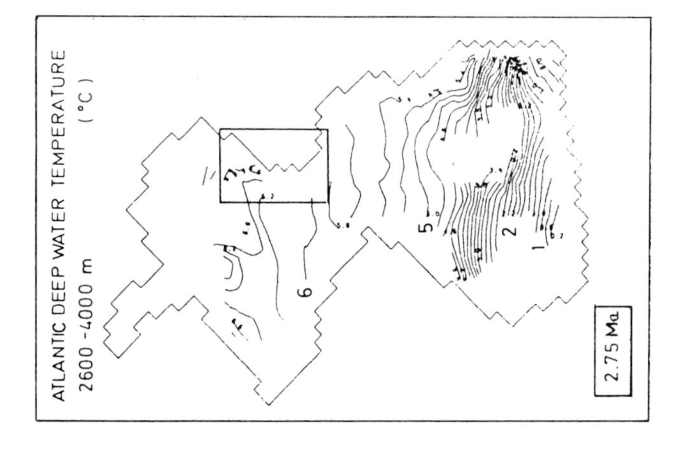

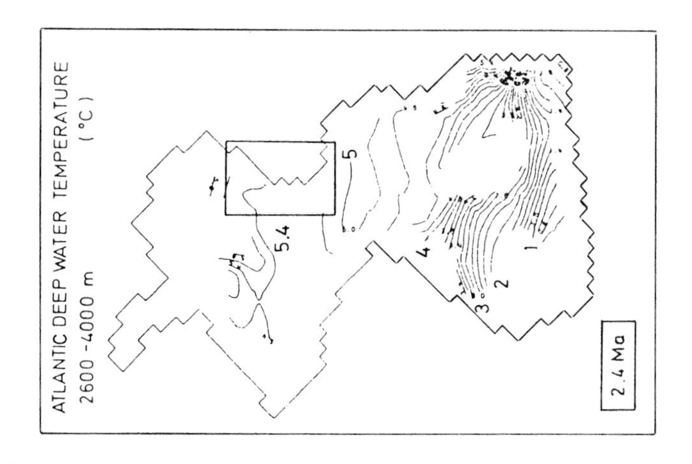

FIG. 4. Results of the dynamic ocean model for 3.5, 2.75, and 2.4 Ma time slices. The box indicates the study area. (a) Palaeotemperatures of Atlantic deep water (2600 m to 4000 m water depth).

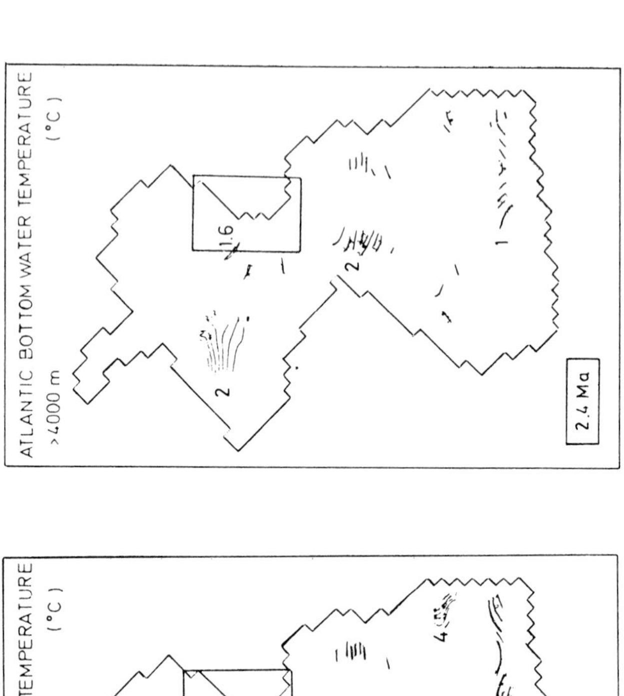

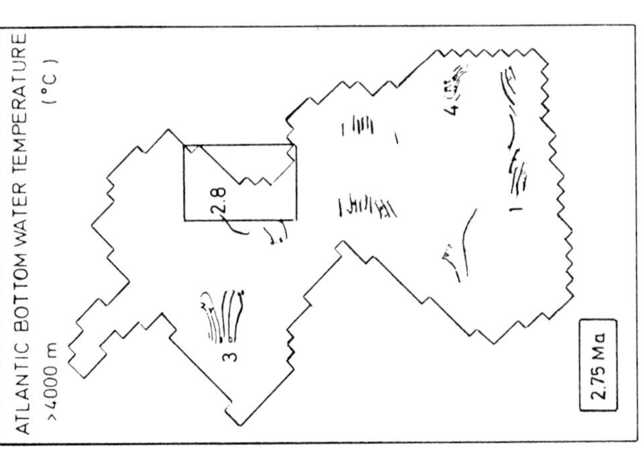

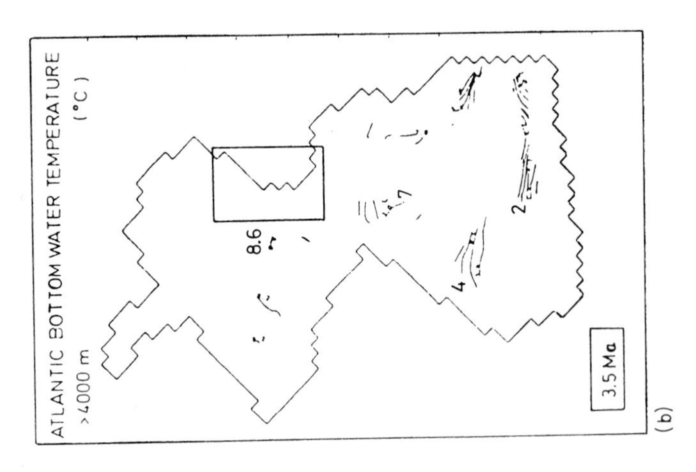

FIG. 4. (b) Palaeotemperatures of Atlantic bottom water (water depth below 4000 m).

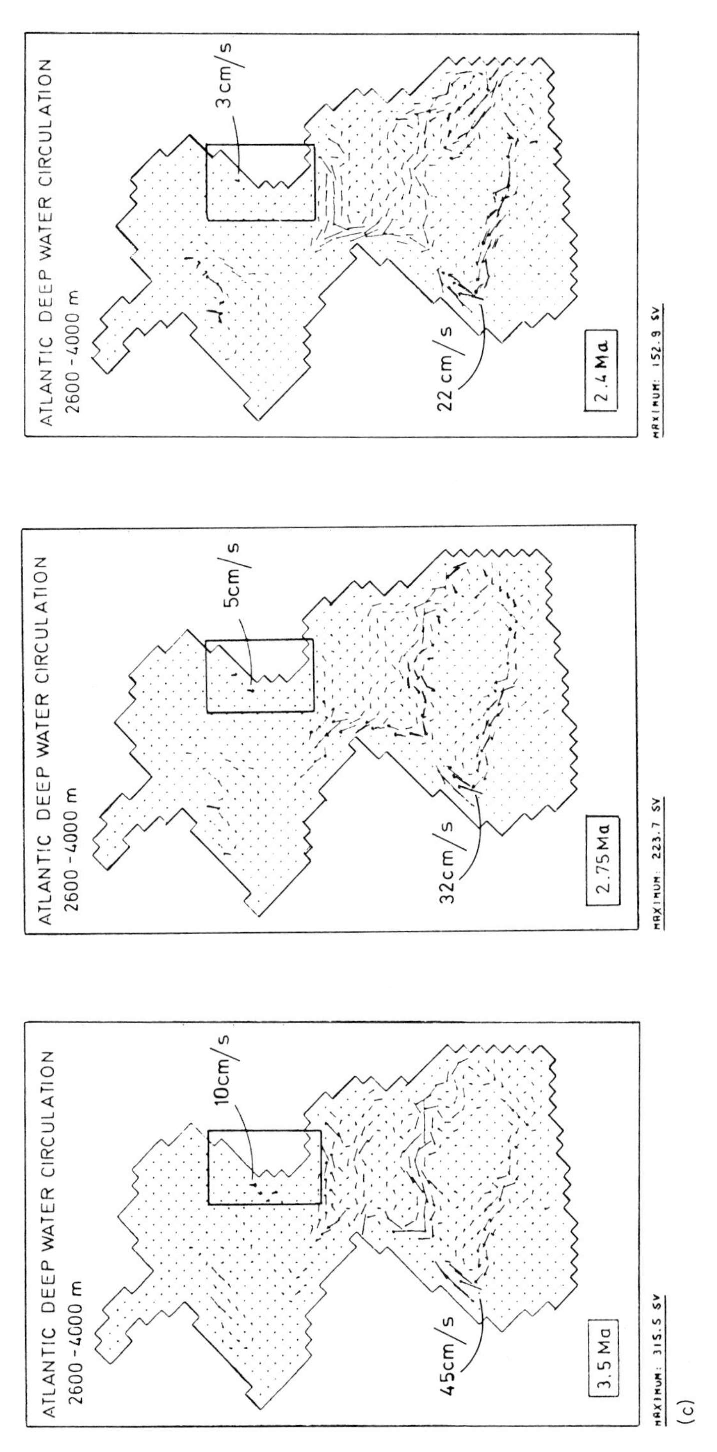

Fig. 4. (c) Speeds and directions of Atlantic palaeo-deep water circulation (2600 m to 4000 m water depth). Maximum speeds are calculated for the region under discussion in this study and for the Vema Channel region (details in Suendermann 1985).

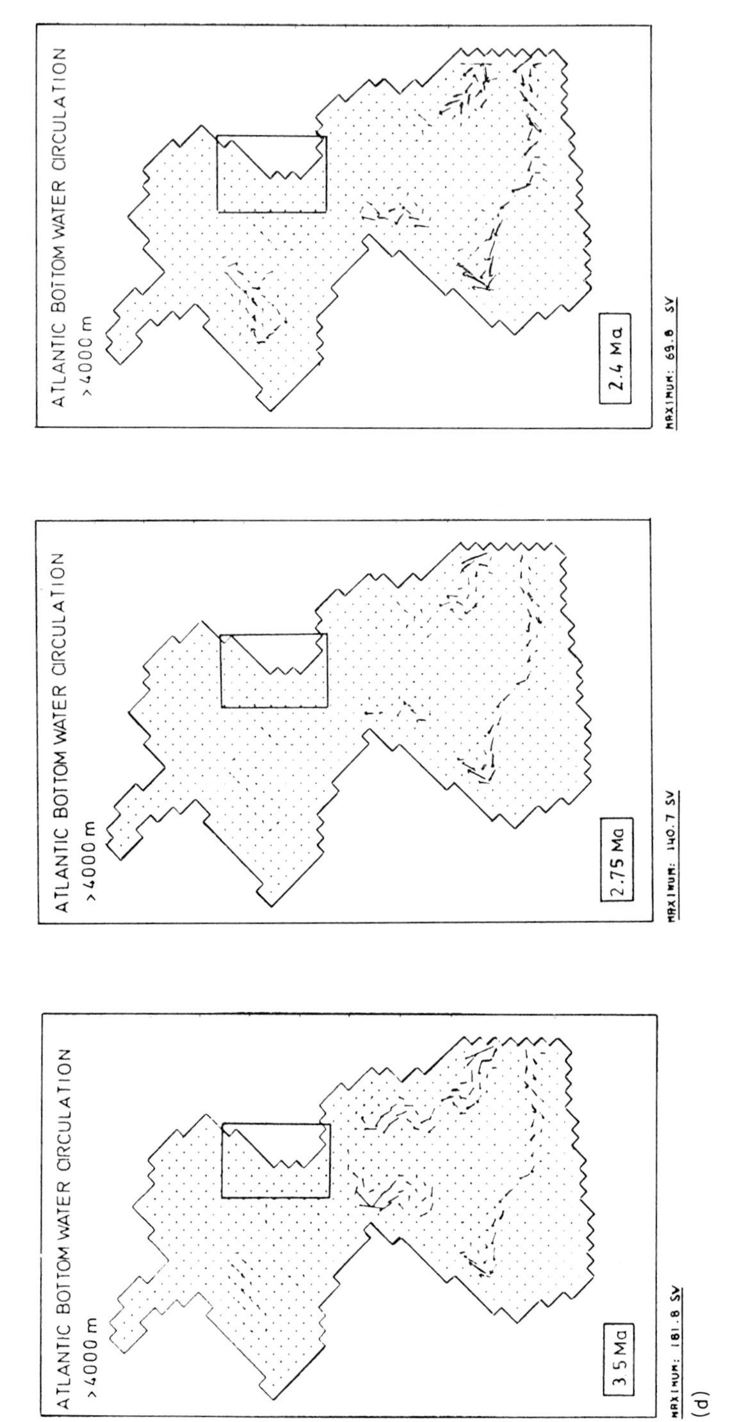

FIG. 4. (d) Speeds and directions of Atlantic palaeo-bottom water circulation (below 4000 m water depth).

(3) 2.4 Ma (time of first extreme glacial event). Although the maps of Fig. 4 show the palaeo-temperature and palaeo-circulation patterns of the whole Atlantic (with special features, such as the deep-water transport through the Vema Channel), only the deep-water circulation off north-west Africa is discussed.

The calculated deep-water temperatures at 2600 m to 4000 m depth in the north-east Atlantic have decreased from about 9° C at 3.5 Ma to about 5.4° C at 2.4 Ma (Fig. 4(a)). The palaeo-temperatures of the bottom water mass below 4000 m are not well documented by the model in the area off north-west Africa (Fig. 4(b)). However, data points further offshore also indicate a distinct decrease from more than 8° C at 3.5 Ma to less than 2° C (i.e. approximately the present value) at about 2.4 Ma. This implies (1) that there was almost no temperature difference between the deep (2600 m–4000 m) and bottom (>4000 m) water masses during Early Pliocene times, (2) that the temperature difference between deep and bottom water recorded at about 2.4 Ma has been already similar to that between the modern NADW and AABW masses in the north-east Atlantic, and (3) that the distinct increase in $\delta\ ^{18}O$ observed at Site 141 (Fig. 3) may have been mainly caused by cooling.

In Fig. 4(c), the calculated palaeo-currents in water depths between 2600 m and 4000 m show an apparent decrease from about 10 cm/s at 3.5 Ma to about 3 cm/s at 2.4 Ma. Below 4000 m water depth, bottom water currents appear to have been negligible off north-west Africa at 3.5 Ma and 2.4 Ma (Fig. 4(d)). These modelled values may be biased, because the boundary conditions applied are still incomplete. For example, the model does not consider the Mediterranean Outflow Water (MOW), which may be an important control on the formation and advection of NADW (Reid 1979). Suendermann (1985) shows that a marked increase of MOW would lead to intensification of the deep and bottom water circulation in the Atlantic, especially off north-west Africa (Fig. 5). Thus the speeds of deep-water currents calculated by the present model (which excludes MOW fluctuations) are probably too low. Similarly, the inflow of polar waters at the northern boundary should be considered.

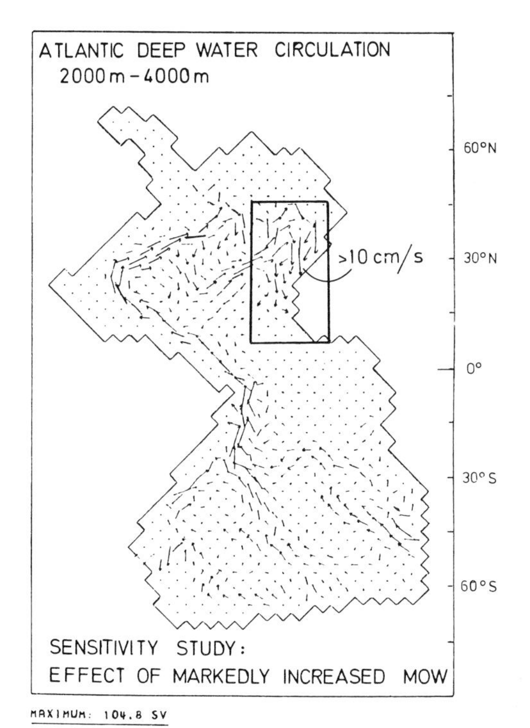

FIG. 5. Results of a sensitivity study of the dynamic ocean model, showing the effect of markedly increased Mediterranean Outflow Water (MOW) on the deep-water circulation. Arrows indicate intensity and direction of deep-water currents, dots indicate 'no currents' (details in Suendermann 1985).

Discussion

The authors consider that three different mechanisms have controlled the formation of hiatuses, as summarized in Table 3 (cf. Moore *et al.* 1978).

In the study area, the $CaCO_3$ dissolution mechanism (Table 3) probably was not very important for hiatus formation. Dissolution cannot be invoked to explain the loss of 30–75 m of sediments in the absence of a dissolution lag sediment (i.e. a distinct sequence of hemipelagic or pelagic clays). Dissolution of $CaCO_3$ may have been an important cause of the reduced sedimentation rate during Miocene and Early Pliocene times at Site 141, i.e. in water depths below 4100 m. Dissolution effects there are inferred from the increased fragmentation of foraminifera tests (Beckman 1972, Diester-Haass & Chamley 1978), and from low mass accumulation rates of $CaCO_3$ (Stein 1984). At Site 544B, there is a distinct increase in the amount of large-sized planktonic foraminifera tests, instead of increased fragmentation, which argues against dissolution. Furthermore, the hiatus at Sites 366, 397, and 544B is associated with an increase in the $\delta\ ^{13}C$ values of benthic foraminifera at Site 141. This isotopic pattern suggests an increase in production and advection of well oxidized NADW, which would not have encouraged dissolution (Blanc & Dup-

TABLE 3. *Possible mechanisms causing hiatuses (e.g. Moore et al. 1978)*

Formation of hiatuses by:	(1) slumping/turbidity currents	(2) bottom water erosion	(3) dissolution
Occurrence:	continental slope or rise	western continental margins of ocean basins; deep-water channels	depths below the lysocline
Trigger:	tectonic events; increased sediment transport to the shelf edge during times of lowered sea-level; under-cutting by increased bottom water circulation	increased production and advection of deep water	increased corrosiveness of bottom water (e.g. due to cooling, reduced circulation, increased input of organic matter)
Extent:	local	regional	local to global
Signals in the sediment along with hiatuses:	shallow-water material, coarse-grained terrigenous matter, disturbed sediment structures, mud flow material	parallel coarsening of both biogenic and terrigenous matter, increased $\delta^{13}C$-values, grain-size sorting	abundant organic matter, reduced $CaCO_3$%, dissolution effects such as foraminifera fragmentation, high per centages of resistant foram and coccolith species

lessy 1982; Stein 1984). While dissolution may have caused short-term fluctuations in $CaCO_3$ in the study area (cf. Diester-Haass 1979; Sarnthein *et al.* 1984; Stein 1984), long-term non-deposition or hiatuses were not caused by persistent dissolution here.

Current erosion is the process most likely to have eroded tens of metres of the north-east Atlantic continental slope/rise (Table 3). At Site 544B, the coarsening of both planktonic foraminifera tests and terrigenous matter, and the reduction in accumulation rates just before the hiatuses, are taken to indicate current-induced winnowing of the fine fraction (Stein & Sarnthein 1984b). At Site 397, the slow accumulation rate associated with the Pliocene hiatus (Stein 1984) is almost contemporaneous with the three sand layers (Diester-Haass 1979), which are possibly related to contour currents. At the same time, increased winnowing of the fine fraction (Stein 1984), and sorting of the sand fraction (Diester-Haass & Chamley 1978), occur at Site 141.

These various data suggest an intensification of deep-water circulation in the north-east Atlantic. Because (1) the hiatuses are confined to water depths between 2800 m and 4000 m, (2) the lost sediment sequence generally decreases in thickness from north to south (Table 2), and (3) the dynamic ocean model suggests southward current directions (Fig. 4), there are good grounds for suggesting that the hiatuses formed in response to enhanced circulation of NADW. Further evidence of large-scale current activity during that time is given by a distinct mid-Pliocene seismic reflector at Site 397 (Shipboard Scientific Party 1979), and in the Kane Gap (Sarnthein *et al.* 1983; Mienert, in prep.). The reflector has the hyperbolic character typical of current activity (Jacobi & Hayes 1982). The near contemporaneous upper limit of hiatuses over a distance of some 4000 km supports the concept of a regional pattern of erosion induced by current activity.

Although the speeds of deep-water currents derived from modelling are too low (see above), the model is probably correct in indicating higher current speeds during the early phases of climatic change (3.5 to 2.75 Ma), than during times of glacial maximum (2.4 Ma) (Fig. 4). This inference is supported by the findings of Ledbetter *et al.* (1978) and Ciesielsky & Weaver (1983), who found that bottom water currents formed hiatuses in the south-west Atlantic between 3.8 and 2.9 Ma. Recent studies of cores from the Vema Channel confirm that maximum deep-water circulation probably corresponded with the onset of ice-growth phases, whereas the circulation of Antarctic Bottom Water was sluggish during glacial times (Johnson *et al.* 1984; Jones & Johnson 1984; Tappa & Thunnel 1984; Ledbetter 1984).

Submarine erosion by enhanced circulation of NADW was probably the major cause of the hiatuses observed. However, at Site 397 another explanation is required for hiatus formation. This Site is situated south of the Canary Islands (Fig.

1), and protected against the direct influence of southward flowing deep water by a ridge less than 2000 m deep between the islands and the continent. Thus, an additional erosion mechanism is needed to explain the hiatus formation. Hiatuses at Site 397 were most probably caused by slumping on the continental rise (Table 3). The slump may have been triggered by turbidity currents, the latest phases of which are preserved as thin sandy turbidites just subsequent to the slumping events (cf. Diester-Haass 1979). The turbidity currents may have been triggered by instability caused (1) by an increased flux of terrigenous sediments at times of lowered sea-level (Stein 1984; 1985), (2) by sea-level fluctuations and/or (3) by increased current erosion along the continental slope.

Conclusions

Distinct hiatuses of mid-Pliocene and Early Quaternary age on the north-east Atlantic continental margin, in water depths between 2800 m and 4000 m, were most probably caused by intensified southward flow in deep-water currents. In some places slumping may have formed hiatuses by eroding sediments from the continental rise, as at Site 397.

The results of a dynamic ocean model simulating deep-water circulation suggest enhanced circulation during times of climatic change rather than during times of glacial maximum. Although the absolute palaeo-speeds of deep-water currents obtained by the present model are probably too low, this study suggests that further use of the model with more realistic boundary conditions would help to understand and explain changes in deep-water circulation.

ACKNOWLEDGEMENTS: The authors thank D. Hellmer and F. Sirocko for computer and technical assistance. The authors thank their reviewers, especially C. Summerhayes, for their numerous constructive suggestions for improvement of the manuscript. This study was financially supported by the 'Deutsche Forschungsgemeinschaft'.

References

ARTHUR, M.A., VON RAD, U., CONRAD, C., MCCOY, F.W. & SARNTHEIN, M. 1979. Evolution and sedimentary history of the Cap Bojador continental margin, Northwest Africa. *In*: VON RAD, U., RYAN, W.B.F. *et al.* (eds). *Init. Repts DSDP* **47** (1). US Govt. Print. Off., Washington, DC. 773–816.

BACKMAN, J. & SHACKLETON, N.J. 1983. Quantitative biochronology of Pliocene and early Pleistocene calcareous nannofossils from the Atlantic, Indian and Pacific Ocean. *Mar. Micropaleont.* **8,** 141–70.

BECKMAN, J.P. 1972. The foraminifera and some associated microfossils of Sites 135 to 144. *In*: HAYES, D.E., PIMM, A.C. *et al.* (eds), *Init. Repts DSDP* **14.** US Govt. Print. Off., Washington, DC. 389–420.

BERGER, W.H. & VON RAD, U. 1972. Cretaceous and Cenozoic sediments from the Atlantic Ocean. *In*: HAYES, D.E., PIMM, A.C. *et al.* (eds), *Init. Repts DSDP* **14.** US Govt. Print. Off., Washington, DC. 787–954.

BLANC, I. & DUPLESSY, J.-C. 1982. The deep water circulation during the Neogene and the impact of the Messinian salinity crisis. *Deep Sea Res.* 1391–414.

CEPEK, P., JOHNSON, D., KRASHENINNIKOV, V.A. & PFLAUMANN, U. 1978. Synthesis of the Leg 41 biostratigraphy and paleontology, Deep Sea Drilling Project. *In*: LANCELOT, Y., SEIBOLD, E. *et al.* (eds), *Init. Repts DSDP* **41.** Govt. Print. Off., Washington, DC. 1181–98.

—— & WIND, F.H. 1979. Neogene and Quaternary calcareous nannoplankton from DSDP Site 397 (Northwest African margin). *In*: VON RAD, U., RYAN, W.B.F. *et al.* (eds), *Init. Repts DSDP* **47** (1) US Govt. Print. Off., Washington, DC. 289–315.

CIESIELSKY, P.F. & WEAVER, F.M. 1983. Neogene and Quaternary paleoenvironmental history of Deep Sea Drilling Project Leg 71 sediments, Southwest Atlantic Ocean. *In*: Ludwig, W.J., Krasheninnikov, V.A. *et al.* (eds), *Init. Repts DSDP* **71.** US Govt. Print. Off., Washington, DC. 461–77.

CITA, M.B. & RYAN, W.B.F. 1979. Late Neogene environmental evolution. *In*: VON RAD, U., RYAN, W.B.F. *et al.* (eds), *Init. Repts DSDP* **47** (1). US Govt. Print. Off., Washington, DC. 447–59.

DIESTER-HAASS, L. 1979. Climatological, sedimentological and oceanographic changes in the Neogene authochthonous sequence. *In*: VON RAD, U., RYAN, W.B.F. *et al.* (eds), *Init. Repts DSDP* **47** (1) US Govt. Print. Off., Washington, DC. 647–70.

DIESTER-HAASS, L. & CHAMLEY, H. 1978. Neogene paleoenvironment off Northwest Africa based on sediments from DSDP Leg 14. *J. sedim. Petrol.* **48,** 879–96.

JACOBI, R.D. & HAYES, D.E. 1982. Bathymetry, microphysiography, and reflectivity characteristics of the West African margin between Sierra Leone and Mauritania. *In*: VON RAD, U., HINZ, K. *et al.* (eds), *Geology of the Northwest African Continental Margin*. Springer-Verlag, Berlin. 182–212.

JOHNSON, D.A., LEDBETTER, M.T., TAPPA, E. & THUNELL, R. 1984. Late Tertiary/Quaternary magnetostratigraphy and biostratigraphy of Vema Channel sediments. *Mar. Geol.* **58,** 89–100.

JONES, G.A. & JOHNSON, D.A. 1984. Displaced Antarctic diatoms in Vema Channel sediments: Late Pleistocene/Holocene fluctuations in AABW flow. *Mar. Geol.* **58,** 165–86.

LEDBETTER, M.T. 1984. Bottom-current speed in the Vema Channel recorded by particle size of sediment fine-fraction. *Mar. Geol.* **58,** 137–49.

——, WILLIAMS, D.F. & ELLWOOD, B.B. 1978. Late Pliocene climate and south-west Atlantic abyssal circulation. *Nature* **272,** 237–9.

MANTYLA, A.W. & REID, J.L. 1983. Abyssal characteristics of the World Oceans waters. *Deep-Sea Res.* **30,** 805–33.

MIENERT, J. 1985. *Akustostratigraphische Untersuchungen zur Klarung der Bodenwasserzirkulation im Aquatorialen Ostatlantric.* Dissertation, Kiel University, in prep.

MOORE, T.C., JR., VAN ANDEL, T.H., SANCETTA, C. & PISIAS, N. 1978. Cenozoic hiatuses in pelagic sediments. *Micropaleontology* **24,** 114–38.

PFLAUMANN, U. & KRASHENINNIKOV, V.A. 1978. Zonal stratigraphy of Neogene deposits of the eastern part of the Atlantic Ocean by means of planktonic foraminifera, Leg 41, Deep Sea Drilling Project. *In*: Lancelot, Y., Seibold, E. *et al.* (eds), *Init. Repts DSPP* **41.** 613–58.

REID, J.L. 1979. On the contribution of the Mediterranean Sea outflow to the Norwegian-Greenland Sea. *Deep-Sea Res.* **26A,** 1199–223.

SARNTHEIN, M., ERLENKEUSER, H., VON GRAFENSTEIN, R. & SCHROEDER, C. 1984. Stable isotope stratigraphy for the last 750,000 years: 'Meteor' core 13519 from the eastern equatorial Atlantic. *'Meteor' Forschungsergebn. C* **38,** 9–24.

——, KOEGLER, F.C. & WERNER, F. 1983. Forschungsschiff 'Meteor', Reise Nr. 65, Aequatorialer Ostatlantik-GEOTROPEX 83, Juni–August 1983, Bericht der wissenschaftlichen Fahrtleiter. *Berichte-Reports* **2.** Geol.-Palaeont. Institut, Universitaet Kiel. 90 pp.

——, THIEDE, J., PFLAUMANN, U., ERLENKEUSER, K., FUETTERER, D., KOOPMANN, B., LANGE, H. & SEIBOLD, E. 1982. Atmospheric and oceanic circulation patterns off Northwest Africa during the past 25 million years. *In*: VON RAD, U., HINZ, K. *et al.* (eds), *Geology of the Northwest African Continental Margin.* Springer Verlag, Berlin. 545–604.

SHACKLETON, N.J., BACKMAN, J., ZIMMERMAN, H., KENT, D.V., HALL, N.A., ROBERTS, D.G., SCHNITKER, D., BALDAUF, J.G., DESPRAIRIES, A., HOMRIGHAUSEN, R., HUDDLESTUN, P., KEENE, J.B., KALTENBACK, A.J., KRUMSIEK, K.A.O., MORTON, A.C., MURRAY, J.W. & WESTBERG-SMITH, J. 1984. Oxygen isotope calibration of the onset of ice-rafting and history of glaciation in the North Atlantic region. *Nature* **307,** 620–3.

—— & CITA, M.B. 1979. Oxygen and carbon isotope stratigraphy of benthic foraminifers at Site 397: Detailed history of climatic change during the Late Neogene. *In*: VON RAD, U., RYAN, W.B.F. *et al.* (eds), *Init. Repts DSDP* **47** (1). US Govt. Print. Off., Washington, DC. 433–45.

SHIPBOARD SCIENTIFIC PARTY 1972. Site 141. In: HAYES, D.E., PIMM, A.C., *et al.* (eds), *Init. Repts DSDP* **14.** US Govt Print. Off., Washington, DC. 217–48.

—— 1978. Site 366, Sierra Leone Rise. *In*: LANCELOT, Y., SEIBOLD, E. *et al.* (eds), *Init. Repts DSDP* **41.** US Govt. Print. Off., Washington, DC. 21–162.

—— 1979. Site 397. *In*: VON RAD, U., RYAN, W.B.F. *et al.* (eds), *Init. Repts DSDP* **47** (1). US Govt. Print. Off., Washington, DC. 17–217.

—— 1984. Site 544. *In*: WINTERER, L., HINZ, K. *et al.* (eds), *Init. Repts DSDP* **79.** US Govt. Print. Off., Washington, DC.

STEIN, R. 1984. Zur neogenen Klimaentwicklung in Nordwest-Afrika und Palaeo-Ozeanographie im Nordost-Atlantik: Ergebnisse von DSDP-Sites 141, 366, 397 und 544B. *Berichte-Reports* **4.** Geol. Palaeont. Institut, Universitaet Kiel. 210 pp.

—— 1985. Late Neogene changes of paleoclimate and paleoproductivity off Northwest Africa (DSDP-Site 397). *Palaeogeogr., Palaeoclimatatol., Palaeoecol.* **49,** 47–59.

—— & SARNTHEIN, M. 1984a. Late Neogene events of atmospheric and oceanic circulation offshore Northwest Africa: High resolution record from deep-sea sediments. *Palaeoecol. Afr.* **16,** 9–36.

—— & SARNTHEIN, M. 1984b. Late Neogene oxygen isotope stratigraphy and terrigenous flux rates at Site 544B off Morocco. *In*: WINTERER, L., HINZ, K. *et al.* (eds), *Init. Repts DSDP* **79.** US Govt. Print. Off., Washington, DC. 385–95.

SUENDERMANN, J. 1985. Late Neogene Atlantic circulation model. *Geol. Rundschau*, in press.

TAPPA, E. & THUNELL, R. 1984. Late Pleistocene glacial/interglacial changes in planktonic foraminiferal biofacies and carbonate dissolution patterns in the Vema Channel. *Mar. Geol.* **58,** 101–22.

R. STEIN* & M. SARNTHEIN, Geologisch—Palaeontologisches Institut Universitaet Kiel, Olshausenstrasse 40, 2300 Kiel, Federal Republic of Germany. * Present address: Geologisch-Palaeontologisches Institut, Universitaet Giessen, Senckenbergstrasse 3, 6300 Giessen, Federal Republic of Germany.

J. SUENDERMANN, Institut fuer Meereskunde, Universitaet Hamburg, Heimhuder Str. 71, 2000 Hamburg 13, Federal Republic of Germany.

Sediment waves in the eastern equatorial Atlantic: sediment record during Late Glacial and Interglacial times

M. Sarnthein & J. Mienert

SUMMARY: Standing and migrating sediment waves along the West African continental rise (3000–4500 m water depth) are characterized by a regular spacing of wave length periods that are multiples of 280 m, and by long-term pelagic sediment composition. Interpretations of their origin must account for the following evidence: (1) The wave crests generally run parallel or oblique to the continental rise and also to prevailing (net) currents, as confirmed by SEA BEAM data; (2) Migrating waves along the West African continental margin are directed only upslope, i.e. towards the East. (3) The waves have been accreting for a long time, (> 1 Ma). (4) Glacial to interglacial differences in the evolution of sediment accumulation rates on crests and troughs may indicate that the accreting regime was intermittent, and that the waves were more levelled prior to 7500 years BP due to increased deposition of siliciclastic matter in the troughs. (5) Absent patterns of grain-size sorting in the fine-grained muds contradict 'conventional' models of current-related sediment transport.

A superposition of large tidal components and slow thermohaline net currents (Lonsdale 1978), secondary current processes (Unsöld, pers. comm.), or a superposition of internal waves and thermohaline currents (Kolla *et al.* 1980) are regarded as the most probable modes of origin.

Sediment waves are elongated sinusoidal mounds of pelagic sediments on the deep-sea floor. They form well layered, arc-shaped echos on 3·5 kHz records; many of them are fairly regular in 'wave length' (λ, distance between 2 troughs) and amplitude (height of the crest) (Fig. 1).

These waves are associated with 'sediment drifts' and are widespread along Atlantic continental margins (Allen 1982). The waves have been interpreted as an expression of contour currents, a continual thermohaline circulation that flows along bathymetric contours (e.g. Heezen & Hollister 1964; Embley *et al.* 1978; Jacobi & Hayes 1982; Mountain & Tucholke 1983). However, a number of problems regarding the origin of sediment waves remain unsolved, especially the origin of 'standing waves' (stationary waves) and of less distinct undulatory sediment layerings associated with sediment waves. Some major problems are:

(1) whether many of these waves are mere projections of a sub-bottom relief, or are hydrodynamically enforced features;
(2) the difficulty in deciphering any hydrodynamic effect on the sediment texture and composition;
(3) the genetic relations between standing and migrating sediment waves;
(4) the mode of hydrodynamic processes actually capable of controlling the formation of (pelagic) sediment waves;
(5) the long-term continuity or discontinuity of such hydrodynamic processes;
(6) if such processes can be defined, the possible implications for palaeoceanography.

The authors will try to contribute to the understanding of sediment waves by presenting some new data on their morphology and, in particular, on the physical properties, the composition and high resolution stratigraphy of the sediments of several sediment waves from the eastern equatorial Atlantic (Fig. 2).

Methods

Most of the results are based on the evaluation of high resolution 3.5 kHz seismic records and cores, large-scale spade box cores (50 cm cubes) and up to 14.5 m long gravity cores with 12 cm diameter collected by the GEOTROPEX '83 cruises to the eastern equatorial Atlantic (Sarnthein *et al.* 1983).

Physical properties and the composition of the (pelagic) sediments were determined by standard procedures; the stratigraphy was defined by oxygen-isotope curves and a few radiocarbon dates (Kassens 1985; Vogelsang 1985). Sound velocity was measured by means of a sounding probe connected to a Krautkrämer Ultrasonic Flow Detector USM2 (Cologne) (Mienert 1985).

Distribution and morphology of sediment waves and their variability

Sediment waves from the eastern equatorial Atlantic have been described recently in detail by Jacobi & Hayes (1982) and correspond to mud-

From SUMMERHAYES, C.P. & SHACKLETON, N.J. (eds), 1986, *North Atlantic Palaeoceanography*, Geological Society Special Publication No. 21, pp. 119–130.

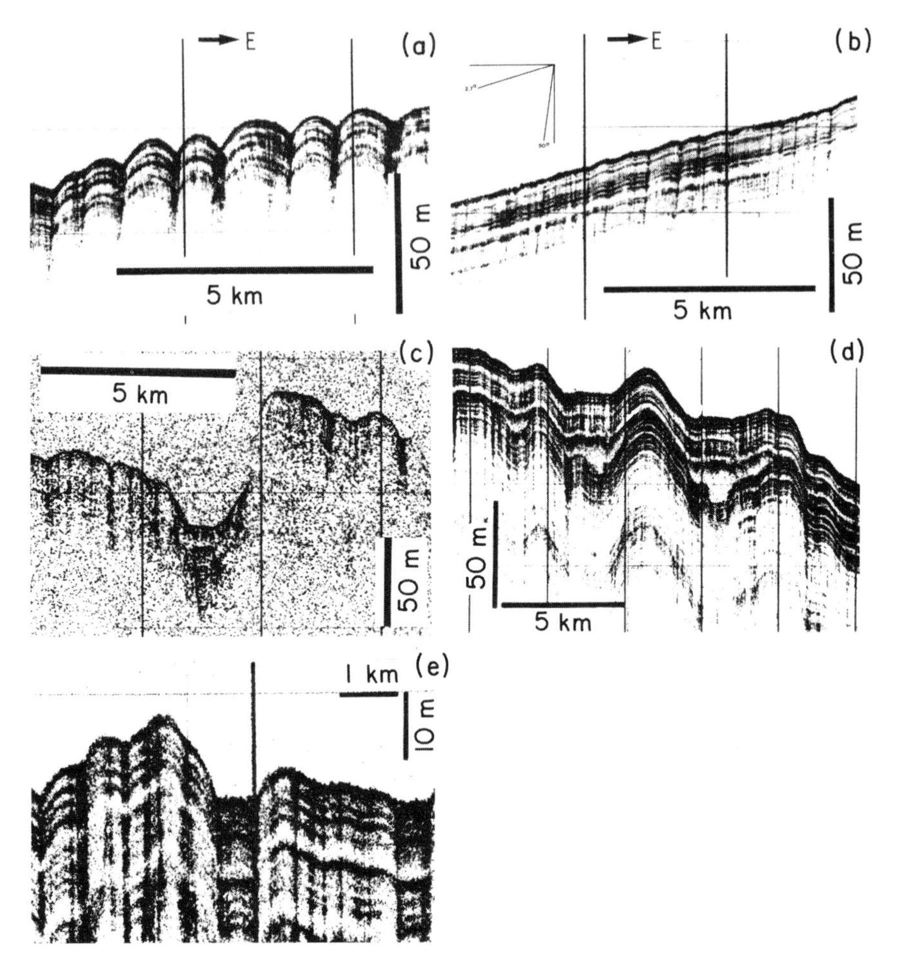

FIG. 1. 3.5 kHz records of sediment waves. (a), (b) Migrating waves at 25°55 N. and 12–13°N. on the north-west African continental rise. Insert diagram in (b) shows actual angles of vertical dark stripe of hyperbolic echos. (c) Unconformable sediment filling of wave trough. (d) Burial of sediment wave relief. (e) Details of small-scale sediment waves. (c)–(e) are from the north-eastern foot region of the Sierra Leone Rise.

wave types 1 and 2 of Allen (1982). The longitudinal form and lateral extension of some very closely spaced narrow sediment waves have been depicted by Lonsdale (1978) (Fig. 3). However, they still remain somewhat obscure because the third, i.e. the lateral dimension, is difficult to register by the two dimensional 3.5 kHz records. A combined SEA-BEAM and GLORIA record from the north-eastern foot region of the Sierra Leone Rise (Fig. 4) suggests that the crests strike obliquely across the bathymetric contours (Fig. 2), as also observed by Lonsdale (1978) from other regions.

Figure 2 shows the widespread distribution of sediment waves in the eastern equatorial North Atlantic. They extend all along the margins of the African continent and the Sierra Leone Rise between approximately 3000 and 4750 m water depth and also on the plateau on top of the Sierra Leone Rise at 2800 m.

Wave length (λ) and the amplitude of sediment waves from the continental rise off Gambia and from the north–north western margin of the Sierra Leone Abyssal Plain show a linear correlation (Fig. 5). At the NNW margin of the Sierra Leone Abyssal Plain, SEA-BEAM data (Fig. 4) have been used to correct the apparent wave lengths of the (two-dimensional!) 3.5 kHz profiles to actual values perpendicular to the wave crest. The resulting correlation factor between amplitude and wave length has been applied to also correct the wave length of the sediment waves

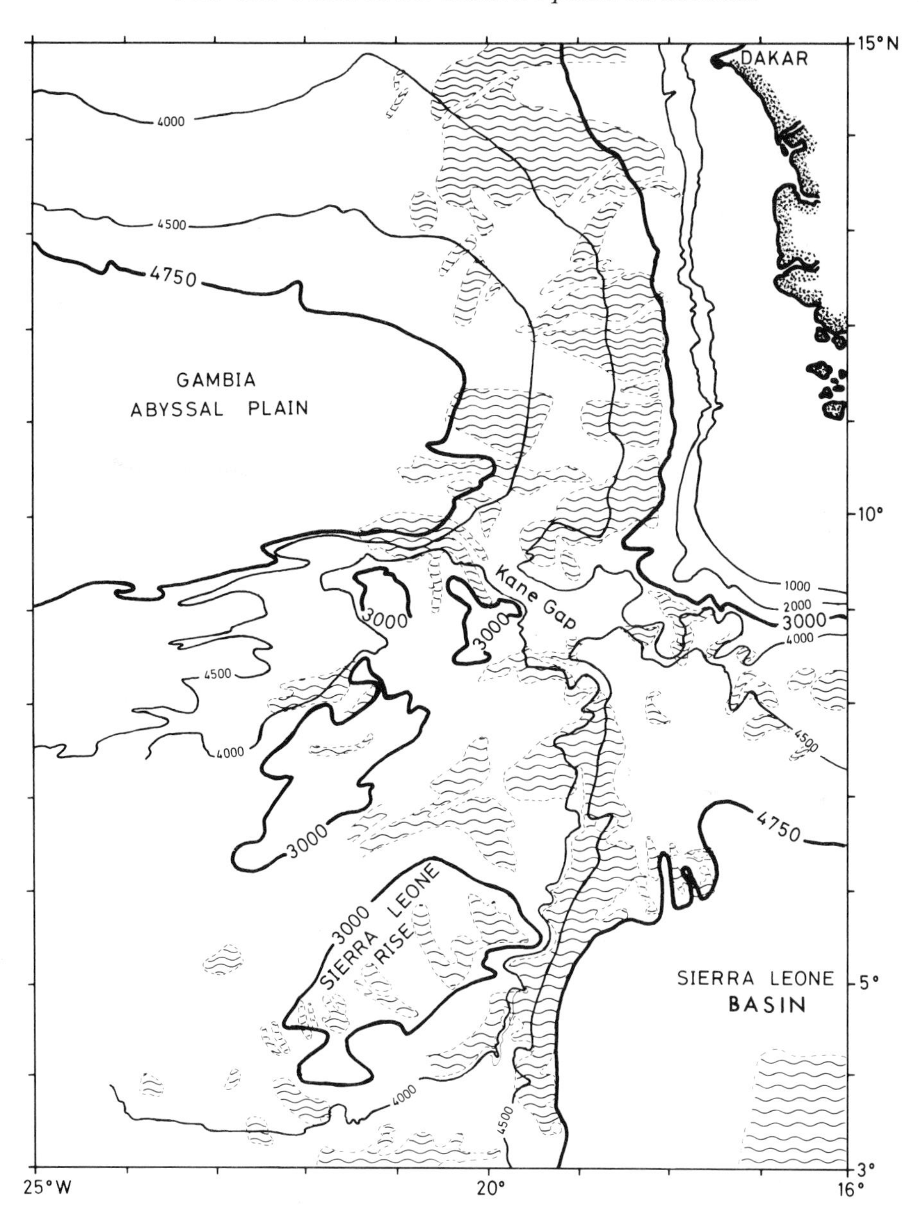

FIG. 2. Distribution of sediment waves along part of the eastern Atlantic continental margin (after Jacobi & Hayes 1982 and the authors' 3.5 kHz echograms).

south of Dakar (Fig. 5 (a)). A crude spectral analysis of the corrected wave lengths reveals a number of distinct and approximately regularly spaced frequency maxima (Fig. 6), many of which occur with the same periods (multiples of 220–280 m) in both regions.

The wave troughs are generally obscured on the 3.5 kHz records by a vertical dark stripe (or two stripes) of hyperbolic echos (Fig. 1; high resolution exposure at site 16416, Fig. 9 (a)). A model for their origin has recently been presented by Flood (1980). He has shown that the stripe increasingly obscures the concave wave troughs with increasing water depth. In harmony with this model, the dark stripe gradually disappears upslope at about 3400 m water depth (Fig. 7). The inclined or vertical dip of the stripe enables one to separate migrating from standing sediment

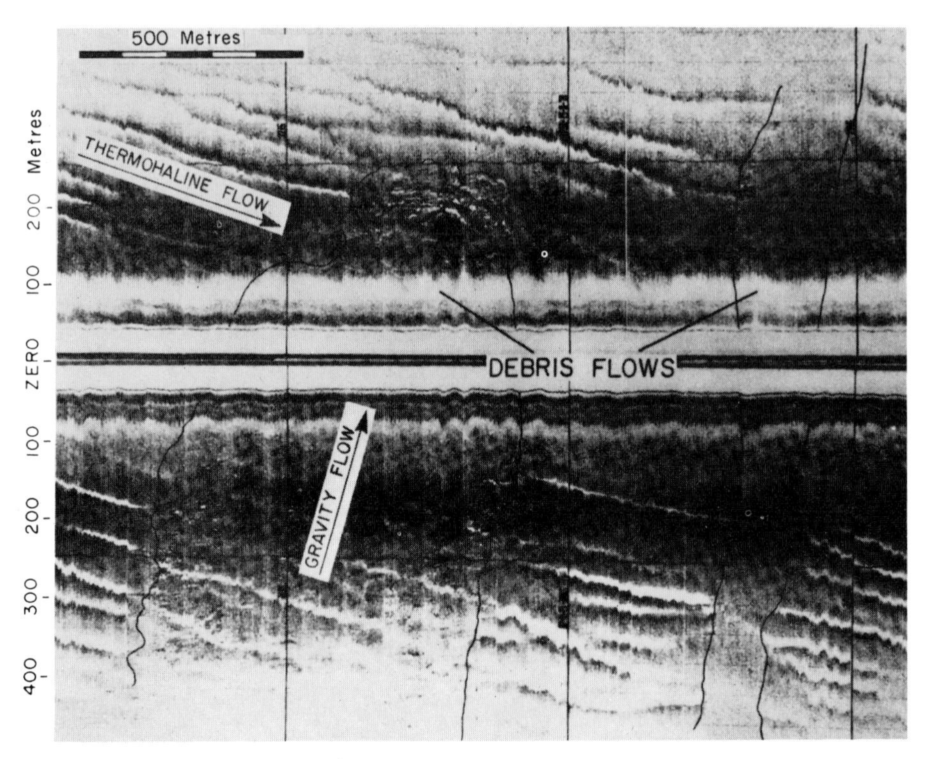

FIG. 3. A pair of side-scan records from the Saharan Rise (25°55′N., 19°50′W.) showing linear along-slope bedforms parallel to measured thermohaline currents, and cross cutting debris flows (from Lonsdale 1978).

waves. Off Gambia, the seismic record of Fig. 7 shows that the transition between the two modes of waves is gradual. Standing waves occur on the upper continental rise between 3100 and 4000 m water depth, where the slope gradient is about 0.5°. Migrating waves appear to be confined to the lower rise between 4000 and 4500 m water depth, where the slope gradient has decreased to about 0.25°. Because of the characteristically high vertical exaggeration of the 3.5 kHz records (1:50–80), the dip of the dark stripe indicative of migrating waves appears to deviate very little (only a few degrees) from the perpendicular, although it is inclined at actual angles of 45° off the perpendicular in many cases. Thus, the migrating nature of the sediment waves may easily remain undetected by the casual observer.

Along the north-west African continental margin, the migration of waves is always directed upslope, i.e. towards the east (see also Fig. 1 (a) and (b)), which may help to identify the driving mechanism of wave formation. At an angle of 45° for the dip of the dark stripes in the wave troughs, the speed of migration must correspond to the average sedimentation rates, amounting to 20–40 m per million years. In some cases, the dip of the stripes, i.e. the migration speed of the sediment waves, has slightly varied with time (Fig. 7). Occasionally, the growth of sediment waves has been simultaneously stopped over large regions because they have been buried (Fig. 1 (d)).

The apparent maximum slope gradient of individual sediment waves on 3.5 kHz records (Fig. 1) ranges from 1.25–6.0°. Flood (1980) has shown that the actual slope angle approximately equals the apparent slope angle. Extreme apparent angles of 8–16° occur at 'mega-waves' along the north-eastern foot region of the Sierra Leone Rise. The vertical exaggeration of the profiles makes some adjoining troughs appear like 20 m deep fissures (Fig. 4, near 04.00 and 04.25 hours), others show an apparently uncomformable sediment fill up to more than 50 m thick (Fig. 1 (c)), corresponding to approximately 2.5 m.y. (Stein & Sarnthein 1984). The long-term persistence of these deep troughs and their uncomformable filling structures suggest hydrodynamic processes as a generally underlying cause.

Sediment composition

The typically pelagic to hemipelagic sediments of the waves are fine-grained and consist of plank-

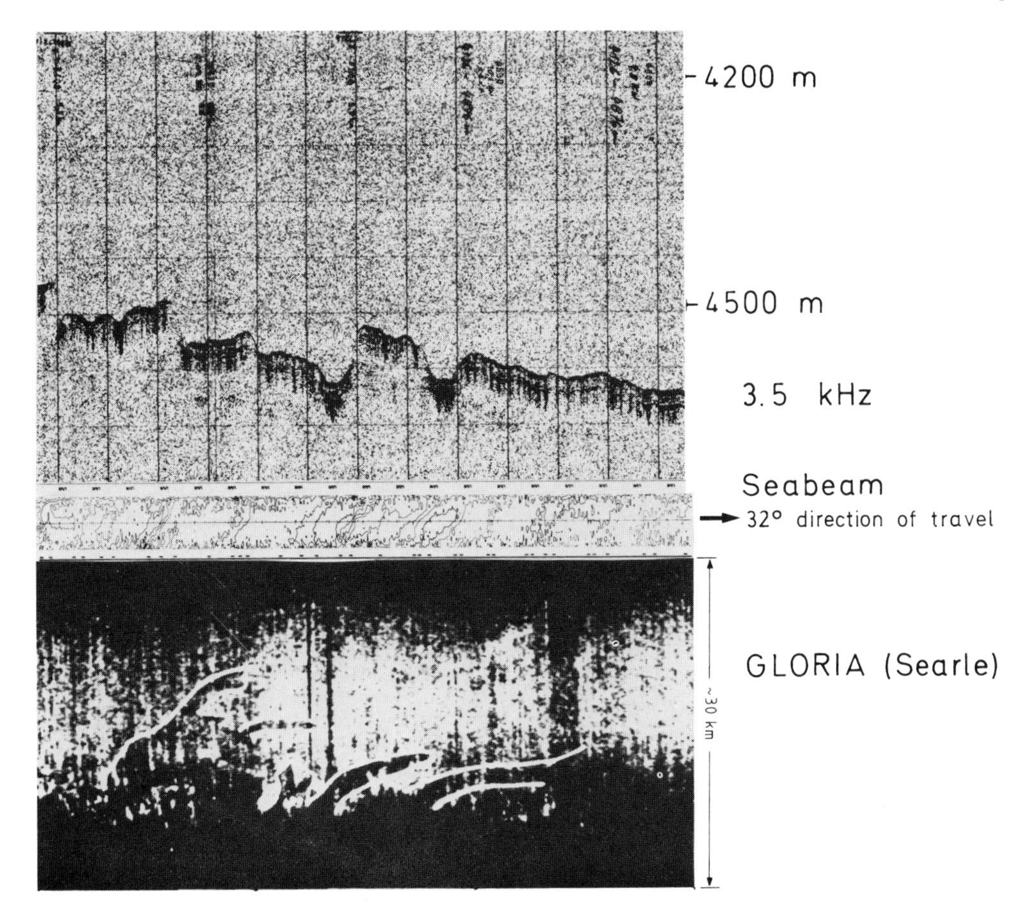

FIG. 4. Combined 3.5 kHz SEA-BEAM and Gloria record of sediment waves south of the Kane Gap (GLORIA record from R. Searle, IOS, UK). Approximate position: 7°45 N. 18°50 W. The SEA-BEAM record shows small false contours at the far ranges.

ton carbonate, clay and (siliciclastic) silt, as in the non-turbidite sediments elsewhere in the region (Sirocco 1985; Vogelsang 1985). Structures or textures directly indicating any processes of hydrodynamic sorting and sediment transport that may cause sediment waves are not observed. Undisturbed pelagic sedimentation has persisted at most sites in the study area during the last 2 million years, apart from the imprint of dissolution cycles, some variations in the input of siliciclastic material and biogenic opal, and minor stratigraphic gaps (Sarnthein *et al.* 1984; Stein & Sarnthein 1984).

At 'Meteor' sites 16402 and 16403 (Fig. 8) the authors succeeded in separately coring the crest and trough of a sediment wave. Only a few of the sediment parameters differ between crest and trough (Fig. 8). Grain sizes are similar at both positions, although the coarse fraction of the crest is slightly higher. Wet bulk density and, especially, shear strength, are much higher on the crest than in the trough. The contents of calcium carbonate also differ; they are equal or higher in trough sediments near to the surface (late Holocene), but much lower in the trough sediments than on the crest during the early Holocene prior to 6000–7500 years BP and during Termination I, the end of the last glaciation. During the whole period, organic matter was equally abundant in the trough and on the crest. From this, one might conclude that calcium carbonate was more strongly diluted by siliciclastic input <63 μm in the troughs prior to some 9000 years BP, and that $CaCO_3$ dissolution apparently did not control the wave patern. This interpretation is corroborated by bulk sedimentation rates as deduced from oxygen isotope and radiocarbon stratigraphy (Fig. 8). The rates are slightly higher on the crest

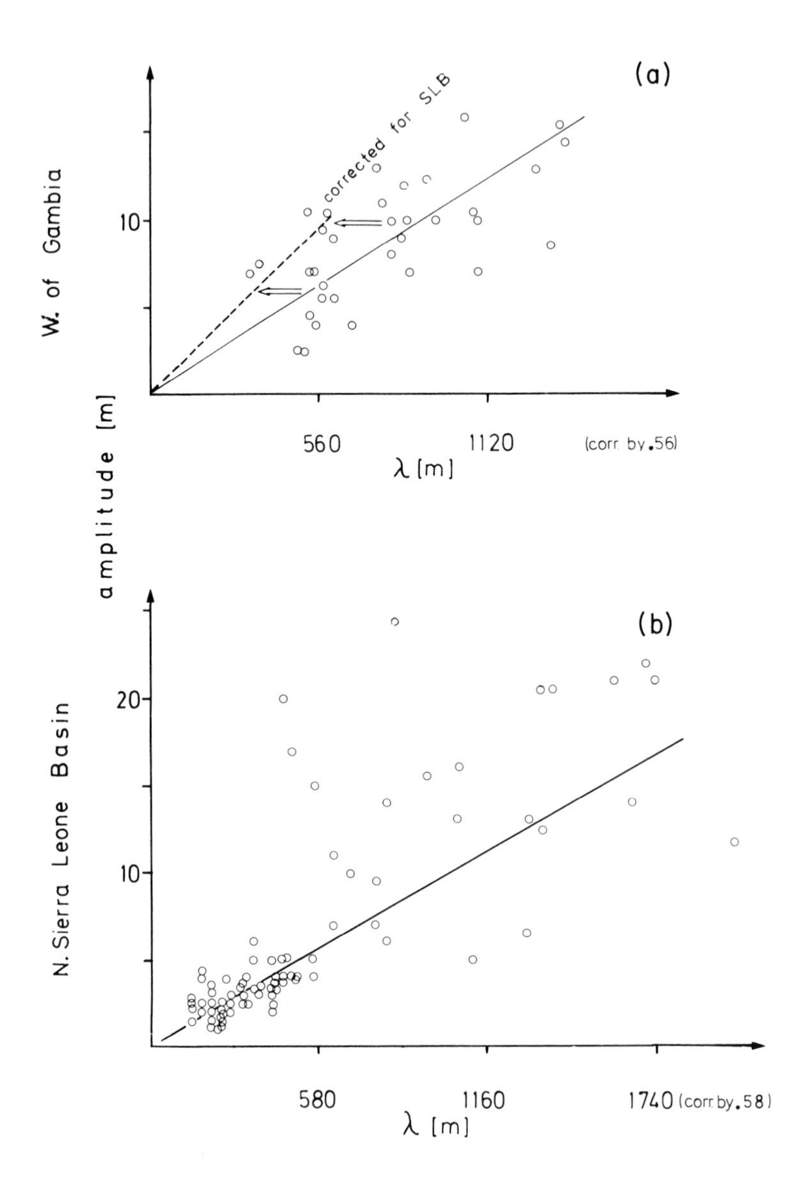

FIG. 5. Correlation of wave lengths and amplitudes of sediment waves. (a) West of Gambia. (b) Northernmost Sierra Leone Basin (=SLB). Correction factors are 0.58 and 0.56 from SEA-BEAM records.

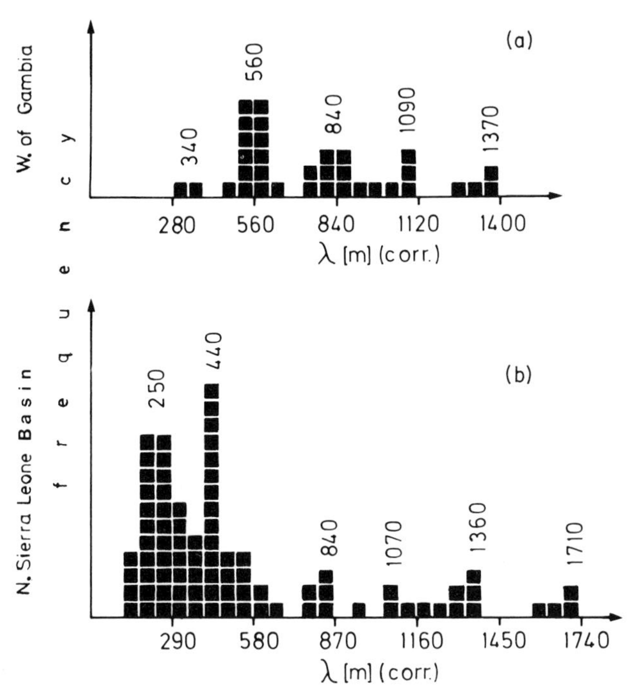

FIG. 6. Frequency distribution of wave lengths (corrected according to Fig. 5). (a) Continental rise west of Gambia. (b) Northernmost Sierra Leone Basin.

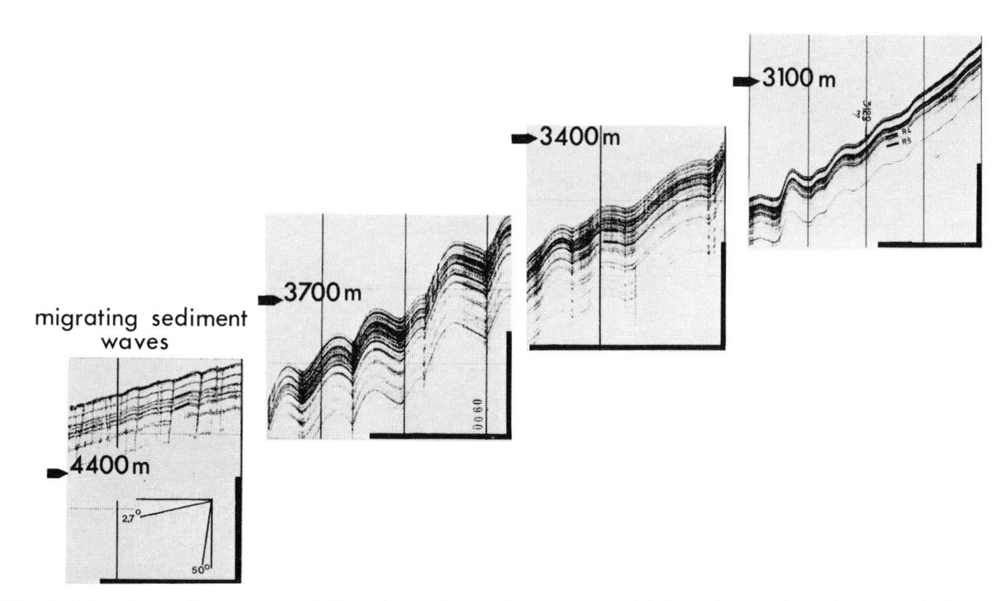

FIG. 7. Migrating sediment waves (left) and standing sediment waves (right) on the continental rise south of Dakar. Note the extreme vertical exaggeration and the gradual disappearance of vertical shadows upslope. Length of horizontal bar = 5000 m; length of vertical bar = 50 m.

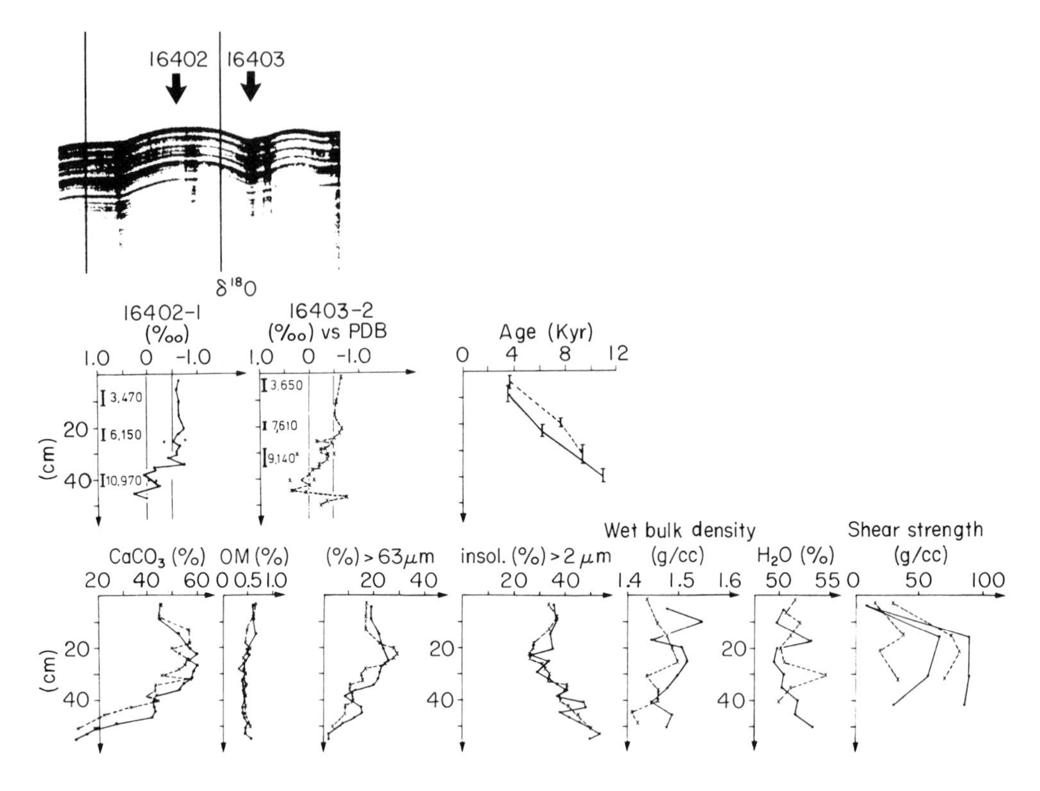

FIG. 8. 3.5 kHz-echograms of sample locations, oxygen-isotope records and ^{14}C-dates (years BP), age-depth curves, and various sediment properties from two cores retrieved from crest (solid line; 16402) and trough (hatched; 16403) of a sediment wave. Note the different sedimentation rates. Data from Vogelsang (1985) and Kassens (1985). OM = organic matter. Insol. = insolubles.

(5.5 cm/1000 year) than in the trough (4.0 cm/1000 year) during late Holocene times. A similar difference has been noted by Normark *et al.* (1980) from sediment waves of California. But the rates of the trough (7.5 cm/year) clearly exceed those of the crest (4.5 cm/1000 year) prior to 7500 years BP.

The most spectacular property recorded from the wave troughs is a largely constant sound velocity that elsewhere is completely unusual (Fig. 9 (a)). The authors do not, at present, see any sediment property that might explain the uniform sound velocity. Wet bulk density (Fig. 9 (b)), carbonate content, and grain sizes, were highly variable through time. The constancy of the sound velocity of core 16416 has only one analogue, that of 2 'Meteor' cores from the bottom of the Kane Gap (16422, 16425) where the sedimentation is probably current controlled (Hobart *et al.* 1975; Sarnthein *et al.* 1983) and where the sediment properties are also strongly variable (Mienert 1985).

Discussion and preliminary physical interpretation

In summary, a number of observations show that despite their 'normal' pelagic sediment composition, the sediment waves are not mere projections of sub-bottom relief, but are controlled by hydrodynamic processes. Major arguments are:

(1) Some of the wave fields are migrating and show a gradual lateral transition to fields of standing waves;

(2) Wave lengths are regularly spaced, as shown by the frequency maxima of wave length in two different wave fields;

(3) Steep slope angles bordering the troughs far exceed the usual range of slope angles of pelagic sediments which otherwise may be subjected to slope instability in such a steep depositional setting;

(4) Shear strength, wet bulk density, calcium carbonate content, and accumulation rates vary between wave crest and troughs;

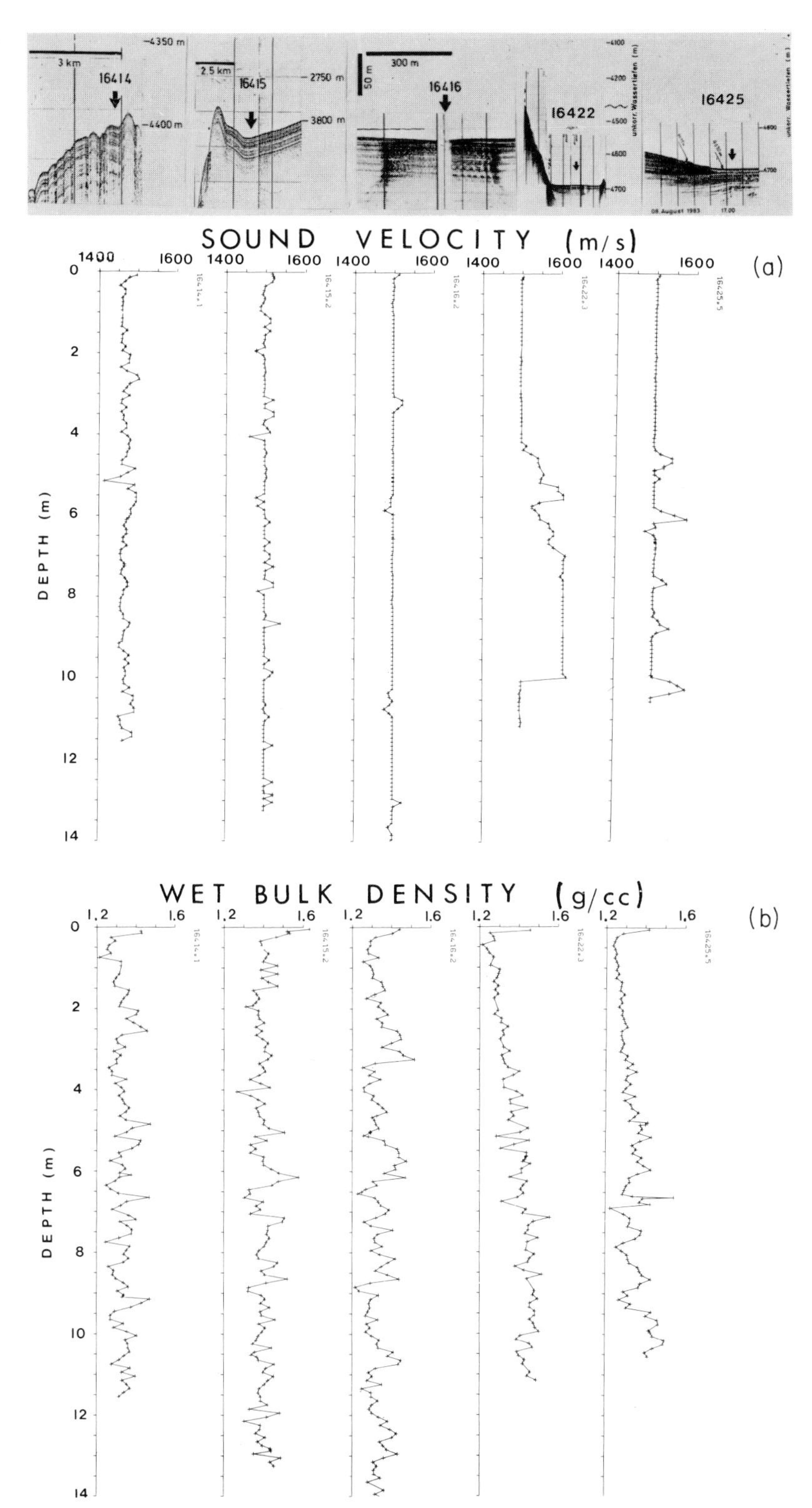

FIG. 9. 3.5 kHz-echograms from sample locations, sound velocities (a), and wet-bulk densities (b) of five gravity cores from the Kane Gap region. Note the unusually constant sound velocity only recorded from the sediment wave trough, core 16416, and from sites 16422 and 16425 in the Kane Gap with current-winnowed sedimentation.

(5) Sound velocities in thick sediment sections from the troughs are largely constant (in contrast to common changes in sediment composition), a feature observed elsewhere only in current-winnowed sediments.

The concepts about how sediment waves form have been rather vague. Partly this is because direct observations of hydrodynamic processes close to the sediment surface are rare in the deep sea (Lonsdale 1978; McCave *et al.* 1980; Hollister & McCave 1984). Interpretations of the origin of sediment waves from the eastern Atlantic, in terms of dynamic and chemical processes, must account for the following findings and conclusions:

(1) The trend of the crests of sediment waves is generally parallel or oblique to, not perpendicular to the trend of the continental rise, and probably also to the direction of prevailing currents (e.g. Lonsdale 1978; Asquith 1979; and the SEA-BEAM data).

(2) Migrating sediment waves along the North African continental margin are directed only upslope, i.e. towards the East.

(3) The sediment waves have been accreting on the average persistently for a significant length of time, most probably more than a million years (e.g. waves shown on Fig. 1 (a), (b), (d)).

(4) Differential variations of sediment accumulation rates in crests and troughs, in glacials and interglacials may indicate that the depositional regime was not uniform over the whole accretion period, but intermittent.

(5) Differential variations of accumulation rates and of calcium carbonate in the wave trough and crest sediments indicate a locally and only temporarily enhanced dilution of calcium carbonate in the troughs by siliciclastics, which has led to a temporary slight reduction of the wave relief during peak interglacial times.

(6) The lack of any recognizable grain-size sorting in the pelagic fine-grained sediment contradicts the assumption of 'conventional' processes of current-related sorting and sediment transport.

The authors conclude that sediment (=mud) waves along the east Atlantic continental margin cannot be explained by any simple forcing mechanism such as direct action of an ocean-boundary current. Many models may provide better interpretation. For example, Allen (1982) has summarized two possible modes of origin. One mode relies on an unstable interaction between a turbid ocean bottom current and a deformable bed form, the other one on large-scale stationary lee waves imposed on such a current behind a suitable bottom irregularity. However, both modes imply a downcurrent migration of the waves, an assumption conflicting with (mostly stationary) wave-crests running parallel or oblique to the current direction.

On the other hand, Hollister & McCave (1984) have shown that oblique oriented sediment waves on the west Atlantic continental margin may result from ephemeral processes, 'deep sea storms'. In this case, the small type of mud waves on the east Atlantic continental rise might reflect distal effects of such storms.

In addition, Lonsdale (1978) and McCave *et al.* (1980) have been able to relate mud-wave formation to a superposition of large tidal components and a low thermohaline net current along continental contours. According to Lonsdale (1978), the resulting current components reach speeds of 17 cm/second on the Saharan continental rise and are sufficient for bimodal branching sediment transport that forms mud waves with crests almost parallel to the net flow direction (Fig. 3).

According to Unsöld (pers. comm.), mud waves may also form by a mechanism analogous to multi-cellular secondary currents, the principles of which have been reviewed by Allen (1982), who has invoked them for the generation of other longitudinal bedforms. These secondary currents, a product of mutual interactions between current flows and longitudinal depositional bedforms parallel to the slow main flow, have been observed for the first time by Nakagawa *et al.* (1983) in laboratory flume experiments.

A completely different mechanism has been suggested by Kolla *et al.* (1980). They have shown that sediment waves can be modelled after a fluvial antidune mechanism in which the air–water interface is replaced by a fluid–fluid interface with internal waves typical of a benthic thermocline in the deep sea.

There is too little evidence in the authors' sediment record to enable one to choose between the last three or four models. The long-term upslope, i.e. eastward migration of the wave fields along the north-west African continental margin indicates persistent Coriolis forcing. This possibly agrees better with the models of persistently northward flowing secondary currents, or with internal waves, than with the model of ephemeral 'storms', although a Coriolis forcing of 'storms' may also be likely.

It may be significant that wave formation appears discontinuous from glacial to interglacial times. Sediments have accumulated faster on the wave crests than in the trough over the last 7500 years, but faster in the trough than on the crest during the Termination I (the end of the last glaciation). This was due to enhanced fine-grained siliciclastic deposition. The change of sediment distribution fits the assumption that Atlantic deep-water circulation was more sluggish during the Termination of the last glacial stage than it is today (Jones & Johnson 1984; Curry & Lohmann 1983) and has occasionally led to a levelling of the waves.

ACKNOWLEDGEMENTS: The authors thank Drs G. Unsöld and F. Werner for stimulating and critical discussions and Drs K. Herterich and J. Willebrand for advice about internal waves. H. Kassens and E. Vogelsang provided unpublished sediment data from cores 16402 and 16403, and the C-14 laboratory of Kiel University (Drs Erlenkeuser and Willkomm) provided the ^{14}C and oxygen-isotope data of Fig. 8. We gratefully acknowledge the technical assistance of M. Baumann, M. Rösler, and A. Walker. This study was financially supported by the Deutsche Forschungsgemeinschaft.

References

ALLEN J.R.L. 1982. *Sedimentary Structures, Their Character and Physical Basis*, Vol. II. Elsevier, Amsterdam. 663 pp.

ASQUITH, S.M. 1979. Nature and origin of the lower continental rise hills off the East Coast of the United States. *Mar. Geol.* **32,** 165–90.

BAGNOLD, R.A. 1965. *The Physics of Blown Sand and Desert Dunes*. Halstead Press, New York.

CURRY, W.B. & LOHMANN, G.P. 1983. Reduced advection into the deep eastern basins of the Atlantic Ocean during the maximum of the last glaciation ($\delta^{18}O$ stage 2). *Nature* **306,** 577–80.

EMBLEY, R.W., RABINOWITZ, P.D. & JACOBI, R.D. 1978. Hyperbolic echo zones in the eastern Atlantic and the structure of the southern Madeira Rise. *Earth planet. Sci. Lett.* **41,** 419–33.

FLOOD, R.D. 1980. Deep-sea sedimentary morphology: modelling and interpretation of echo-sounding profiles. *Mar. Geol.* **38,** 77–82.

HEEZEN, B.C. & HOLLISTER, C. 1964. Turbidity currents and glaciation. *In: Probl. Palaeoclim. Proc. Conf. Newcastle upon Tyne*. Interscience Publishing. 99–108, 111–12.

HOBART, M.A., BUNCE, E.T. & SCLATER, J.G. 1975. Bottom water flow through the Kane Gap, Sierra Leone Rise, Atlantic Ocean. *J. geophys. Res.* **80,** 5083–8.

HOLLISTER, C.D. & MCCAVE, I.N. 1984. Sedimentation under deep-sea storms. *Nature* **309,** 220–5.

JACOBI, R. & HAYES, D. 1982. Bathymetry, microphysiography and reflectivity characteristics of the West African margin between Sierra Leone and Mauritania. *In:* U. VON RAD *et al.* (eds), *Geology of the Northwest African Continental Margin*. Springer Verlag, Berlin. 182–212.

JONES, G.E. & JOHNSON, D.A. 1984. Displaced Antarctic diatoms in VEMA Channel sediments: Late Pleistocene/Holocene fluctuations in AABW flow. *Mar. Geol.* **58,** 165–86.

KASSENS, H. 1985. *Verteilung der physikalischen Sedimenteigenschaften in Oberflächen-Sedimenten des äguatorialen Ostatlantik*. M.Sc. Thesis, University of Kiel.

KOLLA, V., EITTREIM, S., SULLIVAN, L., KOSTECKI, J.A. & BURCKLE, L.H. 1980. Current-controlled, abyssal microtopography and sedimentation in Mozambique Basin, southwest Indian Ocean. *Mar. Geol.* **34,** 171–206.

LONSDALE, P. 1978. Bedforms and the Benthic Boundary Layer in the North Atlantic: A Cruise Report of Indomed Leg II. Scripps Institution of Oceanography, San Diego, California 92152. *SIO Reference 78–30*, 29 December 1978, 1–15.

MCCAVE, I.N., LONSDALE, P.F., HOLLISTER, C.D. & GARDNER, W.D. 1980. Sediment transport over the Hatton and Gardar contourite drifts. *J. sedim. Petrol.* **50** (4), 1049–62.

MIENERT, J. 1985. *Akustostratigraphische Untersuchungen zur Klärung der Bodenwasserzirkulation im äquatorialen Ostatlantik*. Ph.D. Thesis, University of Kiel.

MOUNTAIN, G.S. & TUCHOLKE, B.E. 1983. Abyssal sediment waves. *In:* Bally, A.W. (ed.), *Seismic Expression of Structural Styles*. Vol. 1. AAPG Studies in Geology No. 15, 1.2.5-22/24.

NAKAGAWA, H., NEZU, I. & TOMINAGA, A. 1983. Secondary currents in a straight channel flow and the relation to its aspect ratio. *4th Symp. Turbulent Shear Flows, Sept. 12–14, 1983*, E. Naudascher (ed.) Karlsruhe, 3.8–3.13.

NORMARK, W.R., HESS, G.R., STOW, D.A.V. & BOWEN, A.J. 1980. Sediment waves on the Monterey fan levee: A preliminary physical interpretation. *Mar. Geol.* **37,** 1–8.

SARNTHEIN, M., ERLENKEUSER, H., V. GRAFENSTEIN, R. & SCHRÖDER, C. 1984. Stable-isotope stratigraphy for the last 750,000 years: "Meteor" core 13519 from the eastern equatorial Atlantic. *Meteor Forsch. Ergebn. C,* **38,** 9–24.

——, KÖGLER, F.C. & WERNER, F. 1983. Forschungsschiff Meteor, Reise Nr. 65. Berichte der wissenschaftlichen Leiter. *Geologisch-Paläontologisches Institut der Universität Kiel, Berichte-Reports.* 1–90.

SIROCCO, F. 1985. *Zum Abbild von NE-Passat, SW-Monsun und Oberflächenströmungen in rezenten*

Tiefseesedimenten vor Westafrika. M.Sc. Thesis University of Kiel. 77 p.

STEIN, R. & SARNTHEIN, M. 1984. Late Neogene events of atmospheric and oceanic circulation offshore north-west Africa: high-resolution record from deep-sea sediments. *Palaeoecol. Afr.* **16,** 9–36.

VOGELSANG, E. 1985. *Hochauflösende Zeitreihen von den Sedimenten der Termination I im äquatorialen Ostatlantik*. M.Sc. Thesis, University of Kiel.

M. SARNTHEIN & J. MIENERT, Geologisch Palaeontologisches Institut, Universitaet Kiel, Federal Republic of Germany.

Turbidite deposition and the origin of the Madeira Abyssal Plain

P.P.E. Weaver, R.C. Searle & A. Kuijpers

SUMMARY: Extensive investigations in the Madeira Abyssal Plain have revealed thick sequences of turbidites. Individual turbidites can be correlated over an area of 2° × 2°, and their emplacement can be shown to coincide with glacial onsets and terminations. The turbidites originate on the NW African margin and contain up to 2% organic carbon. They are virtually ungraded massive silts and clays, some with coarser basal layers.

Seismic records through the abyssal plain show distinct changes in sediment character at a few hundred metres depth. Rates of accumulation are calculated from piston cores to be about 7.5 cm/1000 years, suggesting that the turbidites are a recent phenomenon. If the link with glacial onsets and terminations has held throughout, the first major turbidites may have been emplaced at 2.4 Ma, at the onset of Northern Hemisphere glaciation. Some turbidites may also have been emplaced during the supposed late Miocene glaciations. The Madeira Abyssal Plain is therefore a recent feature, and before its initiation the area would have had a ridge-flank type topography.

The Madeira Abyssal Plain (MAP) occupies the central and deepest part of the Canary Basin. It forms a narrow strip of flat sea floor about 200 km wide (Fig. 1). It is bounded to the west by the lower flank of the Mid-Atlantic Ridge, which is characterized by numerous abyssal hills elongated in a NE–SW direction. The western boundary of the abyssal plain trends NNE–SSW along the base of the Ridge flank. Fracture zone valleys cut through these hills at right angles to this general trend and, in the valleys, fingers of the abyssal plain stretch westward into the Ridge flank. To the north, the MAP is bounded by the Azores-Gibraltar Rise and, to the south, by the Cape Verde Rise. The area of study (GME area) lies within the MAP and is bounded to the south by a ridge of hills extending east from the Great Meteor Seamount. To the north it is bounded by a ridge of hills extending east from Cruiser Seamount. To the east there is a distinct break of slope, from the very gentle gradient of the continental rise (about 1:300 to 1:1000) to the virtually flat (< 1:2000) abyssal plain at a water depth of 5440 m. This break of slope occurs at about 24°W (Fig. 2).

Within this area, the Institute of Oceanographic Sciences and the Geological Survey of the Netherlands have obtained over 7000 km of seismic profiles and taken over 50 piston cores (Searle *et al.* 1985; Kuijpers *et al.* 1984). This work has formed part of the OECD/NEA Seabed Working Group's investigations of deep ocean environments for the feasibility of disposal of high-level radioactive waste in the deep ocean (Anonymous 1984). Sediments cored on the abyssal plain reveal thick (up to 5 m) distal turbidite units separated by thin bioturbated pelagic units.

Sedimentology and stratigraphy

Sediments in the Madeira Abyssal Plain area fall into two basic types—those derived from turbidity current transport (turbidites), and those derived from surface water productivity together with fine detrital material (pelagic sediments). These two types alternate, generally with thick turbidite units separated by thin layers of pelagic sediment.

Pelagic sediments

The area today lies within the lysocline but above the CCD (Thunell 1982). Thus, a proportion of the calcium carbonate reaching the sea floor is dissolved before burial, and the resulting sediment appears relatively enriched in non-carbonate components. The amount of dissolution depends on the corrosiveness of the bottom water, and this is greatly increased during glacial periods (Crowley 1983). During the severest glacial periods in the GME area, dissolution exceeds supply of calcium carbonate and the resulting sediment is pelagic clay. Because the clays have resulted from removal of material, they usually form thinner layers than the marls.

Three cores, consisting predominantly of pelagic sediment (D10320, D10323 and 82PCS17) have been taken from the periphery of the area (Figs 2 and 3). These cores show the history of pelagic deposition, generally unaffected by turbidity currents (Weaver & Kuijpers 1983). Cores D10323 and 82PCS17 were located to the south in a slightly shallower region, whereas core D10320 was located to the west in the low abyssal hills. All three show alternations of more marly and less marly units. Core D10320 has the shallowest

From Summerhayes, C.P. & Shackleton, N.J. (eds), 1986, *North Atlantic Palaeoceanography*, Geological Society Special Publication No. 21, pp. 131–143.

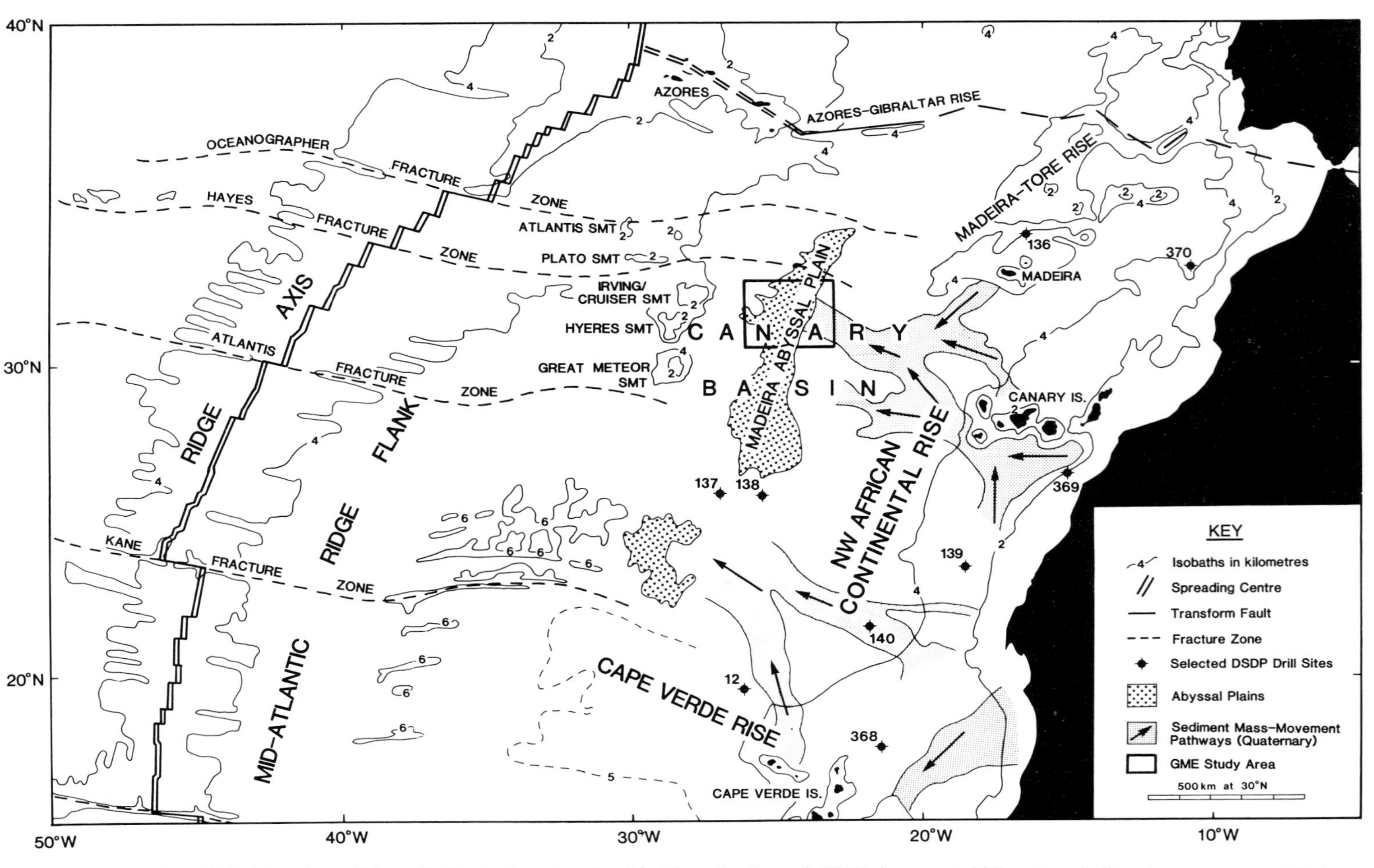

FIG. 1. Regional setting of Madeira Abyssal Plain. Outline bathymetry, simplified from Searle *et al.* (1982), in corrected kilometres. Sediment mass-movement features compiled and simplified from Embley (1975, 1976); Jacobi & Hayes (1982); Kidd & Searle (1984); and Jacobi (unpublished chart). Fracture zones from Searle *et al.* (1982) and Collette *et al.* (1984).

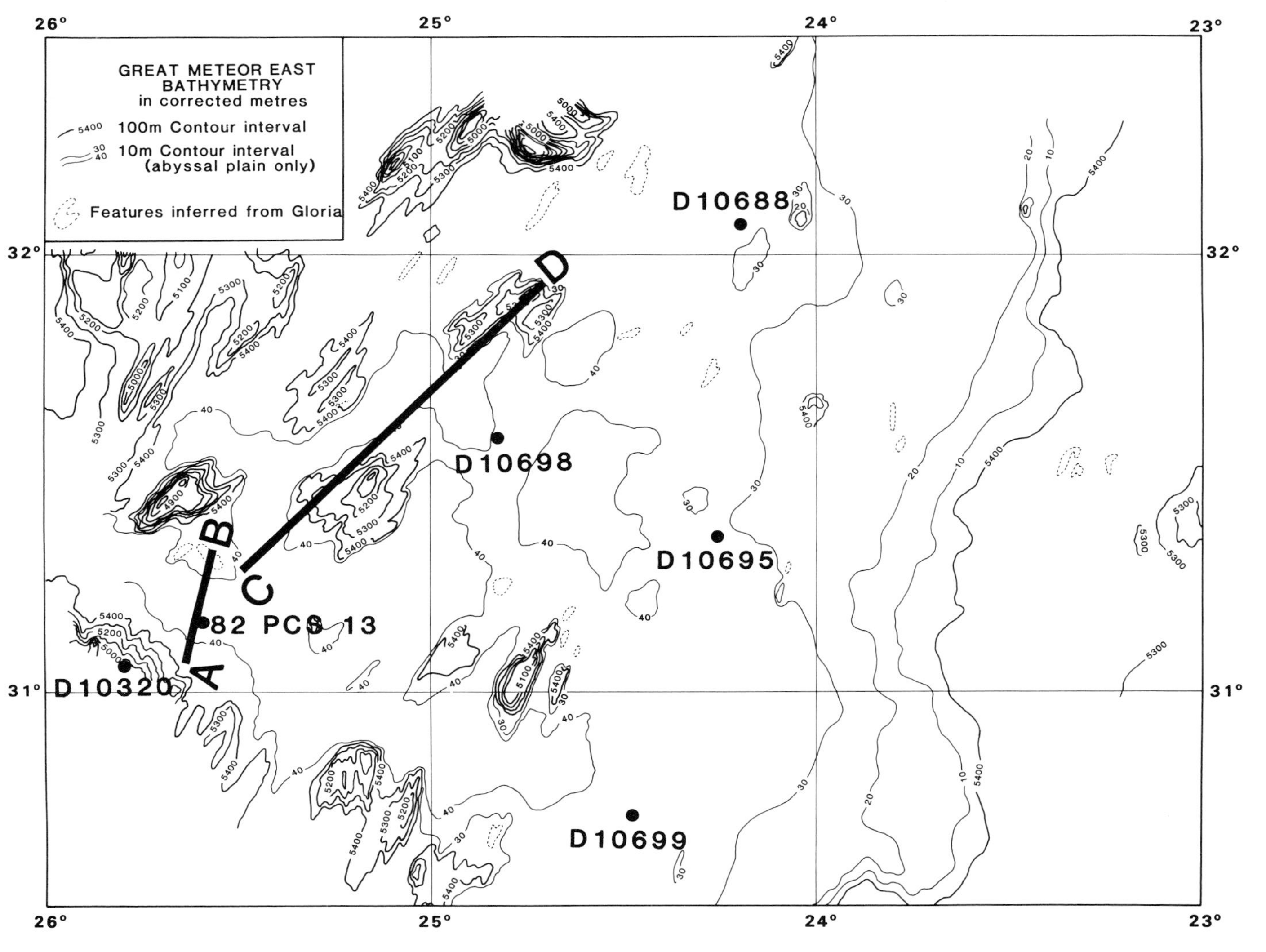

FIG. 2. Bathymetric chart of Great Meteor East, in corrected metres. Based on combined data from IOS and the Geological Survey of the Netherlands with interpolation of contours between tracks guided by GLORIA data. Line A–B = watergun profile in Fig. 7. Line C–D = airgun profile in Fig. 8.

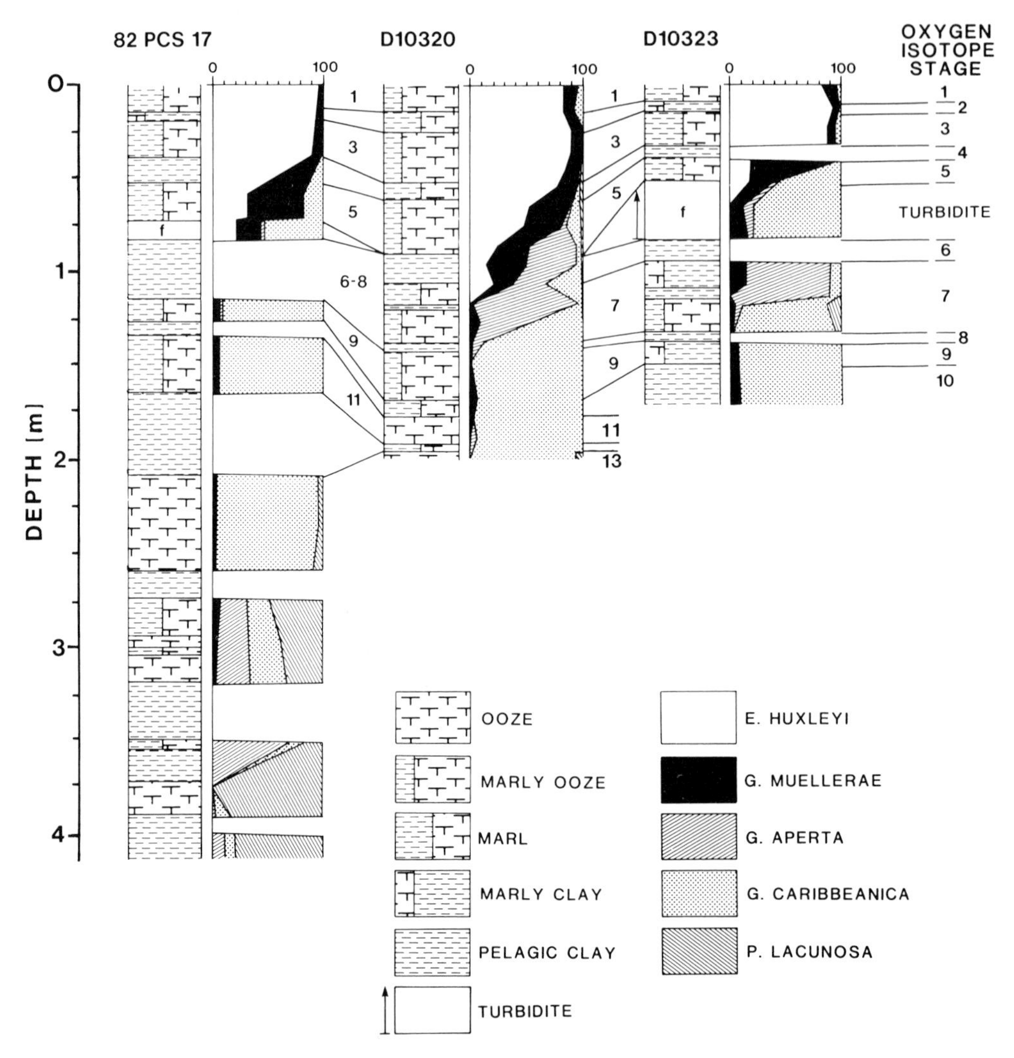

FIG. 3. Stratigraphy and lithology of pelagic cores D10320, D10323 and 82PCS17 (upper part only). Coccolith stratigraphy plotted as cumulative percentages of five coccolith species. Blank sections are due to complete dissolution. Oxygen isotope stage boundaries are placed within the constraints of the coccolith stratigraphy at lithological changes.

location and contains marls and marly oozes, with rare pelagic clay layers. Cores D10323 and the upper part of 82PCS17 show alternations of marl and pelagic clay, while the lower part of 82PCS17 shows alternations of calcareous oozes, marls and pelagic clays. Typically, marl units vary from 10–30 cm thick, pelagic clays from 6–10 cm and oozes from 7–60 cm. Some unusually thick pelagic clay units are found in core 82PCS17. These possibly can be explained by the effects of local topography on bottom flow patterns (Gould *et al.* 1981).

The carbonate fraction of the cores consists exclusively of coccoliths and planktonic foraminifera. The latter are often corroded and broken in the marls, while both are absent in the the pelagic clays.

Median grain sizes of the pelagic units range between fine silt and clay.

Turbidites

Turbidites comprise most of the sediment recovered from the area. The Madeira Abyssal Plain lies at the base of the continental rise and turbidites have become ponded between the base of the rise to the east and the hills of the Mid-Atlantic Ridge to the west.

The turbidites all belong to the E division of Piper (1978). They consist of ungraded massive silts (E3 division), in some cases with graded laminated basal layers (E1 and E2 divisions; Figs 4 and 5). Grain sizes in the silty bases vary across the area and from turbidite to turbidite. Out of cores D10688 and D10695, the coarest sample the authors have measured is the base of turbidite 'b' in core D10695. This has a median grain size of Mdϕ 4.6 (coarse silt). The ungraded parts of all turbidites have a median grain size in the clay range (Md$\phi > 8$). The basal layers contain an assemblage of minerals including quartz grains, glass shards, micas, predominantly basic heavy minerals and foraminfera. The massive silts consist of about 50–60% calcium carbonate, predominantly coccoliths, with clay minerals (illite and montmorillonite). The oldest turbidites recorded consist of over 70% calcium carbonate which may reflect a different source area (Fig. 4).

Many, but not all, of the turbidites consist of a couplet of two distinctly coloured units—a lower deep olive green unit and an upper pale green or grey unit. Mineralogically there is no difference between these two but the dark green unit contains 1–2% of organic carbon, in contrast to the pale units, which contain less than 0.5% organic carbon. A series of bluish laminae above the colour change suggests the presence of chemical fronts (Colley *et al.* 1984; Wilson *et al.* 1985).

The authors believe the turbidites with 50–60% $CaCO_3$ and high organic carbon contents originate on the upper continental slope off NW Africa, north of 20°N or possibly from the Madeira Rise. Sediments enriched in organic carbon occur today along the NW African slope, with the highest concentrations at 1000–2000 m; the shelf itself is depleted in organic material (Summerhayes 1983). During glacial periods the band of organic rich sediments may have been concentrated along the upper slope (Fütterer

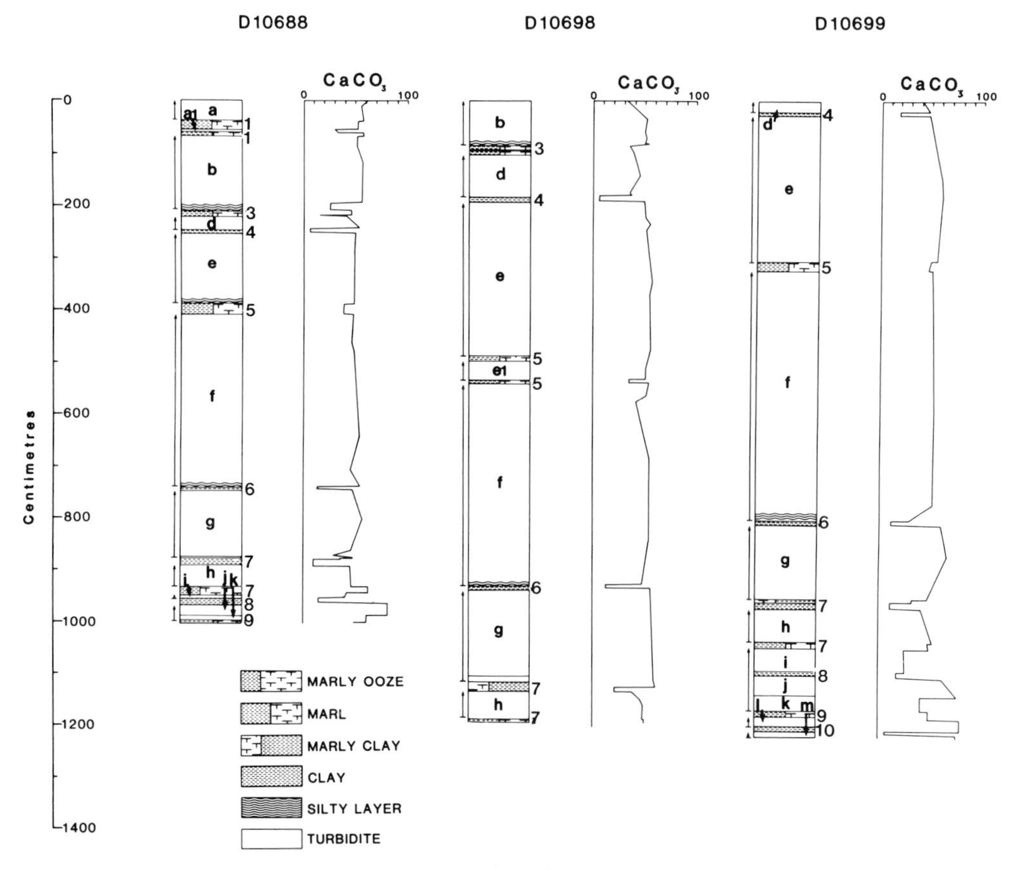

FIG. 4. Lithology and calcium carbonate percentages through three piston cores. Note high carbonate values in turbidites j, l and m. Numbers alongside pelagic intervals denote oxygen isotope stages.

1983). This is the only possible source for the organic carbon content in the turbidites and thus they must be initiated on the slope or upper slope. The silty bases of the turbidites contain benthic foraminifera such as *Brizalina spathulata*, *B. pseudopunctata* and *Gavelinopsis praegeri*, which are typical of the continental shelf/upper slope environment, together with deep-sea species (J. Weston, pers. comm.). The turbidites with calcium carbonate contents over 70% and very low organic carbon contents may have a different source. They do not have silty bases and hence no benthic foraminifera have been found. Further, their composition suggests initial deposition in an area receiving very little terrigenous input but well above the CCD. These turbidites, therefore, may have originated on the steep slopes around the nearby seamounts of Madeira, the Canaries, or even the Great Meteor Seamount group to the west (Fig. 1). Channel systems are known to cross the lower continental rise from the direction of Madeira and the Canaries (Simm & Kidd 1984).

Stratigraphy

Cores D10320, D10323 and 82PCS17, which consist predominantly of pelagic sediment, provide good reference sections for a combined bio-lithostratigraphy (Fig. 3). This stratigraphy can be compared directly to oxygen isotope stages (Emiliani 1955; Shackleton & Opdyke 1973), since the sediments vary according to the corrosiveness of bottom water, which is directly linked to ice production (Crowley 1983). The alternating lithologies, therefore, represent alternating glacial/interglacial conditions, with the more carbonate-rich units reflecting interglacial conditions. Used in isolation, this stratigraphy could not identify missing units, but when it is used in conjunction with the coccolith stratigraphy, the particular glacial and interglacial stages can be recognized. The main elements of the coccolith stratigraphy are: dominance of *Emiliania huxleyi* since late oxygen isotope stage 5 (Thierstein *et al.* 1977), dominance of *Gephyrocapsa aperta* in oxygen isotope stage 7 (Weaver 1983), dominance of *Gephyrocapsa caribbeanica* below oxygen isotope stage 7 (Kidd *et al.* 1983a), and the extinction of *Pseudoemiliania lacunosa* in oxygen isotope stage 12 (Thierstein *et al.* 1977). This stratigraphy can be clearly seen for core D10320 in Fig. 3.

Weaver & Kuijpers (1983) used the above stratigraphy to show that incoming turbidity currents are non-erosive across most of the area, and that their timing is not random. The turbidites lie between different pelagic lithologies, suggesting a strong relationship between climate change (probably via rising and falling sea-level) and turbidite emplacement. Weaver & Kuijpers used only three cores for their analysis, all peripheral to the turbidite province. Data are now available on many more cores (Kuijpers *et al.* 1984; Searle *et al.* 1985; Fig. 5) and it has been possible to correlate individual turbidites across the whole area. Individual turbidites can be recognized by a variety of criteria, including colour (which is partly a function of organic carbon content), $CaCO_3$ content, micropalaeontological composition (see Weaver & Kuijpers 1983), and age of the underlying and overlying pelagic units. The silty bases of some of the turbidites can be identified as reflectors on 3.5 kHz seismic records and correlated by this means (Searle *et al.* 1985).

Weaver & Kuijpers (1983) letter-coded the major turbidites from 'a' to 'g' with turbidite 'g' lying between oxygen isotope stages 6 and 7. Recently-taken cores, D10688 and D10699, have revealed deeper turbidites and enabled the lettering sequence to continue to turbidite 'm' (Fig. 5). As with the more recent turbidites, these deeper ones also lie between pelagic units of different sediment type. The correlation of turbidites with glacial onsets and terminations therefore continues at least to turbidite 'l' which lies between oxygen isotope stages 9 and 10. Turbidites 'j', 'l' and 'm' are high-calcium carbonate turbidites, whereas all the others have about 50–60% calcium carbonate. The high $CaCO_3$ turbidites are only found below oxygen isotope stage 8. The reason for this apparent change in source material at this time may be changes in the channel systems which feed the abyssal plain. Networks of channels are known on the lower continental rise (Fig. 1; Simm & Kidd 1984), but thick debris flow deposits also cross this area. It is possible that individual channels may be periodically blocked by debris flows, thus causing any turbidity current flowing down the channel to overflow and disperse. This system may cause different source areas to be periodically turned on and off. At present the data on the lower rise are insufficient to show if this mechanism is operative or not.

Rate of accumulation

The average rate of accumulation in the abyssal plain over the last 200 000 years varies from about 10 cm/1000 years in the extreme west of the area (core 82PCS13) to about 4 cm/1000 years in the north-east (core D10688). In the centre of the area, the rate of accumulation is between 6 and 9 cm/1000 years. On 3.5 kHz seismic records, strong reflectors can be identified throughout (i.e. down to the maximum depth of resolution at *c.* 100 m; Fig. 6). Such reflectors have been found to

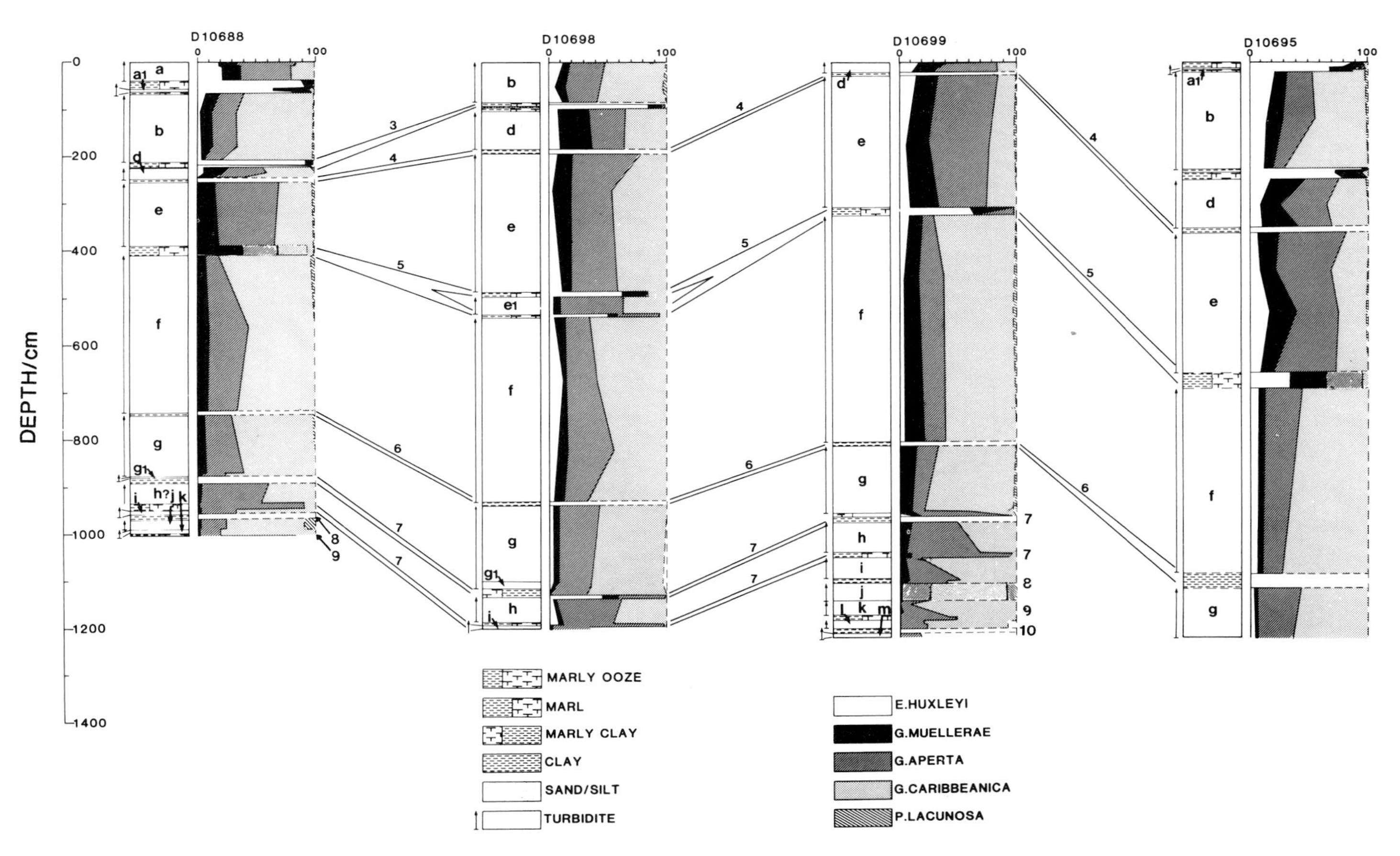

FIG. 5. Stratigraphy and lithology of four turbidite cores showing continuation of the Weaver & Kuijpers (1983) model back to oxygen isotope stage 10.

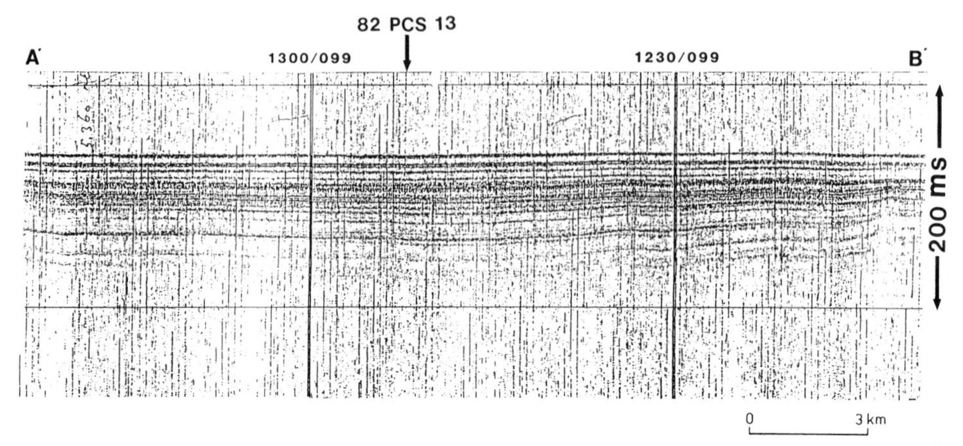

FIG. 6. 3.5 kHz profile from Tyro 82, along centre part of profile shown in Fig. 7. Horizontal (depth) lines spaced 200 ms (approximately 150 m) apart.

correlate with silty layers generally only present at the bases of the thickest turbidites (Searle *et al.* 1985).

The authors therefore expect the same pattern of sedimentation to continue to at least 100 m and probably to the top of the draped sediments identified on the airgun profiles (Fig. 8). The 3.5 kHz records do show occasional bands of weaker reflectors, and these may correlate with bands of turbidites without silty bases, such as the high $CaCO_3$ turbidites. If the volume of material in each turbidity current has remained approximately the same throughout, the early turbidites would have accumulated at higher rates since the depositional area would have been smaller (compare flat areas in Figs 2 and 10(b)).

Initiation of turbidite deposition

If the hypothesis that turbidite deposition is linked to climate change via sea-level fluctuations holds for all of the turbidites on the Madeira Abyssal Plain, then it would be expected that the age of the oldest turbidite would approximate the age of the onset of glaciation. The onset of Northern Hemisphere glaciation on a major scale began 2.4 m.y. ago (Shackleton *et al.* 1984; Kidd *et al.* 1983b), as evidenced by ice rafting and oxygen isotope measurements. However, there is evidence for fluctuating sea-levels before this. Early build-ups of ice have been reported from Iceland at 3.1 Ma (McDougal & Wensink 1966), South America in the latest Miocene (Mercer & Sutter 1982), and Antarctica, also in the latest Miocene (Ciesielski *et al.* 1982). Vail & Mitchum (1979) also show fluctuating sea-levels in the latest Miocene (Messinian). The early Pliocene appears to have been a much quieter interval, both on Vail & Mitchum's sea-level fluctuation curves, and on oxygen isotope evidence (Keigwin 1979). Below the latest Miocene, Vail & Mitchum show much lower frequency sea-level oscillations, these being generally of lower magnitude.

Turbidites initiated by sea-level fluctuations may, therefore, be expected through the 2.4 m.y. to Recent interval, but may also occur slightly earlier than this, and may have occurred during the latest Miocene (e.g. at DSDP site 138, 500 km south of GME (Fig. 1; Hayes, Pimm *et al.* 1972)). If this is the case, then major turbidite input to the Madeira Abyssal Plain is a relatively recent phenomenon.

Seismic evidence

Over 7000 km of airgun and watergun seismic profiles have been examined from the area. The turbidites can be seen on such records as a sequence of horizontal reflectors (Fig. 7). Four units are recognized (Duin & Kok 1984; Searle *et al.* 1985; Fig. 8): an uppermost stratified unit containing many closely-spaced, flat-lying, strong reflectors (unit A); a relatively transparent unit containing only a few parallel reflectors (unit B); a lower stratified unit containing closely-spaced, strong reflectors (unit C); and, in some places, a basal transparent unit which displays no reflectors (unit D). Units A and B are separated by a prominent seismic reflector (Reflector 2). Generally, the reflectors are continuous. They are only locally disrupted where fault-like features occur (Duin *et al.* 1984).

Searle *et al.* suggest that both units A and B represent ponded turbidites and that unit C contains draped pelagic sediment. Their reasons are as follows: first, although unit A contains

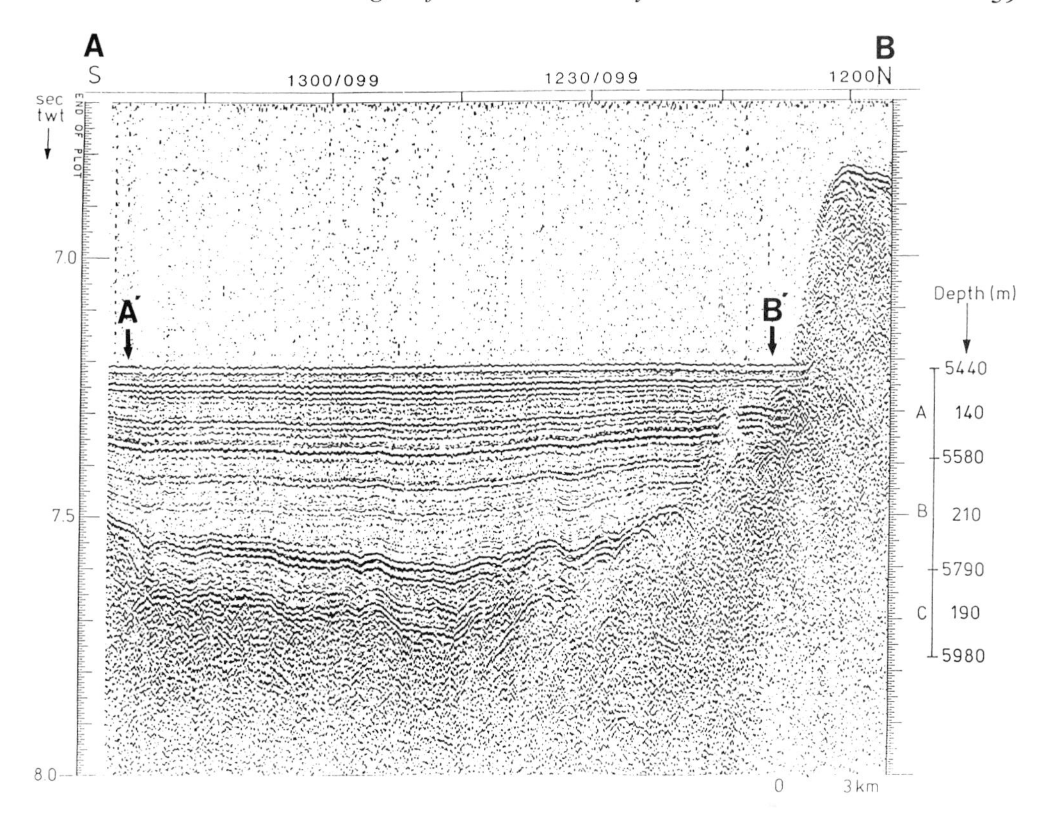

FIG. 7 Watergun seismic profile from Tyro 82. Three interpreted seismic units with estimated thicknesses shown on right. Location = line A–B in Fig. 2.

many more reflectors than unit B, in most places there is no sharp cut-off and the incidence of reflectors dies off gradually with depth. In many places, reflectors occur below Reflector 2 that appear very similar to those immediately above it (Fig. 8). Secondly, no regional unconformity is recognized between units A and B, but there is a clear unconformity between units B and C. This is hard to see in the centres of basins, where all reflectors tend to be flat-lying, but on the sides of the basins one can often recognize a very clear angular unconformity between the sub-horizontal reflectors of unit B and the steeper, draped surface of unit C (Fig. 8). The authors, therefore, prefer to interpret this unconformity as the base of the turbidites. The combined thickness of units A and B ranges from about 150 to 560 ms (125 to 530 m) across the Plain but averages about 350 m.

Duin & Kok suggest that unit B represents pelagic sediments (red clay) but the authors discount that interpretation for the reasons given above.

Searle *et al.* suggest that unit C represents predominantly clay. A considerable thickness of clay would be expected in this area since it has remained below the CCD for much of its history (Van Andel 1975; Tucholke & Vogt 1979; Fig. 9). Predominantly clay sequences were found in the Tertiary to Quaternary section of DSDP hole 137 and the early Tertiary at hole 130 (Hayes, Pimm *et al.* 1972). The seismic data thus suggest that a long period of predominantly pelagic deposition preceded the onset of turbidite deposition. Average thickness of the turbidite unit is 350 m.

Dating the origin of the Abyssal Plain

The average thickness of the turbidite unit (A and B) is 350 m and the average accumulation rate measured from cores is between 6–9 cm/1000 years. If major turbidites in the area have all been initiated during sea-level changes associated with the 2.4 m.y. to Recent glaciations then an average accumulation rate of 14.6 cm/1000 years is required. If the Messinian sea-level changes also caused turbidite initiation between 6.2–5.2 m.y.,

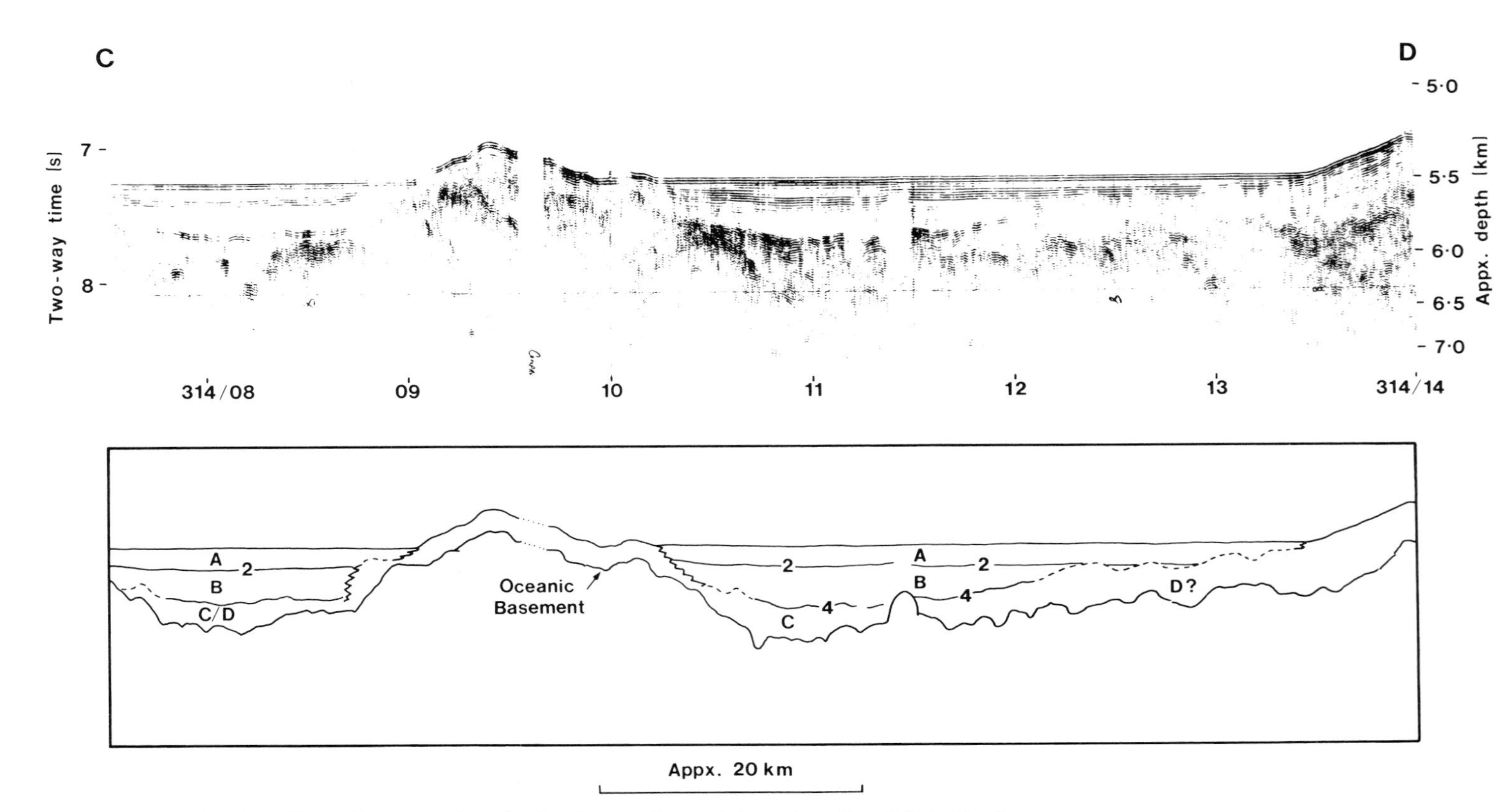

FIG. 8. Airgun seismic profile and interpretation after Searle *et al.* (in press). Location = line C–D in Fig. 2.

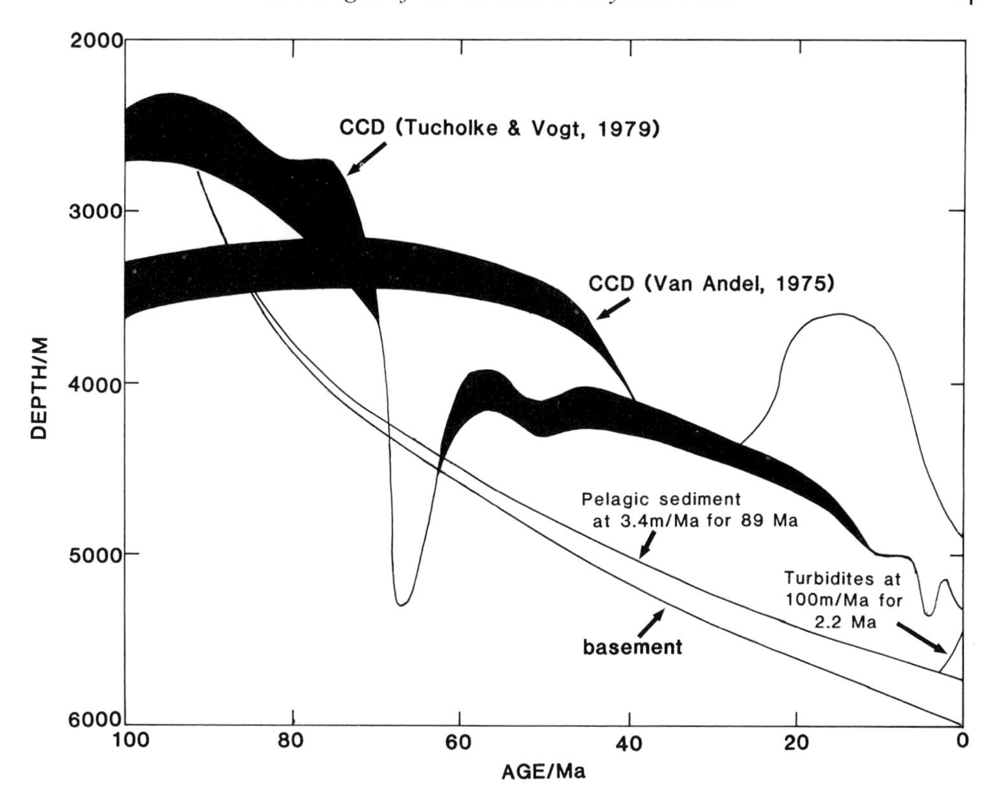

FIG. 9. Variation of carbonate compensation depth (CCD) with time in the North Atlantic, after Van Andel (1975) and Tucholke & Vogt (1979). Also shown is the predicted track for 92 Ma old sea floor, assuming that it was created at the Mid-Atlantic Ridge at a depth of 2600 m and subsequently followed the normal age-depth curve of Parsons & Sclater (1977). The effects of a simple two-stage sedimentation model are also shown, though it is emphasized that the actual history may have been more complex.

then 10.3 cm/1000 years average accumulation rate would be required. Both values are close to the calculated accumulation rates.

The age of the onset of turbidite deposition may therefore be 6.2 or 2.4 m.y. and succeeds a long interval of pelagic clay deposition. As can be seen from Fig. 10, if the turbidite sequence was removed, a typical ridge-flank type topography would be revealed in the area. Thus, before the initiation of turbidite deposition, no abyssal plain existed in this area. The origin of the Madeira Abyssal Plain therefore dates from 2.4 m.y. or possibly 6.2 m.y.

ACKNOWLEDGEMENTS: The authors thank Dr R.B. Kidd for discussions on various aspects of this work. Dr T.J.G. Francis offered helpful comments on the manuscript.

Cores 82PCS-13 and 17 were collected during a cruise of MV *Tyro* in 1982 as part of the Dutch contribution to the Seabed Working Group activities. Cores D10320 and D10323 were collected during RRS *Discovery* Cruise 118, and cores D10688–95, 98 and 99 were collected during RRS *Discovery* Cruise 134.

The Ministry of Economic Affairs of the Netherlands and the Director of the Geological Survey of the Netherlands are acknowledged for their permission to publish data. IOS research has been carried out under contract for the Department of the Environment as part of its radioactive waste management research programme. The results will be used in the formulation of government policy but, at this stage, they do not necessarily represent government policy.

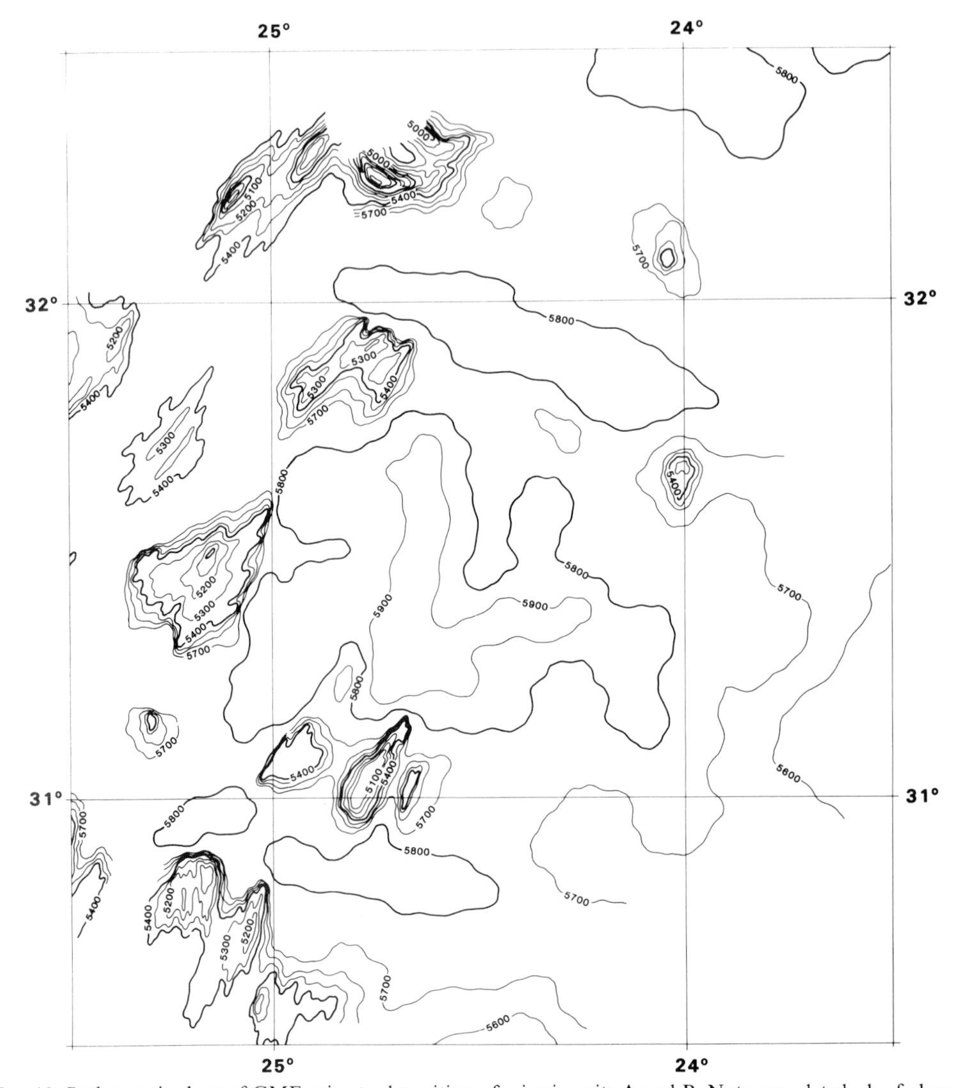

FIG. 10. Bathymetric chart of GME prior to deposition of seismic units A and B. Note complete lack of abyssal plain, and WNW–ESE trending fracture zone valleys with intervening ridge flank topography. Produced by contouring thicknesses of units A and B from seismic profiles across the area, and subtracting these contours from the present-day bathymetry. Depths in metres.

References

ANONYMOUS 1984. *Seabed Disposal of High-level Radioactive Waste.* Nuclear Energy Agency, OECD, Paris. 1–248.

CIESIELKSI, P.F., LEDBETTER, M.T. & ELLWOOD, B.B. 1982. The development of Antarctic glaciation and the Neogene palaeoenvironment of the Maurice Ewing Bank. *Mar. Geol.* **46,** 1–51.

COLLETTE, B.J., SLOOTWEG, A.P., VERHOEF, J. & ROEST, W.R. 1984. Geophysical investigations of the floor of the Atlantic Ocean between 10° and 38°N (Kroonflag-Project). *Proc. K. Nederlandse akad. van wetenschappen*, C, **87** (1), 1–76.

COLLEY, S., THOMSON, J., WILSON, T.R.S. & HIGGS, N.C. 1984. Post-depositional migration of elements during diagenesis in brown clay and turbidite sequences in the North East Atlantic. *Geochim. Cosmochim. Acta* **48,** 1223–35.

CROWLEY, T.J. 1983. Calcium-carbonate preservation patterns in the central North Atlantic during the last 150,000 years. *Mar. Geol.* **51.** 1–14.

DUIN, E.J.T. & KOK, P.T.J. 1984. A geophysical study of the western Madeira Abyssal Plain, eastern North Atlantic. *Meded. Rijks Geol. Dienst.* **38** (2), 67–89.

——, MESDAG, C.S. & KOK, P.T.J. 1984. Faulting in Madeira Abyssal Plain sediments. *Mar. Geol.* **56,** 299–308.

EMBLEY, R.W. 1975. *Studies of Deep-Sea Sedimentation Processes using High-Frequency Seismic Data.* Ph.D. thesis, Columbia University. 332 pp.

—— 1976. New evidence for occurrence of debris flow deposits in the deep sea. *Geology* **4,** 371–4.

EMILIANI, C. 1955. Pleistocene temperatures. *J. Geol.*, **63,** 538–8.

FÜTTERER, D.K. 1983. The modern upwelling record off northwest Africa. *In*: THIEDE, J. & SUESS, E. (eds), *Coastal Upwelling; Its Sediment Record—Part B.* Plenum Press, New York. 105–21.

GOULD, W.J., HENDRY, R. & HUPPERT, H.E. 1981. An abyssal topographic experiment. *Deep Sea Res.* **28A,** 409–40.

HAYES, D.E., PIMM, A.C. *et al.* 1972. *Init. Repts DSDP* **14.** US Govt. Print. Off., Washington, DC. 85–155.

JACOBI, R.D. & HAYES, D.E. 1982. Bathymetry, microphysiography and reflectivity characteristics of the west African margin between Sierra Leone and Mauritania. *In*: VON RAD, U., HINZ, K., SARNTHEIN, M. & SIEBOLD, E. *Geology of the Northwest African Continental Margin.* Springer-Verlag, Berlin. 182–212.

KEIGWIN, L.D. 1979. Late Cenozoic stable isotope stratigraphy and palaeoceanography of DSDP sites from the east equatorial and central North Pacific ocean. *Earth Planet. Sci. Lett.* **45,** 361–82.

KIDD, R.B., SEARLE, R.C., WEAVER, P.P.E., JACOBS, C.L., HUGGETT, Q.J., NOEL, M.J. & SCHULTHEISS, P.J. 1983a. King's Trough Flank: geological and geophysical investigations of its suitability for high-level radioactive waste disposal. *IOS Report* **166,** 93 pp.

—— *et al.* (Shipboard Staff DSDP Leg 94) 1983b. Sediment drifts and intraplate tectonics in the North Atlantic. *Nature* **306,** 532–3.

—— & SEARLE, R.C. 1984. Sedimentation in the southern Cape Verde Basin: regional observations by long-range sidescan sonar. *In*: STOW, D.A.V. & PIPER, D.J.W. (eds) *Fine-grained Sediments.* Spec. Pub. Geol. Soc. London. Blackwell Scientific Publications, Oxford. 145–52

KUIJPERS, A., RISPENS, F.B. & BURGER, A.W. 1984. Late Quaternary sedimentation and sedimentary processes on the Madeira Abyssal Plain, eastern North Atlantic. *Meded. Rijks Geol. Dienst.* **38** (2), 91–118.

MCDOUGAL, I. & WENSINK, H. 1966. Palaeomagnetism and geochronology of the Pliocene-Pleistocene lavas in Iceland. *Earth Plan. Sci. Lett.* **1,** 232–6.

MERCER, J.G. & SUTTER, J.F. 1982. Late Miocene earliest Pliocene glaciation in southern Argentina: implications for global ice-sheet history. *Paleogeogr. Paleoclimatol. Paleoecol.* **38,** 185–206.

PARSONS, B. & SCLATER, J.G. 1977. An analysis of the variation of ocean floor bathymetry and heat flow with age. *J. geophys. Res.* **82** (5) 803–27.

PIPER, D.J.W. 1978. Turbidite muds and silts on deep sea fans and abyssal plains. *In*: STANLEY, D.J. & KELLING, G. (eds), *Sedimentation in Submarine Canyons, Fans and Trenches.* Dowden, Hutchinson & Ross, Stroudsborg, Pa. 163–76.

SEARLE, R.C., MONAHAN, D. & JOHNSON G.L. 1982. *General Bathymetric Chart of the Oceans.* 5th edition., sheet 5.08. Canadian Hydrographic Service, Ottawa. (Scale 1:10,000,000 at the Equator).

——, SCHULTHEISS, P.J., WEAVER, P.P.E., NOEL, M., KIDD, R.B., JACOBS, C.L. & HUGGETT, Q.J. 1985. Great Meteor East (distal Madeira Abyssal Plain): geologic studies of its suitability for disposal of heat-emitting radioactive wastes. *IOS Report* **193,** 161 pp.

SHACKLETON, N.J., BACKMAN, J., ZIMMERMAN, H., KENT, D.V., HALL, M.A., ROBERTS, D.G., SCHNITKER, D., BALDAUF, J.G., DESPRAIRIES, A., HOMRIGHAUSEN, R., HUDDLESTUN, P., KEENE, J.B., KALTENBACK, A.J., KRUMSIEK, K.A.O., MORTON, A.C., MURRAY, J.W. & WESTBERG-SMITH, S. 1984. Oxygen isotope calibration of the onset of ice-rafting and history of glaciation in the North Atlantic region. *Nature* **307,** 5952, 620–3.

—— & OPDYKE, N.D. 1973. Oxygen isotope and palaeomagnetic stratigraphy of equatorial Pacific core, V28-238. Oxygen isotope temperatures and ice volumes on a 10^5 and 10^6 year scale. *J. Quat. Res.* **3** (1) 39–55.

SIMM, R.W. & KIDD, R.B. 1984. Submarine debris flow deposits detected by long-range sidescan sonar 1000 km from source. *Geomar. Lett.* **3,** 13–16.

SUMMERHAYES, C.P. 1983. Sedimentation of organic matter in upwelling regimes. *In*: THIEDE, J. & SUESS, E. (eds), *Coastal Upwelling. Its Sediment record–Part B.* Plenum Press, New York. 29–72.

THIERSTEIN, H.R., GEITZENAUER, K.R., MOLFINO, B. & SHACKLETON, N.J. 1977. Global synchroneity of late Quaternary coccolith datum levels: validation by oxygen isotopes. *Geology* **5,** 404–4.

THUNELL, R.C. 1982. Carbonate dissolution and abyssal hydrography in the Atlantic Ocean. *Mar. Geol.* **47,** 165–80.

TUCHOLKE, B.E. & VOGT, P.R. 1979. Western North Atlantic: sedimentary evolution and aspects of tectonic history. *In*: MURDMAA, I.O. *et al., Init. Repts. DSDP* **43,** 791–825. US Govt. Print. Off., Washington, DC.

VAIL, P.R. & MITCHUM, R.M. 1979. Global cycles of sea-level change and their role in exploration. *In*: *Proceedings of the Tenth World Petroleum Congress, Vol. 2: Exploration Supply and Demand.* Bucharest, Romania. 95–104.

VAN ANDEL, T.H. 1975. Mesozoic-Cenozoic calcite compensation depth and the global distribution of calcareous sediments. *Earth Planet. Sci. Lett.* **26,** 187–94.

WEAVER, P.P.E. 1983. An integrated stratigraphy of the upper Quaternary of the King's Trough Flank area, NE Atlantic. *Oceanol. Acta* **6,** 451–6.

—— & KUIJPERS, A. 1983. Climatic control of turbidite deposition on the Madeira Abyssal Plain. *Nature* **306,** 360–3.

WILSON, T.R.S., THOMSON, J., COLLEY, S., HYDES, D.J., HIGGS, N.C. & SORENSEN, J. 1985. Early organic diagenesis: the significance of progressive subsurface oxidation fronts in pelagic sediments. *Geochim Cosmochim Acta*, in press.

P.P.E. WEAVER & R.C. SEARLE, Institute of Oceanographic Sciences, Brook Road, Wormley, Godalming, Surrey GU8 5UB.

A. KUIJPERS, Geological Survey of the Netherlands, Spaarne 17, Postbus 157, 2000 AD Haarlem, The Netherlands.

Late Weichselian transgression, erosion and sedimentation at Gullfaks, northern North Sea

Randi Carlsen, Tor Løken & Elen Roaldset

SUMMARY: A prograding sequence of near shore sediments is identified at the border of the northern North Sea plateau (water depth 135 m) and the western slope of the Norwegian Trench. In Late Weichselian time (probably about 15 000 years BP) the relative sea level was approximately 180 m lower than today. During this low sea-level stand, the older consolidated glaciomarine sediments in the slope area were exposed to coastal erosion, leading to a prograding unit of sand fining outwards to silt and clay. During rising sea-level the coastal erosion gradually reached a lodgement till deposited at the top of the plateau, resulting in coarser sediments such as stones and gravel, fining outwards to sand down the slope. The plateau area was transgressed shortly after 13 000 years BP.

The Quaternary sediments in the northern North Sea are dominated by stiff to hard clay with varying amounts of silt, sand and gravel, and occasional layers of sand (Skinner & Gregory 1983; Rise & Rokoengen 1984; Rise *et al.* 1984). The thickness varies from a thin cover to several hundred metres, normally 100–200 m. In depressions, like the Norwegian Trench, the upper metres are very soft clay.

Previous investigations indicate several unconformities which most probably are related to events of glacial/periglacial character (e.g. Feyling-Hanssen 1980, 1982; Skinner & Gregory 1983; Read & Løken 1984).

During the Late Weichselian, the absolute sea-level in the North Sea region was much lower than at present, probably 85–90 m (e.g. Mørner 1969; Jansen 1976). This is mainly based on investigations from land or near land areas surrounding the North Sea.

Sea-level variations during the Weichselian have been indicated by foraminiferal analyses in the northern North Sea (Gregory & Harland 1978; Feyling-Hanssen 1980; 1982; Skinner & Gregory 1983). Few low sea-level stands from Late Weichselian are reported. A 'submerged beach' at 90 m water depth, of supposed Late Weichselian age, is interpreted from geophysical records at 58°N (Hovland & Dukefoss 1981). At the Gullfaks area a 'submerged beach' of probably Late Weichselian age was described by Dekko & Rokoengen (1978), later confirmed by ^{14}C-age datings (Rokoengen *et al.* 1982).

The present water current pattern was established at the beginning of the Holocene (less than 10 000 years ago). North Atlantic water flows into the northern North Sea between Scotland, the Orkney Islands, the Faroe Islands and Norway as surficial currents, and as a deeper current along the western part of the Norwegian Trench (Lee 1980).

Site description

The Gullfaks area is located at 61°10′N., 2°15′E. in the Norwegian sector of the North Sea (Fig. 1). The investigated area covers the outermost part of the North Sea Plateau, across the shoulder and down the upper part of the western slope of the Norwegian Trench. The water depth in the area ranges from 135 m to 220 m (below sea-level).

The investigated sediments are mainly of glaciomarine origin with different degrees of consolidation. The sediments are supposedly Weichselian, although the boundary to Pliocene sediments is reported not far below the end of the borings. This paper concentrates upon a coastal progradation during a low sea-level stand in the Late Weichselian. The Weichselian chronology used is according to Mangerud & Berglund (1978).

As a part of the Gullfaks production development several platform locations have been investigated for geotechnical purposes, see Fig. 2. Each site investigation includes 5–10 deep (max. 150 m) boreholes for geotechnical soil sampling and *in situ* testing.

Methods of analyses

This paper presents the results from special geological investigations made on selected soil samples from several of the geotechnical boreholes.

The diameter of the undisturbed soil samples is 7.4 cm. Procedures for drilling, sampling and sample handling are described by Schjetne & Brylawski (1979).

The sedimentological and the stratigraphical results are based on different kinds of analyses.

(1) Geotechnical tests of water content, liquid and plastic limits, grain-size analyses and

From SUMMERHAYES, C.P. & SHACKLETON, N.J. (eds), 1986, *North Atlantic Palaeoceanography*, Geological Society Special Publication No. 21, pp. 145–152.

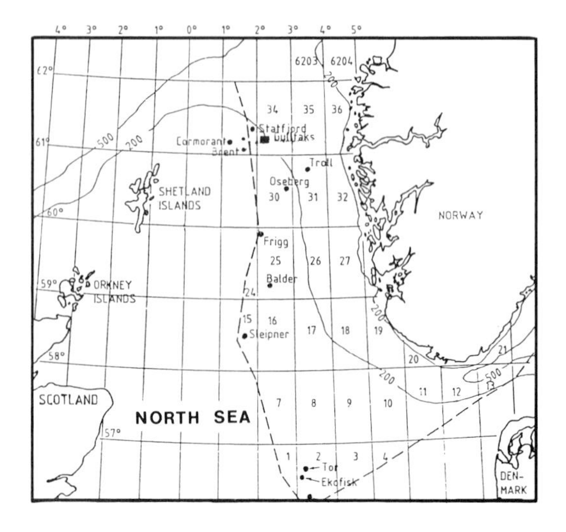

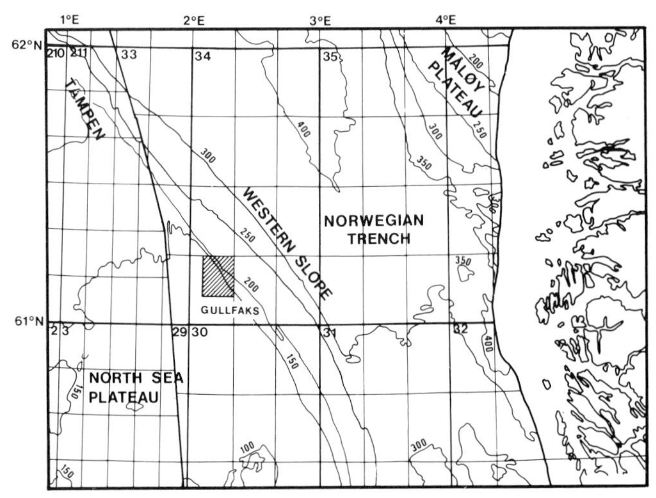

FIG. 1. Location of the Gullfaks area in the North Sea.

different kinds of shear strength measurements were performed at NGI: methods according to Andresen *et al.* (1979).

(2) For mineral identification, X-ray diffraction was used, after standard procedures on oriented slides for the fractions <0.2 μm, 2 μm and 2–75 μm.

(3) For preparing thin sections, 4 cm sediment cubes were air-dried before impregnation with 'Epofix', and then treated as usual for hard rocks.

(4) The foraminiferal analyses were conducted by R.W. Feyling-Hanssen and K.L. Knudsen, Geological Institute, University of Aarhus: using methods according to Feyling-Hanssen (1958) and Meldegaard & Knudsen (1979).

(5) The dinoflagellate cysts were examined by B. Dale, Institute of Geology, University of Oslo, using standard palynological techniques involving HCl, HF and sieving to give the >25 μm fraction. Samples were analysed both qualitatively and quantitatively.

(6) The foraminifera *Elphidium excavatum* was used for oxygen isotope investigations by J.C. Duplessy, Centre des Faibles Radioactivités, CNRS, Gif sur Yvette: methods according to Duplessy (1978).

(7) For amino acid analyses the foraminifera

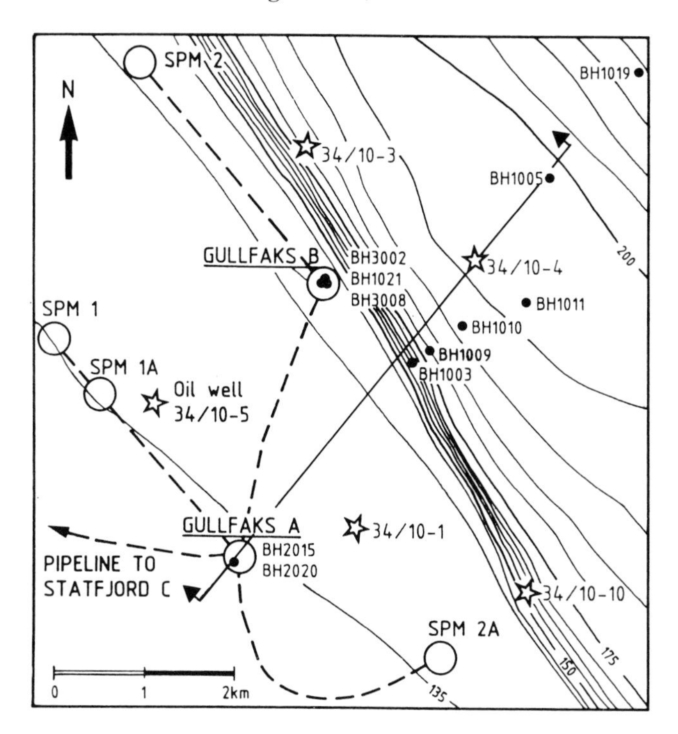

FIG. 2. Location map of the Gullfaks area. The bathymetric contours are water depths in metres.

Elphidium excavatum and *Cibicides lobatolus* were used, along with two shell samples, *Macoma calcarea* and *Arctica islandica*. The analyses were performed by H.P. Sejrup, Geological Institute, Dep. B, University of Bergen, according to Bada & Man (1980). His interpretations are based on the relation between D-allo-isoleucine and L-isoleucine from the total fraction.

(8) ^{14}C-analyses of shell and shell fragments were performed by R. Nydal: methods according to Radiological Dating Laboratory, Norwegian Institute of Technology, Trondheim, including a correction of 440 years for the apparent age of sea water. ^{14}C half-life of 5570 years was used.

(9) Geochemical analyses of ash fragments were performed by H. Furnes, Geological Institute, Dep. A, University of Bergen, using SEM with EDS, according to Mangerud *et al.* (1984).

Results

Figure 3 illustrates the different stages of development of the Gullfaks area. Figure 3(a) illustrates the oldest situation: what was left after the retreat of the last glacier. Figure 3(e) shows the present situation. The schematic profiles illustrate the selected main sediment layers, penetrated by four selected borings. The gradually rising relative sea-level is also sketched. Results of the ^{14}C-datings from the different soil units are presented in Fig. 3(e).

The sediment units in Fig. 3(a) are dominated by clayey glaciomarine and marine sediments. Typical grain-size distribution is shown in Fig. 4(a). Scattered ice-dropped gravel is found in the glaciomarine units.

Both stadial and interstadial climates are recognized. Sedimentological variations are due to changes in sea-level, water current regime and ice distribution. The sedimentation rate seems to have been periodically very high, interrupted by periods of non-deposition or erosion. The investigations of dinoflagellate cysts and foraminifera indicate offshore environment, probably with water current interaction with the Norwegian Sea during the periods of deposition. The water depths were probably ranging from 50 to 100 m.

Several glacial advances have consolidated and eroded the sediments. These erosion surfaces are marked with solid lines in Fig. 3. They probably represent Weichselian ice-shelf advances from Norway. The two uppermost erosion surfaces (2

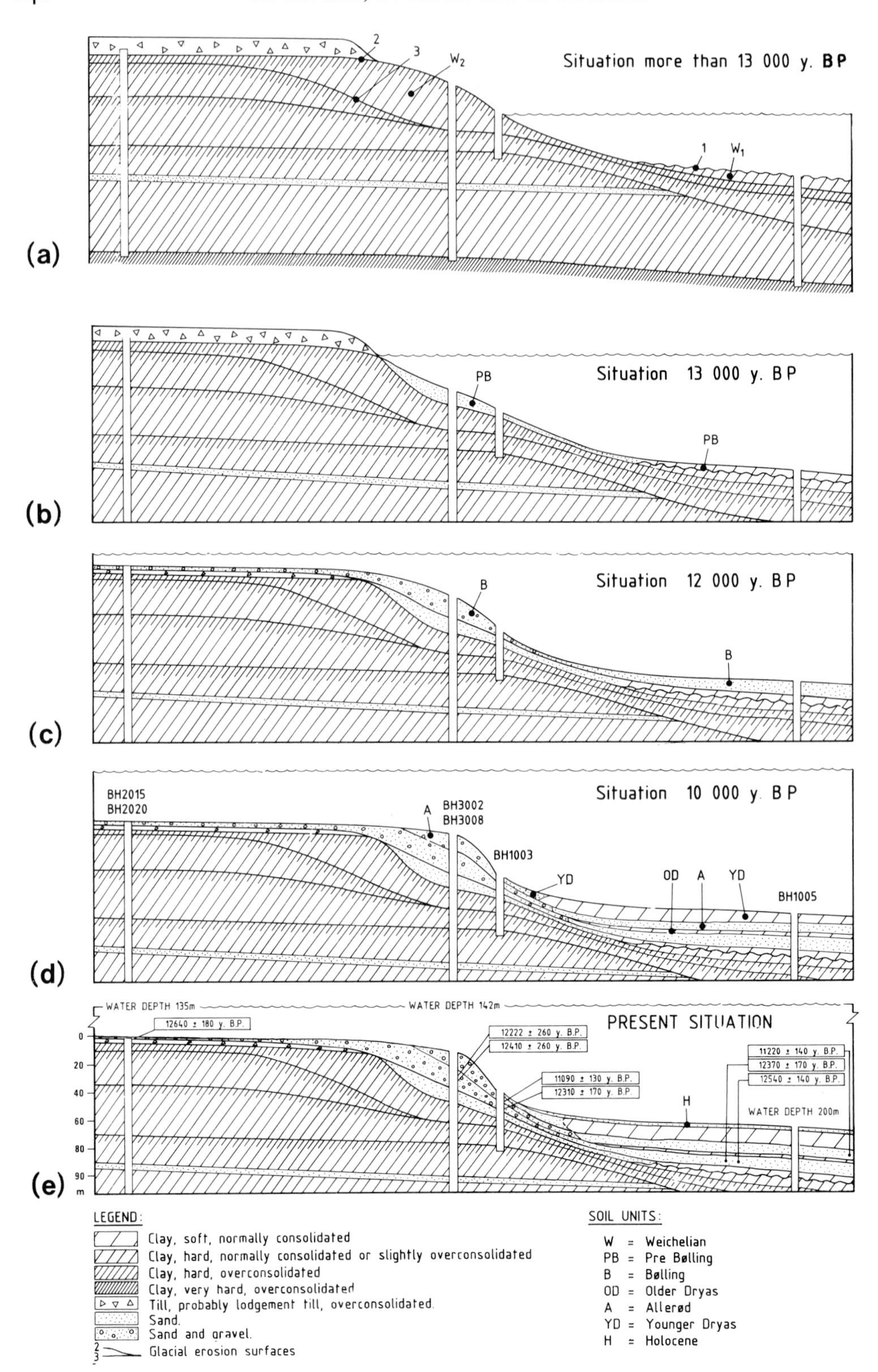

FIG. 3. Schematic section at different stages of development in the Gullfaks area.

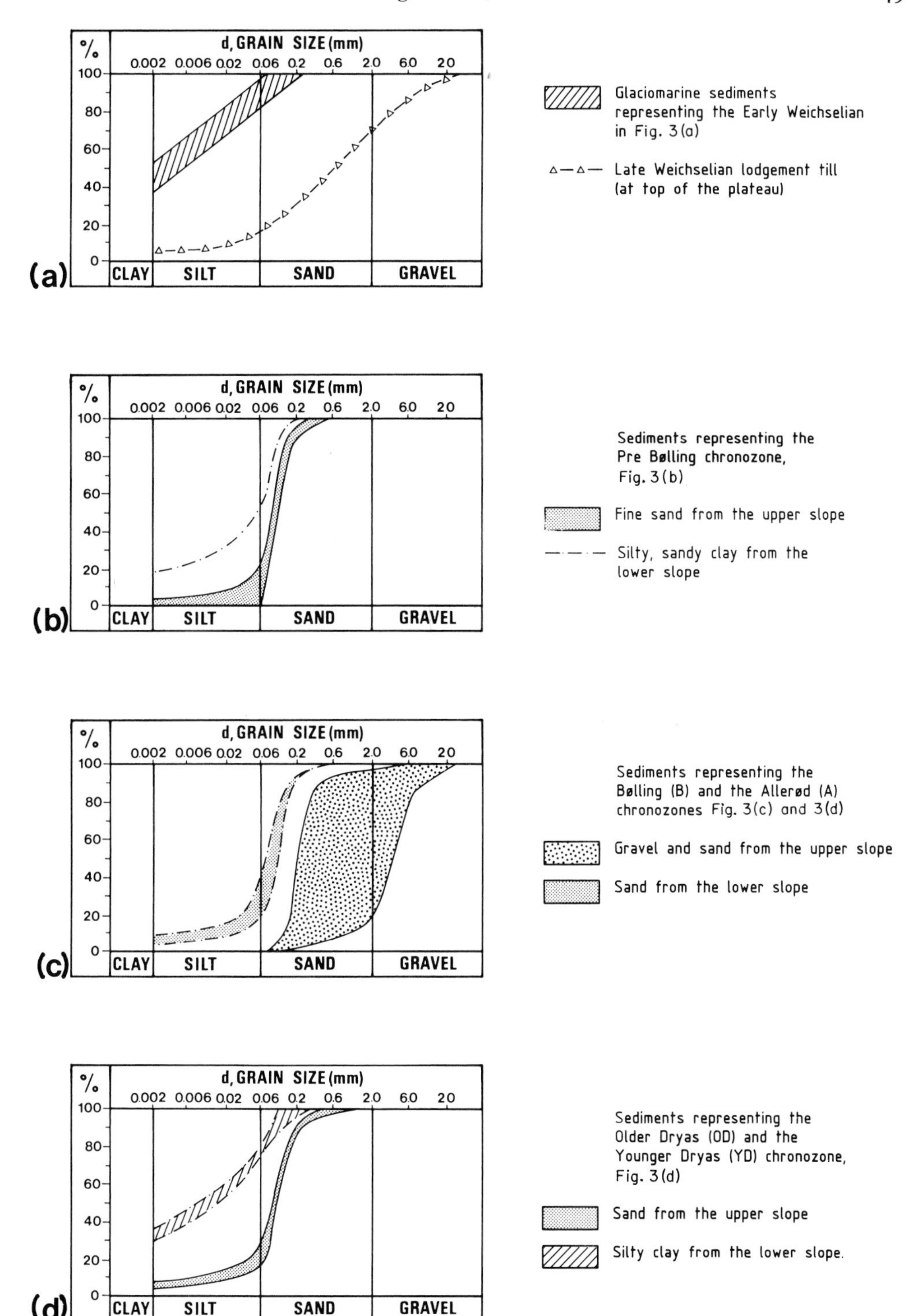

FIG. 4. Representative grain-size distribution curves from the Gullfaks area.

and 3, Fig. 3(a)) are probably the result of the Late Weichselian glacial maximum; the lower, 3, representing the main glacial advance (~20 000 years BP), the upper, 2, may represent a later less pronounced glacial advance (depositing the till on top). The ^{14}C-datings from this till clearly indicate an age younger than 19 000 years BP. The sediments between the two erosion surfaces (sediment unit W_2) were probably deposited very close to an ice front, when the water current interaction with the Norwegian Sea had almost ceased and the planktonic foraminifera disappeared.

After the last advance, glaciomarine sediments (W_1) were deposited, maybe sub-glacially, in the downslope part. These sediments were later eroded (see erosion surface 1 in the lower slope in Fig. 3(a)), but not overconsolidated.

The probable relative sea-level after that event is sketched in Fig. 3(a), as 180–190 m lower than today. The plateau area was above sea-level. The exact age is not known, but is believed to be younger than 16 000 years BP, and is clearly older than 13 000 years BP.

Due to the low sea-level, the waves eroded into the fine-grain overconsolidated glaciomarine sediments (W_2) at the slope. While the sea-level gradually rose, the eroded sediments were reworked and sorted into a sediment unit (PB), fining outwards from sand to silt and clay (PB, Fig. 3(b)).

Figure 3(b) shows the situation at the beginning of the interstadial Bølling chronozone, 13 000 years BP. The relative sea-level had risen to 140–150 m lower than today, and thus the lodgement till at the plateau became exposed to wave erosion and reworking. The plateau area was transgressed during the early part of the Bølling chronozone. The coastal unit deposited during this interstadial is thus characterized by sediments fining outwards from gravel and sand to sand and fine sand (soil unit B, Fig. 3(c)). The grain size distribution is shown in Fig. 4(c). Inclined layers indicate coastal outbuilding. Layers completely dominated by gravel are interpreted as storm layers. The main sediment source is probably the lodgement till, but large amounts of sand were also brought in by long shore currents heading northwards. At the lower slope, soil unit B is ~10 m thick, giving a possible sedimentation rate of about 1 cm/year if the duration of this period were 1000 years.

Figure 3(c) shows the situation at the beginning of the stadial Older Dryas chronozone, 12 000 years BP. The high energy environment which prevailed during the Bølling zone had now become a low energy one. Sedimentation almost ceased at the shoulder and uppermost slope. Downslope, the chronozone is represented by a 1 m thick silty and clayey layer (soil unit OD), indicating a sedimentation rate of 0.5 cm/year, if assuming a duration of 200 years for the Older Dryas zone. The grain size distribution at the lower slope is shown in Fig. 4(d).

The coastal outbuilding continued during the interstadial Allerød chronozone (see soil unit A, Fig. 3(d)). Again, a high energy environment prevailed, and both the grain size distribution and the sedimentation rate were as during the Bølling chronozone (Fig. 4(c)). The sediments were slightly finer, probably due to a higher relative sea-level.

With the stadial starting at about 11 000 years BP (the Younger Dryas chronozone) the sedimentation mode changed again, giving rise to fine-grained deposits as in the Older Dryas chronozone. The sea-level was higher than during the Older Dryas chronozone, so the soil unit is found over greater parts of the slope, fining outwards from sand to silty clay (soil unit YD, Fig. 3(d) and 4(d). The soil units OD and YD are both fine-grained, and probably had the same sedimentation rate. This is also reflected in the fact that YD is thicker than OD, because of the longer duration of the Young Dryas (1000 years).

Figure 3(e) shows the present situation. The Holocene sedimentation rate has been minimal in relation to the rates typical of the Late Weichselian. On the plateau and upper slope only winnowing has taken place. Further downslope the sediments (soil unit H, Fig. 3(e)) are fining outwards from sand to silty sandy clay. Most of the unit was deposited during the first part of the Holocene. At present, winnowing or non-deposition is dominant, leading to a top sand layer. The oceanographic interaction with the Norwegian Sea was re-established in the first part of the Holocene, leading to the reappearance of planktonic foraminifera.

Discussion

According to the findings of this investigation, the relative sea-level was at least 180–190 m lower than today sometime between 16 000 and 13 000 years BP. This coincides with the eustatic sea-level which was probably at its lowest 15 000 years BP (Mørner 1969). Mørner (1979) has also calculated the peripheral subsidence and central uplift resulting from the ice melting and the glacial retreat. The results give a rough figure indicating a subsidence of 50–100 m in this area in response to the Scandinavian isostatic uplift. The authors' investigation shows that if the absolute sea-level lowering at that time was 90 m, an isostatic subsidence of 100 m has taken place during Late

Weichselian and Holocene, which is in accordance with Mørner's data. No geophysical calculations based on the low sea-level stands in the northern North Sea have been made as far as is known. A geophysical calculation could possibly separate the isostatic subsidence due to the central ice melting from the probably Holocene subsidence which seems to have affected the northernmost part of the North Sea, and represents a quite different, long term movement (Clarke 1973).

Conclusions

(1) In the Gullfaks area the 2–5 m thick lodgement till on the plateau area represents a small late Weichselian glacial advance (18–19 000 years BP).

(2) After this advance, the plateau area was above sea-level and glaciomarine sediments were deposited on the downslope area.

(3) The sea-level rose gradually, and the waves eroded the upper part of the slope, resulting in fine-grained sediments on the lower slope.

(4) The sea transgressed the plateau area during the early part of the Bølling chronozone (~13 000 years BP). Now erosion of the lodgement till took place and coarse-grained coastal sediments were built out at the shoulder with deposition of sediments fining outwards down the slope area.

(5) During the alternating stadial and interstadial chronozones (Older Dryas, Allerød and Younger Dryas) a layered sequence of alternating fine-grained and somewhat coarser grained sediments was deposited in the slope area. The plateau remained as an area of erosion and reworking.

(6) The continuous rise in relative sea-level from 180–190 m lower than today at 16–13 000 years BP up to the present, can be explained by a combination of the eustatic sea-level rise with the isostatic subsidence of the Gullfaks area.

ACKNOWLEDGMENTS: The described development of the Late and Post Glacial sediments and the sea-level variations in the Gullfaks area, is extracted from very extensive geotechnical site investigations. The authors want to thank Statoil A/S for permission to publish the results.

References

ANDRESEN, A., BERRE, T., KLEVEN, A. & LUNNE, T. 1979. Procedures used to obtain soil parameters for foundation engineering in the North Sea. *Mar. Geotech.* **3,** 201–66. Also published in Norwegian Geotechnical Institute Publication, 129.

BADA, J.L. & MAN, E.H. 1980. Amino acid diagenesis in Deep Sea Drilling Project cores; kinetics and mechanisms of some reactions and their applications in geochronology and paleotemperature and heat flow determinations. *Earth Sci. Rev.* **16,** 21–55.

CLARKE, R.H. 1973. Cainozoic subsidence in the North Sea. *Earth Planet. Sci. Lett.* **18,** 329–32.

DEKKO, T. & ROKOENGEN, K. 1978. A submerged beach in the northern part of the North Sea. *Norsk Geol. Tidskr.* **58,** 233–6.

DUPLESSY, J.C. 1978. Isotope studies. *In:* Gribbin, J. (ed.), *Climatic Change.* Cambridge University Press, Cambridge, 46–67.

FEYLING-HANSSEN, R.W. 1958. Mikropaleontologiens teknikk. *Norges Geol. Unders.* **203,** 35–48.

—— 1980. Foraminiferal indication of Eemian interglacial in the northern North Sea. *Geol. Soc. Denm. Bull.* **29,** 175–89.

—— 1982. Foraminiferal zonation of a boring in Quaternary deposits of the northern North Sea. *Geol. Soc. Denm. Bull.* **31,** 29–47.

GREGORY, D. & HARLAND, R. 1978. The late Quaternary climatostratigraphy of IGS borehole SLN 75/133 and its application to the palaeoceanography of the north-central North Sea. *Scott. J. Geol.* **14,** 147–55.

HOVLAND, M. & DUKEFOSS, K.M. 1981. A submerged beach between Norway and Ekofisk in the North Sea. *Mar. Geol.* **43,** M19–M28.

JANSEN, J.H.F. 1976. Late Pleistocene and Holocene history of the northern North Sea, based on acoustic reflection records. *Neth. J. Sea Res.* **10,** 1–43.

LEE, A.J. 1980. North Sea: physical oceanography. *In:* BANNER, F.T., COLLINGS, M.B. & MASSIE, K.S. (eds), *The North-West European Shelf Seas: the Sea Bed and the Sea in Motion. II. Physical and Chemical Oceanography, and Physical Resources.* Elsevier, Amsterdam. 467–93.

MANGERUD, J. & BERGLUND, B.E. 1978. The subdivision of the Quaternary of Norden; a discussion. *Boreas* **7,** 179–81.

——, LIE, S.E., FURNES, H., KRISTIANSEN, I.L. & LØMO, L. 1984. A Younger Dryas ash bed in W. Norway, with possible correlations to the Norwegian Sea and the North Atlantic. *Quat. Res.* **21,** 85–104.

MELDEGAARD, S. & KNUDSEN, K.L. 1979. Metoder til indsamling og oparbejdning af prøver foraminiferanalyser. *Dansk Natur—Dansk Skole. Arssk.* **1979,** 48–57.

MØRNER, N.A. 1969. The late Quaternary history of the Kattegatt Sea and the Swedish West Coast. Deglaciation, shore-level displacement, chronology, isostasy and eustasy. *Sver. Geol. Unders. Avhandlingar och uppsatsar. Serie C. Arsbok* **63,** 487.

—— 1979. The Fennoscandian uplift and Late Cenozoic geodynamics; geological evidence. *Geo. J.* **3,** 287–318.

READ, A. & LØKEN, T. 1984. Quaternary stratigraphy at Statfjord B, northern North Sea, correlated with geophysical profiles. *Norw. Geotech. Inst. Rep.* **52002-5.**

RISE, L. & ROKOENGEN, K. 1984. Surficial sediments in the Norwegian sector of the North Sea between 60°31'N and 62°N. *Mar. Geol.* **58,** 287–317.

——, SKINNER, A.C. & LONG, D. 1984. Northern North Sea. Quaternary geology map between 60°30'N and 62°N, east of 1°E. M 1:500000. *Continental Shelf Institute, Trondheim.*

ROKOENGEN, K., LØFALDLI, M., RISE, L., LØKEN, T. & CARLSEN, R. 1982. Description and dating of a submerged beach in the northern North Sea. *Mar. Geol.* **50,** M21–M28.

SCHJETNE, K. & BRYLAWSKI, E. 1979. Offshore soil sampling in the North Sea. International Symposium on Soil Sampling. Singapore 1979. *Proceedings; State of the Art on Current Practice of Soil Sampling*, 139–156. Also published in Norwegian Geotechnical Institute Publication, 130.

SKINNER, A.C. & GREGORY, D.M. 1983. Quaternary stratigraphy in the northern North Sea. *Boreas* **12,** 145–52.

RANDI CARLSEN & TOR LØKEN, Norwegian Geotechnical Institute, P.B. 40 Tåsen, N-0801 Oslo 8, Norway.

ELEN ROALDSET, University of Oslo. Present address: Norsk Hydro A/S, Research Centre, Bergen, Norway.

Neogene deep and surface water palaeoceanography

North Atlantic sea-surface temperatures for the last 1.1 million years

W.F. Ruddiman, N.J. Shackleton & A. McIntyre

SUMMARY: A 1.2 Ma time series of north-east Atlantic sea-surface temperature (SST) has been assembled from the spliced record of piston core K708–7 (0.68–0 Ma) and hydraulic piston cores taken in nearby DSDP Hole 552A (1.2–0.68 Ma). A no-analogue fauna precludes making credible SST estimates in the interval 1.2–1.1 Ma, but the record above 1.1 Ma is suitable for SST estimates and related time-series analysis.

The amplitude of SST variation is considerably higher in the Brunhes than in the upper Matuyama. This is due both to colder glacial SST minima after 0.85 Ma, and to increasingly warm interglacial SST maxima from 0.7 to 0.4 Ma. The dominant periodicity in the SST signal is centred near 95 000 years; it increases in amplitude by a factor of four from the bottom of the record to the top, with the largest increase occurring between 0.7 and 0.4 Ma. This suggests a relatively gradual evolution of '100 000-year' power over half a million years of the late Quaternary, rather than a single abrupt change at 0.9 Ma.

Significant, but progressively smaller, spectral peaks occur at periods of 54 000, 41 000, 31 000, 23 000 and 19 000 years. Both the 54 000-year and 31 000-year signals may be a response to insolation forcing by minor obliquity terms. Long-term $\delta^{18}O$ records show small responses by the ice sheets at or near these two periods, indicating the probable initial response of the climate system to these insolation rhythms. In-phase relationships between $\delta^{18}O$ and SST at these periods suggest that the ice sheets then impose these rhythms on the ocean with no lag. The mechanism by which such small insolation and ice-volume signals become enhanced in the SST record is not clear.

Rationale for the study

Several high-quality records of $\delta^{18}O$ variations spanning over a million years have now been published (Shackleton & Opdyke 1973, 1976; Prell 1983; Shackleton *et al.* 1984). Interpreted as indicators of global ice volume, these records are vital components of the total picture of global climate. Few long-term records are available for other components of the climatic system such as estimated sea-surface temperature (SST).

One of the regions where SST changes most is the subpolar North Atlantic. During the late Brunhes, the warm North Atlantic Drift flow typical of brief interglaciations was periodically replaced by a cold, ice-filled subpolar gyre that extended south to a boundary marked by a strong thermal gradient at 40° to 50° N. (McIntyre *et al.* 1972; Ruddiman & McIntyre 1976). Recent work has shown that this cold gyre developed and disappeared with dominant rhythms of 100 000 and 41 000 years through the late Brunhes, and that the oceanic temperatures north of 50°N. were in phase with, or during brief intervals lagged slightly behind, the $\delta^{18}O$ signals at the same periodicities (Ruddiman & McIntyre 1984). This close link between the ice-volume and surface-ocean responses in the high-latitude North Atlantic during the last 250 000 years raises an obvious question: were the two signals similarly linked during earlier intervals when ice-volume variations were different?

In 1981, Leg 81 of DSDP/IPOD retrieved a record of the last several million years at Site 552, Hole 552A in the North Atlantic (Roberts *et al.* in press). By combining counts of part of this record with previously published data from piston core K708–7 (Ruddiman & McIntyre 1976), the authors have obtained an apparently continuous record of North Atlantic surface-ocean changes back to 1.2 Ma.

Cores examined

Hole 552A is located 130 miles north-north-east of core K708–7 (Fig. 1) and beneath surface waters that today are 1.0°C cooler in summer (14.3°C vs. 13.3°C) and 0.2°C cooler in winter (8.9°C vs. 8.7°C). In addition, Hole 552A lies some 1200 metres shallower in the water column.

Previously published data from K708–7 include: planktonic foraminiferal counts and $CaCO_3$ analyses throughout the entire 0.69 Ma record spanned by the core (Ruddiman & McIntyre 1976), $\delta^{18}O$ analyses of the planktonic foraminifer *Neogloboquadrina pachyderma* (sinistral) through most of the upper 0.50 Ma of the record (Thierstein *et al.* 1977); and estimates of SST over

From SUMMERHAYES, C.P. & SHACKLETON, N.J. (eds), 1986, *North Atlantic Palaeoceanography*, Geological Society Special Publication No. 21, pp. 155–173.

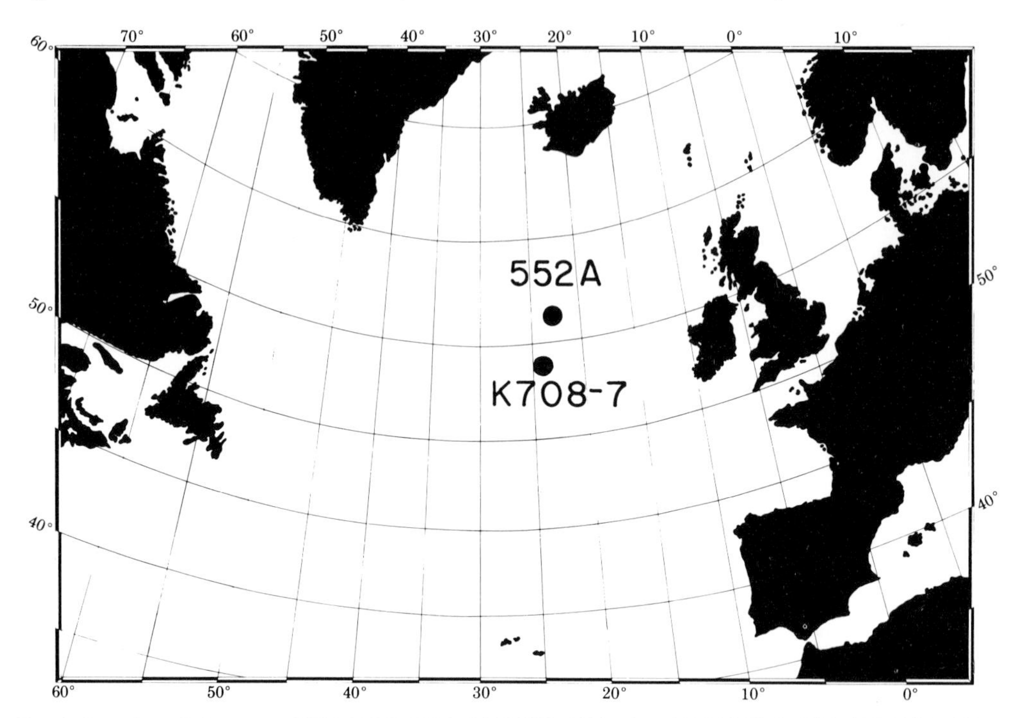

FIG. 1. Location of piston core K708–7 (53° 56′N., 24° 05′W., 3502 m) and DSDP Hole 552A (56° 03′N., 23° 14′W, 2311 m).

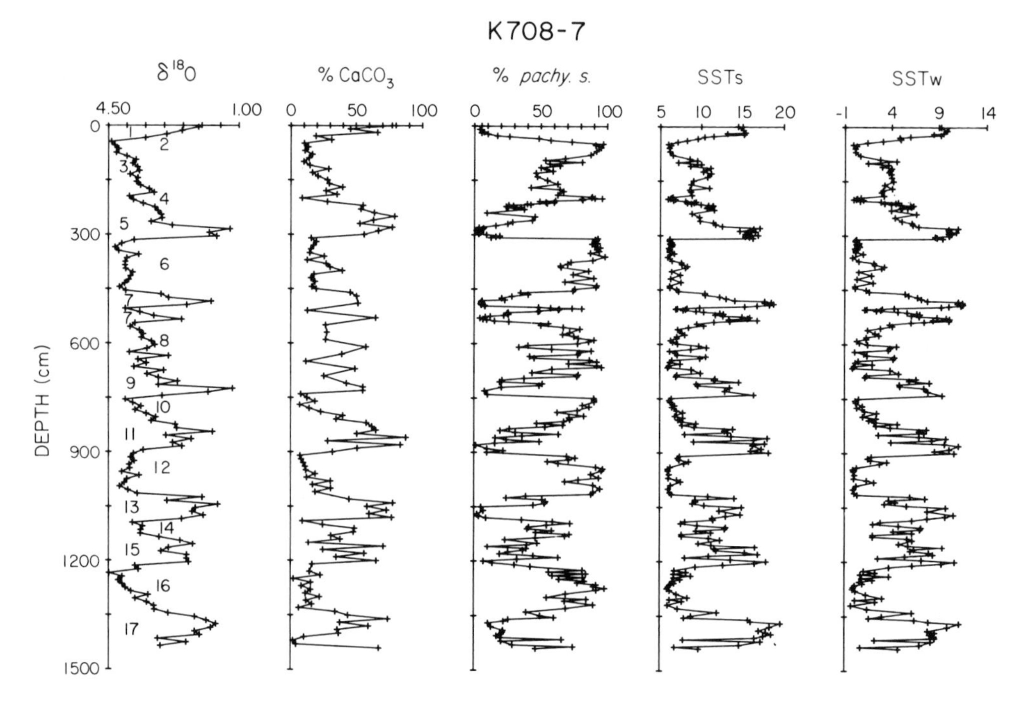

FIG. 2. Climatic signals in core K708–7 plotted against depth in core in cm: oxygen isotopic curve from planktonic foraminifer *N. pachyderma* (s.); % $CaCO_3$, % *N. pachyderma* (s.), and estimated summer and winter sea-surface temperature.

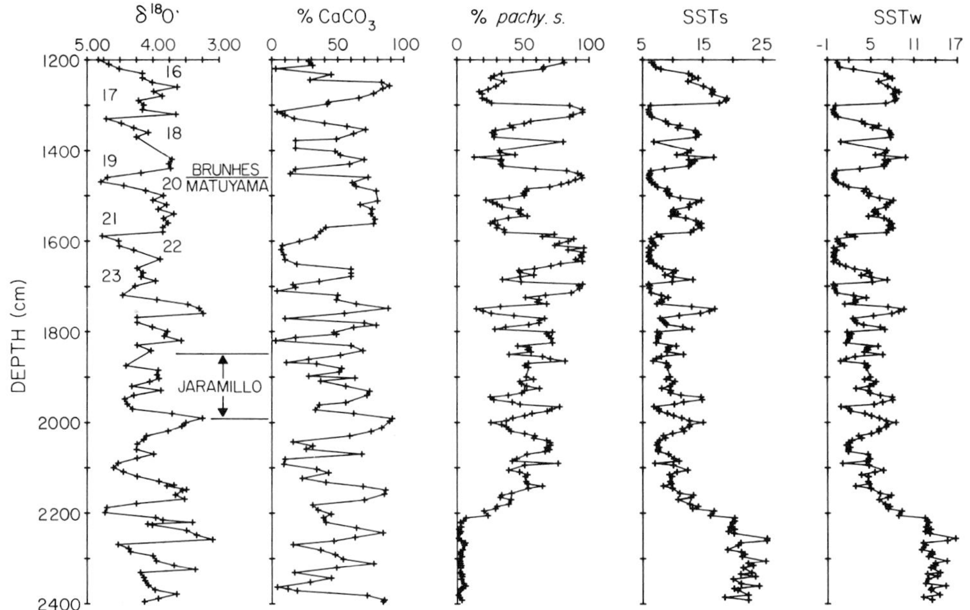

FIG. 3. Climatic signals from DSDP Hole 552A plotted against sub-bottom depth in core in cm: oxygen isotopic curve based on mixed benthic foraminifera, % $CaCO_3$, % *N. pachyderma* (s.), and estimated summer and winter sea-surface temperature.

the last 0.25 Ma (Ruddiman & McIntyre 1984). The data from this core reported here are: percentage abundances of *N. pachyderma* (s.), estimates of winter SST, and $\delta^{18}O$ analyses on *N. pachyderma* (s.), all reported for the entire length of the core (Fig. 2; Appendix 1).

Previously published or in press data from Hole 552A include: an oxygen isotopic curve based on analyses of the benthic foraminifera *Planulina wuellerstorfi*, *Uvigerina peregrina*, and *Globocassidulina subglobosa*; analyses of percentage $CaCO_3$; and palaeomagnetic stratigraphy (Schackleton *et al.* 1984). New data reported here (Appendix 2) are: percentage abundances of *N. pachyderma* (s.) and estimates of winter SST, both shown in Fig. 3 along with the published oxygen isotopic and $CaCO_3$ data.

The data reported from Hole 552A span an interval from the lower Brunhes to the upper–middle part of the Matuyama (about 0.65–1.2 Ma). The upper limit of 0.65 Ma was selected to overlap the lower part of core K708–7. The lower limit was set by the disturbed condition of core 6 at Hole 552A, leaving a gap in the record from about 1.2–1.4 Ma. Counts in Hole 552A were made based on a new transfer function (F13X5) in which four species of foraminifera are counted and all other species are grouped as a fifth category (Ruddiman & Esmay, in press). For this region of the North Atlantic, equation F13X5 gives very similar SST estimates during the upper Quaternary to equations based on full populations (e.g. Kipp 1976).

Time-scale

There are two steps necessary to convert the planktonic foraminiferal data and SST estimates in Figs 2 and 3 into a form useful for detailed analysis: (1) choice of a level at which to splice the two records; and (2) selection of a timescale. Both steps rely primarily on the oxygen isotopic record.

The SST record of K708–7 extends down to the base of oxygen isotopic stage 17 (Fig. 2). The choice of isotopic stages in the $\delta^{18}O$ record of *N. pachyderma* (s.) in core K708–7 is confirmed by the occurrence of the *Pseudoemiliania lacunosa* datum in stage 12 and the *Emiliania huxleyii* datum in stage 8 (Thierstein *et al.* 1977). The 3.5‰ amplitude of the oxygen isotopic curve based on *N. pachyderma* in K708–7 during the Brunhes is 50% larger than that of the benthic foraminiferal record in 552A during the same interval (Fig. 3), apparently because of excursions of the *N. pachyderma* curve to values as light as

1.5‰ during maximum interglaciations. During Brunhes glaciations, the *N. pachyderma* isotopic values in K708–7 are close to the benthic values in 552A.

For the central purpose—splicing the two records—the basic oxygen isotopic stages stand out clearly in both records (Figs 2 and 3). An interval of overlapping faunal counts within stage 17 also permits a direct evaluation of the consequences of splicing SST records from two slightly different geographic locations. The SST values in the middle of interglacial stage 17 from Hole 552A are offset an average of about 1°C colder for summer but agree very closely with K708–7 for winter, giving the same sense of offset as the modern atlas values and validating the splicing of the two records for the winter season.

One problem area exists. In the earliest part of oxygen isotopic stage 17, the SST values at K708–7 show full interglacial warmth, whereas in 552A the earliest part of isotopic stage 17 remains at cold SST values. The $CaCO_3$ values are very low in the early parts of stage 17 in both records, suggesting a colder climate at this time. For this reason, the authors chose for the composite record the SST trends shown by Hole 552A, with cold SST values lingering into the early part of stage 17. In any case, tests showed that all the results that follow were stable for either choice of spliced level. Thus, the two records were spliced in the lower part of stage 17 at 0.685 Ma. Figures 5, 6 and 7 treat this composite record as a single time series 1.2 Ma in length.

The second step is to assign ages to the ^{18}O stages in the composite record. Diagnostic isotopic minima, maxima, and transitions are used as stratigraphic pinning points for the SPECMAP time-scale back to 0.814 Ma (Imbrie *et al.* 1984). All diagnostic features that could be identified are listed in Appendix 3, along with their ages.

Below the 0.814 Ma level reached by the SPECMAP time scale, the best chronological control is that from palaeomagnetic stratigraphy. The authors followed Shackleton *et al.* (1984) and directly interpolated all ages between the oldest isotopic control levels in the SPECMAP time scale, the upper and lower boundaries of the Jaramillo chron (0.91, 0.98 Ma) and the upper boundary of the Olduvai chron (1.66 Ma).

SST record

The time-stretched and spliced SST record is plotted in Fig. 4. Several intervals of differing surface-water response appear in this time series.

Prior to 1.1 Ma and down to the bottom of the record at 1.2 Ma, there was a striking absence of both the cold SST values and the high *N. pachyderma* (s.) percentages typical in this area later in the Pleistocene. Instead, the record through this 100 000-year interval was one of uniformly warm estimated SST. A 'no-analogue' fauna occurred at this time, with *G. inflata* reaching percentages of 40–50% compared to a maximum of just under 40% in modern Atlantic core tops (Kipp 1976). With a no-analogue assemblage present, the palaeoenvironmental estimates in this interval must be considered suspect.

Other species percentages during the interval 1.2–1.1 Ma constrain the possible SST values only broadly. The low values of *G. ruber* and other lower-latitude species indicate that warm subtropical conditions did not develop during this interval near Hole 552A. Selected full counts of the assemblages indicate that much of the 'other' fauna consisted of *N. pachyderma* (dextral), a species which reaches maximum percentages in core tops beneath temperate waters like those overlying this area today. This implies that from 1.2–1.1 Ma the area was occupied by waters not markedly dissimilar from those today.

Between 1.1 and 0.9 Ma, substantial oscillations of SST and percentage *N. pachyderma* (s.) occurred, but the warmest SST values never returned to the extreme values estimated for the preceding interval, nor did the coldest SST values reach the extreme values typical later in the record. The SST oscillations averaged about 6°C in winter and 7°C in summer.

Between 0.9 and 0.7 Ma, the SST oscillations increased due mainly to colder glacial values but also to warmer interglaciations. In fact, the SST amplitudes during both this interval and the rest of the Brunhes may have been larger than shown in Fig. 4, because the glacial SST minima repeatedly bottomed against the effective cold limit of the transfer function (1°C in winter, 6°C in summer). This limit occurs at levels where *N. pachyderma* (s.) percentages reach values of 90% or larger.

The tendency toward increasingly warm interglacial SST values continued from 0.7–0.4 Ma, and most glacial SST minima reached the cold limit of the equation. The two seasons show somewhat different responses. Summer SST maxima became abruptly warmer in stage 17 at 0.68 Ma and consistently reached very warm values after 0.41 Ma. In the winter estimates, there was a more progressive increase in interglacial SST maxima that began around 0.8 Ma but continued up through the middle of the Brunhes at 0.41 Ma (stage 11). The change toward warmer intergla-

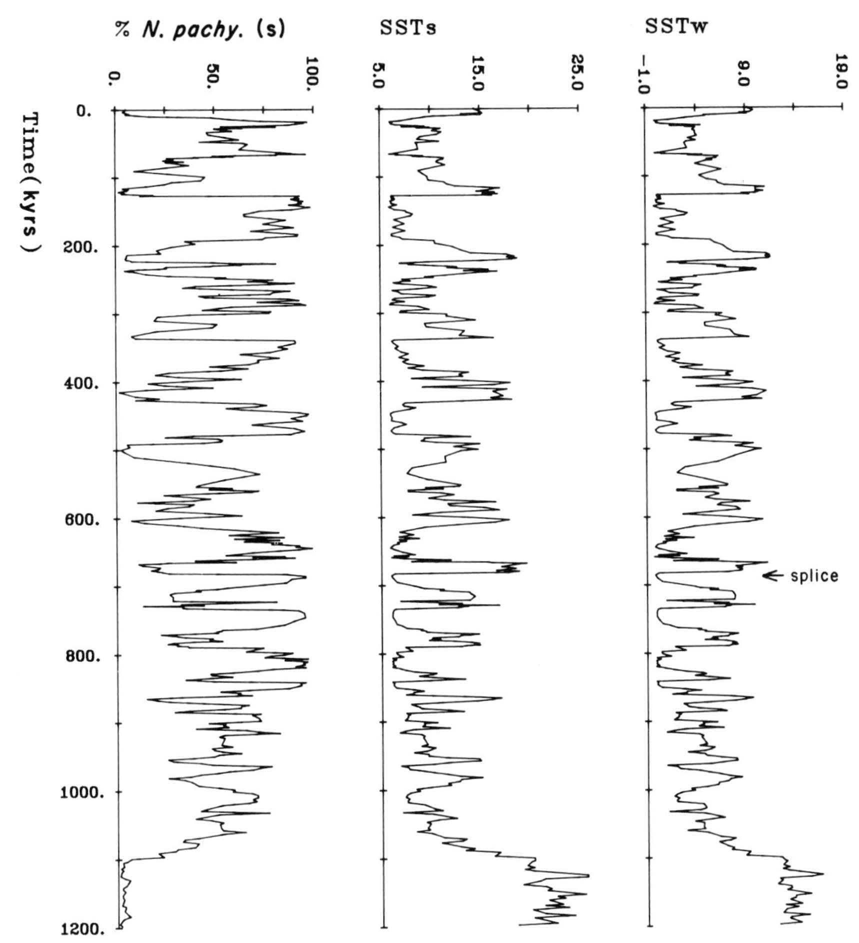

FIG. 4. Time series of % *N. pachyderma* (s.) and estimated sea-surface temperature for the last 1.2 Ma spliced from combined records of cores K708–7 (Fig. 2) and 552A (Fig. 3).

cial winter SST at 0.685 Ma cannot be attributed to splicing the two records, because both records show the same shift to warmer SST in the overlapped section of stage 17.

The upper half of the Brunhes was generally characterized by large-scale SST fluctuations, with a range of about 11°C in summer and 8°C in winter. Again, the actual range was probably larger because of the artificial cold limits to SST estimates.

Signal processing of the SST record

Spectral analysis: the full 1.1 Ma record

The authors first ran a spectral analysis of the 1.1 Ma segment of spliced winter SST record lying above the no-analogue interval. Standard Blackman-Tukey techniques were used (Hays *et al.* 1976) on a time series resampled by interpolation from the orginal data at equal increments of 2500 years (equivalent to the original mean sampling density). The winter season was chosen because the two locations (Hole 552A and core K708–7) have nearly identical winter SST values today and show close agreement in SST estimates across the overlapped interval in stage 17.

The spectra (Fig. 5) reveal concentrations of power at four orbital periodicities (95 000, 40 000, 23 000 and 19 000 years), as well as clearly marked peaks at 54 000 and 31 000 years. The concentration of power is largest under the 95 000-year peak and progressively lower for the five higher-frequency peaks. All but the 19 000-

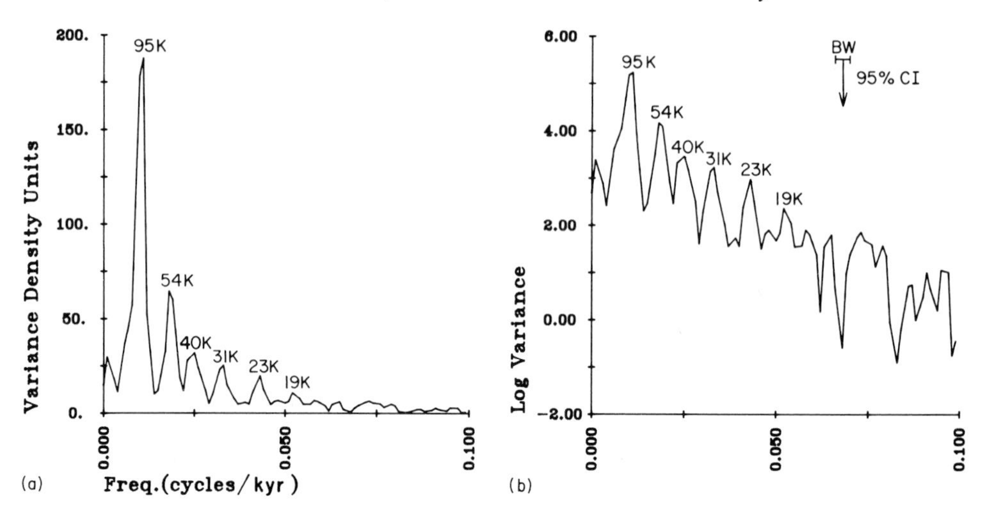

FIG. 5. Spectral analysis of winter SST record (Fig. 4) from 1.1–0 Ma. (a) Area/variance plot. (b) Log/variance plot showing bandwidth and 95% confidence interval.

year peak stand out clearly above the baseline of spectral power and are significant at the 95% confidence interval (Fig. 5). The four orbital periodicities have been previously recognized in North Atlantic SST spectra, and all have been attributed to forcing by fluctuations in northern hemisphere ice sheets at the same periodicities (Ruddiman & McIntyre 1984).

Because of the long record length and short sampling interval, the central frequency of these peaks can be constrained better than in previous studies. The '100 000-year' period is actually centred at 95 000 years, which is one of the two important eccentricity periods in this part of the spectrum, along with that at 123 000 years (Berger 1977). The 41 000-year period is centred around 40 300 years, probably reflecting a combination of the dominant 41 000-year first term in the obliquity signal, with a somewhat weaker second term at 38 900 years. The rhythms at 54 000, 31 000, 23 000 and 19 000 years all have central frequencies within a few hundred years of those rounded-off values.

The periods at 54 000 and 31 000 have not been previously noted in climatic time series, although Imbrie *et al.* (1984) found a 59 000-year response in the global δ^{18}O signal constructed by SPECMAP.

Filtered SST records

To explore the time evolution of these signals, individual time series of each of the five prominent SST rhythms were filtered. Central frequencies for each filter were chosen from the spectral peaks in Fig. 5 and the widest filter bandwidths that could be used without contaminating each filtered signal with substantial power from adjacent peaks were empirically selected. The filtered results from the winter SST record are shown in Fig. 6, alongside the original SST curve. All the signals are plotted at the same SST scale in order to show how important each filtered signal is in the original record.

The 95 000-year signal increases in amplitude by a factor of four from the bottom of the filtered record at 0.97 m.y. to the top at 0.13 Ma (Fig. 6). The largest increase occurs between 0.7 and 0.4 Ma.

The filtered 54 000-year SST signal varies widely in amplitude, ranging over a factor of five from an early maximum at 0.72 Ma to a late minimum at 0.45 Ma. Because the maximum amplitude for this signal is concentrated at 0.8–0.65 Ma, it might be thought to be caused by an error in the splice of the two records at 0.685 Ma. However, it will be shown later that this is not the case.

The three higher-frequency rhythms also vary in amplitude through time, but the credibility of this apparent modulation becomes increasingly dependent on the accuracy of the time-scale and on the record having no small gaps. Viewed with these uncertainties in mind, the 41 000-year SST signal is, on the average, somewhat stronger in the bottom half of the record, and it shows moderate variations in amplitude, with maxima at >0.94 Ma, 0.67 Ma, 0.44 Ma, and <0.15 Ma. The 31 000-year SST signal also varies somewhat in amplitude, with maxima at <0.13 Ma, 0.3 Ma, 0.64 Ma, and 0.87 Ma. The 23 000-year cycle is very strongly modulated at a wavelength of about

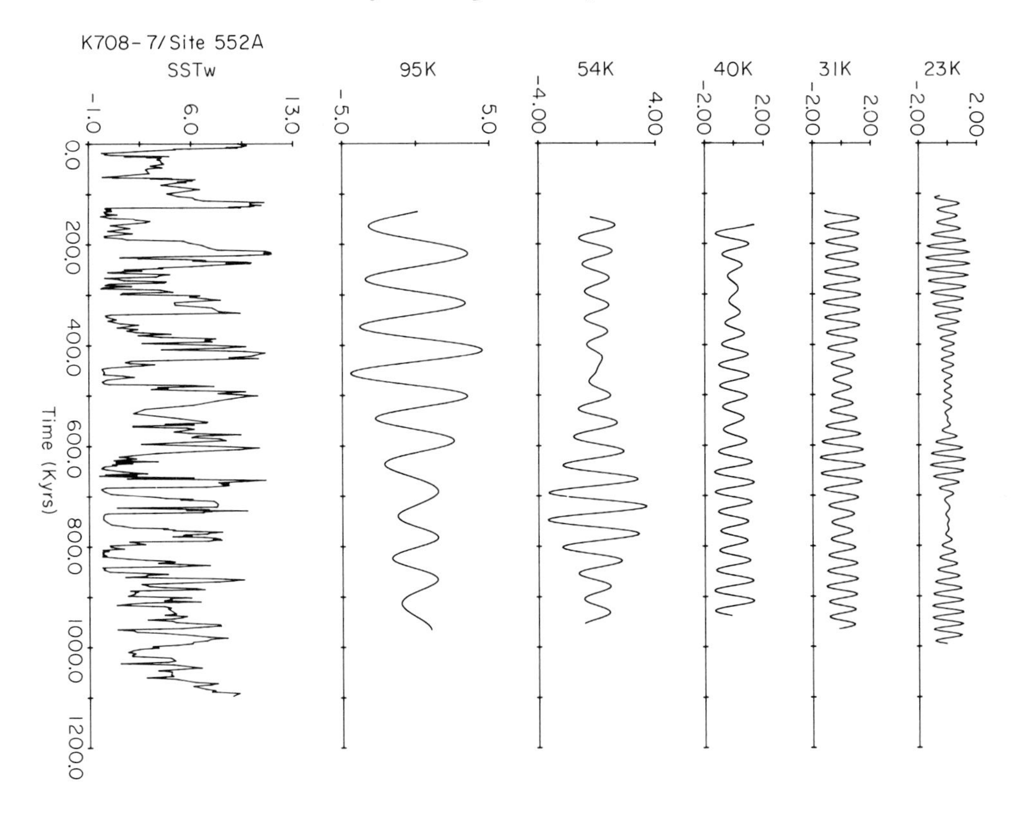

FIG. 6. Filtered records of signals at five periodicities prominent in spectral analysis (Fig. 5) of 1.1–0 Ma record of winter SST shown at the left. Central periods are: 95 000 years, 54 000 years, 40 000 years, 31 000 years and 23 000 years. Filters are Gaussian and are tapered such that zero-point crossings occur at adjoining spectral peaks.

410 000 years, with maxima at 0.22, 0.63, and greater than 0.96 Ma. These closely match maxima in the modulation of the precessional component of insolation by the dominant eccentricity term at 413 000 years (Berger 1977) and similar modulation of the 23 000-year $\delta^{18}O$/ice-volume signal over the last 0.78 Ma (Imbrie *et al.* 1984).

Spectral analysis: record segments

For an alternative view of the evolution of the SST response of the North Atlantic during the last 1.1 Ma, spectral analyses were run of four shorter intervals: 1.1–0.6 Ma, 0.9–4.0 Ma, 0.7–0.2 Ma, and 0.5–0 Ma. Equal time lengths were chosen to give comparable definition of each frequency in each interval, and 0.5 Ma record lengths were picked to give five full repetitions of the longest period, 95 000 years. The results are shown in Fig. 7.

These spectra reflect the evolution of spectral power suggested in the filtered records of Fig. 6. The 95 000-year signal dominated strongly from 0.5–0 and from 0.7–0.2 Ma, was weaker by almost a factor of two but still dominant from 0.9–0.4 Ma, and was weaker by a factor of almost four and clearly sub-dominant from 1.1–0.6 Ma (Fig. 7). In contrast, the 54 000-year rhythm was dominant from 1.1–0.6 Ma, was subdominant but still strongly developed from 0.9–0.4 Ma, and was much weaker from 0.7–0.2 and 0.5–0 Ma.

The other four periods are relatively weak in all four intervals. All of these higher-frequency peaks are more apparent in the spectrum from the full 1.1 Ma record than in the shorter segments.

Discussion

Evolution of 100 000 Year-Power over the Last Million Years

The most striking result of this study is the increase in amplitude of the SST signal between

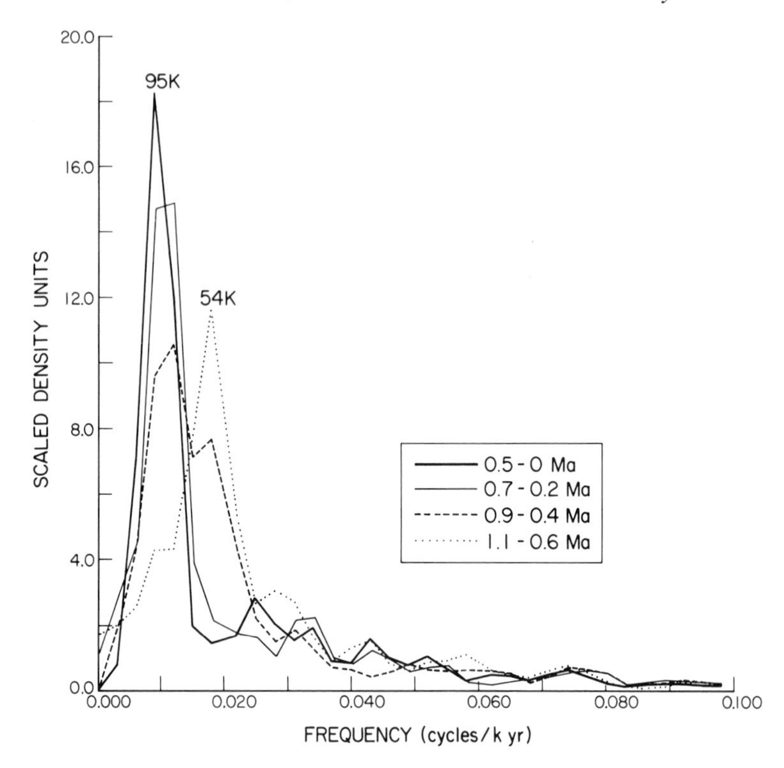

FIG. 7. Area/variance composite of spectral analysis of winter SST record from four record segments.

0.9 and 0.4 Ma, with colder glacial SST minima after 0.85 Ma and increasingly warm interglacial SST maxima from 0.9 Ma until 0.4 Ma. This change is directly associated with the gradual fourfold increase in power at the dominant 95 000-year period from 0.9–0.4 Ma (Figs 6 and 7).

Because high-latitude North Atlantic SST during the late Quaternary responds directly to ice volume (Ruddiman & McIntyre 1984), this longer SST record has been compared to long-term $\delta^{18}O$ results. The shift toward colder SST extremes around 0.85 Ma has a direct counterpart in the oxygen isotopic record. Several studies have found an increase in amplitude between the Brunhes and Jaramillo Chrons, during isotopic stages 23 and 22 at roughly 0.9–0.8 Ma (Shackleton & Opdyke 1973, 1976). This change is also evident in high-quality HPC cores (Prell 1983; Shackleton *et al.* 1984), in DSDP rotary drill cores (Keigwin 1979; Shackleton & Cita 1979; Thunnell & Williams 1983), and even in piston cores with slow deposition rates (van Donk 1976). Shackleton & Opdyke (1976) interpreted this change as marking the first strong northern hemisphere glaciation. Pisias & Moore (1981) concluded that spectral power increased in amplitude after 0.9 Ma for all periods longer than 60 000 years, and particularly for 100 000 years. Prell (1983) suggested that an abrupt change in 'mode of variability' occurred in 50 000 years or less at 0.9 Ma.

Regional surface-ocean responses interpreted as showing a significant shift to cooler planktonic faunal assemblages or increased sea-ice cover also occurred between the Jaramillo and Brunhes chrons in many areas: the Arctic (Hermann & Hopkins 1980); the Mediterranean (Cita *et al.* 1977; Thunell 1979); the south-central North Atlantic (Berggren 1968; but see Briskin & Berggren 1975); the equatorial Atlantic (Ruddiman 1971); and the sub-Antarctic (Hays 1967; Bandy *et al.* 1971; but see Keany & Kennett 1972). No dramatic change is apparent in central European loess sequences at this time, although the thickest loess sequences post-date the Jaramillo (Kukla 1977).

The other component of the increased amplitude in the North Atlantic SST record is the trend toward progressively warmer interglaciations from 0.9–0.4 Ma. (Figs 4 and 6). This also has an oxygen isotopic counterpart in the tendency

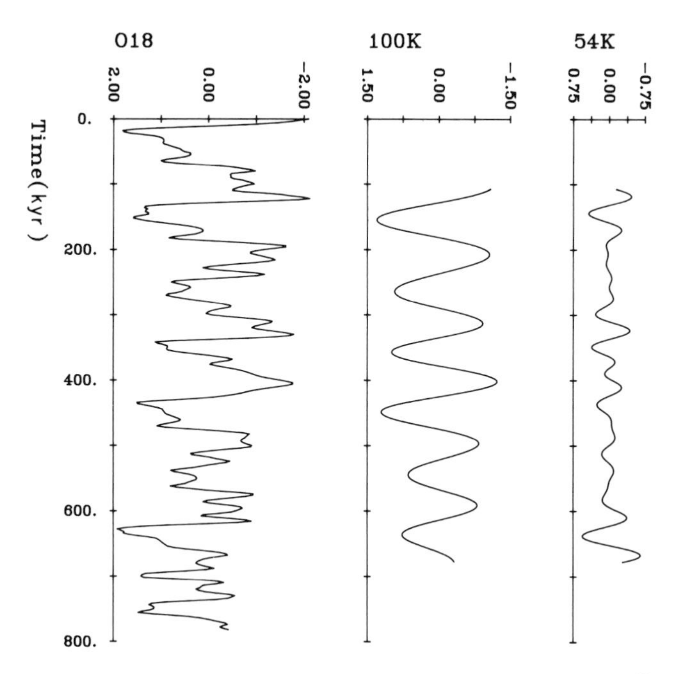

FIG. 8. Filtered signals of the 95 000-year and 54 000-year periods from the SPECMAP $\delta^{18}O$ composite record shown at the left. Filters are identical to those used in Fig. 6.

toward lighter values during interglacial stages during the lower and middle Brunhes. This trend appears in $\delta^{18}O$ records from the Caribbean (Emiliani 1966; Prell 1983) and emerges in the SPECMAP isotopic 'stack' shown in Fig. 8 (Imbrie *et al.* 1984). Prell (1983) interpreted this trend as indicating that Brunhes interglacials were more ice-free than those in the Matuyama. Several of the regional responses noted above also indicate warmer interglaciations during the Brunhes. In addition, warmer faunal assemblages during the last five interglaciations have been noted in the equatorial Atlantic, along with colder glacial assemblages (Ruddiman 1971).

In summary, the SST results indicate a progressive increase in climatic variability beginning after 0.9 Ma and continuing through the succeeding 500 000 years into the middle of the Brunhes. This variability is directly associated with a fourfold increase in the strength of the 95 000-year SST cycle. This view differs from the concept that an increase in mode of climatic variability and/or 100 000-year power occurred very abruptly at 0.9 Ma (Pisias & Moore 1981; Prell 1983). The SST evidence is interpreted to indicate a more gradual evolution of spectral power, particularly the slower increase in power at periods near 100 000 years (Figs 6 and 7).

This apparent disagreement is probably methodological. The isotopic studies used relatively long and non-overlapping segments of record for spectral or statistical analysis, thus precluding fine-scale observations on the evolution of spectral power. This study utilizes both filtered signals and spectral analysis of short overlapped record segments to arrive at a more detailed picture of signal evolution. If examined with similar techniques, isotopic evidence supports the idea of a gradual evolution of power at periods near 100 000 years. In a recent study, Imbrie (in press) divided the SPECMAP $\delta^{18}O$ record into segments and concluded that 100 000-year power was twice as strong from 0–0.4 Ma as from 0.4–0.78 Ma. This matches closely the near-doubling of 95 000-year SST power from the 0.9–0.4 to the 0.5–0 Ma interval (Fig. 7). Furthermore, if the SPECMAP $\delta^{18}O$ 'stack' is processed through the 95 000-year SST filter, it too shows the kind of gradual evolution of 95 000-year power as the SST record (Fig. 8).

It is concluded that the colder glacial and warmer interglacial extremes in the upper

Brunhes part of the SST record are specifically linked to a gradual increase in the 95 000-year SST signal, and that this increase appears to be correlated to a similarly gradual evolution of the 100 000-year $\delta^{18}O$ (ice-volume) signal. Again it is cautioned that this conclusion is partly dependent on the accuracy of the SPECMAP time-scale from 0–0.78 Ma, and especially from 0.6–0.78 Ma.

It is not unreasonable on theoretical grounds that larger Brunhes ice-volume maxima (and associated colder glacial SST) could alternate with smaller ice-volume minima (and warmer interglacial SST) through a causal link involving the 100 000-year signal. Although the origin of 100 000-year power is being debated, it is probably linked to non-linear feedback effects lying within the climate system and probably associated with the ice sheets (Hays *et al.* 1976; Imbrie & Imbrie 1980). The basic ice-volume response recorded by the $\delta^{18}O$ record indicates that the larger the size reached by ice sheets at their periodic late Quaternary maxima, the stronger must have been the subsequent operation of the internal feedbacks in rapidly destroying that ice during the next deglaciation. If sufficiently powerful, these feedbacks might have accelerated the destruction of ice during the late Quaternary to the point that deglaciations were more complete than during the earlier Quaternary, leaving northern hemisphere land masses more ice-free. Ice-free land surfaces might then respond directly to the waning phases of the regime of high summer insolation that initiated the destruction of the ice. Summer insolation heating of the continents could then, through air-sea interaction, warm the ocean.

Origin of power at 54 000 and 31 000 years

The 54 000-year and 31 000-year SST rhythms could originate in several ways. In the following discussion, firstly, the insolation spectrum is searched for forcing at these two periods. It is then considered whether insolation forces the ocean directly at these periods or is mediated by the ice sheets.

Both rhythms can be ruled out as simple harmonics of primary SST rhythms. The 54 000-year signal is clearly not a first harmonic of an SST signal near 100 000 years, because the results constrain the latter to a period of 95 000 years. The 31 000-year signal is also obviously not a first harmonic.

One of three viable possibilities is that these are responses to minor terms in the spectrum of primary insolation. Berger (1977) predicted that both periodicities would appear in geologic records, because both are small but significant terms in time series of obliquity. The moderately strong third term of the obliquity series has a period of 53 616 years, while the considerably weaker fifth, seventh and tenth terms have periods between 28 910 and 30 365 years (Berger 1977). This does not explain why these minor rhythms in the insolation spectrum would stand out so prominently in the SST signal.

A second possibility is that the 54 000-year SST signal is a response to orbital eccentricity, which modulates insolation forcing at the precessional periods, but does not appear as a primary term. There is a very weak term in the series expansion for eccentricity with a period near 56 000 years (Imbrie *et al.* 1984). This explanation has the dual problem of explaining the prominence of a rhythm that not only is weak, but also does not appear in the primary insolation spectrum.

A third possibility is that the 54 000- and 31 000-year rhythms are combination-tone responses to orbital forcing. The summation tone of a 100 000-year and 41 000-year signal has a period of 29 000 years; the difference tone between 100 000-year and 23 000-year signal has a period of 30 000 years; and the difference tone between a 41 000-year and 23 000-year signal has a period of 52 400 years. This explanation is the least likely of the three. Why, out of the large array of possible combination tones, would only these particular tones not just be present in the SST record but strongly enhanced? In addition, because there is effectively no primary power in received insolation at the eccentricity periods, neither combination tone near 31 000 years could be a direct response to insolation forcing.

Whichever of the above explanations of the SST responses is correct, there must be a very marked enhancement of minor insolation forcing at the 54 000- and 31 000-year rhythms. This enhancement can only come from internal components of the climate system. Because high-latitude North Atlantic SST changes are thought to respond primarily to changes in ice volume, as noted earlier, the most likely source of this enhancement is the ice sheets. There are weakly developed spectral peaks near 59 000 and 30 000 years in the stacked $\delta^{18}O$ time series developed by SPECMAP (Imbrie *et al.* 1984). This suggests that the ice sheets have varied at rhythms near the SST responses observed.

For the 54 000-year rhythm, the SPECMAP $\delta^{18}O$ record shows strongest 54 000-year power below 0.65 Ma, in the same region where the SST signal is most prominent (Fig. 8). This demonstrates that the 54 000-year SST signal is not an artifact of splicing the two records; it exists because the SST signal closely tracks an oxygen isotopic signal which, in the SPECMAP time-scale, positions stages 17, 19, and 21 roughly 48 000–55 000 years apart (Imbrie *et al.* 1984). Either the time-scale is in error, or there is a brief interval during which the Northern Hemisphere system responds strongly at a 54 000-year period. In any case, the origin of the 54 000-year SST rhythm appears to be linked to the ice sheets.

Phase analysis of SST against $\delta^{18}O$ for the planktonic foraminferal $\delta^{18}O$ record in K708–7 and for the benthic foraminiferal $\delta^{18}O$ record in Hole 552A shows that SST is in phase with $\delta^{18}O$ at both the 54 000-year and 31 000-year periodicities, at least to within the 25 000-year sampling intervals. This implies transferrence of these signals from the ice sheets to the ocean with negligible lag, just as occurs for the 100 000- and 41 000-year signals in the late Quaternary Atlantic (Ruddiman & McIntyre 1984).

Thus there is a plausible pathway by which the insolation forcing impacts the high-latitude North Atlantic surface ocean. But information necessary to resolve the largest problem is lacking: what mechanism takes minor rhythms in the insolation and $\delta^{18}O$/ice-volume spectra and amplifies them so strongly in the North Atlantic SST changes?

Climate during the no-analogue interval at 1.2–1.1 Ma

The warm North Atlantic record from 1.2 Ma (or earlier) until 1.1 Ma appears somewhat anomalous relative to other evidence. Within this interval at Hole 552A, there are three brief pulses of heavier benthic foraminiferal oxygen isotopic values, suggesting more ice on land. The $CaCO_3$ record also shows synchronous and pronounced minima (2200–2400 cm depth in Fig. 3), implying increased dilution by ice rafting, dissolution on the sea floor, or suppression of productivity in surface waters. Non-carbonate coarse-fraction values in Zimmerman *et al.* (1985) suggest that ice rafting is at least part of the cause of reduced carbonate percentages within this interval. At equivalent levels, the SST curves show barely perceptible trends toward slightly cooler values (Fig. 4).

On the other hand, some evidence supports the possibility of a somewhat warmer Atlantic at this time. Kukla (1977) noted that Czechoslovakian soil records indicate a long interval without strong glacial loesses during the upper Matuyama, from about 1.2–0.95 Ma. This interval appears to correlate with the relatively warm Waalian interval recognized from pollen evidence in the Netherlands and Belgium (van der Hammen *et al.* 1971). Although climatic fluctuations are found during this interval, they are superimposed on a warm trend at a longer wavelength.

There are two possible explanations of these data. One is that several significant coolings of the surface North Atlantic ocean actually did occur, but that the foraminiferal fauna registered them in an anomalous way because of a fundamental change in environmental responses of one or more species. This would in part require a short-term break-down in the response of the cold end-member species, *N. pachyderma* (s.).

Raymo *et al.* (in press) have shown that *N. pachyderma* (s.) did not take on a distinct 'cold-indicator' role in the North Atlantic until 1.7 Ma, suggesting that there may still have been a state of flux in its environmental response that could have persisted until 1.1 Ma. On the other hand, Raymo *et al.* (in press) also showed that apparent northern hemisphere coolings occurring earlier in the Matuyama from 1.7–1.25 Ma (marked by low $CaCO_3$ percentages and moderate influxes of ice-rafted sand) were registered by *N. pachyderma* (s.) in more or less the same sympathetic manner as they have been during the last 1.1 Ma. This would seem to require that *N. pachyderma* (s.) first acquire a set of environmental tolerances at 1.7 Ma, then abandon them from 1.2–1.1 Ma, and finally reacquire them at 1.1 Ma.

A second possibility is that the sea surface west of Great Britain actually was very warm during the 1.2–1.1 Ma interval, but was in some unknown way different from conditions during later interglaciations. This need not require a total decoupling of the North Atlantic response from the ice-sheet response indicated by the $\delta^{18}O$ data; it could occur if the colder response were isolated in the western part of the subpolar gyre near Newfoundland (with a strong east–west temperature gradient) or in the Norwegian Sea (with a strong north-south gradient). Such compartmentalizations of surface-ocean response have occurred during the latest Brunhes, although not for 0.1 Ma or longer.

Further research on sites cored during Leg 94 should help resolve the problems during this time interval.

ACKNOWLEDGEMENTS: The authors thank Joe Morley, George Kukla, and Brian Funnell for critical reviews, Ann Esmay for data processing, and Beatrice Rasmussen for help with the manuscript. This research was funded by grant OCE82–19862 from the Submarine Geology and Geophysics Program of the Ocean Science Section of the National Science Foundation. The authors also acknowledge the help of the DSDP Core Curator in providing samples from Hole 552A. This is LDGO contribution 3822.

References

BANDY, O.L., CASEY, R.E. & WRIGHT, R.C. 1971. Late Neogene planktonic zonation, magnetic reversals, and radiometric dates, Antarctic to tropics. *Am. Geophys. Un. Ant. Res. Ser.* **15,** 1–26.

BERGER, A. 1977. Support for the astronomical theory of climatic change. *Nature (Lond.)* **269,** 44–5.

BERGGREN, W.A. 1968. Micropaleontology and the Plio-Pleistocene Boundary in a deep-sea core from the South-Central North Atlantic. *Giorn. Geol., Ser. 2,* **35** (1967), fasc. 2 (Comm. Mediterranean Neogene Stratigr.), 4th Sess., Proc., Pt. 2, 291–312.

BRISKIN, M. & BERGGREN, W.A. 1975. Pleistocene stratigraphy and quantitative paleo-oceanography of tropical North Atlantic core V16–205. *In*: SAITO, T. & BURCKLE, L. (eds), *Late Neogene Epoch Boundaries.* Micropaleontology Press, Museum of Natural History, New York. 167–98.

CITA, M.B., VERGNAUD-GRAZZINI, C., ROBERT, C., CHAMLEY, H., CIARANFI, N. & D'ONOFRIO, S. 1977. Paleoclimatic record of a long deep-sea core from the eastern Mediterranean: *Quat. Res.* **8,** 205–35.

EMILIANI, C. 1966. Paleotemperature analysis of Caribbean cores P6304–8 and P6304–9 and a generalized temperature curve for the past 425,000 years. *J. Geol.* **74,** 109–24.

HAYS, J.D. 1967. Quaternary sediments of the Antarctic ocean. *In*: SEARS, M. (ed.), *Progress in Oceanography.* Pergamon Press, Oxford. 117–31.

——, IMBRIE, J. & SHACKLETON, N.J. 1976. Variations in the Earth's orbit; pacemaker of the Ice Ages. *Science* **194,** 1121–32.

HERMANN, Y. & HOPKINS, D.M. 1980. Arctic ocean climate in Late Cenozoic time. *Science* **209,** 557–62.

IMBRIE, J., 1985. A theoretical framework for the Pleistocene Ice Ages. *J. geol. Soc. Lond.* **142,** 417–32.

—— & IMBRIE, J.Z. 1980. Modeling the climatic response to orbital variations. *Science* **207,** 943–53.

——, HAYS, J.D., MARTINSON, D.G., MCINTYRE, A., MIX, A.C., MORLEY, J.J., PISIAS, N.G., PRELL, W.L. & SHACKLETON, N.J. 1984. The orbital theory of climate: support from a revised chronology of the marine $\delta^{18}O$ record. *In*: BERGER, A.L. *et al.* (eds), *Milankovitch and Climate.* D. Reidel Publishers. 269–305.

KEANY, J. & KENNETT, J.P. 1972. Pliocene-early Pleistocene paleoclimatic history recorded in Antarctic-subAntarctic deep-sea cores: *Deep-Sea Res.* **19,** 529–48.

KEIGWIN, L.D. 1979. Late Cenozoic stable isotope stratigraphy and paleoceanography of DSDP sites from the east equatorial and north central Pacific ocean. *Earth Planet. Sci. Lett.* **45,** 361–82.

KIPP, N.G. 1976. New transfer function for estimating past sea-surface conditions from sea-bed distributions of planktonic foraminiferal assemblages in the North Atlantic. *In*: CLINE, R.M. & HAYS, J.D. (eds), *Investigation of Late Quaternary Paleoceanography and Paleoclimatology. Geol. Soc. Am. Mem.* **145,** 3–42.

KUKLA, G.J. 1977. Pleistocene Land-Sea Correlations I. Europe. *Earth-Sci. Rev.* **13,** 307–74.

MCINTYRE, A., RUDDIMAN, W.F. & JANTZEN, R. 1972. Southward penetrations of the North Atlantic polar front: Faunal and floral evidence of large-scale surface water mass movements over the last 225,000 years. *Deep-Sea Res.* **19,** 61–77.

PISIAS, N.G. & MOORE, T.C., JR. 1981. The evolution of Pleistocene climate: a time series approach. *Earth Planet. Sci. Lett.* **52,** 450–8.

PRELL, W.L. 1983. Oxygen and carbon isotope stratigraphy for the Quaternary of Hole 502B: evidence for two modes of isotopic variability. *In*: PRELL, W.L., GARDNER, J.V. *et al.* (eds), *Init Repts* DSDP **68.** US Govt. Print. Off., Washington, DC. 269–76.

RAYMO, M., RUDDIMAN, W.F. & CLEMENT, B., in press. Paleonvironmental results from the Late Pliocene and Early Pleistocene of Site 609. *Init. Repts* DSDP **94.** US Govt. Print. Off., Washington, DC.

ROBERTS, D.G., SCHNITKER, D. *et al.*, 1985. *Init. Repts DSDP* **81.** US Govt. Print. Off., Washington, DC.

RUDDIMAN, W.F. 1971. Pleistocene sedimentation in the equatorial Atlantic: stratigraphy and faunal paleoclimatology. *Geol. Soc. Am. Bull.* **82,** 283–302.

—— & ESMAY, A., in press. A streamlined foraminiferal transfer function for the subpolar North Atlantic. *Init. Repts DSDP* **94,** US Govt. Print. Off., Washington, DC.

—— & MCINTYRE, A. 1976. Northeast Atlantic paleoclimatic changes over the last 600 000 years. *In*: CLINE, R.M. & HAYS, J.D. (eds), *Investigation of Late Quaternary Paleoceanography and Paleoclimatology. Geol. Soc. Am. Mem.* **145,** 111–46.

—— & MCINTYRE, A. 1984. Ice-age thermal response and climatic role of the surface Atlantic Ocean, 40° to 63°N. *Geol. Soc. Am. Bull.* **95,** 381–96.

SHACKLETON, N.J., BACKMAN, J., ZIMMERMAN, H., KENT, D.V., HALL, M.A., ROBERTS, D.G., SCHNITKER, D., BALDAUF, J.G. DESPRAIRIES, A., HOMRIGHAUSEN, R., HUDDLESTUN, P., KEENE, J.B., KALTENBACK, A.J., KRUMSIEK, K.A.O., MORTON, A.C., MURRAY, J.W. & WESTBERG-SMITH, J. 1984. Oxygen Isotope calibration of the Onset of Iceraftng in DSDP Site 552A: history of glaciation in the North Atlantic Region. *Nature (Lond.)* **307,** 620–3.

—— & CITA, M.B. 1979. Oxygen and carbon isotope stratigraphy of benthic foraminifers at site 397: detailed history of climatic change during the Neogene. *In*: VON RAD, W.B.F., RYAN, W.B.F. *et al.*, *Init. Repts DSDP* **47.** US Govt. Print. Off., Washington, DC. 433–59.

—— & OPDYKE, N.D. 1973. Oxygen isotope and palaeomagnetic stratigraphy of equatorial Pacific core V28–238: Oxygen isotope temperatures and ice volumes on a 10^5 and 10^6 year scale. *Quat. Res.* **3,** 39–55.

—— & OPDYKE, N.D. 1976. Oxygen-Isotope and Paleomagnetic Stratigraphy of Pacific Core V28–239 Late Pliocene to Latest Pleistocene. *In*: CLINE, R.M. & HAYS, J.D. (eds), *Investigations of Late Quaternary Paleoceanography and Paleoclimatology. Geol. Soc. Am. Mem.* **145,** 449–64.

THIERSTEIN, H.R., GEITZENAUER, K.R., MOLFINO, B. & SHACKLETON, N.J. 1977. Global synchroneity of late Quaternary coccolith datum levels: validation by oxygen isotopes. *Geology* **5,** 400–04.

THUNELL, R.C. 1979. Pliocene-Pleistocene paleotemperature and paleosalinity history of the Mediterranean Sea: results from DSDP sites 125 and 132. *Mar. Micropaleontol.* **4,** 173–87.

—— & WILLIAMS, D.F. 1983. The stepwise development of Pliocene-Pleistocene paleoclimatic and paleoceanographic conditions in the Mediterranean: Oxygen isotopic studies of DSDP sites 125 and 132. *In*: MEULENKAMP, J.E. (ed.), *Reconstruction of Marine Paleoenvironments. Utrecht Micropaleont Bull.* **30,** 111–27.

VAN DONK, J. 1976. O^{18} record of the Atlantic ocean for the entire Pleistocene epoch. *In*: CLINE, R.M. & HAYS, J.D. (eds), *Investigations of Late Quaternary Paleoceanography and Paleoclimatology. Geol. Soc. Am. Mem.* **145,** 147–63.

VAN DER HAMMEN, T., WIJMSTRA, T.A. & ZAGWIJN, W.H. 1971. The floral record of the Late Cenozoic of Europe. *In*: TUREKIAN, K.K. (ed.), *The Late Cenozoic Glacial Ages.* Yale University Press, New Haven, Connecticut, USA. 391–424.

ZIMMERMAN, H.B., SHACKLETON, N.J., BACKMAN, J., KENT, D.V., BALDAUF, J.G., KALTENBACK, A.J. & MORTON, A.C. 1985. History of Plio-Pleistocene Climate in the Northeastern Atlantic Atlantic: DSDP Site 552A. *Init. Repts DSDP* **81.** US Govt. Print. Off., Washington, DC. 861–75.

W.F. RUDDIMAN, Lamont Doherty Geological Observatory and Department of Geological Sciences of Columbia University, Palisades, New York 10964, USA.

N.J. SHACKLETON, Godwin Laboratory, Cambridge University, Cambridge CB2 3RS.

A. MCINTYRE, Lamont Doherty Geological Observatory and Department of Geological Sciences of Columbia University, Palisades, New York 10964, USA. Queens College of the City University of New York, Flushing, New York 11367, USA.

Appendix I: Listing of new data from core K708–7: faunal data

Depth (cm)	% *N. pachyderma* (s)	SSTw	Depth (cm)	% *N. pachyderma* (s)	SSTw	Depth (cm)	% *N. pachyderma* (s)	SSTw
0.0	5.4	9.2	285.0	5.9	10.1	523.0	22.3	6.9
10.0	4.0	9.8	288.0	1.5	10.1	525.0	12.1	8.8
18.0	7.2	9.5	290.0	3.0	10.8	527.0	8.3	9.8
20.0	5.3	8.4	293.0	4.8	10.0	530.0	4.3	10.2
23.0	10.1	9.2	295.0	3.4	10.3	533.0	6.6	8.4
28.0	17.2	6.0	297.0	3.3	10.1	535.0	9.4	10.1
30.0	26.3	4.8	300.0	9.1	10.1	538.0	15.3	7.5
35.0	48.4	5.0	303.0	19.1	8.6	540.0	26.1	6.1
40.0	57.3	3.1	305.0	16.1	8.8	545.0	55.9	3.8
44.0	73.1	1.4	308.0	12.5	9.4	550.0	50.2	4.1
50.0	96.9	−0.1	310.0	93.0	0.2	555.0	67.0	2.8
53.0	91.1	0.2	313.0	90.9	0.1	560.0	79.5	1.4
60.0	95.2	0.1	315.0	92.0	0.2	570.0	71.8	2.2
70.0	91.7	0.1	320.0	90.5	0.5	575.0	66.5	2.7
80.0	87.5	0.6	323.0	92.8	0.2	580.0	75.0	1.3
90.0	68.2	2.8	325.0	88.5	0.5	585.0	78.3	1.3
95.0	53.4	4.5	330.0	90.6	0.6	590.0	90.3	0.3
100.0	81.0	1.5	335.0	94.8	0.1	595.0	86.2	0.6
105.0	56.6	3.6	340.0	92.3	0.1	600.0	77.8	1.5
110.0	64.3	3.0	345.0	94.0	0.2	605.0	40.5	4.6
115.0	50.0	4.0	350.0	87.3	1.0	610.0	33.6	3.8
120.0	54.2	3.9	360.0	98.4	−0.2	615.0	58.4	3.7
125.0	59.0	3.7	370.0	89.1	0.5	620.0	88.2	0.1
130.0	46.3	3.9	375.0	72.7	2.1	625.0	79.1	1.0
140.0	47.2	4.0	380.0	70.2	2.2	630.0	77.9	1.4
150.0	54.8	4.2	385.0	64.7	3.2	635.0	41.8	4.3
160.0	62.6	3.2	390.0	65.2	2.9	640.0	45.2	4.2
170.0	42.5	4.1	400.0	86.0	0.5	645.0	86.9	0.5
175.0	64.2	3.1	410.0	74.3	1.9	650.0	92.5	0.1
180.0	67.0	2.9	420.0	90.0	0.3	655.0	71.1	2.0
190.0	62.6	3.2	430.0	68.1	2.0	660.0	92.4	0.0
195.0	88.7	0.7	440.0	92.2	0.1	665.0	96.0	−0.2
197.0	88.1	0.3	445.0	91.6	0.1	670.0	58.6	3.9
199.0	90.6	0.2	450.0	75.6	1.3	680.0	43.6	4.8
200.0	96.2	−0.1	455.0	74.2	1.6	685.0	78.3	1.3
201.0	80.8	1.0	460.0	34.8	5.5	690.0	77.2	1.2
204.0	81.4	0.6	465.0	40.5	5.8	695.0	37.6	6.6
206.0	60.9	2.8	470.0	28.3	6.8	700.0	21.1	5.9
208.0	49.1	4.7	475.0	20.7	7.2	705.0	19.3	8.0
210.0	53.7	4.4	480.0	23.0	7.8	710.0	51.3	4.9
211.0	59.9	3.2	482.0	5.9	11.1	710.0	51.3	4.9
213.0	54.7	4.0	485.0	5.3	11.5	715.0	49.0	4.9
215.0	39.7	5.1	487.0	5.7	11.0	720.0	20.6	7.4
217.0	42.3	6.3	490.0	4.9	11.6	730.0	7.8	7.9
218.0	25.8	6.0	493.0	5.3	11.4	740.0	10.1	9.4
220.0	29.3	6.1	495.0	7.8	10.1	750.0	90.2	0.5
222.0	23.9	5.1	497.0	8.1	9.3	755.0	90.5	0.1
224.0	34.8	5.7	500.0	48.4	4.3	760.0	90.5	0.2
225.0	25.3	5.0	502.0	57.5	3.0	770.0	81.5	0.9
230.0	37.5	4.0	503.0	73.6	1.4	775.0	86.7	0.3
240.0	9.4	6.6	504.0	80.9	1.1	785.0	77.7	1.1
250.0	45.6	4.4	505.0	64.0	2.8	790.0	62.6	2.4
260.0	43.5	5.1	507.0	63.3	2.4	800.0	82.6	0.9
265.0	28.7	6.1	510.0	49.2	4.7	805.0	71.4	2.4
270.0	25.0	6.3	512.0	48.0	5.7	810.0	72.1	1.6
275.0	16.1	6.9	515.0	25.6	6.7	815.0	67.5	2.2
280.0	3.5	11.0	517.0	24.8	7.0	820.0	47.2	4.7
282.0	6.8	10.1	520.0	25.7	6.1	825.0	66.8	2.4

Depth (cm)	% *N. pachyderma* (s)	SSTw	Depth (cm)	% *N. pachyderma* (s)	SSTw	Depth (cm)	% *N. pachyderma* (s)	SSTw
830.0	53.4	3.9	1040.0	53.0	3.3	1230.0	56.4	2.3
835.0	26.7	7.7	1045.0	44.6	5.6	1235.0	84.4	0.7
840.0	19.7	6.9	1050.0	5.3	9.7	1240.0	59.1	3.8
845.0	31.1	7.5	1060.0	7.0	7.8	1245.0	82.0	0.8
850.0	63.5	2.6	1070.0	2.2	10.6	1250.0	64.1	2.0
855.0	36.2	6.9	1075.0	3.5	9.7	1255.0	83.3	0.6
860.0	16.0	9.8	1080.0	9.0	7.8	1260.0	77.9	0.9
865.0	24.2	8.3	1085.0	35.9	6.2	1265.0	92.1	0.1
870.0	49.2	4.0	1090.0	59.2	2.8	1270.0	90.1	−0.1
875.0	15.7	9.6	1095.0	72.3	2.0	1275.0	98.5	−0.1
880.0	1.4	11.1	1100.0	54.7	4.8	1280.0	92.2	0.2
890.0	9.8	10.1	1105.0	41.3	7.1	1290.0	69.1	1.3
895.0	22.0	8.6	1110.0	40.1	7.0	1300.0	54.8	3.0
900.0	9.7	10.6	1115.0	58.5	4.2	1305.0	84.7	0.1
910.0	69.7	1.9	1120.0	46.6	6.2	1310.0	68.3	2.3
915.0	76.1	1.5	1125.0	71.8	2.0	1320.0	89.9	−0.3
920.0	71.1	1.8	1130.0	68.4	2.0	1330.0	69.1	1.4
925.0	55.3	3.5	1135.0	46.3	5.7	1340.0	39.3	6.2
930.0	63.3	2.7	1140.0	23.5	6.2	1350.0	49.9	3.0
940.0	91.8	0.3	1150.0	47.7	4.8	1355.0	60.2	1.6
945.0	97.0	−0.1	1155.0	39.8	5.8	1360.0	25.9	6.4
950.0	96.0	−0.1	1160.0	10.3	9.4	1365.0	22.3	7.9
960.0	88.0	0.1	1165.0	39.4	6.4	1370.0	10.7	11.1
970.0	94.0	0.0	1170.0	36.8	6.0	1380.0	12.9	9.9
975.0	78.5	1.3	1175.0	25.7	8.0	1390.0	22.0	7.8
980.0	68.3	2.1	1180.0	19.5	8.4	1395.0	21.5	8.6
990.0	89.7	0.2	1185.0	44.7	5.5	1400.0	20.6	8.1
1000.0	95.0	−0.1	1190.0	63.4	2.6	1405.0	17.1	8.5
1010.0	89.7	0.0	1195.0	32.5	7.2	1410.0	20.2	8.6
1015.0	88.2	0.3	1200.0	7.2	10.7	1415.0	66.0	2.2
1020.0	39.0	6.0	1205.0	13.9	9.3	1420.0	20.5	8.2
1025.0	24.3	7.6	1210.0	30.5	6.2	1430.0	29.0	7.0
1030.0	51.3	3.7	1215.0	42.5	3.7	1435.0	74.4	0.7
1035.0	53.8	4.4	1225.0	81.7	1.0	1440.0	46.4	4.7

Appendix I: Listing of new data from core K708–7: oxygen isotope data

Depth (cm)	‰ (PDB)	Depth (cm)	‰ (PDB)	Depth (cm)	‰ (PDB)
5.0	2.09	485.0	1.75	965.0	3.67
15.0	2.53	495.0	2.40	975.0	4.00
25.0	2.94	505.0	4.04	985.0	4.07
35.0	3.50	515.0	3.65	995.0	4.20
45.0	4.42	525.0	3.28	1005.0	3.97
55.0	4.33	535.0	2.54	1015.0	3.72
65.0	4.22	545.0	3.78	1025.0	1.99
75.0	4.29	555.0	3.91	1035.0	2.93
85.0	4.00	565.0	3.66	1045.0	1.57
95.0	3.75	575.0	3.56	1055.0	2.17
105.0	3.84	585.0	3.61	1065.0	2.26
115.0	3.71	595.0	3.34	1075.0	1.96
125.0	3.66	605.0	3.25	1085.0	2.54
135.0	3.91	615.0	3.46	1095.0	3.85
145.0	3.70	625.0	3.93	1105.0	3.58
155.0	3.74	635.0	2.89	1115.0	3.63
165.0	3.63	645.0	3.70	1125.0	3.64
175.0	3.40	655.0	3.48	1135.0	3.14
185.0	3.27	665.0	3.81	1145.0	2.58
195.0	3.93	675.0	3.02	1155.0	2.24
205.0	3.82	685.0	3.46	1165.0	2.89
215.0	3.56	695.0	3.16	1175.0	3.09
225.0	3.25	705.0	2.65	1185.0	2.40
235.0	3.17	715.0	3.16	1195.0	2.41
245.0	3.08	725.0	1.17	1205.0	2.35
255.0	3.06	735.0	1.83	1215.0	3.78
265.0	3.35	745.0	3.06	1225.0	3.69
275.0	2.79	755.0	4.04	1235.0	4.49
285.0	1.24	765.0	3.84	1245.0	4.13
295.0	1.80	775.0	3.62	1255.0	4.24
305.0	1.60	785.0	3.78	1265.0	4.13
315.0	3.80	795.0	3.48	1275.0	4.04
325.0	4.14	805.0	3.23	1285.0	3.89
335.0	4.32	815.0	3.35	1295.0	3.43
345.0	4.22	825.0	2.71	1305.0	3.78
355.0	3.68	835.0	2.67	1315.0	3.47
365.0	4.00	845.0	1.71	1325.0	3.28
375.0	4.04	855.0	2.96	1335.0	3.27
385.0	4.05	865.0	2.28	1345.0	2.90
395.0	4.03	875.0	2.78	1355.0	2.19
405.0	3.84	885.0	2.53	1365.0	1.88
415.0	3.90	895.0	3.56	1375.0	1.62
425.0	3.95	905.0	3.82	1385.0	1.76
435.0	4.06	915.0	3.90	1395.0	2.19
445.0	4.20	925.0	3.80	1405.0	2.06
455.0	4.03	935.0	3.94	1415.0	3.18
465.0	3.09	945.0	3.92	1425.0	2.42
475.0	2.89	955.0	4.14	1435.0	3.11

Appendix II: Listing of new data from DSDP Site 552A: faunal data

Depth (cm)	% *N. pachyderma* (s)	SSTw	Depth (cm)	% *N. pachyderma* (s)	SSTw	Depth (cm)	% *N. pachyderma* (s)	SSTw
1206.5	81.1	0.4	1540.0	48.1	6.0	1825.0	72.2	1.6
1216.5	65.7	0.7	1545.0	53.3	4.7	1832.0	45.8	6.0
1220.5	64.9	2.6	1549.0	39.0	6.8	1835.0	54.4	4.5
1230.5	33.6	6.8	1555.0	28.7	7.6	1840.0	53.3	4.6
1235.5	28.0	7.3	1561.0	25.2	8.0	1845.0	56.0	4.1
1240.5	25.5	8.0	1565.0	30.8	7.5	1850.5	39.3	6.6
1247.5	35.5	6.1	1571.0	30.0	8.1	1855.0	64.7	2.9
1258.5	29.4	7.3	1575.0	35.7	7.4	1860.0	70.6	1.6
1265.5	23.5	8.4	1581.0	36.1	7.2	1865.0	81.8	0.8
1270.5	16.7	8.9	1585.0	73.8	1.7	1870.0	54.0	4.1
1275.5	19.0	8.1	1590.0	64.7	2.9	1875.0	51.1	4.6
1285.0	19.0	8.5	1595.0	88.4	0.4	1879.0	54.0	4.4
1289.5	22.5	8.3	1601.0	84.2	0.5	1900.0	52.3	4.9
1295.5	26.1	7.1	1605.0	77.5	0.8	1905.0	57.9	4.1
1300.5	85.3	0.2	1610.0	74.0	1.4	1910.0	47.9	5.7
1309.5	95.0	−0.2	1615.0	96.1	−0.1	1915.0	48.7	5.4
1315.5	95.1	−0.1	1620.0	84.1	0.3	1920.0	52.7	4.8
1320.5	88.0	0.0	1625.0	95.8	−0.1	1925.0	62.4	3.0
1325.5	85.3	0.5	1631.0	93.6	0.0	1930.0	51.0	4.7
1335.5	55.9	3.8	1635.0	94.4	−0.1	1935.0	48.9	4.9
1340.5	50.9	4.2	1641.0	89.7	0.2	1940.0	38.7	6.3
1344.5	39.9	6.1	1645.0	95.0	−0.1	1944.0	25.0	7.9
1349.5	42.4	5.3	1650.0	78.5	0.7	1950.0	27.7	8.0
1355.5	29.1	7.5	1655.0	71.1	1.6	1955.0	41.7	6.6
1360.5	26.4	7.7	1661.0	58.7	2.8	1960.0	47.6	5.4
1364.5	28.2	7.8	1665.0	46.8	4.5	1965.0	77.6	0.9
1370.5	28.0	7.8	1669.0	47.7	4.9	1970.0	71.8	2.0
1380.5	80.4	0.9	1675.0	58.2	3.7	1975.0	68.2	2.2
1400.0	32.1	7.2	1679.0	47.9	5.1	1980.0	56.7	4.2
1405.0	32.6	7.0	1685.0	34.3	7.3	1985.0	51.1	5.1
1410.0	44.0	5.6	1690.0	48.5	5.2	1990.0	42.1	6.1
1415.5	13.0	9.9	1695.0	95.1	−0.1	1996.0	37.1	7.1
1421.0	33.6	6.9	1699.0	92.2	0.1	2000.0	25.4	8.5
1424.0	33.0	7.4	1705.0	92.8	0.0	2005.0	33.6	7.3
1431.0	33.0	6.8	1715.0	86.5	0.4	2010.0	37.2	7.1
1435.0	34.6	6.9	1720.0	67.0	2.7	2015.0	39.0	6.1
1441.0	59.6	3.5	1725.0	51.8	4.4	2020.0	40.7	6.1
1445.0	83.1	0.3	1730.0	62.1	2.9	2025.0	51.0	5.0
1451.0	91.5	0.1	1735.0	60.7	2.9	2030.0	58.8	3.3
1455.0	94.2	0.0	1739.0	67.9	1.4	2035.0	58.0	3.7
1461.0	94.8	0.0	1745.0	32.8	7.4	2040.0	67.7	2.4
1465.0	89.7	0.2	1750.0	14.6	9.6	2045.0	70.5	1.9
1471.0	84.5	0.5	1755.0	19.6	8.9	2050.5	70.6	1.6
1475.0	78.9	1.3	1760.0	25.8	7.9	2055.5	67.7	2.2
1481.0	70.3	2.0	1765.0	43.0	5.4	2055.5	67.7	2.2
1485.0	53.0	4.4	1771.0	66.3	2.6	2060.5	69.9	1.9
1490.0	52.7	4.1	1775.0	62.4	2.9	2064.5	66.5	1.9
1495.0	50.0	4.9	1781.0	62.1	3.1	2070.5	52.8	4.6
1500.0	51.0	4.7	1785.0	54.3	4.0	2080.5	45.0	4.9
1505.0	39.6	6.2	1790.0	36.8	6.3	2085.5	41.6	4.3
1510.0	21.8	8.1	1795.0	28.5	6.9	2090.5	76.4	1.1
1515.0	26.9	7.7	1801.0	72.2	1.8	2094.5	51.1	4.7
1519.0	30.3	7.2	1805.0	67.7	2.4	2105.5	38.9	6.7
1525.0	34.1	7.1	1810.0	69.5	2.1	2109.5	47.1	5.5
1531.0	48.4	5.6	1815.0	71.1	2.1	2115.5	53.2	3.7
1535.0	46.6	5.4	1825.0	72.2	1.6	2119.5	51.4	4.6

Depth (cm)	% *N. pachyderma* (s)	SSTw	Depth (cm)	% *N. pachyderma* (s)	SSTw	Depth (cm)	% *N. pachyderma* (s)	SSTw
2130.5	52.1	4.6	2220.5	2.4	13.0	2310.5	2.4	13.9
2135.5	53.0	4.7	2230.5	3.1	12.6	2315.5	3.0	13.9
2140.5	64.5	2.9	2235.5	1.3	13.3	2320.5	2.7	13.3
2145.5	53.8	5.0	2240.5	2.7	12.5	2331.5	1.8	14.6
2156.5	41.1	6.2	2244.5	1.8	13.2	2334.5	3.8	12.8
2160.5	33.6	7.8	2255.5	1.1	16.7	2339.5	3.7	14.2
2165.5	32.5	7.3	2260.5	3.0	15.6	2345.5	3.1	13.0
2170.5	40.8	6.1	2265.5	6.1	12.3	2355.5	5.0	12.6
2180.5	39.0	6.9	2270.5	5.4	12.6	2360.5	6.3	15.3
2185.5	30.0	7.6	2280.5	4.1	12.0	2366.5	3.1	13.4
2190.5	28.9	7.3	2285.5	2.3	13.4	2369.5	2.3	13.3
2195.5	20.2	9.3	2290.5	2.4	13.5	2380.5	0.8	14.5
2205.5	23.2	8.8	2295.5	3.6	13.0	2385.5	1.9	12.3
2210.5	6.7	12.5	2305.5	2.2	15.5	2391.5	3.5	13.5
2217.5	4.7	12.7						

Appendix III: Listing of depths (in cm) and ages (in years) of distinctive features in the oxygen isotope records used to plot K708–7 and 552A data to time

Core name	Depth of level (cm)	Age of level (K. yr)	Name of level	Core name	Depth of level (cm)	Age of level (K. yr)	Name of level
K708–7	27.0	9.8	Ash Zone I	552A	1290.0	679.0	Stage 17.2
K708–7	37.0	12.0	Stage 2.0	552A	1320.0	689.0	Stage 17.3
K708–7	45.0	17.8	Stage 2.22	552A	1330.0	700.0	Stage 18.23
K708–7	115.0	28.0	Stage 3.1	552A	1360.0	711.0	Stage 18.3
K708–7	185.0	53.0	Stage 3.3	552A	1370.0	721.0	Stage 18.4
K708–7	190.0	59.0	Stage 4.0	552A	1430.0	731.0	Stage 19.1
K708–7	195.0	65.0	Stage 4.2	552A	1452.0	736.0	Stage 20.0
K708–7	217.0	71.0	Stage 5.0	552A	1470.0	756.0	Stage 20.24
K708–7	250.0	99.0	Stage 5.3	552A	1482.0	763.0	Stage 21.0
K708–7	265.0	107.0	Stage 5.4	552A	1560.0	784.0	Stage 21.3
K708–7	290.0	122.0	Stage 5.5	552A	1585.0	790.0	Stage 22.0
K708–7	313.0	128.0	Stage 6.0	552A	1589.0	795.0	Stage 22.2
K708–7	335.0	135.0	Stage 6.2	552A	1853.0	910.0	Top of Jaramillo
K708–7	380.0	151.0	Stage 6.4				
K708–7	415.0	171.0	Stage 6.5	552A	1998.0	980.0	Bottom of Jaramillo
K708–7	445.0	186.0	Stage 6.6				
K708–7	462.0	194.0	Stage 7.0	552A	3210.0	1660.0	Top of Olduvai
K708–7	485.0	216.0	Stage 7.3				
K708–7	507.0	228.0	Stage 7.4				
K708–7	535.0	238.0	Stage 7.5				
K708–7	541.0	245.0	Stage 8.0				
K708–7	555.0	249.0	Stage 8.2				
K708–7	600.0	257.0	Stage 8.3				
K708–7	625.0	269.0	Stage 8.4				
K708–7	699.0	303.0	Stage 9.0				
K708–7	705.0	310.0	Stage 9.1				
K708–7	715.0	320.0	Stage 9.2				
K708–7	725.0	331.0	Stage 9.3				
K708–7	755.0	341.0	Stage 10.2				
K708–7	795.0	362.0	Stage 11.0				
K708–7	805.0	368.0	Stage 11.1				
K708–7	865.0	405.0	Stage 11.3				
K708–7	893.0	423.0	Stage 12.0				
K708–7	915.0	434.0	Stage 12.2				
K708–7	970.0	458.0	*P. lacunosa* Extinction				
K708–7	995.0	471.0	Stage 12.4				
K708–7	1017.0	478.0	Stage 13.0				
K708–7	1075.0	502.0	Stage 13.13				
K708–7	1105.0	552.0	Stage 14.3				
K708–7	1136.0	564.0	Stage 15.0				
K708–7	1155.0	574.0	Stage 15.1				
K708–7	1175.0	585.0	Stage 15.2				
K708–7	1235.0	628.0	Stage 16.22				
K708–7	1305.0	656.0	Stage 16.4				
K708–7	1375.0	668.0	Stage 17.1				

Meltwater influences and palaeocirculation changes in the North Atlantic during the last glacial termination

Douglas F. Williams & Richard H. Fillon

SUMMARY: Major meltwater plumes during the last glacial termination 13 000 years BP are recorded in the $\delta^{18}O$ anomalies of planktonic foraminifera from North Atlantic and Gulf of Mexico marine sediments. These distinct meltwater plumes represent the discharge of isotopically light freshwater from specific portions of the Northern Hemisphere ice sheets. Plume geometries show that northward incursions of warm, normal saline waters occurred during the last glacial termination, thereby delivering latent and sensible heat to high latitudes of the North Atlantic and enhancing the rate of ice sheet melting.

Learned efforts have been made to reconstruct the climatic conditions of the earth for the last glacial maximum 18 000 years BP when ice sheets covered large portions of the Northern Hemisphere (McIntyre *et al.* 1976). The volume of these ice sheets has been estimated from terrestrial evidence of ice limits (Flint 1947), glaciological modelling (Hughes *et al.* 1981), the $\delta^{18}O$ of foraminifera from deep-sea cores (Emiliani 1966; Fillon & Williams 1984), and sea-level evidence from raised coral terraces (Chappell 1974; Fairbanks & Matthews 1978).

It is likely, however, that the environmental stresses caused by climatic change (maritime and terrestrial) and oceanographic processes (sea surface temperature, salinity, plankton community structure, etc.) were most severe and dramatic during the last deglaciation or termination of ice age conditions. Broecker & van Donk (1970) coined the phrase 'Terminations' to describe the rapid end of glacial conditions and return to interglacial conditions which characterized the climatic changes of the last 700 000 years. These terminations can be clearly seen in the oxygen isotope record of the last 800 000 years and may even extend back into the early Pleistocene (Shackleton & Opdyke 1976).

One of the important events of the last glacial termination was the return to the oceanic reservoir of the isotopically light (^{18}O depleted) freshwater that had been locked in the ice sheets. This return was accompanied by a rapid rise in sea-level that reflooded the continental shelves. Duplessy *et al.* (1981) have shown that the record of the last termination in the eastern North Atlantic occurred in a step-wise fashion. Detailed oxygen isotope studies of planktonic foraminifera from areas like the Gulf of Mexico also support a step-wise change in ice volume and reveal that many areas near to the North American continent received inputs of Laurentide Ice Sheet meltwater during the last termination (Leventer *et al.* 1982, 1983). This meltwater affected thermohaline processes in the North Atlantic (Worthington 1968) and the exchange of CO_2 between the oceans and atmosphere in high latitudes (Berger 1977, 1978). Reconstructing the areal extent of the meltwater may help in understanding the dynamic mechanisms of deglaciation (Fillon & Williams 1984).

Documentation of meltwater distributions during the last termination

Various efforts have been made to estimate the timing and impact of ice sheet meltwater on the surface waters of various ocean basins. Oxygen isotope profiles in deep-sea cores from the Gulf of Mexico and the Labrador Sea/Baffin Bay are often anomalous compared to the global average $\delta^{18}O$ shift, and may reflect the timing of ice sheet decay (Kennett & Shackleton 1975; Emiliani *et al.* 1975, 1978; Fillon & Duplessy 1980). Jones & Ruddiman (1982) mapped the oceanic regions potentially affected by ice sheet meltwater and enhanced freshwater precipitation, but used no direct measure of either phenomenon.

In this study, it is shown that geographically distinct $\delta^{18}O$ gradients can be mapped in planktonic foraminifera from areas of the North Atlantic and Gulf of Mexico likely to be impacted by freshwater runoff from melting ice sheets and associated precipitation. It is proposed that these $\delta^{18}O$ anomaly patterns can be used to understand, (1) the oceanic responses of these areas to meltwater lids, and (2) the dynamics of ice sheet decay as it affected oceanic processes in the North Atlantic and Gulf of Mexico.

From SUMMERHAYES, C.P. & SHACKLETON, N.J. (eds), 1986, *North Atlantic Palaeoceanography*, Geological Society Special Publication No. 21, pp. 175–180.

Selecting and mapping horizons of $\delta^{18}O$ anomalies in deep-sea cores

In using planktonic foraminifera to assess meltwater-induced $\delta^{18}O$ anomalies in sea water, it is assumed that the oxygen isotopic temperature fractionation in the species selected for study (*Globigerinoides ruber*, *Globigerinoides sacculifer*, and *Neogloboquadrina pachyderma* sinistral) parallels the equilibrium fractionation relationship (Epstein *et al.* 1953). This assumption is supported by data from both plankton and core top studies (Williams *et al.* 1979, 1981; Kahn & Williams 1981; Duplessy 1982).

Using this assumption identification of the major features of meltwater-induced isotopic anomalies and selection of the appropriate interval of meltwater discharge to map the anomalies synoptically in the North Atlantic and Gulf of Mexico was attempted (refer to Table 1 in Fillon & Williams (1984) for the pertinent data and references). Cores from the Gulf of Mexico and Labrador Sea/Baffin Bay are excellent for documenting the timing and magnitude of meltwater discharges because they have accumulation rates of 2 to over 30 cm/10^3 years. The core coverage enables us to identify the extent of the meltwater discharge during the last termination from:

(1) the southern margin of the Laurentide ice sheet through the Mississippi drainage system;
(2) the central, northern and eastern portions of the Laurentide ice sheet which must have drained through the Hudson Strait and Lancaster Sound;
(3) the European, Asian and North American ice masses draining into the Arctic Ocean, the north-eastern North Atlantic and the Norwegian Sea.

To select the most appropriate horizon of the last termination to map, detailed stratigraphic comparisons of the cores were made employing a combination of both published and unpublished radiocarbon dating, benthic and planktonic $\delta^{18}O$ records and volcanic ash horizons (as cited in Fillon & Williams 1984). A summary of the down-core chronology for the meltwater anomalies in the Labrador Sea (Core HU75-41) and Gulf of Mexico (core EN32-PC6) indicates that the peak $\delta^{18}O$ anomalies (although not necessar-

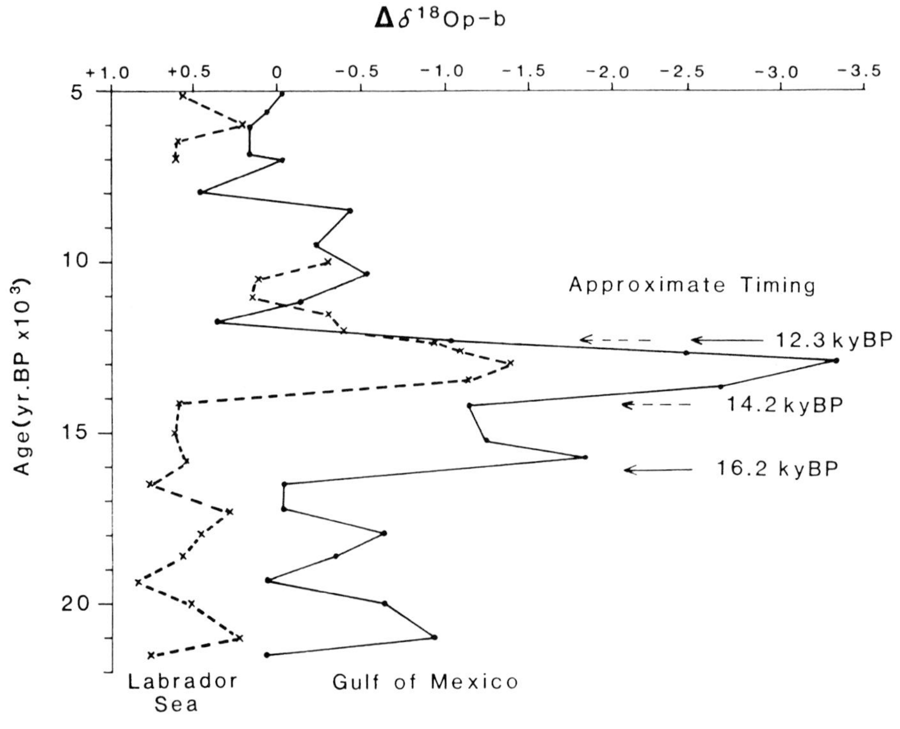

FIG. 1. The timing of the meltwater anomaly during the last termination in the Gulf of Mexico (EN32-PC6) and the Labrador Sea (HU75-41). Each of the planktonic $\delta^{18}O$ records has been normalized to the bottom water conditions at HU75-41 and corrected for global ice volume using the $\delta^{18}O$ records for benthic foraminifera.

ily the peak meltwater discharge) coincide at about 13 000 years BP in the Labrador Sea and Gulf of Mexico (Fig. 1). Discharge from the southern portion of the Laurentide ice sheet began as early as 16 200 years BP and ended by about 12 300 years BP. Melting of the northern portion of the ice sheet began later (approx. 14 200 years BP), peaked at 13 000 years BP and continued until about 7000 years BP. This deep-sea stratigraphic evidence is in agreement with the terrestrial records of deglacial ice limits (Prest 1970; Denton & Hughes 1981).

On the basis of this evidence, the $\delta^{18}O$ values for *N. pachyderma* and *G. ruber* were mapped at the 13 000 years BP horizon at each core location, without applying any corrections for surface temperature or ice volume (Fig. 2). Temperature is likely to be a minor influence except in areas of sharp temperature contrast. The contoured planktonic $\delta^{18}O$ values show:

(1) a zone of steep $\delta^{18}O$ gradients in the Davis Straits region of Baffin Bay;
(2) relatively light $\delta^{18}O$ values corresponding to the East Canadian Current and extending out into the central North Atlantic and;
(3) relatively light $\delta^{18}O$ values off East Greenland and Scandinavia (Fig. 2).

An isotopic anomaly gradient of over 3‰ extends from Baffin Bay, along the Canadian Current system and out into the central North Atlantic. Isotopically light water is also associated with an expanded East Greenland–Norwegian Coastal Current system and run-off from Scandinavia. There is evidence for the incursion of meltwater as far south as Bermuda (Keigwin *et al.*, in press). Water with heavier $\delta^{18}O$ values, similar to modern $\delta^{18}O$ values, extends from the region between Great Britain and Ireland into the eastern Labrador Sea. These anomaly patterns can be used to (1) infer the configuration of surface currents important in distributing heat for the destruction of ice shelves, and (2) distinguish between low salinity water from different sectors of the North American and Eurasian ice sheets (Fig. 3).

Nearly the entire Gulf of Mexico was covered

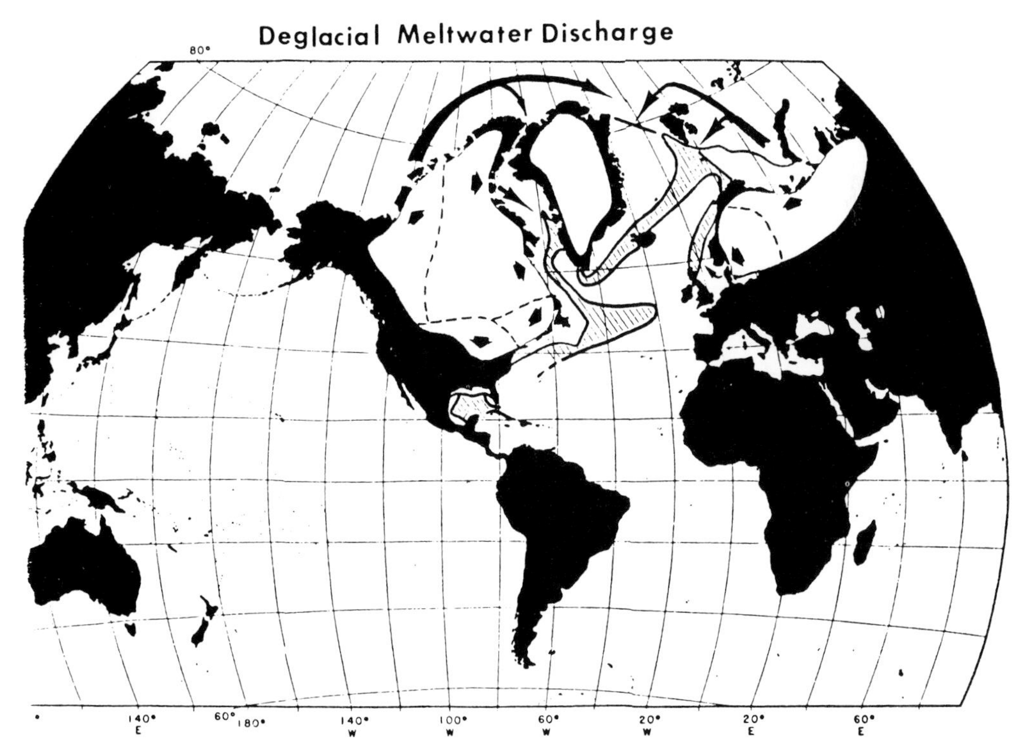

FIG. 2. Planktonic foraminiferal $\delta^{18}O$ values for the deglacial interval centred on about 13 000 years BP representing the maximum of the ^{18}O-depleted waters associated with the last glacial termination. For the North Atlantic, the $\delta^{18}O$ record is based on *Neogloboquadrina pachyderma* sinistral and for the Gulf of Mexico, on *Globigerinoides ruber* or *sacculifer*. Refer to Table 1 found in Fillon & Williams (1984) for exact core designations and locations.

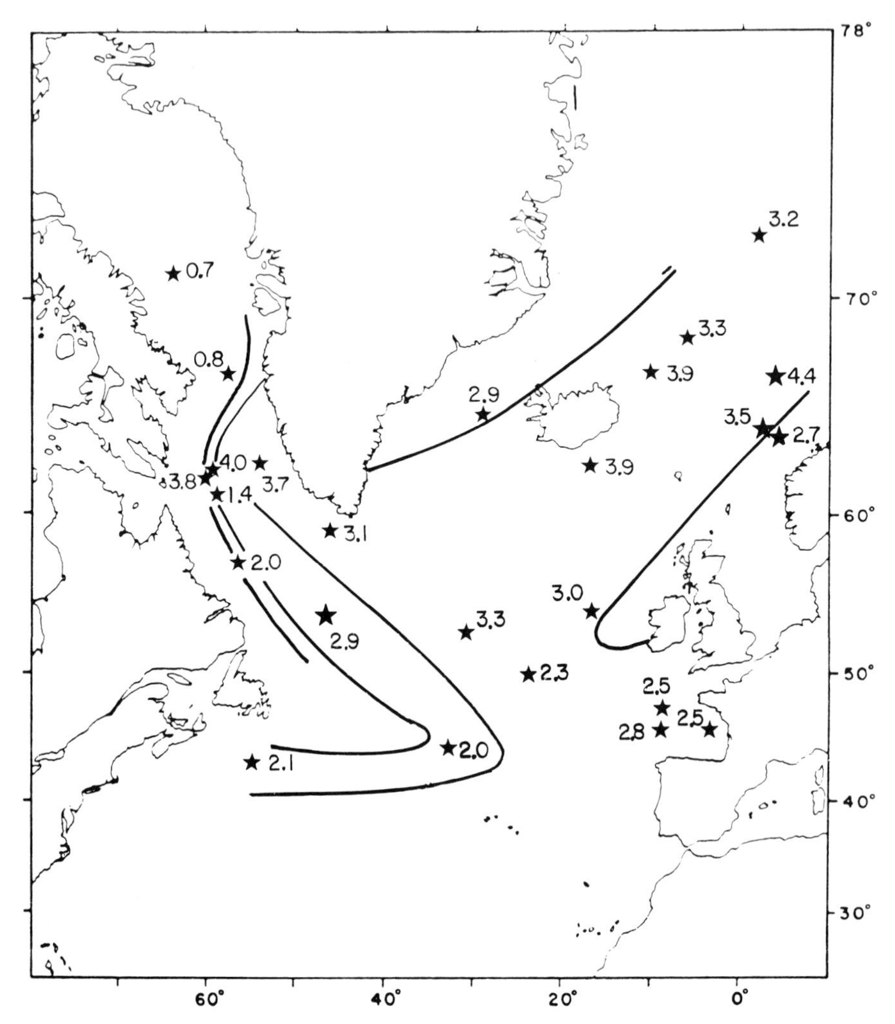

FIG. 3. Schematic representation of the meltwater discharge into the North Atlantic and Gulf of Mexico from various portions of the Northern Hemisphere ice masses.

by isotopically light meltwater, except off the Texas and Mexican coasts (Fig. 2). A lobe of light $\delta^{18}O$ values intrudes from the mouth of the Mississippi River into the south-western Gulf. The $\delta^{18}O$ gradients appear to be particularly steep along the west coast of Florida and in the Florida Straits, although the core coverage could be improved. However, the light $\delta^{18}O$ value found in core GS7603-13 from the Florida Straits suggests that the surface circulation 13 000 years BP may have been very different from that of today, with weakening of both the Loop Current system and the surface exchange between the Caribbean and the Gulf of Mexico. The lack of core coverage in the region of the Yucatan Straits prevents us from delineating the meltwater/normal marine boundary with the Caribbean.

Oceanographic implications of the meltwater reconstructions

Meltwater input from the last glacial termination (spanning the isotope stage 2/1 transition) had a major impact on the North Atlantic Ocean by controlling the distribution of heat and salt to climatically sensitive areas. The geometry of the major $\delta^{18}O$ anomalies at 13 000 years BP provides evidence for three plumes of meltwater in the North Atlantic associated with:

(1) discharge from Baffin Bay and the Labrador Sea via the route of the present day Canadian/Labrador current systems;

(2) discharge from the Greenland Sea via the route of the present day East Greenland Current and;

(3) discharge from the North Sea via the low salinity Norwegian Coastal current (Figs 2 and 3).

Isotopically normal surface water from the subtropical central North Atlantic gyre separated the three meltwater plumes and provided an important meridional component of the surface circulation to the Norwegian and Labrador Seas. These northward intrusions helped to minimize the areal extent and duration of sea ice cover in the deglacial ocean. The latent and sensible heat delivered to high latitudes by these intrusions would have enhanced the rate of ice sheet melting, especially in summer, and thus counteracted the negative impact on ice sheet disintegration produced by the cold iceberg-laden waters of the meltwater plumes.

Northward transport of latent and sensible heat in these incursions would have been important during the initial phase of deglaciation. Ice loss from Northern Hemisphere ice sheets is considered to have been attributable chiefly to marine draw-down (iceberg calving) rather than to surface ablation (Ruddiman & McIntyre 1981; Denton & Hughes 1983). Marine draw-down can proceed rapidly only if stable ice shelves are not present. The stability of ice shelves is strongly dependent on water temperatures. In the Antarctic, for example, ice shelves are absent (and cannot exist) where water temperatures exceed 0°C (Mercer 1968; Thomas *et al.* 1979). Therefore, it is quite possible that extensive marine draw-down would have been aborted as ice shelves reformed against new pinning points without the incursion of relatively warm, normal saline water into the Labrador and Norwegian Seas. The incursion of the warm, salty sea water into high latitudes of the deglacial North Atlantic made readjustment of the ice shelves impossible, and marine deglaciation was able to proceed.

The oceanographic changes caused by meltwater input to the Gulf of Mexico 13 000 years BP were as pronounced as those in the North Atlantic. The geometry of the meltwater plumes (Figs 2 and 3) suggests that the surface water circulation of the Gulf, including the Loop Current system, was significantly depressed. Cores along western Florida and in the Florida Strait suggest that the meltwater was deflected largely to the west and south-west from its entry point and exited into the Atlantic via the southernmost part of the Florida Strait and perhaps into the Caribbean via the Yucatan Strait. More cores are needed, however, from the central Gulf and Yucatan Strait to determine (a) if isotope gradients exist within the Gulf of Mexico plume and (b) exactly how the surface exchange between the Caribbean and Gulf of Mexico was affected.

Conclusions

The assumptions used to interpret the foraminiferal $\delta^{18}O$ gradients in terms of meltwater anomalies are not without problems, but there is good agreement between these results and those from other glaciological and palaeoceanographic studies (Denton & Hughes 1981; Duplessy *et al.* 1981; Ruddiman & McIntyre 1981). The distinct meltwater plumes in the North Atlantic and Gulf of Mexico indicate the extent of the oceanographic changes associated with the last glacial termination. The incursion of warm, normal saline sea water into the high latitude portions of the North Atlantic permitted marine deglaciation to proceed at a rapid pace. The results lead the authors to believe that other potential outlets for Northern Hemisphere meltwater, like the eastern Mediterranean/Black Sea (Williams *et al.* 1978; Rossignol-Strick *et al.* 1982; Jenkins & Williams 1984) and the Bering Strait/Bering Sea (Sancetta & Robinson 1983), were volumetrically minor relative to the volume of meltwater which drained directly into the North Atlantic.

ACKNOWLEDGEMENTS: This research was supported by AFM82-00424 from NSF. Jean-Claude Duplessy, Colin Summerhayes, Nick Shackleton, and B. Funnell are thanked for reviewing the manuscript. Charlotte Brunner and the Lamont-Doherty core storage facility are thanked for supplying core samples.

References

BERGER, W.H. 1977. Carbon dioxide excursions and the deep-sea record: aspects of the problem. *In*: ANDERSON, N.R. & MALAHOFF, A. (eds), *Fate of Fossil Fuel CO_2 in the Oceans*. Plenum Press, New York. 505–42.

BERGER, W.H. 1978. Oxygen-18 stratigraphy in deep-

sea sediments: additional evidence for the deglacial meltwater effect. *Deep-Sea Res.* **25**, 473–80.
BROECKER, W.W. & VAN DONK, J. 1970. Insolation changes, ice volumes and the ^{18}O record in deep sea cores. *Rev. Geophys. Space Phys.* **8,** 169–91.
CHAPPELL, J. 1974. Relationships between sea levels, ^{18}O variations and orbital perturbations during the past 250,000 years. *Nature* **252,** 199–201.
DENTON, G.H. & HUGHES, T. 1981. *The Last Great Ice Sheets.* Wiley-Interscience, New York. 447 pp.
DENTON, G.H. & HUGHES, T.J. 1983. Milankovitch theory of ice ages: hypothesis of ice-sheet linkage between regional insolation and global climate. *Quat. Res.* **20,** 125–44.
DUPLESSY, J.C. 1982. Glacial to interglacial contrasts in the northern Indian Ocean. *Nature* **245,** 494–8.
DUPLESSY, J.C., DELIBRIAS, G., TURON, J.L., PUJOL, C. & DUPRAT, J. 1981. Deglacial warming of the northeastern Atlantic Ocean: correlation with the paleoclimatic evolution of the European continent. *Palaeogeogr., Palaeoclimatol., Palaeoecol.* **35,** 121–44.
EMILIANI, C. 1966. Paleotemperature analysis of Caribbean cores P6304-8 and P6304-9 and generalized temperature curve for the last 425,000 years. *J. Geol.* **74,** 109–26.
EMILIANI, C., GARTNER, S., LIDZ, B., ELDRIDGE, K., ELVEY, D.K., HUANG, T.C., STIPP, L. & SWANSON, M.F. 1975. Paleoclimatological analysis of Late Quaternary cores from the northeastern Gulf of Mexico. *Science* **189,** 1083–8.
EMILIANI, C., ROOTH, C. & STIPP, J.J. 1978. The late Wisconsin flood into the Gulf of Mexico. *Earth Planet. Sci. Lett.* **41,** 154–62.
EPSTEIN, S., BUCHSBAUM, R., LOWENSTAM, H. & UREY, H.C. 1953. Revised carbonate-water isotopic temperature scale. *Geol. Soc. Am. Bull.* **64,** 1315–25.
FAIRBANKS, R.G. & MATTHEWS, R.K. 1978. The marine oxygen isotope record in Pleistocene coral, Barbados, West Indies. *Quat. Res.* **10,** 181–96.
FILLON, R.H. & DUPLESSY, J.C. 1980. Labrador Sea bio-, tephro-, oxygen isotopic stratigraphy and late Quaternary paleoceanographic trends. *Can. J. Earth Sci.* **17,** 831–54.
FILLON, R.H. & WILLIAMS, D.F. 1984. Dynamics of meltwater discharge from Northern Hemisphere ice sheets during the last deglaciation. *Nature* **310,** 674–77.
FLINT, R.F. 1947. *Glacial Geology and the Pleistocene Epoch.* Wiley & Sons, New York. 324 p.
HUGHES, T.J., DENTON, G.H., ANDERSEN, B.G., SCHILLING, D.H., FASTOOK, J.L. & LINGLE, C.S. 1981. *In:* DENTON, G.H. & HUGHES, T.J. (eds), The last great ice sheets. Wiley & Sons, New York. 275–95.
JENKINS, J.A. & WILLIAMS, D.F. 1983/84. Nile water as a cause of eastern Mediterranean sapropel formation: evidence for and against. *Mar. Micropaleontol.* **9,** 521–34.
JONES, G.A. & RUDDIMAN, W.F. 1982. Assessing the global meltwater spike. *Quat. Res.* **17,** 148–72.
KAHN, M.I. & WILLIAMS, D.F. 1981. Oxygen and carbon isotopic composition of living planktonic foraminifera from the Northeast Pacific Ocean. *Paleogeogr., Paleoclimatol., Paleoecol.* **33,** 47–69.
KEIGWIN, L.D., CORLISS, B.H. & DRUFFEL, E.M., in press. High resolution isotope study of the latest deglaciation based on Bermuda Rise cores. *Quat. Res.*
KENNETT, J.P. & SHACKLETON, N.J. 1975. Laurentide Ice Sheet meltwater records in Gulf of Mexico deep-sea cores. *Science* **188,** 147–50.
LEVENTER, A., WILLIAMS, D.F. & KENNETT, J.P. 1983. Relationships between anoxia, glacial meltwater and microfossil preservation in the Orca Basin, Gulf of Mexico. *Mar. Geol.* **53,** 23–40.
LEVENTER, A., WILLIAMS, D.F. & KENNETT, J.P. 1982. Dynamics of Laurentide ice sheet during the last deglaciation: evidence for the Gulf of Mexico. *Earth Planet. Sci. Lett.* **59,** 11–17.
MCINTYRE, A., KIPP, N., BÉ, A.W.H., CROWLEY, T., GARDNER, J.V., PRELL, W.L. & RUDDIMAN, W.F. 1976. Glacial North Atlantic 18,000 years ago: A CLIMAP reconstruction. *In:* CLINE, R.M. & HAYS, J.D. (eds), *Investigations of Late Quaternary Paleoceanography and Paleoclimatology. Geol. Soc. Am. Mem.* **145,** 43–76.
MERCER, J.H. 1968. Antarctic ice and Sangamon Sea level. *Int. Ass. Sci. Hydrol.* **79,** 217–29.
PREST, V.K. 1970. Quaternary geology of Canada. *In:* DOUGLAS, R.J.W. (ed.), *Geology and Economic Minerals of Canada.* Department of Energy, Mine, and Resources, Canada. 675–764.
ROSSIGNOL-STRICK, M., NESTEROFF, W., OLIVE, P. & VERGNAUD-GRAZZINI, C. 1982. After the deluge: Mediterranean stagnation and sapropel formation. *Nature* **295,** 105–10.
RUDDIMAN, W.F. & MCINTYRE, A. 1981. Oceanic mechanisms for amplification of the 23,000-year Ice-Volume Cycle. *Science* **212,** 617–27.
SANCETTA, C. & ROBINSON, W.S. 1983. Diatom evidence on Wisconsin and Holocene events in the Bering Sea. *Quat. Res.* **20,** 232–45.
SHACKLETON, N.J. & OPDYKE, N.D. 1976. Oxygen-isotope and paleomagnetic evidence for early Northern Hemisphere glaciation. *Nature* **270,** 216–19.
THOMAS, R.H., SANDERSON, T.J.O. & ROXE, K.E. 1979. Effect of climatic warming on the West Antarctic ice sheet. *Nature* **277,** 385–8.
WILLIAMS, D.F., BÉ, A.W.H. & FAIRBANKS, R.G. 1981. Seasonal stable isotopic variations in living planktonic foraminifera from Bermuda plankton tows. *Palaeogeogr., Palaeoclimatol., Palaeoecol.* **33,** 71–102.
WILLIAMS, D.F., BÉ, A.W.H. & FAIRBANKS, R. 1979. Seasonal oxygen isotopic variations in living planktonic foraminifera off Bermuda. *Science* **206,** 447–9.
WILLIAMS, D.F., THUNELL, R.C. & KENNETT, J.P. 1978. Periodic freshwater flooding and stagnation of the eastern Mediterranean Sea during the late Quaternary. *Science* **201,** 252–4.
WORTHINGTON, L.V. 1968. Genesis and evolution of water masses. *Meteorol. Monogr.* **8,** 63–7.

DOUGLAS F. WILLIAMS & RICHARD N. FILLON*, Department of Geology and Belle W. Baruch Institute for Marine Biology and Coastal Research, University of South Carolina, Columbia, South Carolina 29208, USA. *Present address: Texaco Inc., PO Box 60252, New Orleans, Louisiana 70160, USA.

Late Pliocene to Recent planktonic foraminifera from the North Atlantic (DSDP Site 552A): quantitative palaeotemperature analysis

P.W.P. Hooper & B.M. Funnell

SUMMARY: Hydraulic piston cores from Hole 552A (56′02.56′N., 23′13.88′W.), recovered during Leg 81 of the Deep Sea Drilling Project, provide a near-perfect record of the glacial history of the North Atlantic. Oxygen isotope data gathered by Dr N.J. Shackleton were used to select 40 samples representing maxima of glacial and interglacial events, from the initiation of ice-rafting at 2.5 Ma to the present. For each sample the planktonic foraminiferal assemblage was analysed, enabling distinctive warm and cold water assemblages to be identified.

The relative abundance distributions of individual species were compared with the isotope data to assess temperature preferences, including those of two extinct species: *Globorotalia puncticulata* (Deshayes) and *Neogloboquadrina atlantica* (Berggren). A similar correlation for the coiling direction ratio in *Neogloboquadrina pachyderma* (Ehrenberg) gave a coefficient greater than 0.9, showing this to be an especially sensitive indicator of temperature at this latitude.

From these data a preliminary quantitative palaeotemperature analysis, using a simple transfer function, enabled a sea-surface temperature curve to be calculated for the Late Pliocene and Pleistocene to Recent, which correlates well with both the isotopic data and ecological data for extant species.

It is some eight years now since the results of the CLIMAP project were published (Cline & Hays 1976). The bringing together of climatic reconstructions from the world's oceans and continents was a major stimulus to palaeoclimatic and palaeoceanographic research. MacIntyre and others produced a classic paper on climatic reconstructions in the North Atlantic (MacIntyre *et al.* 1976) which looked in detail at the last major glacial (18 000 years BP). They were able to calculate palaeotemperatures for this interval by utilizing a transfer function developed by Imbrie & Kipp (1971) and further refined by Imbrie *et al.* 1973 and Kipp (1976), using planktonic foraminiferal data from a large number of North Atlantic core tops. This temperature reconstruction was for 18 000 years BP—extrapolation beyond this was made difficult by two factors: a lack of long, undisturbed cores in the North Atlantic, and changes in the faunal record through time (i.e. extinctions and evolutions) which prohibit the use of the more sophisticated transfer functions. However, Thunell (1979) has used the Imbrie & Kipp (1971) model to calculate palaeotemperatures for the last 5.0 million years in the Mediterranean Sea, making the basic assumption that evolving lineages of planktonic foraminifera retain the same ecological requirements throughout their history.

The purpose of this research project was to attempt to extend CLIMAP-style interpretations back into the Pleistocene and if possible back into the Pliocene, using a similar but less rigorous transfer function developed by Hecht (1973), and attempting to quantify, rather than assume, the ecological niche of extinct species. This was made possible by the acquisition of a near-perfect sedimentary record of the Late Neogene in the North Atlantic by Leg 81 of the Deep Sea Drilling Project (Shackleton *et al.* 1984). Using a hydraulic piston corer, a long sequence of relatively undisturbed cores were collected at Site 552 (Hole 552A), situated south-west of Rockall (56′02.56′N., 23′13.88′W.) (Fig. 1) and lying near the boundary of the polar and subpolar water-masses (Kipp 1976). The oxygen isotope curve, lithological log and carbonate content of these cores clearly show glacial and interglacial cycles through the Pleistocene and Late Pliocene (Shackleton *et al.* 1984), with the first incidence of ice-rafting at 2.5 Ma. DSDP Leg 12 (Hole 116) was drilled in a very similar location (57′30′N., 15′55′W.), but because of drilling disturbance the resolution of the glacial events in the area was very poor (Poore & Berggren 1975).

Methods

Using the oxygen isotope curve published by Shackleton *et al.* (1984), 40 samples representing glacial and interglacial extremes were selected for analysis. The actual samples used in the isotope analyses were made available by Dr N.J. Shackle-

From SUMMERHAYES, C.P. & SHACKLETON, N.J. (eds), 1986, *North Atlantic Palaeoceanography*, Geological Society Special Publication No. 21, pp. 181–190.

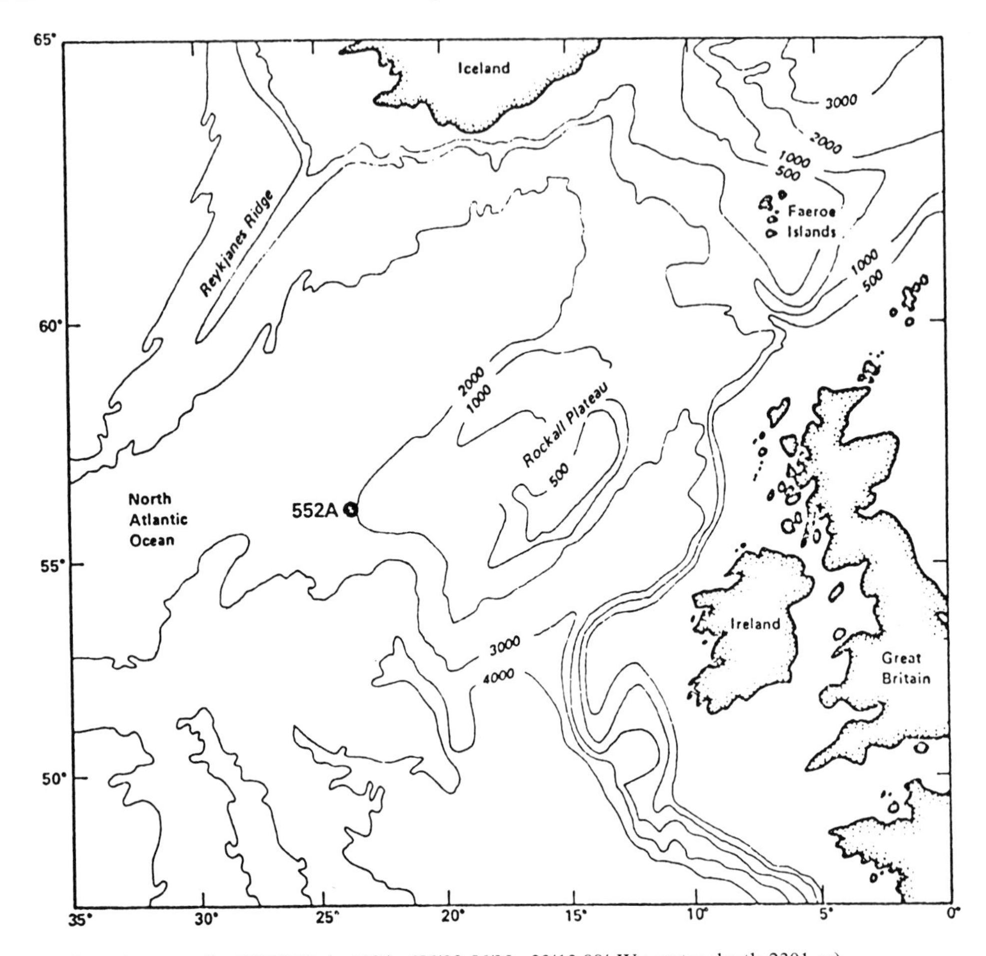

FIG. 1. Location map for DSDP Hole 552A. (56′02.56′N., 23′13.88′ W.; water depth 2301 m).

ton for this study. They had been previously washed and sieved (to 150 μm) and were ready for splitting into fractions containing 300–400 specimens, prior to picking, sorting and counting.

The distribution of the samples from Hole 552A is shown in Fig. 2 and Table 1, and it can be seen that between two and four samples were chosen from each of the most extreme glacial and interglacial intervals in the Pleistocene, whilst the 2.4 Ma ice-rafting event was treated in more detail, having eleven closely-spaced samples through the interval to monitor faunal changes in more detail.

The taxonomy followed during counting was that specified by Kipp (1976), with the addition of four species: *Neogloboquadrina atlantica* (Berggren), *Globorotalia puncticulata* (Deshayes), *Globorotalia crassula* (Cushman and Stewart) and *Globigerina woodi* (Jenkins), all of which occur in the Late Pliocene but not in the Pleistocene.

The raw counts obtained from the samples were converted into percentage abundances for each species and correlated with the oxygen isotope data (Table 2); the most important species were then plotted up as graphs and visually compared with the isotope curve (Fig. 3).

The methods used in determining palaeotemperatures in the Pleistocene and Late Pliocene are briefly summarized in a later section, but for a full explanation of the derivation and use of the relevant transfer function consult Hecht (1973).

Results

Pleistocene

One of the most striking results to come out of the study was the almost complete faunal change

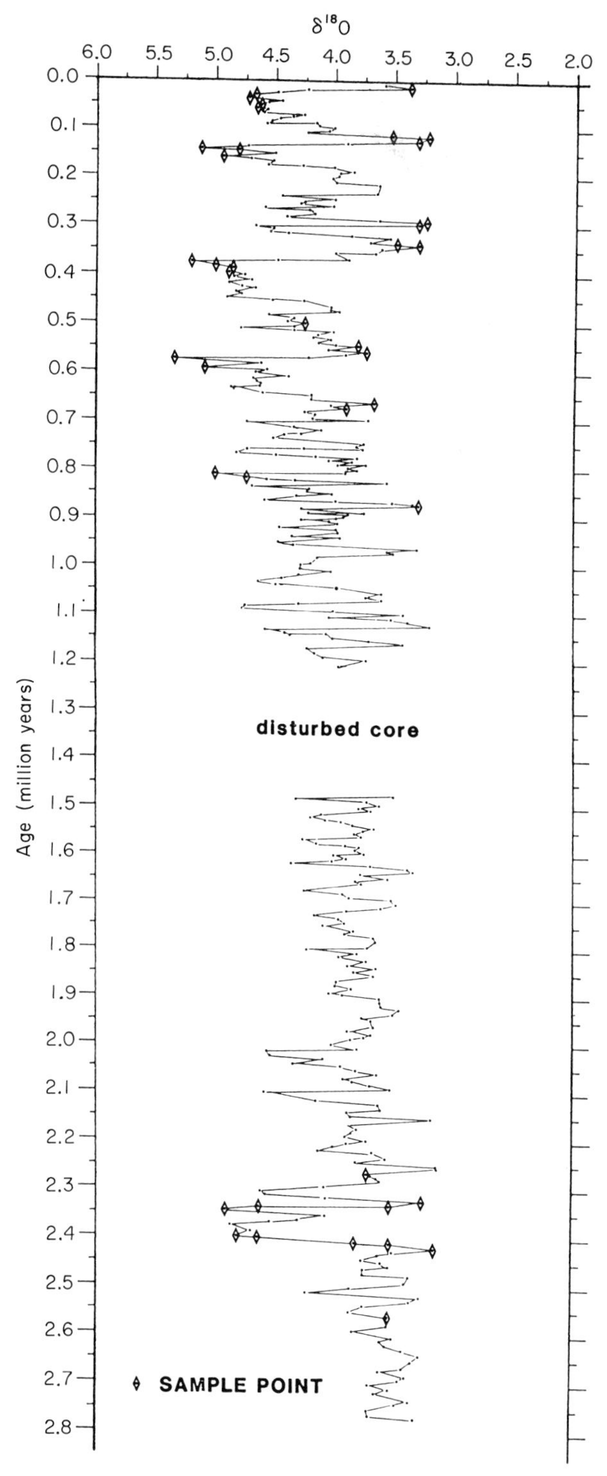

FIG. 2. Oxygen isotope curve for 552A showing sample distribution downhole (isotope data from Shackleton *et al.* 1984).

TABLE 1. *Sample data for Hole 552A.*

Sample no.	Sample	Depth down-hole (M)	Age (Ma)	σ¹⁸O	Isotope stage
1	1/1/20	0.20	0.011	3.35	1
2	1/1/51	0.51	0.027	4.63	2*
3	1/1/62	0.62	0.033	4.72	2*
4	1/1/89	0.89	0.047	4.61	2*
5	1/1/100	1.00	0.053	4.64	2*
6	1/2/61	2.11	0.113	3.48	5E
7	1/2/69	2.19	0.117	3.19	5E
8	1/2/78	2.28	0.122	3.27	5E
9	1/2/110	2.60	0.141	5.12	6*
10	1/2/120	2.70	0.148	4.84	6*
11	1/2/140	2.90	0.161	4.93	6*
12	2/2/1	5.51	0.326	3.18	9
13	2/2/10	5.60	0.332	3.24	9
14	2/2/90	6.40	0.383	3.46	11
15	2/2/96	6.46	0.388	3.27	11
16	2/3/10	7.10	0.429	5.18	12*
17	2/3/30	7.30	0.434	4.96	12*
18	2/3/39	7.39	0.436	4.83	12*
19	2/3/49	7.49	0.438	4.85	12*
20	3/1/45	9.45	0.519	4.21	14*
21	3/1/140	10.40	0.588	3.75	15
22	3/2/10	10.60	0.601	3.67	15
23	3/2/40	10.90	0.618	5.29	16*
24	3/2/70	11.20	0.629	5.05	16*
25	3/3/60	12.60	0.678	3.63	17
26	3/3/80	12.80	0.685	3.86	17
27	4/2/30	15.89	0.797	4.95	22*
28	4/2/39	16.00	0.801	4.75	22*
29	4/3/60	17·60	0.859	3.23	
30	9/1/40	40.40	2.287	3.60	
31	9/2/60	41.10	2.334	3.19	
32	9/2/70	41.20	2.342	3.49	
33	9/2/80	41.30	2.350	4.56	*
34	9/2/90	41.40	2.358	4.85	*
35	9/3/0	42.00	2.401	4.75	*
36	9/3/10	42.10	2.408	4.57	*
37	9/3/20	42.20	2.415	3.78	*
38	9/3/30	42.30	2.422	3.48	
39	9/3/40	42.40	2.429	3.09	
40	10/1/0	44.00	2.579	3.48	

(* = Glacial horizon)

between glacial and interglacial intervals. A comparison of the graphs in Fig. 3 shows that glacial faunas are completely dominated by *N. pachyderma* (sinistral)—the heavily encrusted and compact morphotype described by Kennett (1968), Bandy (1972), Kennett & Srinivasan (1980), Arikawa (1983), Kennett & Srinivasan (1983) and Huddleston (1984)—with only very minor proportions of *N. pachyderma* (dextral) and other elements of the interglacial fauna. Sinistral *N. pachyderma* reaches a peak of abundance in sample 16 (7.10 m downcore, oxygen isotope stage 12), where it forms 95% of the fauna.

In interglacial intervals a much more diverse fauna is recorded, though it is still dominated by *N. pachyderma* (the dextral, lobate form), which forms around 50% of the fauna. Other species present in some abundance are *Globigerina bulloides*, *Globigerina quinqueloba*, *Globorotalia scitula*, *Globorotalia inflata* and *Globigerinata glutinata*; also present are *Orbulina universa*, *Globorotalia crassaformis*, *Globorotalia truncatulinoides* and *Globigerinelloides calida*, with rare *Globigerinoides ruber*.

A plot of species diversity through time at this site clearly reflects the changing environmental

TABLE 2. *Correlation coefficients of selected species with the isotope curve.*

Species	Correlation coefficient
G. inflata	−0.4847
G. puncticulata	−0.6626
N. pachyderma (sinistral)	+0.7208
N. pachyderma (dextral)	−0.7633
N. pachyderma (s:d ratio)	+0.9143
N. atlantica (sinistral)	+0.8338

NB Critical value (99.5% significance level) = 0.4026
+ = polar water
− = subpolar/temperate water

conditions (Fig. 4), but by far the most sensitive environmental indicator is the coiling ratio in *N. pachyderma*, which gave a correlation coefficient of 0.91 with the oxygen isotope curve (the critical value at 99.5% significance level is 0.4026) (Fig. 5).

Pliocene

The heavily encrusted morphotype of sinistral *N. pachyderma* is absent from this interval, since it makes its first appearance at 1.6 Ma, at the base of the Pleistocene (Huddleston 1984; Weaver, in press). The dominant sinistral neogloboquadrinid at this level at Site 552A is *N. atlantica*, which seems to effectively replace sinistral *N. pachyderma* in terms of abundance and also correlates well with the isotope curve through the 2.4 Ma ice-rafting event, showing a distinct preference for cold (polar) water (Fig. 3). Warmer water species absent at this time include *G. inflata* and *G. truncatulinoides*. The transition between *G. inflata* and its immediate ancestor, *G. puncticulata* (Malmgren & Kennett 1982; Kennett & Srinivasan 1983) occurs just above the ice-rafting event. Through the 2.4 Ma cold interval *G. puncticulata* shows a marked decrease in abundance, thus exhibiting a similar temperature preference to *G. inflata*. Other species present in the warmer water assemblage in the Pliocene but extinct in the Pleistocene include *G. woodi* and *G. crassula*, though these are both comparatively scarce.

N. pachyderma (dextral) is the only numerically important species that occurs both in the Pleistocene and Pliocene at this site. It shows an identical relationship to the isotope curve through both periods, virtually disappearing from the assemblage during the 2.4 Ma cold interval. Other warmer water indicators such as *G. bulloides*, *G. quinqueloba*, *G. calida*, *O. universa* and *G. crassaformis* also occur with dextral *N. pachyderma*; the ubiquitous *G. glutinata* is present in virtually every sample in small numbers.

Discussion

The results above clearly indicate the marked differences between the faunas of the polar and subpolar watermasses (Kipp 1976) and reveal that Site 552 must have lain very close to the Polar Front throughout the late Pliocene and Pleistocene. The almost monospecific assemblages of encrusted sinistral *N. pachyderma* are indicative of polar water (Kennett 1968), and thus the samples analysed show the repeated migration of the polar front across the site as glacial episodes cooled the North Atlantic (cf. Ruddiman & MacIntyre 1976). A corollary of this is that the distribution of individual species downhole closely matches the oxygen isotope record, though the extent to which the coiling ratio in *N. pachyderma* correlates with the isotope data is exceptional and shows this parameter to be a particularly sensitive indicator of surface-water temperature at this latitude. This exact correlation also confirms that the Polar Front migrates backwards and forwards in phase with ice-sheet growth and decay (Ruddiman, this volume) and not as a direct result of temperature changes.

Correlating the downhole distribution of extant species (with known temperature preferences) with the oxygen isotope curve indicates that over the time-scale considered here (2.5 m.y.) the species' ecological requirements do not change appreciably. It follows that the correlation of extinct species with the isotope curve should give some indication of their temperature preference, and this is what is assumed here. From the results *N. atlantica* is taken as a polar-water form, since it is during a glacial interval that it shows highest abundances. However, Poore & Berggren (1975) regard this species as a sub-polar form. A possible explanation of this apparent difference in opinion is that *N. atlantica* has a rather broad geographical range in the North Atlantic (Berggren 1981) and appears abundant at high latitudes merely because other less tolerant species are absent from such assemblages.

From the above assumption, the relationship between the downhole abundance of *G. puncticulata* and the oxygen isotope curve suggests that this is a sub-polar or transitional water-mass species like its descendant *G. inflata* (Kipp 1976; Malmgren & Kennett 1982). The apparent and very rapid transition between these two species, at around 2.3 Ma at this site, is interesting since it

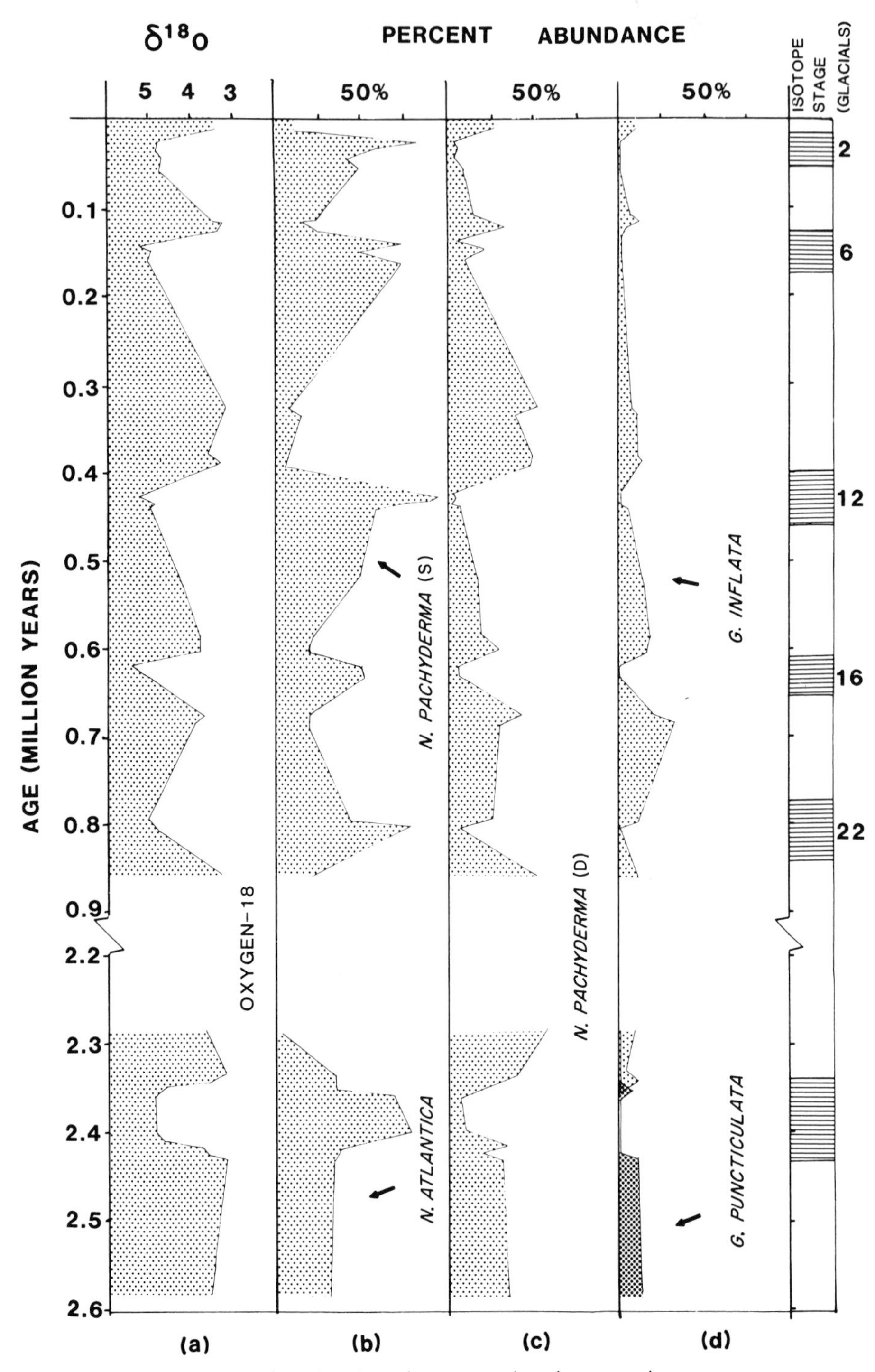

FIG. 3. Percent abundance curves for selected species compared to the oxygen isotope curve.

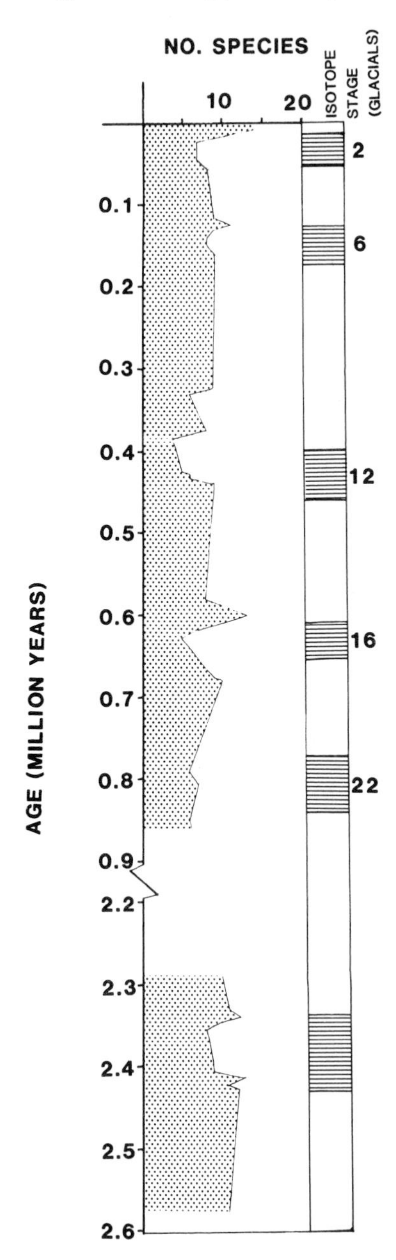

FIG. 4. Plot of species diversity through time.

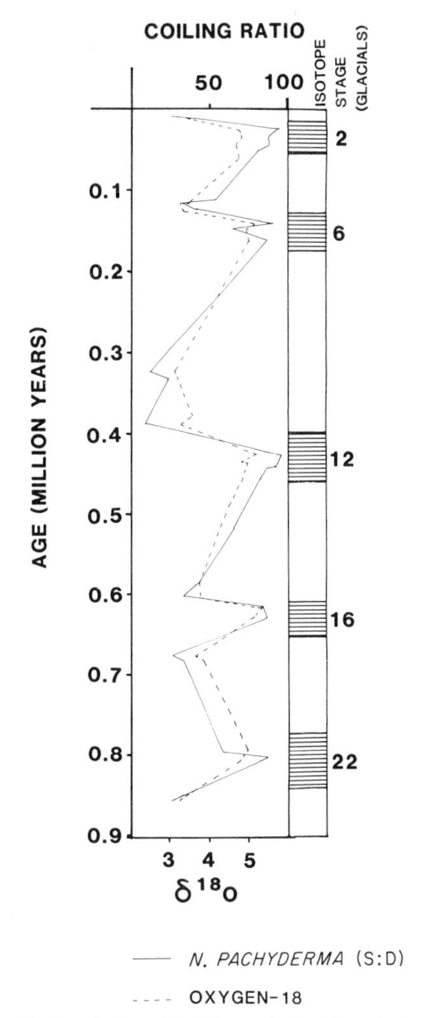

FIG. 5. Correlation of coiling ratio in *N. pachyderma* with oxygen isotope curve (the isotope curve is plotted on a reverse scale to facilitate the comparison).

occurs considerably later than the gradational transition documented by Malmgren & Kennett (1982) in the Pacific, where it occurs at 2.9 Ma. Since a single species is unlikely to evolve independently in separate oceans at different times (unless this can be regarded as a proof of gene migration and directed evolution) it seems likely that the first appearance of *G. inflata* in the North Atlantic is a migratory one (Weaver, in press; Kennett, pers. comm.), though this in itself raises problems of migration across water-mass boundaries.

Palaeoclimatic interpretation

Planktonic foraminifera have been used in a variety of ways to determine palaeoclimate and palaeoceanographic conditions (see Berger & Gardner (1975) for discussion). A number of transfer functions have now been developed to calculate palaeotemperatures from fossil planktonic foraminiferal assemblages (Imbrie & Kipp 1971; Hecht 1973; Kipp 1976; Ruddiman, this

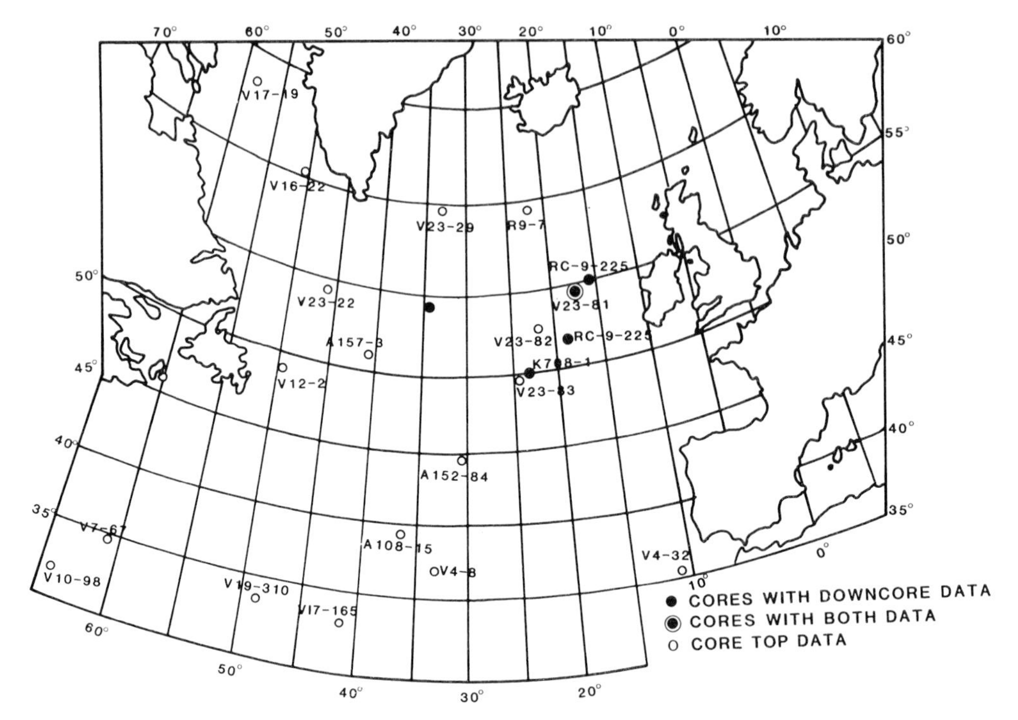

FIG. 6. North Atlantic core locations, used in the derivation of the transfer function formula of Hecht (1973).

volume), and such models have not only reached a high level of sophistication, but have also been adapted to use a very limited number of species (Ruddiman, this volume). Since the purpose of this study was to attempt to calculate palaeotemperatures for both the late Pleistocene and the late Pliocene, from foraminiferal assemblages containing extinct species, it seemed expedient to use a simple model, at least in the first instance, and therefore that of Hecht (1973) was adopted.

The transfer function developed by Hecht (1973) has five basic steps, as follows:

(1) Percentage abundances of planktonic foraminifera in core-top samples from a wide range of sites (Fig. 6) are obtained, in this case from the CLIMAP data base.

(2) An index sample is chosen, to which all other coretops will be compared; one with a high species diversity from a low-latitude site is preferable. This is then statistically compared to the other sites using a distance coefficient, so that each site has a numerical value placed on the difference between that site and the index (this is why a mid-latitude site cannot be used as the index, since sites north and south would be equally 'different' and would be given identical distance coefficients). The coefficient used was Parks' Distance Coefficient, which is defined as follows:

$$D = \sum_{i=1}^{N} \frac{(X_{ij} - X_{ik})^2}{N}$$

where X equals the percent of the species in samples j and k, and N is the number of species.

(3) The distance coefficients are regressed against sea-surface temperature data (winter and summer) for the sites to obtain a regression equation (the transfer function).

(4) Downhole samples from the particular core being studied (552A in this case) are compared to the original index sample, so that a distance coefficient is obtained for each sample.

(5) The distance coefficient for each sample is substituted into the regression equation obtained in (3), which 'transfers' it into sea-surface temperature; the result is a palaeotemperature curve such as that in Fig. 7. The precision of this method is +/− 1.18°C (Hecht 1973).

In order that a fair comparison can be made between core-tops and downhole samples a consistent taxonomy must be followed. This is obviously not possible in this case, with extinct

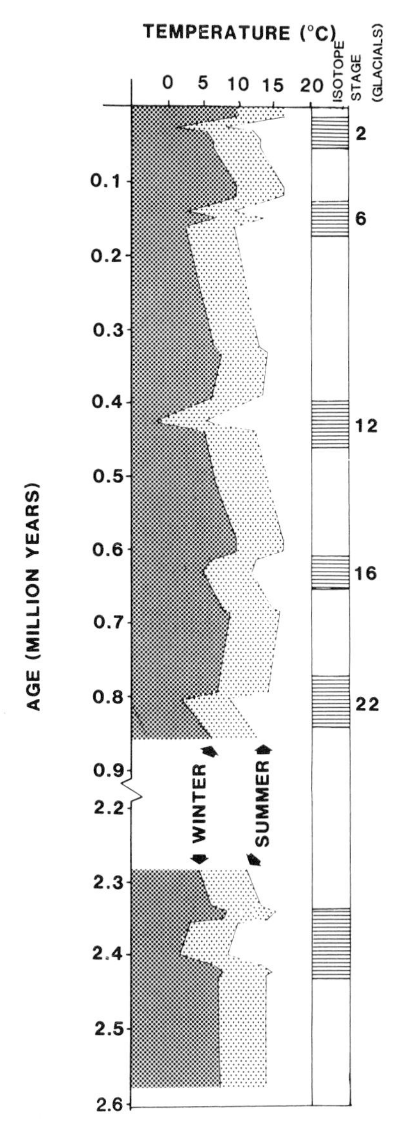

FIG. 7. Palaeotemperature reconstruction for Hole 552A, for the Late Pliocene and Pleistocene.

species such as *N. atlantica* to consider in the lower part of the section, but from the results obtained from the initial part of this project it was felt justifiable to substitute data for these species into the transfer function under their ecological analogues. A similar approach was adopted by Cita & Columbo (1979) for their studies on DSDP Site 397. They studied the correlation of planktonic foraminifera (both extant and extinct species) with physical parameters such as stable isotope composition and carbonate content, and came to the conclusion that 'it is possible to apply conventional techniques to extend the climatic record of the late Pleistocene back in time by using substitute species for climatic indicators, which had not yet appeared in the fossil record.'

Hence, for the purpose of this study, *N. atlantica* was substituted for sinistral *N. pachyderma* and *G. puncticulata* for *G. inflata*, to enable palaeotemperatures to be calculated for the Pliocene. The results of this analysis are shown in Fig. 7. In the Pleistocene, interglacial temperatures varied from around 16°C in the summer to 9°C in the winter, whilst glacial temperatures were appreciably lower, varying from 9°C (summer) to 0°C (winter). These figures agree well with previously published data, e.g. Ruddiman & MacIntyre (1976), and with recent work by Ruddiman (pers. comm.). The Late Pliocene ice-rafting event is of very similar magnitude to a Pleistocene glacial, showing summer and winter minima of 8°C and 2°C respectively, representing a cooling of approximately 6°C.

Conclusions

A Late Pliocene cooling in the North Atlantic has been postulated for some time (e.g. Poore & Berggren 1975; Shackleton & Kennett 1975; Thunell & Belyea 1982) but results from DSDP Leg 81 were finally able to show that major continental glacial cycles were initiated at about 2.4 Ma, when intense ice-rafting is first recorded in the North Atlantic.

Inferred palaeoecological data for two extinct species, *N. atlantica* and *G. puncticulata*, have been obtained from a comparison of their downhole distributions with oxygen isotope data, and this has enabled a palaeotemperature curve to be calculated back to the Late Pliocene. The temperature profile of the Late Pliocene glaciation closely matches that of a Pleistocene glacial interval and shows a drop of about 6°C in the sea-surface temperature to a minimum of 2°C in winter and 8°C in summer.

ACKNOWLEDGEMENTS: The authors would like to thank Dr T. Atkinson of the University of East Anglia and Dr P.P.E. Weaver of the Institute of Oceanographic Sciences for their help and advice during this research. Dr N.J. Shackleton of the University of Cambridge provided the samples and much useful guidance. The authors also thank Professor W.F. Ruddiman of the Lamont-Doherty Geological Observatory for his comments on the results.

This research was undertaken whilst the first author was on a Natural Environment Research Council/Institute of Oceanographic Sciences CASE studentship.

References

ARIKAWA, R., 1983. Distribution and taxonomy of *Globigerina pachyderma* (Ehrenberg) off the Sanriku Coast, Northeast Honshu, Japan. *Tohoku University, Science Reports, 2nd Ser. (Geology)* **53** (2), 103–57.

BANDY, O.L. 1972. Origin and development of *Globorotalia (Turborotalia) pachyderma* (Ehrenberg). *Micropaleontology* **18** (3), 294–318.

BERGER, W.H. & GARDNER, J.V. 1975. On the determination of Pleistocene temperatures from planktonic foraminifera. *J. Foram. Res.* **5** (2), 102–13.

BERGGREN, W.A. 1981. Correlation of Atlantic, Mediterranean and Indo-Pacific stratigraphies: geochronology and chronostratigraphy. *Proc. Int. Geol. Correlation Prog. – 114. Int. Workshop Pacific Neogene Biostratigraphy*. 29–60.

CLINE, R.M. & HAYS, J.D. 1976. Investigation of late Quaternary paleoceanography and paleoclimatology. *Geol. Soc. Am. Mem.* 145.

CITA, M.B. & COLUMBO, M.R. 1979. Late Neogene paleoenvironment: quantitative micropaleontology. *In:* VON RAD, U., RYAN, W.B.F. *et al.* (eds), *Init. Repts DSDP* **47** (1). part 1: US Govt Print. Off., Washington, DC. 391–417.

HECHT, A.D. 1973. A model for determining Pleistocene paleotemperatures from planktonic foraminiferal assemblages. *Micropaleontology* **19,** 68–77.

HUDDLESTON, P.F. 1984. Planktonic foraminiferal biostratigraphy, Deep Sea Drilling Project leg 81. *In:* ROBERTS, D.G., SCHNITKER, D. *et al.* (eds) *Init. Repts DSDP,* **81.** US Govt Print. Off.; Washington, DC. 429–38.

IMBRIE, J. & KIPP, N.G. 1971. A new micropaleontological method for quantitative paleoclimatology: application to a late Pleistocene Caribbean core. *In:* TUREKIAN, K.K. (ed.), *Late Cenozoic Glacial Ages.* Yale University Press, Hartford. 71–181.

——, VAN DONK, J. & KIPP, N.G. 1973. Paleoclimatic investigation of a late Pleistocene Caribbean deep-sea core: comparison of isotopic and faunal methods. *Quat. Res.* **3,** 10–38.

KENNETT, J.P. 1968. Latitudinal variation in *Globigerina pachyderma* (Ehrenberg) in surface sediments of the southwest Pacific Ocean. *Micropaleontology* **14** (3), 305–18.

—— & SRINIVASAN, M.S. 1980. Surface ultrastructural variation in *Neogloboquadrina pachyderma* (Ehrenberg): Phenotypic variation and phylogeny in the Late Cenozoic. *In:* SLITER, W.V. (ed.), *Studies in Marine Micropaleontology: A Memorial Volume to Orville L. Bandy. Cushman Found. Foram. Res., Spec. Publ.* **19,** 134–62.

—— & SRINIVASAN, M.S. 1983. *Neogene Planktonic Foraminifera: A Phylogenetic Atlas.* Dowden, Hutchinson & Ross, Stroudsburg, Pa. 265pp.

KIPP, N.G. 1976. New transfer function for estimating past sea-surface conditions from sea-bed distribution of planktonic foraminiferal assemblages in the North Atlantic. *Geol. Soc. Am. Mem.* **145,** 3–41.

MACINTYRE, A., KIPP, N.G., BE, A.W.H., CROWLEY, T., GARDNER, J.V., KELLOGG, T., PRELL, W. & RUDDIMAN, W.F. 1976. Glacial North Atlantic 18,000 years ago: a CLIMAP reconstruction. *Geol. Soc. Am. Mem.* **145,** 43–76.

MALMGREN, B.A. & KENNETT, J.P. 1982. The potential of morphometrically based phylo-zonation: application of a Late Cenozoic planktonic foraminiferal lineage. *Mar. Micropaleontol.* **7,** 285–96.

POORE, R.Z. & BERGGREN, W.A. 1975. Late Cenozoic planktonic foraminiferal biostratigraphy and paleoclimatology of the Hatton-Rockall Basin: DSDP Site 116. *J. Foram. Res.* **5** (4), 270–93.

RUDDIMAN, W.F. & MACINTYRE, A. 1976. Northeast Atlantic paleoclimatic changes over the past 600,000 years. *Geol. Soc. Am. Mem.* **145,** 111–46.

SHACKLETON, N.J., BACKMAN, J., ZIMMERMAN, H., KENT, D.V., HALL, M.A., ROBERTS, D.G., SCHNITKER, D., BALDAUF, J.G., DESPRAIRIES, A., HOMRIGHAUSEN, R., HUDDLESTUN, P.F., KEENE, J.B., KALTENBACK, A.J., KRUMSIEK, K.A.O., MORTON, A.C., MURRAY, J.W. & WESTBERG-SMITH, J. 1984. Oxygen isotope calibration of the onset of ice-rafting and history of glaciation in the North Atlantic region. *Nature* **307,** 620–3.

—— & KENNETT, J.P. 1975. Paleotemperature history of the Cenozoic and initiation of Antarctic glaciation: Oxygen and carbon isotope studies in DSDP Sites 277, 279 and 281. *In:* KENNETT, J.P., HOUTZ, R.E., *et al.* (eds), *Init. Repts. DSDP* **29.** US Govt Print. Off., Washington, DC. 743–55.

THUNELL, R.C. 1979. Climatic evolution of the Mediterranean Sea during the last 5.0 million years. *Sed. Geol.* **23,** 67–79.

—— & BELYEA, P. 1982. Neogene planktonic foraminiferal biogeography of the Atlantic Ocean. *Micropaleontology* **28** (4), 381–98.

WEAVER, P.P.E., in press. Late Miocene to Recent planktonic foraminifera from the North Atlantic: DSDP Leg 94. *In:* KIDD, R.B., RUDDIMAN, W.F. *et al.* (eds), *Init. Repts DSDP* **94.** US Govt Print. Off., Washington, DC.

P.W.P. HOOPER, Institute of Oceanographic Sciences, Brook Road, Wormley, Godalming, Surrey GV8 5UB and School of Environmental Sciences, University of East Anglia, Norwich, Norfolk NR4 7TJ.

B.M. FUNNELL, School of Environmental Sciences, University of East Anglia, Norwich, Norfolk NR4 7TJ.

North-east Atlantic Neogene benthic foraminiferal faunas: tracers of deep-water palaeoceanography

Detmar Schnitker

SUMMARY: The qualitative and quantitative composition of deep-sea benthic foraminiferal faunas from DSDP Sites 400 (4399 m), 552 (2301 m), 553 (2329 m), 554 (2576 m), and 555 (1659 m) in the north-eastern North Atlantic have undergone strong changes during the past 27 Ma, very probably reflecting changes in the deep oceanic environment. Throughout this time interval the faunas of the deepest site (400) were distinctly different from the faunas of the shallower sites, indicating a persistent depth stratification of the deep water masses. The strongest qualitative faunal change occurred during the middle Miocene, about 15–13 Ma ago, when at all depths several species that had persisted since the Oligocene became extinct, several new species appeared for the first time, and the quantitative faunal composition of all sites changed drastically. This change is thought to be indicative of the (re-)initiation of North Atlantic Deep Water formation, and occurred at the same time as the rapid growth of the Antarctic ice sheet and a strong change in the carbon and oxygen isotopic composition of deep sea foraminifers worldwide. At the shallower and intermediate sites, the composition of benthic faunas changed again about 6.5–7 Ma ago, a change that also affected several key species at the abyssal site as well. This change was probably caused by the closure of the Mediterranean, the Messinian Crisis. The introduction of a new 'shallow water' species into the bathyal and upper abyssal sites, as well as a change in dominance in the deep abyss at 2.5 Ma, signifies the onset of modern glacial conditions.

Many studies have documented a distinct relationship between the distribution of deep-sea benthic foraminifers and deep ocean hydrography. Since tests of deep-sea benthic foraminifers are often the only remains of bathyal and abyssal life that are incorporated into the sediments in significant numbers, they are potentially a powerful tool for deciphering the deep environments of the past. Because of the great diversity of deep-sea environments and the high degree of covariance between most of the deep water attributes, a clear delineation of the relationship between the various environmental attributes and the composition of deep-sea benthic foraminiferal faunas has not yet been established. However, the nature of major environmental differentiations is clearly indicated by several key species, as has recently been corroborated geochemically (carbon isotopes and the Cd/Ce ratio of foraminiferal calcite) (Boyle & Keigwin 1982; Curry & Lohmann 1982). At the very least, compositional changes in fossil deep-sea foraminiferal faunas reflect the timing, magnitude, and general nature of changes in the deep-sea environment and thus of deep-sea circulation. At present, the North Atlantic is a very unusual ocean in that it allows penetration of subtropical surface water into arctic latitudes, thereby strongly modifying the northern hemisphere energy balance and climate. The return flow of this water, now cooled, supplies perhaps as much as one half of the volume of the global deep water circulation. It is therefore desirable to know when these conditions became established, and how they might have affected or caused oceanographic and climatic changes elsewhere.

Geographic and oceanographic setting

Site 400 was drilled at 4399 metres depth at the base of the continental slope of the Armorican Margin (Fig. 1). It lies under the lower North Atlantic Deep Water (NADW) but away from the influence of strong geostrophic bottom currents. Sites 552, 553, and 554 were drilled on or near the Hatton Drift, one of several North Atlantic sediment bodies thought to have been formed by strong geostrophic bottom currents (Jones *et al.* 1970; Ruddiman 1972; Roberts 1975). The shaping of these sedimentary features, and thus the existence of geostrophic contour currents, dates back to at least the early or middle Miocene (Shor & Poore 1979). The water mass at these sites originates in the overflow from the Norwegian Sea, through Faeroe Bank Channel, following the base of Rockall Plateau as a contour current in a clockwise direction. During its descent into Rockall Trough and along its flow path, this water entrains warmer and less saline deep water of Labrador Sea origin (Lonsdale & Hollister 1979), so that it has a salinity of about 34.96‰, a potential temperature of 3.3°C, and an oxygen content of about 6.1 ml/l (Worthington &

From Summerhayes, C.P. & Shackleton, N.J. (eds), 1986, *North Atlantic Palaeoceanography*, Geological Society Special Publication No. 21, pp. 191–203.

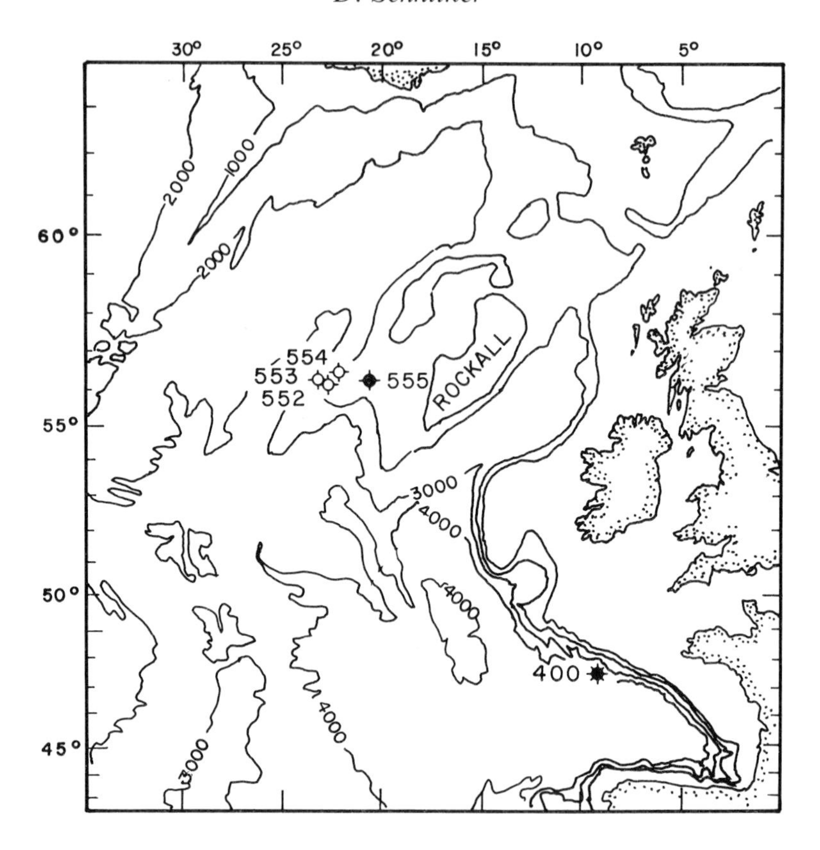

FIG. 1. Location of DSDP drilling sites investigated in this study.

Wright 1970). Site 555 was drilled near the top of Rockall Bank at 1659 metres depth. The water at this depth, the upper NADW, is largely of Labrador Sea origin, with possibly a fair contribution from the lower portion of the Mediterranean Outflow water.

Materials and methods

The goal of this study was to sample the Neogene sections of these drill sites at intervals of approximately one million years, and more closely at sections of specific interest where rapid faunal or sedimentary changes were detected. For significant portions of the record this has been accomplished, but gaps occur, mainly because of hiatuses within the sections. The time control for this study is based on sedimentation rate curves constructed from the combined biostratigraphies and magnetochronologies from the Initial Reports of Deep Sea Drilling Legs 48 and 81 (Montadert *et al.* 1979; Roberts *et al.* 1984). Samples of approximately 20 cc's wet volume were dried, weighed, disaggregated in a weak Calgon solution and wet sieved on a 125 micrometre mesh screen. Usually these samples yielded more than the 300 benthic specimens needed for significant quantitative analysis and were subdivided with a rotary splitting device. Preservation of the benthic foraminifers is excellent or very good in the late Pliocene and Pleistocene sections, but deteriorates to only fair in the Miocene material of all sites. The effects of preservation upon the quantitative species composition has not yet been assessed. The faunal tallies were subjected to a hierarchial cluster analysis (Sneath & Sokal 1973), which sorted 'Canberra Metric' similarity coefficients (Lance & Williams 1967) into groups of decreasing faunal similarity. A separate Q-mode varimax principal component analysis was used to reduce the diversity of the 77 taxonomic categories employed in this study to a lesser number of covariant faunal groupings.

Results

A total of 152 species of benthic foraminifers were identified from the material of the five DSDP sites, but only 76 were sufficiently abundant and/or recurring that they were incorporated into the faunal census matrix. The remainder were carried as 'others'.

Inspection of the faunal data clearly shows that the deep water faunas of the different sites have been distinctly differentiated and that their composition has not remained steady through the time interval studied. Several of the long ranging species that evolved sometime in the early or mid-Palaeogene are common in nearly all samples. They comprise an ubiquitous component of the Cenozoic deep sea community. *Globocassidulina subglobosa* and *Oridorsalis umbonatus* are two of these species, each comprising, on average, between 5% and 10% of the fauna. Depth differentiation is best illustrated by several species of *Stilostomella*, which are most abundant in the intermediate and shallower sites (Fig. 2), and *Osangularia umbonifera* (Fig. 3) or *Melonis pompilioides*, which are decidedly more abundant in the deepest site. Strong systematic abundance changes over time are exhibited by many species, either at one, two, or all sites. For example, *Stilostomella* spp. (Fig. 2) and *Gyroidinoides neosoldanii* decreased strongly in relative abun-

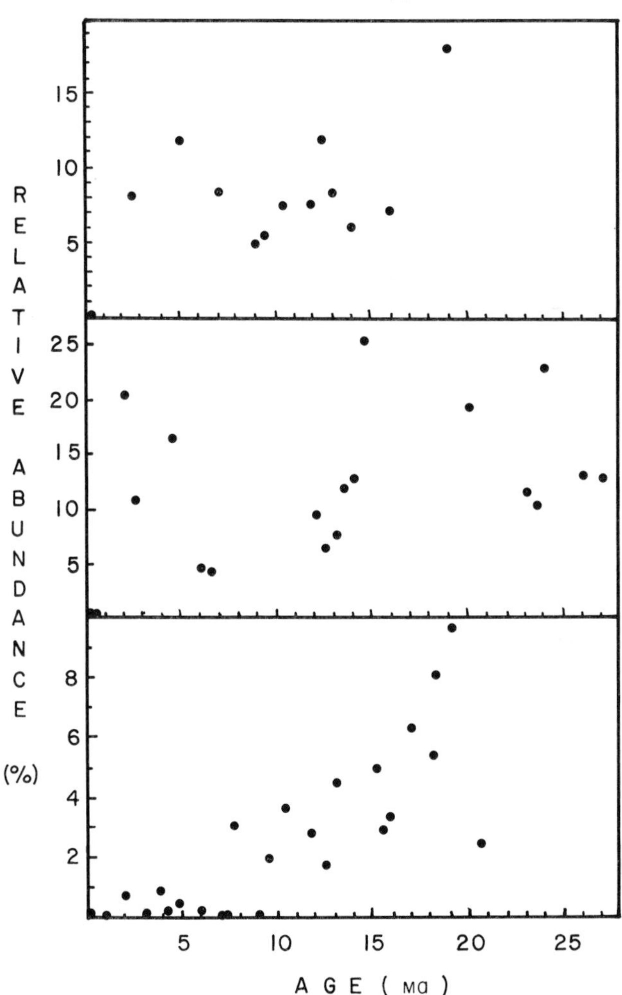

FIG. 2. Relative abundance plots of *Stilostomella* spp. Upper panel Site 555, centre panel Sites 552, 553, and 554, lower panel Site 400.

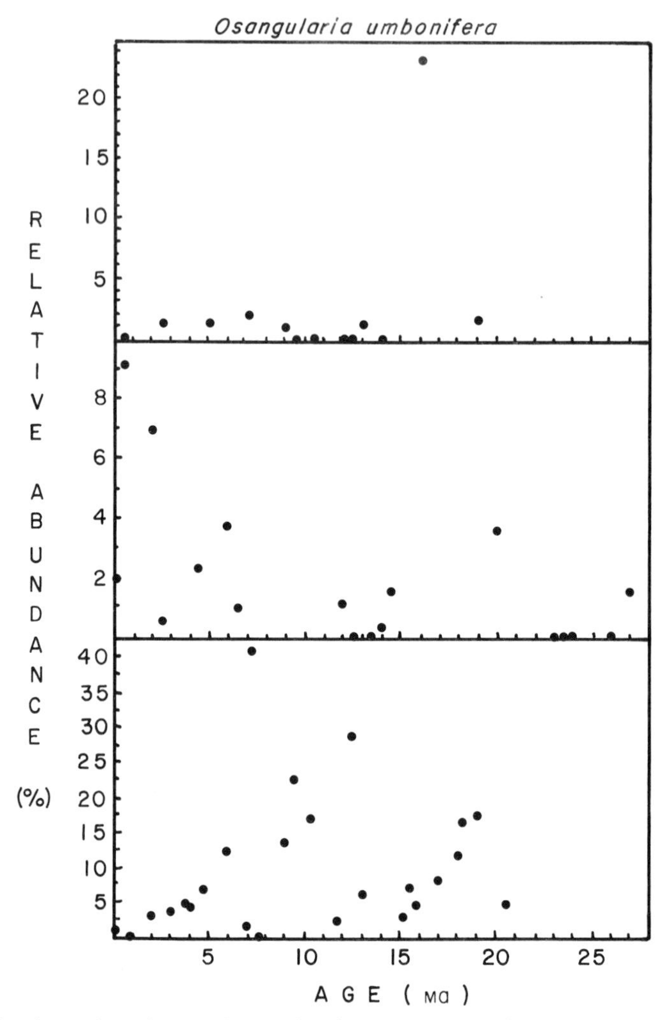

FIG. 3. Relative abundance plots of *Osangularia umbonifera*. Upper panel Site 555, centre panel Sites 552, 553, and 554, lower panel Site 400.

dance during the past 20 Ma in Site 400 samples but less so in samples from the shallower sites. *Pullenia quinqueloba* and *Parrelloides bradyi* display very similar tendencies but are not as abundant as *Stilostomella* spp. or *G. neosoldanii*. In contrast, species such as *Gyroidinoides orbicularis* (Fig. 4), *Siphotextularia* spp., *Oridorsalis tener*, or *Epistominella exigua* increase strongly in abundance during the Neogene, some at the deeper, others at the shallower sites. A faunal turnover of several species occurs throughout the time interval studied, but most events are centred in the middle Miocene interval (Fig. 5). Most of the Palaeogene species that come to the end of their range have been minor constituents of their faunas, their termination therefore does not change the overall appearance of the faunas to a great degree. The newly appearing species, such as *C. wuellerstorfi* (Fig. 6), soon become ubiquitous and abundant, essentially 'modernizing' the deep-sea faunas.

The resolution of six significant clusters of faunal similarity (Fig. 7), as well as of nine principal components (Fig. 9) from the faunal census date, summarizes and categorizes the earlier observations on individual species records. The occurrences of the individual cluster and factor sequences indicate that the deep faunas, and thus the deep environments, have not been homogenous throughout the past 27 Ma. As illustrated in Figs 8 and 9, the differentiation was both vertical and temporal:

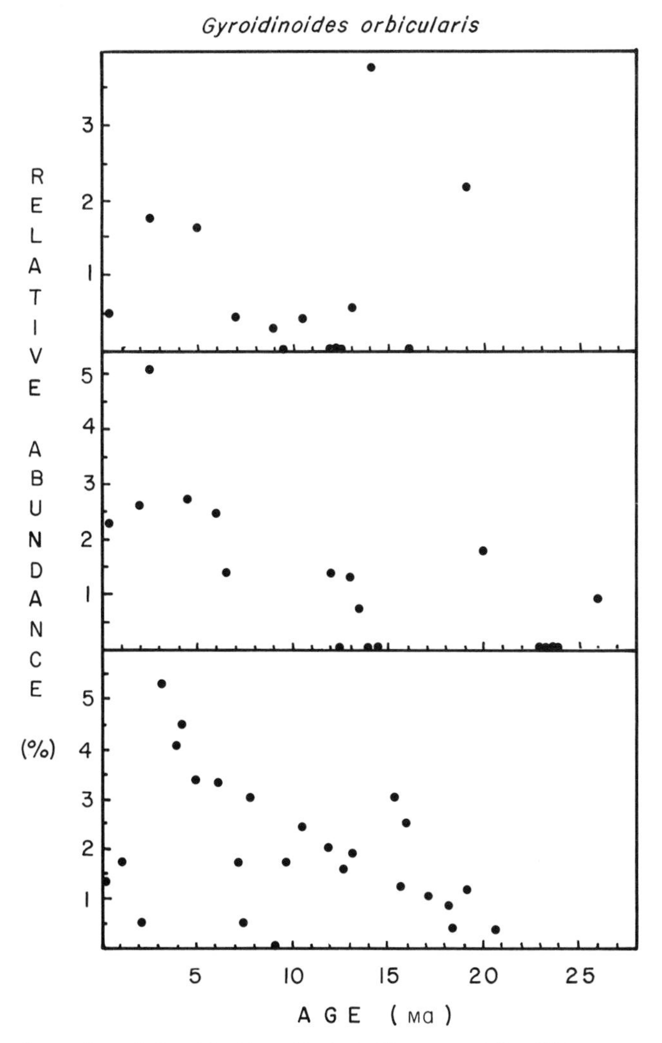

FIG. 4. Relative abundance plots of *Gyroidinoides orbicularis*. Upper panel Site 555, centre panel sites 552, 553, and 554, lower panel Site 400.

(1) The faunas of Site 400 (clusters 2, 3 and 4; principal components 6, 1, and 2) have continuously been different from those of the shallower sites.

(2) The temporal faunal divisions revealed by the principal components analysis are coincident in the deep and the intermediate waters only once, at about 15 Ma in the middle Miocene.

The nine principal component groupings represent faunal assemblages that succeeded each other as follows:

(1) A late Oligocene assemblage (principal component 3) (three samples only), occurring at intermediate depth, is characterized by the presence of two species of *Alabamina*, a species of *Epistominella* (probably the ancestor of *Epistominella exigua*), *Cibicidoides pseudoungerianus*, and abundant ubiquitous deep water species *Globocassidulina subglobosa* and *Oridorsalis umbonatus*. No samples of Oligocene age from the deep and the shallow sites were incorporated into this analysis.

(2) An early Miocene to early mid Miocene assemblage (principal component 4) at the shallow and intermediate sites is characterized by great abundances of *Stilostomella* spp. (particularly at the intermediate depth sites) and relatively high numbers of

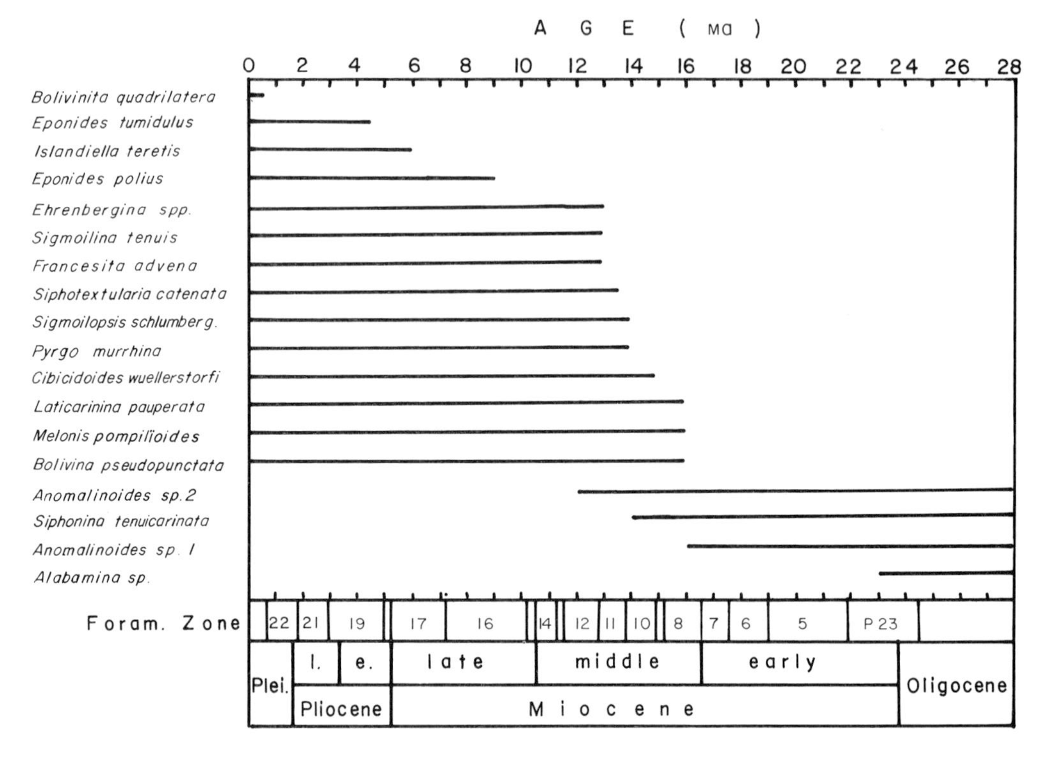

FIG. 5. Chart of species whose range begins or ends within the time interval of this study.

Bulimina alazanensis. *Globocassidulina subglobosa* and *Oridorsalis umbonatus* are again significant accessory species. The samples from the intermediate sites also contain *Melonis barleeanum* in fair abundance.

(3) The contemporary deep water counterpart of assemblage 2 is expressed as principal component 6, which is also dominated by *Stilostomella* spp. It differs in the absence of *Bulimina alazanensis* and the significant presence of *Parrelloides bradyi*, *Pleurostomella* spp., *Gyroidinoides neosoldanii*, and *Osangularia umbonifera*.

(4) A mid-Miocene to early late Miocene intermediate assemblage (principal component 5) is homogenous in both the intermediate and shallow water sites. This assemblage is characterized by the dominance or exclusive presence of *Pullenia bulloides*, *Cibicidoides wuellerstorfi*, *Melonis barleeanum*, and *Oridorsalis umbonatus*. The upper temporal limits of this assemblage cannot be clearly defined because of a very large sampling gap in the range from 11–7 Ma at the intermediate sites and a gap from 8–6.5 Ma at the shallower site.

(5) A mid late Miocene and early Pliocene assemblage (principal component 7) occurs at the intermediate and shallow water sites. Again, because of fairly wide sampling gaps, the lower and upper temporal boundaries of this assemblage cannot be clearly established. The five samples that are characterized by this assemblage are from about 6.5–4 Ma old, centred on the time interval of the Messinian event. The fauna is characterized by the first abundant occurrence of *Epistominella exigua*, with *Cibicidoides wuellerstorfi*, *Ehrenbergina* spp., and *Globocassidulina subglobosa* as a prominent accessory species.

(6) The (partial) temporal deep water equivalent of the last two assemblages is given by principal component 1, which ranged from 15 Ma to about 5 Ma before present. This assemblage consists mainly of *Osangularia umbonifera*, *Gyroidinoides neosoldanii*, *Melonis pompilioides*, *Pullenia quinqueloba*, and, especially in the most recent samples of this group, *Ehrenbergina* spp. Until the late Pliocene the depth stratification of benthic foraminiferal faunas has been twofold, but about 2.5 Ma ago, with the onset

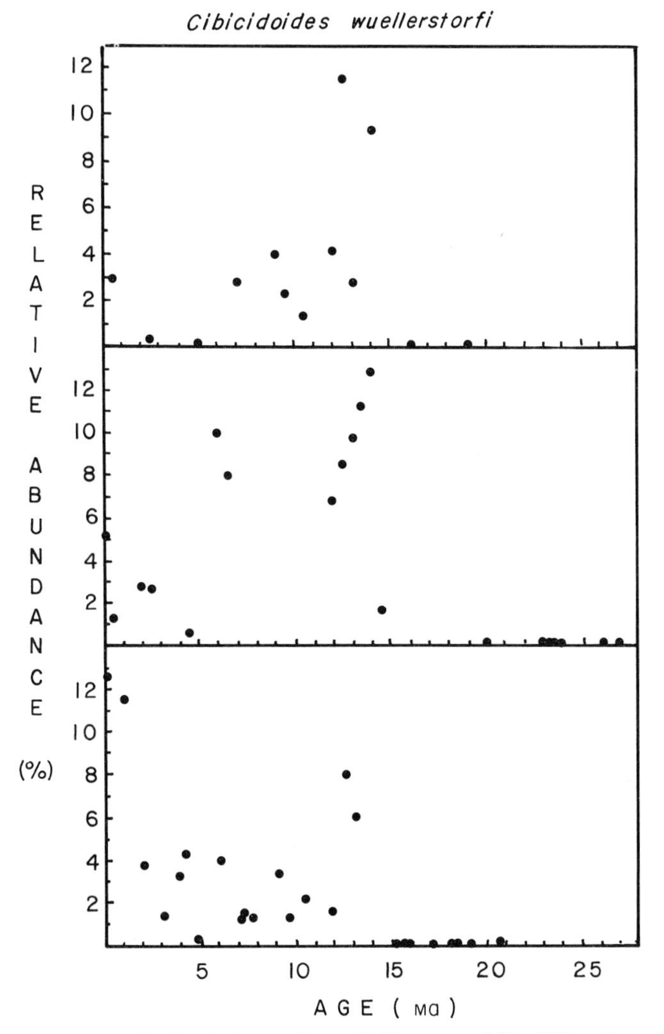

FIG. 6. Relative abundance plots of *Cibicidoides wuellerstorfi*. Upper panel Site 555, centre panel Sites 552, 553, and 554, lower panel Site 400.

of Northern Hemisphere glaciations, the stratification becomes more intense, as indicated by a clear separation of the faunas at the shallow site (555) and the intermediate sites (552, 553, 554).

(7) At Site 555 a faunal association (principal component 9) of Pleistocene age is characterized by *Triloculina frigida*, *Islandiella teretis*, *Epistominella exigua*, and in particular, by *Uvigerina peregrina*.

(8) At Sites 552, 553, and 554 a faunal association (principal component 8) occurs during the latest Pliocene and Pleistocene that is typified by *Triloculina frigida*, *Islandiella teretis*, *Osangularia tener*, *Pullenia quinqueloba*, and *Melonis pompilioides*.

(9) The Pliocene and Pleistocene deep water succession is grouped into principal component 2. This assemblage consists of *Cibicidoides wuellerstorfi*, *Melonis pompilioides*, a hispid *Uvigerina*, *Gyroidinoides orbicularis*, *Textularia* spp., and *Siphotextularia catenata*. It is somewhat surprising that this group did not split into two separate entities as did its shallow and intermediate temporal equivalents. The break at 2.5 Ma that divides the shallower faunas is clearly seen in the deep fauna when inspecting the census data or the abundance plots of individual species (see Figs 2, 3, 4, 6 and 10).

The cluster analysis, as interpreted in Fig. 8,

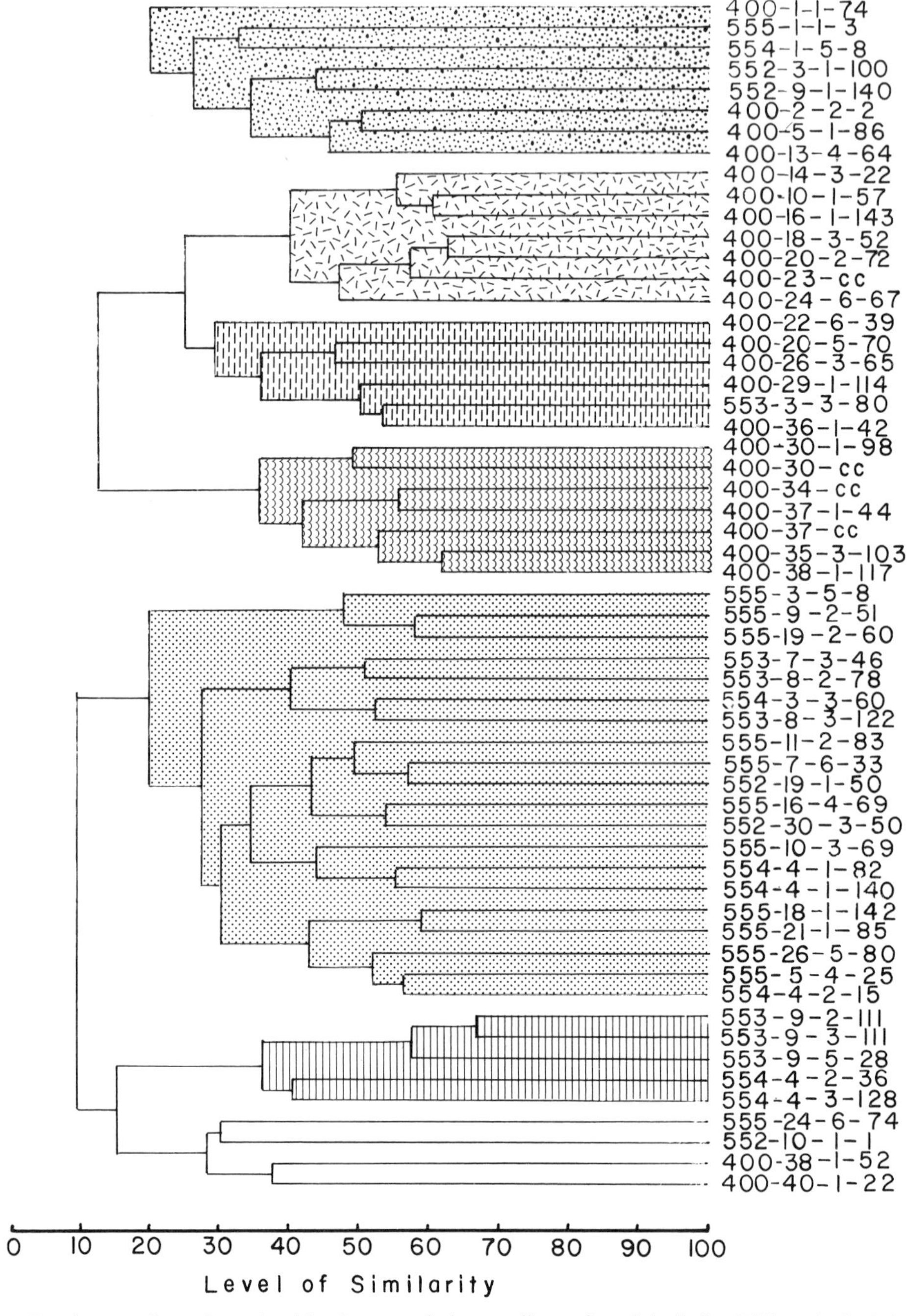

Fig. 7. Dendogram of samples ordered by cluster analysis according to faunal similarity. Differently shaded clusters denote assemblages discussed in text.

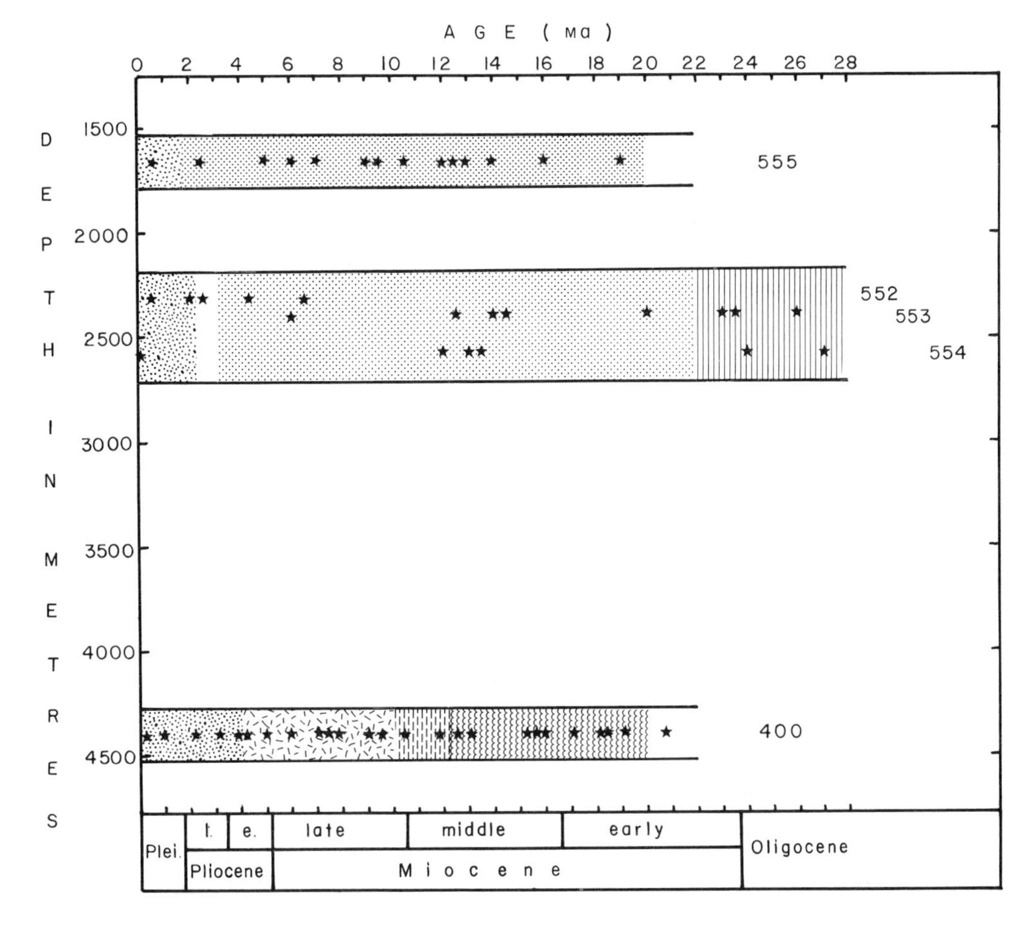

FIG. 8. Distribution of faunal clusters in time and space. Shading of assemblages same as in Fig. 7.

shows less complexity than the principal component analysis. This difference is due in part to the different sensitivity of the methods of analysis to relative abundances of species and to the appearance and disappearance of species. Most of the difference, however, is due to operator choice of accepted clusters or principal components. On the one hand, the large Miocene cluster of the intermediate and shallow sites consists of four subclusters that divide the Miocene faunas of the intermediate sites into early, middle, and late Miocene groups, and the Miocene/early Pliocene of the shallow Site 555 into two groups, divided in the middle mid-Miocene. On the other hand, allowing the principal component analysis to resolve nine groups resulted in several groups that, although ecologically very interesting, are quantitatively not very significant. Forcing the reduction of data into progressively fewer principal components results in the merger of principal components 9 and 8, 7 and 5, thereby making the results of the principal components analysis very similar to the results of the cluster analysis. The only lasting difference is that the division within the middle Miocene at about 15 Ma, so very clear in the principal component analysis, is not brought out very well in the cluster analysis.

Discussion and conclusion

Many workers have shown that the ecological success or failure of deep water benthic foraminiferal species is dependent upon their environment, of which the physical and chemical characteristics of the deep water masses are a very important, if not dominant, aspect (Streeter 1973; Schnitker 1974, 1980a; Corliss 1979, 1983; Lohmann 1978; Douglas & Woodruff 1981; Weston & Murray 1984). Faunal differentiation, both

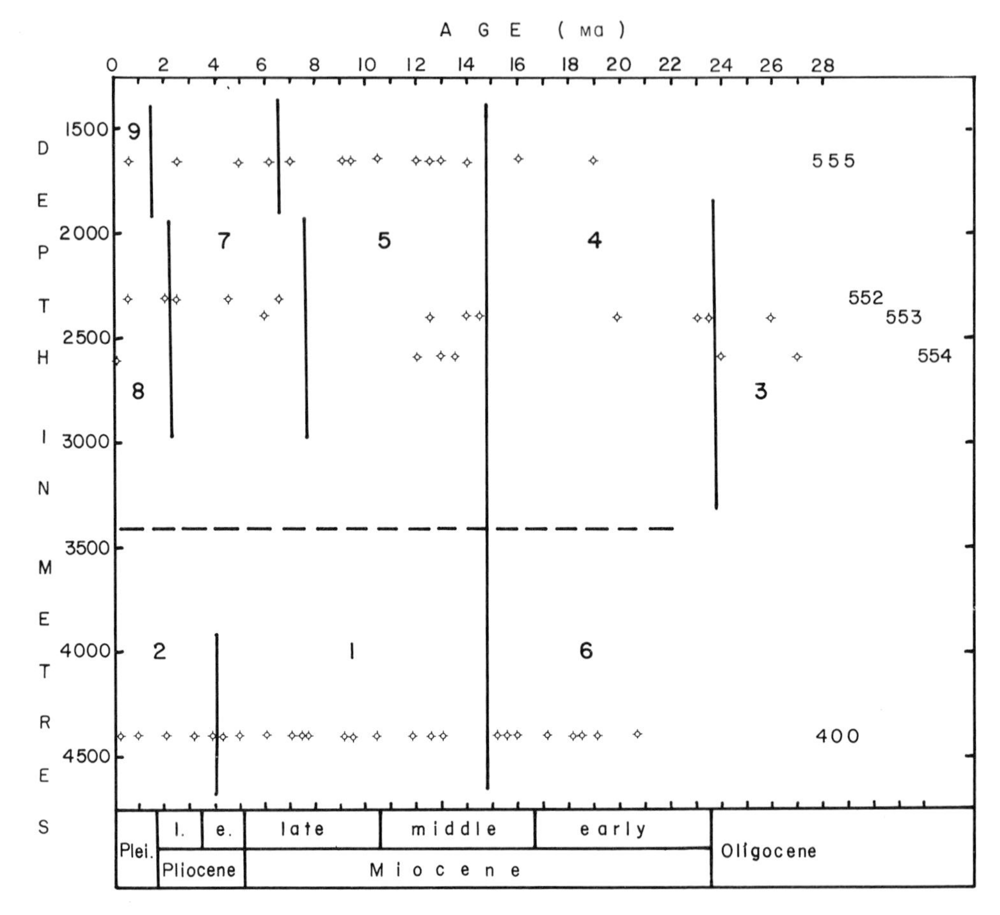

FIG. 9. Distribution of faunal principal components in time and space. Numbers correspond approximately to the ranking of relative quantitative significance of each principal component.

temporal and spatial, therefore reflects in large part the differentiation of the deep oceanic water masses at these sites.

The results of both the cluster and principal component analyses clearly indicate strong and persistent vertical faunal differentiation between the deepest site (400, 4399 m), the intermediate (552, 2301 m; 553, 2329 m; and 554, 2576 m), and shallow (555, 1659 m) sites. This suggests that the water masses of the eastern North Atlantic were stratified into abyssal and deep/intermediate water masses throughout the Neogene, much as they were in the Palaeogene (Schnitker 1979). The faunal analyses suggest that the water mass differentiation was stronger in the past than it is now because cluster analysis shows that the late Pliocene–Quaternary faunas at all levels are nearly the same. The principal component analysis distinguishes three faunas at three depth levels (Fig. 9: PC 9, PC 8, and PC 2) but the 'weight' of each of these three principal components is not very high. The relative similarity of faunas thus reflects clearly the relatively minor differentiation and stratification of North Atlantic Deep Water (NADW). These results closely parallel those of Weston & Murray (1984), who also found three faunal components associated with three levels of the North Atlantic Deep Water.

This modern situation became established only in the fairly recent geological past: it started nearly 4 Ma ago at the abyssal site, at about 2.5 Ma at intermediate depth, and slightly later at the shallow site (Fig. 9). The initial appearance of these faunas at Site 400 is perhaps related to the resumption of NADW production, as postulated by Blanc & Duplessy (1982), replacing there an antecedent Antarctic Bottom Water (AABW). While the first NADW filled the deep eastern North Atlantic Basin, the intermediate water masses did not appear until after the cooling

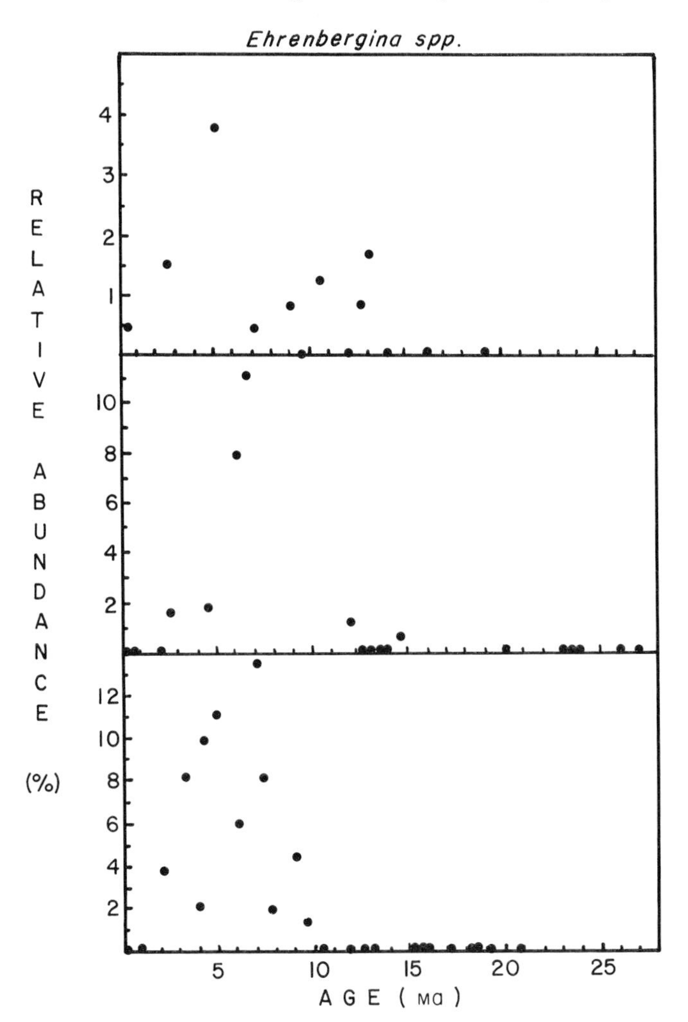

FIG. 10. Relative abundance plots of *Ehrenbergina* spp. Upper panel Site 555, centre panel Sites 552, 553, and 554, lower panel Site 400.

about 3.5 Ma ago (Berggren & Schnitker 1983) of the Labrador Sea (which is now a principal source of intermediate NADW) and the initiation of northern hemisphere glacial conditions at 2.4 Ma (Shackleton *et al.* 1984b).

None of the faunal assemblages older than the late Pliocene–Quaternary have exact modern analogues in the North Atlantic and cannot be directly interpreted to represent a particular water mass of a type found in the present North Atlantic. However, their similarity to modern faunas is close enough to enable us to make inferences about the deep water conditions of the past.

The abundance of *Osangularia umbonifera* at Site 400 during the Miocene and early Pliocene (Fig. 3) strongly suggests that water very much like Antarctic Bottom Water penetrated this far into the eastern North Atlantic at that time. The differentiation of this sequence into two or three faunas, by the appearance/disappearance of various accessory species, indicates that the characteristics of the early AABW were not constant through time, with the strongest changes occurring in the middle Miocene.

At the intermediate and shallow sites the late Miocene and early Pliocene faunas are persistently subdivided only temporally, which can be taken as evidence that the water masses were not as strongly differentiated and stratified as they are today (Fig. 9). The late Miocene and early Pliocene faunas are perhaps closest to modern deep NADW fauna of Weston & Murray (1984) but with considerably fewer *Epistominella exigua*, which may indicate somewhat lower levels of dissolved oxygen than in the present Norwegian Sea Overflow.

The faunal break during the middle Miocene, coincident with a similar faunal break at Site 400, signals the (re-) establishment of the North Atlantic Deep Water formation through overflow from the Norwegian Sea (Schnitker 1980b). Very high mid Miocene δ^{13}C values elsewhere in the Atlantic (Boersma & Shackleton 1977; Savin *et al.* 1981; McKenzie *et al.* 1981; Shackleton *et al.* 1984a) corroborate this faunally derived notion of new North Atlantic Deep Water formation. A change in the character of the Antarctic Bottom Water fauna, and thus of the AABW itself, is a necessary consequence of the new North Atlantic Deep Water formation: AABW is largely derived from Circumpolar Deep Water, itself a product of much NADW.

The early Miocene and early middle Miocene faunas, although containing most species of the modern faunas, were nevertheless so dominated by species which are now very rare (f. ex. *Stilostomella* spp., Fig. 2) that the deep waters at that time were probably not strict analogues of the modern waters at these sites.

An interesting phenomenon is the sharply increased abundance of *Ehrenbergina* spp. during the latest Miocene (Fig. 10). This genus has been associated with relatively low oxygen water in the South Atlantic (Lohmann 1978) and may here perhaps indicate a diminution of 'NADW' formation in response to the Messinian salinity crisis. During this period the Mediterranean became isolated from the North Atlantic and ceased to supply the highly saline outflow that apparently is necessary for the NADW formation. Blanc & Duplessy (1982) also noted this latest Miocene cessation of NADW formation.

ACKNOWLEDGEMENTS: The author thanks Drs Berggren, Corliss, Murray, and Summerhayes for reading the manuscript and for their considered criticism. Mr A. Hillyard helped with the computer data analysis. Financial support for this study came from a consortium of oil companies for the study of Cenozoic Benthic Foraminifers.

References

BERGGREN, W.A. & SCHNITKER, D. 1983. Cenozoic marine environments in the North Atlantic and Norwegian-Greenland Sea. *In:* BOTT, M.H.P., SAXOV, S., TALWANI, M. & THIEDE, J. (eds), *Structure and development of the Greenland-Scotland Ridge*. Plenum Press, New York. 495–548.

BLANC, P.L. & DUPLESSY, J.C. 1982. The deep water circulation during the Neogene and the impact of the Messinian salinity crisis. *Deep-Sea Res.* **29,** 1391–414.

BOERSMA, A. & SHACKLETON, N.J. 1977. Tertiary oxygen and carbon isotope stratigraphy, Site 356 (Mid latitude south Atlantic). *In: Init. Repts DSDP* **39,** US Govt. Print. Off., Washington, DC. 911–24.

BOYLE, E.A. & KEIGWIN, L.D. 1982. Deep circulation of the North Atlantic: geochemical evidence. *Science* **218,** 784–7.

CORLISS, B.H. 1979. Recent deep-sea benthonic foraminiferal distributions in the southeast Indian Ocean: inferred bottom-water routes and ecological implications. *Mar. Geol.* **31,** 115–38.

—— 1983. Distribution of Holocene deep-sea benthonic foraminifera in the southwest Indian Ocean. *Deep-Sea Res.* **30,** 95–117.

CURRY, W.B. & LOHMANN, G.P. 1982. Reduced advection into Atlantic Ocean deep eastern basins during last glacial maximum. *Nature* **308,** 377–80.

DOUGLAS, R.G. & WOODRUFF, F. 1981. Deep sea benthic foraminifera. *In:* C. EMILIANI (ed.), *The Ocean Lithosphere. The Sea,* **7**. Wiley Interscience, New York. 1233–327.

JONES, E.J.W., EWING, J.I. & EITTREIM, S.L. 1970. Influences of Norwegian overflow water on sedimentation in the northern North Atlantic and Labrador Sea. *J. geophys. Res.* **75,** 1655–80.

LANCE, G.N. & WILLIAMS, W.T. 1967. Mixed-data classificatory programs. I. Agglomerative systems. *Aust. Computer J.* **1,** 15–20.

LOHMANN, G.P. 1978. Abyssal benthonic foraminifera as hydrographic indicators in the western south Atlantic Ocean. *J. Foram. Res.* **8,** 6–34.

LONSDALE, P. & HOLLISTER, C.D. 1979. A near-bottom traverse of Rockall Trough: hydrographic and geologic inferences. *Oceanol. Acta* **2,** 91–105.

MCKENZIE, J.A., WEISSERT, H., POORE, R.Z., WRIGHT, R.C., PERCIVAL, S.F., OBERHANSLI, H. & CASEY, M. 1984. Paleoceanographic implications of the stable-isotope data from the upper Miocene–lower Pliocene sediments from the southeast Atlantic. *Init. Repts. DSDP* **74.** US Govt. Print. Off., Washington, DC. 717–24.

MONTADERT, L., ROBERTS, D.G. *et al.* 1979. *Init. Repts. DSDP* **48.** US Govt. Print. Off., Washington, DC. 1183 pp.

ROBERTS, D.G. 1975. Marine geology of the Rockall Plateau. *Phil Trans. R. Soc. Lond, Ser. A* **278,** 447–509.

——, SCHNITKER, D. *et al.* 1984. *Init. Repts. DSDP* **81.** US Govt. Print. Off., Washington, DC. 923 pp.

RUDDIMAN, W.F. 1972. Sediment redistribution on the Reykjanes Ridge: seismic evidence. *Geol. Soc. Am. Bull.* **83,** 2039–62.

SAVIN, S.M., DOUGLAS, R.G., KELLER, G., KILLINGLEY, J.S., SHAUGHNESSY, L., SOMMER, M.A., VINCENT, E.

& Woodruff, F. 1981. Miocene benthic foraminiferal isotope records: a synthesis. *Mar. Micropaleontol.* **6,** 423–50.

——, Abel, L., Barrera, E., Hodell, D., Keller, G., Kennett, J.P., Killingley, J., Murphy, M., Vincent, E. & Woodruff, F., in press. The evolution of Miocene surface and near-surface oceanography: oxygen isotope evidence. *Geol. Soc. Am. Mem.*

Schnitker, D. 1974. West Atlantic abyssal circulation during the past 120,000 years. *Nature* **248,** 385–7.

—— 1979. Cenozoic deep water benthic foraminifers, Bay of Biscay. *In: Init. Repts DSDP* **48.** US Govt. Print. Off., Washington, DC. 377–413.

—— 1980a. Quaternary deep-sea benthic foraminifers and bottom water masses. *Ann. Rev. Earth Planet. Sci.*, **8,** 343–70.

—— 1980b. North Atlantic oceanography as possible cause of Antarctic glaciation and eutrophication. *Nature*, **284,** 615–16.

Shackleton, N.J., Hall, M.A. & Boersma, A. 1984a. Oxygen and carbon isotope data from Leg 74 foraminifers. *In: Init. Repts DSDP* **74.** US Govt. Print. Off., Washington, DC. 599–612.

——, Backman, J., Zimmerman, H.B., Kent, D.V., Hall, M.A., Roberts, D.G., Schnitker, D., Baldauf, J.G., Desprairies, A., Homrighausen, H., Huddlestun, P., Keene, J.B., Kaltenback, A.J., Krumsiek, K.A., Morton, A.C., Murray, J.W. & Westberg-Smith, J. 1984b. Oxygen isotope calibration of the onset of ice-rafting and history of glaciation in the North Atlantic region. *Nature* **307,** 620–3.

Shor, A.N. & Poore, R.Z. 1979. Bottom currents and ice rafting in the North Atlantic: Interpretation of Neogene depositional environments of Leg 49 cores. *In: Init. Repts DSDP* **49,** US Govt. Print. Off, Washington, DC. 859–72.

Sneath, P.H.A. & Sokal, R.R. 1973. *Numerical Taxonomy.* W.H. Freeman & Co., San Francisco. 573 pp.

Streeter, S.S. 1973. Bottom water and benthonic foraminifera in the North Atlantic: glacial-interglacial contrasts. *Quat. Res.* **3,** 131–41.

Weston, J.F. & Murray, J.W. 1984. Benthic foraminifera as deep-sea water-mass indicators. *In:* Oertli, H.J. (ed.), *Benthos '83; Second International Symposium on Benthic Foraminifera (Pau, 1983).* Elf-Aquitaine, Esso REP and Total CFP, Pau. 607–10.

Worthington, L.V. & Wright, W.R. 1970. *North Atlantic Ocean Atlas (V).* Woods Hole Oceanographic Institution, Woods Hole, Massachusetts.

Detmar Schnitker, Department of Geological Sciences and Program in Oceanography, University of Maine at Orono. Mailing address, Ira C. Darling Center, Walpole, Maine 04573, USA.

Early to Middle Miocene benthic foraminiferal faunas from DSDP Sites 608 and 610, North Atlantic

Ellen Thomas

SUMMARY: Benthic foraminifera were studied in lower and middle Miocene cores from DSDP Site 608 (3534 m, 42°50′N., 23°05′W.) and Site 610 (2427 m, 53°13′N., 18°53′W.), north-eastern North Atlantic Ocean. There were extremely high relative abundances of *Bolivina spp.* between about 19.5 and 17 Ma at both sites. These high abundances were probably caused by oxygen deficiency in the bottom waters, which is suggestive of sluggish circulation.

Some changes in the faunal composition occurred at the beginning of the period of sluggish circulation (20–19 Ma), and between about 15 and 14 Ma, at which time there was a global increase in oxygen isotopic ratios in benthic foraminiferal tests. The changes in faunal composition involved less than 20% of the total fauna. Most first and last appearances of benthic species were not coeval at the two sites. The first appearance of *Cibicidoides wuellerstorfi* is an exception and occurred at the same time as the middle Miocene increase in oxygen isotopic ratios.

The early to middle Miocene was a period of major changes in the ocean-atmosphere system. Studies of oxygen isotopes in tests of benthic and planktonic foraminifera indicate that a general cooling and an increase in temperature gradients occurred at about 15–14 Ma. This increase is observed in lateral (pole-to-equator) gradients in surface waters, and in vertical (depth) gradients at low latitudes (Shackleton & Kennett 1975; Matthews and Poore 1980; Berger 1981; Berger *et al.*, 1981; Savin *et al.* 1981; Savin & Yeh 1981; Woodruff & Douglas 1981; Kennett 1983; Loutit *et al.* 1983).

There is discussion on how much of the isotopic signal was caused by initiation or growth of the Antarctic ice caps, and how much was a temperature effect. The importance of plate tectonic processes as initiators of the general cooling is also under discussion (e.g. the opening of the Drake Passage, the sinking of the Greenland-Faeroe Ridge).

Recent deep-sea benthic foraminiferal assemblages have been correlated with the water masses in which they live (see, e.g. Douglas & Woodruff 1981, for a review). It is to be expected, therefore, that deep-sea benthic foraminiferal faunas in the early to middle Miocene reacted to environmental changes in the deep waters at that time. Until recently, however, only one study on a detailed correlation between the faunal composition of benthic foraminifera and the isotopic record had been published (Woodruff & Douglas 1981). The authors concluded that there was a direct correlation between faunal composition and the increase in oxygen isotopic values in the western Pacific (DSDP Site 289, Ontong-Java Plateau), although the faunal event appears to last longer than the main isotopic change. Woodruff (1985) studied many more sites in the Pacific Ocean and concluded that the main phase in benthic evolution occurred at 16–13 Ma, correlated with the increase in oxygen isotopic values.

In the author's study of three sites in the central equatorial Pacific Ocean, however, it was concluded that the changes in the benthic foraminifera started at about 18.5 Ma and lasted until about 13 Ma (Thomas 1985). These changes affected only part of the total benthic foraminiferal faunas (less than 40%). A global increase in $\delta^{13}C$ values of benthic and planktonic foraminifera also began after the beginning of faunal changes in benthic foraminifera; this increase was centred at about 17.5 Ma and lasted about 1 m.y. (Vincent & Killingley 1985; Vincent *et al.* 1985). In the author's opinion there was no simple correlation between the faunal changes and the isotopic events, since the former started earlier. The faunal changes in benthic foraminifera in the equatorial Pacific were more closely related with a decrease in the $CaCO_3$-content of the sediments and an increase in dissolution (Thomas 1985; Vincent & Killingley 1985).

Atlantic early to middle Miocene benthic foraminiferal faunas changed in composition in the middle Miocene (Berggren 1972; DSDP Site 116, North Atlantic, and DSDP Sites 118, 119, Bay of Biscay). The time control at these sites is not precise, and isotopic data were not available. The exact level of first and last appearances of species at the different sites was not specified.

In contrast with the studies mentioned above are the conclusions of Boltovskoy (e.g. 1980), who observed only minor changes between Oligocene and Recent deep-sea benthic foraminiferal

From SUMMERHAYES, C.P. & SHACKLETON, N.J. (eds), 1986, *North Atlantic Palaeoceanography*, Geological Society Special Publication No. 21, pp. 205–218.

faunas from the South Pacific, South Atlantic, and Indian Oceans.

In this paper, data are presented on deep-sea benthic foraminiferal faunas from the early to middle Miocene at DSDP Sites 608 and 610 in the North Atlantic Ocean. The material recovered at Site 608 is ideal for studying lower to middle Miocene benthic foraminiferal assemblages: excellent time-control is provided by palaeomagnetic data (Clement & Robinson, in press), there are no appreciable hiatuses in the lower to middle Miocene part of the section, and recovery and preservation of calcareous microfossils are good. The record at Site 610 is not so well suited, because of intermittent coring; also, recovery and preservation were not so good.

Material and methods

On DSDP Leg 94 sediments older than upper Miocene were recovered at Sites 608 (present water depth 3534 m, 42°50′N., 23°05′W.) and 610 (2427 m, 53°13′N., 18°53′W.; Fig. 1). Site 608 is on the southern flank of the King's Trough tectonic complex. Major debris flows are recognized in the sediments at 369–75 m sub-bottom depth (site chapter in Kidd *et al.*, in press). Site 610 is on a major North Atlantic sediment drift, the Feni Drift. A regional reflector ('R2'; Miller & Tucholke 1983) was recognized at Site 610 at about 0.75 seconds two-way travel time, corresponding to about 640–60 m sub-bottom depth (site chapter in Kidd *et al.*, in press). Site 608 was cored continuously with the extended core barrel (XCB) to basement (middle Eocene) at about 530 m sub-bottom. Site 610 was also cored with the XCB, but intermittently (1 core every 50 m) in middle Miocene sediments (450–635 m sub-bottom). In the lowermost middle Miocene and lower Miocene coring was continuous (635–720 m sub-bottom).

At Site 608 the recovered record is good and time-control is satisfactory because of the overall good quality of the palaeomagnetic data (Clement & Robinson, in press). The author used the time-scale of Berggren *et al.* (in press), including a correlation of Anomaly 5 to Chron 11. Core recovery and micro-fossil preservation are good, in contrast with Site 610, where recrystallization is severe below about 640 m sub-bottom. Many specimens, especially of planktonic foraminifera, are crushed and flattened. The age control at Site 610 is poor, because of discontinuous coring and the generally poor recovery.

From Sites 608 and 610 all core catcher samples were studied. For a discussion of the faunas other than lower–middle Miocene see Thomas (in press, b). The most detailed research was done at Site 608, where additional samples were studied from cores 31 through 41. The author tried to obtain a sample distance equivalent to 100 000 to 200 000 years, which corresponds to 2–6 samples per core. The sedimentation rates were variable, and averaged about 17 m/m.y. (site chapter in Kidd *et al.*, in press). At Site 610 additional samples were studied from cores 19 through 27. It was intended to study one sample per section of 1.5 m, but several samples in cores 25, 26 and 27 were too indurated to be processed.

Samples were washed through a 63 μm sieve. Samples of strongly indurated chalks (cores 19–27, Site 610) were dried, soaked in kerosene, and heated in water. Splits were made of such a size that they contained about 200 specimens; all specimens counted were picked and mounted in slides. About 200 specimens were counted, since the species-versus-specimens curve for each sample is almost parallel to the specimen-axis after about 180 specimens, i.e. further counting gives very few extra species (Thomas 1985, in press, b). Samples containing less than 180 specimens were not used; these occurred only in cores 610–24–26. The lack of correlation between number of species and number of specimens in a sample is demonstrated in Fig. 2. Thus the number of species in a sample can be used as a measure of the species diversity.

Results and discussion

The diversity of the benthic assemblages as expressed in number of species per sample (200 specimens) is similar at Sites 608 and 610 (Figs 2 and 3), except for an interval with unusually low diversity at Site 608 between 345 and 355 m sub-bottom depth. This interval is called 'Interval A' and will be discussed below. The low diversity is caused by the extremely high relative abundance of species of the genus *Bolivina*, especially of *Bolivina spathulata*. Otherwise the number of species is usually between 50 and 65; the average is 53 ($\pm$11) at Site 608, 56 ($\pm$7) at Site 610. This diversity closely resembles the diversity at deep equatorial Pacific sites (Thomas 1985; in press, a). No obvious correlation exists between the number of species and sub-bottom depth, or between the number of species and the induration of the sediments as indicated by the level of penetration of the hydraulic piston corer (HPC), or the level of the first core to be split by the saw (Fig. 3).

The relative abundances of the most common species and species-groups are plotted versus sub-

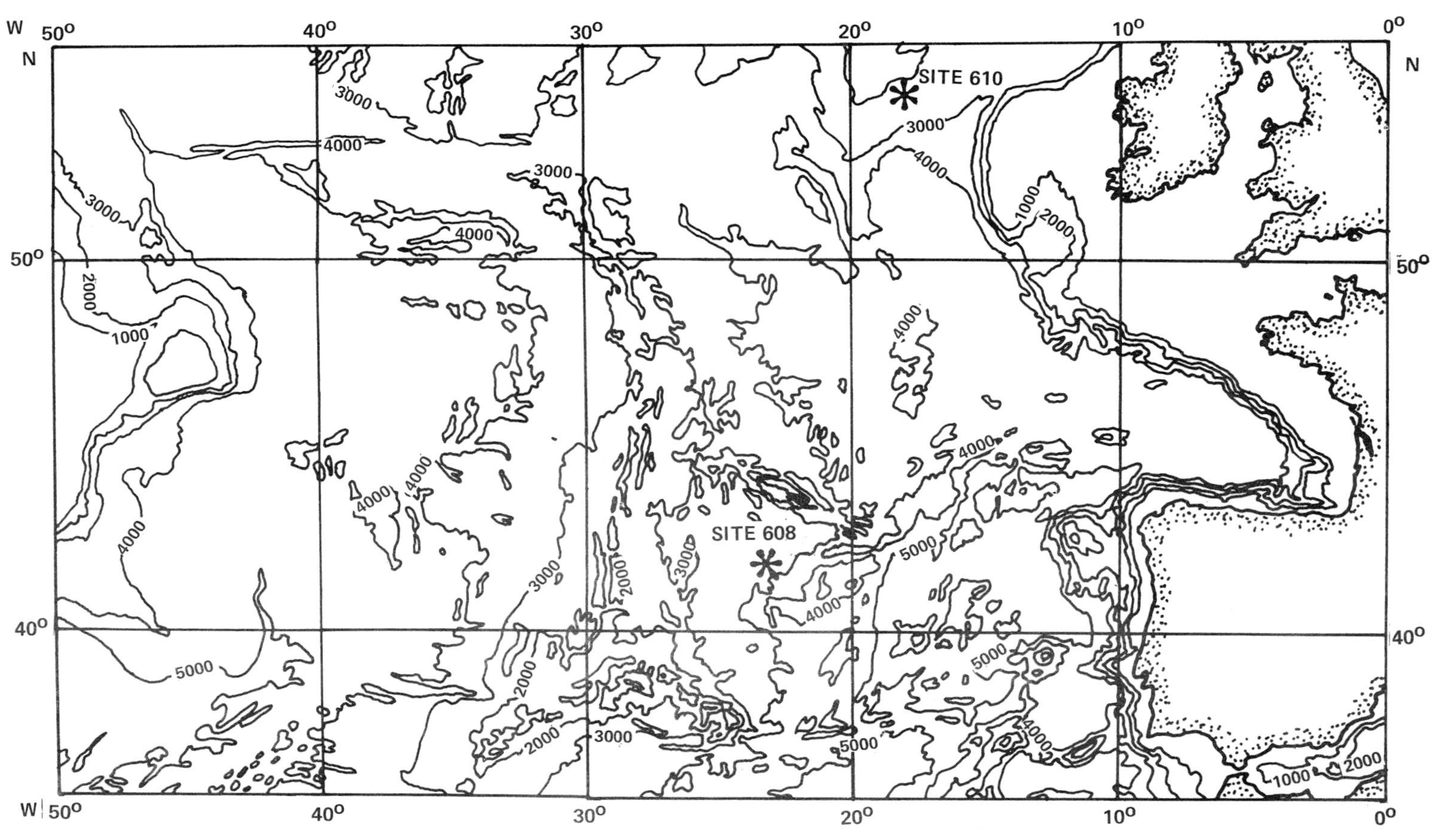

Fig. 1. Location of Sites 608 and 610, depth in metres, contour interval 1000 m.

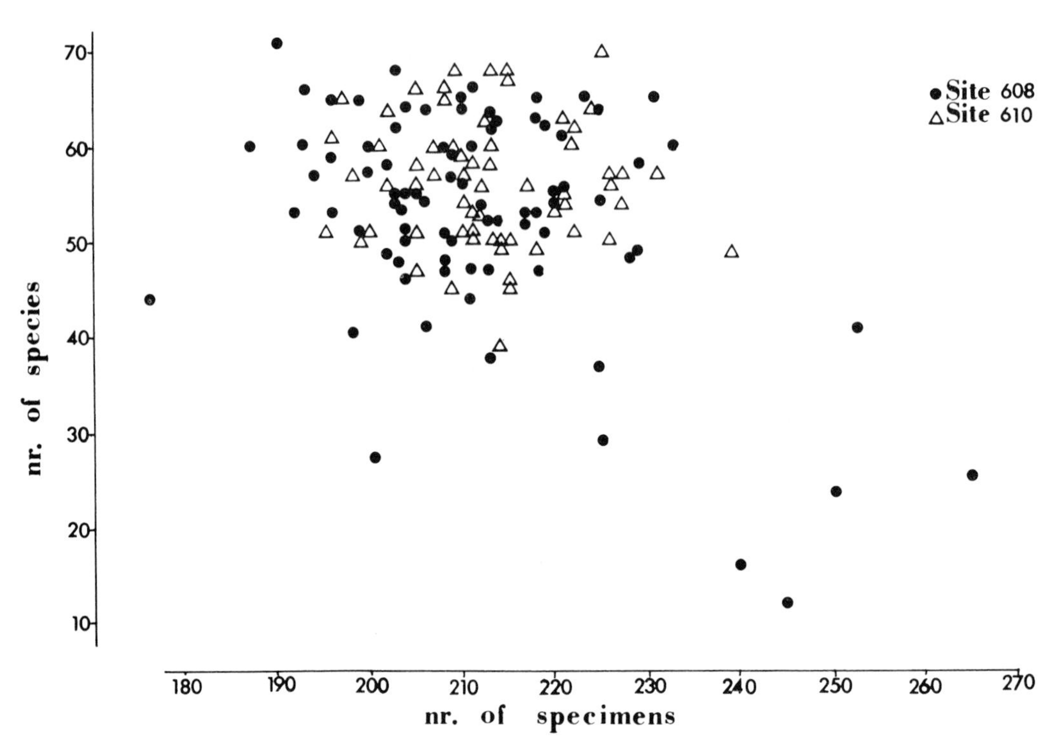

FIG. 2. Plot of numbers of species versus numbers of specimens for all samples from Sites 608 and 610. Note the absence of a positive correlation.

bottom depth in Figs 4 and 5. For a discussion of the taxonomy and a complete listing of the counts see Thomas (in press, b). Core numbers, core recovery, palaeomagnetic chrons (Clement & Robinson, in press), calcareous nannofossil zonation (Takayama & Sato, in press) and diatom zonation (Baldauf, in press, a) are shown. Note that calcareous nannofossil zone NN4 could not be recognized at Site 608, because of the absence of *Helicosphaera ampliaperta*.

Although the present water depth of the sites is quite different (Site 608 is in water about 1000 m deeper) many taxa occur at both sites. Overall, *Oridorsalis umbonatus*, *Globocassidulina subglobosa*, *Pullenia* spp., *Gyroidinoides* spp., and *Cibicidoides* spp. are the most common taxa, and all somewhat more abundant at Site 608 than at Site 610. Buliminids and *Fursenkoina* spp. are more common at Site 610. At the latter site *Bolivina* and *Fursenkoina* have the highest relative abundance below core 20. Relative abundances fluctuate, with maximum values of 43% (*Bolivina* spp.) and 21% (*Fursenkoina* spp.) (Fig. 5). The most common species of *Bolivina* are *B. spathulata* (highest relative abundance in cores 21–23) and *B. pseudoplicata* (core 24) and *B. striatula* (core 27).

At Site 608 *Bolivina* is most abundant in cores 37, 38 and the upper part of core 39 (340–65 m sub-bottom); the genus reaches a much higher maximum relative abundance than at Site 610 (Figs 4 and 5). In cores 608–37 and –38 *B. spathulata* has a relative abundance of more than 90%. In these samples the total number of benthic foraminifera as a fraction of the larger than 63 μm fraction is much larger than in samples with a more normal faunal composition; this might suggest higher population densities of benthic foraminifera during periods when *Bolivina* species had high relative abundances. Different species of *Bolivina* (*B. spathulata*, *B. striatula*, and *B. pseudoplicata*) have their highest abundance at different levels, as do *Uvigerina* spp. and *Bulimina* spp. (Fig. 6). *B. striatula* and *B. pseudoplicata* are large species with an ornamented test (the former has costae, the latter a reticulate ornamentation); *B. spathulata* is a small, thin-walled, smooth species.

The times of the highest relative abundance of *Bolivina* spp. at both sites appear to be coeval within the time-resolution of this study. Interval A can be correlated with magnetic Chron C5C through the topmost part of Chron C6 at Site 608 (Fig. 4). Time control is less precise at Site 610, but interval A appears to occur in the same

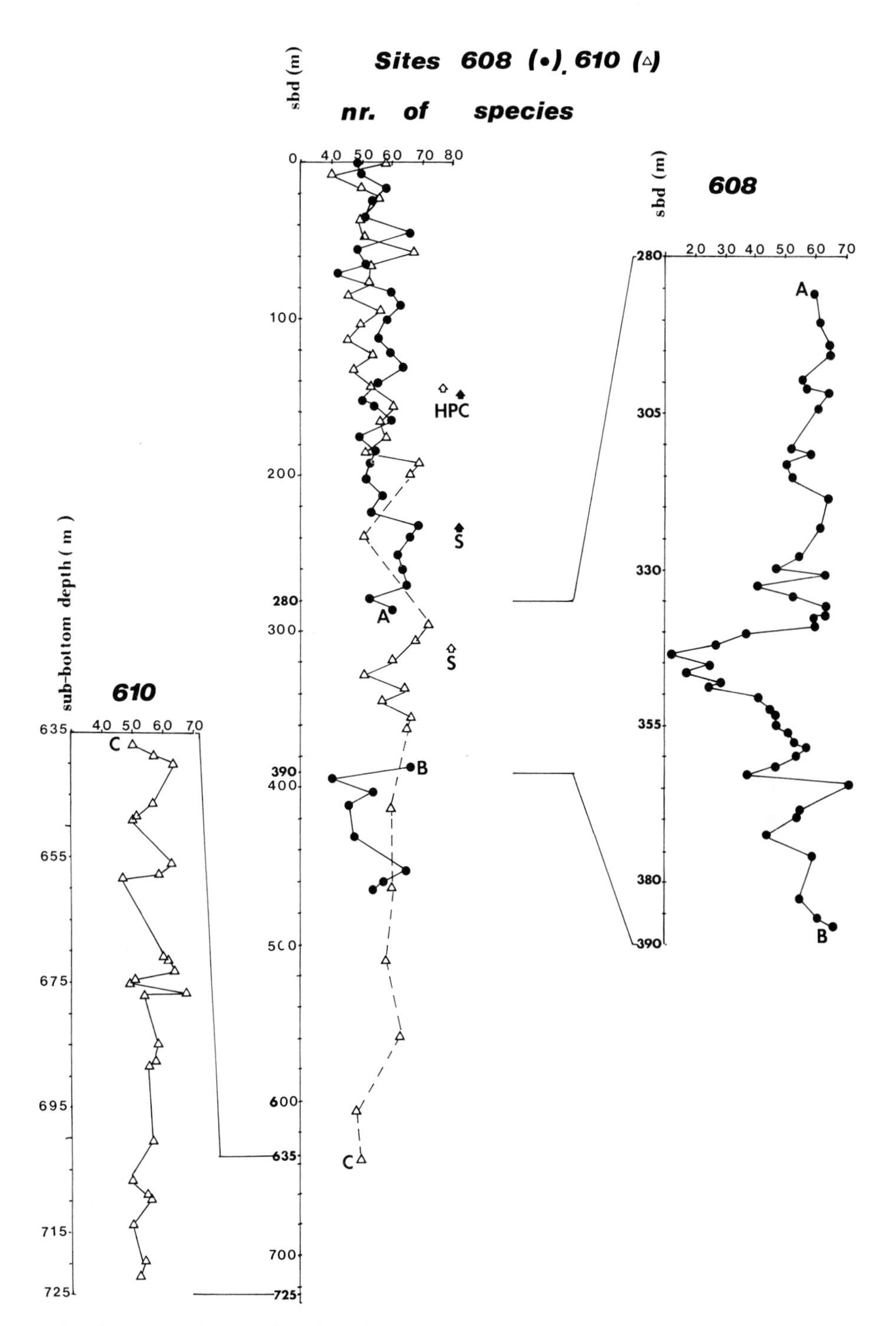

FIG. 3. Diversity expressed as number of species per sample, plotted versus sub-bottom depth for Sites 608 and 610. ◆ HPC: lower level of HPC penetration at Site 608. ◇ HPC: lower level of HPC penetration at Site 610. ◆ S: level of first core split by saw, Site 608. ◇ S: level of first core split by saw, Site 610.

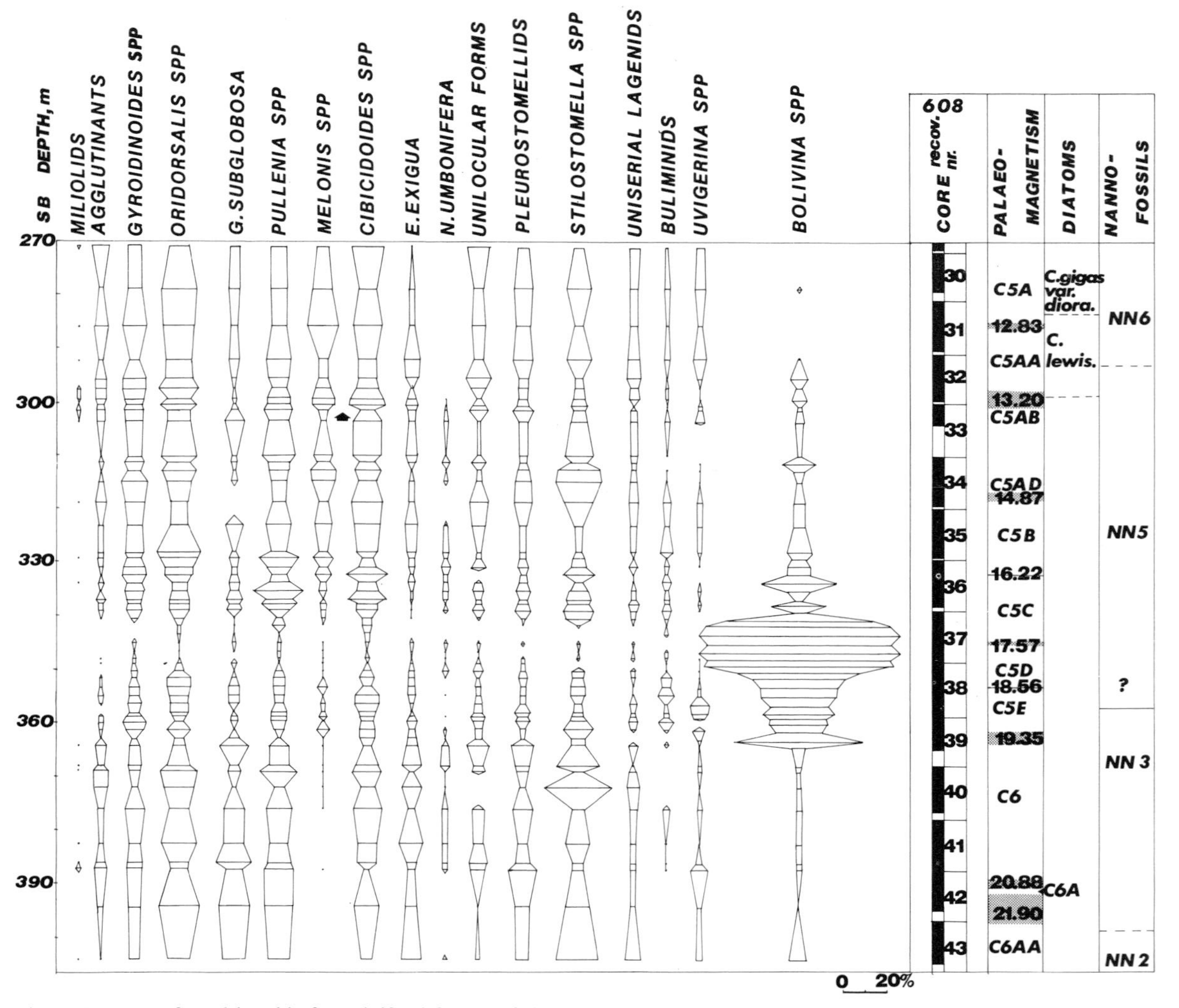

FIG. 4. Relative abundance (percent of total benthic foraminiferal fauna) of the most common species and species groups plotted versus sub-bottom depth for Site 608. Arrow indicates the FA of *Cibicidoides wuellerstorfi*. Indicated are core number, core recovery (black = recovered), palaeomagnetic data (Clement & Robinson in press), calcareous nannofossil zonation (Takayama & Sato, in press), and diatom zonation (Baldauf, in press, a).

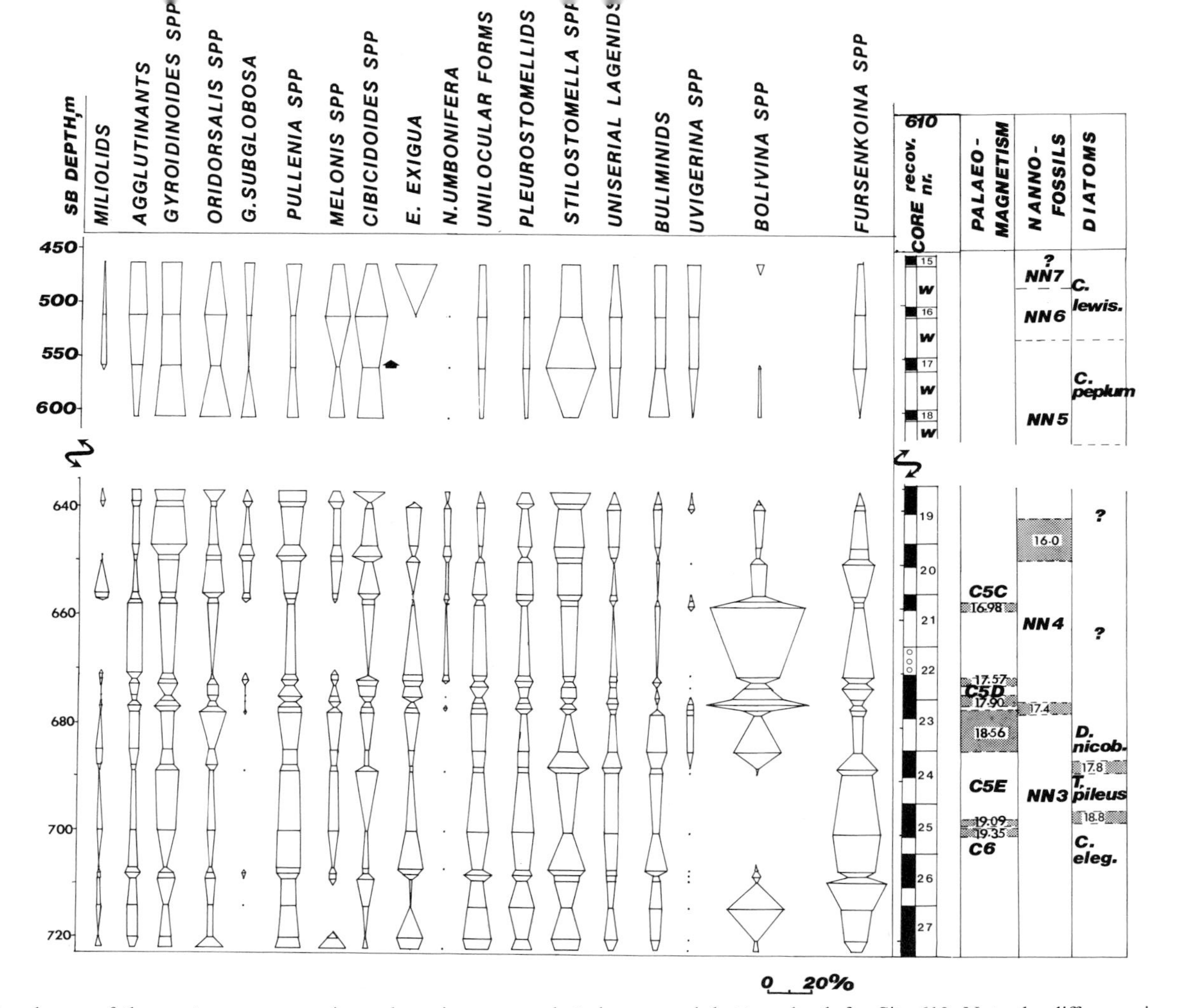

FIG. 5. Relative abundances of the most common species and species groups plotted versus sub-bottom depth for Site 610. Note the difference in scale in the upper part of the figure. Arrow indicates the FA of *Cibicidoides wuellerstorfi*. Indicated are core number, core recovery (black = recovered), palaeomagnetic data (Clement & press), calcareous nannofossil zonation (Takayama & Sato, in press), and diatom zonation (Baldauf, in press, a).

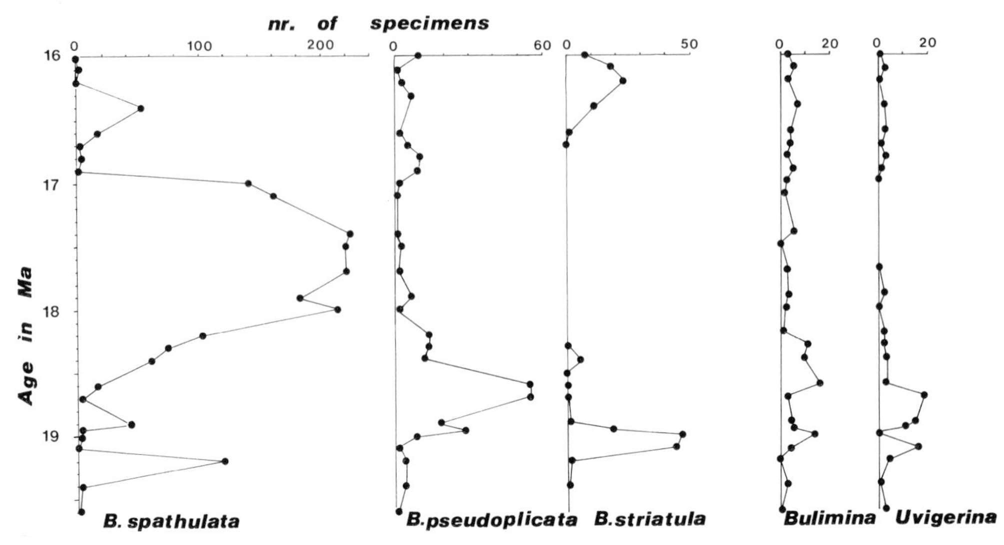

FIG. 6. Numbers of specimens of different species of *Bolivina*, *Bulimina spp.*, and *Uvigerina spp.* plotted versus time for Site 608. Note that the scale for *Bolivina spathulata* is different.

interval of the magnetic polarity time-scale, i.e. chron C5C-top C6, about 19.5–17 Ma (Figs 4 and 5).

High relative and total abundances of *Bolivina* species (and of species of the related genera *Brizalina* and *Fursenkoina*) are known to occur in areas where the bottom waters are deficient in oxygen, and the sediments are rich in organic matter (Boltovskoy & Wright 1976). Such circumstances exist in basins with sluggish circulation, e.g. the southern California basins (Harman 1964; Phleger & Soutar 1973; Douglas 1981). In those basins small, smooth species of *Bolivina* (similar to *B. spathulata*) are associated with the lowest concentrations of oxygen. Oxygen deficiencies in bottom waters also exist in shelf and slope areas in the oxygen-minimum zone (generally between 200 and 1500 m depth), and are caused by the accumulation of large amounts of organic matter in zones of high surface productivity (upwelling) (Boltovskoy & Wright 1976; Sen Gupta *et al.* 1981; Poag & Low 1985). Recently high abundances of *Bolivina* species have also been noted where the abundant organic matter is supplied by oil spills (Casey *et al.* 1980, 1981).

The pattern of successive species dominance noted at Site 608 (Fig. 6) resembles faunal patterns in Mediterranean cores with sapropel layers, where different species alternate in abundance at different levels in or close to a sapropel layer (e.g. Van der Zwaan 1980; Mullineaux & Lohmann 1981; Cita & Podenzani 1982; Ross & Kennett 1983): as conditions become more extreme (less oxygen available), species with somewhat different environmental tolerances survive. *B. spathulata* and species of similar morphology are among the species that tolerate oxygen deficiencies well.

It is surprising to find *Bolivina* abundances as high as at Site 608 in the deep open Atlantic, not a shelf-slope area or an enclosed basin. The sediments at Site 608 show some evidence that oxygen levels may have been somewhat lower than usual: cores 37 through 40 are light brown, in contrast with white chalks above and below. *Zoophycus* burrows are rare to absent in these cores. There is no evidence for high productivity in the surface waters: diatoms are not found in cores 608–32 and lower (site 608 chapter, Kidd *et al.*, in press). At Site 610 there is no sedimentological evidence for oxygen deficiency. At this site fragments of the diatom *Ethmodiscus* are abundant in cores 24 and 25, and diatoms are absent or badly dissolved in the cores with the highest *Bolivina* abundance. *Ethmodiscus* oozes have been explained as resulting from phytoplankton blooms, or from differential dissolution (Baldauf, in press b).

It is not probable that the *Bolivina*-rich faunas are the result of down-slope transport: the specimens are fragile and thin-walled, and do not show evidence of abrasion. The faunal pattern (Fig. 6) also does not suggest transport. In addition, level A at both sites does contain chalk-breccias, thought to represent debris-flows (Hill, in press). These debris-flows do contain faunas that are not rich in *Bolivina* species.

In the author's opinion the high *Bolivina*

abundances cannot be fully explained. The geographic setting of the sites and the nature of the sediments are not similar to those of known areas with high *Bolivina* abundances (enclosed basins or shelf-slope areas with high surface water productivity). At about the time of the high *Bolivina* abundances there was a worldwide increase in $\delta^{13}C$ values in benthic and planktonic foraminifera, which suggests that there was an increase in global organic productivity (Vincent & Killingley 1985; Vincent & Berger 1985; Miller & Fairbanks 1985). This increase in $\delta^{13}C$ values does occur at Site 608, but seems to be at the top of the *Bolivina*-rich interval (Miller *et al.*, in press). Interval A may have been deposited in a time when the local topography at Site 608 had been rejuvenated and was rugged, as indicated by the debris-flows. Probably the circulation in the north-east North Atlantic in this period became more sluggish, which caused low oxygen contents in the bottom waters. At Site 608 the effects of sluggish bottom water circulation may have been enhanced by the local topography. At Site 610 the *Bolivina* abundance data suggest that the sluggish circulation ended relatively suddenly (core 20), just below the level of a regional reflector ('R2', Miller & Tucholke 1983). This reflector has generally been interpreted as the result of a period of more vigorous circulation.

The author studied first appearances (FA's) and last appearances (LA's) of species of benthic foraminifera to determine whether the high relative abundances of *Bolivina* spp. are related to major, irreversible changes in faunal composition. For deep-sea benthic foraminifera there are presently not enough data available to evaluate if FA's and LA's are synchronous on a global, basin-wide or local scale. Therefore a 'faunal event' was defined as a first or last appearance of a taxon, whether this FA or LA is evolutionary or migratory. In the author's opinion it is reasonable to use this combination of migratory and evolutionary events: when there are important changes in the environment, benthic faunas can react by migrating (geographic or depth migration) or evolving. If faunal events turn out to be more numerous in certain periods, then we can assume that these were times of important environmental changes. We have to look at the number of faunal events per time unit over long periods of time to determine whether there are significant fluctuations in the event-rate. The number of faunal events at Sites 608 and 610 is plotted cumulatively versus time in Fig. 7.

There was no exceptionally large increase in the number of faunal events per time-unit correlated with the high relative abundance of *Bolivina sp.*, although there are considerable fluctuations in the event-rate during the last 25 m.y. There appears to be a small increase in the number of events at Sites 608 and 610 at about the beginning of interval A, and a somewhat larger increase at about the level of the increase in oxygen isotopic values. The increase in oxygen isotopic values at Site 608 was observed in cores 34–32 (Miller *et al.*, in press), with the larger part of the increase between about 301 and 311 m sub-bottom. This increase in oxygen isotopic ratios has been corre-

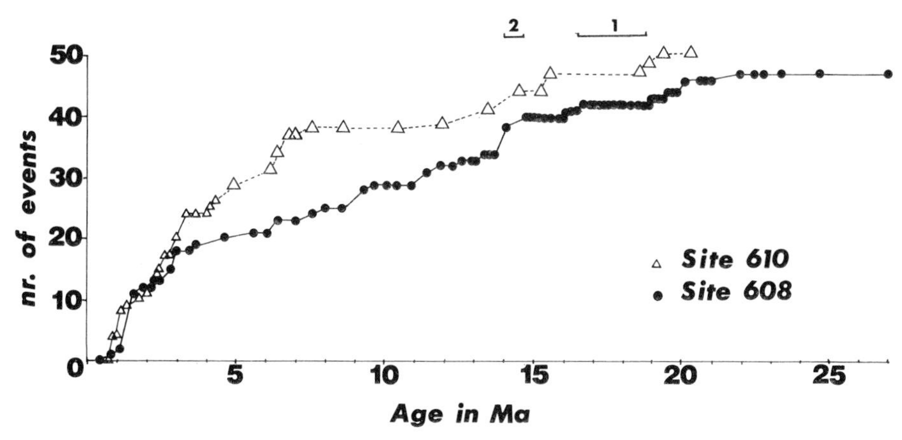

FIG. 7. Number of faunal events (first and last appearances, evolutionary and migratory) plotted cumulatively versus time for Sites 608 and 610. Absolute ages from extrapolation of palaeomagnetic data (Clement & Robinson, in press). Time control is poor at Site 610 for the dotted part of the curve. Figure after Thomas (in press.). 1: time of increase of oxygen isotopic ratios in benthic foraminiferal tests (*Cibicidoides* spp.) at Site 608; 2: time of increase in carbon isotopic ratios in benthic foraminiferal tests (*Cibicidoides* spp.) at Site 608. Isotopic data are after Miller *et al.* (in press).

TABLE 1. *Faunal events at Site 608*

Faunal event	Sample	Assigned age (Ma)**
1. FA *Cibicidoides laurisae*	29,CC	11.4–11.9
2. FA *Melonis pompilioides*	31–4	12.6–12.9
3. FA *Quinqueloculina compta*	32,CC	13.4–13.5
4. LA *Cibicidoides havanensis*	33,CC	13.7–14.1
5. FA *Cibicidoides wuellerstorfi*	33,CC	14.1–14.8*
6. FA *Eilohedra weddellensis*	33,CC	14.1–14.8
7. FA *Ophthalmidium pusillum*	33,CC	14.1–14.8
8. LA *Cibicidoides perlucidus*	34–1	14.1–14.8
9. LA *Gyroidinoides girardanus*	34–1	14.1–14.8
10. LA *Bolivina spathulata*	36–2	16.0–16.1
11. FA *Cibicidoides cicatricosus*	36–5	16.6–16.7
12. FA *Ehrenbergina caribbea*	38,CC	18.8–18.9*
13. FA *Bolivina spathulata*	40–1	19.6–19.7
14. FA *Melonis barleeanus* group	40,CC	20.1–20.6
15. FA *Uvigerina peregrina*	41,CC	22.0–22.4

* Faunal event coeval at Site 610
** Extrapolated from palaeomagnetic data, Clement & Robinson (in press)

TABLE 2. *Faunal events at Site 610*

Faunal event	Sample	Assigned age (Ma)**
1. LA *Anomalina spissiformis*	16,CC	±13.5
2. FA *Francesita advena*	16,CC	±13.5
3. LA *Bolivinopsis cubensis*	17,CC	±14.5
4. FA *Cibicidoides wuellerstorfi*	17,CC	±14.5*
5. FA *Sigmoilopsis schlumbergeri*	17,CC	±14.5
6. LA *Fursenkoina mexicana*	19–2	±15.3
7. FA *Bolivina translucens*	19,CC	±15.5
8. FA *Bulimina elongata*	19,CC	±15.5
9. FA *Ehrenbergina caribbea*	24–3	±18.8*
10. FA *Stilostomella annulifera*	24–3	±18.8
11. LA *Bigenerina nodosaria*	26–4	±19.5?

* Faunal event coeval at Site 608
** Age control at Site 610 not precise

lated with palaeomagnetic Chron C5AD (Barron *et al.* 1985; Miller *et al.* 1985). The age of this isotopic event is estimated at between 14.6 and 14.0 Ma at Site 608, which is in excellent agreement with earlier estimates (Woodruff & Douglas 1981).

At Site 608 the faunal event rate was somewhat higher after about 14.5 Ma than before that time; at Site 610 the event rate was low after 14.5 Ma, and similar to the rate before that time. The FA's and LA's that occurred during the early and middle Miocene are listed in Table 1 (Site 608) and Table 2 (Site 610). Only two events are synchronous at the two sites: the FA of *Cibicidoides wuellerstorfi* (14.1–14.8 Ma) and the FA of *Ehrenbergina caribbea* (18.8–18.9 Ma; ages from extrapolation of palaeomagnetic data). More events were coeval at a deep Atlantic site (Site 608) and deep sites in the equatorial Pacific (Table 3) than at a deep (Site 608) and a shallower (Site 610) Atlantic site. According to Murray (1984) different watermasses overlie these sites presently: North-east Atlantic Deep Water (NEADW) Site 610, North Atlantic Deep Water Site 608. A

TABLE 3. *Comparison of faunal events at Site 608 (North Atlantic Ocean) and Sites 573, 574 and 575 (equatorial Pacific, depth ±4.5 km)*

Faunal event	Sites 573–575	Site 608
1. FA *Melonis pompilioides*	11.8–13.1	12.6–12.9
2. FA *Cibicidoides wuellerstorfi*	15.0–15.6	14.1–14.8
3. LA *Cibicidoides perlucidus*	12.0–15.0	14.1–14.8
4. FA *Ehrenbergina caribbea*	17.5–20.5	18.8–18.9
5. FA *Melonis barleeanus* group	20.1–20.6	18.2–20.1

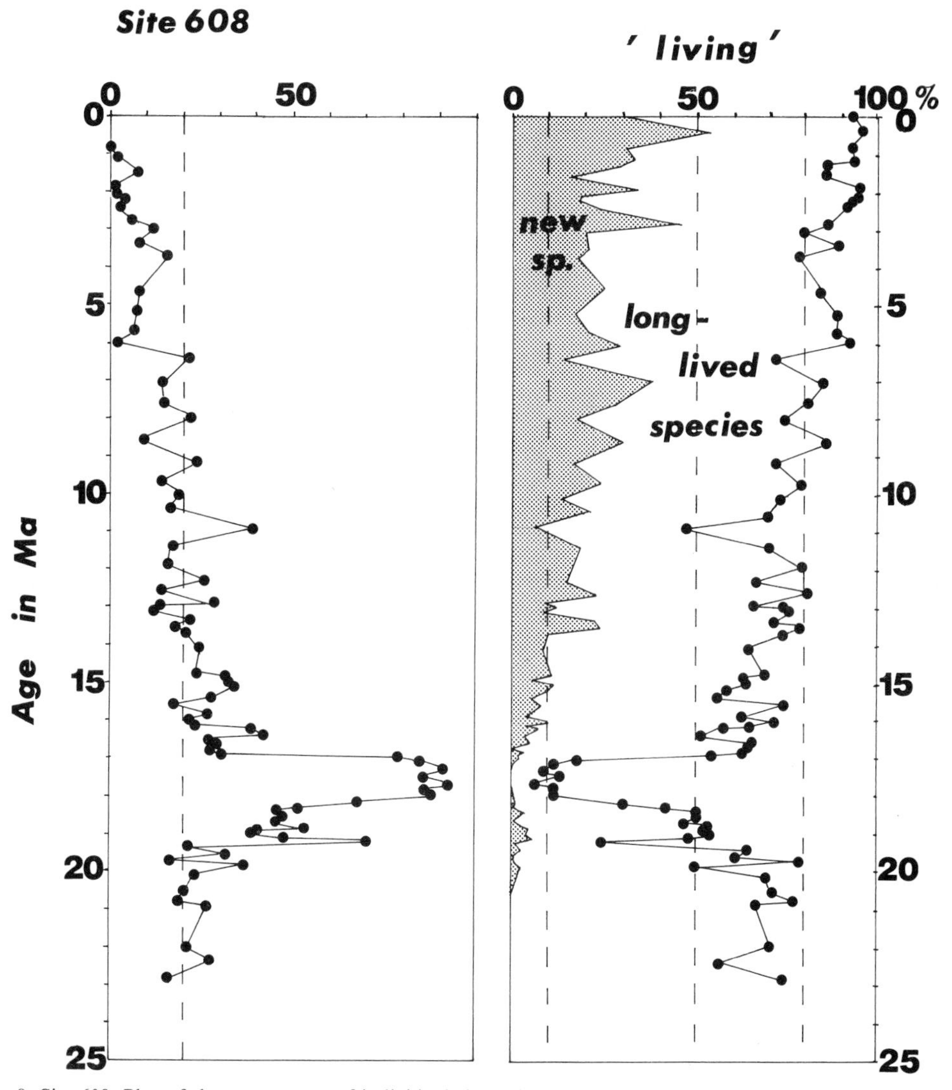

FIG. 8. Site 608. Plot of the percentages of individuals in each sample, belonging to species that are not present in a sample from the mud-line at Site 608 on the left side. On the right side are plotted the percentages of specimens in each sample that belong to a species present in a mud-line sample from Site 608, divided into long-lived species (found throughout the section, 0–25 Ma) and 'new species', which have a first appearance during the last 25 Ma (shaded area).

similar difference in watermasses in the Miocene could explain the difference in benthic foraminiferal faunas at the sites. The FA of *Cibicidoides wuellerstorfi* occurred at about the same time in Atlantic and Pacific, deeper and shallower sites; this FA is also coeval with the shift in oxygen isotopic ratios. It should be emphasized that the faunal changes in the early and middle Miocene in the North Atlantic appear to be less important than later faunal changes: at both sites important changes occurred at about 3 Ma (probably associated with the initiation of northern hemisphere glaciation), and at Site 610 a major period of change was at about 6 Ma (possibly associated with the desiccation of the Mediterranean in the Messinian; see Thomas, in press, b for a more detailed discussion of these periods).

Data on first and last appearances emphasize the rare species; Fig. 8 shows the proportion of the fauna that belonged to species that have either an FA or a LA at Site 608. This figure emphasizes the more common species: it is obvious that less than 20% of the total benthic foraminiferal fauna at about 14 Ma belonged to species that had appeared at the site since the early Miocene.

Conclusions

(1) There were strong fluctuations in the number of faunal events (first apperances or last appearances) of benthic foraminifera per time unit during the last 25 m.y.

(2) There were changes in the faunal composition of deep-sea benthic foraminifera in the early to middle Miocene at Sites 608 and 610 in the North Atlantic. A first, relatively minor change occurred at about 20–19 Ma, followed by a somewhat larger change at 15–14 Ma. Species that have first or last appearances at those times constitute less than 20% of the total fauna.

(3) Individual events were generally not coeval at Sites 608 and 610, with the first appearance of *Cibicidoides wuellerstorfi* and of *Ehrenbergina caribbea* (at about 14.5 and 18.8 Ma respectively) as exceptions.

(4) The first appearance of *Cibicidoides wuellerstorfi* is coeval with the major increase in oxygen isotopic values in the early middle Miocene; both events might be related to environmental changes at that time.

(5) There was a peak in the relative abundance of *Bolivina spp.* at Sites 608 and 610, but more extreme at Site 608 (more than 90% versus 45% at Site 610). This high relative abundance can be correlated with the top of palaeomagnetic Chron C6 through Chron C5C (19.5–17 Ma); it may indicate that at that time the bottom waters in the north-eastern North Atlantic were deficient in oxygen, probably as a result of sluggish circulation.

ACKNOWLEDGEMENTS: I thank the shipboard scientific party and the DSDP technicians on Leg 94 for making this Leg so successful, and Joop Varekamp and Drew Carey at the Department of Earth and Environmental Sciences at Wesleyan University (Middletown, CT) for the use of facilities. The manuscript has benefitted from reviews by K. G. Miller, E. Boltovskoy, J. W. Murray and C. Summerhayes. This research has been partially funded by NSF grant OCE 83-10518.

References

BALDAUF, J.G., in press a. Diatom biostratigraphy of the middle and high latitude North Atlantic Ocean, DSDP Leg 94. *In:* KIDD, R.B., RUDDIMAN, W.F. *et al.* (eds), *Init. Repts DSDP* **94.** US Govt. Print. Off., Washington, DC.

——, in press b. Biostratigraphic and paleoceanographic interpretations of diatom preservation in lower and middle Miocene sediments from the Rockall Plateau Region, North Atlantic Ocean. *In:* KIDD, R.B., RUDDIMAN, W.F., *et al.* (eds), *Init. Repts DSDP* **94.** US Govt. Print. Off., Washington, DC.

BARRON, J.A., KELLER, G. & DUNN, D.A., 1985. A multiple microfossil biochronology for the Miocene. *In:* KENNETT, J.P. (ed.), *The Miocene Ocean: Paleoceanography and Biogeography. Mem. Geol. Soc. Am.* **163,** 21–35.

BERGER, W.H. 1981. Paleoceanography: the deep sea record. *In:* EMILIANI, C. (ed.), *The Oceanic Lithosphere, The Sea,* **7.** Wiley Interscience, New York 489–504.

BERGGREN, W.A. 1972. Cenozoic biostratigraphy and paleobiogeography of the North Atlantic. *In:* LAUGHTON, A.S., BERGGREN, W.A. *et al.* (eds), *Init. Repts DSDP* **12.** US Govt. Print. Off., Washington, DC. 965–1002.

——, KENT, D.V. & VAN COUVERING, J. 1985. Neogene geochronology and chronostratigraphy. *In:* SNELLING N.J. (ed.), *The Chronology of the Geological Record,* Spec. Publ. Geol. Soc. London. Blackwell Scientific Publications, Oxford.

BOLTOVSKOY, E. 1980. On the benthonic bathyal zone foraminifera as stratigraphic guide fossils. *J. Foram. Res.* **10,** 163–72.

—— & WRIGHT, R. 1976. *Recent Foraminifera.* W. Junk, The Hague.

CASEY, R., AMOS, R., ANDERSON, J., KOEHLER, R., SCHWARZER, R. & SLOAN, J. 1980. A preliminary report of the microplankton and microbenthon responses to the 1979 Gulf of Mexico oil spills (Ixtoc I and Burmah Agate), with comments on avenues of oil to the sediments and the fate of oil in the column and on the bottom. *Gulf Coast Ass. Geol. Soc., Trans.* **30,** 273–81.

——, HUENI, C. & LEAVESLEY, A. 1981. *Brizalina lohmani*, a meroplankton foraminiferan useful as an indicator of shelfal circulation and eutrophication (with comments on biostratigraphy and evolution). *Gulf Coast Ass. Geol. Soc., Trans.* **31,** 249–55.

CITA, M.B. & PODENZANI, M. 1980. Destructive effect of oxygen starvation and ash falls on benthic life: a pilot study. *Quat. Res.* **13,** 230–41.

CLEMENT, B.M. & ROBINSON, F., in press. The magnetostratigraphy of Leg 94 sediments. *In:* KIDD, R.B., RUDDIMAN, W.F. *et al.* (eds), *Init. Repts DSDP,* **94.** US Govt. Print. Off., Washington, DC.

DOUGLAS, R.G. 1981. Paleoecology of continental margin basins: a modern case history from the borderland of Southern California. *In: Depositional Systems of Active Continental Margin Basins. Soc. econ. Paleont. Miner., Pacific Section, Short Course,* 1–165.

—— & WOODRUFF, F. 1981. Deep sea benthic foraminifera. *In:* EMILIANI, C. (ed.), *The Oceanic Lithosphere, The Sea,* **7.** Wiley Interscience, New York. 1233–327.

HARMAN, R.A. 1964. Distribution of foraminifera in the Santa Barbara Basin, California. *Micropaleontology* **10,** 81–96.

HILL, P.R., in press. Chalk solution structures in cores from DSDP Leg 94. *In:* KIDD, R.B., RUDDIMAN, W. F. *et al.* (eds), *Init. Repts DSDP* **94.** US Govt. Print. Off., Washington, DC.

KENNETT, J.P. 1983. Paleo-oceanography: global ocean evolution. *Rev. Geophys. Space Phys.* **21,** 1258–74.

KIDD, R.B., RUDDIMAN, W.F., BALDAUF, J.G., CLEMENT, B.M., DOLAN, J., EGGERS, M.R., HILL, P.R., KEIGWIN, L.D., MITCHELL, M., PHILIPPS, I., ROBINSON, F., SALEHIPOUR, S., TAKAYAMA, T., THOMAS, E., UNSOLD, G., WEAVER, P.P.E., in press. *Init. Repts DSDP* **94.** US Govt. Print. Off., Washington, DC.

LOUTIT, T.S., KENNETT, J.P. & SAVIN, S.M. 1983. Miocene equatorial and southwest Pacific paleoceanography from stable isotope evidence. *Marine Micropaleontol* **8,** 215–33.

MATTHEWS, R.K. & POORE, R.Z. 1980. Tertiary delta-^{18}O record and glacioeustatic sea-level fluctuations. *Geology* **8,** 501–4.

MILLER, K.G., AUBRY, M.P., KHAN, M.J., MELILLO, A.J., KENT, D.V. & BERGGREN, W.A. 1985. Oligocene to Miocene biostratigraphy, magnetostratigraphy, and isotopic stratigraphy of the western North Atlantic. *Geology* **13,** 257–61.

—— & FAIRBANKS, R.G. 1985. Oligocene to Miocene global carbon isotope cycles and abyssal circulation changes. *In:* SUNDQUIST, E.T. & BROECKER, W.S. (eds), *The Carbon Cycle and Atmospheric CO_2: Natural Variations, Archean to Present. Geophysical Monographs* **32,** 469–86.

——, FAIRBANKS, R.G. & THOMAS, E., in press. Benthic foraminiferal carbon isotopic records and the development of abyssal circulation in the eastern North Atlantic. *In:* KIDD, R.G., RUDDIMAN, W.F. *et al., Init. Repts. DSDP* **94.** US Govt. Print. Off., Washington, DC.

—— & TUCHOLKE, B.E. 1983. Development of Cenozoic abyssal circulation south of the Greenland-Scotland Ridge. *In:* BOTT, M.H.P., SAVOX, S., TALWANI, M. & THIEDE, J. (eds), *Structure and development of the Greenland-Scotland Ridge.* NATO Conf. Series. Plenum Press, New York. 549–89.

MULLINEAUX, L.S. & LOHMANN, G.P. 1981. Late Quaternary stagnation and recirculation of the eastern Mediterranean: changes in the deep water recorded by fossil benthic foraminifera. *J. Foram. Res.* **11,** 20–39.

MURRAY, J.W. 1984. Paleogene and Neogene benthic foraminifers from Rockall Plateau. *In:* ROBERTS, D. G., SCHNITKER, D. *et al. Init. Repts DSDP* **81.** 503–34.

PHLEGER, F.B. & SOUTAR, A. 1973. Production of benthic foraminifera in three East Pacific oxygen minima. *Micropaleontology* **19,** 110–15.

POAG, C.W. & LOW, D. 1985. Paleoenvironmental trends among Neogene benthic foraminifers at DSDP Site 549, Irish continental margin. *In:* DE GRACIANSKY, P.C., POAG, C.W. *et al., Init. Repts DSDP* **80.** US Govt. Print. Off., Washington, DC. 489–504.

ROSS, C.R. & KENNETT, J.P. 1983. Late Quaternary paleoceanography as recorded by benthonic foraminifera in Strait of Sicily sediment sequences. *Mar. Micropaleontol.* **8,** 315–36.

—— & YEH, H. 1981. Stable isotopes in ocean sediments. *In:* EMILIANI, C. (ed.), *The Oceanic Lithosphere, The Sea* **7.** Wiley-Interscience, New York. 1521–54.

SAVIN, S.M., DOUGLAS, R.G., KELLER, G., KILLINGLEY, J.S., SHAUGHNESSY, L., SOMMER, M.A., VINCENT, E. & WOODRUFF, F. 1981. Miocene benthic foraminiferal isotope record: a synthesis. *Mar. Micropaleontol* **6,** 423–50.

SEN GUPTA, B.K., LEE, R.F. & MAY, M.S. 1981. Upwelling and unusual assemblages of benthic foraminifera on the northern Florida continental slope. *J. Paleontol* **55,** 853–7.

SHACKLETON, N.J. & KENNETT, J.P. 1975. Paleotemperature history of the Cenozoic and the initiation of Antarctic glaciation: oxygen and carbon isotope analysis in DSDP Sites 277, 279, and 281. *In:* KENNETT, J.P., HOUTZ, R.E. *et al.* (eds), *Init. Repts DSDP* **29,** 743–55.

TAKAYAMA, T. & SATO, T., in press. Coccolith biostratigraphy, DSDP Leg 94. *In:* KIDD, R.B., RUDDIMAN, W.F. *et al.* (eds), *Init. Repts DSDP* **94.** US Govt. Print. Off., Washington, DC.

THOMAS, E. 1985 Late Eocene to Recent deep-sea benthic foraminifera from the central equatorial Pacific Ocean. *In:* MAYER, L.A., THEYER, F. *et al.* (eds), *Init. Repts DSDP* **85.** US Govt. Print. Off., Washington, DC. 655–94.

——, in press, a. Changes in composition of Neogene

benthic foraminiferal faunas in equatorial Pacific and north Atlantic. *Palaeogeogr., Palaeoclimatol., Palaeoecol.*, **53.**

THOMAS, E. in press,b. Oligocene to Recent benthic foraminifera from the northeastern North Atlantic Ocean. *In:* KIDD, R.B., RUDDIMAN, W.F. *et al.* (eds), *Init. Repts DSDP* **94.** US Govt. Print. Off., Washington, DC.

VAN DER ZWAAN, G.J. 1980. The impact of climatic changes on deep sea benthos. *Kon. Nederlandse Akad. Wetensch., Proc., Series B,* **83,** 379–97.

VINCENT, E. & BERGER, W.H., 1985. Carbon dioxide and Antarctic ice build-up in the Miocene: the Monterey hypothesis. *In:* SUNDQUIST, E.T. & BROECKER, W.S. (eds), *The Carbon Cycle and Atmospheric CO_2: Natural Variations Archean to Present. Geophysical Monographs* **32,** 455–68.

—— & KILLINGLEY, J.S., 1985. Oxygen and carbon isotope record for the early and middle Miocene in the central equatorial Pacific, DSDP Leg 85, and paleoceanographic implications. *In:* MAYER, L.A., THEYER, F. *et al., Init. Repts DSDP* **85.** US Govt. Print. Off., Washington, DC. 749–70.

——, KILLINGLEY, J.S. & BERGER, W.H., 1985. Miocene oxygen and carbon isotope stratigraphy of the tropical Indian Ocean. *In:* KENNETT, J.P. (ed.), *The Miocene Ocean: Paleoceanography and Biogeography. Mem. Geol. Soc. Am.* **163,** 131–75.

WOODRUFF, F. 1985. Changes in Miocene deep-sea benthic foraminiferal distribution in the Pacific Ocean: relationship to paleoceanography. *In:* KENNETT, J.P. (ed.), *The Miocene Ocean: Paleoceanography and Biogeography. Mem. Geol. Soc. Am.* **163,** 131–75.

—— & DOUGLAS, R.G. 1981. Response of deep sea benthic foraminifera to Miocene paleoclimatic events, DSDP Site 289. *Mar. Micropaleont.* **6,** 617–32.

ELLEN THOMAS, Scripps Institution of Oceanography, La Jolla, California 92093, USA. Present address: Lamont-Doherty Geological Observatory, Palisades, New York 10964, USA.

Miocene to Recent bottom water masses of the north-east Atlantic: an analysis of benthic foraminifera

J.W. Murray, J.F. Weston, C.A. Haddon & A.D.J. Powell

SUMMARY: The modern benthic foraminifera of the north-east Atlantic Ocean show a distribution pattern related to bottom water masses. Varimax factor analysis of data from depths >2000 m gives factors which correlate with North-East Atlantic Deep Water (NEADW), North Atlantic Deep Water (NADW) and Antarctic Bottom Water (AABW); a fourth factor corresponds with the basal part of the Mediterranean Water.

Miocene to Pleistocene benthic foraminiferal assemblages have been compared with the modern assemblages. On the assumption that species have not changed their environmental preferences through time, the fossil assemblages can be used to infer the past existence of water masses similar to those of today. The results presented here show that in the late Miocene, AABW penetrated to the southern flank of Rockall Plateau at the time of build-up of the East Antarctic ice sheet. During the early Pliocene the rate of production of NADW increased, perhaps as a result of a decrease in size of the Antarctic ice sheet, which in turn led to the southward retreat of the limit of penetration of AABW. Throughout much of the Pleistocene, as now, AABW has not penetrated north of the Azores Ridge. NEADW was the most extensive bottom water mass in the late Pliocene and Pleistocene.

The earliest study of the modern benthic foraminiferal assemblages from the North Atlantic was that of Phleger *et al.* (1953). Streeter (1973) reinterpreted their data, using Q-mode factor analysis, and recognized three main assemblages whose distribution correlated with that of the bottom water masses. Subsequent studies have confirmed this relationship and have used comparisons between modern and fossil assemblages to infer the existence of similar water masses in the Quaternary (Streeter 1973; Schnitker 1974, 1976, 1979, 1980, 1982; Streeter & Shackleton 1979).

The Recent benthic foraminifera of the north-east Atlantic have been studied in detail by Weston (1982). This author carried out varimax factor analysis and showed that each water mass has a characteristic benthic fauna. Weston's data were used for comparison with the Neogene assemblages of DSDP Hole 119 (Weston & Murray 1984), Holes 552A–555 (Murray 1984) and Holes 609B, 610 and 611 (Murray, in press) in order to recognize the distribution of past water masses. In this paper the study is extended to interpret the regional changes in the distribution of water masses during the Miocene to Pleistocene. The authors have contributed as follows: J.W. Murray, DSDP Holes 400A, 552A, 555, 609B, 610, 611; J.F. Weston, Recent data base and DSDP Hole 119; C.A. Haddon, Pleistocene; A.D.J Powell, DSDP Holes 116, 116A

Material and methods

Assemblages of ~200 benthic individuals were picked from the >125 μm fraction of the sediment. Most of the species discussed are illustrated in Murray (1984). The biostratigraphic data for DSDP samples are taken from the *Initial Reports*, with additional data for Site 116 from Backman (1979). The Pleistocene successions have been correlated by their isotope stratigraphies (using data supplied by the Institute of Oceanographic Sciences). The borehole and core sites are shown on Fig. 1.

Interpretation of fossil assemblages

The fossil assemblages have been interpreted by subjective comparison with the modern assemblages recognized by Weston (1982) and also by Q-mode varimax factor analysis. The latter was carried out using a program which expresses the fossil assemblages in terms of the modern data set with which it is being compared. With two exceptions all the fossil material came from sites having a present water depth >2000 m, and these were compared with modern data from >2000 m (open ocean data of Weston 1982). Sites 116 and 555 are from 1151 m and 1666 m respectively, and these have been compared with a large data set that includes samples from a depth range ~200 to ~5200 m (Weston & Murray 1984). A summary of the Recent data and water masses is given in Table 1.

The foraminiferal assemblages, in the source areas of AABW and in the South Atlantic where AABW is close to source, are dominated by *Osangularia umbonifera* (Anderson 1975; Lohmann 1978). In the north-east Atlantic the so

From Summerhayes, C.P. & Shackleton, N.J. (eds), 1986, *North Atlantic Palaeoceanography*, Geological Society Special Publication No. 21, pp. 219–230.

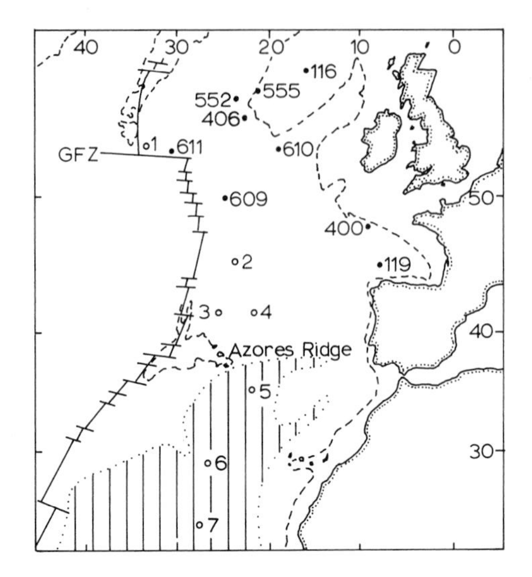

FIG. 1. Position of the boreholes: ● = DSDP Sites; ○ = IOS cores, 1 = D10611, 2 = D9812-1, 3 = 82PCS01, 4 = 58-79-3, 5 = D10330, 6 = D10320, 7 = D10316. Dashed line is 2000 m isobath. GFZ = Gibbs Fracture Zone. Shaded area = region where AABW for ms > 5% of the total bottom water. (From Heezen & Hollister 1971, p. 377.)

called AABW is made up of 5–10% true AABW mixed with waters of northern origin (Heezan & Hollister 1971). *O. umbonifera* makes up <20% of the foraminiferal assemblages beneath such water but, nevertheless, this is strongly picked up in the varimax factor analysis.

The fossil data are summarized in Table 2. These results are considered to be statistically significant because the sample variance explained is > 50%, and the loading on a factor is 0.5–1.0 or −0.5– −1.0. These samples can be interpreted in terms of past water masses similar to those observed today, namely NEADW, NADW and AABW. Because of its importance as an indicator of AABW, the abundances of *O. umbonifera* are listed in Table 3.

In order to show sequential changes in the distribution of past bottom water masses, the results are presented as a series of time-slices.

Miocene–Pliocene

The data are summarized in Tables 2 and 3, and the interpretations are shown in Figs 2–4.

The earliest record of AABW (defined by the abundance of *O. umbonifera*, factor 2, Tables 1–3) is from the middle Miocene (NN5) of Sites 119

TABLE 1. *Q-mode varimax factor analysis of open ocean Recent benthic foraminiferal assemblages from depths >2000 m in the north-east Atlantic (from Weston 1982). S = salinity. T = temperature*

Factor	Principal species	Varimax factor score	Water mass
			North Atlantic Deep Water =
1	*Planulina wuellerstorfi*	0.51	NADW
	Globocassidulina subglobosa	0.49	S > 34·90–35·0‰
	Cibicidoides kullenbergi	0·40	T 2.5°–4°C
	Oridorsalis umbonatus	0.38	O_2 5.2–5.6 ml/l
			Antarctic Bottom Water =
2	*Osangularia umbonifera*	−0.91	AABW
			S <34·90‰
			T <2°C
			O_2 >6 ml/l
			North-East Atlantic Deep Water =
3	*Epistominella exigua*	−0.80	NEADW or upper
			NADW
			S 34.92–34.97‰
			T 3°–4°C
			O_2 6 ml/l
			Mediterranean Water =
4	*Cassidulina obtusa*	0.67	MW
	Globocassidulina subglobosa	0.56	S variable >35.0‰
			T >4°C
			O_2 ~4 ml/l

TABLE 2. *Fossil assemblages. Distribution of varimax factors by nanno zone, based on samples with >50% variance accounted for and factor scores of 0.5–1.0 or −0.5– −1.0. 0 = no definitive factor; ( ) = factor score 0.45–0.49; * = affected by dissolution; / = no data. 1 = NADW, 2 = AABW, 3 = NEADW = factors from Recent data base >2000 m; I = NADW, II, VII, IX = MW = factors from Recent data for all depths (Weston 1982)*

Site	611	609	400	119	610	406	552	555	116
NN 18	(3)	3	3	/	0	/	3	/	VII
NN 17	/	/	/	/	/	/	0	/	/
NN 16 top	3	3	3	1,2,3	1	/	0	/	II
NN 16 bottom	2	2	1	/	1	/	0	/	I
NN 15	1+(2)	2	/	/	0	/	0	/	0
NN 14	1	/	1	/	/	/	0	/	/
NN 13	2	1/2	(1)	/	/	/	?1	/	0
NN 12	2	1	/	/	/	2	1	I	IX
NN 11 top	1+3	2	3	/	1	1	1	I	0
NN 11 mid	1	/	3	2	0	1	/	I/VII	I
NN 11 bottom	0	/	2	/	0	1	0	/	0
NN 10	0	/	/	/	0	/	0	/	/
NN 9	0	/	/	/	0	/	0	/	/
NN 8	/	/	0	/	/	/	/	/	/
NN 7	/	/	0	/	3	/	0	/	/
NN 6	/	/	/	2/1	0	/	/	/	?I
NN 5	/	/	2*	/	1/3	/	/	/	/

TABLE 3. *Per cent abundance of* Osangularia umbonifera *in the fossil assemblages. *Sample has undergone dissolution*

Site	611	609	400	119	406
NN 16 top				6–22	
NN 16 bottom	8	11			
NN 13	12	15			
NN 12					22
NN 11 top		18			
NN 11 middle		26*		6–14	
NN 11 bottom			15		
NN 5			20*	7–21	

and 400. From the late Miocene to the late Pliocene there were progressive changes in the geographic extent of AABW. In the late Miocene (NN11 bottom part, Fig. 2(a), (b)) it was present at Site 400, and in the middle and upper parts of NN11 it was present at Sites 119 and 609 (Fig. 2(c)–(f)). By the Miocene–Pliocene boundary AABW had reached its northernmost extent (Fig. 3(a), (b); no data for Sites 119 and 400). In the early Pliocene the northern limit of AABW retreated to the south (Fig. 3(c)–(f)), by the late Pliocene (NN16 top, Fig. 4(a), (b)) it was present only at Site 119, and then it withdrew from the area altogether (Fig. 4(c), (d)). These changes are summarized in Fig. 4(e), (f), each line representing AABW distribution at the times given in millions of years. NADW (defined by factor 1, Tables 1 and 2) was the dominant northern water

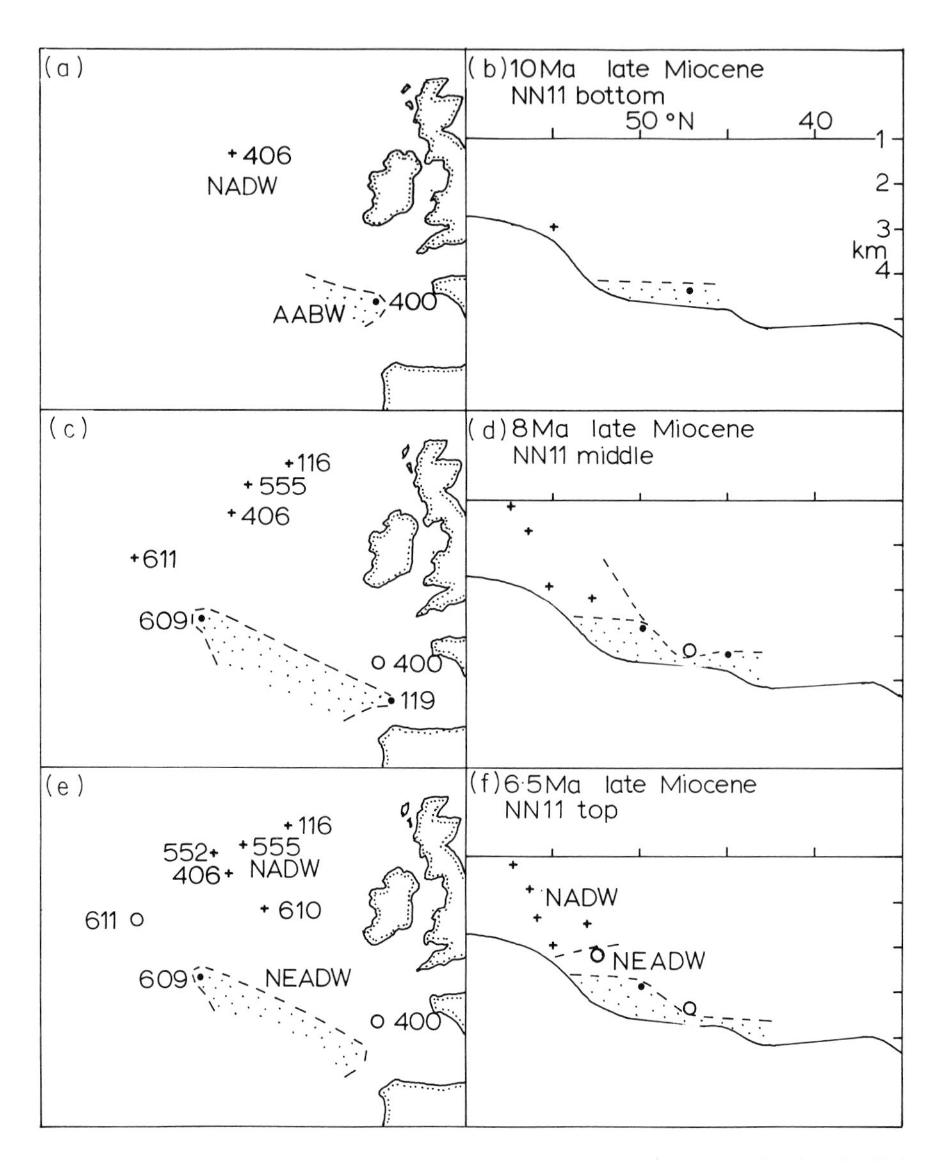

FIG. 2. Distribution of late Miocene factors and inferred bottom water masses. The profiles (b), (d), (f) show the bathymetry of the deepest areas at each latitude but excluding closed basins. ○ = factor 3 = NEADW; + = factor 1 = NADW; ● = factor 2 = AABW. Area of occurrence of AABW stippled.

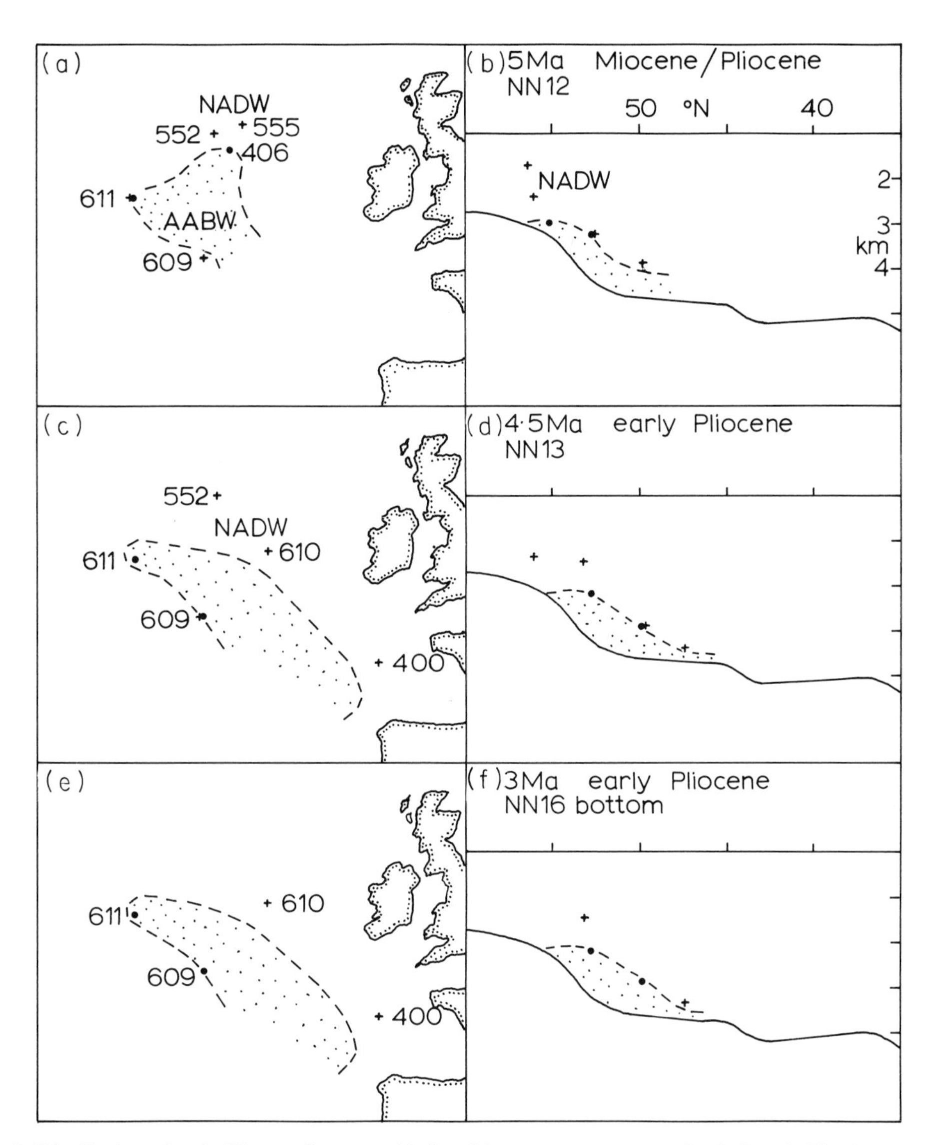

FIG. 3. Distribution of early Pliocene factors and inferred bottom water masses. Symbols as in Fig. 2.

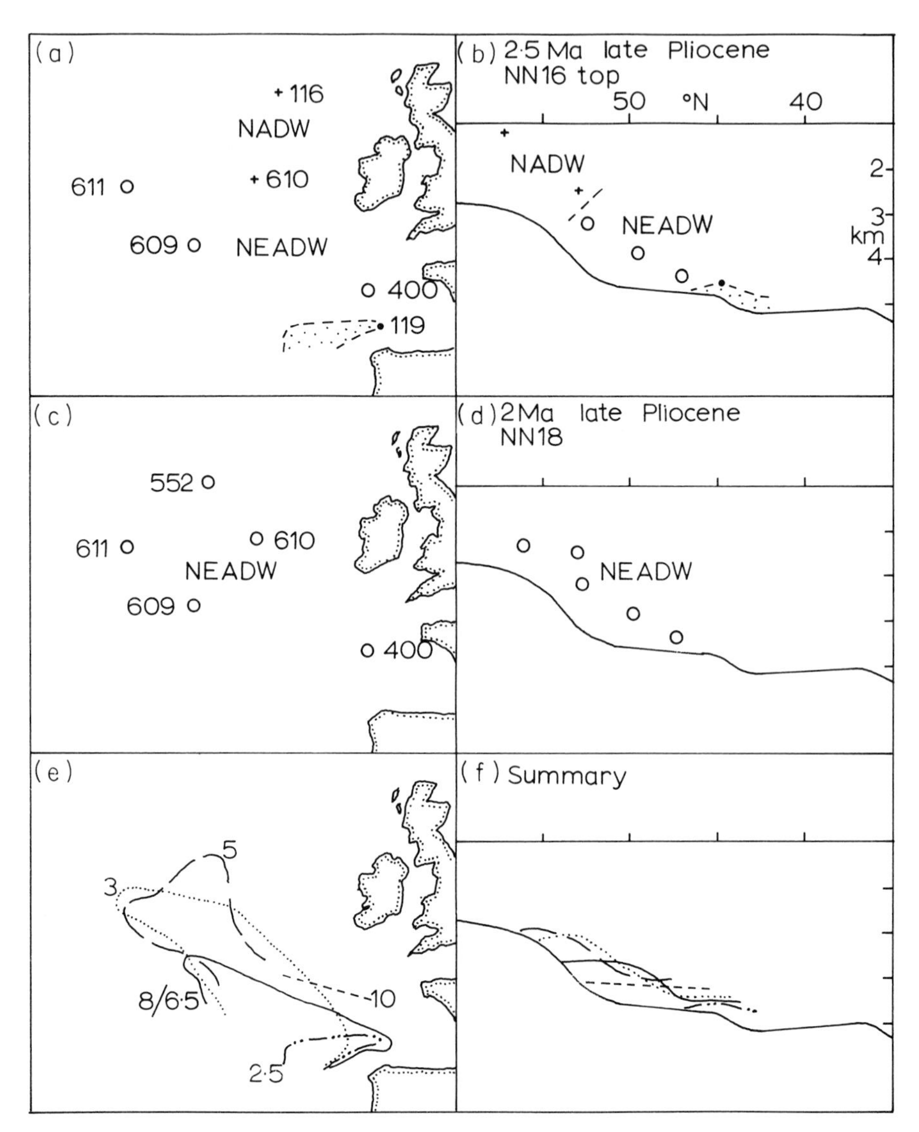

FIG. 4. (a)–(d). Distribution of late Pliocene factors and inferred bottom water masses. Symbols as in Fig. 2. (e)–(f). Summary of the northern limit of AABW. Numbers represent millions of years: 10, 8, 6.5 = late Miocene; 5 = Miocene–Pliocene boundary; 3 = early Pliocene; 2.5 = late Pliocene.

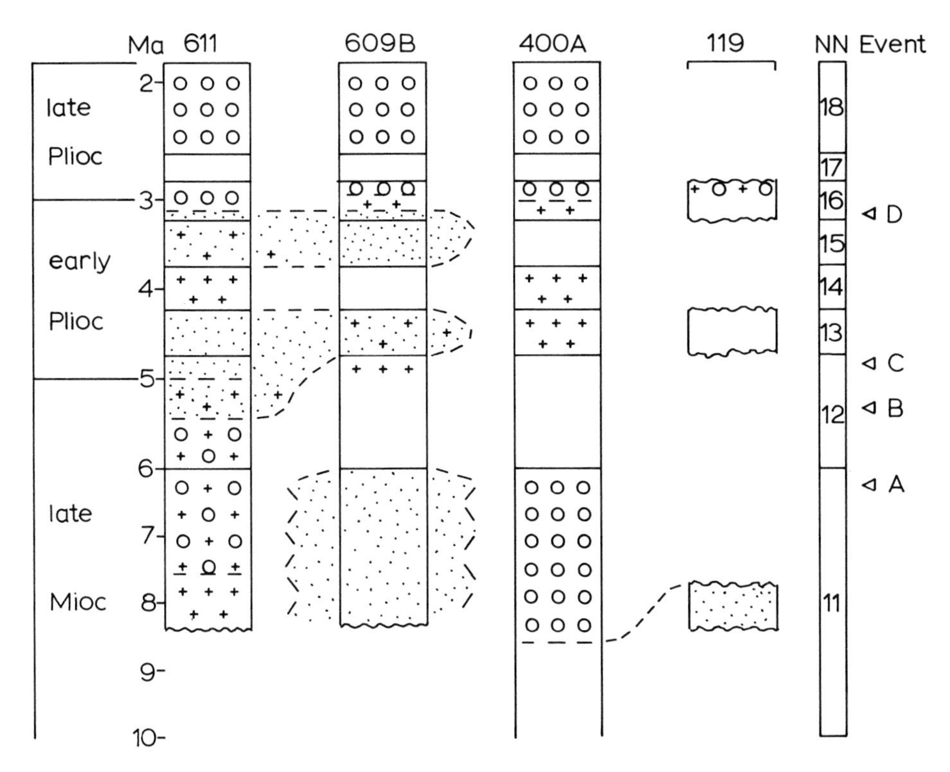

FIG. 5. Comparison of the Neogene for four DSDP sites. Symbols as in Fig. 2. Mixed ornament indicates two equally dominant factors. Blank = no data. Event A = maximum build up of ice in East Antarctica; B = Messinian salinity crisis; C = resumption of formation of NADW; D = build up of continental ice in the northern hemisphere.

mass from the late Miocene to early Pliocene (Figs 2 and 3) but in the late Pliocene NEADW (defined by factor 3, Tables 1 and 2) became increasingly important (compare Figs 4(a), (b) and 4(c), (d)).

These changes in bottom water mass distribution through time are also shown in a vertical sequence along a NW–SE transect from Site 611, by the Gibbs Fracture Zone, to Site 119, in the Bay of Biscay (Fig. 5). The interplay between AABW and bottom waters of northern origin is clearly seen in the period from the late Miocene to the early Pliocene, but in the late Pliocene NEADW was the major bottom water mass.

Pleistocene

Data are available for the period from isotope Stage 6 to the present (Stage 1), (Figs 6 and 7). During full glacial episodes, such as Stage 6 (Fig. 6(a), (b)) and late Stage 2 (Fig. 7(c), (d)), NEADW reached its most southerly extent and NADW was absent. In the full interglacial conditions of Substage 5e (Fig. 6(c), (d)) NADW was present and NEADW was absent. These events are thought to correlate with the increase and decrease respectively of the northern hemisphere ice. Other bottom water mass distributions, intermediate between the glacial and interglacial distribution patterns, can be seen in Stages 5A, 4, 3 and 1 (Figs 6(e)–(g), 7(a), (b), (g), (h)). The stage 3 pattern most closely resembles that of the present.

Throughout the period from isotope Stage 6 to the present, AABW has been confined to the deeper basin south of the Azores Ridge (Figs 6 and 7).

Discussion

If the premises are correct that the ecological preferences of the benthic foraminifera have not changed significantly with time, and that different assemblages characterize different bottom water masses, then the preliminary results presented here have some palaeoceanographic significance.

It has already been emphasized that the pres-

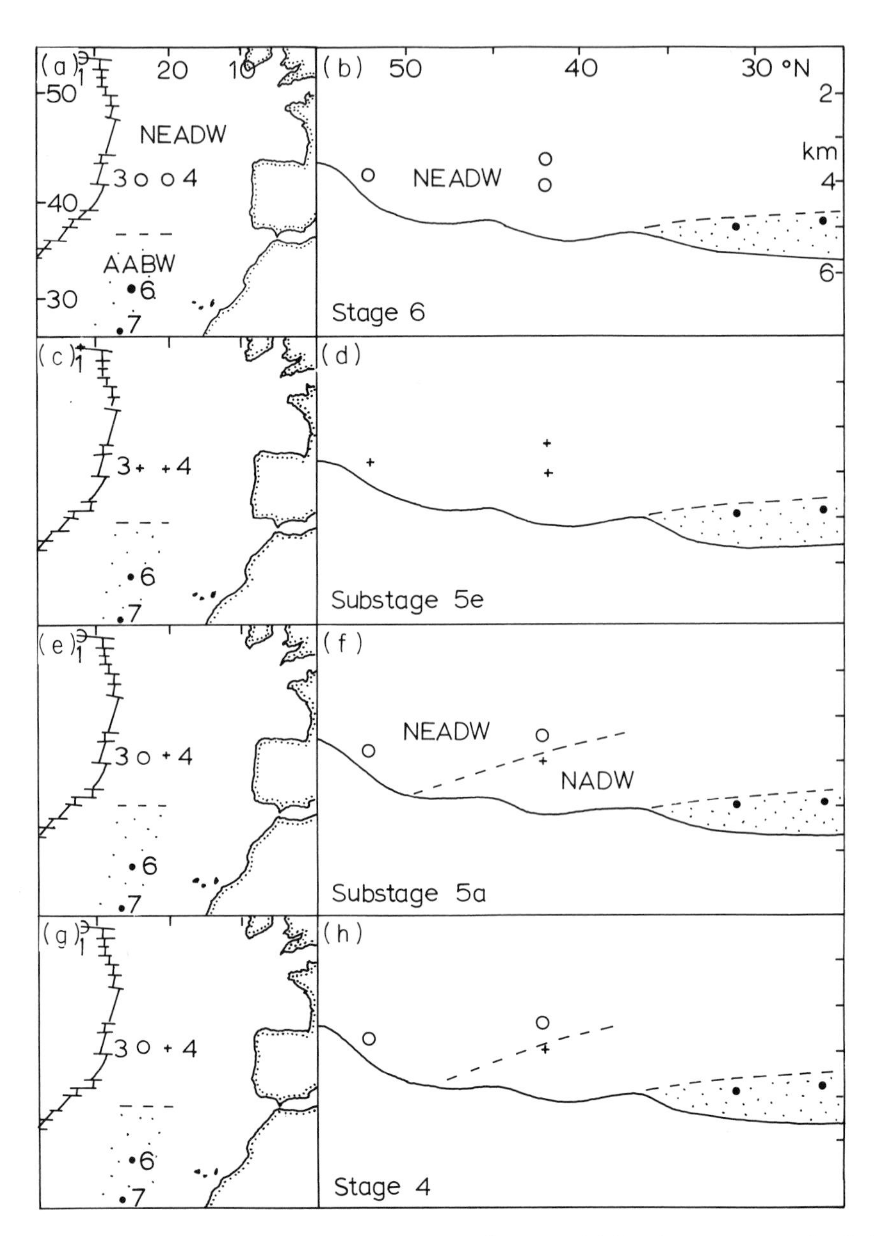

Fig. 6. Distribution of Pleistocene (isotope stages 6 to 4) factors and inferred water masses. Symbols as in Fig. 2.

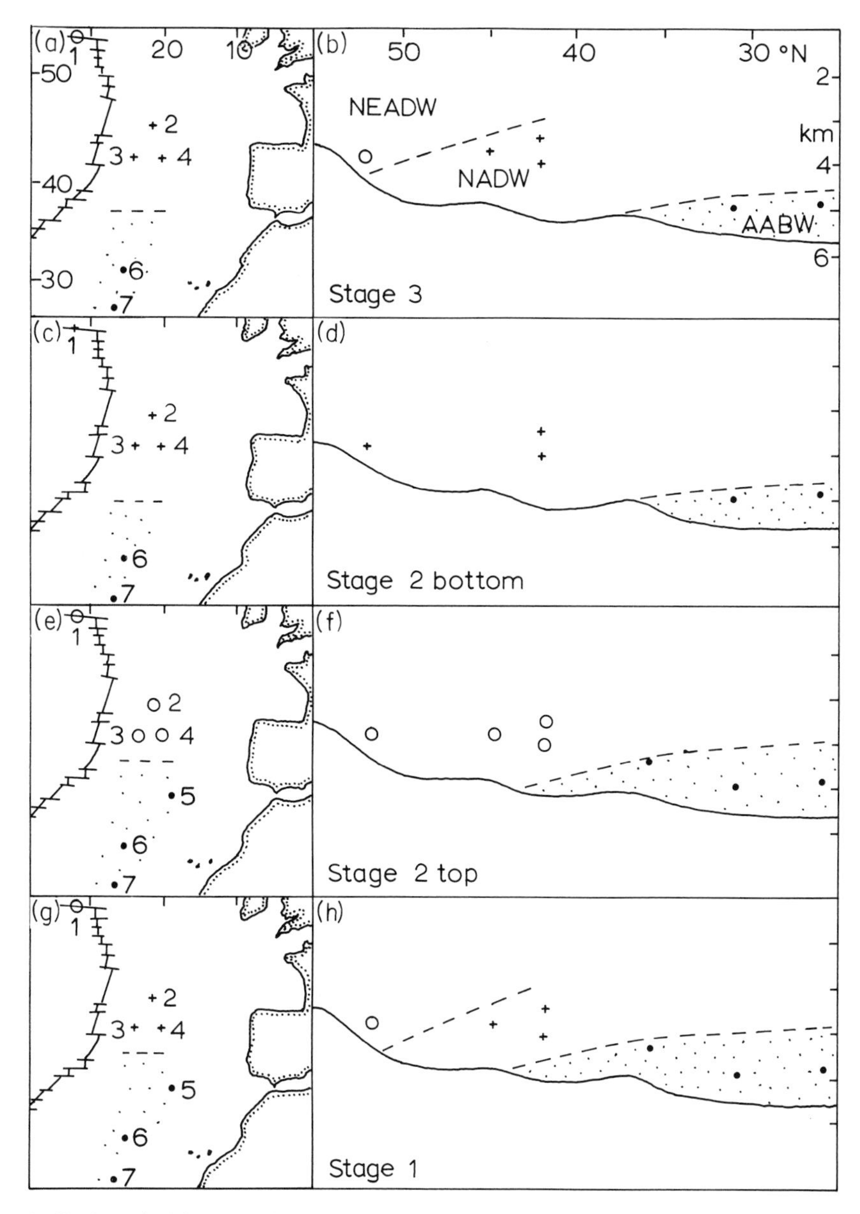

FIG. 7. Distribution of Pleistocene (isotope stages 3 to 1) factors and inferred water masses. Symbols as in Fig. 2.

ence of even small amounts of AABW is strongly-mirrored in the modern benthic assemblages (especially by *O. umbonifera*). Johnson (1982) described the zone of contact of AABW and NADW as a 'wedging' teleconnection, i.e. it is nearly horizontal or only slightly inclined, up to several hundred metres in thickness, and extends over several thousand kilometres. In the modern north-east Atlantic, the Azores Ridge forms a topographic barrier to AABW moving up from the south. According to Heezen & Hollister (1971) AABW forms ($>5\%$) of the total bottom water immediately south of the Azores Ridge (see Fig. 1). For AABW to extend further north, the AABW layer would have to thicken to overtop the sills which lie at ~ 5000 m on the Azores Ridge. The extent of the northward movement would depend on two factors: the relative rates of production of AABW and the density difference between these two water masses.

There is foraminiferal evidence that even in the middle Miocene (NN5) AABW extended as far north as Sites 400 and 119. There is more detailed evidence beginning with the late Miocene. The maximum extent of ice build-up in East Antarctica (Hayes & Frakes 1975) took place during zone NN12 (Fig. 5, Event A). Shackleton & Kennett (1975a) believe that a rapid and major build-up of continental ice took place in the early middle Miocene and led to the formation of the East Antarctic ice cap. Furthermore, oxygen isotope evidence suggests that, during the latest Miocene, this ice cap was 50% larger than at present, and that a decrease in size took place during the Pliocene (Shackleton & Kennett 1975b). The results presented here show that the maximum northward event of AABW (to Site 406) took place in NN12, close to the Miocene–Pliocene boundary, in good agreement with the inferred build-up of the East Antarctic icecap.

A second major event (Fig. 5, Event B) was the Messinian 'salinity crisis' (Hsu *et al.* 1973), during which the Mediterranean was isolated from the Atlantic Ocean. As Mediterranean Water plays an important role in the formation of modern NADW, this crisis should have had a major effect on the distribution of the bottom water masses. Blanc & Duplessy (1982) used the evidence from carbon stable isotopes to infer that the supply of young northern water to the NE Atlantic ceased during the Messinian. The deep waters had a longer residence time and were derived from AABW and other waters having a southerly source. Unfortunately only Site 611 has yielded information for this period and here the factors suggest mixed AABW and NADW, which is rather inconclusive. According to Blanc & Duplessy (1982) NADW production was resumed at the beginning of the Pliocene and NADW is recorded at Site 609 (Fig. 5, Event C). The Labrador Current was initiated at this time (Berggren & Schnitker 1983) and this has since played a part in the formation of the upper layers of NADW (=NEADW). However, NEADW did not become widespread until the late Pliocene (Figs 4(a)–(d) and 5) as already noted by Weston & Murray (1984).

The build-up of continental ice in the northern hemisphere reached a critical level at ~3.2 Ma (Shackleton & Opdyke 1977). With the cooling of northern waters, true modern NADW must then have been formed (Schnitker 1980).

Bender & Graham (1981) found little evidence of the 'carbon shift' at ~3.1 Ma at Site 116. Shackleton *et al.* (1964), in a study of Hole 552A, concluded that, for the period 1.5–3.6 Ma, there was a consistently large isotopic difference between the C^{13} of the North Atlantic and Pacific, as at the present day. They interpreted this as showing bottom water mass characteristics similar to those of the present.

Changes in benthic foraminiferal assemblages and their inferred water masses have continued from the Miocene into the Pleistocene. The influence of AABW has remained throughout th latter period. This result is in good agreement with the studies of Corliss (1979) in the Indian Ocean, where AABW production occurred during both glacial and interglacial episodes. It is in conflict with Weyl's (1968) suggestion that the formation of AABW was cut off in colder periods by stable density stratification in the Southern Ocean. At present AABW is confined to the deeper basin south of the Azores Ridge. With the exception of a record at an unspecified time in the Pleistocene of Site 119 (Weston & Murray 1984) AABW has not extended north of the Azores Ridge since the late Pliocene. North of the Azores Ridge the influence of both NEADW and NADW is widespread. However, their distribution undergoes a major change between glacial and interglacial modes (McIntyre *et al.* 1976). The southern limit of NEADW lay north of 50°N during interglacials, yet moved southward to 45°N during glacial maxima, wedging out NADW.

These advances and retreats of NADW and NEADW demonstrate the response of the deep ocean circulation to the dramatic Pleistocene climatic variations. Weyl (1968) and Newell (1974) have suggested on theoretical grounds that major changes were to be expected. Duplessy *et al.* (1975) produced isotopic evidence suggesting that the Norwegian Sea was not a source of cold, deep water formation during the most recent glacial maximum, agreeing with the observations of Streeter (1973) and Schnitker (1974). More recently Duplessy *et al.* (1980) have suggested that the North Atlantic Ocean has been a source of deep water during the past 75 000 years, even during the coolest periods. However, the presence of certain benthic foraminiferal species (i.e. *U. peregrina*), suggests that the net production of NADW was substantially diminished or had even ceased during the last glaciation, and that the present deep water circulation, with the production of large volumes of NADW, is typical only of a limited portion of the past 150 000 years. The role of NADW in the Pleistocene is, therefore, much in dispute. From the benthic foraminiferal evidence, it has not been possible to determine whether there has been continuation, reduction or cessation of NEADW and NADW production during glacial times. However, the authors conclude that the pattern of distribution of AABW and NADW in the modern north-east Atlantic, has a long history of development throughout the Neogene and Quaternary.

ACKNOWLEDGEMENTS: The authors are grateful for samples from the following sources: Deep Sea Drilling Project, Institute of Oceanographic Sciences, British Museum (Natural History), Marine Biological Association. Dr J.F. Weston and Miss C.A. Haddon are grateful to the Natural Environment Research Council for studentships and Miss A.D.J. Powell to British Petroleum for a studentship. Miss J. Eggins kindly typed the manuscript.

References

ANDERSON, J.B. 1975. Ecology and distribution of foraminifera in the Weddell Sea of Antarctica. *Micropaleontology* **21,** 69–96.

BACKMAN, J. 1979. Pliocene biostratigraphy of DSDP sites 111 and 116 from the North Atlantic Ocean and the age of the Northern Hemisphere glaciation. *Stockh. Contr. Geol.* **32,** 115–37.

BENDER, M.L. & GRAHAM, D.W. 1981. On Late Miocene abyssal hydrography. *Mar. Micropaleont.* **6,** 451–64.

BERGGREN, W.A. & SCHNITKER, D. 1983. Cenozoic marine environments in the North Atlantic and Norwegian-Greenland Sea. *In:* BOTT, M.H.P., SAXOV, S., TALWANI, M. & THIEDE, J. (eds) *Structure and Development of the Greenland-Scotland Ridge.* Plenum Press, New York. 495–548.

BLANC, P.L. & DUPLESSY. V.C. 1982. The deep-water circulation during the Neogene and the impact of the Messinian salinity crisis. *Deep-Sea Res.* **29** 1391–1414.

CORLISS, B.H. 1979. Quaternary bottom water history: deep sea benthonic foraminiferal evidence from the Southeast Indian Ocean. *Quat. Res.* **12,** 271–89.

DUPLESSY, J.C., CHENOUARD, L. & VILA, F. 1975. Weyl's theory of glaciation supported by isotopic study of Norwegian core K11. *Science* **188,** 1208–9.

——, MOYES, J. & PUJOL, C. 1980. Deep water foraminifera in the North Atlantic during the last Ice Age. *Nature (Lond.)* **286,** 479–82.

HAYES, D.E. & FRAKES, L.A. 1975. General synthesis, Deep Sea Drilling Project Leg 28. *In:* HAYES, D.E., FRAKES, L.A. *et al.* (eds), *Init. Repts DSDP* **28.** US Govt. Print. Off., Washington, DC. 919–42.

HEEZEN, B.C. & HOLLISTER, C.D. 1971. *The Face of the Deep.* Oxford University Press, New York.

HSU, K.J., CITA, M.B. & RYAN, W.B.F. 1973. The origin of Mediterranean evaporites. *In:* RYAN, W.B.F., HSU, K.J. *et al.* (eds), *Init. Repts. DSDP* **13.** US Govt. Print. Off., Washington, DC. 1203–31.

JOHNSON, D.A. 1982. Abyssal Teleconnections: interactive dynamics of the deep ocean circulation. *Palaeogeogr., Palaeoclimatol., Palaeoecol.* **38,** 93–128.

LOHMANN, G.P. 1978. Abyssal benthonic foraminifera as hydrographic indicators in the western South Atlantic Ocean. *J. Foramin. Res.* **8,** 6–34.

MCINTYRE, A., KIPP, N.C., BE, A.W.H., CROWLEY, T., KELLOGG, T., GARDNER, J.V., PRELL, W. & RUDDIMAN, W.F. 1976. Glacial North Atlantic 18,000 years ago: a CLIMAP reconstruction. *Mem. Geol. Soc. Am.* **145,** 43–76.

MURRAY, J.W. 1984. Paleogene and Neogene benthic foraminifers from Rockall Plateau. *In:* ROBERTS, D.G., SCHNITKER, D. *et al.* (eds), *Init. Repts DSDP* **81.** US Govt. Print. Off., Washington, DC. 503–34.

—— in press. Benthic foraminifera and Neogene bottom water masses as DSDP Leg 94 North Atlantic sites. *In:* RUDDIMANN, W.F., KIDD, R. *et al.* (eds), *Init. Repts DSDP* **94.** US Govt. Print. Off., Washington, DC.

NEWELL, E. 1974. Changes in the poleward energy flux by the atmosphere and ocean as a possible cause for ice ages. *Quat. Res.* **4,** 117–27.

PHLEGER, F.B., PARKER, F.L. & PEIRSON, J.F. 1953. North Atlantic foraminifera. *Rept. Swedish Deep Sea Exped.* **7,** 1–122.

ROBERTS, D.G. 1975. Marine geology of the Rockall Plateau and Trough. *Phil. Trans. R. Soc. Lond. (Ser. A)* **278,** 447–509.

SCHNITKER, D. 1974. West Atlantic abyssal circulation during the past 120,000 years. *Nature (Lond.)* **248,** 385–7.

—— 1976. Structure and cycles of the western North Atlantic bottom water, 24,000 years BP to present. *EOS* **57,** 257–8.

—— 1979. The deep waters of the western North Atlantic during the past 24,000 years, and the re-initiation of the Western Boundary Undercurrent. *Mar. Micropaleontol.* **4,** 265–80.

—— 1980. Quaternary deep-sea benthic foraminifers and bottom water masses. *Ann. Rev. Earth Planet. Sci.* **8,** 343–70.

—— 1982. Climatic variability and deep ocean circulation: evidence from the North Atlantic. *Palaeogeogr., Palaeoclimatol., Palaeoecol.* **40,** 213–34.

SHACKLETON, N.J., BACKMAN, J., ZIMMERMAN, H., KENT, D.V., HALL, M.A., ROBERTS, D.G., SCHNITKER, D., BALDAUF, J.G., DESPAIRIES, A., HOMRIGHAUSEN, R., HUDDLESTUN, P., KEENE, J.B., KALTENBACK, A.J., KRUMSIEK, K.A.O., MORTON, A.C., MURRAY, J.W. & WESTBERG-SMITH, J. 1984. Oxygen isotope calibration of the onset of ice-rafting and history of glaciation in the North Atlantic region. *Nature (Lond.)* **307,** 620–3.

—— & KENNETT, J.P. 1975a. Paleotemperature history of the Cenozoic and the initiation of Antarctic glaciation: oxygen and carbon isotope analyses in DSDP sites 277, 279 and 281. *In:* KENNETT, J.P., HOUTZ, R.E. *et al.* (eds) *Init. Repts DSDP* **29.** US Govt. Print. Off., Washington, DC. 743–55.

—— & KENNETT, J.P. 1975b. Late Cenozoic oxygen and carbon isotopic changes at DSDP site 284: implications for glacial history of the Northern Hemisphere and Antarctica. *In:* KENNETT, J.P., HOUTZ, R.E. *et al.* (eds), *Init. Repts DSDP* **29.** US Govt. Print. Off., Washington, DC. 801–7.

—— & OPDYKE, N.D. 1977. Oxygen isotope and paleomagnetic evidence for early northern hemisphere glaciation. *Nature (Lond.)* **270,** 216–19.

STREETER, S.S. 1973. Bottom water and benthonic foraminifera in the North Atlantic—glacial–interglacial cycles. *Quat. Res.* **3,** 131–41.

—— & SHACKLETON, N.J. 1979. Paleocirculation of the deep North Atlantic: 150,000 yr record of benthic foraminifera and oxygen—18. *Science* **203,** 168–71.

WESTON, J.F. 1982. *Distribution and Ecology of Recent Deep Sea Benthic Foraminifera in the Northeast Atlantic Ocean.* Unpublished Ph.D. Thesis, University of Exeter.

—— & MURRAY, J.W. 1984. Benthic Foraminifera as deep-sea water-mass indicators. *In:* OERTLI, H.J. (ed.), *Benthos 1983; Second International Symposium on benthic foraminifera (Pau, 1983).* Elf-Aquitaine, Esso REP and Total CFP, Pau. 605–10.

WEYL, P.K. 1968. The role of the oceans in climatic change: a theory of the Ice Ages. *Meteorol. Monogr.* **8,** 37–62.

J.W. MURRAY, C.A. HADDON & A.D.J. POWELL, Department of Geology, University of Exeter, Exeter, Devon, EX4 4QE.

J.F. WESTON, Stratigraphic Services International (UK) Ltd., Chancellor Court, 20 Priestley Rd., Guildford, Surrey, GU2 5YL.

Palaeoclimatic and palaeoceanographic development in the Pliocene North Atlantic: *Discoaster* accumulation and coarse fraction data

Jan Backman, Pierre Pestiaux, Herman Zimmerman & Otto Hermelin

SUMMARY: Accumulation variations of the calcareous nannofossil genus *Discoaster* have been determined from a high latitude North Atlantic sediment sequence (DSDP Hole 552A) of late Pliocene age. Spectral analysis for the preglacial Pliocene reveals a dominant quasiperiodicity associated with obliquity-induced temperature variations in surface water. Spectral peaks corresponding to the 100 Kyr eccentricity and the 21 Kyr precession periodicities are also detected.

Coarse fraction analysis of core V26-145 (Blake Plateau) illustrates the Pliocene development of an increasingly intense subtropical circulation in the North Atlantic. A trend towards increased current velocity occurs through the early Pliocene and probably reflects the progressive emergence of the Panamanian Isthmus, and concomitant intensification of surface water circulation of the North Atlantic subtropical gyre system. In the Rockall Plateau area (Hole 552A), however, no evidence is seen for a related surface water warming. *Discoaster* accumulation data indicate that the preglacial Pliocene development of surface water temperature in the high latitude North Atlantic reflects a progressive climatic deterioration.

Examination of the Pliocene onset of glacial/interglacial cycles in the high latitude North Atlantic has, until recently, been prevented by the lack of suitable study material. The breakthrough came when Shackleton *et al.* (1984) demonstrated that the first Pliocene glaciation was an intense and abrupt event occurring close to 2.4 Ma ago. They achieved this by using the virtually complete Neogene section recovered at DSDP Hole 552A (56°N, 23°W. 2,301 m water depth) on the south-west flank of the Rockall Plateau, and also were able to show that considerable climatic variability existed during the million years (2.4–3.5 Ma) prior to the initial glacial event. Results from previous studies (e.g. Shackleton & Opdyke 1977; Ledbetter *et al.* 1978; Backman 1979; Keigwin & Thunell 1979; Hodell *et al.* 1983) indicate that the time interval between 2.4 and 3.2 Ma is of special interest in this context, although it has been difficult to es[illegible]lish casual relationships for the immediately preglacial palaeoclimatic pattern.

One line of reasoning is that the initiation of North Atlantic glaciation was linked with the closing of the Central American seaway. This would intensify the North Atlantic Drift/Gulf Stream system, bringing warm surface waters and an increased rate of evaporation to the northern North Atlantic and Norwegian Sea. The resulting precipitation over the surrounding continents would thus trigger the onset of glacial/interglacial cycles (Emiliani *et al.* 1972; Luyendyk *et al.* 1972; Berggren & Hollister 1974; 1977). There is, however, a lengthy gap in timing between the first glacial build-up (2.4 Ma) and the suggested closing of the Central American seaway (3.1–3.6 Ma; Saito 1976; Keigwin 1978).

In the light of recent improvements in Pliocene bio- and chronostratigraphy and the conceptual framework discussed above, the purpose of this paper is to present data which are likely to reveal possible relationships between palaeoclimatic development in the high latitude North Atlantic and the closing of the Central American seaway. In order to achieve this the authors have analysed Pliocene sediment sequences from two different geographic areas: where palaeoclimatic change can be recorded in the high latitude North Atlantic and where flow intensity variation of the Gulf Stream/North Atlantic Drift system can be recorded. For this purpose the authors have investigated accumulation fluctuations of the calcareous nannofossil genus *Discoaster*, taken from the identical samples used by Shackleton *et al.* (1984) from DSDP Hole 552A, and sediment coarse fraction percentages from core V26-145 located on the Blake Plateau. Age assignments are based on the marine magnetic anomaly time-scale of Berggren *et al.* (1985).

Late Pliocene *Discoaster* data (DSDP Hole 552A)

From the emergence of the first species in the Palaeocene to the extinction of the last represen-

From SUMMERHAYES, C.P. & SHACKLETON, N.J. (eds), 1986, *North Atlantic Palaeoceanography*, Geological Society Special Publication No. 21, pp. 231–242.

tative in the Pliocene, *Discoaster* spp. showed an ecological preference for tropical and subtropical environments, that is, for warm water masses (e.g. Haq *et al.* 1976; Haq & Lohmann 1976; Backman & Shackleton 1983). The latitudinal decrease in abundance of *Discoaster* spp., however, does not occur linearly. At 46°S in the South Atlantic (DSDP Site 514), for example, Pliocene *Discoaster* spp. occur only at a few sample levels, which are interpreted on the basis of radiolarian biofacies to represent warm water maxima (Ludwig *et al.* 1983). This is in marked contrast to the continuous presence of Pliocene *Discoaster* spp. at 56°N. (DSDP Hole 552A) in the North Atlantic. Rather than simple latitudinal position, the distribution and abundance of *Discoaster* spp. in high latitudes, therefore, to some extent reflects water mass boundary position. Accordingly, the authors consider that gross fluctuations in *Discoaster* abundance at higher latitudes reflect relative changes in surface water palaeotemperature; a greater abundance representing warmer waters.

Because its relatively shallow depth (2301 m) makes dissolution negligible, DSDP Hole 552A provides a good opportunity to study the high latitude behaviour of *Discoaster* spp. Approximately 26 m of the sedimentary column was investigated at 10 cm intervals, using the counting procedure described by Backman & Shackleton (1983). Due to the low abundance of Pliocene *Discoaster* spp. in the area of Hole 552A, one hundred viewfields in the light microscope were scanned in each sample at a viewfield diameter of 0.3 mm; roughly 250 nannofossils were present in each viewfield. The *Discoaster* spp. were counted at the species level, and the results are shown in Fig. 1. The biostratigraphical implications of the results are discussed by Backman (1984).

The abundance plots (Fig. 1) are characterized by marked short-term changes and a dominance of *D. surculus* between approximately 44 and 51 m. Bukry (1978) observed that 'even within the warm-water genus *Discoaster*, relative abundance of species . . . suggest temperature distinctions', and that *D. surculus* dominated in cool areas.

Backman & Shackleton (1983) showed that the counting procedure employed here gives abundance patterns coherent to those derived from counts of absolute abundances (i.e. number of specimens per unit weight of sediment). This implies a proportional relationship between the authors' *Discoaster* abundance curves and the accumulation of *Discoaster* spp. However, in order to account for changes in sediment accumulation rates (Fig. 2), and thus the accumulation rate of *Discoaster* spp. per unit time, the total abundance in each sample has been multiplied with the sediment accumulation rate in the pertinent interval. Accumulation plots where each plot represents the added abundance of all *Discoaster* spp. in each sample are shown in Fig. 3. The sum is used for two reasons:

(1) a possible bias induced by the ecological preference of a single species is avoided. In this data set this is of special importance, because *D. brouweri* is the only species ranging throughout the investigated interval.

(2) In view of the low relative abundance of *Discoaster* spp. in the 522A sequence, conceivably less than 0.1% of the total assemblage in most samples, the sum of all species had to be used in order to reach statistically meaningful numbers.

Marked variability in *Discoaster* abundance is the most prominent character of the curve shown in Fig. 3. The preglacial/glacial transition occurs at about 2.4 Ma, and the low accumulation after that time partly reflects the influence of more severe climatic conditions and the fact that only one species is present.

Spectral characteristics

Using the age model in Fig. 2, the four different segments of the accumulation rate curve were used to transform the 228 sample depths into non-equidistant time-dependent values. These original data were interpolated by third degree polynomials (cubic splines) at the constant time-interval of 5 Kyrs, corresponding to the average sample resolution.

Each spectral analysis technique is based on an *a priori* model which fits the observations. The performance of the various spectral techniques, therefore, is attributed to how well the assumed model matches the process under analysis. This implies that different models may not necessarily yield similar results. After having compared advantages and disadvantages of the main existing techniques (Pestiaux & Berger 1984), it was decided to use two of them for their complementary characteristics.

The classical lagged product spectral analysis is used for its statistical properties and its accuracy in the spectral amplitude estimation. This method is applied to the *Discoaster* accumulation data in order to select the statistically significant parts of the spectrum which give the mean length of the cycles (quasiperiodicities) as contained in the whole set. Figure 4(a) shows four broad peaks which are significant at the 90% confidence level. These peaks correspond to mean periodicities of 230 Kyr, 52 Kyr, 42 Kyr and 23 Kyr. The most

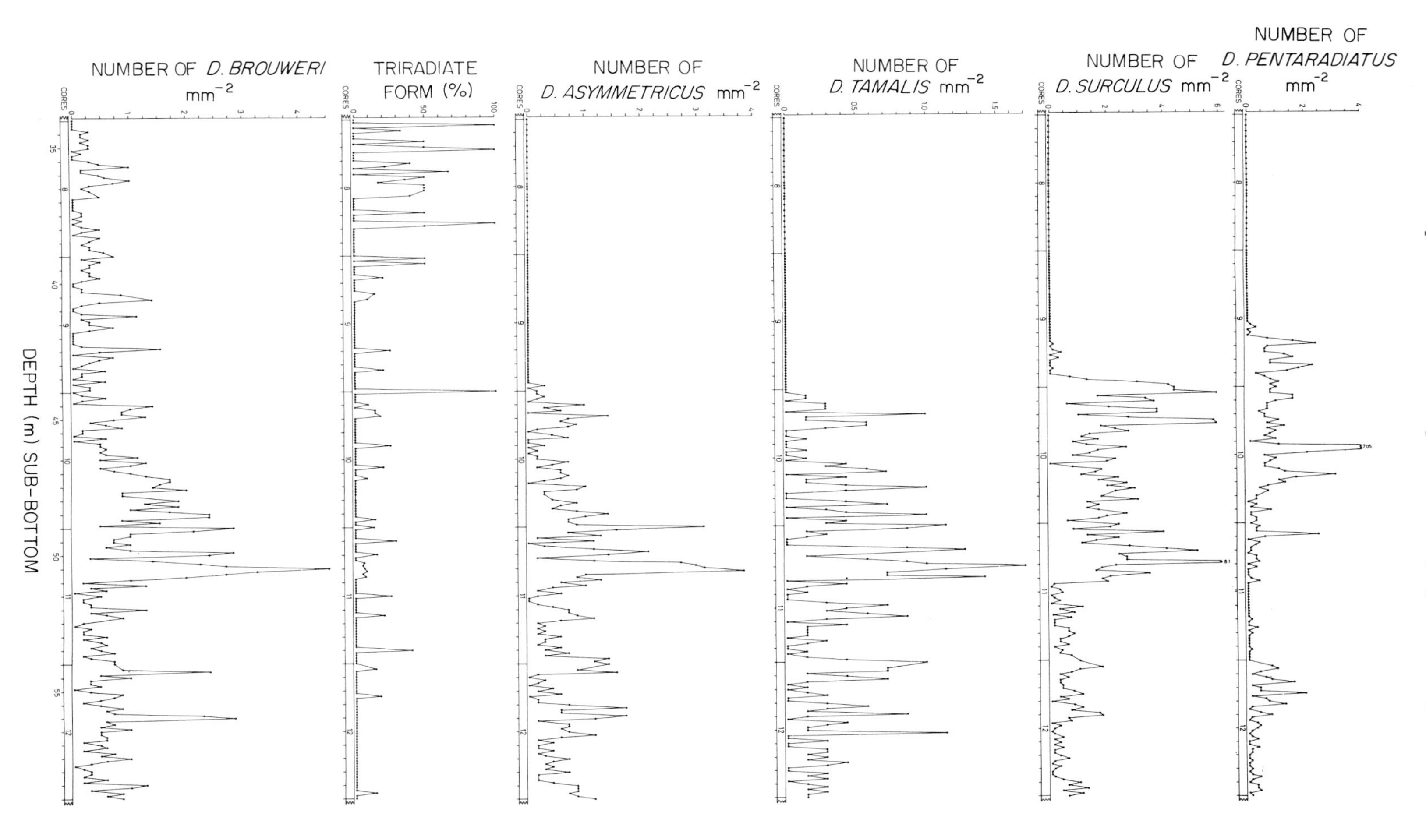

FIG. 1. Plots of *Discoaster* abundance data at the species level versus depth in DSDP Hole 552A. The percentage of the triradiate form is calculated relative to *D. brouweri*. The plots represent the time-interval from about 1.9–3.5 Ma.

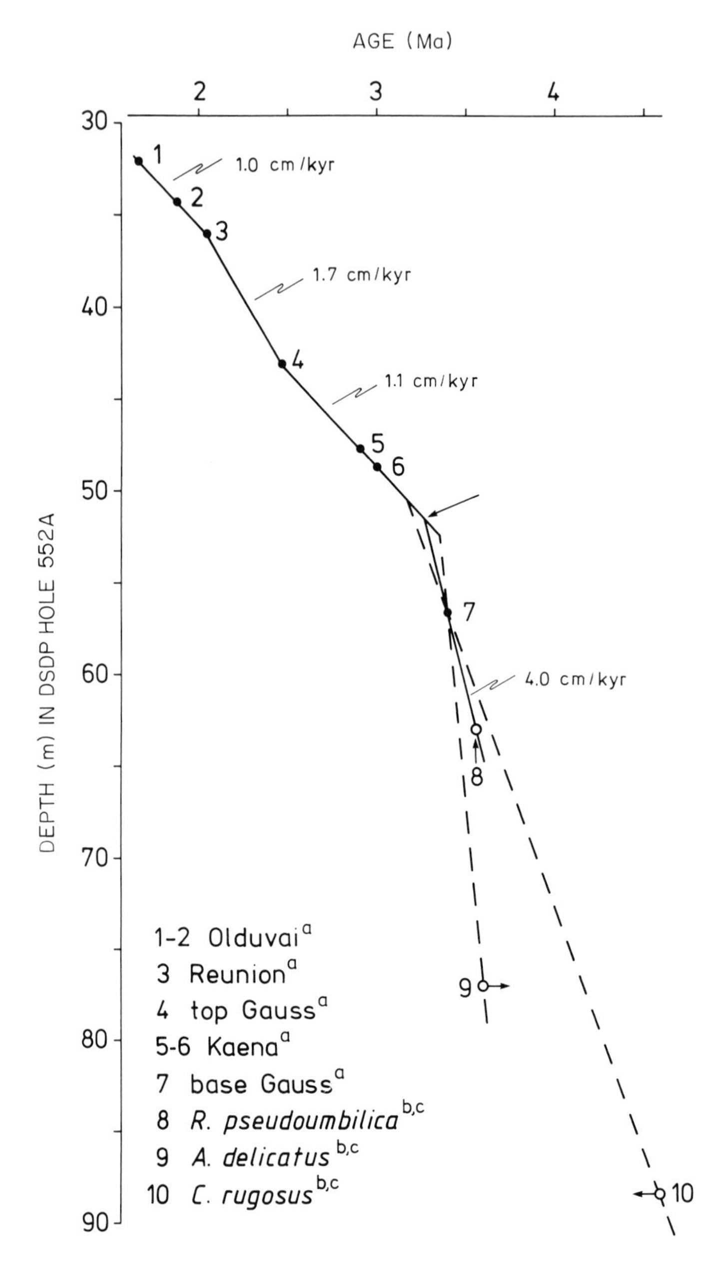

FIG. 2. Age/depth relationships of magnetostratigraphic and biochronologic control points in DSDP Hole 552A. The species *A. delicatus* occurs sporadically in the sequence, indicating that its highest occurrence (point 9) has an age equivalent to or older than its extinction age of 3.6 Ma. The species *C. rugosus* also occurs very sporadically, indicating that its lowest occurrence (point 10) has an age equivalent to or younger than its first appearance age of 4.6 Ma. These two points and the Gauss/Gilbert boundary (point 7) give the limits for possible sediment accumulation rates below that boundary. The chosen rate is one determined by the extinction of *R. pseudoumbilica* (point 8) and the Gauss/Gilbert boundary. Extrapolated above this boundary, the suggested sediment accumulation rate intercepts the rate extrapolated from Kaena at 3.28 Ma (arrow). Superfixes: (a) Shackleton *et al.* (1984); (b) Backman (1984); (c) Backman & Shackleton (1983).

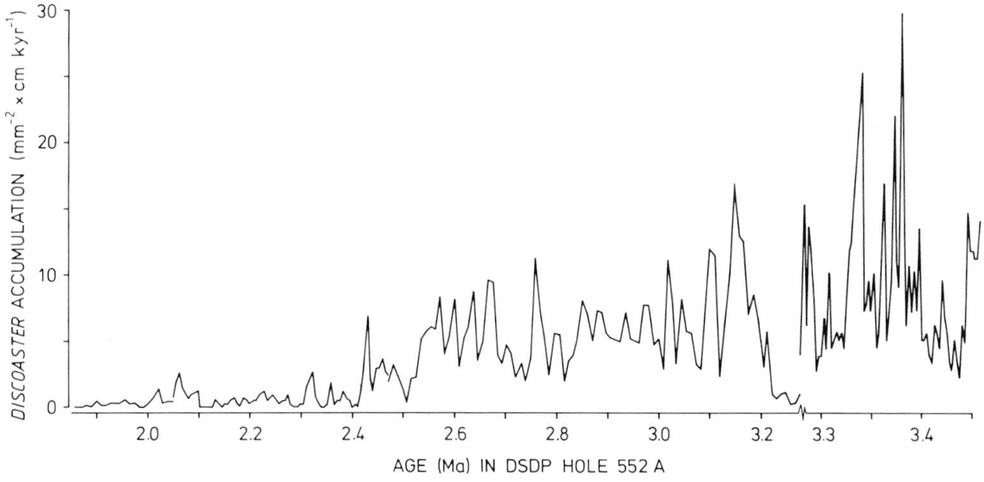

FIG. 3. Plots of the late Pliocene accumulation of *Discoasters* in DSDP Hole 552A. The three gaps in accumulation at 2.05 Ma, 2.47 Ma and 3.28 Ma represent places of change in sediment accumulation rate (Fig. 2). Each sample at the place of rate transition is given two values, one based on the sediment accumulation rate above the change and one on the rate below the change, hence the gaps. The oldest part of the record (> 3.28 Ma) is characterized by a higher sediment accumulation rate (Fig. 2), making a scale change necessary. The preglacial/glacial transition occurs at about 2.4 Ma. The average accumulation of *Discoaster* spp. is roughly 7 units in the preglacial part (compare with Fig. 9).

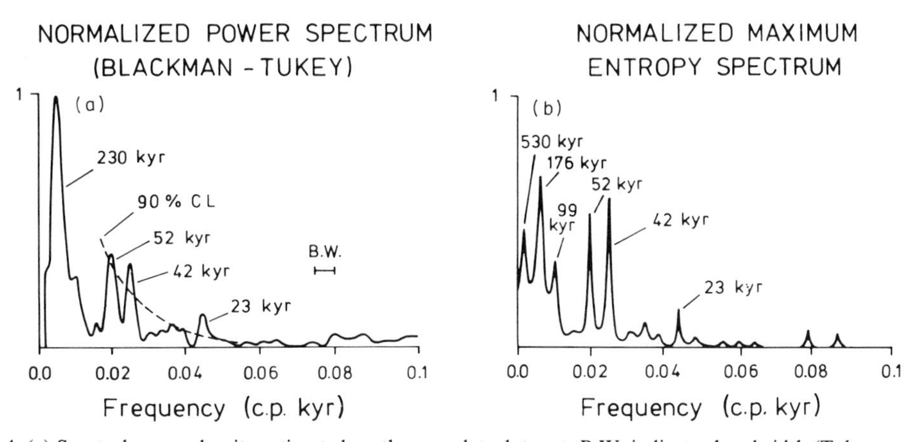

FIG. 4. (a) Spectral power density estimated on the complete data set. B.W. indicates bandwidth (Tukey window) and 90% C.L. the upper red noise confidence limit around the spectral peaks. (b) shows the maximum entropy spectrum for the same record.

important disadvantage of the lagged product spectral analysis is its lack of frequency resolution. In order to investigate how far it is possible to sharpen the peaks and split the broader ones, the maximum entropy spectral analysis was used.

Figure 4(b) shows this analysis performed on the whole data set. The broad peak at 230 Kyr splits into three, corresponding to periodicities of 530 Kyr, 176 Kyr and 99 Kyr. The three peaks at 99 Kyr, 42 Kyr and 23 Kyr are in agreement with the well known astronomical periodicities of 100 Kyr, 41 Kyr and 21 Kyr associated with eccentricity, obliquity and precession. The obliquity peak is the dominant one in the authors' record.

Implications of the spectral analyses

Numerous Pliocene and Pleistocene palaeoclimatic records demonstrate variability at periodicities corresponding to the orbital elements (e.g. Hays *et al.* 1976; Berger 1977; Kominz *et al.* 1979; Pestiaux & Berger 1984). Improvements in climate modelling permit a better understanding of the relationships between insolation forcing and its imprint in the geological record (Imbrie & Imbrie 1980; Berger *et al.* 1981; Ruddiman & McIntyre 1984).

Spectral analyses of monthly insolations at different latitudes clearly indicate that the variability of insolation associated with obliquity (41 Kyr periodicity) increases with increasing latitude (Berger & Pestiaux 1984). The spectra of the entire data set shown in Fig. 4 have to be interpreted as a mixture of the spectra characterizing the three time-increments shown in Fig. 5. The middle time-increment (2.4–3.3 Ma) is considered to represent the best part of the record in terms of time-control. The spectral peaks of 106 Kyr and 42 Kyr are particularly well defined in that interval, and because a sharper peak reflects a more regular oscillation, these two peaks are considered to reflect the periodicities of 100 Kyr and 41 Kyr generally associated with astronomical frequencies.

Peaks which cannot directly be linked with astronomical frequencies may be explained in two ways. Firstly, they may be due to varying sedimentation rates between the chronological control points. Secondly, the climate system is non-linear and complex, which may generate combined tones of the forcing frequencies of the insolation (Le Trent & Ghil 1983). For example, the 52 Kyr periodicity could result from a combination of the 42 Kyr and the 23 Kyr periodicities (see Fig. 4), i.e. a combination of the obliquity and precessional effects. Within the bandwith of the spectral analysis used, the spectral peaks detected between about 50 Kyr and 35 Kyr are considered to be related to the quasiperiodicity of obliquity at 41 Kyr, which is present in the insolation spectrum that characterizes the latitude of Hole 552A (56°N). If *Discoaster* accumu-

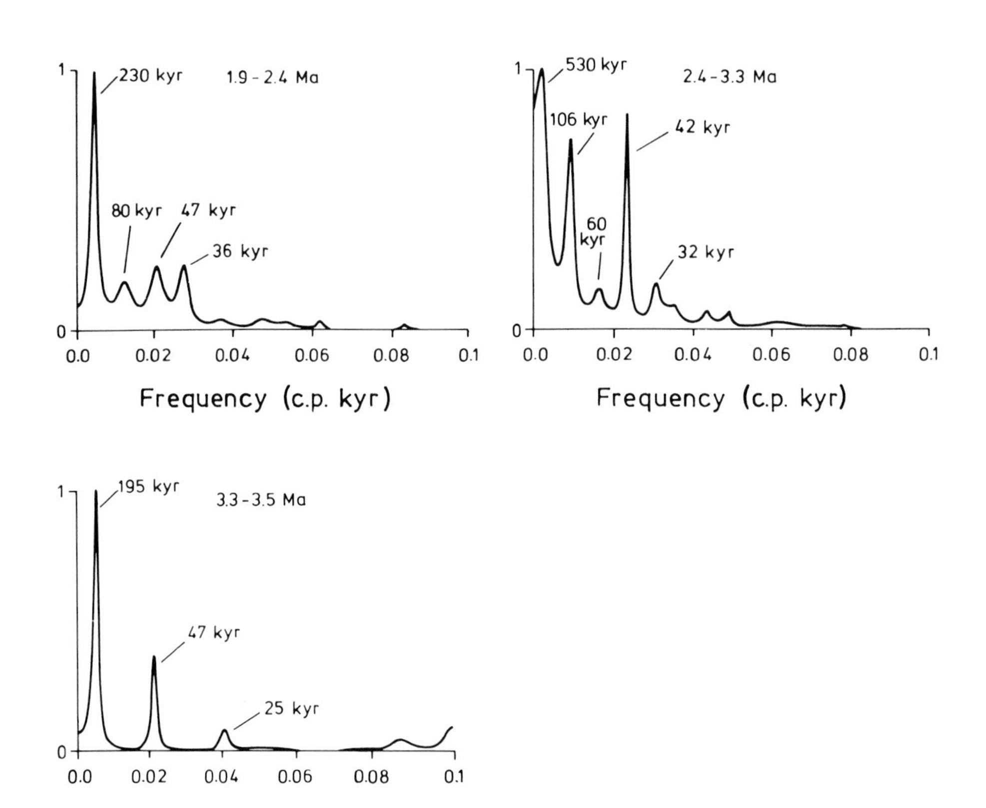

FIG. 5. Maximum entropy spectrum estimated on three time-increments of the record.

lation also reflects water mass boundaries, then the variation recorded here may reflect, at least in part, orbital-induced changes in Gulf Stream/North Atlantic Drift position in the preglacial North Atlantic.

The 23/19 Kyr quasiperiodicities are expected in the three time-increments shown in Fig. 5, since the precessional bimodal peak is also present in the insolation spectrum of high latitudes. However, a quasiperiodicity of 25 Kyrs is observed only in the increment showing the highest sediment accumulation rate. Apparently the combined effect of the sampling intervals and comparatively low sediment accumulation rates in the glacial and immediate preglacial interval (1.9–3.3 Ma) probably has prevented the identification of the precessional peak. Moreover, the fact that quasiperiodicities are less clear in the glacial record (post 2.4 Ma) is probably due to the low accumulation of a single species and to the influence of ice rafted deposition, which is likely to have caused short-term variations in sediment accumulation rates.

Pliocene coarse fraction data from the Blake Plateau (V26–145)

A stratigraphical succession of coarse fraction (CF-dry weight) data may be viewed in terms of current velocities influencing the depositional environment; a greater velocity resulting in higher CF percentages (Kaneps 1979; Brunner 1984). The Blake Plateau is ideally located for a study of variations in Gulf Stream velocity, because it is affected by the Gulf Stream and associated currents (Pinet & Popenoe 1982). Kaneps (1979) presented CF data for late Miocene through Pleistocene times from a Blake Plateau core (V26-145). The biostratigraphy of the foraminiferal datums has been re-examined and those provided by the nannofossils have been added (Fig. 6).

The results of the CF determination and cumulative percent of sediment size are shown in Fig. 7, together with age assignments at levels characterized by changes in total CF proportions. The plots of the total CF content are in substantial agreement with Kaneps (1979), although the curves presented here are based on a closer sample interval.

Kaneps (1979) discussed three possible causes for the observed changes in his CF data: (1) adjustments of Gulf Stream velocity to tectonic modifications; (2) shifts in the position of the Gulf Stream axis; and (3) velocity perturbations in response to global climate events. Kaneps dismissed the second possibility on hydrodynamic grounds, and argued that the long-term velocity trend represents a response to a tectonic event (the closing of the Central American seaway), while the short-term changes reflect variability in global climate. Pinet & Popenoe (1982), however, suggested that the path of the Gulf Stream may indeed shift—but in response to changes in sea-level. It should be pointed out that the position of core V26-145, as well as the other cores studied by Kaneps (1979), is significantly eastward of the present-day axis of the Gulf Stream. The position of core V26-145, on the south-eastern Blake Plateau, is actually beneath the Antilles Current and thus reflects the flow intensity of the entire subtropical gyre circulation system.

The progressive increase in CF, which began in the late Miocene (Kaneps 1979), continues up to about 4.1 Ma where the curve levels out (Fig. 7). Keigwin (1982a; 1982b), based his history of the Central American seaway on the initiation of Atlantic/Pacific provincialism among planktonic foraminifers at about 4.2 Ma and on the interpretation of salinity contrast at about the same time (from oxygen isotope measurements of planktonic foraminifers from Caribbean and Pacific drill sites). The timing of these events (4.0–4.2 Ma) is in substantial agreement with the permanent shift in CF percentages observed in core V26-145 (Fig. 7). The authors consider this correlation to represent a sedimentological response to an increase in the North Atlantic surface circulation generated by the increasingly restricted flow through the emerging isthmus.

It is difficult to assign specific causes to the short-term oscillations in CF content. However, the event at 2.4 Ma (Fig. 7), coincides in time with the onset of large scale North Atlantic glaciation (Shackleton *et al.* 1984), and CF oscillations further up-section may be similarly related to intense climatic cyclicity. Moreover, since the subtropical gyre system is likely to have also responded to preglacial climatic events, such events at 3.1 Ma and 3.5 Ma (e.g. Shackleton & Opdyke 1977; Kennett *et al.* 1979; Keigwin & Thunell 1979; Hodell *et al.* 1983; Backman & Pestiaux, in press) possibly may be related to the CF oscillations at these times.

Comparison of CF-data: core V26-145 and DSDP Hole 552A

DSDP Hole 552A was drilled on the Hatton Drift. The formation of the northern North Atlantic sediment drifts was strongly influenced by abyssal circulation (Miller & Tucholke 1983), and hence by the production rates of the deep and bottom waters which are formed in the northernmost North Atlantic and Norwegian Sea. If the Gulf Stream input has been significant with

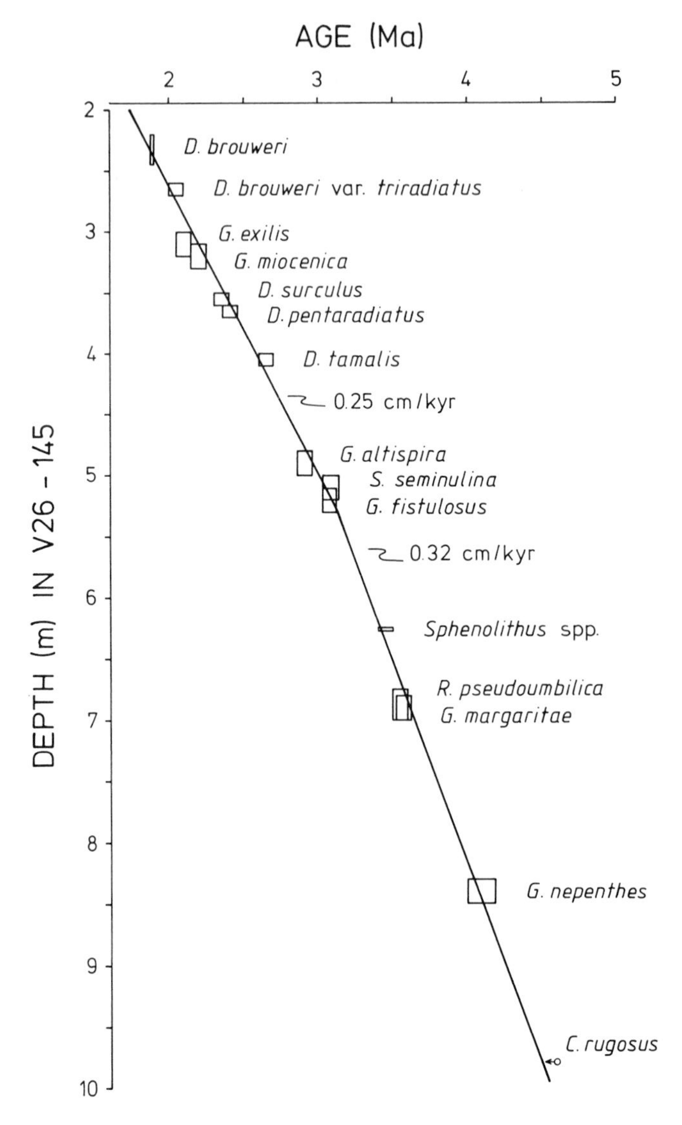

FIG. 6. Age/depth relationships of quantitatively determined planktonic foraminiferal and nannofossil datums in core V26-145. The abundance of the foraminiferal species has been quantified by evenly spreading 1000–2000 specimens on a picking tray, using a microsplitter for obtaining the appropriate sample size. Species of interest were counted relative to the remaining forms, and percentages were calculated. The 125–250 μm fraction was used, except in the case of *G. nepenthes* (> 250 μm). Both groups indicated that one sample, the 4.9 m level, is characterized by reworking. Keigwin's (1982a) data from DSDP Site 502 were used for calibration of the foraminiferal ages, except for *G. exilis* and *G. miocenica* (Berggren *et al.*, 1985). Backman & Shackleton's (1983) data was used for the nannofossil ages. In the lowermost sample, *C. acutus* occurs together with *C. rugasus*, indicating that this sample is close to the first appearance level of the latter (Backman & Shackleton 1983).

respect to the production rates of the deep waters and velocity of the abyssal circulation, then a comparison of CF curve patterns from Hole 552A and core V26-145 might reveal at least gross similarities.

The CF data from Hole 552A have been 'smoothed' (N = 3) (Zimmerman *et al.* 1984), and the resulting plots are shown in Fig. 8. The curve represents the time interval between approximately 2.5 and 3.5 Ma. The lower part of the record

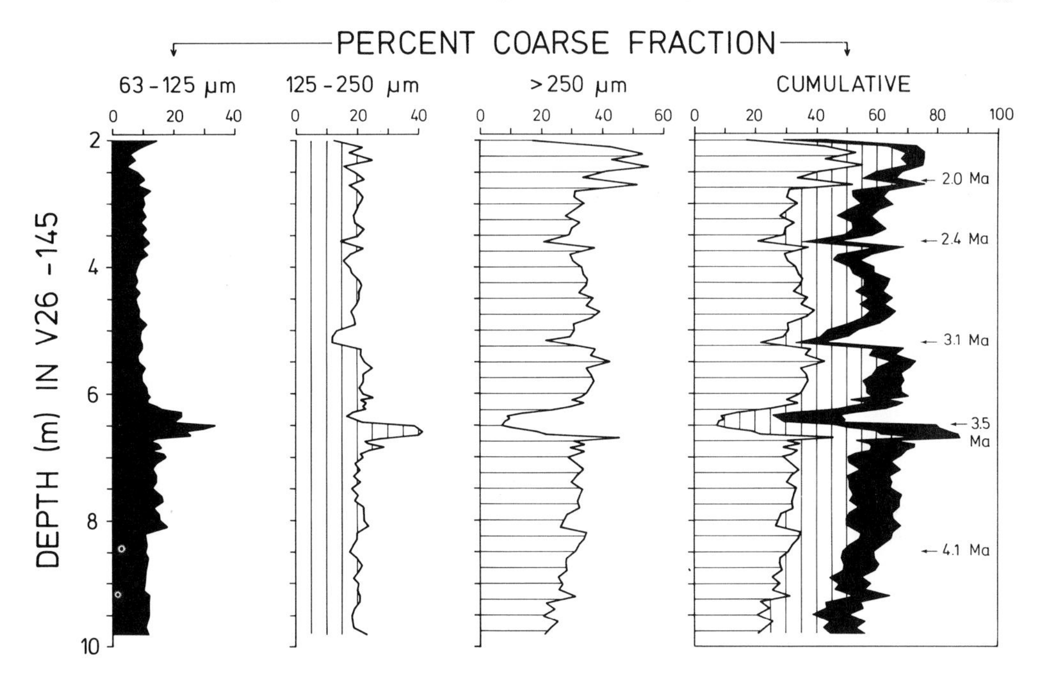

Fig. 7. Plots of sediment coarse fraction percentages in core V26-145. The unfilled part in the cumulative plots represents the size fraction <63 µm. The ages indicated are derived from Fig. 6.

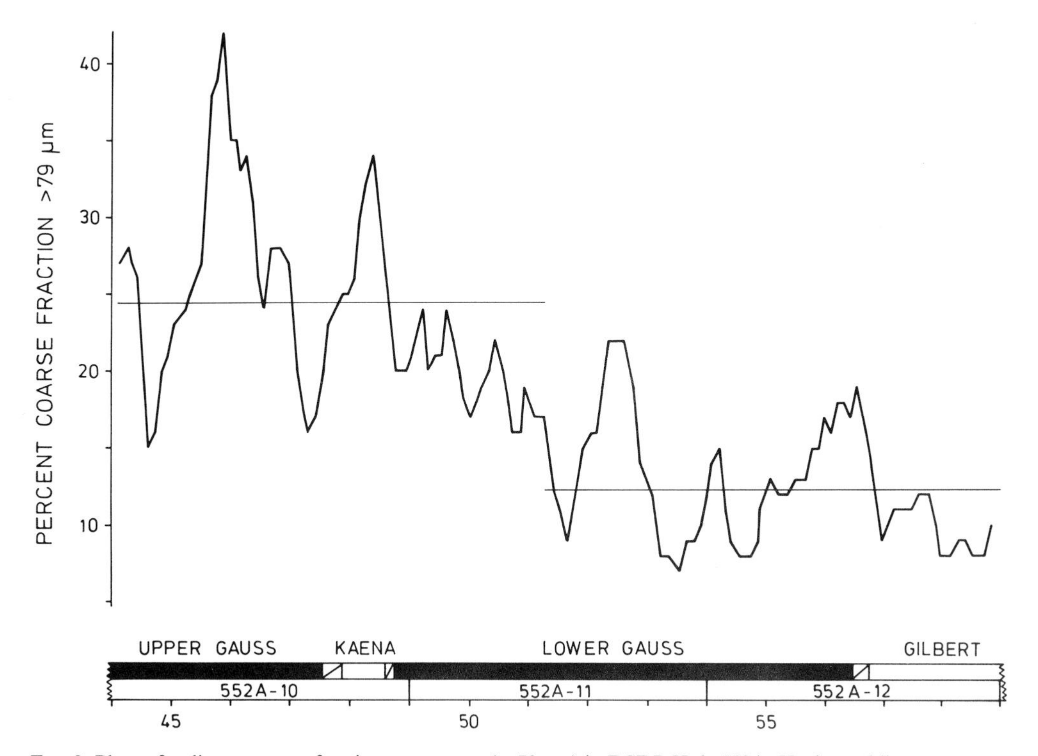

Fig. 8. Plots of sediment coarse fraction percentage (>79 µm) in DSDP Hole 552A. Horizontal lines represent mean values for the two segments of the record.

shows an average CF content (12.2%), which is considerably less than the upper part (24.4%). The transition from lower to higher values occurs in a depth interval centred on approximately 51 m (about 3.2 Ma), which also is characterized by a major change in sediment accumulation rate (Fig. 2). There is little similarity in detail between this and the CF curve of core V26–145 (Fig. 7).

The preglacial CF of Site 552A consists almost entirely of biogenous carbonate material (Zimmerman *et al.* 1984). The increase in the proportion of CF (Fig. 8) in concert with the decrease in the rate of sediment accumulation (Fig. 2) at 3.2. Ma, suggests an increase in the winnowing of the fine sediment. The authors infer, therefore, higher abyssal current velocity and an increase in the production of deep and bottom water at 3.2 Ma in the high latitude North Atlantic.

Early Pliocene *Discoaster* data (DSDP Hole 552A)

If the general increase in velocity of the Gulf Stream system during early Pliocene times had a significant effect on the high latitude North Atlantic surface water temperatures, this should be seen as a corresponding shift in *Discoaster* accumulation. The results of *Discoaster* counts from Hole 552A which represent the time interval between 3.8 and 5.0 Ma are shown in Fig. 9. The species are identical to those previously used, and the age model stems from combined nannofossil (Backman 1984) and diatom data (Baldauf 1984).

Figure 9 reveals pronounced short-term variation in *Discoaster* accumulation. Omitting the single peak value of 262 units, the average *Discoaster* accumulation in this time interval is four times the value in the time interval between 2.4 and 3.5 Ma (Fig. 3). The implication is that the surface temperatures were warmer in the Rockall area during early Pliocene times, as compared to preglacial late Pliocene times. The development of Pliocene surface water temperature in the high latitude North Atlantic thus appears to reflect a progressive climatic deterioration unaffected by an influx of warm water related to the closing of the Central American seaway. Similarly, the preglacial obliquity-induced climatic forcing observed in the *Discoaster* data is regarded as an overprint on a broader climatic trend, resulting in full-scale glacial conditions at about 2.4 Ma.

Conclusions

The cause-and-effect relationship between insola-

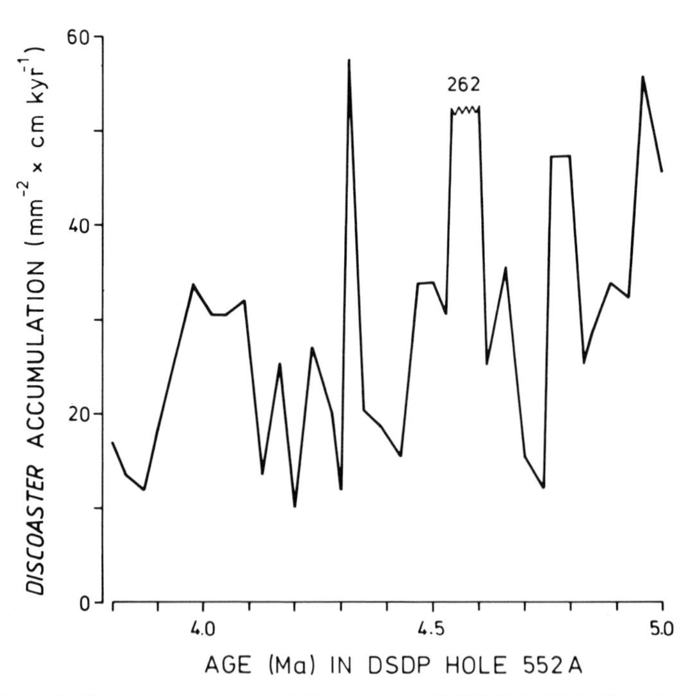

FIG. 9. Plots of the early Pliocene accumulation of *Discoasters* in DSDP Hole 552A. Omitting the single peak value of 262 units, the average accumulation value is about 28 units (compare with Fig. 3; 7 units).

tion and palaeoclimatic response has been derived mostly from oxygen isotope analyses of Pleistocene sequences and, therefore, interpreted in terms of glacial/interglacial cycles. The record from DSDP Hole 552A, reflecting the total accumulation of *Discoaster* spp., has been found to be sensitive to sea surface temperature variations at high latitudes during a period of time prior to the onset of North Atlantic glacial effects. The accumulation variations expressed in the dominant spectral peak are in the range of the 41 Kyrs periodicity, which is attributed to insolation changes due to obliquity.

Shoaling of the Central American seaway intensified the flow of the subtropical gyre in the North Atlantic. However, the constriction of the Central American seaway and the concomitant introduction of warm water into the North Atlantic did not appreciably alter the overall deterioration of Northern Hemisphere climate during preglacial Pliocene times.

ACKNOWLEDGMENTS: J.B. and H.Z. thank the staff of the Deep Sea Drilling Project for the opportunity to participate on Leg 81. Office of Naval Research grant ONR-N 00014-75-CO210 and NSF grant OCE 78-25448 supported the Lamont-Doherty Geological Observatory core collection. Financial support for J.B. was provided by the Swedish Natural Science Research Council and for H.Z. by the US National Science Foundation (OCE 82-07164). The authors thank B. Funnell, A. McIntyre, N.J. Shackleton and C. Summerhayes for their constructive criticism.

References

BACKMAN, J. 1979. Pliocene biostratigraphy of DSDP Sites 111 and 116 from the North Atlantic Ocean and the age of Northern Hemisphere glaciation. *Stockholm Contr. Geol.* **35,** 115–37.

—— 1984. Cenozoic calcareous nannofossil biostratigraphy from the notheastern Atlantic Ocean—Deep Sea Drilling Project Leg 81. *Init. Repts DSDP* **81,** 403–27.

—— & SHACKLETON, N.J. 1983. Quantitative biochronology of Pliocene and early Pleistocene calcareous nannofossils from the Atlantic, Indian and Pacific oceans. *Mar. Micropaleontol* **8,** 141–70.

—— & PESTIAUX, P. in press. Pliocene *Discoaster* abundance variations from DSDP Site 606: biochronology and paleoenvironmental implications. *Init. Repts. DSDP* **94.**

BALDAUF, J.G. 1984. Cenozoic diatom biostratigraphy and paleoceanography of the Rockall Plateau region, North Atlantic, Deep Sea Drilling Project Leg 81. *Init. Repts. DSDP* **81,** 439–78.

BERGER, A. 1977. Support for the astronomical theory of climatic change. *Nature* **269,** 44–5.

BERGER, A., GUIOT, J., KUKLA, G. & PESTIAUX, P. 1981. *Geol. Rundsch.* **70,** 748–58.

—— & PESTIAUX, P. 1984. Accuracy and stability of the Quaternary terrestrial insolation. *In*: *Milankovitch and Climate.* I. Riedel Publishing Company, Bordrecht. 83–113.

BERGGREN, W.A. & HOLLISTER, C.D. 1974. Paleogeography, paleobiogeography and the history of circulation in the Atlantic Ocean. *Soc. econ. Paleont. Min. Spec. Publ.* **20,** 126–86.

—— & HOLLISTER, C.D. 1977. Plate tectonics and paleocirculation—commotion in the ocean. *Tectonophysics* **38,** 11–48.

——, KENT D.V., FLYNN, J.J. & VAN COUVERING, J.A. 1985. Cenozoic geochronology. *Bull. Geol. Soc. Am.* **96,** 1407–18.

BRUNNER, C.A. 1984. Evidence for increased volume transport of the Florida Current in the Pliocene and Pleistocene. *Mar. Geol.***54,** 223–35.

BUKRY, D. 1978. Biostratigraphy of Cenozoic marine sediment by calcareous nannofossils. *Micropaleontology* **24,** 44–60.

EMILIANI, C., GARTNER, S. & LIDZ, B. 1972. Neogene sedimentation on the Blake Plateau and the emergence of the Central American isthmus. *Paleogeogr., Paleoclimatol., Paleoecol.* **11,** 1–10.

HAQ, B.U. & LOHMANN, G.P. 1976. Early Cenozoic calcareous nannoplankton biogeography of the Atlantic Ocean. *Mar. Micropaleontol.* **1,** 119–94.

——, LOHMANN, G.P. & WISE, S.W. 1976. Calcareous nannoplankton biogeography and its paleoclimatic implications: Cenozoic of the Falkland Plateau (DSDP Leg 36) and Miocene of the Atlantic Ocean. *Init. Repts DSDP* **36,** 745–59.

HAYS, J.D., IMBRIE, J. & SHACKLETON, N.J. 1976. Variations in the Earth's orbit: pacemaker of the ice ages. *Science* **194,** 1121–32.

HODELL, D.A., KENNETT, J.P. & LEONARD, J.P. 1983. Climatically induced changes in vertical water mass structure of the Vema Channel during the Pliocene: evidence from Deep Sea Drilling Project Holes 516A, 517, and 518. *Init. Repts. DSDP* **72,** 907–19.

IMBRIE, J. & IMBRIE, J.Z. 1980. Modelling the climatic response to orbital variations. *Science* **207,** 943–53.

KANEPS, A.G. 1979. Gulf Stream: velocity fluctuations during the late Cenozoic. *Science* **204,** 297–301.

KEIGWIN, L.D. 1978. Pliocene closing of the Isthmus of Panama, based on biostratigraphic evidence from nearby Pacific Ocean and Caribbean Sea cores. *Geology* **6,** 630–4.

—— 1982a. Neogene planktonic foraminifers from Deep Sea Drilling Project Sites 502 and 503. *Init. Repts DSDP* **68,** 269–77.

—— 1982b. Stable isotope stratigraphy and paleoceanography of Sites 502 and 503. *Init. Repts DSDP* **68,** 445–53.

—— & THUNELL, R.C. 1979. Middle Pliocene climatic change in the Western Mediterranean from faunal and oxygen isotopic trends. *Nature* **282,** 294–6.

KENNETT, J.P., SHACKLETON, N.J., MARGOLIS, S.V., GOODNEY, D.E., DUDLEY, W.C. & KROOPNICK, P.M. 1979. Late Cenozoic oxygen and carbon isotopic history and volcanic ash history: DSDP Site 284, South Pacific. *Am. J. Sci.* **279,** 52–69.

KOMINZ, M.A., HEATH, G.R., KU, T.L. & PISIAS, N.G. 1979. Bruhnes time scales and the interpretation of climatic change. *Earth Planet. Sci. Lett.* **45,** 3394–410.

LEDBETTER, M.T., WILLIAMS, D.F. & ELLWOOD, B.B. 1978. Late Pliocene climate and south-west Atlantic abyssal circulation. *Nature* **272,** 237–9.

LE TRENT, H. & GHIL, M. 1983. Orbital forcing, climatic interactions and glaciation cycles. *J. geophys. Res.* **88,** 5167–90.

LUDWIG, W.J., KRASHENINNIKOV, V.A. *et al.* 1983. *Init. Repts DSDP* **71**(I), 1–477.

LUYENDYK, B.P., FORSYTH, D. & PHILLIPS, J.D. 1972. An experimental approach to the paleocirculation of the ocean surface water. *Geol. Soc. Am. Bull.* **83,** 2649–64.

MILLER, K.G. & TUCHOLKE, B.E. 1983. Development of Cenozoic abyssal circulation south of the Greenland-Scotland Ridge. *In: Structure and Development of the Greenland-Scotland Ridge.* Plenum Press, New York. 549–89.

PESTIAUX, P. & BERGER, A. 1984. An optimal approach to the spectral characteristics of deep-sea climatic records. *In: Milankovitch and Climate.* Riedel Publishing Company, Bordrecht. 417–46.

PINET, P.R. & POPENOE, P. 1982. Blake Plateau: Control of Miocene sedimentation patterns by large-scale shifts of the Gulf Stream axis. *Geology* **10,** 257–9.

RUDDIMAN, W.F. & MCINTYRE, A. 1984. Ice-age thermal response and climatic role of the surface Atlantic Ocean, 40°N to 63°N. *Geol. Soc. Am. Bull.* **95,** 381–96.

SAITO, T. 1976. Geologic significance of coiling direction in the planktonic foraminifera *Pulleniatina. Geology* **4,** 305–9.

SHACKLETON, N.J. & OPDYKE, N.D. 1977. Oxygen isotope and paleomagnetic evidence for early Northern Hemisphere glaciation. *Nature* **270,** 216–19.

—— *et al.* 1984. Oxygen isotope calibration of the onset of ice-rafting and history of glaciation in the North Atlantic region. *Nature* **307,** 620–3.

ZIMMERMAN, H.B. *et al.* 1984. History of Plio-Pleistocene climate in the northeastern Atlantic: DSDP Site 552A. *Init. Repts DSDP* **81,** 861–75.

JAN BACKMAN & OTTO HERMELIN, Department of Geology, University of Stockholm, S–106 91 Stockholm, Sweden.

PIERRE PESTIAUX, Institut d'Astronomie et de Geophysiqué, Universite Catholique de Louvain, B–1348 Louvain-La-Neuve, Belgium.

HERMAN ZIMMERMAN, Department of Geology, Union College, Schenectady, New York 12308, USA.

Diatom biostratigraphic and palaeoceanographic interpretations for the middle to high latitude North Atlantic Ocean

Jack G. Baldauf

SUMMARY: Diatoms are useful for biostratigraphic correlation of Neogene sediment sections in the North Atlantic Ocean. The warm–temperate assemblage observed in cores from the middle and high latitude Atlantic enables partial application of the diatom zonations defined in the eastern equatorial Pacific. Sixteen of the equatorial Pacific diatom zones are recognized in cores from the middle latitudes of the North Atlantic.

The diatom assemblage implies that surface waters were warm during the early Miocene in the Rockall Plateau region. A decrease in the quality of diatom preservation, and the occurrence of numerous hiatuses at approximately the early–middle Miocene boundary, suggest a major change in oceanic circulation. An increase in deep water overflow from the Norwegian Sea may have been a contributing factor. Although middle Miocene sediments at both middle and high latitudes are generally enriched with silica, a possible restriction of siliceous sedimentation towards the high latitudes is observed during this time. The interval representing the early late Miocene and earliest Pliocene is generally devoid of diatoms.

Conditions favouring siliceous sedimentation continued at high latitudes and returned to middle latitudes of the North Atlantic during the late Pliocene, maybe in response to increased intensity of the Gulf Stream. The abundance and preservation of diatoms in high latitude Quaternary sediments is partially controlled by climatic fluctuations. The occurrence of the *Denticulopsis seminae* group suggest that cool, low saline waters may have occurred in the North Atlantic between 1.1 and about 0.70 Ma.

The North Atlantic has been the setting for numerous studies concerned with oceanic responses to climatic fluctuations and regional tectonics. Some of these studies used either calcareous nannofossils or foraminifera for stratigraphic control and/or palaeoceanographic interpretations, because these are the most common microfossil remains found in North Atlantic sediment. Because siliceous microfossils such as diatoms are less common in the sediment, they have rarely been used for such studies (see Baldauf 1984). This study will show that diatoms have great potential as biostratigraphic and palaeoceanographic indicators in the North Atlantic region.

Diatoms were examined from Neogene sediment recovered during DSDP Legs 81 (Baldauf 1984) and 94 (Baldauf, in press, a, b). The significant results of these studies are synthesized here. The geographic locations of the 13 DSDP sites discussed in this paper are shown in Fig. 1. With the exception of a 10 cm sampling interval used for the upper 50 metres of Hole 552A, generally one sample was examined from each core-section.

Because the stratigraphic usefulness of diatoms has not been rigorously tested previously in the North Atlantic, it is important that the diatom stratigraphy be tied to other stratigraphies. Stratigraphic control is also based on the calcareous nannofossil stratigraphies of Backman (1984), Takayama (in press), and Muller (1979), and the palaeomagnetostratigraphies of Kent (in Shackleton *et al.* 1984) and Clement & Robinson (in press). Sample preparation techniques are described in Baldauf (1984, in press, a).

Biostratigraphy

The Neogene diatom species observed in sediment recovered from the DSDP sites compose a warm–temperate assemblage similar to that observed in the eastern equatorial Pacific. The occurrence in North Atlantic sediment of diatom species which are stratigraphically useful in eastern equatorial Pacific sediment has enabled the recognition and use of most biozones defined by Burckle (1972, 1977) and Barron (1983, 1985a).

The diatom zonations defined for the equatorial Pacific are based on the first or last occurrence of selected species. Figure 2 compares the diatom datums used in the equatorial Pacific to those used in the North Atlantic. Thirty-three of the 47 equatorial Pacific biostratigraphic events are recognized in the same sequence in the North Atlantic. The greatest difference between records from the two oceans occurs in the middle and late Miocene. Only nine of the 14 middle Miocene and three of the nine late Miocene

From SUMMERHAYES, C.P. & SHACKLETON, N.J. (EDS), 1986, *North Atlantic Palaeoceanography*, Geological Society Special Publication No. 21, pp. 243–252.

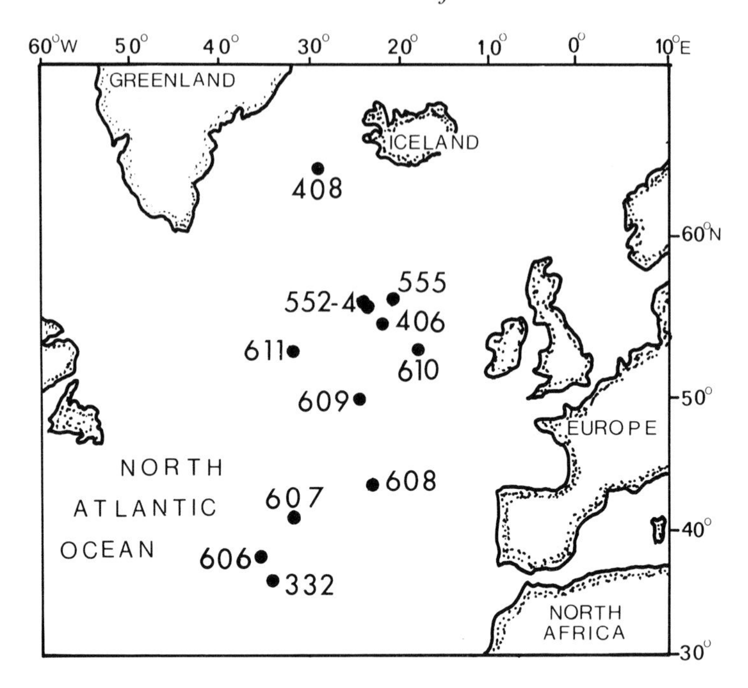

FIG. 1. Geographic location of Deep Sea Drilling Sites discussed in this study.

equatorial Pacific datums are recognized in the North Atlantic. This is because certain species are absent from the North Atlantic sediment.

Datums which are only useful in the equatorial Pacific generally are those of species restricted to tropical waters. These include species such as *Coscinodiscus temperi* var. *delicata*, *C. turberculatus*, *Nitzschia porteri*, *Thalassiosira burckliana*, and *T. bukryi*. Although other species such as *Rhizoselenia praebergonii* and *Nitzschia miocenica* are most abundant at low latitudes, they do occur sporadically at middle latitudes but are not stratigraphically useful.

The biostratigraphic use of equatorial Pacific datums at middle and high latitudes of the North Atlantic assumes that biostratigraphic events are isochronous between the North Atlantic and eastern equatorial Pacific. Baldauf (1984, in press, a), correlated the late Pliocene through Quaternary biostratigraphic events with the palaeomagnetostratigraphies available from DSDP Site 522 (Kent, *in*: Shackleton *et al.* 1984) and Sites 606 through 611 (Clement & Robinson, in press). Comparing these results to similar correlations in the equatorial Pacific indicates that while some events are isochronous, others are diachronous (Table 1). For example, the first occurrence of *Pseudoeunotia doliolus* is assigned an age of 1.8 Ma by Barron (1985a) for the equatorial Pacific and 1.84 by Baldauf (in press, a) for the North Atlantic. On the other hand, Barron (1985a) assigned an age of 0.65 Ma for the last occurrence of *Nitzschia reinholdii* in the equatorial Pacific, whereas this datum has an approximate age of 0.44 Ma (Baldauf, in press, a) in the North Atlantic.

The isochroneity of Miocene and early Pliocene biostratigraphic events cannot be determined at this time because palaeomagnetic or isotope stratigraphies have either not yet been completed, or the results are inadequate. The stratigraphic sequence of datums observed in the North Atlantic, however, agrees with that observed in the equatorial Pacific (Fig. 2). This suggests that some events are likely to be synchonous between the two oceans. This is supported by the correlation between diatom and calcareous nannofossil stratigraphies (Baldauf 1984, in press, a), although palaeomagnetic or isotope data would be required for confirmation of this assumption. This assumption is adopted for the Miocene and early Pliocene events in this paper.

The Neogene diatom zonation applied in the North Atlantic consists of 16 zones (Fig. 3). The early Miocene through early late Miocene zones are assumed to be the same as defined by Barron (1983) for the equatorial Pacific. Secondary datums and/or substitute zones were defined by Baldauf (1984, in press, a) for the intervals representing the *Coscinodiscus gigas* var. *diorma*

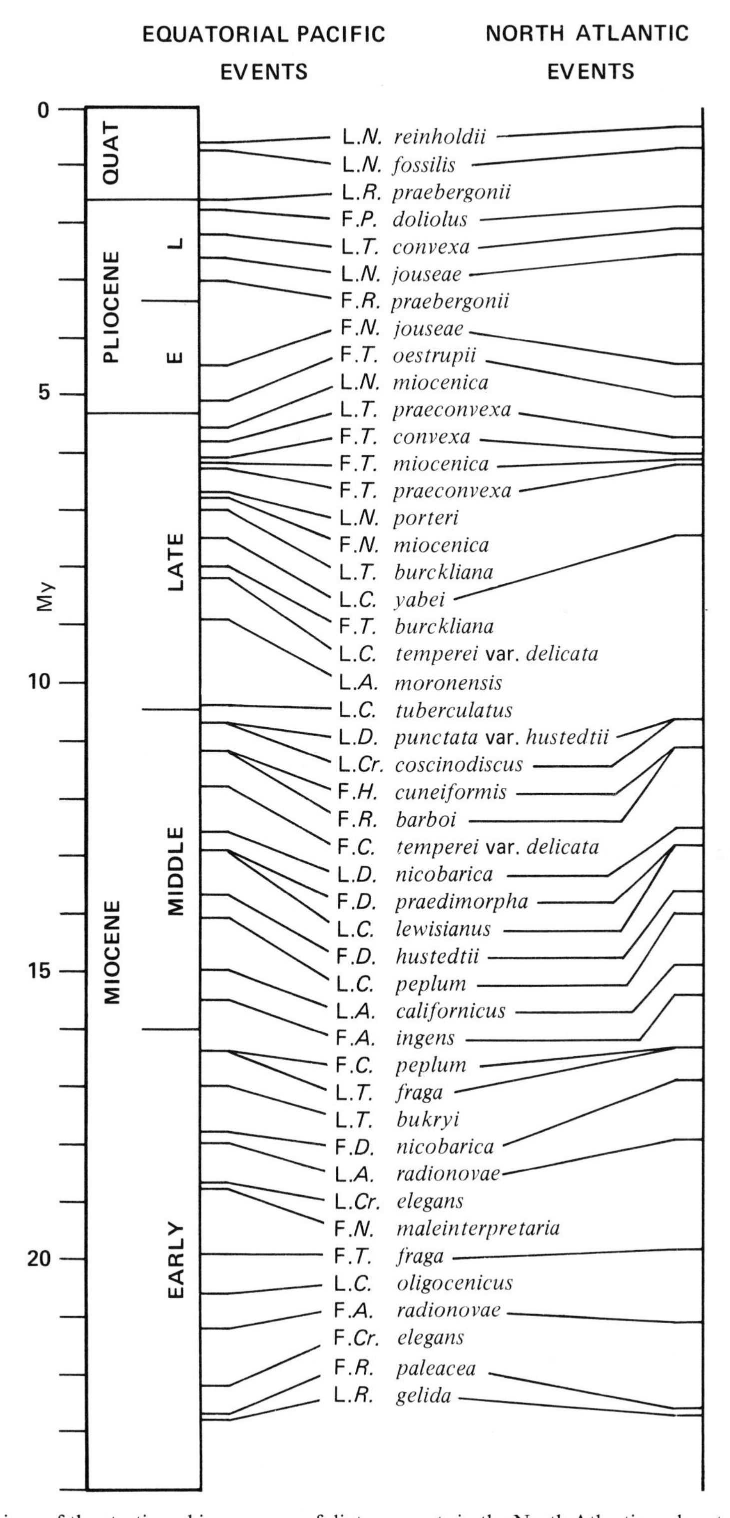

FIG. 2. Comparison of the stratigraphic sequence of diatom events in the North Atlantic and eastern equatorial Pacific Oceans. F = first stratigraphic occurrence. L = last stratigraphic occurrence. Age assignments are from Barron (1985a) and Baldauf (in press, a).

TABLE 1. *Comparison of the assigned ages for the late Pliocene and Quaternary datums between the Atlantic and Pacific Oceans*

Datum	Atlantic Ocean	Pacific Ocean
(L) *N. reinholdii*	0.44 Ma (7)	0.65 Ma (1)
(L) *N. fossils*	0.51–0.64 Ma (7)	0.79 Ma (2)
(L) *M. quadrangula**	0.64–0.74 Ma (7)	0.79 Ma (1)
(L) *R. praebergonii* var. *robusta*	not observed	1.55 Ma (3,4,5)
(F) *P. doliolus*	1.74–1.86 Ma (7)	1.80 Ma (1,3–5)
(L) *R. praebergonii*	not observed	1.85 Ma (3,8)
(L) *T. convexa*	2.20 Ma (6)	2.20 Ma (3)
(L) *N. jouseae*	2.60 Ma (6)	2.60 Ma (3)

* = silicoflagellate. (1) = Burckle 1977. (2) = Kozumi & Kanaya 1976. 3 = Burckle & Trainer 1979. (4) = Barron 1980b. (5) = Barron, 1985b. (6) = Baldauf, 1984. (7) = Baldauf 1985b. (8) = Baldauf, 1985. (L) = last stratigraphic occurrence. (F) = first stratigraphic occurrence.

and *Actinocyclus moronensis* zones (upper middle Miocene), because the primary taxa are excluded from the high-latitude North Atlantic (Sites 408 and 555) at this time.

The diatom species observed by Baldauf (1984) include *Hemidiscus cuneiformis*, *Rhizosolenia barboi*, *Denticulopsis punctata* var. *hustedtii*, and *D. praedimorpha*. These species are characteristic of the high latitude North Atlantic and are absent, or have inconsistent stratigraphic ranges, at middle latitude sites. As a result, the late middle Miocene diatom zones defined by Baldauf (1984) are not applicable at middle latitudes of the North Atlantic.

With the following exceptions, the latest Miocene through Quaternary equatorial Pacific zones as defined by Burckle (1977) are applied here:

(1) Due to the rare occurrence of *Nitzschia miocenica* and the preservational or ecological exclusion of *Nitzschia porteri* from North Atlantic sediments, the *Nitzschia porteri* and *N. miocenica* Zones of Burckle (1972, 1977) cannot be differentiated in the North Atlantic but are termed the *Nitzschia porteri-N. miocenica* interval.

(2) The defintion of the top of the *Nitzchia jouseae* Zone of Burckle (1972) is also modified due to sporadic occurrence of *Rhizosolenia praebergonii*, and is defined in the North Atlantic at the last occurrence of *Nitzschia jouseae*. Consequently, the *Nitzschia marina* zone is defined as the interval between the last occurrence of *N. jouseae* and the first occurrence of *P. doliolus*. This zone partially replaces the *Rhizosolenia praebergonii* Zone of Burckle (1977) (see Baldauf, 1984, in press, a, for a detailed discussion).

Results

The stratigraphic occurrence of diatoms at DSDP sites from the middle and high latitude North Atlantic is shown in Fig. 4. Stratigraphic intervals enriched with diatoms occur in the lower Miocene at DSDP Sites 406 and 610, the middle Miocene at Sites 406, 408, 553, 608, and 610, and the upper Pliocene through Quaternary at most sites. The lower upper Miocene (portion of the *N. porteri-N. miocenica* interval) is generally barren of diatoms.

With the exception of a stratigraphically short sediment section recovered from Site 553, diatoms are generally absent from the lowest Miocene. Samples at Site 553 from this time interval contain rare specimens of *Rocella gelida*. This species occurs in material from the Antarctic (Weaver & Gombos 1981), Pacific (Barron 1983) and the Bering Sea (Baldauf, unpublished data). Diatoms are present in the upper portion of the lower Miocene at Sites 406 and 610.

Three stratigraphic intervals defined by different states of diatom preservation are observed at middle latitudes of the North Atlantic during the middle Miocene. The interval approximating the lower–middle Miocene boundary is either barren of diatoms or contains rare, poorly preserved specimens. The middle Miocene interval is enriched with diatoms and the upper middle Miocene interval is generally barren of diatoms. A similar sequence of preservation is observed at

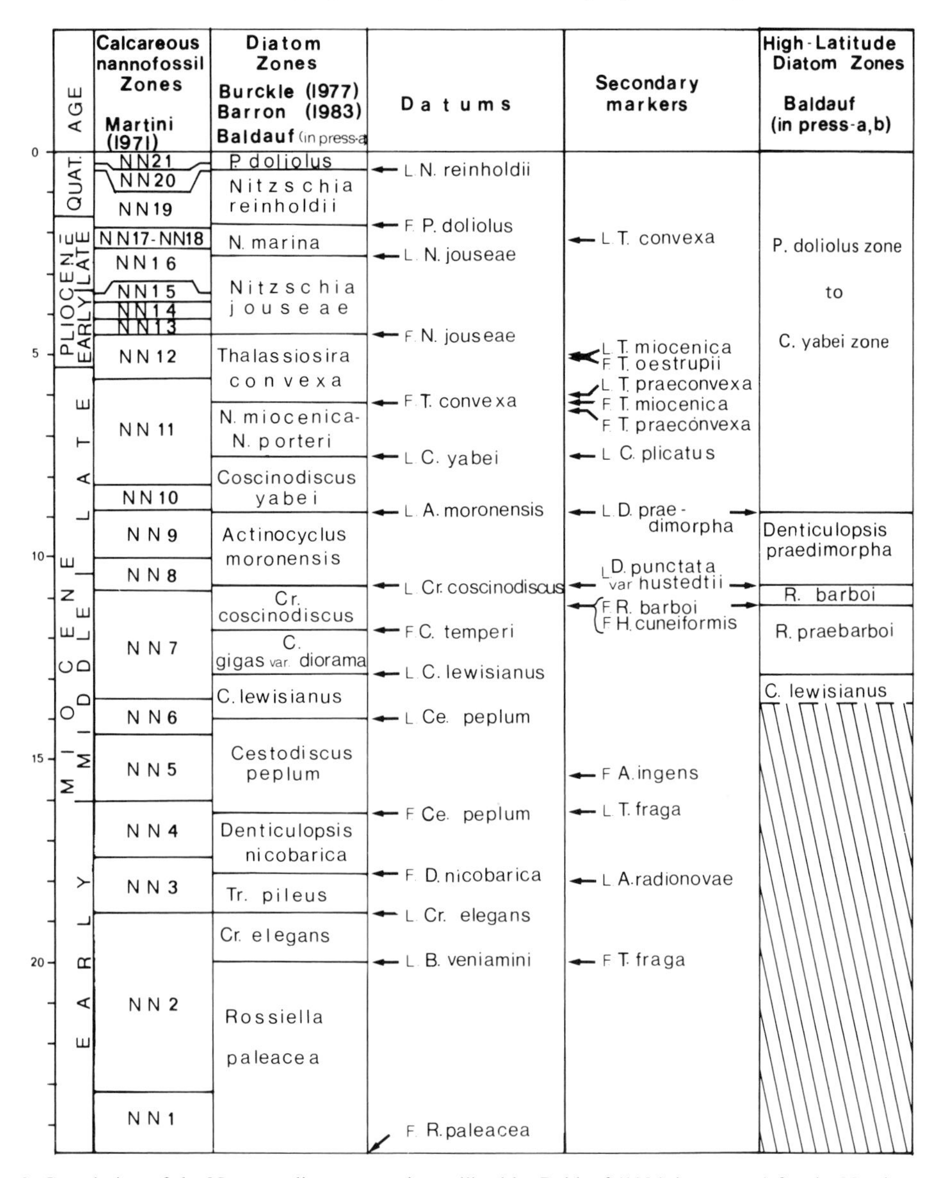

FIG. 3. Correlation of the Neogene diatom zonation utilized by Baldauf (1984, in press, a) for the North Atlantic to the calcareous nannofossil zonation of Martini (1971). The diatom zonation of Burckle (1977) and Barron (1983, 1985a) modified by Baldauf (1984) are utilized for the low and middle latitudes of the North Atlantic. Zonal and secondary markers are shown for this zonal scheme. Several zones cannot be recognized at high latitude sites of the North Atlantic during the late middle and late Miocene. An alternative high latitude zonation (Baldauf 1984, in press, a) is shown in the right column.

high latitudes of the North Atlantic, but the sequence is diachronous between the two regions (Fig. 4).

The stratigraphic interval approximating the lower–middle Miocene boundary (upper portion of the *Denticulopsis nicobarica* and the lower portion of the *Cestodiscus peplum* zones, 17–15.5 Ma) is characterized by poor diatom preservation at Sites 406 and 610. This interval of poor preservation occurs stratigraphically above an interval containing common to abundant *Ethmodiscus* fragments. A hiatus may occur within

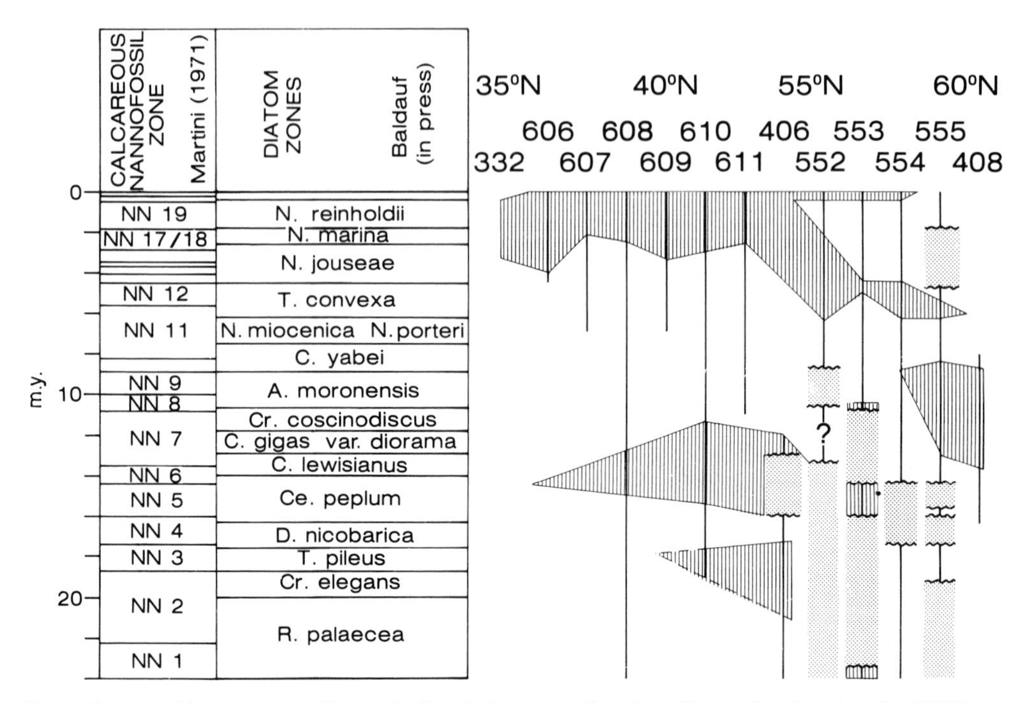

FIG. 4. Stratigraphic occurrence of biogenic silica (striped area) (based on diatom abundance) at the DSDP sites examined in this study. Stippled area represents hiatuses. See Baldauf (1984, in press, a) for detailed stratigraphic discussion.

the earliest middle Miocene at Site 406. The stratigraphic occurrence of the hiatus is based on diatom and silicoflagellate analysis and correlations between these groups and calcareous nannofossils (Baldauf, in press, b). The sub-bottom depth of the hiatus corresponds to the placement of the '2' seismic reflector (Montadert *et al.* 1979), and to an abrupt sedimentological change from calcareous diatomite (lower Miocene) to nannofossil ooze (middle Miocene). At Site 610, a similar change in diatom preservation corresponds to a lithification of the sediment and an increase in sonic velocity (Baldauf, in press, b).

Diatoms are present in the short sedimentary sequence occurring in the lowermost middle Miocene at Sites 553 and 555. Stratigraphic placement of these intervals is based on the nannofossil biostratigraphy of Backman (1984). Although diatoms are present, they are rare and poorly preserved. The sedimentary sequences at both Sites 553 and 555 are interrupted by hiatuses. Diatoms are absent from the stratigraphic interval approximating the lower–middle Miocene boundary at Site 608.

Sediments were generally enriched with respect to biogenic silica (based on diatom abundance) during the middle and late middle Miocene (upper *Cestodiscus peplum* through lower *Craspedodiscus coscinodiscus* zones). The age of this siliceous interval differs between middle and high latitudes of the North Atlantic (Fig. 4). At Site 608 (42°N.) diatoms occur in the upper portion of the *Cestodiscus peplum* zone (about 15 Ma) through the upper portion of the *Coscinodiscus lewisianus* zone (12.9 Ma). At Site 610 (53°N.), diatoms also occur in the upper portion of the *C. peplum* zone, but disappear later (lower *Craspedodiscus coscinodiscus* zone, 11.4 Ma). At Site 555 (56°N.), diatoms occur in the uppermost portion of the *Coscinodiscus lewisianus* zone through the lower portion of the late Miocene *Coscinodiscus yabei* zone (8.5 Ma).

Diatoms are generally absent at DSDP sites located in middle latitudes of the North Atlantic during the interval representing the late middle Miocene (top of the *Coscinodiscus lewisianus* zone) through early Pliocene (lower portion of the *Nitzschia jouseae* zone). At high latitudes, diatoms are present during this time at Sites 552, 553, 554, and 555 (Rockall Plateau Region).

In Hole 552A, few, well-preserved diatoms are observed in the interval approximating the base of the late Miocene *Thalassiosira convexa* zone. Late Miocene and early Pliocene diatoms are not as well represented at Sites 553–555. At these sites, diatoms occur sporadically. The occurrence

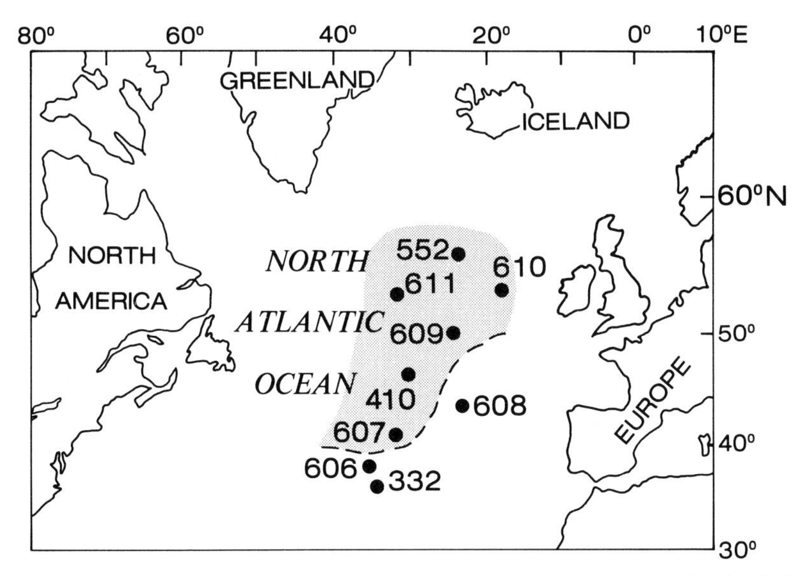

FIG. 5. Geographic distribution of the *Denticulopsis seminae* group (shaded area) in North Atlantic Ocean sites during the interval from about 1.1 to about 0.70 Ma.

of a hiatus at Site 555 suggests an increase in oceanic circulation at middle water depths (Site 555 was cored in 1659 metres of water) causing erosion or non-deposition. In addition, fewer samples were examined from this interval as a result of the poor core recovery.

Sediments enriched with diatoms increased at middle North Atlantic latitudes at about 4 Ma, and are commonly present above this initial Pliocene occurrence. As previously mentioned, diatom rich sediments increased in the Rockall region by about 6 Ma. At high latitudes, diatoms are not consistently present above this 6 Ma occurrence. In Hole 552A, the interval from about 1.8–0.44 Ma is generally devoid of diatoms (Baldauf 1984). At present, the reason for the absence of diatoms from this interval is unknown. Baldauf (1984) also shows that during the late Quaternary in Hole 552A, changes in the abundance of diatoms corresponds somewhat to fluctuations of the oxygen isotope curve. In general, diatoms are present in Hole 552A during intervals corresponding to interglacials and absent during intervals corresponding to glacial stages.

The late Quaternary biogeography of *Denticulopsis seminae* and *Denticulopsis seminae* var. *fossils* (*D. semeinae* group) in the North Atlantic is shown in Fig. 5. This group occurs at Sites 607, 609, 610, 611, and 552. In addition, Schrader (1977) recorded *Denticula seminae* (at present referred to as *Denticulopsis seminae*) at Site 410. The first occurrence of the *D. seminae* group, at Sites 607, 609, and 611 occurs at about 1.1 Ma. The last occurrence of this group is slightly diachronous between Site 607 (about 0.8 Ma) and Sites 609 and 611 (about 0.70 Ma). The absence of the *D. seminae* group from DSDP Sites 332, 606, and 608, and its rarity at Sites 552 and 610, indicates that these forms were most common in middle and high latitudes of the central North Atlantic.

Denticulopsis seminae occurs at present in the Bering Sea (Baldauf 1982) and at middle and high latitudes of the North Pacific (Barron 1980a, 1981). The first stratigraphic occurrence of this group is within the middle Gauss palaeomagnetic Chron (about 3.1 Ma) at middle latitudes of the North Pacific (Barron 1980a, 1981), and between the b and c events of the Gilbert Chron (about 4.2 Ma) at the high latitudes of the North Pacific (Burckle & Opdyke 1977). The *D. seminae* group does not occur in low latitude sediments of the North Atlantic.

Discussion

Surface water productivity and silica dissolution control the abundance and preservation of diatoms within North Atlantic sediment. Within the Rockall Plateau region, these processes are influenced by palaeoceanographic changes resulting from tectonics and climatic fluctuations. The diatoms present at Sites 406 and 610 with the earliest Miocene include warm–temperate species such as *Synedra jouseana* and *Raphidodiscus marylandicus*, and warm-water species such as *Cestodiscus spp.*, *Coscinodiscus rhombicus*, *Tri-*

ceratium pileus, and a species similar to *Ethmodiscus rex*. The occurrence of these species suggests that warm surface waters were present in the Rockall Plateau region during the earliest Miocene (about 22–19 Ma).

The abundance of *Ethmodiscus* fragments increases during the stratigraphic interval representing the *Triceratium pileus* and the lower portion of the *Denticulopsis nicobarica* zones (about 18.7–17.0 Ma) at Sites 406 and 610 (Rockall Plateau). Samples from this interval have common to abundant fragments of *Ethmodiscus* that suggests surface waters were tropical in character and possibly low in phosphate (Belyayeve 1968), that middle and bottom water circulation was intensified, or that a combination of these two conditions occurred during this time in the Rockall Plateau region (Baldauf, in press, b). Hiatuses approximating the early–middle Miocene boundary are observed at DSDP Site 406 and 610 (Baldauf, in press, b), Sites 552, 553, 554, and 555 (Site reports, Roberts *et al.* 1984), and Sites 336 and 352 (Iceland-Faroe Ridge), 114 (Gardner Drift), 403–405 (Rockall Plateau), 407 (Reykjanes Ridge), and 112 (Labrador Sea) (Shor & Poore 1979). The stratigraphic correspondence between changes in diatom preservation, sedimentology, numerous hiatuses, and a major seismic reflector (Sites 406 and 610) suggest that these hiatuses result from major increases in deep water circulation.

Sea water is typically undersaturated with respect to silica (Lisitzin 1972; Heath 1974). As a result, siliceous frustules dissolve as they settle through the water column towards the ocean floor. Dissolution of the frustules normally continues as long as the frustules remain in contact with sea water or pore water which is undersaturated with respect to silica. The biogenic silica which is dissolved may be incorporated into middle and bottom water masses. Younger bottom water contains less biogenic silica and is more corrosive to biogenic silica than older bottom water which is enriched in dissolved silica (Heath 1974). Increasing the volume of young water entering the North Atlantic basin during the latest and early middle Miocene could explain the abrupt decline in the diatom preservation at Sites 406 and 610. An alternative is that the burial rate is decreased, exposing the diatom frustules to undersaturated water for prolonged periods, thereby increasing frustule dissolution. As a result of the geographically restricted data base, it is uncertain whether this abrupt preservational change is a local event restricted to the Rockall Plateau region, or is characteristic of the middle and high latitude North Atlantic.

The results of this study suggest that siliceous sedimentation during the middle and late Miocene of the North Atlantic possibly retreated to higher latitudes and a major switch from siliceous to calcareous sedimentation occurred during the late middle and early late Miocene. Keller & Barron (1983) reached a similar conclusion during their cursory examination of 7 DSDP Atlantic sites. They indicate that siliceous sedimentation declined abruptly in the Caribbean region during the late early Miocene and more gradually at middle and high latitudes of the North Atlantic Ocean.

The decline of siliceous sedimentation from the North Atlantic corresponds to an increase in siliceous sedimentation along the North Pacific margin (Keller & Barron 1983). This interval of declined siliceous productivity in the North Atlantic may, therefore, be controlled in part by the global silica budget.

Increased siliceous sedimentation at middle latitudes of the North Atlantic during the late Pliocene could have occurred as a result of increased intensity of the Gulf Stream, resulting from the closing of the Pan-American Seaway. Intensification of the Gulf Stream would transport greater volumes of warm, saline, nutrient rich surface water to middle and high latitudes of the North Atlantic. A second possibility is that silica was preserved because of a decrease in the intensity of deep water circulation at the middle and high-latitude North Atlantic. A third possibility is that an increase in the silica or nutrient concentration of North Atlantic surface waters resulted in an increase in diatom productivity, thereby increasing sedimentation and burial rates and diatom preservation.

The most likely factor responsible is the latitudinal migration of the polar front and Gulf Stream, which have been shown to migrate latitudinally during climatic cycles (Ruddiman & McIntyre 1976). During interglacials the high latitude North Atlantic Ocean is free of ice cover, allowing increased surface productivity and the transport of warm saline water northward into the high latitudes. During glacials, the high latitude North Atlantic is ice-covered and the Polar front is restricted southward of 40°N. (maximum glacial stage, Ruddiman & McIntyre 1976) limiting diatom productivity in the high latitude North Atlantic.

Preliminary studies indicate that diatoms are present in glacial intervals at both Sites 606 and 607 (Baldauf, in press, a). The general lack of diatoms at Site 552 for the interval from 1.8–0.44 Ma suggest that other factors are also responsible for the abundance and preservation of diatoms in North Atlantic sediments.

The occurrence of the *Denticulopsis seminae*

group in both the North American and western European Basins of the North Atlantic suggest that a cool, low salinity surface water mass may have occurred in the North Atlantic during the interval from 1.1 to about 0.70 Ma. A possible origin of this cool, low saline water is the Arctic Ocean. If this hypothesis is correct then the occurrence of this group also suggests that the Arctic Ocean was partially ice free during the middle Quaternary in order to allow transportation of this group between the Bering Sea and the North Atlantic. The possibility that the *D. seminae* group occurred in the surface waters but was not recorded in the sedimentary record should not be dismissed at this time. Additional studies are required to further examine the occurrence of this group in North Atlantic sediment.

Conclusion

Neogene diatoms were examined from the thirteen North Atlantic DSDP sites. The presence of warm–temperate species at these sites has allowed the development of a North Atlantic diatom zonation which is partially correlated with the diatom zonation defined for the eastern equatorial Pacific Ocean.

Palaeoceanographic interpretations based on changes in the abundance and preservation of the diatom assemblage are also discussed. Warm temperate surface waters may have occurred in the Rockall Plateau region of the North Atlantic during the early Miocene. An abrupt decrease in the quality of diatom preservation at approximately the early–middle Miocene boundary suggests a major reorganization of oceanic circulation during this time. The correspondence of this decline in preservation with numerous hiatuses suggests an increase in bottom water circulation resulting from increased Norwegian Sea overflow.

Diatom enriched sediments are widespread in the middle Miocene of the North Atlantic (30–65°N). These sediments may be gradually restricted to the higher latitudes. At low latitudes, upper middle Miocene through lower Pliocene sediments are generally devoid of diatoms. At high latitudes only lower upper Miocene sediments are barren of diatoms.

Siliceous sedimentation increases in the North Atlantic during the late Pliocene. The abundance of late Pliocene and Quaternary diatoms in the high latitude North Atlantic is controlled partially by climatic fluctuation. The occurrence of the *Denticulopsis seminae* group in the North Atlantic Ocean from about 1.1–0.70 Ma may indicate the occurrence of cool, low saline surface waters in the central North Atlantic during the early Quaternary.

ACKNOWLEDGMENTS: The manuscript was reviewed by John Barron, Charles Blome and Charlotte Brunner. Additional comments by Constance Sancetta and Audrey Meyer also improved this manuscript. Special thanks are given to the Geological Society of London for travel support to attend the meeting and to present this paper. This study was completed while employed at the US Geological Survey and while enrolled at the University of California, Berkeley, USA.

References

BACKMAN, J. 1984. Cenozoic calcareous nannofossil biostratigraphy from the northeast Atlantic Ocean, Deep Sea Drilling Project, Leg 81. *In*: ROBERTS, D., SCHNITKER, D. *et al.*, *Init. Repts DSDP* **81.** US Govt. Print. Off., Washington, DC. 403–8.

BALDAUF, J.G. 1982. Identification of the Holocene–Pleistocene boundary in the Bering Sea by diatoms. *Boreas* **11,** 113–8.

—— 1984. Cenozoic diatom biostratigraphy and paleoceanography of the Rockall Plateau Region of the North Atlantic, Deep Sea Drilling Project leg 81. *In*: ROBERTS, D., SCHNITKER, D. *et al.*, *Init. Repts DSDP* **81.** US Govt. Print. Off., Washington, DC. 439–78.

——, 1985. A high resolution late Miocene-Pliocene diatom biostratigraphy for the eastern equatorial Pacific. *In*: MAYER, L., THEYER, F. *et al.*, *Init. Repts DSDP* **85.** US Govt. Print. Off., Washington, DC, 457–475.

——, in press, a. Diatom biostratigraphy of the middle and high latitude North Atlantic Ocean, Deep Sea Drilling Project Leg 94. *In*: RUDDIMAN, W., KIDD, R. *et al.*, *Init. Repts DSDP* **94.** US Govt. Print. Off., Washington, DC.

——, in press, b. Biostratigraphic and paleoceanographic interpretations of diatom preservation in lower and middle Miocene sediments examined from the Rockall Plateau region of the North Atlantic Ocean. *In*: RUDDIMAN, W., KIDD. R. *et al.*, *Init. Repts DSDP* **94.** US Govt. Print. Off., Washington, DC.

BARRON, J.A. 1980a. Lower Miocene to Quaternary diatom biostratigraphy of Leg 57, off Northeastern Japan, Deep Sea Drilling Project. *In*: HONZA, E. *et al.*, *Init. Repts DSDP* **56** and **57.** US Govt. Print. Off., Washington, D.C. 641–85.

BARRON, J.A. 1980b. Upper Pliocene and Quaternary diatom biostratigraphy of the Deep Sea Drilling Project Leg 54, Tropical Eastern Pacific. *In*: ROSENDAHL, B.R., HEKINIAN, R. *et al.*, *Init. Repts DSDP* **54.** US Govt. Print. Off., Washington, DC. 455–62.

—— 1981. Late Cenozoic diatom biostratigraphy and paleoceanography of the middle-latitude eastern North Pacific, Deep Sea Drilling Project Leg 63. *In*: YEATS, R.S., HAQ, B. *et al.*, *Init. Repts DSDP* **63.** US Govt. Print. Off., Washington, DC. 507–38.

—— 1983. Latest Oligocene through early middle Miocene diatom biostratigraphy of the eastern tropical Pacific. *Mar. Micropaleont.* **7,** 487–15.

——, 1985a. Late Eocene to Holocene diatom biostratigraphy of the equatorial Pacific Ocean, DSDP Leg 85. *In*: MAYER, L., THEYER, F. *et al.*, *Init. Repts DSDP* **85.** US Govt. Print. Off., Washington, DC.

——, 1985b. Miocene to Holocene planktic diatoms. *In*: BOLLI, H.M., SAUNDERS, J.B. & PERZH-NIELSEN, K. (eds), *Plankton Stratigraphy*. Cambridge University Press, Cambridge. 763–809.

BELYAYEVE, T.V. 1968, Distribution and numbers of diatoms *Ethmodiscus* Castr. in plankton and in bottom sediments of the Pacific Ocean. *Okeanologiya* **8** (1).

BURCKLE, L.H. 1972. Late Cenozoic plankton diatom zones from the eastern equatorial Pacific. *Nova Hedwidiga, Beihft.* **39,** 217–50.

—— 1977. Pliocene and Pleistocene diatom datums from the equatorial Pacific. Republic of Indonesia. *Geol. Res. Dev. Centre, Spec. Publ.* **1,** 25–44.

—— & OPDYKE, N.D. 1977. Late Neogene diatom correlations in the circum-Pacific. *Proceedings of the First International Congress on Pacific Neogene Stratigraphy, Tokyo, 1976*. 255–84.

—— & TRAINER, J. 1979. Middle and late Pliocene diatom datum levels from the central Pacific. *Micropaleontology* **25** (3), 281–93.

CLEMENT, B. & ROBINSON, F., in press. Magnetostratigraphy of Leg 94 sediments. *In*: RUDDIMAN, W., KIDD, R. *et al.*, *Init. Repts DSDP* **94.** US Govt. Print. Off., Washington, DC.

HEATH, G.R. 1974. Dissolved silica in deep-sea sediments. *Soc. Econ. Paleont. Min. Spec. Pub.* **20.** 77–93.

KELLER, G. & BARRON, J.A. 1983. Paleoceanographic implications of Miocene deep-sea hiatuses. *Geol. Soc. Am. Bull.* **94,** 590–613.

KOIZUMI, I. & KANAYA, T. 1976. Late Cenozoic marine diatom sequence from Chosi district, Pacific coast, central Japan. *In*: TAKAYANAGI, Y. & SAITO, T. (eds), *Progress in Micropalaeontology*. Micropalaeontology Press, New York. 144–59.

LISITZIN, A.P. 1972. Sedimentation in the world ocean. *Soc. Econ. Paleont. Min. Spec. Pub.* **17.**

MARTINI, E. 1971. Standard Tertiary and Quaternary calcareous nannoplankton zonation. *In*: FARINACCI, A. (ed.), *Proceedings of the second planktonic conference, Roma, 1970, vol. 2*. 739–77.

MONTADERT, L., ROBERTS, D.G. *et al.* 1979. *Init. Repts DSDP*, **48.** US Govt. Print. Off., Washington, DC.

MULLER, C. 1979. Calcareous nannofossils from the North Atlantic (LEG 48). *In*: MONTADERT. L., ROBERTS, D.G. *et al.*, *Init. Repts DSDP.* **48.** US Govt. Print. Off., Washington, DC. 589–39.

ROBERTS, D.G., SCHINTKER, D. *et al.* 1984. *Init. Repts DSDP* **81.** US Govt. Print. Off., Washington, DC.

RUDDIMAN, W.F. & MCINTYRE, A. 1976. Northeast Atlantic Paleoclimatic changes over the past 600,000 years. *Geol. Soc. Am. Mem.* **145,** 111–46.

SCHRADER, H.J. 1977. Opal phytoplankton in DSDP Leg 49 samples. *In*: LUYENDYK, B.P., CANN, J.R. *et al.*, *Init. Repts DSDP* **49.** US Govt. Print. Off., Washington, DC. 589–93.

SHACKLETON, N.J., BACKMAN, J., ZIMMERMAN, H., KENT, D.V., HALL, M.A., ROBERTS, D.G., SCHNITKER, D., BALDAUF, J.G., DESPRAIRIES, A., HOMRIGHAUSEN, R., HUDDLESTUN, P., KEENE, J.B., KALTENBACK, A.J., KRUMSIEK, K.A.O., MORTON, A.C., MURRAY, J.W. & J. WESTBURG-SMITH. 1984. Oxygen isotope calibration of the onset of ice-rafting and history of glaciation in the North Atlantic region. *Nature*. **307,** 620–3.

SHOR, A.N. & POORE, R.Z. 1979. Bottom currents and ice rafting in the North Atlantic: Interpretation of Neogene depositional environments of Leg 49 cores. *In*: LUYENDYK, B., CANN, J. *et al.*, *Init. Repts DSDP* **49.** US Govt. Print. Off., Washington, DC. 859–72.

TAKAYAMA, T., in press. Calcareous nannofossil biostratigraphy. *In*: RUDDIMAN, W., KIDD, R. *et al.*, *Init. Repts DSDP* **94.** US Govt. Print. Off., Washington, DC.

WEAVER, F. & GOMBOS., A.J. JR. 1981. Southern high-latitude diatom biostratigraphy. *In*: WARME, J.E., DOUGLAS, R.G. & WINTERER, E.L. (eds), *The Deep Sea Drilling Project: A Decade of Progress. Soc. Econ. Paleont. Min. Spec. Pub.* **32,** 445–70.

JACK G. BALDAUF, Ocean Drilling Program, Texas A & M University, College Station, Texas 77843, USA.

Mesozoic palaeoceanography and black shales

Pulsation tectonics as the control of North Atlantic palaeoceanography

Robert E. Sheridan

SUMMARY: A new theory of pulsation tectonics involving the cyclic eruption of thermal plumes from the D″ layer in the lowermost mantle is proposed as a control of the first-order cycles of palaeocean changes. Plumes carry heat rapidly away from the core/mantle boundary, changing conditions in a way that results in the quiescence of the earth's magnetic field. The rising plumes, after a phase lag, induce faster plate spreading and widespread intraplate volcanism. Subsequently, colder rock returning to the lower mantle re-establishes the D″ thermal boundary layer and sets up conditions for more turbulent core convection, which in turn allows frequent magnetic reversals. Meanwhile, lower mantle plumes cease erupting, and the plates spread more slowly. The cycles of fast and slow plate spreading and quiet and reversing magnetic field have a period of 60–100 Ma.

Changes in spreading rate on a global scale are probably a major cause of large scale (100–300 m) eustatic sea-level changes. Major global sea-level high-stands in middle to late Jurassic and in middle to late Cretaceous times coincide with fast spreading. Consequential changes in albedo, global temperature, humidity, forestation, atmospheric and oceanic CO_2, oceanic acidity and oxygenation lead to variations in the preservation of black shales in the oceans, and in the level of the calcite compensation depth (CCD). In the North Atlantic, the climatic factors forced by rising sea-level caused increased black shale preservation and a shallowing of the CCD. With falling sea-levels, less black shale is preserved and the CCD is deeper.

As part of Leg 76 of the Deep Sea Drilling Project (DSDP), drilling was completed at Site 534 in the Blake-Bahama Basin (Fig. 1). All the formations and sedimentary units of the western North Atlantic and basaltic basement were penetrated (Fig. 2). Site 534 also penetrated the major seismic reflection Horizons A, β, C, and D, the deepest of which had never been penetrated before (Fig. 3) (Bryan *et al.* 1980; Sheridan *et al.* 1983). The prominent reflection Horizon D is correlated with an early Oxfordian sequence of horizontally bedded, grey carbonate turbidites that overlie buried contourites (Sheridan *et al.* 1983). Site 534 is the only site thus far to penetrate basaltic oceanic basement in this position of the Jurassic magnetic quiet zone. Coring across the contact between sediment and basalt reveals typical basaltic basement features and no evidence of a hiatus. The middle Callovian age for the sediments immediately above the basalt is probably the age of the basement (Gradstein & Sheridan 1983; Sheridan, Gradstein *et al.* 1983; Sheridan, Bates *et al.* 1983).

This paper is about the implications of these new findings on our understanding of the origin of magnetic quiet zones through the theory of pulsation tectonics (Sheridan 1983). This theory also provides explanations of the non-random time variations in plate tectonics that control the long-term cycles of palaeoceanography.

Sedimentary cycles—West North Atlantic

The sedimentary units penetrated at Site 534 are widespread (Sheridan *et al.* 1978) and most have been elevated to formal formation and member status (Jansa *et al.* 1979). Unit 7 of Site 534 is only documented at this single site and so far is not defined formally as a formation. However, it is mappable seismically as the D to basement interval over a wide area.

The western Atlantic sedimentary facies consist of the following units (Fig. 2) (Sheridan, Gradstein *et al.* 1983):

Unit 7: Green, maroon, and black carbonaceous claystones, with low angle cross beds and ripples reflecting deposition from nepheloid layers as contourites. At the top interbedding with grey turbiditic limestones increases. Middle Callovian to early Oxfordian in age.

Cat Gap Formation: Red, variegated claystone and shaly limestone with interbedded grey turbiditic limestone. Deposited by normal pelagic. hemipelagic, and turbiditic processes as draping and basin-levelling sequences. Oxfordian to early Tithonian in age.

Blake–Bahama Formation: White and grey nannofossil and turbiditic limestone interbedded with black shale and quartz sands in its upper part. Deposited by normal pelagic and turbiditic

From Summerhayes, C.P. & Shackleton, N.J. (eds), 1986, *North Atlantic Palaeoceanography*, Geological Society Special Publication No. 21, pp. 255–275.

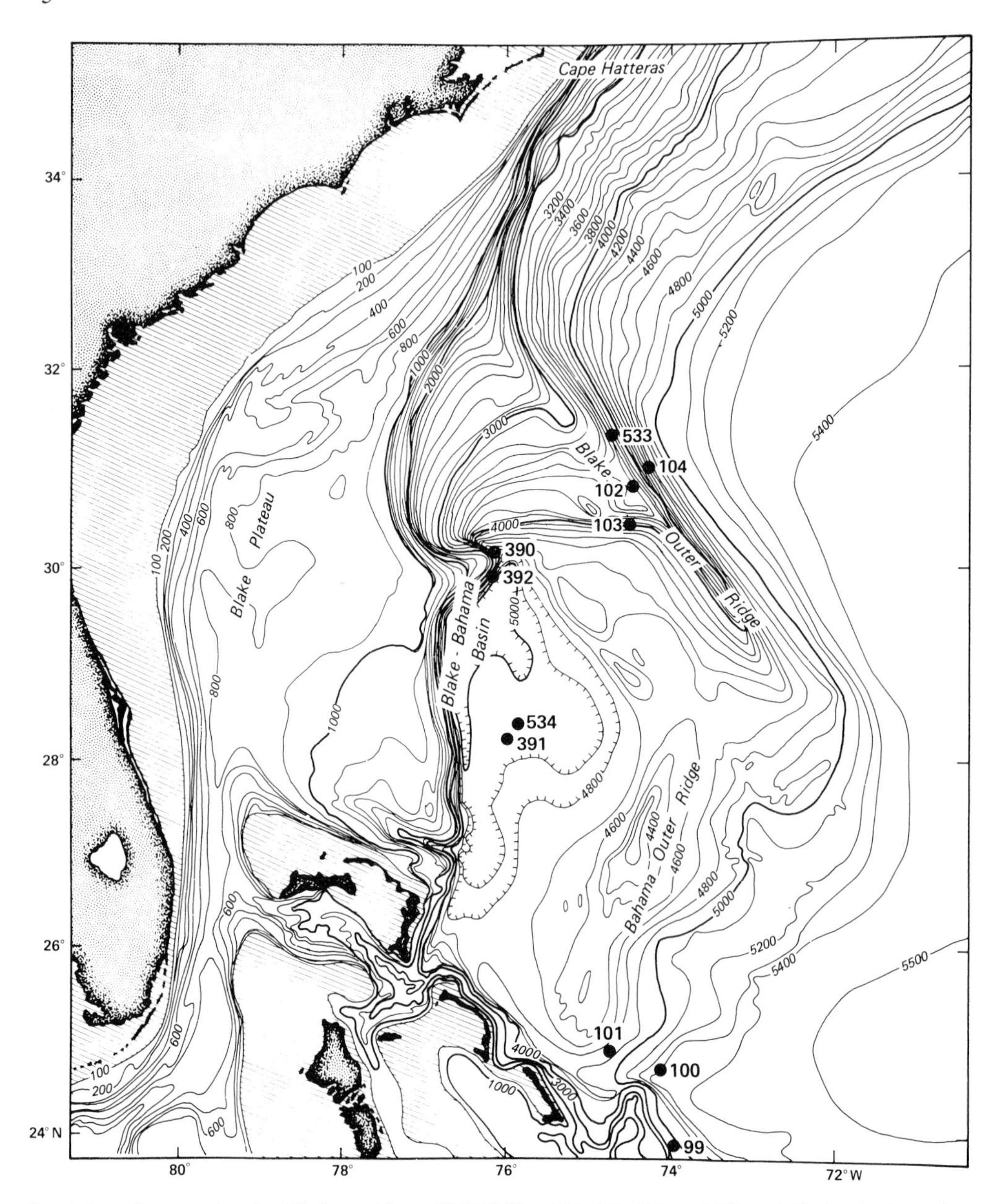

FIG. 1. Location map showing (1) the position of DSDP Sites 391, 534, 101, and 100 and (2) the physiography of the Blake-Bahama Basin (from Sheridan, Gradstein *et al.* 1983).

processes as drape and basin-levelling carbonates; shales and sands are distal portions of margin fan facies. Late Tithonian to Barremian in age.

Hatteras Formation: Black and green carbonaceous claystone and shale with pyrite and radiolarian silts. Deposited as normal pelagic clays and hemipelagic clays in nepheloid layers, and as clay turbidites. Barremian to Cenomanian in age.

Plantagenet Formation: Red, orange, variegated claystone. Deposited as pelagic clays of oxidized mineral matter in a sediment starved basin, forming drape sequences. Cenomanian to Maestrichtian in age.

Bermuda Rise Formation: Green, white, and yellow siliceous ooze, porcellanite, and cherts. Palaeocene to Oligocene in age.

Great Abaco Member: Intraclastic chalk of

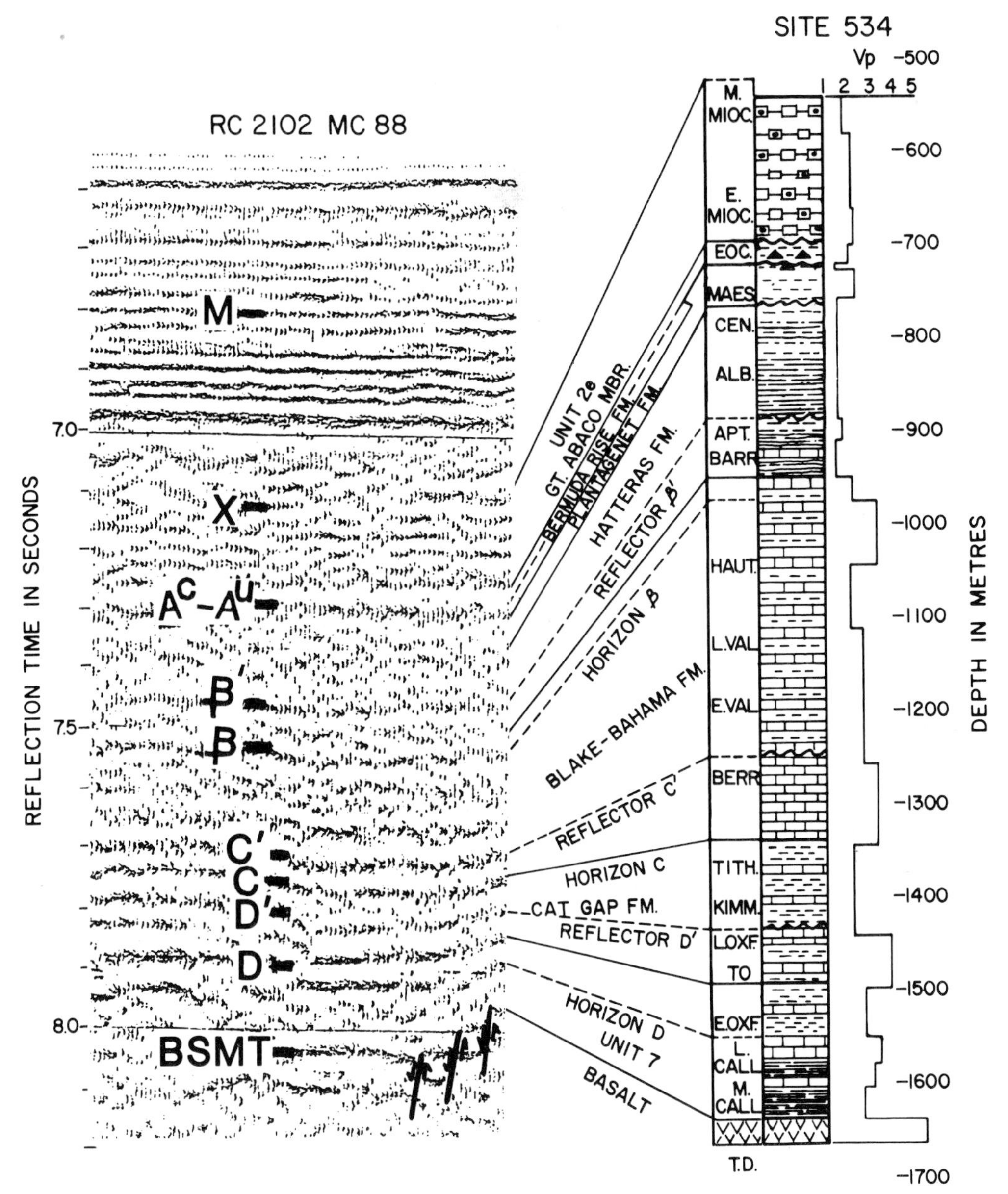

FIG. 2. Results of drilling at Site 534 showing correlation of age, lithologies and compressional wave velocities (V_p) with the seismic reflection profile at the site (from Sheridan 1983). The formation nomenclature used is from Jansa *et al.* (1979).

debris flows, and turbidites of extrabasinal limestone clasts and nannofossil ooze, mixed with intrabasinal clasts of siliceous claystones. Early to late Miocene in age.

Blake Outer Ridge Formation: Grey and green hemipelagic clay and claystone deposited from suspension and as contourites. Miocene to Holocene in age.

Some units are similar to one another, and these sediments may belong to long-term cycles. For example, both the Hatteras Formation and Unit 7 contain layers of highly carbonaceous clays with more than 2% total organic carbon (TOC) (Fig. 4) (Sheridan, Gradstein *et al.* 1983). Reducing conditions in the sediments preserved the carbon, but the cause of these reducing

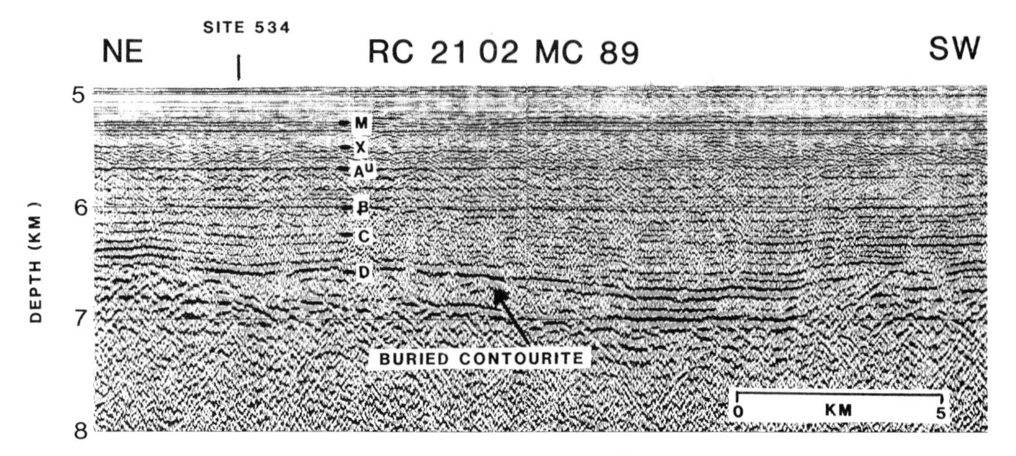

FIG. 3. Multichannel seismic reflection profile across Site 534 in the Blake–Bahama Basin showing the prominent seismic Horizons M, X, A^u, β, C, and D (from Sheridan 1983).

conditions is still controversial (Katz 1983; Habib 1983; Summerhayes & Masran 1983). Whether there were cycles of anoxic ocean bottom, or of influxes of organic matter that overwhelmed oxidized ocean bottoms is still debated. Similar types of organic matter, terrestrial plant material and marine fossil remains, occur in both units.

Another similarity between the Hatteras Formation and Unit 7 is that both are claystones with zero percent calcium carbonate. Both units were deposited below the calcite compensation depth (CCD) (Fig. 4) (Sheridan, Gradstein *et al.* 1983; Sheridan, Bates *et al.* 1983). Geohistory diagrams for Site 534 (Sheridan 1983) and previous DSDP

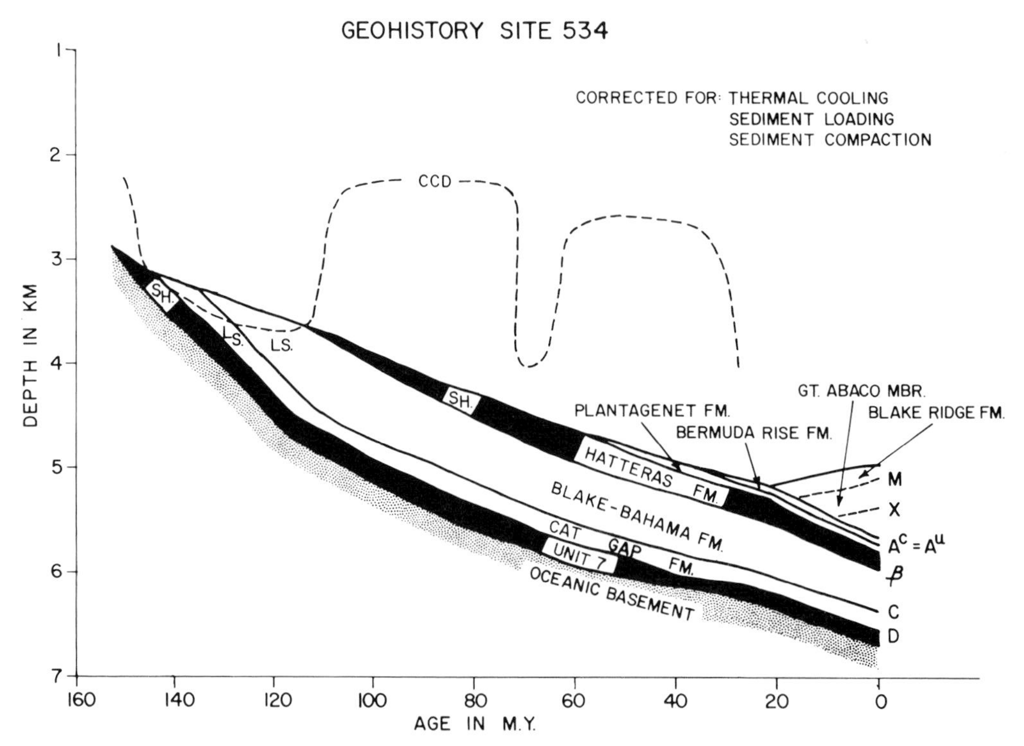

FIG. 4. Geohistory of Site 534 taking into account the effects of thermal cooling, sediment loading, and sediment compaction. The 'black shales' of Unit 7 and the Hatteras Formation are indicated (from Sheridan 1983).

sites in the western North Atlantic (Jansa *et al.* 1979) confirm that the CCD has varied with time.

The Cat Gap Formation and the Plantagenet Formation overlying the carbonaceous claystones are poor in organic carbon. Both are brightly coloured and well oxidized. Also, both contain chalks and limestones suggesting a deepening of the CCD (Fig. 4) (Jansa *et al.* 1978; Sheridan, Gradstein *et al.* 1983; Sheridan, Bates *et al.* 1983).

Thus, in the western North Atlantic there are two similar cycles of carbon-rich to carbon-poor sediments, with two corresponding cycles of calcite-poor to calcite-rich sediments. Unit 7/Cat Gap represents one cycle, while Hatteras/Plantagenet represents a second similar cycle of carbon-rich to carbon-poor sediments. Unit 7/Blake-Bahama represents one cycle of calcite-poor to calcite-rich sediments, while Hatteras/Great Abaco represents a similar younger cycle. Only two cycles are recorded in the North Atlantic, one from Middle Jurassic to Early Cretaceous, and a younger one from Middle Cretaceous to Miocene.

These changes in sedimentary facies extend into the eastern Atlantic (Jansa *et al.* 1979, 1983), and the Ligurian and eastern Tethys (Baumgartner 1983). Because of the coeval nature of these cycles over such a wide geographic area, and in basins that were physically isolated, it is presumed that the cycles were controlled by first-order climatic factors.

Based on the drilling results at Site 534, it appears that the central North Atlantic broke up, and a narrow 'proto-Atlantic' spread rapidly in the Bathonian (Figs 5 and 6). Rapid spreading in the Bathonian created the oceanic crust in the Gulf of Mexico as a continuation of the proto-Atlantic (Sheridan 1983) (Fig. 5). During the early Callovian, a spreading centre shift isolated the proto-Atlantic/Gulf of Mexico oceanic crusts on the North American plate, and the modern central North Atlantic Ocean began to form. Rapid spreading ensued from Callovian to Kimmeridgian (Fig. 6), during which the lower half of the first sedimentary cycle was deposited as Unit 7 and Cat Gap Formation (Sheridan 1983).

Palaeogeographic reconstruction of the North Atlantic, and palaeoenvironmental interpretations based on evidence such as Site 534 (Figs 5 and 6), indicate that the sediments were deposited by typical hemipelagic and pelagic processes. The cycles of high and low carbon content and high and low calcite content are superimposed on these

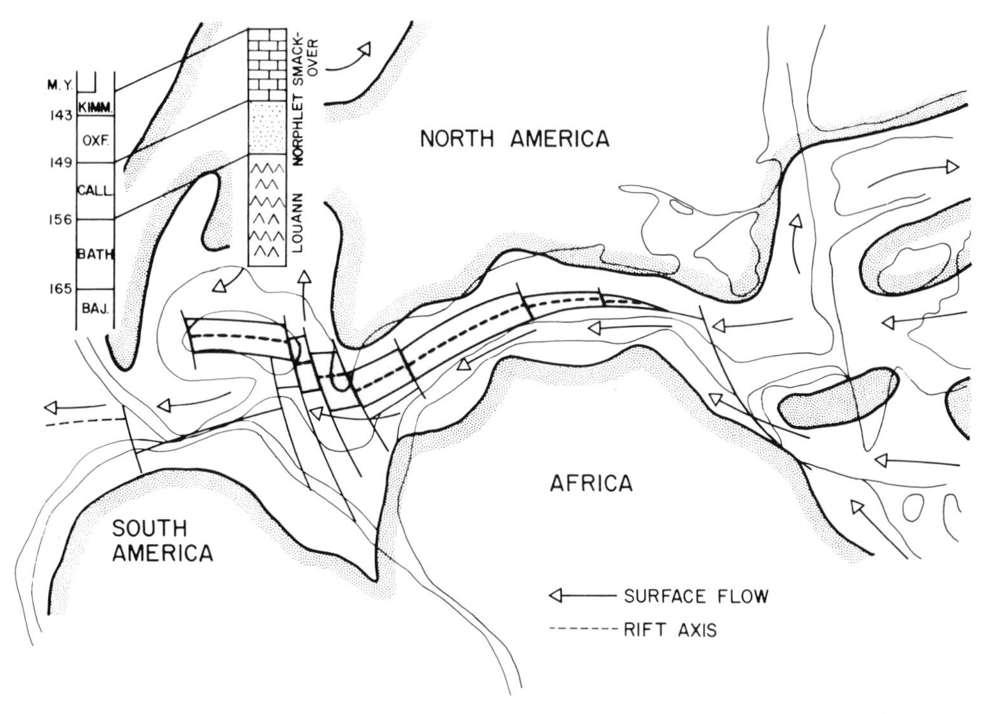

FIG. 5. Continental reconstruction and palaeogeography during the earliest Callovian at the time of formation of the Blake Spur magnetic anomaly. The shaded crustal zone with a dashed rift axis is the Bathonian age proto-Atlantic and Gulf of Mexico oceanic crust (from Sheridan 1983).

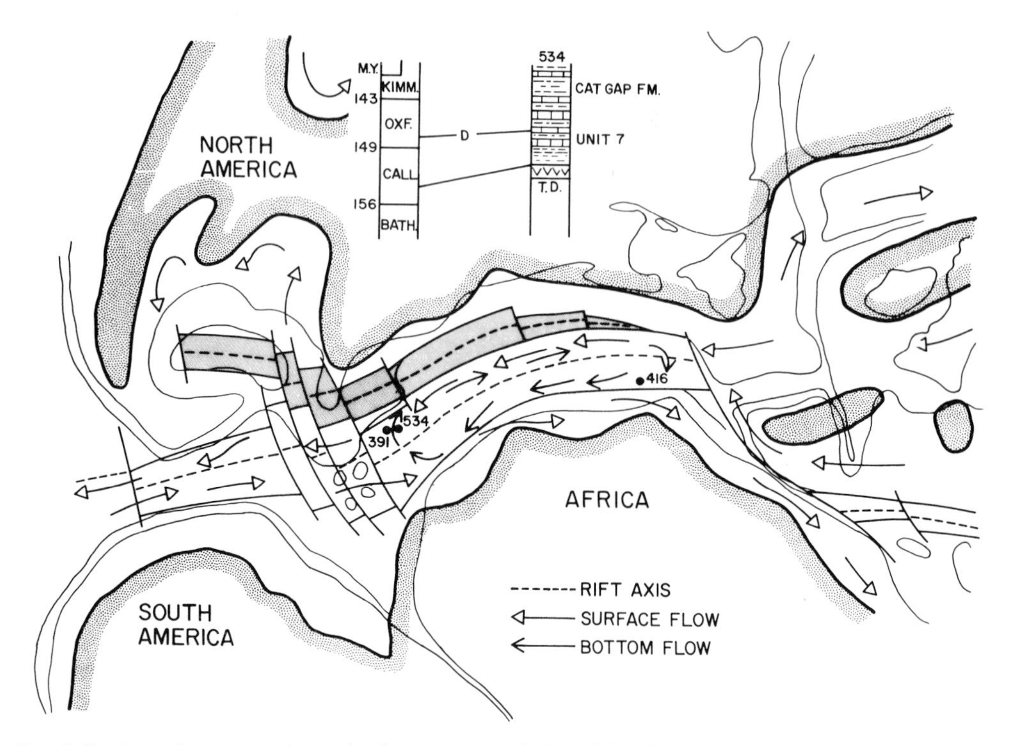

FIG. 6. Continental reconstruction and palaeogeography during mid-Oxfordian time, the age of magnetic anomaly M26. Note that the proto-Atlantic-Gulf of Mexico spreading centres (shaded) are extinct at this time and isolated on the North American plate (from Sheridan 1983).

otherwise normal oceanic sedimentation processes. This is similar to what Stow & Dean (1984) have found for the South Atlantic.

Spreading rate–magnetic quiet zone cycles

Site 534 is located precisely relative to magnetic anomaly M25 on the south-east, and to the Blake Spur anomaly on the north-west. Between anomaly M25 and the Blake Spur anomaly is the Jurassic magnetic quiet zone. Within the quiet zone are weakly defined, small amplitude north-east-striking negative anomalies M26 to M28 (Bryan *et al.* 1980). M26 is probably a true reversal anomaly in that it is found as a series of reversals in the Pacific Jurassic quiet zone (Cande *et al.* 1978). M27 and M28 might only be linear magnetic anomalies following linear basement topography. In the Pacific, M26 through M29 are equivalent to M26 in the Blake Bahama Basin. Site 534 is located near the south-east boundary of anomaly M28.

Spreading rates across the Jurassic quiet zone of the western North Atlantic had been estimated previously by several techniques (Fig. 7) (Vogt & Einwich 1979; Bryan *et al.* 1980; Sheridan 1983). The drilling at Site 534 dates basement as no older than middle Callovian (154 Ma on the van Hinte (1976) scale or 165 Ma on the Kent & Gradstein (1984) scale) and Horizon D as early Oxfordian (149 Ma on the van Hinte (1976) scale or 163 Ma on the Kent & Gradstein (1984) scale). These points plot on a spreading rate of 3.8 cm/yr for the Jurassic quiet zone. This rate agrees with the previous drilling on anomalies M25 (Sites 100 and 105) and M16 (Site 387) but there must have been a drastic change in spreading rate at about M22 time. This spreading rate change within the M sequence is consistent for both the van Hinte (1976) and the Kent & Gradstein (1984) time-scales. Moreover, the ages for basement at M28 and at the Horizon D pinchout yield projected ages for M22 and M25 that agree well with the ages determined independently by Ogg (1980) by palaeomagnetic measurements in the Tethyan stratigraphic section. Now M25 is taken as early Kimmeridgian in age rather than as its oldest possible age, from drilling, of early Oxfordian (Fig. 7).

Thus drilling at Site 534 supports a revision of

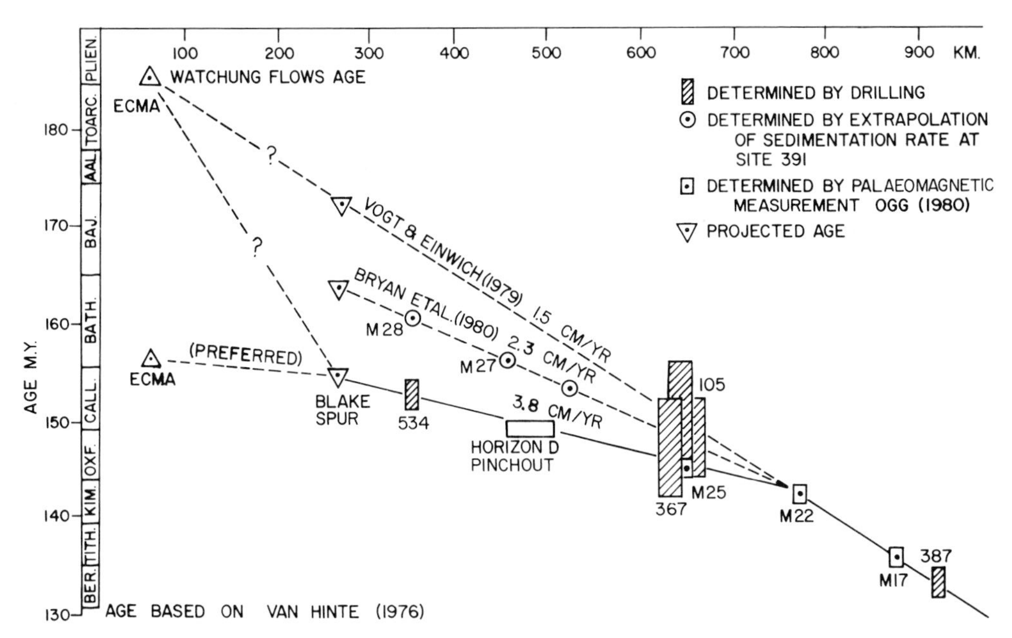

FIG. 7. Plot of horizontal position of key anomalies vs. geologic age. Numerical age from van Hinte (1976). Extrapolation of constant spreading rate by Vogt & Einwich (1979) from DSDP Sites 100, 105, 387 yields a Bajocian age for Blake Spur and Pliensbachian for East Coast magnetic anomaly. Age of basement at Site 534 and age of Horizon D yield younger ages for the Blake Spur (early Callovian) and ECMA (late Bathonian) and require higher (>3 cm/yr) spreading across the Jurassic magnetic quiet zone (from Sheridan 1983).

the palaeomagnetic time-scale for the Jurassic, as proposed by Ogg (1980) and adopted by Kent & Gradstein (1984). The new time-scale implies high spreading rates worldwide for the Jurassic quiet zones.

The new date of the basement at Site 534 now gives an age for the Blake Spur anomaly of basal Callovian. This is some 20 Ma younger than previously thought (Fig. 7). The Blake Spur anomaly is thought to represent extrusions of basalt along the rift that signalled the initiation of the modern central North Atlantic. Prior to Blake Spur time, spreading was confined to the Blake Spur–East Coast magnetic anomaly (ECMA) corridor (Figs 5 and 6). When the spreading-centre shifted to the Blake Spur position, the proto-Atlantic and Gulf of Mexico oceanic crusts were isolated on the North American plate.

The age of the proto-Atlantic crust in the Blake Spur–ECMA corridor is determined by extrapolating Jurassic quiet zone spreading rates landward of the Blake Spur, assuming a two-sided spreading system (Fig. 7) (Sheridan 1983). This gives an age of middle to late Bathonian for the ECMA. The proto-Atlantic and Gulf of Mexico spread in only a few million years at a rate of 7.6 cm/yr.

These new ages for the ECMA and Blake Spur magnetic anomaly, and thus the breakup of the Eastern North American margin, are supported by independent evidence. Biogeographic evidence suggested to Hallam (1977) that the central North Atlantic did not open until after the Callovian. A similarly late opening is likely for the Gulf of Mexico. Anderson & Schmidt (1983) used stratigraphic data to conclude that a small ocean crust formed there in the Bathonian, and that a major spreading centre shift to the south-east margin of Yucatan occurred in the Callovian. The author believes these are the same events as the breakup of the proto-Atlantic and the Blake Spur spreading centre shift.

This new timing for the breakup, and the interpretation of fast spreading on a global scale, are compatible with continental stratigraphy, which records an abrupt widespread transgression in latest Bathonian and earliest Callovian through Oxfordian times. The Cornbrash Formation in England, the Sundance Formation in Wyoming, and the Fernie Formation in Alberta were deposited in these widespread Jurassic transgressions. Hallam (1975) notes this worldwide late Jurassic sea-level rise, as do Vail *et al.* (1977). This eustatic event was caused by the opening of the central North Atlantic.

In the Atlantic, the fast spreading rates of more

than 3 cm/yr for the Jurassic quiet zone are comparable to those values determined for the Middle Cretaceous, when fast spreading also occurred on a global scale (Lowrie *et al.* 1980). The latest revisions of the Cretaceous palaeomagnetic time-scale reaffirm that the Middle Cretaceous magnetic quiet zone was associated with fast sea-floor spreading on a global scale. The Cretaceous quiet zone is a 30 Ma interval between 80 and 110 MA, when the earth's magnetic field was of mostly constant normal polarity (Lowrie *et al.* 1980) (Fig. 8). There is some suggestion that the Cretaceous was also a time of weaker magnetic field (Keating, pers. comm.).

As for the Jurassic magnetic quiet zones, the occurrence of these at approximately the same time in the North Atlantic, the Pacific, and Indian Oceans, in crusts with vastly different absolute spreading rates, seems proof enough that this Jurassic magnetic quiet period was also a global field phenomenon. Palaeomagnetic studies indicate that much of the Oxfordian and Callovian stages was of constant normal polarity, and that this normal polarity could extend into the upper Bathonian for a total interval of 15–17 Ma (Channell *et al.* 1982). Steiner (1980) also indicates that there was a decrease in strength of the Jurassic magnetic field. Other explanations of the Jurassic quiet zones, such as post-facto thermal alteration of the magnetic minerals, or unusually viscous thermal remanent magnetization, have been reviewed by Taylor *et al.* (1968) and Barrett & Keen (1975). The basalt recovered at Site 534, however, was typical oceanic pillow basalt without thick sediment intercalations. It is well magnetized with a normal Jurassic polarization (Sheridan, Gradstein *et al.* 1983).

With the new documentation of the age of the Jurassic quiet zone and its short duration of 15–17 Ma, it is clear that the wide areas of quiet zone crust in the Atlantic, Pacific, and Indian Oceans must have spread at rates relatively high for all these oceans. Thus the fast plate spreading in the Jurassic was indeed a global phenomenon, as it was in the Middle Cretaceous. In Fig. 8 the spreading rate for the western Atlantic plate for a

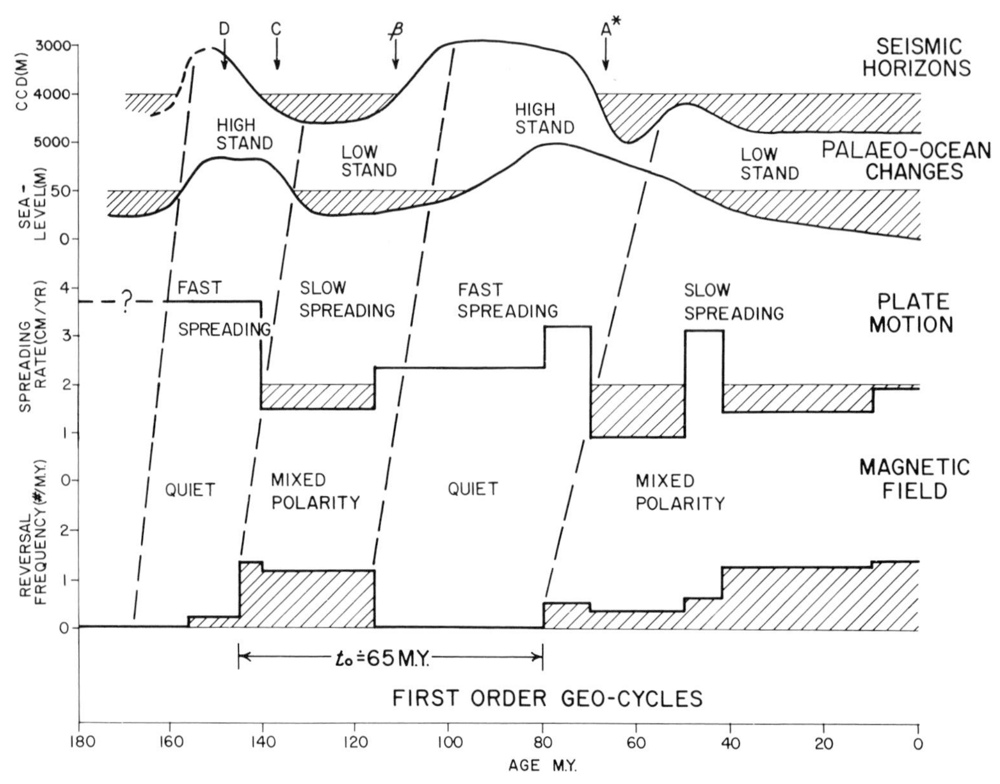

FIG. 8. Correlations of magnetic anomaly reversal frequency and plate spreading rate for the North Atlantic on the flow line through Site 534. Sea-level curve is based on Watts & Steckler (1979) and Hallam (1975); the calcite compensation depth (CCD) is from Jansa *et al.* (1979). Dashed lines imply correlation with a phase lag between magnetic field variation, plate motions, and sea-level changes. Also shown are the ages of the major seismic horizons A*, *β*, C, and D (from Sheridan 1983).

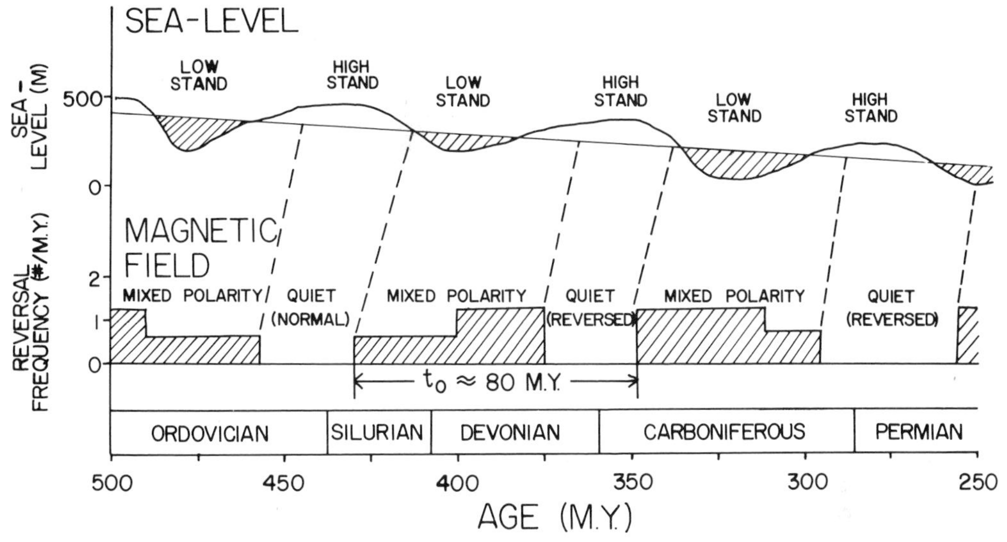

FIG. 9. Palaeozoic pulsation tectonic cycles as revealed in the quiet and reversing magnetic field (Harland *et al.* 1983) and the long-term envelope of the sea-level curve (Vail *et al.* 1977). The same phase lag and approximate period are found for the Palaeozoic as for the Mesozoic and Cenozoic (Fig. 8).

flow line through Site 534 is plotted in comparison with the time of magnetic quiet zones. There are two correlative occurrences of fast sea-floor spreading and relative quiescence of the earth's magnetic field; one in the Middle to Late Cretaceous and one in the Middle to Late Jurassic. These cycles of fast and slow spreading and quiet and reversing magnetic field have a period of approximately 65 Ma, but their period length seems to vary, in that the Jurassic cycle is shorter than the Cretaceous cycle (Fig. 8).

Older magnetic quiet times are documented by palaeomagnetic measurements on land (Harland *et al.* 1983). The Kiaman interval in the late Carboniferous and Permian has a constant reversed polarity; another reversed quiet zone occurs in the late Devonian–early Carboniferous; and a normal quiet zone occurs in the Late Ordovician–Silurian (Fig. 9). The magnetic quiet zone phenomena is probabilistic, and there is no significant difference in terms of magnetic energy or state of the core between normal and reversed polarity (McFadden & Merrill 1984). The period of the Palaeozoic cycles of reversing and quiet magnetic field is approximately 80 Ma, which is about the same as that for the Jurassic and Cretaceous cycles (Fig. 8). The differences between the Palaeozoic and younger periods of these cycles is not significant. The cycles in the magnetic field have periods generally of 60–100 Ma.

Because the Palaeozoic oceanic crust has all been destroyed in subduction, we can only infer that during times of Palaeozoic magnetic quiet zones spreading was relatively faster than in times of reversing magnetic fields. Eustatic sea-level changes in the Palaeozoic suggest that this inference is correct (Figs 8 and 9).

Eustatic sea-level cycles

Land stratigraphers have noted globally synchronous trangressions and regressions (Stille 1924; Grabau 1934; Umbgrove 1947), and advocated a tectonic origin for them. Modern stratigraphers have adopted the term sequence for these cycles (e.g. Sloss 1963), and Vail *et al.* (1977) have used seismic stratigraphic techniques to detect these cycles on continental margins. One modern sea-level curve, that of Vail *et al.* (1977), with later modifications by Hardenbol *et al.* (1981), shows long-term and short-term cycles. The long-term cycles have periods of 60–100 Ma, like those found for the magnetic field and plate motion cycles (Figs 8 and 9).

Most researchers feel that the long-term cycles, 60–100 Ma, are caused by global changes in plate spreading rates (Hays & Pitman 1973; Rona 1973; Vail *et al.* 1977; Pitman 1978). Faster plate spreading makes shallower, broader ridges, which cause the overflow of ocean water onto the shelves and continental interiors.

The absolute magnitudes of the long-term eustatic changes are uncertain. Hays & Pitman (1973) and Pitman (1978) estimate amplitudes of approximately 500 m and 350 m, respectively, for the Cretaceous sea-level highstand based on subjective estimates of ridge volumes at that time. Watts & Steckler (1979), on the other hand, estimate an amplitude of about 150 m for the same Cretaceous highstand, based on the analysis of several wells from the eastern North American continental shelf (Fig. 8). The author favours the estimates of Watts & Steckler (1979) because their technique is more objective and involves fewer assumptions.

The Watts & Steckler (1979) technique determines only the long-term sea-level curve. The seismic technique of Vail *et al.* (1977) can determine the higher-order short-term cycles. The envelope of the Vail *et al.* (1977) sea-level curve gives a long-term curve that is comparable to the Watts & Steckler (1979) curve. These researchers agree that there is a long-term highstand of sea-level in the Middle to Late Jurassic and another in the Middle to late Cretaceous (Fig. 8). The long-term envelope of the Vail *et al.* (1977) curve shows relative highstands in the late Ordovician–early Silurian, in the late Devonian–early Carboniferous, and in the late Carboniferous–Permian (Fig. 9).

The correlations of the Jurassic and Cretaceous highstands of sea-level with the fast spreading episodes is reasonably good (Fig. 8). There is evidence that the fast plate spreading episodes and highstands of sea-level lag the episodes of quiet magnetic field by some 10 Ma or so. High spreading rates and higher sea-levels persist into the end of the Cretaceous and earliest Tertiary after the earth's magnetic field begins to reverse in Campanian time (Fig. 8).

The long-term highstands of the Vail *et al.* (1977) sea-level curve for the Palaeozoic have phase lags of about 10 Ma after the independently determined palaeomagnetic quiet times (Harland *et al.* 1983) (Fig. 9). This consistency in phase lags and periods for the Palaeozoic and Mesozoic show that common geologic processes were operating throughout the Phanerozoic.

Pulsation tectonics

The origin of the magnetic field and the processes causing its reversal are found in the liquid outer core, whereas the plate motions and their spreading velocities are controlled by the properties and stresses in the lithosphere and asthenosphere. Yet these two phenomena are related in some way. Vogt (1975) hypothesized that plume eruptions from the lower mantle could affect both the core/mantle boundary and the overlying asthenosphere (Fig. 10). He noted that the phase lag, Δt, between changes in the magnetic field, which occur first, and the subsequent plate motion effects, was evidence that lower mantle processes such as plumes were involved. The observational data (Figs 8 and 9) appear to document this phase lag where the high spreading rate and higher sea-level episodes occur as much as 10 Ma after magnetic reversals begin to increase in frequency.

Theoretical considerations support this hypothesis. For example, Doell & Cox (1972) liken the core/mantle boundary to the surface of the earth, with solid rock in contact with a hot convecting fluid. If the temperature of the core is hotter relative to the mantle, and the mantle is cooler, core convection will be more turbulent, or 'stormier' (Fig. 10). If the temperature of the core is cooler relative to the mantle, and the mantle is hotter, core convection will be less turbulent. In the former, 'stormier' case, the earth's magnetic field measured at the surface should be more dipolar and strong, but inherently less stable and with more frequent reversals. A more stable, but weaker, toroidal field should result in the less turbulent core convection case (Strangway 1970). The interplay between the toroidal and dipolar components of the magnetic field are critical to the self sustaining dynamo concept and the maintenance of the earth's magnetic field at the surface. Turbulent core convection is necessary, but the energy for this convection is still little understood.

Jones (1977) and later authors (McFadden & Merrill 1984) note that changes in temperature at the core/mantle boundary may cause changes in the long-term reversal frequency of the magnetic field. Convective heat transfer via plumes could cause geologically rapid temperature changes, if the velocity of convection is great enough.

The time for convective circulation of plumes is determined by the equation:

$$t_{cv} \approx \frac{L}{v} \tag{1}$$

where L is the path length of material moving in the convection cell and v is the convection velocity. For plumes erupting from the lowermost mantle and rising to the asthenosphere (Fig. 11), with cooler mantle material sinking to the lower mantle, the size of the convection cell path, L, would be about π times half the diameter of the cell, the diameter being approximately 2200 km. Therefore, $L \approx 3000$ km in equation (1). If the velocity were known, then the time for plume circulation could be calculated.

Plume convection velocities are controlled by the temperature difference relative to the ambient temperature in the lower mantle and by the viscosity of the mantle rock (Yuen & Schubert 1976). This relationship is derived from the basic

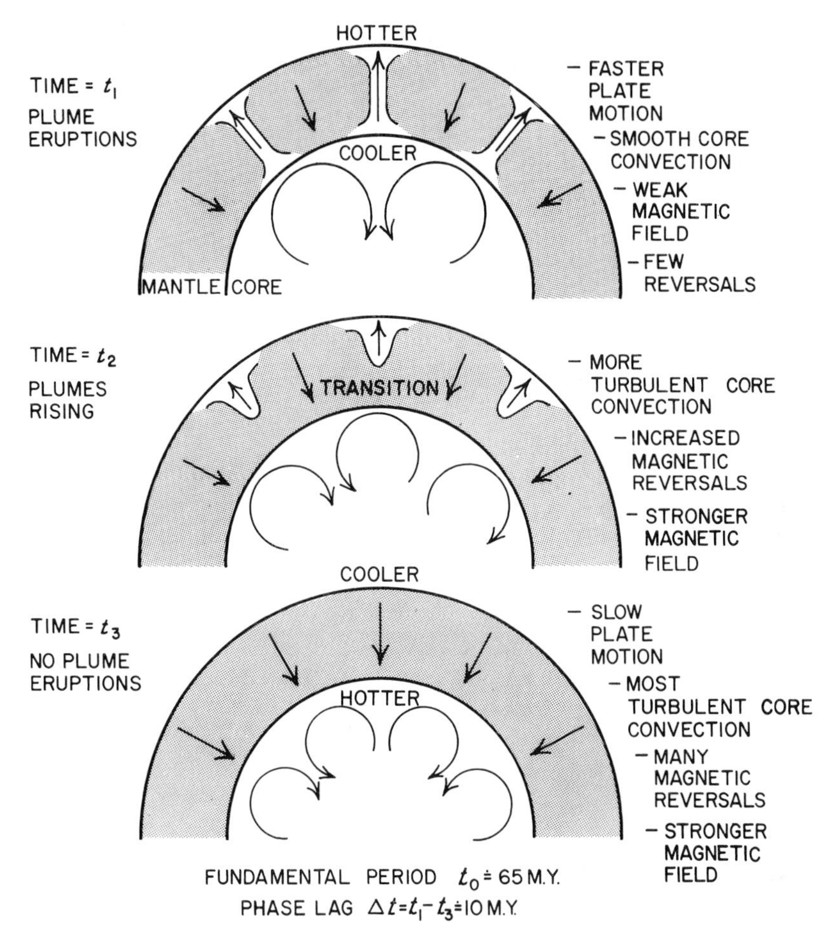

FIG. 10. Diagrammatic model of pulsation tectonics showing the cyclic eruption of plumes from the core/mantle boundary. The period of the cycle (t_o) is controlled by the time it takes for plumes to circulate (convective part of the cycle) and the time it takes for the lowermost D″ layer of the mantle to heat to instability (conductive part of the cycle) (from Sheridan 1983).

equation balancing the viscous forces that resist flow with the buoyancy forces, which cause the plume to rise. Yuen & Schubert (1976) conclude that:

'. . . the rheological behavior of the mantle material is such that narrow plumes can rise hundreds of kilometres through the mantle at velocities between one and several tens of centimetres per year . . .'

Also, in analyses of plume behaviour, Yuen and Peltier (1980) find that:

'Because of the strong temperature dependence of mantle viscosity . . . (the lowermost D″ layer of the mantle) could be strongly unstable against secondary convective instability.'

Such plume eruptions will take heat rapidly from the lower mantle to the asthenosphere, causing partial melting above the broad rising limb of the lower mantle plume (Figs 10 and 11). Local intraplate volcanism through the lithosphere (hot spots) will result, such as happened in the Pacific in the mid-Cretaceous (Larson & Schlanger 1981). Other lower mantle plumes might accidently rise under subduction zones (Fig. 11), lowering the viscosity of the asthenosphere. The subducting slab of oceanic lithosphere would then sink more rapidly. Plume eruptions in the lower mantle will speed up plate motions by generally decreasing the asthenosphere's viscosity, and thus decreasing the resistance of plate motion to continuing boundary forces.

The phase lag, Δt, between the time of initiation or cessation of plume eruptions from the D″ layer and the later time when plate motions speed up or slow down, is caused by the rise time of the plumes (Fig. 10). This creates a basic asymmetry

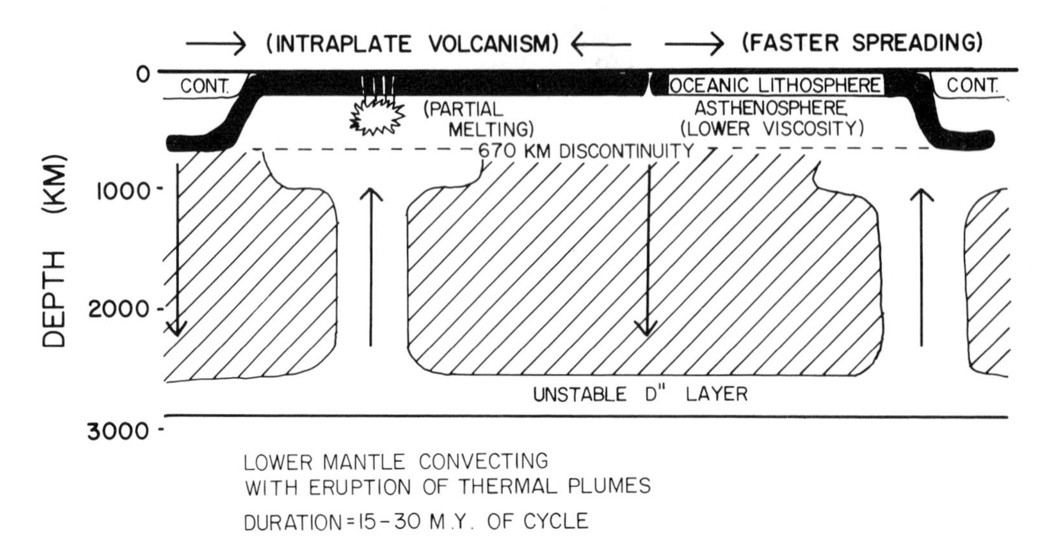

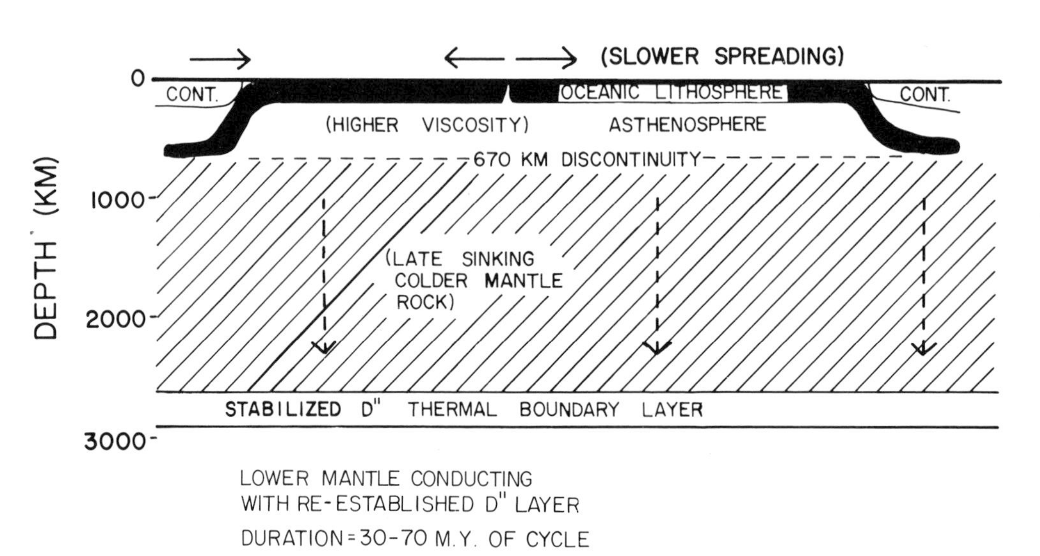

FIG. 11. Diagrammatic model showing the two states of the pulsation tectonic cycle: lower mantle convection and lower mantle conduction. The effects of lower mantle plume eruptions (convection) on the lithosphere/asthenosphere include lowering viscosity, intraplate volcanism, and faster spreading.

within a cycle. During a time in the cycle (t_2 in Fig. 10) the asthenosphere is being heated by still rising plumes while the core/mantle boundary is being cooled by descending colder rock reestablishing the D″ thermal boundary layer. During this time magnetic reversals begin to increase in frequency while the plates are still spreading fast. At a later time (t_3 in Fig. 10) the rising plumes have ceased totally, and the sinking mantle rock has reestablished the D″ thermal boundary layer so that conditions of slow spreading and frequent magnetic reversals occur.

The phase lag of the order of $\Delta t \approx 10$ Ma (Figs 8 and 9) agrees with convection velocities of about 10 cm/yr, which is the same order of magnitude that Yuen & Schubert (1976) determined for the velocities of mantle plumes. When applying the velocities of 10 cm/yr in equation (1), the time, t_{cv}, for convection of the lower mantle is about 30 Ma, which agrees well with the observed lengths of magnetic quiet zones (Figs 8, 9, 10, and 11).

After the plume eruptions, the descending cooler mantle material stabilizes the lower mantle and the D″ thermal boundary layer (Fig. 11). The

D″ layer acts as an insulating layer and confines the heat to the molten core, which increases its turbulent convection (Fig. 10). Meanwhile, via conduction, the D″ layer is being heated and its temperature begins to rise (Jones 1977). The time required for this temperature build-up and for the eventual instability of the D″ layer can be calculated from the conduction equation. For a layer like the D″ layer, of thickness H, the thermal transfer time, t_{cd}, to reach instability is:

$$t_{cd} \approx \frac{H^2}{\pi^2}\frac{\rho c}{K} \qquad (2)$$

For reasonable values of density, ρ, heat capacity, c, and thermal conductivity, K, for a layer the order of 100 km thick, the thermal transfer time would be $t_{cd} \approx 40$ Ma. This time, t_{cd}, will control the time of the cycle in which the lower mantle remains in the conducting state (Fig. 11) and is generally in agreement with the observed lengths of the magnetically reversing episodes (Figs 8 and 9).

These theoretical considerations indicate that lower-mantle-wide plume convection is possible. There appear to be intermittent eruptions of plumes, and an alternation between a convective state of the lower mantle (plumes erupting) and a conductive state (no plumes erupting) (Figs 10 and 11). The complete period of the cycles, t_o, would be:

$$t_o = t_{cv} + t_{cd} \qquad (3)$$

The periods, t_o, of these fundamental geocycles are generally 60 to 100 Ma (Figs 8, 9, 10, and 11).

Other theoretical and observational considerations support this theory, called *pulsation tectonics* by Sheridan (1983). Foster (1971) points out that in intermittent convection, the convective part of the cycle is about one-half the duration of the conductive part, which agrees with the observations on the magnetic field/plate motion cycles (Figs 8 and 9). McFadden & Merrill (1984) have shown statistically that the frequency of reversals after the Cretaceous quiet interval increases gradually. This reflects the gradual build-up of the thermal boundary layer and the longer interval of the conductive part of the cycle.

Other evidence supporting the theory of pulsation tectonics is the history of breakup of the Pangaean supercontinent. Sheridan (1983) points out that most of the breakup occurred with the high spreading rate pulses (Fig 12). For example, more of the length of Pangaea continental margins was created during the mid-Jurassic and mid-Cretaceous fast spreading pulses. Less of the breakup of Pangaea occurred in the Early Cretaceous and mid-Tertiary times (Fig. 12). Apparently the mechanisms for the breakup of continents, intraplate volcanism and slab pull, were more efficient during the time of plume eruptions and global fast plate spreading.

Energy considerations support this theory of pulsation tectonics. Turcotte & Schubert (1982) calculate that the geothermal heat flow during the fast spreading episode in the Cretaceous increased by some 5% relative to the present heat flow. Such variations in heat flow probably occurred throughout geologic time (Figs 8 and 9) with each fast spreading episode. However, over the Phanerozoic the total energy of the earth has decreased by far less than this 5% variation, and certainly during one 60–100 Ma cycle, such a great percentage of the earth's energy cannot be lost.

The increase and decrease in heat flow is compensated for by changes in other forms of energy. It is the author's belief that there is compensation of the variations in thermal energy output by the variation in magnetic energy output. The earth's magnetic field strength measured at the surface varies in geologic time (Strangway 1970; Steiner 1980; Keating, pers. comm.). Weaker field strengths are postulated in the Cretaceous, Jurassic, Permian, and during various times in the Palaeozoic (Keating, pers. comm.). Magnetic energy, proportional to the strength of the field squared, was less during times of quiet intervals when fast plate spreading released more geothermal energy. The earth is pulsating thermally just out of phase with its magnetic pulsation.

Pulsation tectonics control of palaeoenvironment

In several complicated ways pulsation tectonics exerts control on the long-term changes of the global palaeoenvironment. Obviously the more immediate controls on the palaeoenvironment are the influences of solar radiation and resultant atmospheric effects. On a long-term (60–100 Ma) scale, the tendencies of these surficial phenomena are influenced by tectonic events in many ways.

One of the pulsation tectonic controls of the palaeoenvironment is the long-term eustatic sea-level change. Fast spreading episodes produce higher eustatic sea-levels, while slow spreading episodes produce lower eustatic sea-levels (Fig. 8). These major sea-level fluctuations are associated with fluctuations in the calcite compensation depth (CCD) as noted by Fischer & Arthur (1977), Berger (1979) and Sclater *et al.* (1979). For the western North Atlantic basin, the CCD is shallowest in the Middle Jurassic and the Middle Cretaceous, and it deepens in the Early Creta-

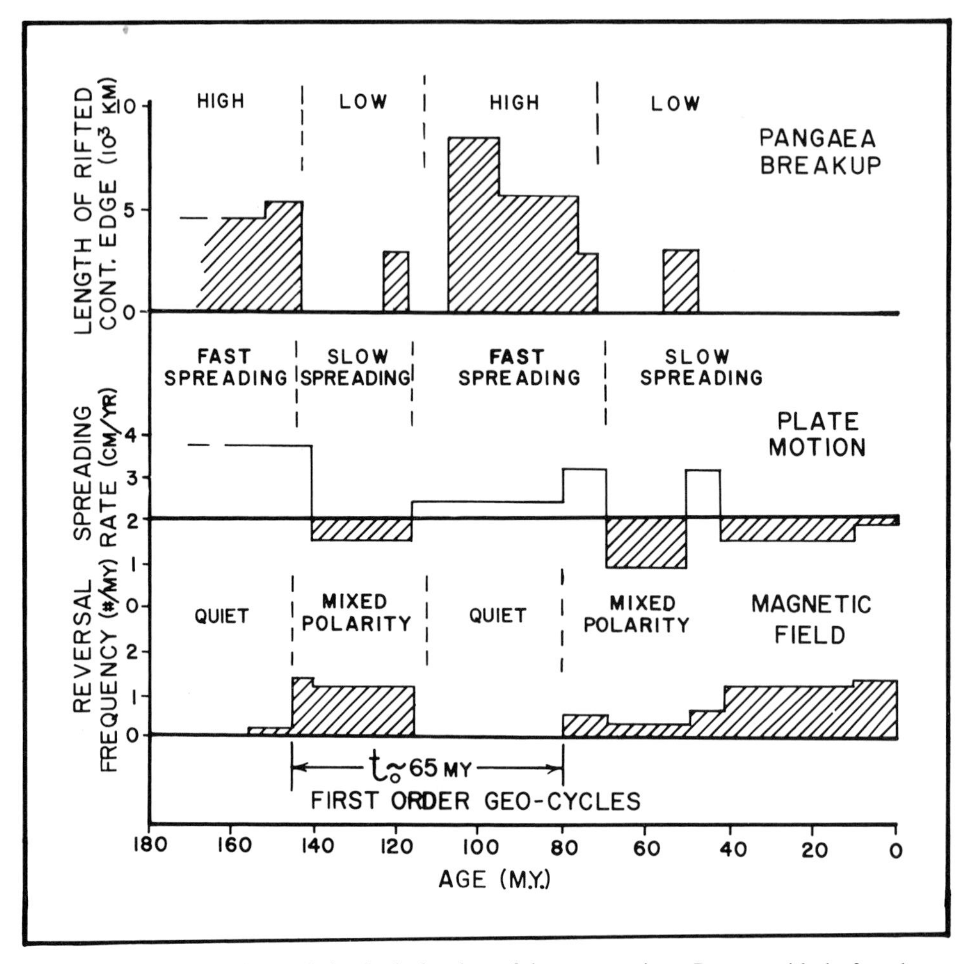

FIG. 12. Correlation of the time variation in the breakup of the supercontinent Pangaea with the fast plate spreading and quiet magnetic field episodes. The plot of length of rifted continental edges of Pangaea versus age reveals greater breakup in Jurassic and Middle and Late Cretaceous (from Sheridan 1983).

ceous and Tertiary (Sheridan 1983). Higher sea-level generally correlates with a shallower depth of the CCD (Figs 4 and 8).

The western North Atlantic stratigraphy (Figs 2 and 4) reveals two carbon rich and poor cycles, with 'black shales' being deposited in the Middle Jurassic (Unit 7) and Middle Cretaceous (Hatteras Formation). Black shales occur in global cycles and have been associated with global 'oceanic anoxic events' (OAE) (Fischer & Arthur 1977; Arthur 1979; Arthur & Schlanger 1979). In the western North Atlantic there is a correlation of the amount and kind of Cretaceous black shales with the fluctuation in sea-level (Habib 1983; Summerhayes & Masran 1983). Generally, more black shale is deposited when there is rising sea-level than when there is falling sea-level. Tissot (1979) observed that on a global scale 95% of all oil source rocks (marine black shales) originate during times of high eustatic sea-level, while 95% of all coal deposits originate during low eustatic sea-levels. Variations in sea-level, controlled by pulsation tectonics, have a fundamental influence on the geological occurrence of energy resources. The oil source rocks and coal deposits are continental platform facies, indicating that the OAE cycles also affect shallow seas and shelves as well as deep basins.

Two lithologic variations, calcite content and organic carbon content, occur in pelagites, turbidites, and contourites without differentiation. This implies that controls of these lithologic variations are in the source areas of the sediment particles and the overall biogeochemical condi-

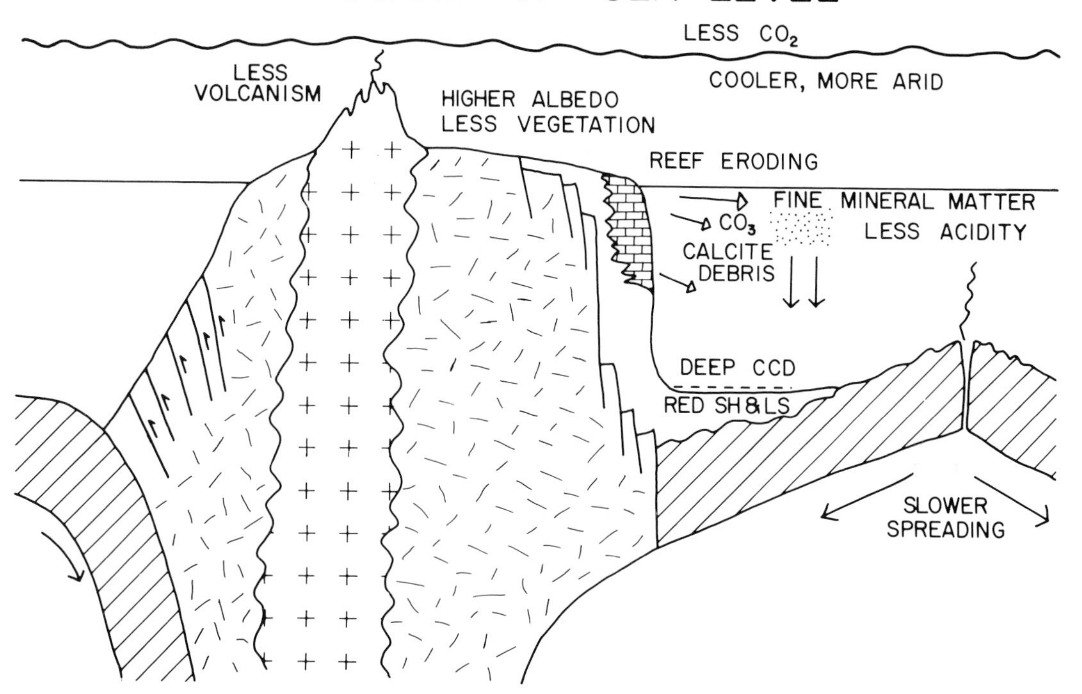

FIG. 13. Palaeoenvironmental effects of slower spreading episodes of the pulsation tectonic cycle shown on a generalized, schematic model. Variations of volcanism, atmospheric CO_2, albedo, land vegetation, global temperature and aridity, oceanic temperature, acidity and sediments are indicated.

tions of the basin. The variations of calcite and organic carbon are similar in that their preservation in the basinal sediments is a matter of supply versus dissolution or diagenetic degradation. Consequently, the models for the variations of these components appeal to changes in provenance and hinterland conditions (Figs 13, 14, and 15). On a first-order, simplifed scale, these critical changes in condition of hinterland and basin can be related to eustatic sea-level fluctuations and plate tectonic changes.

Three basic states prevail in the North Atlantic basin palaeoenvironment: (1) Low eustatic sea-level (Fig. 13). (2) High eustatic sea-level (Fig. 14). (3) High eustatic sea-level plus orogeny in the hinterland (Fig. 15).

Low eustatic sea-level

This condition prevails with slow plate spreading (Fig. 13). Slower plate spreading results in decreased subduction on a global scale, leading to lesser volcanism and uplift in orogenic belts, and less submarine volcanism. Less submarine volcanism contributes less juvenile CO_2 to the oceans, leaving the oceans less acidic and more suitable for the preservation of calcite on the sea-floor. This will contribute to the deepening of the CCD.

With reduced land volcanism, less juvenile CO_2 will enter the atmosphere, resulting in less insulation and a cooler surface temperature. In sympathy with this effect, the increased exposure of land surface with lower sea-level will increase the albedo of the earth and lead to greater loss of solar radiation. This too will cool the surface and oceanic environment.

Lower surface temperatures, and lower humidity because of fewer coastal estuarine environments, less local evaporation, and less moisture-holding capacity of the air, will lead to less land vegetation (Fig. 13). In general, cooler global climates are associated with more arid conditions on a global scale (Dott & Batten 1981). This implies that input of organic carbon to the basin will be reduced during eustatic low sea-level stands.

Sediment input to the basins would be largely mineral matter eroded from the hinterlands, which, because of slower spreading, are undergoing less tectonism and uplift. Consequently the

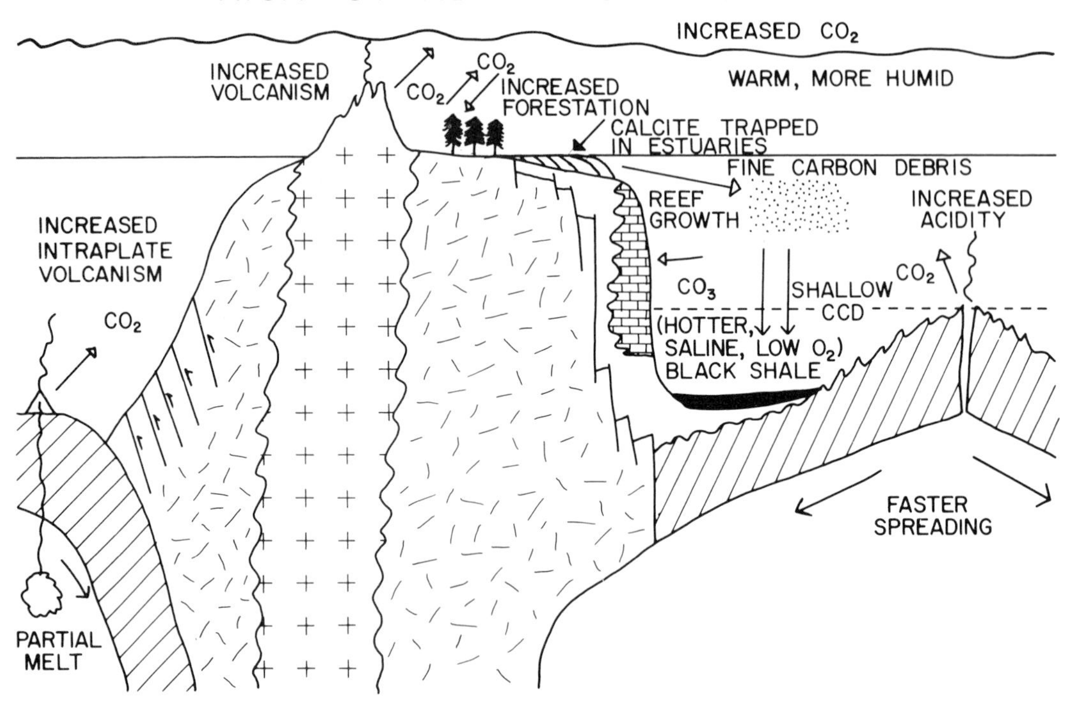

FIG. 14. Palaeoenvironmental effects of faster spreading episodes of the pulsation tectonic cycle shown on a generalized, schematic model. Tendencies in vegetation, volcanism, atmospheric CO_2, surface temperature and humidity, oceanic temperature, O_2 content, acidity and sediments are shown.

mineral matter input would tend to be a smaller amount of finer grain size. The ocean basin would tend to become sediment starved. This leads to deposition of red shale facies in the North Atlantic basin.

Other sediment input would include calcite debris from submarine (wave) erosion of exposed shelf carbonate platforms. These turbiditic limestones deposited in the basin are interbedded with the red shale facies. Also, subaerial exposure of the shelf carbonate platforms would contribute CO_3 in solution to the oceans, which could be utilized by calcite secreting organisms, and thus increase the deposition of calcite and the deepening of the CCD.

Finally, the cooler conditions would favour higher oxygen content in the oceans. Whatever small input of organic carbon enters the basin, there is ample oxygen available for the degradation of the organic carbon, and well-developed bottom fauna for further carbon consumption. Thus, a lack of black shales would be predicted during low eustatic sea-levels.

High eustatic sea-level

This condition prevails with global fast plate spreading (Fig. 14). Associated with faster spreading is increased submarine volcanism, both at the ridge axes and intraplate. Increased input of juvenile CO_2 to the oceans will raise its acidity and make corrosion of calcite more possible leading to a shallowing of the CCD.

On land, increased subduction leads to increased volcanism in orogenic belts, leading to increased CO_2 in the atmosphere. The 'greenhouse' effect takes over to warm the earth. In sympathy with this, the increased water area and decreased land surface creates more absorption of solar heat and further warming of the surface and oceans.

Greater areas of shallow seas and estuaries, with greater evaporation and moisture-holding capacity of the warm air leads to greater global humidity. This warm, humid environment with greater amounts of atmospheric CO_2 favours increased land vegetation and widespread fores-

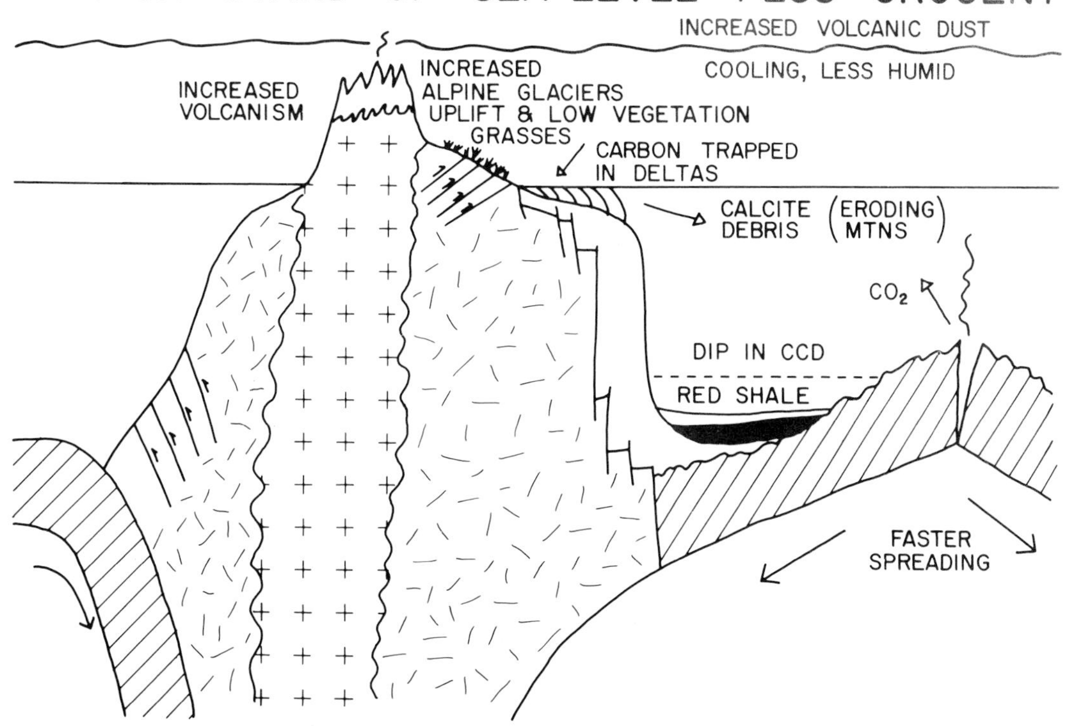

FIG. 15. Palaeoenvironmental effects of faster spreading episodes of the pulsation tectonic cycle with the complication of ubiquitous continental orogenies. The ubiquitous regressions and uplifts change the global temperature, humidity, and oceanic sediments.

tation. Because carbon detritus from these sources is light, it will be carried out into the Atlantic basin, while the mineral matter and calcite from the hinterland will be deposited well inland in estuaries. Thus the mineral matter is filtered out at the source, while the fine carbon debris increases in the basin. This leads to extensive black shale deposition.

Rising sea-levels and warm temperatures are conducive to widespread reef growth and carbonate platform development. This is another sink for carbonate and the deposition of calcite, which is effectively taken out of the oceans. This contributes to a lower supply of carbonate for calcite secreting organisms, less production of calcite in the ocean, and a further shallowing of the CCD.

Meanwhile, the warmer oceans become more thermally stratified, with warmer, more saline bottom water becoming depleted in oxygen. The oxygen minimum zone may expand (Arthur & Schlanger 1979), and low oxygen conditions may touch the bottom of the deep Atlantic basin. Alternatively, just the increased supply of carbon debris, both plant debris and marine organisms included in fecal pellets, could consume enough oxygen to make the bottom sediments anoxic and conducive to preservation of black shales (Habib 1983; Dean *et al.* 1984). In either case, black shales are deposited, and the sediments are anoxic.

High eustatic sea-level plus orogeny

As a general rule during the fast spreading episodes, there is at least one orogeny recognized as a culminating paroxysmal intrusive event over several continents. The mid-Jurassic fast spreading pulse is accompanied by the Late Jurassic–Early Cretaceous Nevadan orogeny in the Cordillera, while the mid-Cretaceous fast spreading episode is accompanied by the Laramide–Alpine events in Late Cretaceous–Early Tertiary. It appears that the orogenies occur late in the interval of fast spreading. Perhaps during global fast spreading it is more likely that microcontinents and aseismic ridges might collide and 'board' the continents, creating extensive granitization in the collision zones.

An orogeny in the hinterland during fast spreading and high eustatic sea-level would add several key elements for North Atlantic palaeoceanography (Fig. 15). Increased uplift of orogenic belts could result in increased alpine glaciation on a global scale, thus cooling the surface climate. Increased eruptions of explosive volcanics in the orogenic belts could increase volcanic dust in the atmosphere, shutting out solar radiation and cooling the earth. Generally this global cooling will cause decreased humidity, and decreased forests and vegetation. Less vegetation will lead to less carbonaceous detritus reaching the deep basin and a return to a red shale facies.

The newly uplifted mountains in the orogenic belts will shed more mineral detritus, but it will be trapped in inland seas and estuaries as deltaic complexes. Because of the eustatic high-stand, more estuaries are available to fill in. Carbon debris, what little is available, will also be trapped in these estuarine deltas as coal deposits.

The newly uplifted orogenic belts will shed more calcite debris as ancient shelf limestones are exposed in overthrust belts. Calcite and carbonate will reach the deep ocean providing carbonate for calcite secreting organisms, and this will contribute to deepening of the CCD. Such a dip is noted in the western North Atlantic in the Maestrichtian when the chalk of the Crescent Peaks Member of the Plantagenet Formation was deposited (Jansa *et al.* 1979).

Generally the cooler global climate during this stage will cool the oceans and create a more oxygen-rich state. The little organic carbon that reaches the basin will most likely not be preserved in the more oxygenated environment.

These generalized models may explain the carbon and calcite cycles. Although they are speculative, and may be wrong in detail, they show how pulsation tectonics, which originates in the core of the earth, has a first-order long-term impact on palaeoceanography. Palaeobathymetric changes occur with global spreading rate changes, new oceans open with widespread hot spots, sea-level rises and falls as mid-ocean ridges

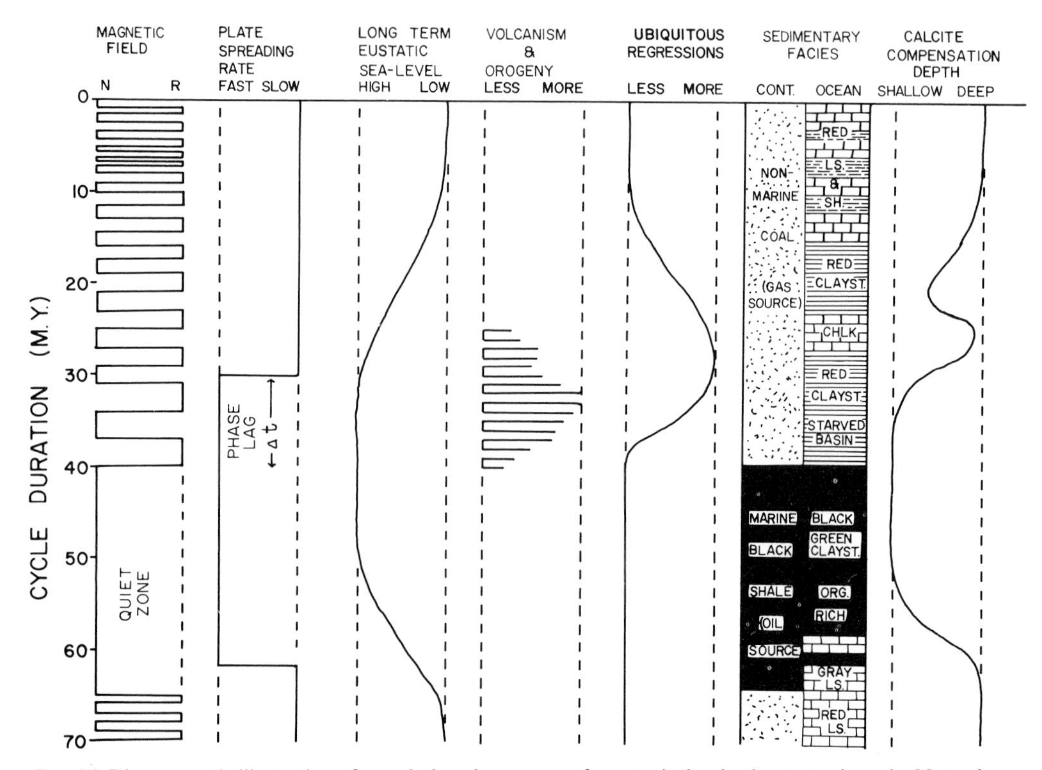

FIG. 16. Diagrammatic illustration of correlative phenomena of one typical pulsation tectonic cycle. Note phase lag, Δt, between plate spreading pulse and change in reversal frequency. Ubiquitous orogenies and regressions change the continental and oceanic sedimentary facies and the CCD variation. The 'black shale' facies correlate with earlier stages of long-term rising sea-level and magnetic quiet zones; red shales in the later stages of eustatic high stands and in low stands.

grow faster and slower, and finally the biogeochemistry of the oceans is influenced. This gross systems analysis of the earth provides a working model to understand the long period cycles, $t_o = 60$–100 Ma, of the entire planetary environment. It forms a strong predictive tool with which to go backward in time to help understand how the earth worked even when the oceanic record has been completely destroyed in subduction, as in the Palaeozoic (Fig. 9).

The unifying theory of pulsation tectonics explains the variation in the magnetic field, plate motions, orogenies, eustatic sea-level, and finally the palaeoenvironment and resulting sedimentary cycles of calcite and carbon (Fig. 16). The magnetic field is quiet for approximately one-third of the pulsation tectonic cycle, with gradually increasing reversal frequency spanning nearly two-thirds of the cycle interval. Fast sea-floor spreading episodes occur with a phase lag of about 10 Ma after the magnetic quiet interval, and high eustatic sea-levels follow closely in time the fast plate spreading. Toward the end of the fast spreading interval, widespread culminating, intrusive orogenies occur. These shed deltas that cause ubiquitous regressions in spite of a eustatic high stand (Fig. 16). These regressions have the effect of limiting the marine black shale facies in epicontinental seas to the early part of the eustatic-high stand, when sea-level was rising. Prior to the orogenic part of the cycle, conditions are right for preservation of black shale in the oceanic basins, but the orogenic effects create red shales during the latter part of the eustatic high-stand. Thus the 'black shale' intervals are apparently more correlated with the magnetic quiet interval than with the total interval of eustatic high-stand (Fig. 16). Such a correlation of black shales was noted by Force (1984). The complex linkages between these diverse factors lead to this otherwise unexplainable correlation. Finally, the variation in CCD is superimposed on the black shale/red shale cycles to yield grey limestones and red limestones. On a local western North Atlantic basin scale, these alternations of limestones and shales, controlled by the fluctuations in the CCD, are the locations of the major seismic horizons such as A*, β, C, and D (Figs 2, 3, 4, and 8). So the theory of pulsation tectonics is applicable to seismic stratigraphy as well.

Appropriateness of pulsation tectonic title

The word pulsation was used by Grabau (1934). He defined a pulsation as a long term rhythm, nearly equal to the length of a geological period, in which there were widespread transgressions and regressions of seas over whole continents. The pulsation tectonic theory predicts eustatic cycles on a 60–100 Ma period, which is about the length of common geological Periods, and these would be equivalent to the pulsations of Grabau (1934). In fact, the Vail *et al.* (1977) long-term sea-level fluctuations are equivalent to the pulsations of Grabau (1934). The work of Vail *et al.* (1977) demonstrating the global synchroneity of long-term sea-level changes is just what Grabau (1934) had attempted to demonstrate in his historic work. Grabau (1934) declared that if a global sea-level curve like that later produced by Vail *et al.* (1977) could be demonstrated, then he would be convinced that his 'pulsation theory' was correct.

In a discussion of his pulsation theory with Stille, Grabau (1934) pointed out that he viewed the cause of pulsations to be in the tectonics of the oceans, in contrast to what he considered the more local effects of continental orogenies. He was almost prophetic in this intuition, although he did not explain the underlying causes of oceanic tectonics.

The cyclic character of crustal movements called for in the theory of pulsation tectonics is what Umbgrove (1947) called the 'pulse of the earth'. We can see now that the pulsating character of orogenies recognized by Umbgrove (1947), and Stille (1924) before him, reflects the pulsations of fast spreading in the oceanic crust. The new theory of pulsation tectonics revives many of the ideas of past advocates of non-steady state tectonics, and provides modern mechanisms that explain plate motions and magnetic field data, phenomena which were totally unknown to these past authors.

References

ANDERSON, T.H. & SCHMIDT, V.A. 1983. The evolution of Middle American and The Gulf of Mexico–Caribbean Sea region during Mesozoic time. *Geol. Soc. Am. Bull.* **94,** 941–66.

ARTHUR, M.A. 1979. Paleoceanographic events; Recognition, resolution, and reconsideration. *Rev. Geophys. Space Phys.* **17,** 1474–94.

—— & SCHLANGER, S.O. 1979. Cretaceous "oceanic anoxic events" as causal factors in development of reef-reservoired giant oil fields. *Am. Ass. Petrol. Geol. Bull.* **63,** 870–85.

BARRETT, D.L. & KEEN, C.E. 1975. Mesozoic magnetic lineations, the magnetic quiet zone, and sea floor spreading in the Northwest Atlantic. *J. Geophys. Res.* **81,** 4875–84.

BAUMGARTNER, P.O. 1983. Summary of Middle Jurassic–Early Cretaceous radiolarian biostratigraphy of Site 534 (Blake-Bahama Basin) and correlation to the Tethyan sections. *In:* SHERIDAN, R.E., GRADSTEIN, F.M. *et al.* (eds), *Init. Repts DSDP* **76.** US Govt Print. Off., Washington, DC. 569–71.

BERGER, W.H. 1979. Impact of Deep Sea Drilling on paleoceanography. *In:* TALWANI, M., HAY, W. & RYAN, W.B.F. (eds), *Deep Drilling Results in the Atlantic Ocean: Continental Margins and Paleoenvironment*, Am. Geophys. Union, Maurice Ewing Series 3. 297–314.

BRYAN, G.M., MARKL, R.G. & SHERIDAN, R.E. 1980. IPOD site surveys in the Blake-Bahama Basin. *Mar. Geol.* **35,** 43–63.

CANDE, S., LARSON, R. & LA BRECQUE, J. 1978. Magnetic lineations in the Pacific Jurassic quiet zone. *Earth Planet Sci. Lett.* **41,** 434–40.

CHANNELL, J.E.T., OGG, J.C. & LOWRIE, W. 1982. Geomagnetic polarity in the Early Cretaceous and Jurassic. *Phil. Trans. R. Soc. Lond.* **A306,** 137–46.

DEAN, W.E., ARTHUR, M.A. & STOW, D.A.V. 1984. Origin and geochemistry of Cretaceous deep sea black shales and multicolored claystones, with emphasis on Deep Sea Drilling Project Site 530, southern Angola Basin. *In:* HAY, W.W., SIBUET, J.C. *et al.* (eds), *Init. Repts DSDP*. US Govt Print Off., Washington, DC. 819–44.

DOELL, R.E. & COX, A. 1972. The Pacific geomagnetic secular variation anomaly and the question of lateral uniformity in the lower mantle. *In:* ROBERTSON, E.C., HAYS, J.F. & KNOPOFF, L. (eds), *The Nature of the Solid Earth*. McGraw Hill, New York. 245–84.

DOTT, R.H., JR & BATTEN, R.L. 1981. *Evolution of the Earth*. McGraw Hill, New York. 113 pp.

FISCHER, A.G. & ARTHUR, M.A. 1977. Secular variations in the pelagic realm. *In:* COOK, H.E. & ENOS, P. (eds), *Deep Water Carbonate Environments. Soc. Econ. Paleontol. Mineral. Spec. Publ.* **25,** 19–50.

FORCE, E. 1984. A relation among geomagnetic reversals, sea floor spreading rate, paleoclimate, and black shales. *EOS Trans. Am. Geophys. Union* **65,** 18–19.

FOSTER, T.D. 1971. Intermittent convection. *Geophys. Fluid Dynamics* **2,** 201–17.

GRABAU, A.W. 1934. *Oscillation or Pulsation*. Report of XVI International Geological Congress, Washington, D.C. 1933. 1–15.

GRADSTEIN, F.M. & SHERIDAN, R.E. 1983. On the Jurassic Atlantic Ocrean and a synthesis of results of Deep Sea Drilling Project Leg 76. *In:* SHERIDAN, R.E., GRADSTEIN, F.M. *et al.* (eds), *Init. Repts DSDP* **76.** US Govt. Print. Off., Washington, DC. 913–43.

HABIB, D. 1983. Sedimentation-rate-dependent distribution of organic matter in the North Atlantic Jurassic-Cretaceous. *In:* SHERIDAN, R.E., GRADSTEIN, F.M. *et al.* (eds), *Init. Repts DSDP* **76.** US Govt Print. Off., Washington, DC. 781–94.

HALLAM, A. 1975. *Jurassic Environments*. Cambridge University Press, New York. 269 pp.

—— 1977. Biogeographic evidence bearing on the creation of Atlantic seaways in the Jurassic. *In:* WEST, R.M. (ed.), *Paleontology and Plate Tectonics.* Spec. Publ. in Biology and Geology, Milwaukee Public Museum. 23–39.

HARDENBOL, J., VAIL, P.R. & FERRER, J. 1981. Interpreting paleoenvironments, subsidence history, and sea-level changes of passive margins from seismic and biostratigraphy. *Oceanol. Acta, Proceedings 26th International Geol. Cong.*, **4,** 33–44.

HARLAND, W.B., COX, A., LLEWELLYN, P.G., PICKTON, C.A.S., SMITH, A.G. & WALTERS, R. 1983. *A Geologic Time Scale*. Cambridge University Press, New York. 112 pp.

HAYS, J.D. & PITMAN, W.C. III 1973. Lithospheric plate motion, sea-level changes and climatic and ecological consequences. *Nature* **246,** 18–22.

JANSA, L., ENOS, P., TUCHOLKE, B.E., GRADSTEIN, F.M. & SHERIDAN, R.E. 1979. Mesozoic and Cenozoic sedimentary formations of the North American Basin; western North Atlantic. *In:* TALWANI, M., HAY, W.W. & RYAN, W.B.F. (eds), *Deep Drilling Results in the Atlantic Ocean: Continental Margins and Paleoenvironment*. Am. Geophys. Union, Maurice Ewing Series 3. 1–57.

JONES, G.M. 1977. Thermal interaction of the core and mantle and long-term behavior of the geomagnetic field. *J. geophys. Res.* **82,** 1703–9.

KATZ, B.J. 1983. Organic geochemical character of some Deep Sea Drilling Project cores from Legs 76 and 44. *In:* SHERIDAN, R.E., GRADSTEIN, F.M. *et al.* (eds), *Init. Repts DSDP* **76.** US Govt Print. Off., Washington, DC. 463–8.

KENT, D.V. & GRADSTEIN, F.M. 1984. A Jurassic to Recent Chronology. *In:* TUCHOLKE, B.E., VOGT, P.R. (eds), *The Western Atlantic Region: The Geology of North America. Geol. Soc. Am.* DNAG Series.

LARSON, R.L. & SCHLANGER, S.O. 1981. Cretaceous volcanism and Jurassic magnetic anomalies in the Nauru Basin, western Pacific Ocean. *Geology* **9,** 480–4.

LOWRIE, W., CHANNEL, J.E.T. & ALVAREZ, W. 1980. A review of magnetic stratigraphy investigations in Cretaceous pelagic carbonate rocks. *J. geophys. Res.* **85,** 3597–605.

MCFADDEN, P.L. & MERRILL, R.T. 1984. Lower mantle convection and geomagnetism. *J. geophys. Res.* **89,** 3354–605.

OGG, J. 1980. Upper Jurassic magnetostratigraphy from northern Italy (abs). *EOS Trans. Am. Geophys. Union* **61,** 216.

PITMAN, W.C. III 1978. Relationship between eustatic and stratigraphic sequences of passive margins. *Geol. Soc. Am. Bull.* **89.** 1389–403.

RONA, P.H. 1973. Relations between rates of sediment accumulation on continental shelves, sea-floor spreading, and eustasy inferred from the central North Atlantic. *Geol. Soc. Am. Bull.* **84,** 2851–72.

SCLATER, J.G., BOYLE, E. & EDMOND, J.M. 1979. A quantitative analysis of some factors affecting carbonate sedimentation in the oceans. *In:* TALWANI, M., HAY, W.W. & RYAN, W.B.F. (eds), *Deep Drilling Results in the Atlantic Ocean: Continental Margins and Paleoenvironments*. Am. Geophys. Union, Maurice Ewing Series 3. 235–48.

SHERIDAN, R.E. 1983. Phenomena of pulsation tectonics related to the breakup of the eastern North American continental margin. *In:* SHERIDAN, R.E., GRADSTEIN, F.M. *et al.* (eds) *Init. Repts DSDP* **76.** US Govt Print. Off., Washington, DC. 897–909.

——, ENOS, P., GRADSTEIN, F.M. & BENSON, W.E. 1978. Mesozoic and Cenozoic environments of the western North Atlantic. *In:* BENSON, W.E., SHERIDAN, R.E. *et al.* (eds), *Init Repts DSDP* **44.** US Govt Print. Off., Washington, DC. 971–80.

——, GRADSTEIN, F.M. *et al.* 1983. Site 534: Blake-Bahama Basin. *In:* SHERIDAN, R.E., GRADSTEIN, F.M. *et al.* (eds), *Init. Repts. DSDP* **76.** US Govt Print. Off., Washington, DC. 141–340.

——, BATES, L.G., SHIPLEY, T.H. & CROSBY, J.T. 1983. Seismic stratigraphy in the Blake-Bahama Basin and the Origin of Horizon D. *In:* SHERIDAN, R.E., GRADSTEIN, F.M. *et al.* (eds), *Init. Repts DSDP* **76.** US Govt Print. Off., Washington, DC. 667–83.

SLOSS, L.L. 1963. Sequences in the cratonic interior of North America. *Geol. Soc. Am. Bull.* **74,** 93–113.

STEINER, M.B. 1980. Investigation of the geomagnetic field polarity during the Jurassic. *J. Geophys. Res.* **85,** 3572–86.

STILLE, H. 1924. *Grundfragen der Vergleichenden Tektonik.* Borntraeger, Berlin.

STOW, D.A.V. & DEAN, W.E. 1984. Middle Cretaceous black shales at Site 530 in the southeastern Angola Basin. *In:* HAY, W.W., SIBUET, J.C. *et al.* (eds), *Init. Repts DSDP* **75.** US Govt Print. Off., Washington, DC. 809–17.

STRANGWAY, D.W. 1970. *History of the Earth's Magnetic Field.* McGraw Hill, New York. 163 pp.

SUMMERHAYES, C.P. & MASRAN, T.C. 1983. Organic facies of Cretaceous and Jurassic sediments from Deep Sea Drilling Project Site 534 in the Blake-Bahama Basin, western North Atlantic. *In: Init. Repts DSDP* **76.** US Govt Print. Off., Washington, DC. 469–80.

TAYLOR, P.T., ZIETZ, I. & DENNIS, L.S. 1968. Geologic implications of aeromagnetic data for the eastern continental margin of the United States. *Geophysics* **33,** 755–80.

TISSOT, B. 1979. Effects on profilic petroleum source rocks and major coal deposits caused by sea-level changes. *Nature* **277,** 463–5.

TURCOTTE, D.L. & SCHUBERT, G. 1982. *Geodynamics, Application of Continuum Physics to Geological Problems.* John Wiley & Sons, New York. 450 pp.

UMBGROVE, J.H.F. 1947. *The Pulse of the Earth.* Nijoff, The Hague. 358 pp.

VAIL, P.R., MITCHUM, R.M. JR. & THOMPSON, S. III 1977. Global cycles of relative changes in sea-level. *Am. Ass. Petrol. Geol. Mem.* **26,** 83–97.

VAN HINTE, J.E. 1976. A Jurassic time scale. *Am. Ass. Petrol. Geol. Bull.* **60,** 489–97.

VOGT, P.R. 1975. Changes in geomagnetic reversal frequency at times of tectonic change. *Earth Planet Sci. Lett.* **25,** 313–21.

—— & EINWICH, A.M. 1979. Magnetic anomalies and sea-floor spreading in the western North Atlantic, and a revised calibration of the Keathley (M) geomagnetic reversal chronology. *In:* TUCHOLKE, B.E., VOGT, P.R. *et al.* (eds), *Init. Repts DSDP* **43.** US Govt Print. Off., Washington, DC. 971–4.

WATTS, A.B. & STECKLER, M.S. 1979. Subsidence and eustasy at the continental margin of eastern North America. *In:* TALWANI, M., HAY, W.W. & RYAN, W.B.F. (eds) *Deep Drilling Results in the Atlantic Ocean: Continental Margins and Paleoenvironment.* Am. Geophys. Union, Maurice Ewing Series 3. 218–34.

YUEN, D.S. & PELTIER, W.R. 1980. Mantle plumes and the thermal stability of the D″ layer. *EOS Trans. Am. Geophys. Union* **61,** 637.

—— & SCHUBERT, G. 1976. Mantle plumes: a boundary layer approach for Newtonian and non-Newtonian temperature dependent rheologies. *J. geophys. Res.* **81,** 2499–510.

ROBERT E. SHERIDAN, Department of Geology, University of Delaware, Newark, Delaware 19716, USA.

Role of climate in affecting late Jurassic and early Cretaceous sedimentation in the North Atlantic

A. Hallam

SUMMARY: Both regional and temporal changes in the influx of terrestrial organic matter, siliciclastic sediments and kaolinite are related to climate over the continents adjacent to the opening North Atlantic, reflecting a change from humid to arid and back to humid in North America and Europe, and more persistent aridity in Africa. Small-scale sedimentary cycles of laminated organic-rich and non-laminated organic-poor claystones have previously been interpreted in terms either of periodic influx of detrital plant material, or variations in sedimentation rate. They are here interpreted as the result of periodic stirring of bottom waters related possibly to Milankovich cycles.

A multiplicity of factors have been invoked to account for regional and temporal variations in sedimentation in the early North Atlantic. Attention has been concentrated on regional tectonics (e.g. Jansa & Wade 1975; Chamley *et al.* 1983; Ogg *et al.* 1983), changes in the level of the sea (Ogg *et al.* 1983; Summerhayes & Masran 1983) or CCD (Herbin *et al.* 1983; Ogg *et al.* 1983) and organic productivity (Habib 1982; Ogg *et al.* 1983; Summerhayes & Masran 1983; Summerhayes, in press). By comparison, the role of climate has received scant attention. It is proposed here that climate might have had a significant influence in a number of respects.

It is necessary initially to outline the major patterns of climatic change that can be inferred to have existed on the adjacent continents, from a study of the distribution of coals and evaporites, as well as other climatically significant deposits such as bauxites, ironstone and kaolinite-bearing clay. In addition, the proportion of siliciclastic to carbonate deposits gives a measure of the amount of continental runoff, which is directly related to precipitation (Hallam 1984).

As the climate within the region in question remained consistently warm during the Mesozoic, the only notable change through the course of time was in the amount of precipitation. The best record comes from Europe, where the evidence clearly points to a change from a wetter to a drier climate in the late Jurassic, with the most arid phase being in the late Tithonian. Thereafter the climate became humid again, shortly after the beginning of the Cretaceous (Valanginian), and remained so for the rest of the period. The most striking manifestation of this change is the sudden replacement over wide areas of Europe of shallow marine and paralic carbonates and evaporites, with a clay mineral content devoid of kaolinite, by arenaceous and argillaceous 'Wealden' deposits rich in both kaolinite and siderite. That a similar climatic change took place on the eastern seaboard of North America north of the Blake Plateau is indicated by a comparable and contemporary facies change under the continental shelf (Jansa & Wade 1975; Owens 1983). In contrast to these more northerly areas, north-west Africa had a drier climate throughout the late Jurassic and early Cretaceous, though not so arid as to preclude an increase in continental runoff in the early Cretaceous giving rise to a Wealden-type facies (de Klasz 1978).

The explanation tentatively advanced for this drastic change across the Jurassic-Cretaceous boundary involves the expansion through sea-floor spreading of a young, narrow ocean within an arid region beyond a certain critical point necessary for the establishment of a strong, monsoonal effect on the adjacent continents, giving pronounced seasonal rainfall (Hallam 1984). This implies that until the early Cretaceous the incipient late Jurassic North Atlantic was, like the present Red Sea, too narrow to exert such an effect. A Red Sea analogue has previously been invoked for the early Jurassic because of the occurrence of extensive evaporite deposits (Kinsman 1975), but it is now apparent that the opening of the North Atlantic took place well after the evaporites were deposited, with true ocean crust not being formed until the late mid Jurassic (Hallam 1983; Sheridan 1983). A further implication of this comparatively late opening is that the Neotethys also opened later than generally accepted. Stöcklin's (1984) suggestion that it did not open before the Cretaceous is supported by the occurrence of major late Jurassic salt deposits in the Caucasus, which are unlikely to have formed adjacent to a major ocean (Hallam 1984).

From SUMMERHAYES, C.P. & SHACKLETON, N.J. (eds), 1986, *North Atlantic Palaeoceanography*, Geological Society Special Publication No. 21, pp. 277–281.

The deep ocean record

Large-scale regional and temporal changes

The most notable regional change in the deep ocean record is the content of terrestrial organic matter. Whereas the early and mid Cretaceous deposits off North America and western Europe (Iberia) have a high proportion of terrestrial matter in the organic-rich mudrocks, the corresponding deposits off north-west Africa have only a low proportion (Summerhayes, in press). As Summerhayes recognizes, this is evidently a consequence of the higher run-off from the first-named continents because of the more humid climate, and the lower run-off from north-west Africa because of the more arid climate. The contrast is enhanced by what appears to have been higher marine productivity on the south-eastern side of the young ocean, as marked by radiolarian concentrations indicative of a zone of upwelling (Ogg *et al.* 1983).

A climate-related change in the organic matter can also be recognized through the course of time. This is apparent in the most extensive Jurassic–Cretaceous sequence yet explored, based on Leg 76 in the Blake-Bahama Basin. According to the data of Habib (1983) the oldest, Callovian, deposits contain a mixture of terrestrial and phytoplankton-derived organic matter, but the Oxfordian to Berriasian deposits contain only the latter. Terrestrial matter returns in abundance in the Valanginian and continues through the younger Cretaceous. The presence of terrestrial matter in the Callovian would appear to relate (a) to the narrowness of the incipient ocean and hence close proximity of land and (b) to the humid climate of the time. The late Jurassic–Berriasian disappearance of terrestrial organic matter and its subsequent reappearance, corresponds well to the change to a more arid climate and reversion to a seasonally humid climate (Hallam 1984).

Clay mineralogy also records the late Jurassic arid phase, in the deep ocean as in the onshore outcrops of shallow-marine deposits of western Europe. With regard to the latter, illite and subordinate kaolinite are the dominant clay minerals in the Jurassic, with smectite important only in the Cretaceous (Sellwod & Sladen 1981; Sladen 1983). Much of this smectite can be shown to be of volcanic origin (Jeans *et al.* 1982; Pacey 1984). The late Tithonian–early Berriasian disappearance of kaolinite reflects the peak of aridity in western Europe (Hallam 1984). In contrast to the shallow marine environment, smectite is the dominant clay mineral in the deep ocean, with other types being very subordinate. The smectite is believed to be pedogenic in origin, derived from weathering in a terrestrial regime of warm climate with strong seasonal variation in precipitation (Chamley & Robert 1982; Chamley *et al.* 1983). This climatic inference is broadly consistent with what can be inferred on independent grounds for some of the late Mesozoic, but Chamley and colleagues appear to underestimate the role of differential settling of land-derived clay minerals. As in the present-day oceans (Griffin *et al.* 1968) most kaolinite is deposited in shallow nearshore areas of humid tropical regimes. Thus kaolinite is an index both of climate and proximity to land. As most or all of the marine illite is probably land-derived, there may also be a nearshore concentration of this mineral as a result of differential settling. A further point to bear in mind is that there is probably a volcanic component in the deep ocean smectite, although the amount of this is likely to be difficult to assess.

Turning to the excellent, comprehensive record of Jurassic and Cretaceous clay mineralogy obtained from leg 76 (Chamley *et al.* 1983) it is notable that the abundance of illite and kaolinite in the Callovian, with minimal smectite, compared most closely with the shallow marine record of western Europe, and clearly relates to (a) proximity of land to the incipient ocean, and (b) a humid climate on that land. Thereafter kaolinite becomes rare in the sequence and is absent altogether in the Kimmeridgian and Tithonian, a feature consistent with the inferred late Jurassic arid phase. Scattered 'illite events' in the sequence record, according to Chamley *et al.* (1983), pulses of tectonically-induced terrigenous sedimentation. The sparsity of kaolinite in the Cretaceous, despite the humid climate of the adjacent continent, evidently reflects the increased width of the ocean during the period, and hence the greater distance from land.

With regard to the sedimentary record as a whole, the change up the Upper Jurassic sequence from less to more calcareous facies in the Leg 76 sites was tentatively interpreted by Ogg *et al.* (1983) to be the consequence of a lowering of the CCD, perhaps associated with increased production of calcareous plankton. It could be, however, at least partly a consequence of the reduced run-off associated with increased aridity on the continental sediment-sourcelands. Calcareous facies in the form of marls and limestones is indeed widespread throughout the late Jurassic and earliest Cretaceous (Berriasian) of the North Atlantic (Jansa *et al.* 1977; Ogg *et al.* 1983; Jansa *et al.* 1984). An abrupt change took place from Valanginian times onward, with the sudden introduction of coarse and fine terrigenous siliciclastic turbidites into the pelagic sequence. The lower continental rise flysch off Morocco is Valanginian to Hauterivian in age (Lancelot & Winterer 1980) while the corresponding deposits north-west of Bermuda, at site 603 (Leg 93) are mainly Hauterivian and Barremian. The latter deposits have been interpreted as belonging to a deep-sea fan and help to disprove the older, widely accepted idea of a continuous Lower Cretaceous reef barrier rimming the North American continent and preventing the passage of coarse terrigenous sediments to the deep sea (JOIDES Journal 1983; Wise, this volume). Further south, in the Blake–Bahama Basin, redeposited siliciclastics consisting of siltstone and sandstone turbidites first appear in the Lower Valanginian and become dominant in the Upper Valanginian and Hauterivian. Similar successions were found off the Bermuda Rise at sites 387 and 391 (Leg 44) (Robertson & Bliefnick 1983).

The parallel between these facies changes on both sides of the early North Atlantic with those on the adjacent continents is striking, with the massive turbidite influx correlating with the 'Wealden' paralic sediments and reflecting a significant phase of comparative humidity. Although the relative sparsity of land-derived plant material in the south-east suggests more arid conditions on the African side, there was evidently sufficient seasonal rainfall in the early Cretaceous to effect the transport of significant quantities of sediment offshore. As the ocean widened and sea-level rose, the proportion of coarse terrigenous siliciclastics in the deep-sea record diminished, although the climate of the continental sourcelands remained relatively humid.

Small-scale cyclicity

In the Jurassic and Cretaceous mudrock deposits of the North Atlantic there are numerous small, decimetre-scale alternations of finely laminated units with a few percent of organic matter, and *Chondrites*—bioturbated units largely devoid of organics (Dean *et al.* 1977; McCave 1979, Cool 1982; Herbin *et al.* 1983; Ogg *et al.* 1983; Robertson & Bliefnick 1983; Summerhayes & Masran 1983). They compare closely with alternations of anoxic and oxic bottom-water neritic deposits known in mudrocks of the English Jurassic, such as the Blue Lias and Kimmeridge Clay. Based on sedimentation-rate estimates, cycle durations of ~ 2 to 5×10^4 years have been inferred for the deep Atlantic deposits (Summerhayes, in press) and ~ 1 to 10×10^4 years for the English Jurassic (Hallam, in press). These periods of time have suggested to several authors some sort of control on sedimentation by 'Milankovich' precession, obliquity and/or eccentricity cycles (McCave 1979; Hallam, in press; Summerhayes, in press).

A number of widely differing interpretations have been put forward to account for such cycles.

(1) Habib (1982, 1983) sees the critical controlling factor as sedimentation rate. A high rate of sedimentation leads to the development of anoxia in buried sediments and perhaps also in the overlying water column, whereas a low rate leads to oxidizing conditions in the sediments. The water column could have been in oxic condition throughout the Jurassic and Cretaceous. Although Habib cites evidence supporting the correlation with sedimentation rate, this only refers to large-scale stratigraphic units. While it is indeed true that there is an overall correlation between organic content and sedimentation burial rate (Summerhayes, in press) there is nevertheless no evidence that the organic-rich cyclic units were deposited more rapidly than the intervening organic-poor units. Habib's hypothesis fails to explain the intimate correlation between fine lamination and high organic content, and bioturbation and low organic content, and disregards the well established fact that by far the most important factor controlling the amount of organic matter preserved in the sediment is oxygen content. In the presence of oxygen, and hence benthic life which effects bioturbation, all but the most refractory humic material is destroyed (Demaison & Moore 1980).

(2) According to Dean *et al.* (1977) the late Jurassic and Cretaceous cycles recorded from the region off the north-west African margin relate to periodic influxes of detrital organic matter from the continent, as a result of Milankovich-forced humid-arid climatic cycles affecting run-off. The presence of large quantities of organic matter in the sediment is thought to create anoxic conditions in the overlying water. This inference confuses cause and effect. Organic matter is in general preserved *because of* anoxic conditions in the bottom water; its presence is unlikely to have created such a condition except perhaps locally and on a small scale (cf. Demaison & Moore 1980). Another serious, indeed critical, objection is that organic geochemical analysis indicates that the organic matter in question is largely of marine origin with a low content of humic compounds (Deroo *et al.* 1977).

In contrast to the situation off north-west Africa, there is much terrestrial organic matter in the Lower Cretaceous deep-sea sediments off North America, and Robertson & Bliefnick (1983) invoke periodic influxes of such material from the land, as a result of variations in precipitation. This interpretation is similar to that of Dean *et al.* (1977) and therefore open to the same objection concerning cause and effect. Furthermore, a significant proportion of the organic content of these sediments is in fact marine (Summerhayes & Masran 1983).

(3) McCave (1979), describing mid Cretaceous sediments collected on Leg 43, observes a correlation in some instances between organic content and abundance of radiolarians. This leads him to propose, in partial explanation of the cyclicity, periodic increases in plankton productivity, which he calls 'large blooms', causing an increased consumption of oxygen and hence generating a condition of anoxia in sluggish, slowly renewed bottom waters. The increased productivity could relate to episodes of warmer climate.

(4) There are no doubts, shared by McCave himself, of the plausibility of his productivity model as a *general* explanation, and he puts forward an alternative based on periodic variations in the strength of bottom water circulation. A comparable interpretation was put forward by Hallam (in press) to account for similar types of cycles in both shallow- and deep-water organic-rich sequences of the North Atlantic region. There has been much dispute, reviewed by Summerhayes (in press) about whether or not the late Mesozoic North Atlantic Ocean generally tended towards a state of anoxia. The existence of these numerous sedimentary cycles, and the occurrence of both planktic and nektic elements, appear to favour the view that much or most of the water column contained dissolved oxygen but that circulation was sluggish and in consequence oxygen content was low near the bottom, with conditions at the sediment-water interface being in a fine state of balance between oxic and anoxic. The range of content of dissolved oxygen was probably in the order of 0.1 to 1.0 ml/l in the deeper water (Hallam, in press). Given such a delicate condition, quite subtle 'Milankovich'-induced climatic changes such as cooler periods causing slight increases in atmospheric and consequently oceanic turbulence, might have been sufficient to promote the critical changes required in bottom-water oxygen content. Thus the increased oceanic turbulence might have been expressed by, among other things, more active bottom currents, which would have served to restrict anoxic conditions to below the sediment-water interface. In this connection it is noteworthy that Quaternary sapropel horizons in the Black Sea apparently formed at times of relatively warm climate (Degens & Stoffers 1980) but perhaps not too much should be made of this comparison, because the Black Sea is a small, restricted marine basin and the Quaternary climate was very different from that of the Mesozoic.

References

Chamley, H. & Robert, C. 1982. Paleoenvironmental significance of clay deposits in Atlantic black shales. *In*: Schlanger, S.O. & Cita, M.B. (eds), *Nature and Origin of Cretaceous Carbon-rich Facies*. Academic Press, London. 101–12.

——, Debrabant, P., Candillier, A.-M. & Foulon, J. 1983. Clay mineralogy and inorganic geochemical stratigraphy of Blake-Bahama Basin since the Callovian, site 534, Deep Sea Drilling Project leg 76. *Init. Repts DSDP* **76,** 437–51.

Cool, T.E. 1982. Sedimentological evidence concerning the paleoceanography of the Cretaceous western North Atlantic Ocean. *Palaeogeog., Palaeoclimatol., Palaeoecol.* **39,** 1–36.

Dean, W.E., Gardner, J.V., Jansa, L.F., Cepek, P. & Seibold, E. 1977. Cyclic sedimentation along the continental margin of northwest Africa. *Init. Repts. DSDP* **41,** 965–86.

Degens, E.T. & Stoffers, P. 1980. Environmental events recorded in Quaternary sediments of the Black Sea. *J. geol. Soc.* **137,** 131–8.

De Klasz, I. 1978. The West African sedimentary basins. *In*: Moullade, M. & Nairn, A.E.M. (eds), *The Phanerozoic Geology of the World, II. The Mesozoic, A*. Elsevier, Amsterdam. 371–400.

Demaison, G.J. & Moore, G.T. 1980. Anoxic environments and oil source bed genesis. *Bull. Am. Ass. Petrol. Geol.* **64,** 1179–209.

Deroo, G., Herbin, J.P., Rouchache, J., Tissot, B., Albrecht, P. & Schaeffle, J. 1977. Organic geochemistry of some Cretaceous black shales from sites 367 and 363; Leg 41, eastern North Atlantic. *Init. Repts DSDP* **41,** 865–73.

Griffin, J.J., Windom, H. & Goldberg, E.D. 1968. Distribution of clay minerals in the world ocean. *Deep-Sea Res.* **15,** 433–61.

Habib, D. 1982. Sedimentary supply origin of Cretaceous black shales. *In*: Schlanger, S.O. & Cita, M.B. (eds), *Nature and Origin of Cretaceous Carbon-rich Facies*. Academic Press, London. 113–27.

—— 1983. Sedimentation-rate-dependent distribution of organic matter in the North Atlantic Jurassic-Cretaceous. *Init. Repts DSDP* **76,** 781–94.

Hallam, A. 1983. Early and mid Jurassic molluscan biogeography and the establishment of the central Atlantic seaway. *Palaeogeog., Palaeoclimatol., Palaeoecol.* **43,** 181–93.

—— 1984. Continental humid and arid zones during the Jurassic and Cretaceous. *Palaeogeog., Palaeoclimatol., Palaeoecol.* **46,** 195–223.

—— in press. Mesozoic marine organic-rich shales. *In*: Brooks, J.R.V. & Fleet, A.J. (eds), *Marine petroleum Source Rocks*. Blackwell Scientific Publications, Oxford.

Herbin, J.P., Deroo, G. & Rouchache, J. 1983. Organic geochemistry in the Mesozoic and Cenozoic formations of site 534, Leg 76, Blake-Bahama Basin, and comparison with Site 391, Leg 44. *Init. Repts DSDP* **76,** 481–93.

Jansa, L.F. & Wade, J.A. 1975. Geology of the continental margin off Nova Scotia and Newfoundland. *Geol. Surv. Can. Paper* **74–30,** 51–105.

——, Gardner, J.V. & Dean, W.E. 1977. Mesozoic sequences of the Central North Atlantic. *Init. Repts DSDP* **41,** 991–1031.

Jansa, L.F., Steiger, T.H. & Bradshaw, M. 1984. Mesozoic carbonate deposition on the outer continental margin off Morocco. *Init. Repts DSDP* **79,** 857–891.

Jeans, C.V., Merriman, R.J., Mitchell, J.G. & Bland, D.J. 1982. Volcanic clays in the Cretaceous of southern England and northern Ireland. *Clay Miner.* **17,** 105–56.

Joides Journal 1983. *Glomar Challenger Operations, Leg 93, US East Coast Continental Rise.* Vol. 9, no. 3 (October), 20–26.

Kinsman, D.J. 1975. Salt floors to geosynclines. *Nature* **255,** 375–8.

Lancelot, Y. & Winterer, E.L. 1980. Evolution of the Moroccan oceanic basin and adjacent continental margin—a synthesis. *Init. Repts DSDP* **50,** 801–21.

McCave, I.N. 1979. Depositional features of organic carbon-rich black and green mudstones at DSDP sites 386 and 387, western North Atlantic. *Init. Repts DSDP* **43,** 411–16.

Ogg, J.G., Robertson, A.H.F & Jansa, L.F. 1983. Jurassic sedimentation history of site 534 (western North Atlantic) and of the Atlantic-Tethys seaway. *Init. Repts DSDP* **76,** 829–84.

Owens, J.P. 1983. The northwestern Atlantic Ocean margin. *In*: Moullade, M. & Nairn, A.E.M. (eds), *The Phanerozoic Geology of the World, II. The Mesozoic, B.* 33–60.

Pacey, N.R. 1984. Bentonites in the Chalk of central eastern England and their relation to the opening of the northeast Atlantic. *Earth Planet. Sci. Lett.* **67,** 48–60.

Robertson, A.H.F. & Bliefnick, D.M. 1983. Sedimentology and origin of Lower Cretaceous pelagic carbonates and redeposited clastics, Blake-Bahama Formation, Deep Sea Drilling Project site 534, western equatorial Atlantic. *Init. Repts DSDP* **76,** 795–828.

Sellwood, B.W. & Sladen, C.P. 1981. Mesozoic and Tertiary argillaceous units: distribution and composition. *Q. J. eng. Geol. Lond.* **14,** 263–75.

Sheridan, R.E. 1983. Phenomena of pulsation tectonics related to the breakup of the eastern North American continental margin. *Init. Repts DSDP* **76,** 897–909.

Sladen, C.P. 1983. Trends in Early Cretaceous clay mineralogy in Europe. *Zitteliana* **10,** 349–57.

Stöcklin, J. 1984. Orogeny and Tethys evolution in the Middle East. An appraisal of current concepts. *Rep. 27th Int. geol. Congr.* **5,** 65–84.

Summerhayes, C.P., in press. Organic-rich Cretaceous sediments from the North Atlantic. *In*: Brooks, J.R.V. & Fleet, A.J. (eds), *Marine Petroleum Source Rocks.* Blackwell Scientific Publications, Oxford.

—— & Masran, T.C. 1983. Organic facies of Cretaceous and Jurassic sediments from DSDP site 534 in the Blake-Bahama Basin, western North Atlantic. *Init. Repts DSDP* **76,** 469–80.

A. Hallam, Department of Geological Sciences, University of Birmingham, P.O. Box 363, Birmingham B15 2TT.

Palaeoceanographic setting of the Callovian North Atlantic

A.H.F. Robertson & J.G. Ogg

SUMMARY: DSDP/IPOD Site 534 located in the Blake Bahama Basin adjacent to the Florida coast off eastern USA exemplifies the early opening stages of the North Atlantic in Callovian (mid-Jurassic) time.

The earliest sediments accumulated on a rugged ocean floor located near the edge of a WNW–ESE-trending fracture zone, not far from the spreading axis. At that time the ocean was *c*.500 km wide with a borderland experiencing a tropical climate. Input of both inorganic and organic material was mostly of marine origin, but also included quartz, illite, chlorite and large quantities of fine plant material. Carbonate banks flourished during a time of raised sea-level, shedding large quantities of sediment, mostly peri-platform ooze, which was redeposited by turbidity currents to Site 534 after a marginal ridge (Blake Spur ridge) was finally breached during later Callovian time. The site subsided *c*.300 m during the Callovian and for a time lay below the CCD which was still shallower than 3000 m. Distinctive hummocky seismic reflections and sedimentary structures in the cores suggest that sluggish bottom water circulation existed, giving rise to sediment-drifts ('muddy contourites'). Hydrothermal precipitates drifted westwards from the ridge. Inferred trade wind patterns favoured extensive coastal upwelling along the African borderland, the closest margin to Site 534 during the Callovian. The favoured origin of numerous, but volumetrically minor, black shale intervals is that marine organic matter was concentrated where the oxygen-minimum-zone intersected the shelf-continental rise and that this material was subsequently redeposited by currents and/or gravity flows to form layers which readily became black and de-oxygenated during diagenesis.

The early opening history of the Atlantic Ocean has intrigued many geologists and oceanographers ever since the continental drift hypothesis gained wide acceptance. There were tantalizing glimpses during the early legs of the Deep Sea Drilling Project, but a more comprehensive picture has only begun to emerge with the first drilling to reach mid-Jurassic oceanic basement. The objectives here are to synthesize the palaeoceanographical and sedimentological evidence from Site 534 (DSDP/IPOD Leg 76) in relation to the early opening history of the central North Atlantic in Callovian time, to discuss alternative interpretations and to highlight remaining problems. Particular reference will be made to the possible origin of the Callovian black shales, which is still subject to controversy. As a starting point, the authors wish to stress that only if the sedimentology of the cores is taken fully into account is there much hope of disentangling the effects of variables, often inter-related, which include sediment source, climate, sea-level, marine productivity, sea water circulation, tectonics and diagenesis.

Geological setting

From a combination of regional geology, seismic surveys, and deep drilling—both on land and at sea—it is now widely accepted that the North Atlantic rifted in the Triassic–early Jurassic, followed by the first creation of oceanic crust in mid-Jurassic time (e.g. Sheridan *et al.* 1983b). Only off the south-eastern USA, adjacent to the Blake Plateau, was the sedimentary succession sufficiently thin enough for the Glomar Challenger drillstring to penetrate mid-Jurassic basement. Site 534 (Fig. 1) is located not far below the crest of a basement high on the north flank of a trough which probably originated as a minor transform fault. The location is within the Outer Magnetic Quiet Zone, which extends from marine magnetic anomaly M25, 370 km to the east, to the Blake Spur anomaly some 110 km to the west. The Blake Spur anomaly represents a major north-east-trending basement ridge, which influenced early sedimentation history (Fig. 2). Its origin is attributed to a reorganization of the spreading axis, which involved a jump in the ridge crest to the east, thus isolating a strip of the oldest North Atlantic oceanic crust between the Blake Spur and the East Coast magnetic anomalies (Vogt 1973).

South of the Blake Spur the basement ridge is buried below the edge of the Blake Plateau and there is no definite evidence of oceanic crust having existed further west. One of the implications of re-dating the relevant magnetic anomalies (M25 to M28) is that early spreading was relatively fast (*c*.3.1 cm/yr). The first sediments at Site 534 thus accumulated near the edge of a

From SUMMERHAYES, C.P. & SHACKLETON, N.J. (eds), 1986, *North Atlantic Palaeoceanography*, Geological Society Special Publication No. 21, pp. 283–298.

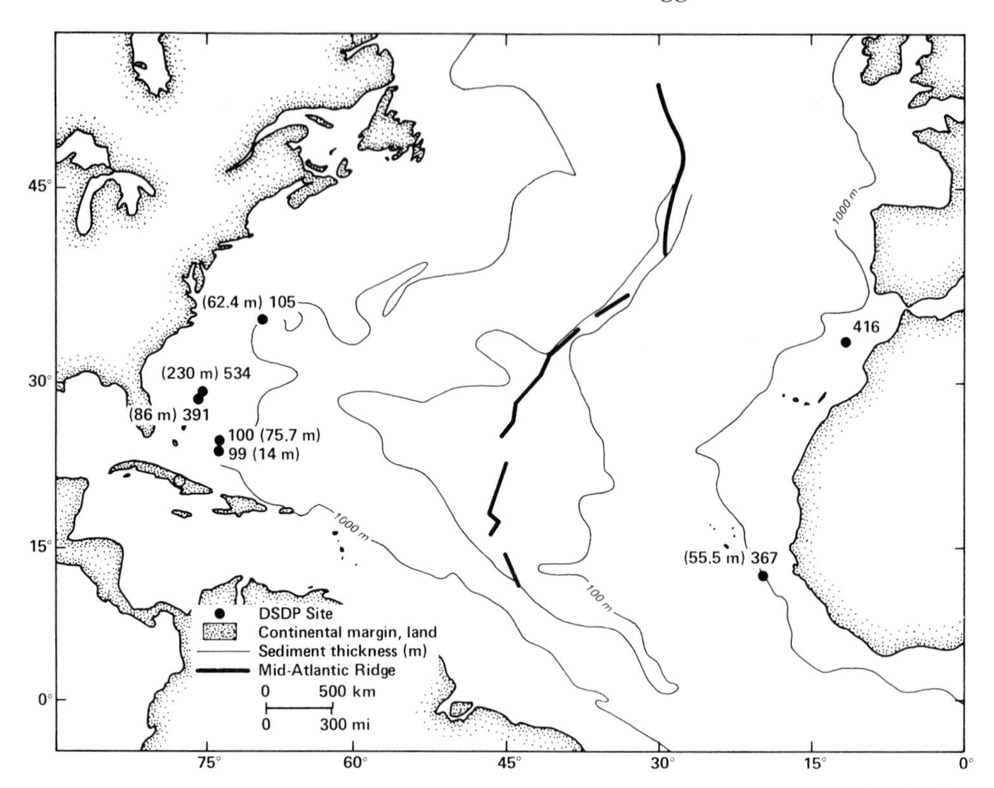

FIG. 1. DSDP/IPOD holes which penetrated Middle and Upper Jurassic successions in the North Atlantic. Note the location of Site 534 in the Blake-Bahama basin.

fracture zone within an asymmetrical ocean basin already some 500 km wide. The following discussion highlights only specific points of particular palaeoceanographical interest; for a more comprehensive treatment the reader is referred to the more detailed descriptions in the Leg 76 Initial Report (Ogg *et al.* 1983).

Mid-Callovian ocean floor

The oceanic basement at Site 534 (Fig. 3) comprises alternations of massive flows and basaltic breccias. Palaeomagnetic inclinations measured for most of the flows are steeper than expected for the palaeolatitude of the site (Steiner 1983), which is in keeping with the evidence, discussed below, that tectonic tilting has occurred. Clasts in the lava breccias are angular and moderately to highly vesicular, set in a matrix of quartz, calcite and green clay. Individual lava clasts are coated with fibrous calcite cement of probable hydrothermal origin. There are also several thin interbeds of laminated reddish-brown marly limestone and siliceous claystone, which resemble the immediately overlying basal sediments, and this tends to favour the onset of sedimentation above with no significant time break.

The Callovian basalts were erupted at a spreading axis close to a minor fracture zone, where slopes were sufficiently steep to give rise to occasional basaltic breccias, in addition to mostly pillowed flows. In the modern oceans strong faulting dominates the slow- and intermediate-rate spreading axes (e.g. Mid-Atlantic Ridge; Ballard & Van Andel 1977). Only exceptionally is there significant eruption away from the median valley. The basal and interlava sediments thus probably accumulated very close to the spreading axis, raised well above the surrounding areas of older oceanic crust. After burial, the interlava sediment was variably calcified and silicified and the basalt was altered with the formation of well crystallized smectite and Na-feldspar (Chamley *et al.* 1983). Such alteration effects extend no more than 9 cm into the overlying (preserved) sediment succession and reflect slow long-term alteration of oceanic crust below a thermal blanket of impermeable fine-grained argillaceous sediment.

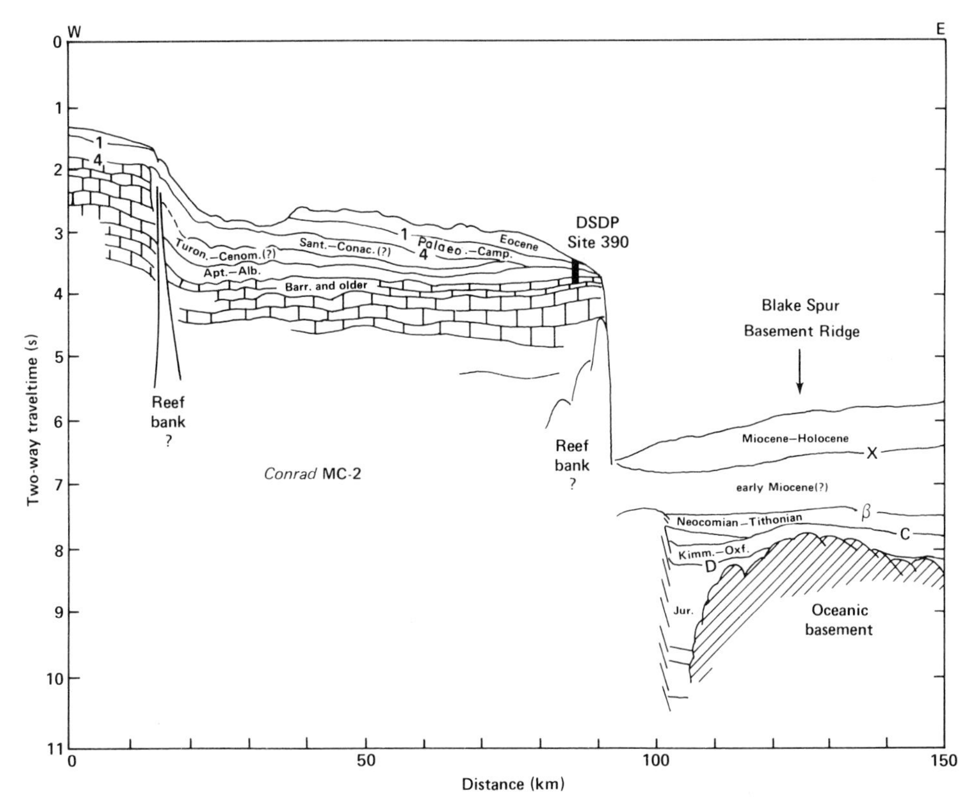

FIG. 2. Interpretation of seismic profiles across the Blake Escarpment and Blake Spur basement ridge (after Sheridan *et al.* 1979). Note the deep sediment basin between the Blake Escarpment and the Blake Spur. The Blake Spur effectively dammed sediment dispersal until later Callovian time.

Callovian succession

The basalt is overlain by 63 m of calcareous claystones (up to 23% $CaCO_3$) with subordinate calciturbidite intercalations (Fig. 4). The basal 10 m of the succession comprise mostly dusky red calcareous silty marls while much of the overlying Callovian sediment interval is made up of alternations of non-calcareous claystones, marly limestones, 'black shales', thin radiolarites, silts and sands. The claystone intervals (above Core 126–2) are mostly greenish or greyish and variably bioturbated (*Chondrites*). The 'black shales' are laminated black or greenish-black nannofossil claystones, present as thin (up to several centimetres thick) interbeds within Cores 122–2 to 125–2; they are especially abundant in Cores 125–3 to 125–6 (Sheridan *et al.* 1983b; Tyson 1984). Grey-green marly limestone interbeds become numerous in the upper part of the Callovian succession (Cores 122–1 through 125–4) and are clearly of calciturbidite origin; above Core 125–4 they are bioclastic. The claystones show extensive soft-sediment disturbance, including slumping, debris-flow, and possible cross-lamination, which is most evident from Cores 125–4 to 127–1. Compositionally, the claystones contain variable amounts of quartzose silt, claystone intraclasts, biotite-chlorite flakes, phosphatic concretions, glauconite and possible minor volcaniclastic material. Total organic carbon values reach 3.1% in the black shale and 1.4% in the claystones.

Dividing the whole Callovian sediment thickness (decompacted) by the biostratigraphically-determined age span suggests average accumulation rates of 1.4 to 2.5 cm/10^3 yr for the Callovian (Gradstein & Sheridan 1983). Ogg *et al.* (1983), however, point out that actual background claystone depositional rates could have been significantly higher, in the order of 6–10 m/Ma (assuming that the recovered proportions of lithologies reflect the original succession).

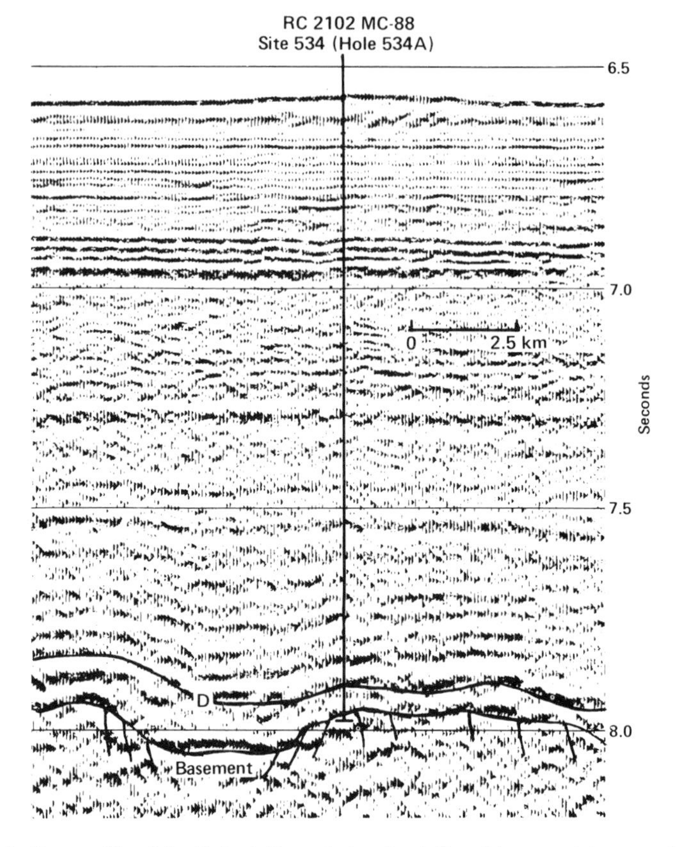

FIG. 3. Seismic reflection profile of the Robert Conrad showing inferred basement topography and approximate location of Site 534 (Sheridan, Gradstein *et al.* 1983b). A major basement trough is located 10 km south-west.

Theoretical circulation patterns

A key question relevant to much of the following discussion is whether there was a global atmospheric circulation similar to the present, particularly the existence of a belt of trade winds. While there is no evidence of major differences in the past, most physical oceanographers recognize that the applicability of present circulation patterns rests on an assumption (e.g. Dean H. Roemmich, pers. comm., 1985). As shown in Fig. 5, the reconstructed palaeogeography of the Atlantic-Tethys area as westward-narrowing gulf implies that a hydrostatic 'head' could have formed in the easternmost Tethys, driving net east to west surface currents through the rest of the Tethys-Atlantic seaway, but only if an adequate surface opening into the Pacific Ocean to the west allowed unrestricted current flow (e.g. Luyendyk *et al.* 1972). There was minimal exchange of western Tethyan and Eastern Pacific faunal zones during Toarcian to Bajocian time, and apparently no significant connection until the early Callovian (Hallam 1983). From mid-Callovian to late Oxfordian time there was a cosmopolitan ammonite assemblage, indicating free exchange of Tethyan, Atlantic and Pacific faunal zones (Hallam 1975; Westermann & Riccardi 1975). In this situation *surface* currents would have primarily flowed from the Atlantic to the Pacific. If, on the other hand, any 'downstream' constriction existed between the Atlantic and the Pacific (or between the eastern and western Tethyan areas) then some back-flow would have occurred (i.e. like the Gulf Stream), but it is unclear whether such western intensification would have developed given the geometry and latitude of the early Atlantic (cf. Thiede 1979; Gradstein & Sheridan 1983).

Assuming the existence of trade winds, these would have caused frictional drag on the surface waters, with a resulting flow at 45° to the right of the wind direction at the surface and 90° to the right below the surface in the northern hemi-

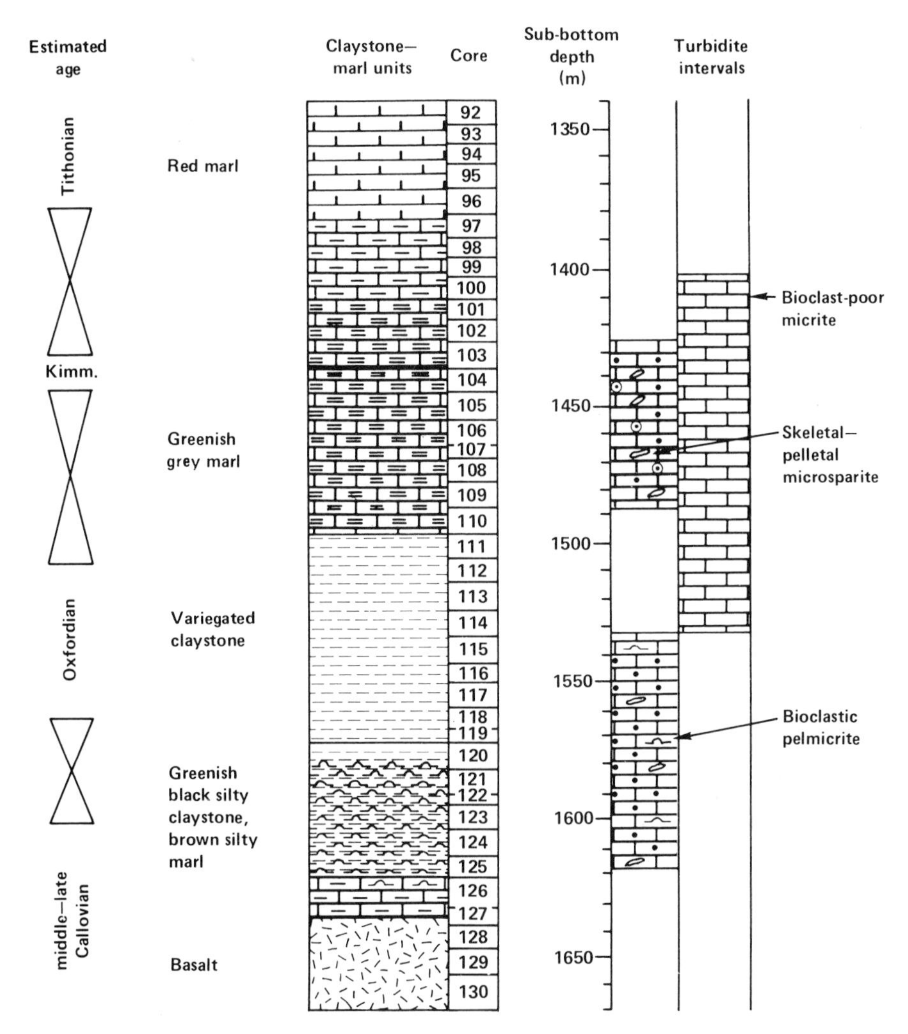

FIG. 4. Jurassic stratigraphy of Site 534. The claystone units and turbidite intervals are shown separately. For biostratigraphy see Sheridan *et al.* (1983b). Only the Callovian interval is discussed here.

sphere ('Ekman flow'). Net flow away from any coast facing north, north-west or west, or away from the equator (equatorial divergence) would then have been compensated by upwelling in all of these areas. The nutrient content of these waters controlled productivity. Decomposition products of any life at the surface would have tended to enrich the upwelling waters, thus helping to maintain fertility. The width of the high fertility zones would thus have depended on the rate of surface flow away from upwelling areas relative to the loss of nutrients from the system.

Deep water circulation patterns are difficult, if not impossible, to predict since they depend less on physical principles than on the total palaeoceanographic setting (e.g. geography, climate, surface circulation). One possibility is that the early Atlantic was deeper than either the Tethys channel and/or the Gulf of Mexico. If so, there could have been denser water inflow over a sill from the Pacific and/or Atlantic oceanic reservoirs, or saline flow from epicontinental areas. Any major influx of fresh water from the bordering tropical continents could conceivably have caused episodes of stratification. Circulation in the below-sill portions of the basin could then have been very sluggish tending to lead to periodical oxygen-depletion. However, predictions of the deep water circulation are very uncertain and one is forced back on the some-

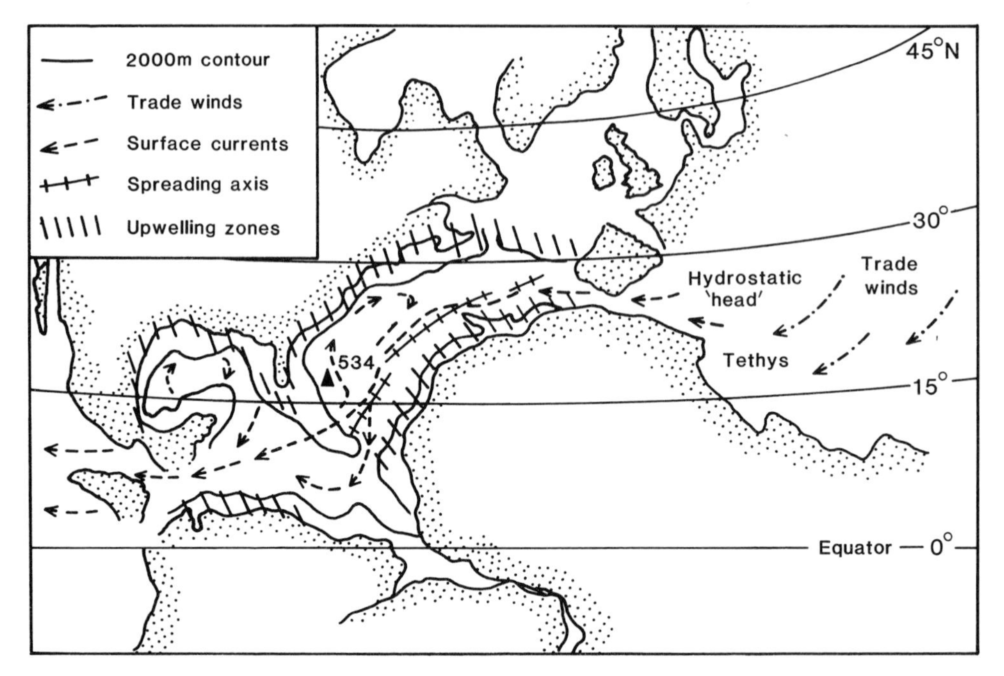

FIG. 5. Reconstruction of the Northern Atlantic in late Callovian time, to show inferred surface and possible deep circulation (reconstruction modified from Sclater *et al.* 1977).

times ambiguous geological evidence of the cores, which is now examined.

Sediment sources

At the time of basal sediment deposition Site 534 was located midway between the African continental margin and the Blake Spur ridge. The North American continental margin was at least another 300 km away across the Blake Plateau carbonate platform and the extinct corridor west of the Blake Spur anomaly (Fig. 1). The closest known exposed continental landmass would thus have been the Palaeozoic fold belts and platform of the Liberian segment of the African margin.

In keeping with this setting (Fig. 5), the major and trace element composition points to mainly terrigenous sources. During the Callovian, illite, chlorite and kaolinite dominate the clay minerals, while in the Oxfordian succession smectites become relatively much more abundant (Chamley *et al.* 1983). The high kaolinite in the basal sediment reflects the hot tropical climate of the adjacent continent and the tendency for kaolinite to be more widely dispersed than illite before settling. Young basins are expected to be bordered by areas of marked relief which formed by differential uplift during the earlier rift phases and this favours a dominant input of illite and chlorite at this stage (Fig. 6).

On modern spreading axes hydrothermal activity produces metal-enriched precipitates which may be widely dispersed by currents. At Site 534 the claystones are enriched in Fe, and trace metals including Zn and Cu. Interbedded limestones are often markedly enriched in Mn. Ratios of Fe/Al and Mn/Al suggest that up to half the metal content is excess to normal deep sea sedimentation and thus is probably of hydrothermal origin (Fig. 6). From the average claystone sedimentation rate it can be calculated that this excess Fe accumulated at *c.*70 mg/cm^2/1000 yr and the excess Mn at 2 mg/cm^2/1000 yr, close to the values on the modern East Pacific Rise (Ogg *et al.* 1983). Many of the claystones are organic-rich (see below). Manganese has apparently been diagenetically mobilized and precipitated in the interbedded calciturbidites, which contain lower organic matter and are considerably more porous.

During the Callovian, input from the North American shelf was dammed behind the Blake Spur ridge (Figs 2 and 6). The prominent seismic reflector, Horizon D, can be traced from a short distance east of Site 534 to the base of the Blake Spur ridge, pinching out across the ridge in topographical lows (Sheridan *et al.* 1983a). The

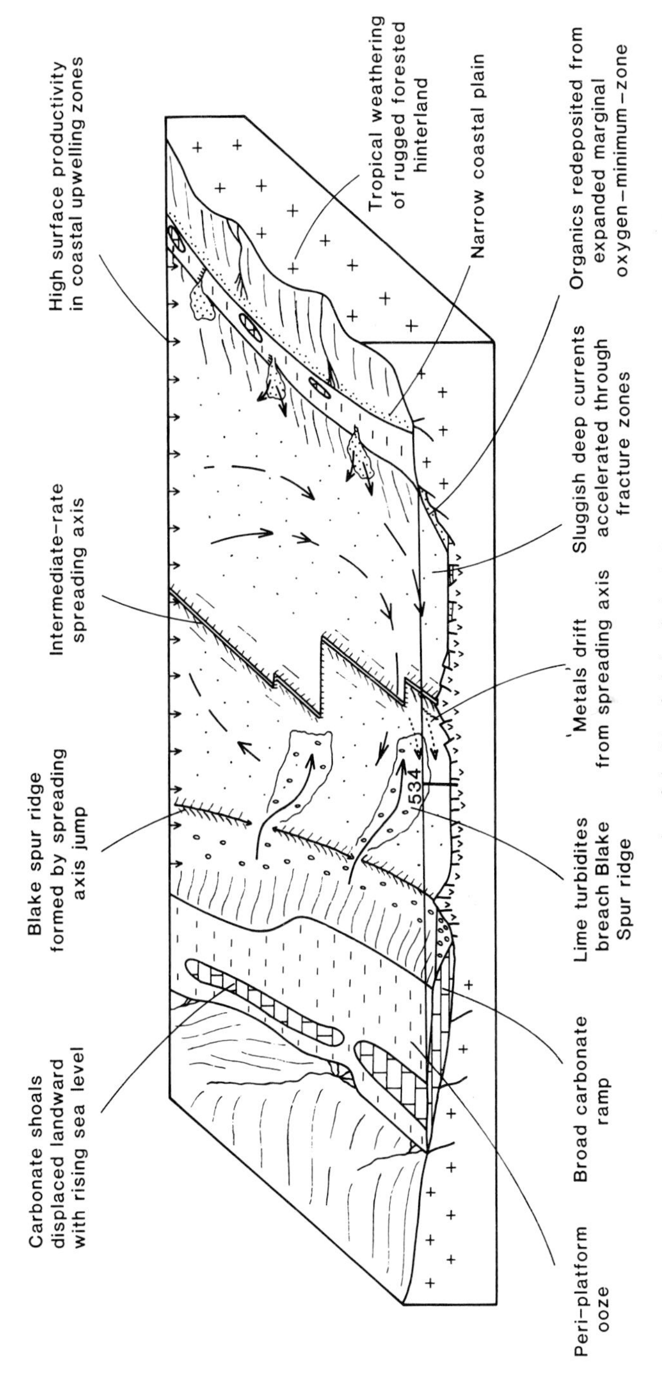

FIG. 6. Block diagram to illustrate the palaeoceanography of the North Atlantic in Callovian time. See text for explanation.

ridge was breached for the first time only in mid-late Callovian time, allowing lime-rich turbidites to flow onto the Blake-Bahama basin floor. By this time the site had already subsided several hundred metres from the spreading axis (see below) and this also facilitated calciturbidite deposition. The bioclasts within the marly calciturbidites are mostly radiolarians and fine pelecypod fragments ('filaments') together with rare shells, echinoderm fragments and benthic foraminifera, but neritic carbonate material is notably absent (e.g. pelecypod shells). The obvious source is the Blake Plateau now located 150 km to the west (Fig. 6). In the Jurassic the margins of the Blake Plateau were quite different from the present steeply sloping Blake Escarpment (Van Buren & Mullins 1983). Seismic and well data (e.g. Jansa & Weidmann 1982) from both the North American and African margin suggest that the Blake platform was then a broad gently-sloping carbonate ramp. During a time of high sea-level, in the early Callovian (see below), carbonate shoals would have retreated landward, increasing the area of peri-platform ooze accumulation above the CCD. This material was then redeposited onto the ocean floor, exploiting fracture zone troughs. Seismic interpretation (Sheridan *et al.* 1983a) suggests that more than 100 m of sediment were banked up against the Blake Spur ridge before it was finally overwhelmed.

Surface productivity and currents

While high productivity related to equatorial divergence is anticipated in the palaeo-Pacific Ocean, at this time the equator still lay south of the area of significant Atlantic opening; Site 534 was located around 15°N in late Callovian time (Fig. 5). In a young narrow equatorial ocean, divergence probably played little role in productivity anyway. Instead, the main influence was important coastal upwelling along the closest (African) landmass to the south and east (Fig. 5).

Elevated productivity could be inferred in the Callovian from:

(1) the abundance of radiolarians, which in modern and ancient oceans are associated with high fertility zones (e.g. Jenkyns & Winterer 1982);
(2) the abundance of marine organic matter and phosphatic pellets (Cores 121–124);
(3) the palaeoceanographical setting next to upwelling regions created by the trade winds.

Several other points, however, need to be considered:

(1) the preservation of marine organic matter, which appears to predominate (see below), depends largely on whether or not the organic matter was combusted on the sea floor and during early diagenesis;
(2) phosphate is not particularly enriched in the black shales, as expected if these are *in situ* below an upwelling zone;
(3) organic matter (e.g. phosphatic pellets) could have been locally concentrated by redeposition, for example, from a coastal upwelling zone (see below).

On balance, although it is clear that preferential preservation probably has much to do with the organic matter content, the authors feel that enhanced productivity related to coastal upwelling (with or without redeposition) constitutes the most attractive model (Fig. 6). Significantly, when, in the Early Cretaceous, there was a return to very similar diagenetic conditions of black shale accumulations (Habib 1983; Robertson & Bliefnick 1983) this was not associated with similar concentrations of marine organic matter and phosphatic material, possibly because the site was by then outside the influence of African coastal upwelling.

Subsidence and the CCD

The basal calcareous claystones give way upwards over 25 m to effectively non-calcareous claystone (Cores 120–122). By this time a projection based on the initial estimated depth of 2500–2700 m for the ridge suggests that the site could have subsided some 300 m from 2800–3100 m (Ogg *et al.* 1983), possibly taking the sea floor below the CCD, or else the CCD could have risen. A combination of subsidence *and* a rising CCD would explain the *progressive* disappearance of pelagic bivalve shells, *then* of calcareous nannofossils within the claystones. There is little textural or mineralogical evidence that the decrease in calcium carbonate content was due to swamping by terrigenous input. However, redox is another factor to be taken into account. As noted by Tyson (pers. comm. 1985) organic-rich lithofacies are liable to be carbonate-poor because of intense CO_2 production by organic matter degradation which is taking place in a chemically-open (bioturbated) pore water system. Only when bioturbation ceases does the balance shift to carbonate preservation which can then be excellent, for example, in the case of the nannoplankton preserved in the black shales (Roth 1983).

The redox factor is not, however, considered to have greatly affected the conclusion of a raised CCD in the Callovian since both the organic-rich and organic-poor facies alike show low calcium carbonate contents above the basal dusky red claystones. In this case there is agreement from both seismic identification of coastal onlap (Vail & Todd 1980) and the recognition of transgressive-regressive events on the continental shelves (Hallam 1975, 1978) that the Callovian was a time of rising sea-level. This could have considerably increased the volume of $CaCO_3$ locked up by shallow shelf production and so depleted the oceanic reservoir, resulting in a rise in the CCD (Fig. 6).

Bottom topography

Seismic reflectors below Horizon D show that Callovian sediments must have progressively filled hollows in an initially rugged sea floor surface (Fig. 2). Thicker Callovian sediments, perhaps as much as 200 m, accumulated along the axis of a fracture zone trough *c.*10 km south-west of Site 534.

The apparent inclination of bedding is 5° to 15° more than can be accounted for either by drill-string deviation (*c.*2.5°) or by angles of deposition (at least for the fine-grained plane-laminated claystones). The bedding returns to horizontal above Core 116. It thus appears that the inclination could have arisen by bulk fault-rotation around the end of the period of Callovian deposition, to be consistent with the palaeomagnetic inclinations determined in the basaltic basement (see above). Alternatively, the fault-rotation could have been progressive, contributing to the evidence of gravitational instability in the cores. In this context it may be significant that fracture zones are normally largely inactive outside the zone of transform fault offset. Fault activity could, however, have persisted in this case, due to continuing accommodation of the spreading axis to its new position following the earlier ridge jump.

Sediment drifts: Evidence of bottom currents

Seismic-reflection-profiling in the vicinity of Site 534 (Fig. 7(a)) has revealed mounded acoustically-transparent bodies above the faulted oceanic basement. By contrast, reflectors above horizon D (post-Callovian) are parallel and basin-levelling. Hummocky seismically-transparent reflectors independent of bottom topography are believed by many to be distinctive of sediment-drift deposits ('contourites') (Sheridan *et al.* 1983a; Fig. 7(b)). Both recent and ancient exam-

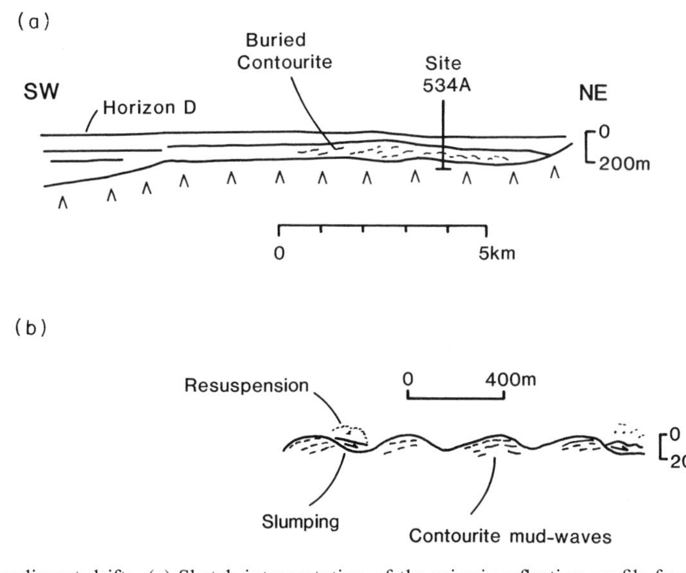

FIG. 7. Buried sediment drifts. (a) Sketch interpretation of the seismic reflection profile from the Robert Conrad near DSDP/IPOD Site 534 (Sheridan *et al.* 1983b). Key seismic horizons are indicated. Note the distinctive hummocky reflectors which are interpreted as a distinctive 'contourite feature' by these authors. Post Horizon-D reflectors are basin-levelling. (b) Sketch of the detailed morphology of the 'contourite' sediment-drifts, as inferred by R.E. Sheridan (pers. comm., 1985). Slumping and resuspension takes place off the mud-waves to produce sedimentary structures seen in the cores.

ples of current-drift deposits have recently been reviewed by Stow & Holbrook (1984) who identify the existence of both distinctive sandy and muddy contourite facies. A fine example drilled by DSDP is located on the east flank of the Rekjanes ridge in the North Atlantic where elongate ridges of rapidly deposited (?) lower Pliocene sediment are believed to have resulted from deep water outflow from the Norwegian Sea (Laughton *et al.* 1972). Similar deposits have recently been recognized in fracture zones, based on evidence of seismic profiles alone (Scrutton & Stow 1984). In both of these cases deep current flow was focused and accelerated by the bottom topography. By contrast, the bottom currents which gave rise to the classic mud-drifts of the Blake Outer ridge at Site 533 (Sheridan *et al.* 1983c) accumulated from more diffuse sluggish currents less constrained by bottom topography. Unlike the Callovian deposits these nannofossil muds are virtually featureless, with no clearly visible current structures in the cores. It follows, thus, that an absence of visually obvious textural features in cores (e.g. lamination, grading, scours) does not necessarily exclude an origin as sediment-drift deposits ('contourites'). If, however, seismic character is to be the sole recognition criterion in certain cases, then it must be noted that similar hummocky reflectors can also be generated from multiple-overlapping sediment lobes, as found in shallow water pro-delta and interdelta settings (Mitchum *et al.* 1977). Similar reflections have recently also been noted in deep turbidites off the Brazilian continental margin (north of the Vema channel; Gamboa *et al.* 1983), and hence the recognition of 'contourites' on seismic criteria alone could be misleading. There is no reason, in principle, why diffuse turbidity currents should not be deflected by sea floor topography to finally accumulate in mounds or ridges away from the accessible basin depocentre.

One has to ask, then, whether the redeposited Callovian sediments at Site 534 are current-drift deposits ('contourites'), debris-flows, slumps, turbidites, or some combination of these features (Fig. 8(a)–(d)). Here the authors prefer to define a current-drift deposit ('contourite') as 'sediment reworked or redeposited by bottom currents, which are commonly geostrophic and contour-flowing, but which may be either fast or sluggish'. The significance is increased since positive identification of current-deposited features would have a considerable bearing on the controversial origin of the intercalated Callovian black shales, discussed below. Consistent with his view that these organic-rich shales accumulated in local sea floor pools of anoxic water, Tyson (1984 and pers. comm., 1985) favours the view that the sediments are essentially slumps, debris-flows and turbidites redeposited off local basement highs, and thus that only gravitational processes, rather than current deposition, need be involved.

Ogg *et al.* (1983) note that thin intervals with changed inclinations could be parts of ripples, cross-bedding or scour-structures, although they concluded that the evidence is ambiguous (Fig. 8(a), (b)). They particularly suggested that concentrations of small flattened intraclasts (clast-supported) could be lag deposits on scoured erosion surfaces (Fig. 8(c)). Tyson (pers. comm. 1985) prefers to interpret these as the result of gravitational downslope redeposition (i.e. debris-flow, slumps, creep effects).

Critical in this respect is the origin of numerous (63), mostly less than 1 cm thick, greenish radiolarian bands, which are either flat-bedded (60%) or lenticular (40%) (Sheridan *et al.* 1983b; Tyson 1984; Fig. 8(b)). Scattered radiolarians within the host claystones are mostly pyritized, in at least one case showing evidence of fragmentation by reworking. Tyson (1984; pers. comm., 1985) interpreted the radiolarites as probably turbiditic, suggesting that the sharp tops could be due to a 'hydraulic jump' between the current carrying the radiolarite tests and that carrying the clay, together with slight winnowing by residual flow during the waning of the turbidity current which introduced the radiolarians. A purely turbiditic origin does not, however, easily explain why the radiolarites are often highly lenticular (cf. dilute turbidite facies model of Stow & Piper 1984) and why in each case the immediately overlying clay appears to be depositionally unrelated. By contrast, many typical on-land Tethyan radiolarian chert turbidites (e.g. Neraida chert, Othris, Greece; Nisbet & Price 1974) typically exhibit a structureless (often ungraded) basal interval overlain by a laminated middle part (sometimes, cross-laminated), then an upper, again structureless, interval. On the other hand, concentrations of other coarse material (e.g. silt, phosphate pellets) would be expected in a purely winnowed origin. The observations can perhaps be reconciled if the radiolarians first accumulated on local basement highs below the CCD, from which they were gravity redeposited then variably reworked by currents, *in situ*. In this context it is interesting that the Jurassic Neraida cherts of Othris, Greece, are believed to have been deposited from low-density turbidites derived from intra-oceanic highs, possibly the spreading ridge axis (Nisbet & Price 1974).

One other problem with the current-drift origin is that the hummocky seismic reflectors are present in the *whole* Callovian interval below seismic Horizon D, yet throughout much of it

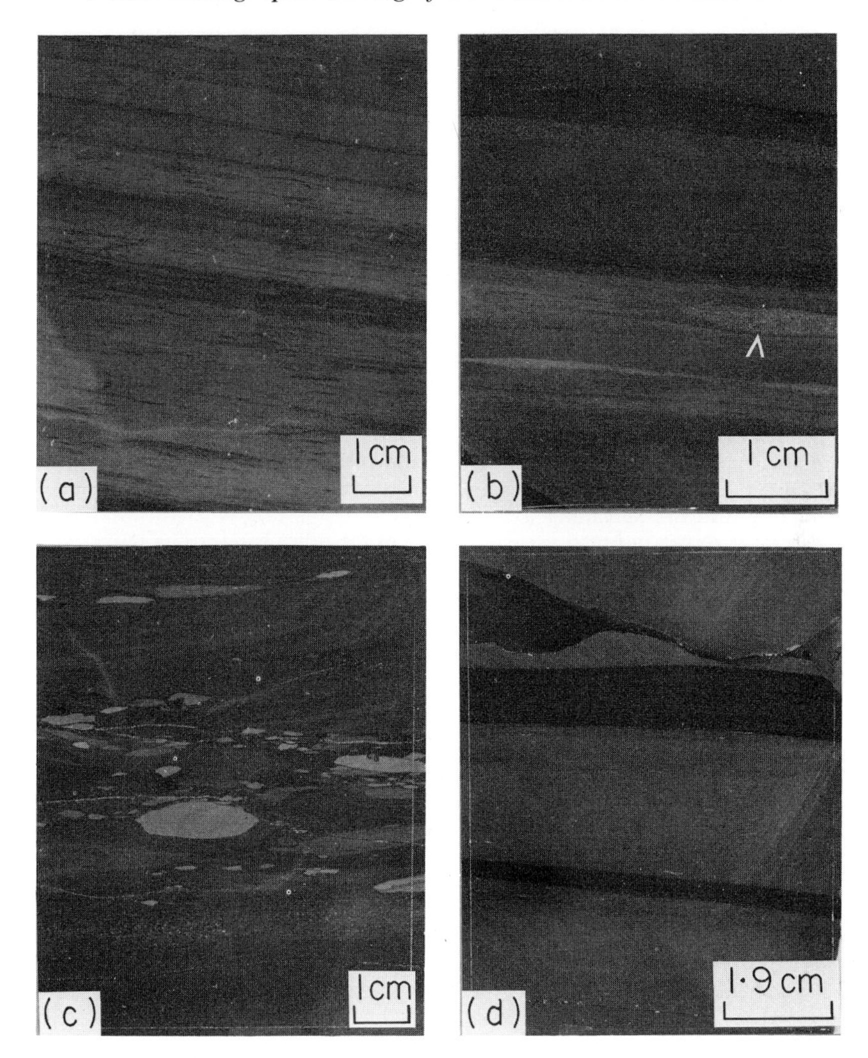

FIG. 8. Photographs of cores of Callovian sediments at Site 534. (a) Note streaky laminations, which may be due to current winnowing. Formation entirely due to slumping is unlikely as graded claystone laminites (upper) are not deformed. The tilt is post-depositional prior to Oxfordian time (Core 122, Section 1, 134–141 cm). (b) Lenticular bands of greenish radiolarian sand in greenish-black silty claystone. The lenticularity is considered influenced by current winnowing. Note also the streaky lamination (lower) (Core 125, Section 2, 33–39 cm). (c) Dark greenish-black claystone with numerous flattened claystone intraclasts, also phosphatic and glauconite concentrates. Could be either a bottom current or slump effect; note the oversteepened cross-lamination (upper right) (Core 125, Section 5d, 47–52 cm). (d) Finely laminated unburrowed black shale in greenish-grey silty claystone. Note fine lamination and sharp cut-off top and bottom (Core 125, Section 116–122 cm).

(above Core 125) the presence of calciturbidites points to a flat abyssal plain, rather than persistences of mud waves, as in the seismic interpretation (see Fig. 7(b)).

Summarizing, while the sedimentological criteria of active bottom currents are apparently, individually, ambiguous, the complete assemblage of sedimentary structures corresponds closely to the model of muddy contourite facies, as proposed by Stow & Piper (1984). The seismic evidence indicates that the sediment mounds occur locally, independent of basement topography (Fig. 7), and thus, on balance, the authors conclude in favour of an essentially current-drift origin, but this need not imply that high velocity currents were involved.

Bottom currents could have funnelled along fracture zones in the vicinity of Site 534. In this case fine terrigenous material of African origin could have drifted across the spreading axis to be deposited at Site 534. Alternatively, the currents could have been salinity-driven, caused by sinking of denser surface water derived from a broad evaporitic continental shelf and marginal environments. By far the largest area would have been the Blake Plateau and the Bahama Banks. The sediments are banked up towards the northern sides of deep troughs orientated ENE–WSW, possibly suggesting a Coriolis effect on a westward-flowing current (Sheridan *et al.* 1983a). Westward rather than eastward bottom current flow would explain the abundance of hydrothermally-precipitated Fe, Mn and trace metals which are easily envisaged as having been carried by currents flowing across the spreading axis in fracture zone lows, as shown in Fig. 6.

Origin of the Callovian black shales

Considerable controversy persists as to the palaeoceanographical implications of black shale occurrences. In essence, do the black shales indicate accumulation under a static de-oxygenated water column, or are differential rates of supply and selective diagenesis capable of producing black shales in a normally circulating oxygenated ocean? Elucidation of the origin of the Callovian black shales at Site 534 would contribute considerably to an understanding of early Atlantic palaeoceanography. Three main lines of evidence are available: (1) analyses of total organic content and kerogen type; (2) palynological studies; (3) the general depositional and diagenetic setting. Black shales form a major component of the late Callovian interval (Cores 125, Section 3 to Core 126, Section 6), and occur more widely as thin interbeds (Core 122, Section 2 to Core 125, Section 2; Fig. 8(d)). Reported total organic carbon analyses ranges from 1.8–3.1%, and in the associated claystones from 0.2–1.3%.

Few organic geochemical data for the Callovian black shales are currently available. Samples of olive black claystone with, respectively, 3.1% (Herbin *et al.* 1983) and 2.8% organic carbon (Rullkotter, this volume) gave hydrogen indices of 221 and 238; another claystone yielded a similar value of 200 to Summerhayes & Masran (1983), but sediment lower in organic carbon (0.4%) gave a hydrogen index of only 58 (Herbin *et al.* 1983). High hydrogen indices normally indicate the supply and/or preservation of a substantial marine-derived organic matter component, but Rullkotter (this volume) cautions against this interpretation, *a priori*, since the organic matter has been attacked by aerobic organisms and partly degraded, a process liable to lower hydrogen indices.

Assessment of the palynological evidence is complicated by the different preparation techniques and nomenclature still used by different workers. In particular, the analytical methods used by Habib (e.g. 1983) may destroy much of the fine-grained marine-derived organic matter leaving the residue preferentially enriched in the more resistant terrestrial material. Habib (1983) concluded in his Leg 76 paper that Cores 125–127 had predominantly terrestrial carbon (cuticle, ligno-cellulosic and carbonized material), but that the amount of marine carbon (mainly in the form of pellets) became increasingly important and dominant above Core 124, in late Callovian time.

On the other hand, using more detailed sedimentological sampling, and a different preparation technique, Tyson (1984) showed that the individual black shale horizons, concentrated in Core 125, were dominated by marine organic matter. He concluded that in all but the calciturbidite interbeds, the main control on organic carbon values was the abundance of amorphous marine organic matter, with plant debris as a low background component. Ultraviolet fluorescence microscopy indicates that even in the black shales the amorphous organic matter has, however, undergone significant degradation, which would not be anticipated in an extremely de-oxygenated deep water column, as in the modern Black Sea. The palynomorph species distributions further suggest that the black shales are the most pelagic of the sediments sampled and show the least obvious signs of redeposition.

That the organic matter *type*, and not just total *abundance*, is considered important stems from knowledge that marine organic matter is much more reactive than most terrestrial material. High concentrations of marine organic matter, thus, tend to imply transport and deposition under at least dysaerobic conditions, since degradation would take place rapidly under more oxidizing conditions. Actual organic-rich intervals are numerous, but thin, implying that if bottom water de-oxygenation was critical, then conditions must have been frequently changing. Tyson (1984; pers. comm., 1985) feels that episodic oxygen-deficiency may have been caused by periodically reduced circulation. If in a restricted Callovian ocean the oxygen content of the bottom water was generally low—probably aerobic-dysaerobic rather than anaerobic—then the circulation could have been further reduced in local sea floor hollows in which bottom waters became periodically dysaerobic to anaerobic. The fact that the black shales apparently occur between the main phase of slumping in the greenish and greyish claystones and the onset of infilling by the marly turbidites would be consistent with existence of small stagnant pockets (Tyson 1984). In support of this view, it may be noted from the evidence of numerous DSDP cores, that organic carbon accumulation rates are often inversely correlated with clastic accumulation rates (Summerhayes 1986).

In the alternative hypothesis, supply of organic matter rather than deoxygenation of the bottom waters plays the key role. Ogg *et al.* (1983) cited frequent mottling in the claystones as evidence of mostly *Chondrites* burrowing in the less organic-rich claystones. Gradstein (1983) also records the presence of a rare small fauna of agglutinating foraminifera which is thought to be indigenous, and thus also indicative of aerobic conditions. It is accepted that *in situ* high productivity is an unlikely cause of the organic-rich accumulations, mainly because the site was by then already removed from marginal upwelling areas. It is, however, notable that the black shales appear nearly simultaneously with the micritic turbidites at the top of the highest obvious slump interval and then continue upwards interspersed with the Callovian calciturbidites. The black shale horizons, individually, exhibit sharp contacts with the host claystones and are often very finely laminated (Fig. 8(d)). An obvious possibility, which would be in keeping with the inferred origin of black shales in the overlying Early Cretaceous interval (Robertson & Bliefnick 1983; Robertson 1984), is that organic-rich muds first accumulated within the oxygen-minimum zone on the continental slope below the upwelling areas and were later transported by low-density turbidity currents and/or nepheloid flow to the present site. In this context, it should be pointed out that once the first definite micrite turbidites began to accumulate, the surrounding sea floor area was essentially basin-levelling and thus gravity deposits could have come from a variety of sites, including the shelf (micrite turbidites), slope (black shales) and from highs within the basin (radiolarites?). Since the ridge was not elevated far above the abyssal plain at this time, turbidites could possibly have been derived from the African margin through fracture zone lows. That, as Tyson (1984) points out, terrestrial organic matter is minimal in the black shales, may well reflect damming of land-derived sediment on the shelf during a time of elevated sea-level. Certainly, the Blake Spur ridge still existed to block any major clastic input from the American side until late Callovian. In this model, the mostly marine organic matter was gravity redeposited off the outer shelf-slope and ponded in topographical lows. Below the surface the use of oxygen by the decomposing organic matter-rich layers exceeded the supply of oxygen from the surface and the more organic-rich layers were turned black and anoxic early during diagenesis.

In summary, any choice between the competing models cannot at present be definitive. There is in any case relatively little effective difference between Tyson's (1984) localized stagnation of circulating aerobic, or dysaerobic, waters, and the author's favoured explanation involving circulating, also aerobic or dysaerobic waters, with periodically increased organic matter input (gravity redeposited) tending to turn the sediment black and anoxic sub-surface. From the scanty on land evidence there is certainly no evidence of a Callovian 'oceanic anoxic event'. In much of the western Tethys area the Callovian interval is represented by a hardground or is un-fossiliferous (and thus not well documented).

Transition to the Oxfordian Atlantic

The transition to the Oxfordian Stage was marked by incoming of more variegated claystones without microfossils. The basal levels exhibit very low $CaCO_3$ contents, as in the middle and upper levels of the Callovian succession. Calciturbidites continued from the Callovian into the Oxfordian without a break, but the black shales become thinner and less abundant. According to Habib (1983), the Oxfordian kerogens contain low contents of terrigenous organic matter, while the amorphous organic matter is considerably more oxidized than in the Callovian interval. These facts, and the distribution of *Classopollis* pollen, as reported by Tyson (1984), point to a switch to a markedly more arid climate. The seas apparently regressed, favouring creation of more pedogenic smectite (Chamley *et al.* 1983) in broad coastal plain settings. Shelf carbonate production decreased, and the CCD dropped to produce a more calcareous pelagic sediment. Basement highs were progressively blanketed by sediment, with redeposition of thin-bedded clay turbidites into lows. No longer constrained, any pre-existing bottom currents became more sluggish and diffuse, so as not to be recognizable in seismic profiles, or as sedimentary structures. With a more arid climate (Hallam 1984) the flux of terrestrial organic matter decreased and the sediment became more oxic, reflected in an often reddish hue and increased *Chondrites* burrowing. By this time the site had moved northwards and away from any African source of coastal upwelling sediment. Marine productivity decreased and radiolarians are no longer present. Post mid-Oxfordian palaeoceanography is from then on much better documented from a number of earlier DSDP sites in both the western and eastern North Atlantic (see Jansa *et al.* 1978; Jansa *et al.* 1979; Ogg *et al.* 1983 for summaries).

Conclusions: a working hypothesis

The authors conclude with their favoured working hypothesis of Callovian palaeoceanography which, it is hoped will be tested by future deep drilling discoveries (Fig. 6). The Callovian interval of DSDP/IPOD Site 534 is of particular interest in its exemplification of the depositional and diagenetic processes in a young, narrow (*c.*500 km), continentally-bordered, ocean basin located near the equator. During the early spreading stages, when the continental borderland topography remained rugged related to earlier rifting, the dominant influence on the clay minerals was terrestrial; large quantities of illite and chlorite accumulated, with kaolinite formed by tropical weathering. Smectites became more abundant later as the topography was smoothed by erosion and this favoured genesis of large volumes of pedogenic smectite and mixed-layer clays in a coastal plain setting (Chamley *et al.* 1983).

Marginal platforms were constructed, especially in the Blake Plateau area, but large volumes of carbonate platform-derived calciturbidites only reached the ocean floor after the Blake Spur ridge to the west was finally breached later in Callovian time. An effect of Callovian high sea-level was also to dam terrigenous sediment in coastal and shelf areas, explaining why terrigenous clastic sediment input was not more extensive at this stage.

In response to inferred trade winds, coastal upwelling was well established especially along the African margin. Marine organic matter was preferentially preserved in an expanded marginal oxygen-minimum-zone, as along the present west-facing continental borderlands. This organic matter was then re-suspended and carried by dilute turbidity currents and/or nepheloid flow onto the abyssal plain. At this time the ocean ridge protruded only slightly above the surrounding abyssal plain, and gravity deposits were able to find their way through fracture zone lows to reach the present site. Circulation was sluggish and the bottom waters dysaerobic (but not anoxic), and this helped to carry fine sediment westwards, taking with it significant volumes of Fe, Mn and trace metals (Cu, Zn), related to ridge crest hydrothermal activity. Assuming that hummocky seismic reflectors below Horizon D have been correctly interpreted, sluggish geostropic currents gave rise to sediment-drifts ('muddy contourites'). Continuing seismic activity, possibly related to block-rotation of the site, triggered slumping and deposition by debris-flow. However finally deposited, the localized concentrations of reactive organic matter were great enough to fully utilize any available oxygen during early diagenesis and thus turn the sediment black and anoxic.

ACKNOWLEDGEMENTS: Many of the basic observations were initially made by our shipboard sedimentological colleagues whom we thank. The manuscript was reviewed by B. Funnell, R.E. Sheridan and R. Tyson. Richard Tyson's detailed comments particularly helped focus our attention on possible alternative interpretations. Mrs D. Baty assisted with drafting and the manuscript was typed by Mrs Marcia Wright.

References

BALLARD, R.O. & VAN ANDEL, TJ.H. 1977. Morphology and tectonics of the inner rift valley at 37° 50′N and the Mid-Atlantic Ridge. *Geol. Soc. Am. Bull.* **88,** 507–30.

CHAMLEY, H., DEBRABANT, P., CANDILLIER, A.M. & FOURON, J. 1983. Clay mineralogical and inorganic geochemical stratigraphy of Blake-Bahama Basin since the Callovian, Site 534, Deep Sea Drilling Project Leg 76 1983. *In:* SHERIDAN, R.E., GRADSTEIN *et al.* (eds), *Init. Repts DSDP* **76.** US Govt. Print. Off., Washington, DC. 437–53.

GAMBOA, L.A., BUFFLER, R.T. & BAKER, P.F. 1983. Seismic stratigraphy and geologic history of the Rio Grande Gap and Southern Brazil Basin. *In:* BAKER, R.F. CARLSON, R.L. & JOHNSON, D.A. *et al.* (eds), *Init. Repts DSDP* **72,** 481–93.

GRADSTEIN, F.M. 1983. Paleoecology and stratigraphy of Jurassic abyssal foraminifera in the Blake-Bahama Basin, Deep Sea Drilling Project Site 534. *In:* SHERIDAN, R.E., GRADSTEIN, R.E. *et al.* (eds), *Init. Repts DSDP* **76.** US Govt. Print. Off., Washington, DC. 511–37.

—— & SHERIDAN, R.E. 1983. On the Jurassic Atlantic ocean and a synthesis of results of Deep Sea Drilling

Project Leg 76. *In:* SHERIDAN, R.E. & GRADSTEIN, F.M. *et al.* (eds), *Init. Repts DSDP* **76.** US Govt. Print. Off., Washington, DC. 913–45.
HABIB, D. 1983. Sedimentation-rate-dependent distribution of organic matter in the North Atlantic Jurassic-Cretaceous. *In:* SHERIDAN, R.E., GRADSTEIN, F.M. *et al.* (eds), *Init. Repts DSDP* **76.** US Govt. Print. Off., 781–95.
HALLAM, A. 1975. *Jurassic Environments.* Cambridge University Press Cambridge. 291pp.
—— 1978. Eustatic cycles in the Jurassic. *Palaeogeogr. Palaeoclimatol. Palaeoecol.* **23,** 1–32.
—— 1983. Early and mid-Jurassic Molluscan biogeography and the establishment of the Central Atlantic Seaway. *Palaeogeogr., Palaeoclimatol. Palaeoecol.* **43,** 181–93.
—— 1984. Continental humid and arid zones during the Jurassic and Cretaceous. *Palaeogeogr., Palaeoclimatol., Palaeoecol.* **47,** 195–223.
HERBIN, J.P., DEROO, G. & ROUCACHE, J. 1983. Organic geochemistry in the Mesozoic and Cenozoic formations of Site 534, Leg 76, Blake-Bahama Basin and comparison with Site 391, Leg 44. *In:* SHERIDAN, R.E., GRADSTEIN, F.M. *et al.* (eds), *Init. Repts DSDP* **76.** US Govt. Print. Off., Washington, DC. 481–97.
JANSA, L.F. & WIEDMANN, J. 1982. Mesozoic-Cenozoic development of the Eastern-North American and Northwestern African continental margins: a comparison. *In:* VON RAD, U., HINZ, K., SARNTHEIN, M. & SIEBOLD, E. (eds), *Geology of the Northwestern African Continental Margin.* Springer-Verlag, Berlin. 212–69.
——, GARDNER, J.V. & DEAN, W.E. 1978. Mesozoic sequences of the Central North Atlantic. *In:* LANCELOT, T., SEIBOLD, E. *et al.* (eds), *Init. Repts DSDP* **41.** US Govt. Print. Off., Washington. 991–1031.
——, ENOS, P., TUCHOLKE, B.E., GRADSTEIN, F.M. & SHERIDAN, R.E. 1979. Mesozoic sedimentary formations of the North American Basin, western North Atlantic. *In:* TALWANI, M., HAY, M. & RYAN, W.B.F. (eds), *Deep Drilling Results in the Atlantic Ocean: Continental Margins and Paleoenvironment.* Am. Geophys. Union, Maurice Ewing Series 3. 1–58.
JENKYNS, H.C. & WINTERER, E.L. 1982. Palaeo-oceanography of Mesozoic ribbon radiolarites. *Earth Planet. Sci. Lett.* **60,** 351–75.
LAUGHTON, A.S., BERGGREN, W.A. *et al.* 1972. *Init. Repts. DSDP* **12.** US Govt. Print. Off., Washington, DC.
LUYENDYK, B.P., FORSYTH, D. & PHILLIPS, J.D. 1972. Experimental approach to the paleocirculation of the oceanic surface waters. *Geol. Soc. Am. Bull.* **83,** 2694–64.
MITCHUM, R.M. JR., VAIL, P.R. & SANGREE, J.B. 1977. Seismic stratigraphy and global changes in sea level. Part 6. Stratigraphic interpretation of seismic reflection profiles in depositional sequences. *In:* PAYTON, C. E. (ed.), *Seismic Stratigraphy Application to Hydrocarbon Exploration. Am. Ass. Petrol Geol. Mem.* **26,** 117–33.
NISBET, E.G. & PRICE, I. 1974. Siliceous turbidites: bedded cherts as redeposited ocean ridge-derived sediment. *In:* HSÜ, K.J. & JENKYNS, H.C. (eds), *Pelagic Sediments: On Land and Under the Sea. Spec. Publ. Int. Ass. Sedimentol.* **1,** 351–67.
OGG, J.G., ROBERTSON, A.H.F. & JANSA, L.F. 1983. Jurassic sedimentation history of Site 534 (Western North Atlantic) and of the Atlantic-Tethys seaway. *In:* SHERIDAN, R.E., GRADSTEIN, F.M. *et al* (eds), *Init. Repts DSDP* **76.** US Govt. Print. Off., Washington, DC. 829–85.
ROBERTSON, A.H.F. 1984. Origin of varve-type lamination, graded claystones and limestone-shale 'couplets' in the Lower Cretaceous of the Western North Atlantic. *In:* STOW, D.A.V. & PIPER, D.W.J. (eds), *Fine-grained Sediments: Deep-Water Sediments and Facies. Spec. Publ. Geol. Soc.* **15,** 437–51. Published for the Geological Society of London by Blackwell Scientific Publications, Oxford.
—— & BLIEFNICK, D.M. 1983. Sedimentology and origin of Lower Cretaceous pelagic carbonates and redeposited clastics. Blake-Bahama Formation, Deep Sea Drilling Project Site 534, Western Equatorial Atlantic. *In:* SHERIDAN, R.E., GRADSTEIN, F.M. *et al., Init. Repts DSDP* **76.** US Govt Print. Off., Washington, DC. 795–828.
ROTH, P.H. 1983. Jurassic and Lower Cretaceous calcareous nannofossils in the Western North Atlantic (Site 534): Biostratigraphy, preservation and some observations on biogeography and paleoceanography. *In:* SHERIDAN, R.E., GRADSTEIN, F.M. *et al., Init. Repts. DSDP* **76.** US Govt Print. Off., Washington, DC. 581–7.
SCRUTTON, R.A. & STOW, D.A.V. 1984. Seismic evidence for early Tertiary bottom-current controlled deposition in the Charlie Gibbs fracture zone. *Mar. Geol.* **56,** 325–34.
SHERIDAN, R.E., BATES, L.G. & CROSBY, J.T. 1983a. Seismic stratigraphy in the Blake-Bahama Basin and the origin of Horizon D. *In:* SHERIDAN, R.E. & GRADSTEIN, F.M. *et al.* (eds), *Init. Repts. DSDP* **76.** US Govt. Print. Off., Washington, DC. 667–85.
——, GRADSTEIN, F.M. *et al.* 1983b. *Init. Repts. DSDP* **76.** US Govt Print. Off., Washington, DC. 947pp.
——, GRADSTEIN, F.M. *et al.* 1983c. Site 543: Blake Outer Ridge. *In:* SHERIDAN, R.E. & GRADSTEIN, F.M. *et al., Init. Repts DSDP* **76.** US Govt Print. Off., Washington, DC. 35–141.
SCLATER, J.G., HELLINGER, S. & TAPSCOTT, C. 1977. The paleobathymetry of the Atlantic Ocean from the Jurassic to the present. *J. Geol.* **85,** 509–22.
STEINER, M.B. 1983. Paleomagnetism of Middle Jurassic basalts, Deep Sea Drilling Project leg 76. *In:* SHERIDAN, R.E., GRADSTEIN, F.M. *et al. Init. Repts DSDP* **76.** US Govt Print. Off., Washington, DC. 705–12.
STOW, D.A.V. & HOLBROOK, J.A. 1984. North Atlantic contourites: an overview. *In:* STOW D.A.V. & PIPER D.W.J. (eds), *Fine-Grained Sediments: Deep Water Processes and Facies. Spec. Publ. Geol. Soc.* **15,** 245–57. Published for the Geological Society of London by Blackwell Scientific Publications, Oxford.
—— & PIPER, D.W.J. 1984. Deep-water fine-grained sediments: facies models. *In:* STOW D.A.V. & PIPER D.W.J. (eds), *Fine-Grained Sediments: Deep-Water Processes and Facies. Spec. Publ. Geol. Soc.* **15,** 611–

46. Published for the Geological Society of London by Blackwell Scientific Publications, Oxford.
SUMMERHAYES, C.P. (1986) Organic-rich sediments from the North Atlantic. *In:* BROOKS, J. & FLEET, A.J. (eds), *Marine Petroleum Source Rocks. Spec. Publ. Geol. Soc.* (in press).
—— & MASRAN, T.C. 1983. Organic facies of Cretaceous and Jurassic sediments from Deep Sea Drilling Project Site 534 in the Blake-Bahama Basin, Western North Atlantic. *In:* SHERIDAN, R.E. & GRADSTEIN, F.M. (eds), *Init. Repts DSDP* **76.** US Govt. Print. Off., Washington, DC. 469–81.
THIEDE, J. 1979. History of the North Atlantic Ocean: evolution of an asymmetrical zonal paleo-environment in a latitudinal ocean basin. *In:* TALWANI, M. *et al.* (eds), *Deep Drilling Results in the Atlantic Ocean: Continental Margins and Paleoenvironment.* Am. Geophys. Union, Maurice Ewing Series **3.** 275–97.
TYSON, R.V. 1984. Palynofacies investigation of Callovian (Middle Jurassic) sediments from DSDP Site 534, Blake-Bahama Basin, western central Atlantic. *Mar. Petrol. Geol.* **1,** 3–13.
VAIL, P.R. & TODD, R.G. 1980. Northern North Sea Jurassic unconformities, chronostratigraphy and sea-level changes from seismic stratigraphy. *Proceedings Petroleum Geology of the Continental Shelf of Northwest Europe Conference, London, March 4–7, 1980.*
VAN BUREN, H.M. & MULLINS, H.T. 1983. Seismic stratigraphy and geological development of an open-ocean carbonate slope: the Little Bahama Bank. *In:* SHERIDAN, R.E. & GRADSTEIN, F.M. *et al.* (eds), *Init. Repts DSDP* **76.** US Govt. Print. Off., Washington, DC. 749–63.
VOGT, P.R. 1973. Early events in the opening of the North Atlantic. *In:* TARLING, D.H. & RUNCORN, S.K. (eds), *Implications of Continental Drift to the Earth Sciences (Vol. 2).* Academic Press, New York. 693–712.
WESTERMANN, G.E.C. & RICCARDI, A.C. 1975. Middle Jurassic ammonite distribution and the affinities of the Andean faunas. *Primer. Congr. Geol. Chileno* **1,** 23–39.

A.H.F. ROBERTSON, Department of Geology, West Mains Road, Edinburgh EH9 3JW.
J.G. OGG, Scripps Institution of Oceanography, La Jolla, A–012, California–92093, USA.

Mesozoic palaeoceanography of the North Atlantic and Tethys Oceans

Peter H. Roth

SUMMARY: The biogeography and preservation of calcareous nannofossil assemblages allow mapping of surface water fertility patterns and put constraints on carbonate dissolution in the global oceans of the middle Cretaceous. Cyclic changes in carbonate dissolution and surface water biologic productivity in the Cretaceous oceans were linked to oceanographic and climatic changes that had frequencies close to the Milankovitch cycles (tens of thousands to hundreds of thousand years). Evolution of calcareous nannoplankton during the Jurassic and Cretaceous periods occurred in two primary cycles of about 75 Ma duration that were punctuated by major diversification events, occurring every 15–30 Ma, that marked the beginning of periods of increased marine organic matter production and preservation (stagnation events). Calcareous nannoplankton evolved largely in shallow seas during the Rhaetian to Kimmeridgian. In the latest Jurassic, calcareous nannoplankton conquered the oceanic realm, became major producers of pelagic carbonates, and are thus responsible for the shift of carbonate deposition from shallow seas to the deep ocean. In the late Cretaceous chalk seas, nannofossils replaced shallow water benthos as major carbonate producers. Warm climates, due to high atmospheric CO_2 concentrations and increased albedo during high sea-level stands caused by global tectonic processes, made these tiny creatures flourish during the late Mesozoic.

This paper describes an attempt to combine calcareous nannofossil data and sedimentological and geochemical information to find out something about Mesozoic climates and ocean circulation. The era is unusual having been much warmer than today. Present theoretical models based on today's climate fail to explain palaeocirculation. The author has used samples from deep-sea cores and land sections from the major ocean basins and adjacent epicontinental sedimentary basins. In the first part of this paper the spatial distribution and temporal variation of middle Cretaceous calcareous nannofossils is discussed using mainly samples from deep-sea cores of the Atlantic, Indian and Pacific Oceans to provide information on surface and deep water circulation and its temporal fluctuations. In the second part of this paper the evolution of calcareous nannofossils and pelagic facies during the Jurassic and Cretaceous is traced. The author shows that the evolutionary rates of nannoplankton are controlled by changes in ocean fertility and ventilation.

Cretaceous palaeoceanography

The recovery of Mesozoic oceanic sedimentary rocks during the Deep Sea Drilling Project (DSDP/IPOD) has been small compared to Cenozoic sedimentary rocks; it is adequate for detailed palaeoecological studies for the Cretaceous, but not yet for the Jurassic. Sufficient core coverage for the middle Cretaceous has made detailed quantitative studies of biogeographic patterns and cyclic fluctuations of nannofossils feasible.

Nannofossil biogeography

Quantitative global distribution patterns of calcareous nannofossil assemblages are available for the late Barremian to early Cenomanian (Roth & Bowdler 1981; Roth & Krumbach, in press) and for the Campanian to Maestrichtian (Thierstein 1981). As in the Cenozoic, nannofossil assemblages from the late Cretaceous show zonal distribution patterns indicating that ocean surface temperature was the major controlling factor on the distribution pattern during this time (Fig. 1).

Substantially different biogeographic distribution patterns characterize the middle Cretaceous, where surface water fertility had a major influence (Fig. 2). Biogeographic distribution patterns of about twenty species or groups of closely related taxa were determined for the Atlantic and Indian Oceans for the Aptian to early Cenomanian, which was a time of major deposition of black shales. Factor analyses were performed on the census data for four time slices. Sample coverage is best for the late Albian to lower Cenomanian time interval (about 103–94 Ma), and thus shows the clearest patterns. A high latitude assemblage with *Seribiscutum primitivum* and *Lithastrinus floralis* as the two most important species is found on the Falkland Plateau and

From Summerhayes, C.P. & Shackleton, N.J. (eds), *North Atlantic Palaeoceanography*, Geological Society Special Publication No. 21, pp. 299–320.

in the Indian Ocean at palaeolatitudes of more than 50° S. (Fig. 2). *Seribiscutum primitivum* is also found in the boreal middle Cretaceous of the North Sea region, but is absent from tropical waters during the middle Cretaceous.

Assemblages enriched in *Zygodiscus erectus* and *Biscutum constans* occur along the eastern margin of the North Atlantic off the Iberian Peninsula and North Africa (Fig. 3(a), (b)). These two species are also relatively enriched in the equatorial Pacific (Roth 1981). Assemblages enriched in *Zygodiscus erectus* and *Biscutum constans* are considered indicative of high surface water nutrient concentration, most likely caused

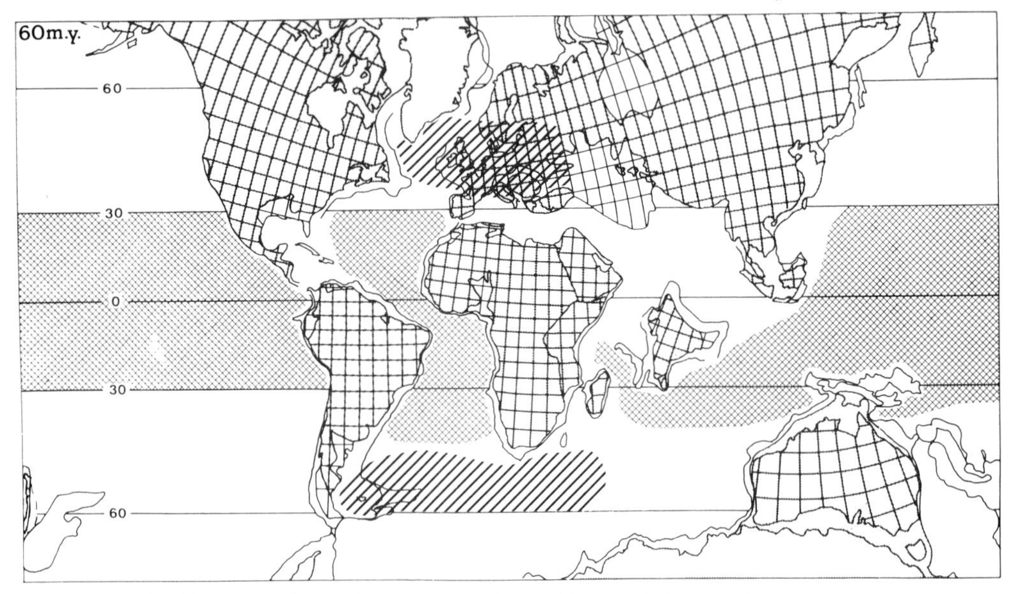

Fig. 1. Nannofossil biogeography for the late Campanian and Maestrichtian, based on Thierstein (1981). Cross-hatched: tropical assemblages. Diagonal line pattern: high latitude assemblages.

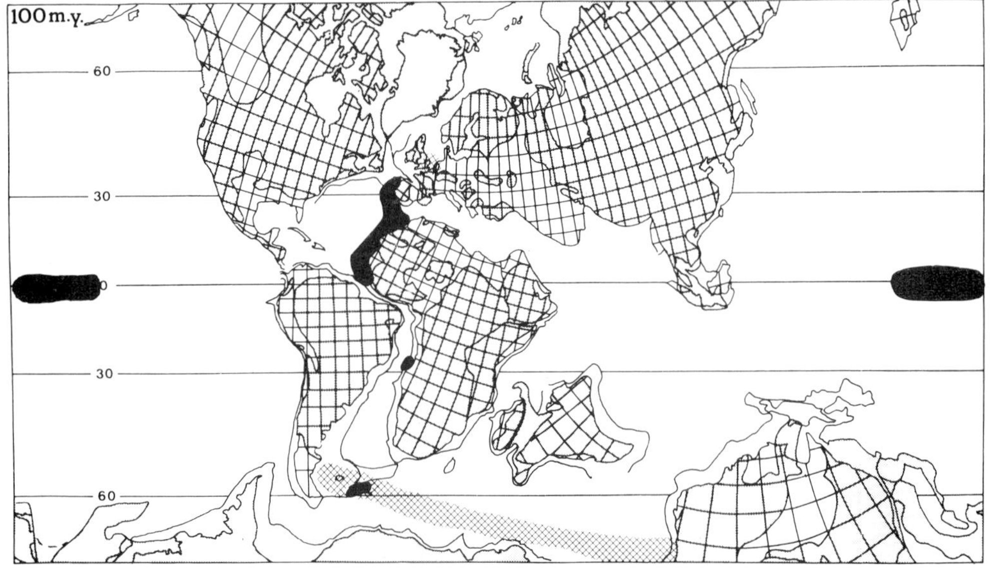

Fig. 2. Biogeography of middle Cretaceous nannofossil assemblages. Cross-hatched: high latitude assemblage with *Seribiscutum primitivum* and abundant *Lithastrinus floralis*. Solid pattern: high fertility assemblages contain common *Zygodiscus erectus* and *Biscutum constans*.

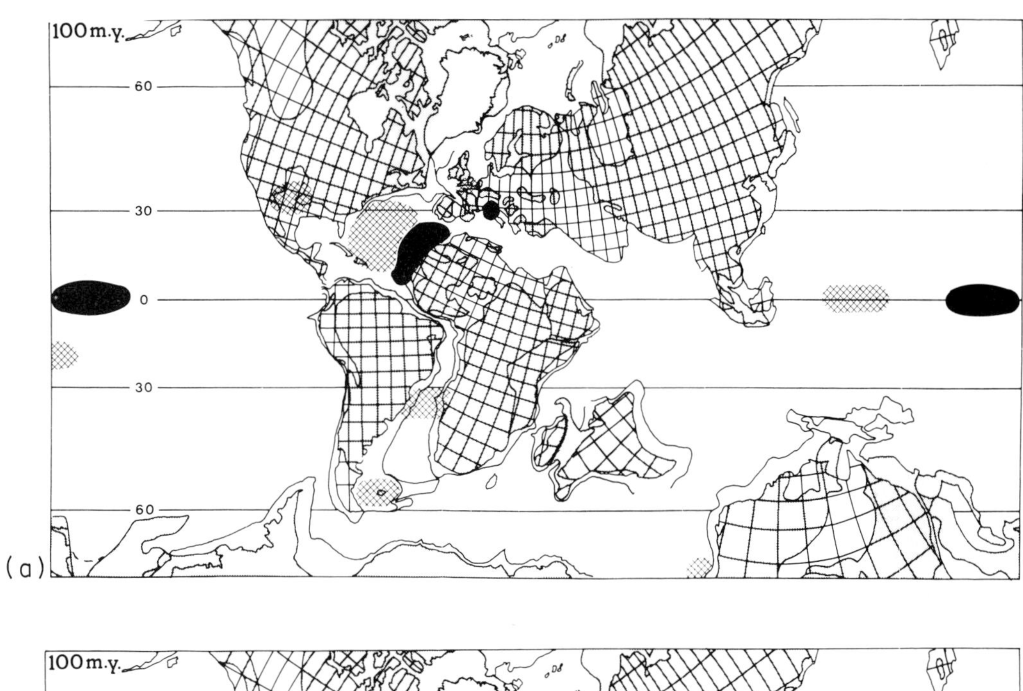

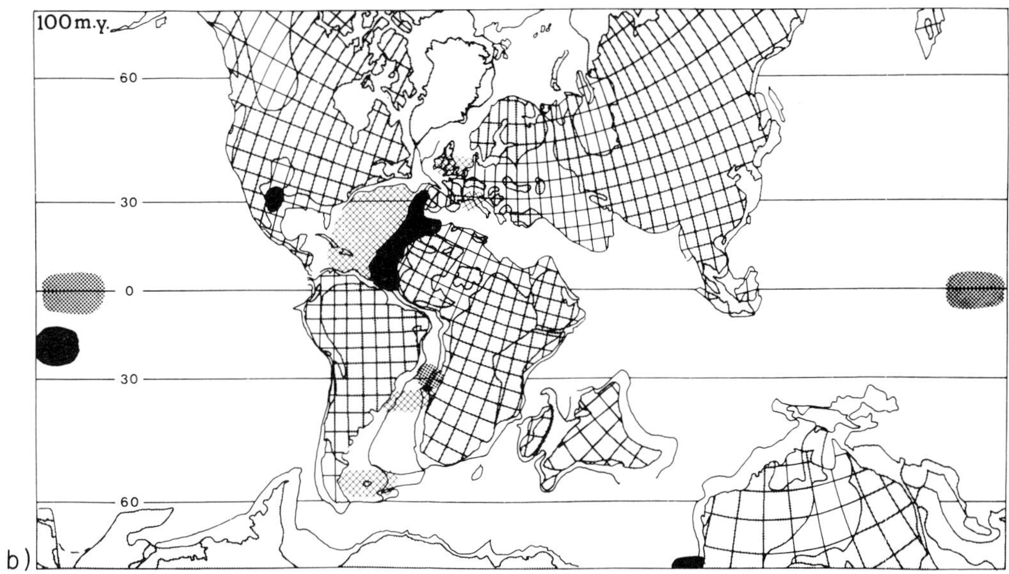

FIG. 3(a). Relative abundance of *Zygodiscus erectus* in the late Albian. Solid pattern: 14–30% of total assemblage. Cross-hatched: less than 14% of total assemblage. (b). Relative abundance of *Biscutum constans* in the late Albian. Coarsely stippled: 35–59% of total assemblage. Solid pattern: 20–30% of total assemblages. Cross-hatched: less than 20% of total assemblage.

by upwelling conditions. Studies on the origin of organic matter by Tissot *et al.* (1979, 1980) and Summerhayes (1981), and empirical climate models by Parrish & Curtis (1982), lend support to this hypothesis.

Oceanic assemblages tend to be dominated by *Watznaueria barnesae* a species that is also among the species that are most resistant to dissolution. A plot of the two surface water fertility indicators versus the oceanic species clearly separates sites of

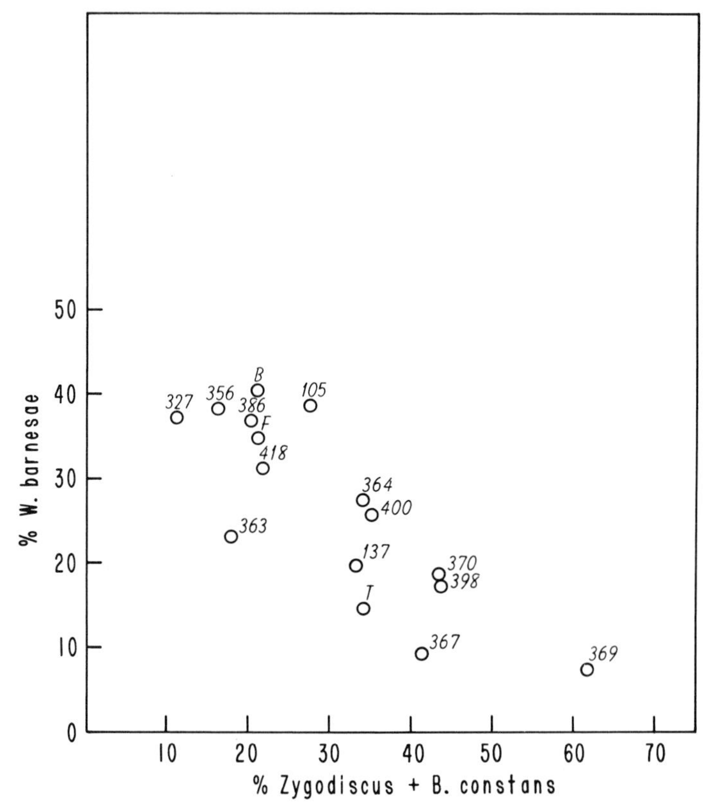

FIG. 4. Plot of the relative abundance of *Watznaueria barnesae* versus the sum of the relative abundances of *Zygodiscus erectus* and *Biscutum constans*. Sites from the eastern North Atlantic (369, 367, 370, 398, 400, 137) and Site 364 in the South Atlantic, both areas of high marine organic productivity, cluster very nicely. Sites underlying low-fertility waters in the western North Atlantic also form well defined clusters. T: Texas; F: France, two samples from land sections.

higher palaeofertility from low fertility sites in the mid-Cretaceous ocean (Fig. 4).

In the shallow margin and epicontinental sea settings *Broinsonia* and *Nannoconus* are most common (Roth & Bowdler 1981; Roth & Krumbach, in press). These two genera are thus considered indicators of neritic conditions.

Biogeographic patterns of mid-Cretaceous calcareous nannofossil assemblages clearly show that temperature gradients were low. High-latitude assemblages are generally restricted to latitudes greater than 45° (Fig. 2). In a wide tropical belt, surface water differences in the concentration of limiting nutrients, rather than temperature, determined the composition of nannofossil assemblages. Calcareous nannofossil assemblages enriched in high fertility indicators can be used to identify coastal upwelling and equatorial divergences (Fig. 5); thus these nannofossil data lend support to model predictions of Parrish & Curtis (1982).

Nannofossil Preservation and Water Mass Structure

Calcareous nannofossil preservation patterns are complex and thus difficult to interpret, because it is difficult to determine if dissolution occurred at the sediment-water interface or within the sediment during oxidation of organic matter, and because of cyclic changes in organic carbon and carbonate production and preservation during the middle Cretaceous. Studies of nannofossil assemblages and stable isotopes (Thierstein & Roth 1980; Roth 1983; Thierstein 1983) indicate that diagenetic alteration of fine-fraction carbonate in middle Cretaceous sediments was considerable in intensity. This is also shown by a positive correlation between the relative abundance of the most dissolution resistant species (*Watznaueria barnesae*) and percent organic carbon in the sediment (Roth & Krumbach, in press). Average relative abundance of the most

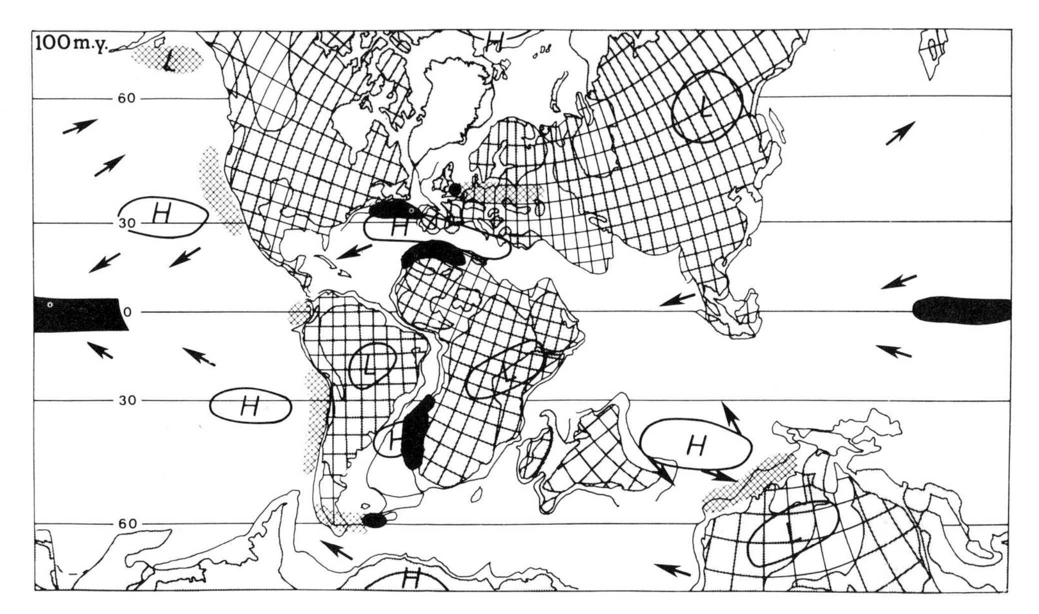

FIG. 5. Atmospheric circulation and upwelling regions predicted by Parrish & Curtis (1982), shown cross hatched, and upwelling regions also predicted and documented by nannofloral patterns (solid pattern), Northern hemisphere winter for the late Albian.

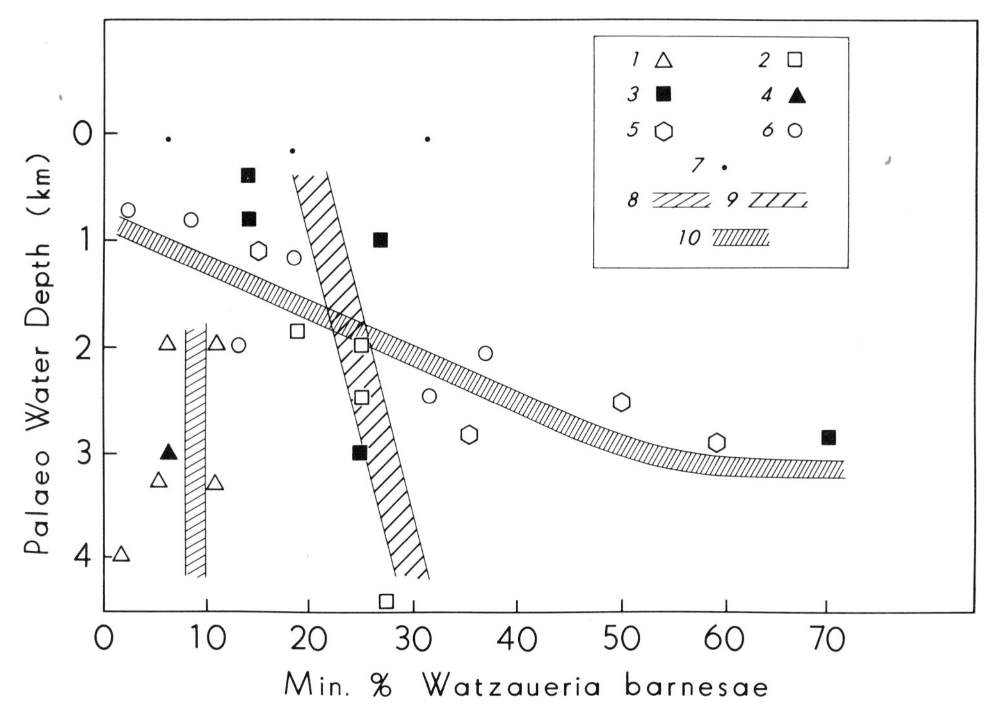

FIG. 6. Minimum abundance of *Watznaueria barnesae* plotted against palaeowater depth for the late Albian of the Atlantic, Indian, and Pacific Oceans. (1) Upwelling samples (DSDP Sites 137, 367, 369, 370, 398; Eastern Atlantic). (2) Low-fertility samples (DSDP Sites 386 and 418, western North Atlantic; DSDP 400 eastern North Atlantic). (3) Poorly preserved South Atlantic samples, Rio Grande-Walvis Ridge (DSDP Sites 327, 356, 363) and poorly preserved deep basin samples (DSDP Sites 364, 531). (4) Sample from DSDP Site 364, Angola Basin, South Atlantic during upwelling conditions. (5) Indian Ocean samples (DSDP Sites 258, 259, 256). DSDP Site 260 has over 90% *Watznaueria barnesae* and thus would fall outside this graph. (6) Pacific Ocean samples (DSDP Sites 305, 306, 363, 464, 465 and 466). (7) Samples from land sections, Texas, Southeastern France and England. (8) Carbonate dissolution trend with depth for high-fertility regions in the Atlantic Ocean. (9) Dissolution trend for low-fertility regions in the Atlantic Ocean. (10) Dissolution trend for the Indian and Pacific Oceans.

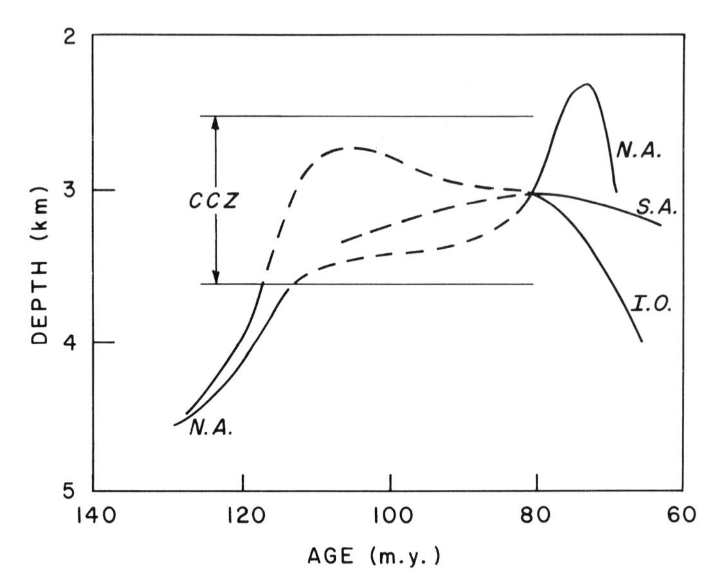

FIG. 7. Calcite compensation depth for the Mesozoic. CCZ: Calcite compensation Zone, a fairly broad transition zone rather than a sharp boundary for the middle Cretaceous. N.A.: North Atlantic. S.A.: South Atlantic. I.O.: Indian Ocean (After Roth & Krumbach, in press).

dissolution resistant species (*Watznaueria barnesae*) for the Albian shows a slight increase with water depth if all sites are combined (Fig. 6). However, if only sites from various oceanic regions that underlie waters of similar organic productivity are compared, this slight increase in dissolution intensity with depth applies only to the Atlantic Ocean (Fig. 6). A much more rapid increase of carbonate dissolution with depth can be inferred from the relative abundance of *Watz-*

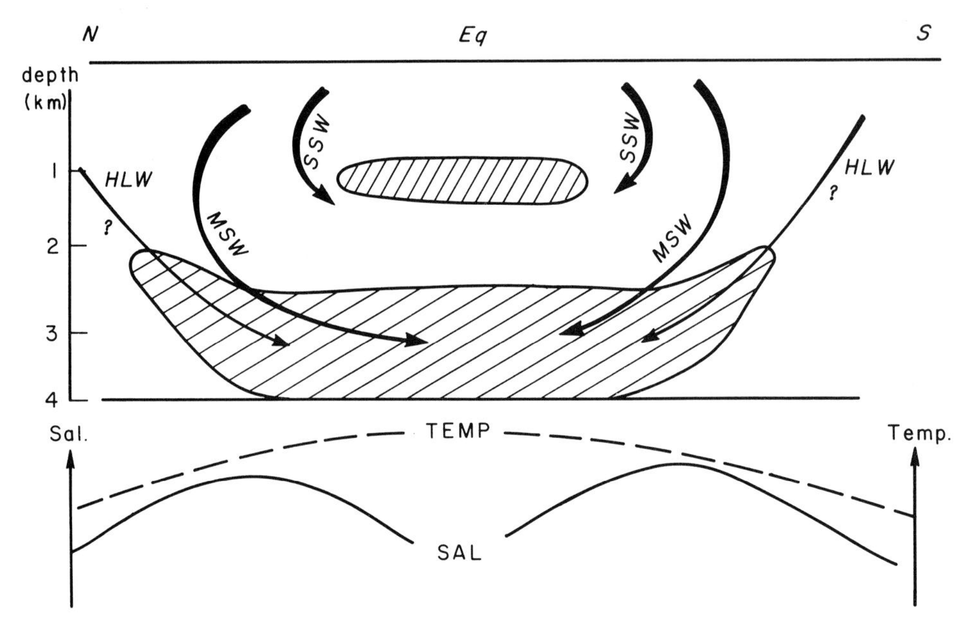

FIG. 8. Model of deep and intermediate circulation in the middle Cretaceous, after Wilde & Berry (1982). HLW: high latitude water. SSW: subtropical saline waters. MSW: marginal sea saline waters. Waters with low dissolved oxygen contents are shown by hatched pattern. A simplified temperature and salinity diagram for surface water is shown on lower graph.

naueria barnesae in the Indian and Pacific Oceans. The lack of a strong dissolution gradient in the Atlantic Ocean compared to the Pacific and Indian Oceans is due to poor deep ocean ventilation, which resulted in the accumulation of increased amounts of detrital and marine organic matter. Processes of downslope sediment transport on the continental slopes and flanks of submarine highs were also important. Calcium carbonate dissolution occurred largely within the sediment, and thus organic carbon concentration and oxygenation of interstitial waters had a greater influence on carbonate dissolution than did water-depth-dependent carbonate-under saturation in the water column. This explains the presence of an indistinct calcite compensation zone, rather than a sharp calcite compensation depth during the middle Cretaceous (Fig. 7). A slight indication of a mid-depth relative abundance peak of *Watznaueria barnesae* (Fig. 6) is perhaps indicative of a mid-water oxygen minimum zone, with increased organic carbon input and thus increased dissolution of carbonate within the sediments of the South Atlantic.

Warm saline waters derived from shallow marginal basins, rather than cold waters from high latitudes, formed the densest waters (Roth 1978; Brass *et al.* 1982). Thus circulation in middle Cretaceous oceans was largely saline-driven, although some high latitude component could have been added to high salinity waters to form the densest bottom waters (Fig. 8) as postulated by Wilde & Berry (1982). Oxygen isotope data by Saltzman & Barron (1982) provide evidence for warm bottom water temperatures in the late Cretaceous South Atlantic Ocean.

Thus, nannofossil preservation patterns lend some support to models of deep water circulation and ventilation.

Cyclic fluctuations

Cyclic alternation of black to green, laminated, organic rich claystones and grey to light green to red, often burrowed and more carbonate-rich layers are widespread in the Barremian to Cenomanian sequences of the Atlantic. Locally, especially in the eastern basin of the North Atlantic, green and red claystones occur. In the underlying Neocomian, white limestones alternate with more organic rich black shales. Organic carbon accumulation rates were fairly high in the Neocomian and increased drastically in the Aptian and Albian (Summerhayes 1986).

Estimates of the periodicity of the lithologic alternations, although poorly constrained, appear to fall into the range of 20 000 to 50 000 years. Various hypotheses have been proposed for the origin of these rhythmic sequences, including climatic changes, changes in terrestrial organic carbon input, turbidity current transport, changing surface water productivity, and changes in redox potential in the sediments (Arthur 1979; McCave 1979; Dean & Gardner 1982; Graciansky *et al.* 1982; Habib 1983; Robertson & Bliefnick 1983).

In order to understand the nature of increased organic carbon sedimentation in the Atlantic during the early and mid-Cretaceous, it is important to consider large-scale regional patterns and the cyclic nature of the rock sequences. Sources, flux rates and burial rates of organic matter need to be determined. Much progress has been made, although there is still some disagreement about the relative importance of terrestrial and marine organic matter (Habib 1979, 1983; Tissot *et al.* 1979 1980; Summerhayes 1981; Simoneit & Stuermer 1982).

Careful sedimentological analyses are helpful in determining modes of deposition, including redeposition by currents and dilution of pelagic components with lithogenous components. Analyses of oxygen and carbon isotope ratios of fine-fraction carbonates put some constraints on temperatures, salinities and productivity of surface waters (Thierstein & Roth 1980). Carbon isotope ratios could possibly reflect high marine organic productivity and upwelling conditions. However, light stable isotope ratios predominantly record diagenetic changes in carbonate dissolution and early diagenetic secondary calcite precipitation. As a result the original palaeoceanographic signals have often been lost (Thierstein 1983).

Surprisingly the overall composition of calcareous nannofossil assemblages is not very much affected unless dissolution or diagenesis is quite severe. A comparison of the fluctuation of the relative abundances of common species of calcareous nannofossils with calcium carbonate percentages in black shale cycles demonstrates that quite clearly (Roth 1983).

Various qualitative and semiquantitative estimates of the amount of carbonate dissolution and its effect on nannofossil assemblages have been made (Roth & Bowdler 1981; Roth & Krumbach, in press). More elaborate estimates of nannofossil preservation, such as dissolution rankings and indices, have not proved to be very informative for the middle Cretaceous. Total species diversity, or the relative abundance of *Watznaueria barnesae*, are the best measure of coccolith preservation (Roth & Bowdler 1981; Roth & Krumbach, in press). Any assemblage that contains over 40% *Watznaueria barnesae* has been greatly affected by dissolution, and primary palaeoenvironmental

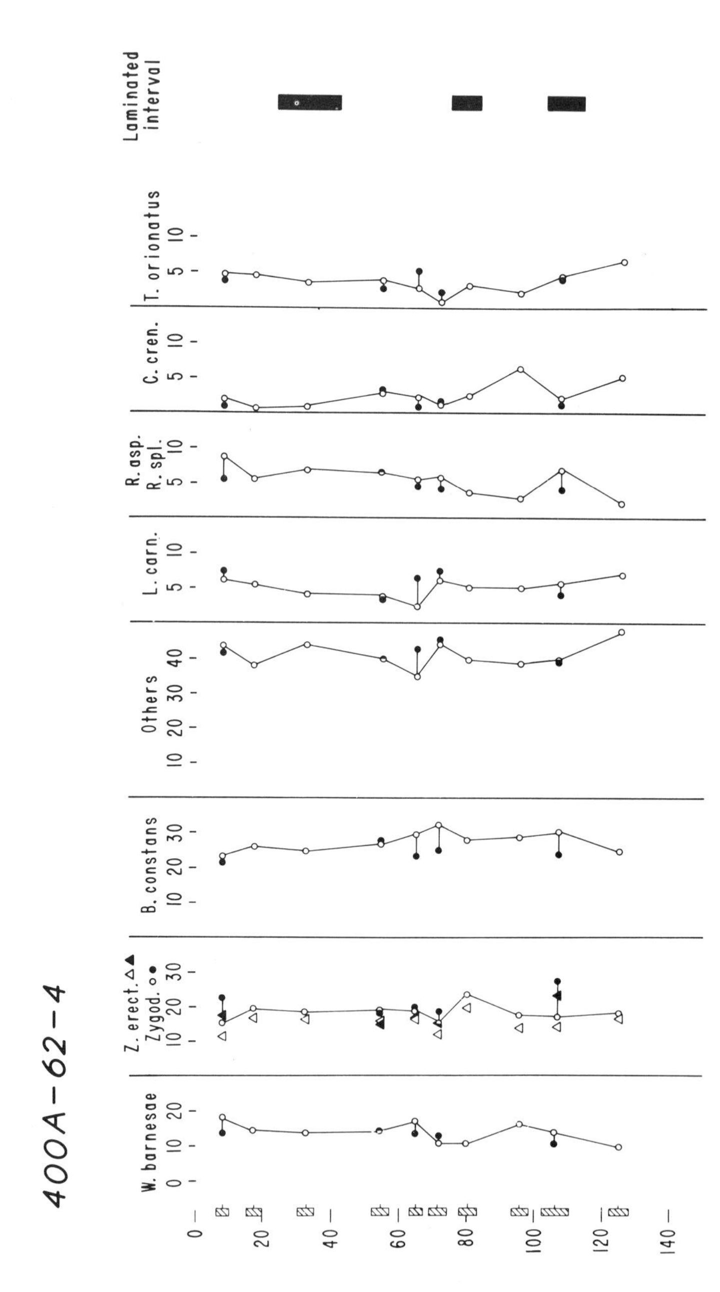

FIG. 9. Relative abundances of important nannofossil species from cyclic black shale section at DSDP Site 400, late Albian. Black bars: laminated organic-rich intervals, cross-hatched bars: sampling intervals.

signals recorded by such assemblages have been greatly distorted. Careful observation of the preservation of individual coccoliths to determine the degree of etching and secondary calcite overgrowth, using methods introduced by Roth (1973), also help to determine the degree of secondary alteration of coccolith assemblages, and allow the exclusion of poorly preserved assemblages.

Biogeographic distribution patterns clearly show that high relative abundances of *Zygodiscus erectus* and *Biscutum constans* are indicative of increased surface water productivity (Fig 3(a), (b)).

To demonstrate temporal changes in palaeoproductivity and palaeocirculation, a nannofossil census was conducted on three rhythmic black shale sequences of late Albian to Aptian age from the Atlantic, and on a homogenous carbonate sequence without much organic carbon from the equatorial Pacific.

A sequence of upper Albian grey marly nannofossil chalks and carbonaceous chalks from DSDP Site 400 contains well preserved coccoliths throughout (Fig. 9). Calcium carbonate contents do not fluctuate very much (41–51%), nor do organic carbon contents (0.4–1.1%). Entire coccoliths far outweigh fragments (about 80% whole coccoliths). *Watznaueria barnesae* is most abundant between the laminated intervals. The two high fertility indicators (*Zygodiscus erectus* and *Biscutum constans*) show peak abundances in laminated intervals. Cyclic changes in surface water organic productivity, caused by more intensive upwelling during the formation of these black shale intervals, best explain the observed nannofossil abundance patterns.

The next example is an early Aptian sequence with distinctive laminated intervals from DSDP Site 367 (Fig. 10). Both calcium carbonate (41–90%) and organic carbon (0–4.7%) are more variable than at Site 400. Relatively constant abundances of *Watznaueria barnesae* indicate only moderate fluctuations in carbonate dissolution. The coincidence of peak abundances of *Zygodiscus erectus* and *Biscutum constans* with laminated intervals and high organic carbon contents is indicative of fluctuations in surface water productivity. Large fluctuations in the relative abundance of *Nannoconus* are difficult to explain because they have not been observed in other cyclic sequences. Redeposition of this neritic species by periodic currents from a shallower source appears most likely.

Figure 11 shows an example of dissolution-dominated upper Albian black shale cycles from DSDP Site 364 in the South Atlantic. Calcium carbonate fluctuates from 0–46% and organic carbon from 0.2–9%. Moderate to severe etching has affected the coccoliths. Coccoliths are totally absent in black laminated layers. If barren layers are ignored, relative abundances of *Watznaueria barnesae*, *Zygodiscus* spp. and *Biscutum constans* are fairly constant. Fluctuating dissolution, most likely within the sediment during break-down of organic matter, destroyed any primary palaeoceanographic signal that calcareous nannofossil assemblages might have originally recorded.

An upper Albian sequence of nannofossil limestones and chalk from Site 466 in the central Pacific, (which lay close to the equator at that time) shows the effects of changes in surface water fertility even in rocks that do not display sedimentary rhythms (Fig. 12). Carbonate contents are high (71–81%) and organic carbon contents are very low. Nannofossil preservation is good to moderate. Although the studied interval is short, a peak of *Zygodiscus* spp. and *Biscutum constans* in the middle of the studied interval is interpreted as an indication of increased surface water productivity (Fig. 12). Higher relative abundances of *Watznaueria barnesae* at the top and at the base of the studied interval are caused by increased dissolution.

The middle Cretaceous cycles discussed here clearly show fluctuations in intensity of dissolution and secondary calcite precipitation. Neither process was sufficiently strong to destroy a primary surface water fertility signal except at Site 364. Peak abundances of high fertility species (*Zygodiscus* spp., *Biscutum constans*) occur in laminated intervals in the upwelling regions in the eastern North Atlantic. In the South Atlantic, dissolution was largely responsible for fluctuations in relative abundances of nannofossil species. In the Pacific, dissolution intensity also fluctuated, but a fertility signal is preserved at Site 466.

Laminated layers generally were produced during periods of increased surface water fertility, although the overall fertility of middle Cretaceous oceans was probably rather low (Bralower & Thierstein 1984). Dissolution and diagenesis generally have only a minor effect on the composition of nannofossil assemblages.

Circulation of the mid-Cretaceous Atlantic

The North Atlantic had reached over half its present width by middle Cretaceous time. A central gyre with upwelling along the eastern margin had probably developed by that time (Fig. 13). Deep-water connections to the north and south did not exist until the end of the middle Cretaceous. It is generally accepted that the connection between the North Atlantic and the

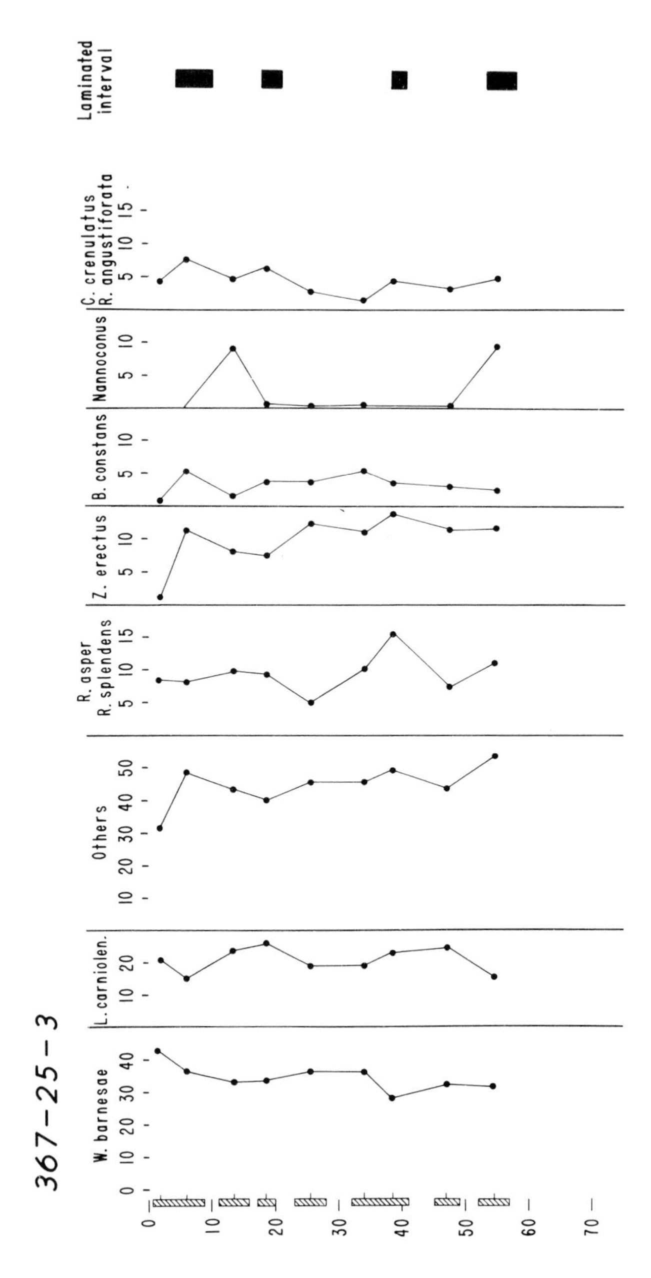

FIG. 10. Relative abundance of nannofossil species in cyclic sediments at DSDP Site 367, late Barremian. Black bars: laminated organic-rich intervals. Cross-hatched bars: sampling intervals.

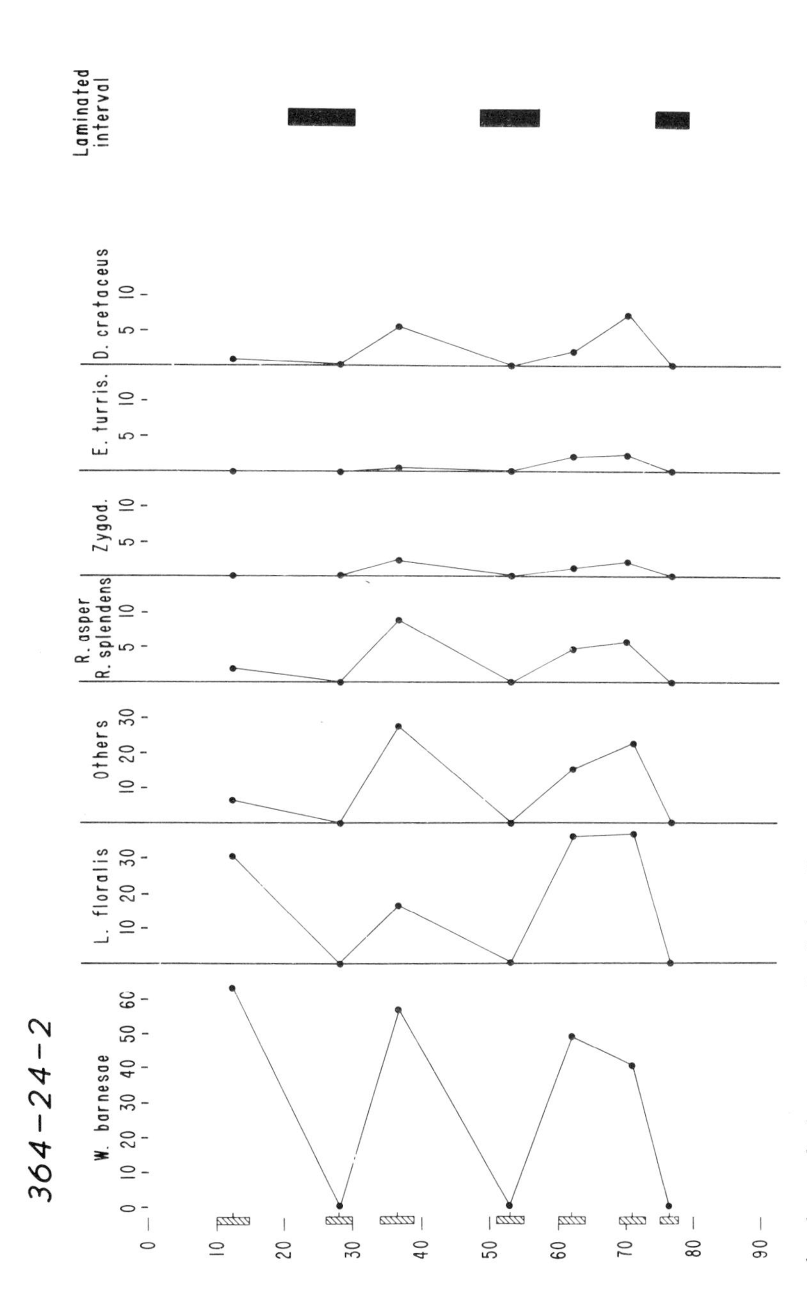

FIG. 11. Relative abundance of calcareous nannofossils in cyclic sediments of late Albian age, DSDP Site 364. Angola Basin, South Atlantic. Black bars: laminated organic-rich intervals. Cross-hatched bars: sampling intervals.

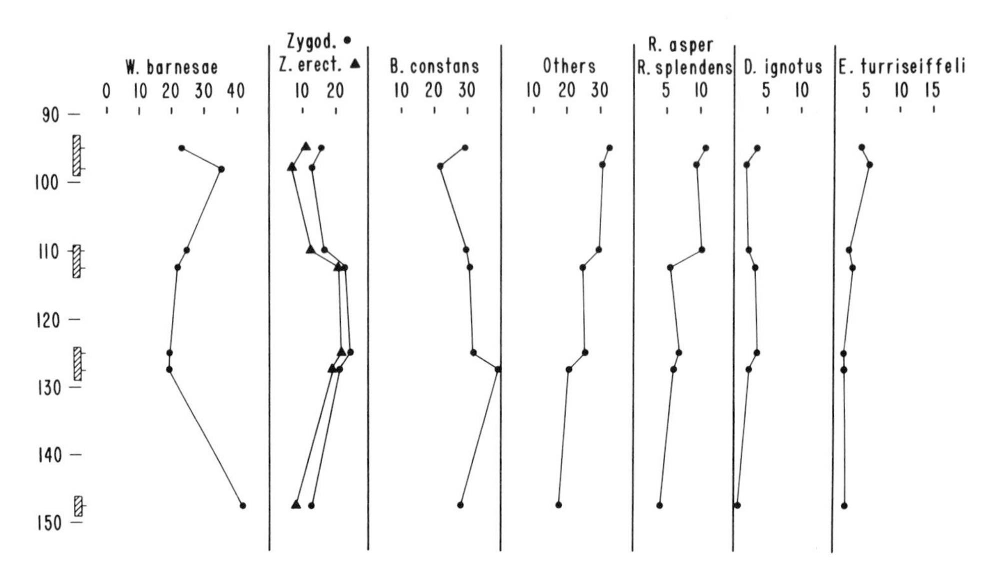

FIG. 12. Relative abundance of calcareous nannofossil species in a chalk section of late Albian age, DSDP Site 466, Central Pacific. Cross-hatched bars: sampling interval.

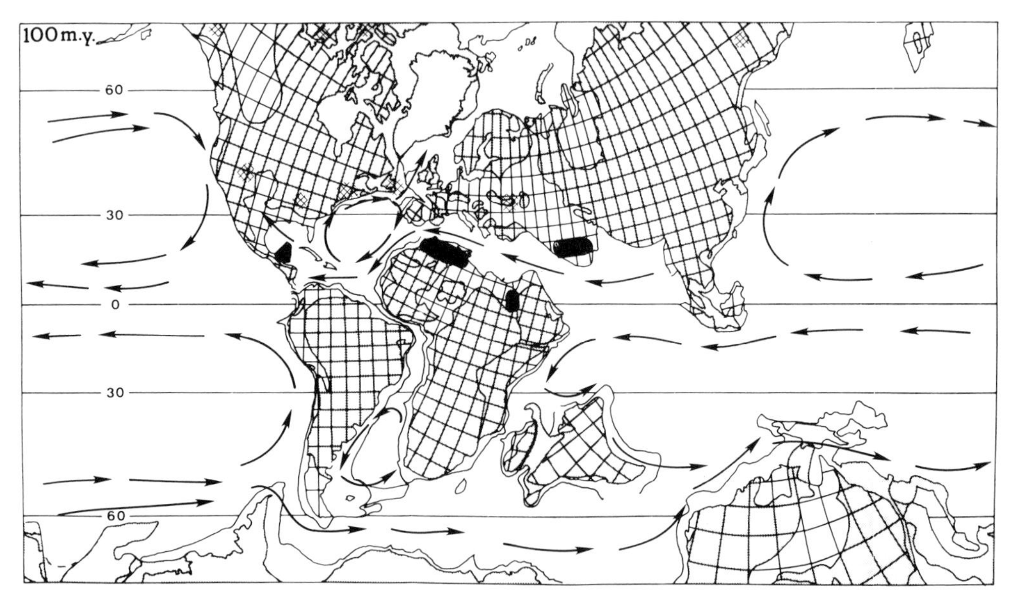

FIG. 13. Surface currents in middle Cretaceous oceans. Distribution of evaporite basins, marked by a solid pattern, may be the source for warm saline bottom waters. Coal deposits, shown by a cross-hatched pattern, may be indicative of high rainfall at mid and high latitudes. (After Roth & Krumbach, in press.)

Pacific oceans through the Caribbean and across Central America was an open seaway of at least several hundred metres depth that would allow exchange of surface waters during high and low sea-level stands and intermediate waters during high sea-level stands (Summerhayes & Masran 1983; Summerhayes, in press). The North Atlantic was surrounded by large continental land masses and received much river inflow (Ryan & Cita 1977; Jenkyns 1980). It is thus entirely possible that surface waters had slightly lower salinities than the tropical Pacific, and that the Atlantic had a estuarine circulation, much like the present-day North Pacific (Berger 1970), with surface water outflow to the Pacific and intermediate water inflow from the Pacific. Deep and bottom waters were derived from tropical to subtropical marginal evaporitic basins and shallow reef-rimmed basins in the Gulf of Mexico–Caribbean–Florida Platform region, northern South America, Tethys-western and southern Europe (Fig. 13; see also Roth 1978; Roth & Krumbach, in press). Some admixture of colder waters seems likely if polar temperature were as low as 5° C to 10° C, and if polar waters had salinities greater than 33.4‰. Such high-latitude surface waters could be as dense as lower latitude surface waters in shallow seas with temperatures of 15 to 20° C and salinities of 36 to 37‰ (Wilde & Berry 1982). However, the author feels that the contribution of cold high latitude waters to the Atlantic would have been very small at best, because of poor connections with boreal seas. Furthermore, high rainfalls at higher latitudes, attested by widespread coal deposits (Fig. 13) would have resulted in low salinity surface waters at high latitudes. Bottom water temperatures based on oxygen isotope ratios in benthic foraminifera would preclude surface temperatures in polar seas much below 15° C (Savin 1977). Thus, tropical warm saline waters were the major source of deep and bottom waters in the Atlantic (Roth 1978; Saltzman & Barron 1982).

During the mid-Cretaceous stagnation event, the North Atlantic seems to have been an ocean basin with poor deep connection to other ocean basins, relatively low surface water salinities, and high surface water temperatures, surrounded by shallow tropical marginal seas with high salinity waters. These marginal basins may have furnished warm high salinity waters that filled the deep ocean basins. Overall marine organic productivity was probably low (Bralower & Thierstein 1986) because the stability of the water column was high. The oxidation of falling marine organic particles and detrital organic matter used up most of the available oxygen, so that bottom waters had very low oxygen concentrations. Influx of large amounts of waters from an oxygen minimum layer at intermediate depths in the Pacific across the Central American Sill during high sea-level stands in the Barremian to early Aptian would explain low oxygen concentrations at intermediate depths in the North Atlantic (Summerhayes & Masran 1983). Reduced influx of intermediate water during the low sea-level stand in the late Aptian, lower evaporation rate in the tropics and thus a lowered flux of warm saline bottom waters to the bottom of the Atlantic basin could have been responsible for more oxidizing conditions during that time period.

Finally, during the Cenomanian, the influx of warm salty water from the South Atlantic (Thierstein & Berger 1978) led to density stratification in the North Atlantic, and resulted in exceptional preservation of organic matter. Wide-spread shallow seas during the initial phase of the great late Cretaceous transgression contributed additional warm saline bottom water. Deposition of large quantities of pelagic carbonates in the vast expanses of the late Cretaceous epicontinental chalk-seas resulted in a shallow calcite compensation depth in the North Atlantic during the late Cretaceous, and overall low sedimentation of biogenous pelagic sediments in the world oceans.

During the initial phases of the late Cretaceous transgression, many shallow depressions of the flooded craton contained pools of anoxic waters that led to the preservation of organic-rich sediments (e.g. the Black Band in Cenomanian chalk of England and Northern France). During the Maestrichtian regression, sedimentation in shelf seas, and thus sequestering of nutrients and calcium carbonate, were globally reduced. The oceans became more productive and the CCD dropped rapidly in all oceans from a depth of about 3 km to 5 km. A general cooling trend during the late Cretaceous caused more vigorous stirring of the oceans and thus more efficient ventilation and nutrient recycling.

During the mid-Cretaceous, the South Atlantic was still a narrow sea enclosed by arid continents, surrounded by evaporite basins and divided into silled basins by the Rio Grande-Walvis Ridge and the Falkland Plateau. The Angola and Cape basins did not exchange much deep and intermediate waters with the world's oceans until (1) connections with the North Atlantic were established, (2) the Falkland Plateau cleared the southern tip of Africa, and (3) the Rio Grande-Walvis Ridge subsided to greater depths. Poor ventilation, especially of the Angola Basin, persisted well into the late Cretaceous. By Campanian time the entire Atlantic had become a better stirred and ventilated ocean.

Superimposed on these large cycles of Cretaceous ocean ventilation, with frequencies in the millions of years, were higher frequency cycles with periods of roughly tens of thousands to one hundred thousand years. These short term cycles are probably responses to climatic changes caused by astronomical forcing functions, as proposed by Milankovitch for the Pleistocene. Such climatic fluctuations affected rates of ocean mixing (and thus biological productivity), rates of warm saline bottom water production, and (most likely) rainfall and fresh-water inflow. These fluctuations account for the formation of mid-Cretaceous black shale cycles, and of the rhythmic sedimentation of late Cretaceous epicontinental chalks.

The late Jurassic carbonate shift and nannofossil evolution

Much of the sea floor older than 120 million years has been subducted or obducted to form structurally complex mountain ranges. To analyse this period it is necessary to obtain information from sections deposited in the ever shifting epicontinental seas (now exposed on land) and to analyse facies patterns from mountain ranges thrust up from the ancient Tethys. Nannofossil data are not precise, derived mostly from a few biostratigraphic studies. Taxonomic problems are vexing, as few researchers have studied these small early calcareous nannofossils. Nevertheless trends in nannofossil production and evolution in the latest Triassic, Jurassic and Cretaceous can be determined, and combined with major facies patterns to deduce palaeoceanographic conditions. This analysis is the basis for speculations on the driving mechanism responsible for the evolution both of the calcareous plankton and the ocean circulation and climate. Spatial resolution is still poor, and temporal resolution limited to millions rather than thousands of years, because of increasing errors in radiometric ages with increasing age.

Pelagic facies patterns

Sediment mass balance calculations for the Phanerozoic show that in mid-Mesozoic time the major sink for calcium carbonate shifted from shallow epicontinental seas to the deep-sea (Kuenen 1941). The author compiled some of the more important and best described sedimentary sections of late Triassic to Maestrichtian age from the Atlantic Ocean, the Tethys, the Indopacific, and the Central American regions in order to determine the timing and cause of this carbonate shift (Fig. 14). Shallow water carbonates, evaporites, and terrestrial or shallow marine siliciclastics dominate the Triassic and lower Jurassic in the region under consideration. Some deep-water micritic limestones formed mostly on submarine highs and benches in small ocean basins that were surrounded by carbonate platforms and thus shielded from much river-born siliciclastic deritus. These fine-grained carbonates often contain *Schizosphaerella*, a nannolith of unknown affinity, and *Bositra*, a planktonic bivalve. Much of the fine carbonate mud was contributed by the shallow platforms and transported by currents. Radiolarian ribbon cherts were widespread in the middle Jurassic of the Mediterranean, and also occurred in the Middle East and the Indopacific Region. These cherts formed in small ocean basins with high fertility, elevated calcite compensation depth, and low terrigenous input (Jenkyns & Winterer 1982; Baltuck 1982, 1983). Where such conditions were met, as in Turkey and Timor, the deposition of radiolarites began before the middle Jurassic and persisted until the late Cretaceous. The presence of these cherts is a reflection of the geometry and bathymetry of the sedimentary basins and surrounding areas, and of fertility. It does not show globally elevated radiolarian productivity.

FIG. 14. Mesozoic pelagic sections. North Atlantic-Tethys-Indopacific-Central America, Northern Europe. Section numbers: 1: DSDP Site 534. 2: DSDP Site 100. 3: DSDP Site 105. 4: DSDP Site 387. 5: DSDP Site 446. 6: DSDP Site 547. 7: DSDP Site 367. 8: DSDP Site 330. 9: Scotian Margin, Georges Bank. 10: Essaouira/ Agadir Basin, Morocco. 11: Tarfaya Basin, Morocco. 12: Subbetic zone, SE Spain. 13: Tunisia. 14: Sicily. 15: Apennines, T. Bosso, Ital. 16: Lombardy Basin, Italy. 17; Salzburg Basin, Austria. 18: Briançonais, French Alps. 19: Piedmont, Southern Alps. 20: Pennine nappes, Swiss Alps. 21: Pienini Klippen, Polish Carpathians. 22: Western Greece. 23: Pelagonia/Argolis. 24: Cyprus. 25: SW Turkey. 26: Syria. 27: Oman. 28: DSDP Site 261, off W Australia. 29: Timor. 30: DSDP Site 267, Magellan Plateau. 31: DSDP 305/306. 32: Sierra Madre Oriental, Mexico. 33: Cuba. 34: England/N France. From many different sources.

Explanation of lithologic symbols: A: bedded cherts, radiolarites, ribbon cherts. B: common occurrence of the nannolith *Schizosphaerella punctulata*. C: coarse siliciclastics, including sandstones, conglomerates and breccias. D: sandstones. E: black shales (laminated, enriched in organic carbon). F: pelagic clays, rich in zeolites (upper left) or volcanic ash (lower right). G: shales and clays. H: silts and muds. I: marls. J: marly and other impure fine-grained limestones. K: nodular pelagic limestones. L: chert nodules (small nodules, upper left; large nodules and slabs, lower right). M: fine-grained limestones, micrites, calcareous oozes and chalks. N: sandy limestones. O: shallow water limestones, bioclastic, biohermal, or oolitic. P: dolostones. Q: evaporites. R: volcanics.

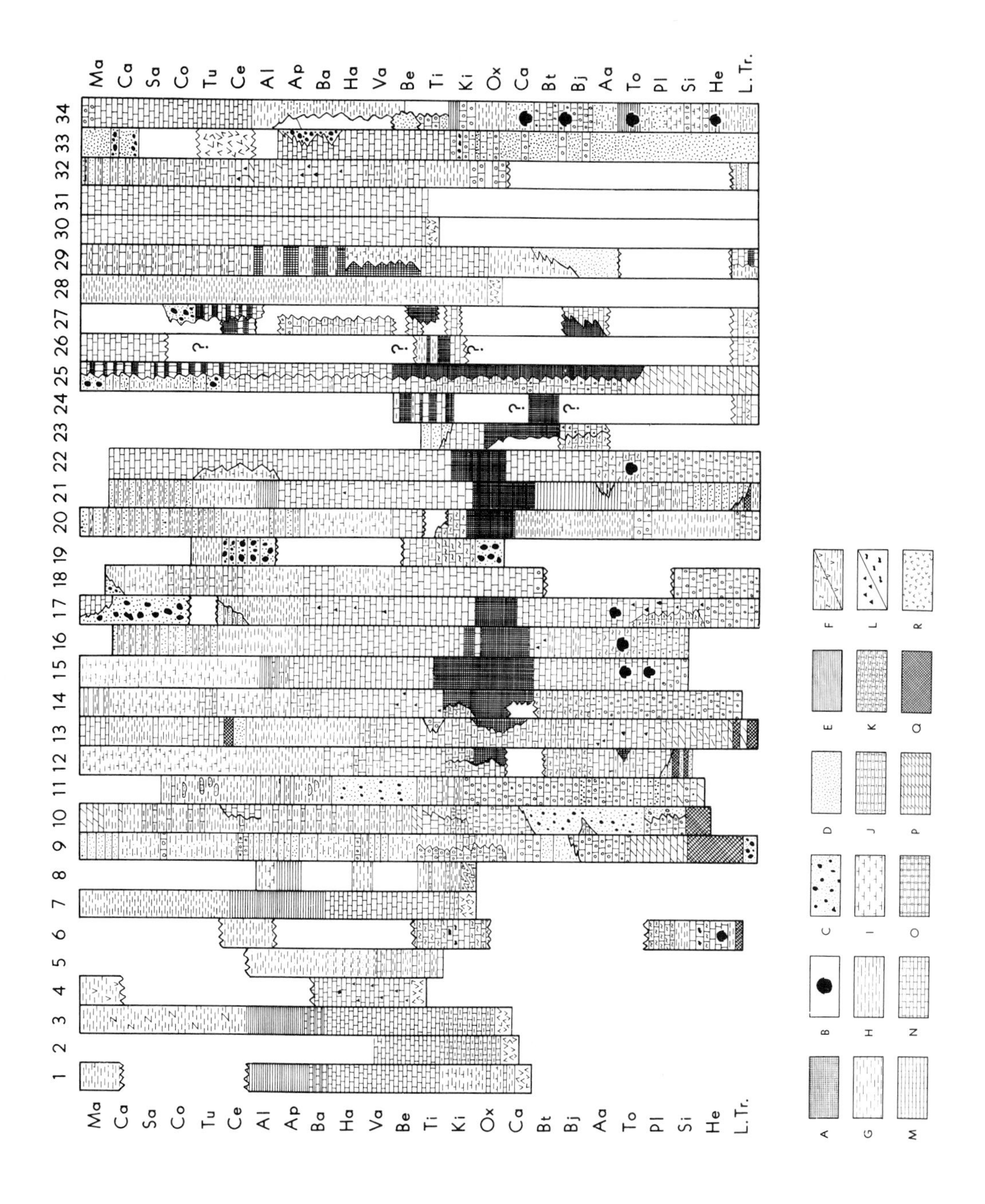
Ma
Ca
Sa
Co
Tu
Ce
Al
Ap
Ba
Ha
Va
Be
Ti
Ki
Ox
Ca
Bt
Bj
Aa
To
Pl
Si
He
L.Tr.
1 2 3 4 5 6 7 8 9 10 11 12 13 14 15 16 17 18 19 20 21 22 23 24 25 26 27 28 29 30 31 32 33 34
A B C D E F
G H I J K L
M N O P Q R

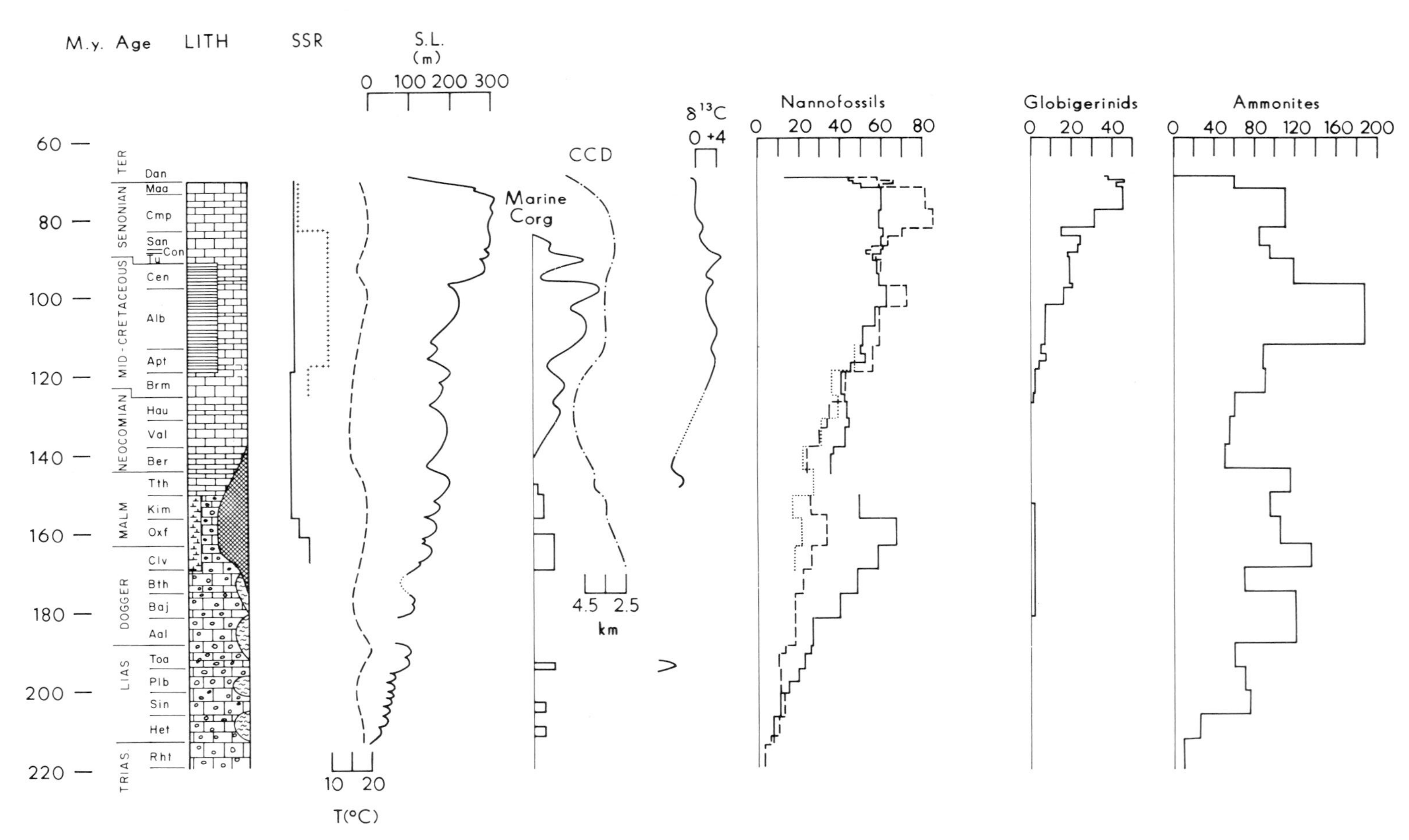

FIG. 15. Mesozoic facies, sea floor spreading rates (SSR). T: palaeotemperature estimates, sea level (S. L.). Marine organic carbon input (Corg.), CCD, ^{13}C. Nannofossil diversity. Jurassic: solid line (Medd 1982); dashed (Hamilton 1983); dotted (Roth 1983). Cretaceous: solid line (Thierstein, pers. comm. 1984); dashed (Perch-Nielsen, in Thierstein 1983). Globigerinid diversity (Thierstein 1983) and ammonite diversity (Kennedy 1977). For lithologic symbols, see Fig. 14.

In the late Jurassic, about late Kimmeridgian to early Tithonian time, there was drastic change in facies (Fig. 14). Ribbon cherts, platform carbonates, or hemipelagic rocks gave way to nannofossil limestones with nannoconids and planktonic crinoids (*Saccocoma*). This facies persisted into the Neocomian but increasing amounts of organic carbon became incorporated, especially in the North Atlantic and western Tethys (Maiolica Facies). This is an indication of decreasing ventilation and increasing stability of the North Atlantic and Western Tethys, culminating in the middle Cretaceous Black Shale Event. The CCD shallowed, and thus carbonate sedimentation decreased during the late Barremian to Cenomainan. The beginning of the Alpine Orogeny in the late Cretaceous led to increased siliciclastic detrital sedimentation in the western Tethys Region. The Pacific Ocean was affected only slightly by these events. Pelagic carbonate sedimentation persisted in the Pacific from the late Jurassic to the end of the Cretaceous except in the deepest basins where pelagic clays with variable admixures of biogenic silica accumulated. During the late Cretaceous, pelagic carbonate sedimentation spread to the epicontinental seas of the European platform, the Middle East and North America. The CCD remained shallow in all ocean basins throughout the late Cretaceous, because calcium carbonate was sequestered in shallow seas, rather than deposited in deep ocean basins. Open-ocean primary productivity was low and carbonate dissolution rates were high.

Evolution of calcareous plankton and ocean ventilation

The geographically widespread carbonate shift in the late Jurassic calls for a global rather than regional explanation. Although planktonic foraminifera originated in the middle Jurassic, they did not become major sediment producers until the Neogene. Thus, it must be the evolution of calcareous nannoplankton and related calcareous algae that was of crucial importance for pelagic carbonate deposition (Fig. 15). A compilation of the best available data on calcareous nannofossil assemblages (Medd 1982; Hamilton 1983) shows some ambiguities but reveals general trends (Fig. 15). Cretaceous nannofossil assemblages are better known and diversity trends are thus more firmly established.

Calcareous nannofossils originated in epicontinental seas during the latest Triassic (Rhaetian). Standing diversity and production of Rhaetian and lower Liassic calcareous nannofosils were low. Small increases in standing diversity occurred in the Hettangian and the Sinemurian. Evolutionary rates increased in the Toarcian. Black shales rich in marine organic matter accumulated in shallow epicontinental seas during all these three Liassic stages but are particularly well developed in the Toarcian (Hallam 1975). A shift to heavier carbon isotope ratios lead Jenkyns (1983) to propose a Jurassic oceanic anoxic event (OAE). Stable isotope data for the Hettangian and Sinemurian are not available to determine if earlier global anoxic events occurred during these stages. Trends of standing diversity of calcareous nannoplankton in the middle Jurassic (Dogger) are ambiguous. Data of Medd (1982) support a significant evolutionary event in the late Aalenian to Bajocian which is not apparent in the data of Hamilton (1983). However, both sets of data show that evolutionary rates of calcareous nannofossils increased significantly in the middle part of the Dogger. The author suspects that the Bajocian evolutionary event among calcareous nannofossils will prove to be significant when more information becomes available. Like other anoxic events it coincided with a major high sea-level stand (Hallam 1981; Vail *et al.* 1984).

The next diversification event coincides with the more pronounced Callovo-Oxfordian OAE (see Fig. 16, and Arthur 1982; Holser 1984). In the Anglo-Paris Basin nannofossil diversity peaked in the Oxfordian and declined in the Kimmeridgian and Portlandian (Tithonian). The terminal Jurassic regression in the North Sea Region lead to shallow, often hypersaline conditions that are poorly tolerated by calcareous nannoplankton. Oceanic nannofossils show a peak in standing diversity in the Tithonian with assemblages characterized by the appearance of the first nannoconids and other nannofossil species that flourished in the early Cretaceous (Roth 1983). Whereas calcareous nannofossil diversity increased steadily during the middle Jurassic and the early part of the late Jurassic, nannofossil production was low; this is shown by the scarcity of pure pelagic carbonates in sequences of this age. A fairly abrupt shift from impure limestones, marls, or cherts to pure nannofossil carbonates in the uppermost Kimmeridgian to Tithonian marks the migration of calcareous nannoplankton from shelf and epicontinental seas to the open ocean (Fig. 16). The production of carbonate by calcareous nannofossils was sufficiently high in the pelagic realm during the latest Jurassic to depress the CCD in the Tethyan Ocean (Fig. 15). High phytoplankton productivity in the boreal seas of northern Eurasia resulted in the accumulation of oil shales (e.g. Kimmeridgian of the North Sea region, and Volgian of the Russian platform). Thus, the boreal seas of Eurasia were poorly

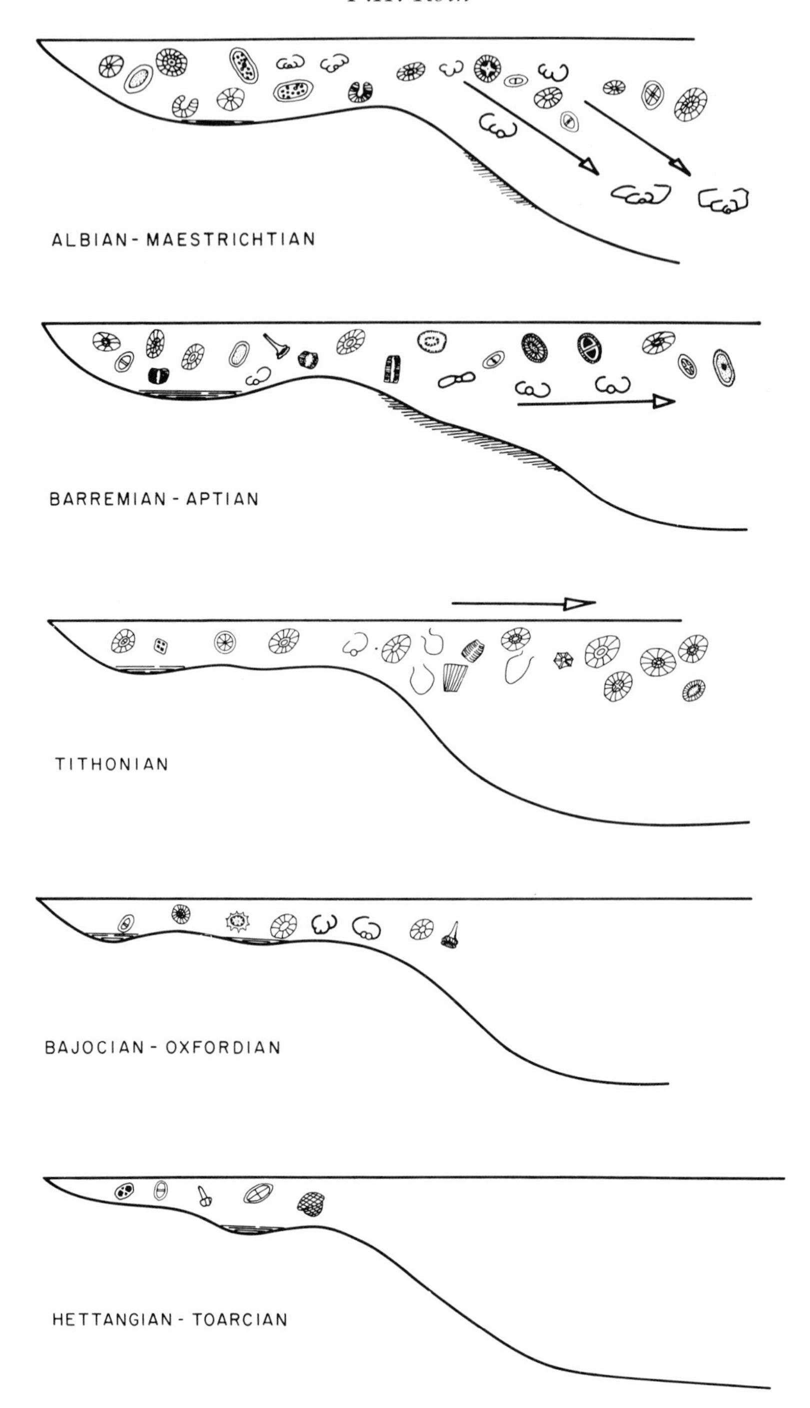

FIG. 16. Model of Jurassic and Cretaceous evolution of calcareous plankton. Hettangian to Toarcian. Nannofossils mostly in shelf seas; diversity low. Bajocian to Oxfordian: nannoplankton still neritic, first globigerinids. Tithonian: invasion of oceanic realm by calcareous nannofossils, first nannoconids, common calpionellids; globigerinids rare and small. Barremian to Aptian: diverse nannofossil assemblages. Globigerinids migrate to open ocean. Albian-Maestrichtian: planktonic foraminifera select different depth habitats.

oxygenated and probably well stratified while the deeper parts of the North Atlantic remained largely oxygenated (Ogg *et al.* 1983). A small but well-defined positive excursion in stable carbon isotope ratios in the upper Tithonian of northern Mexico (Scholle & Arthur 1980) and the occurrence of major upper Jurassic phosphate deposits (Cook 1984) indicate that shallow and intermediate waters in the oceans may have been poorly oxygenated while deep ocean basins were well ventilated.

The end of the Jurassic is marked by a slight decrease in nannofossil diversity in the oceanic realm. Neither extinction rates, nor origination rates of calcareous nannoplankton were particularly high across the Jurassic–Cretaceous boundary, and many Jurassic species survived well into the Neocomian, some even reaching into the middle Cretaceous (Roth, in prep.). A minor evolutionary event occurred in the early Valanginian, about when organic carbon accumulation rates started to increase in the North Atlantic Ocean. The major diversity increase in the Cretaceous coincides with the beginning of wide-spread black shale deposition in the Aptian in the North Atlantic (Roth 1978). Nannofossil diversity remained high during the late Cretaceous, until a decline began in the middle Maestrichtian shortly before the terminal Cretaceous event.

Discussion

Two first-order cycles of calcareous nannoplankton evolution are apparent for the latest Triassic to Cretaceous time period: (1) A Jurassic megacycle (Rhaetian to Tithonian) and (2) A Cretaceous megacycle (Beriasian to Maestrichtian). These two first-order cycles are punctuated by short pulses of increased evolution that resulted in high standing diversities. Primary and secondary cyclic diversity fluctuations of ammonites in the Jurassic and Cretaceous and globigerinids in the Cretaceous are similar to fluctuations of nannofossil diversity (Fig. 16). Monographic effects add bias to ammonite diversities (Kennedy 1977) especially for the middle Jurassic and middle Cretaceous; they explain some of the divergences among the different diversity curves. The Jurassic and Cretaceous megacycles of diversity are in phase with cycles of carbon dioxide concentration in the atmosphere as estimated by Budyko & Ronov (1979) and shown on Fig. 17. Times of high pCO_2 in the atmosphere were characterized by warm equable climates which were conducive to high production of calcareous nannoplankton. During the late Jurassic atmospheric CO_2 maximum, calcareous nannoplankton invaded the open ocean and became highly productive (Tithonian Carbonate Shift). During the late Creta-

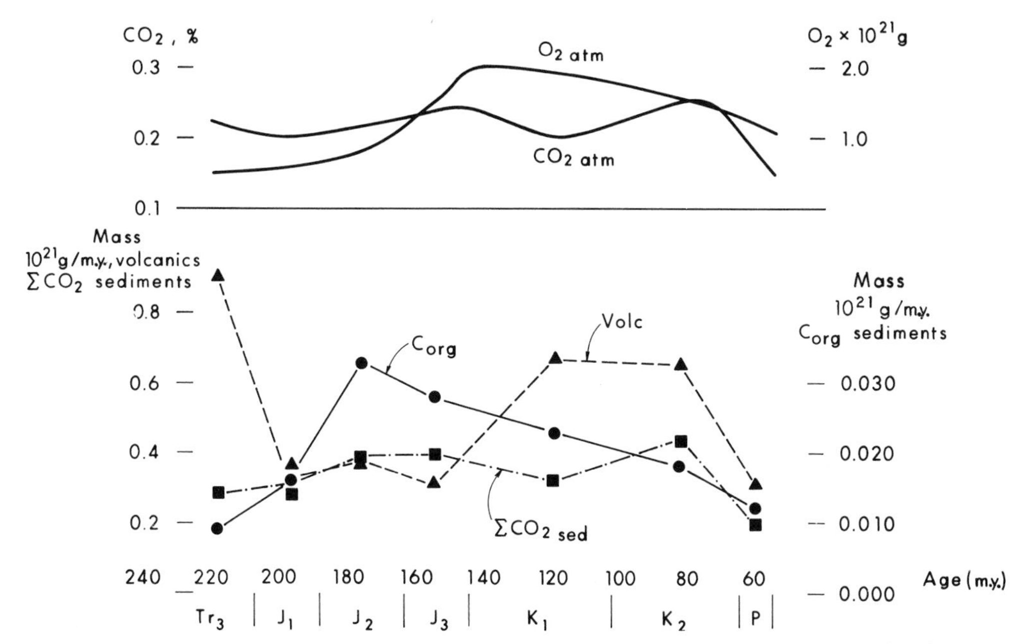

FIG. 17. Oxygen and carbon content of the atmosphere, organic carbon (Corg.) in sediments, total carbon dioxide (CO_2) in sedimentary rocks and the mass of volcanic rocks during the Jurassic and Cretaceous after Budyko & Ronov (1979).

ceous CO_2 peak, widespread nannofossil carbonates (chalk) accumulated in epicontinental seas of Eurasia and North America.

Major increases of standing diversities of calcareous plankton and nektobenthos appear to coincide with OAEs which occurred during eustatic sea-level highs. This coincidence is not fortuitous; both processes are related to the carbon cycle which is affected by sea-floor spreading and climate.

The carbon cycle is very complex and not fully understood (Arthur 1982). On a geologic timescale (Ma) the carbon dioxide concentration in the ocean-atmosphere system is affected mostly by: (1) input rates of volcanic CO_2; (2) burial rates of marine and terrestrial organic matter; (3) burial rate of carbonates. Additional feedback mechanisms are provided by the sulphur cycle (Garrels & Perry 1974; Garrels & Lerman 1984; Holser 1984) and possibly by the calcium cycle (sequestering of Ca in evaporites). Over timespans of more than 100 ky, volcanic input of CO_2, mostly related to rates of formation and destruction of oceanic lithosphere, appears to be most important (Berner *et al.* 1983). High atmospheric CO_2 concentration and reduced albedo occur during periods of rapid sea-floor spreading.

Rapid sea-floor spreading appears to have occurred in the middle Jurassic and middle Cretaceous (Force 1984); highs in atmospheric pCO_2, eustatic sea-level, and average global temperatures occurred during or slighly after these two time periods (Fig. 17). Minor fluctuations in spreading rates and basin configurations affected the burial rates of carbon and thus pCO_2 in the atmosphere during the Jurassic and Cretaceous.

The relative importance of warm saline bottom water production and freshening of surface waters for ocean stability and thus burial of marine organic carbon can be inferred from the temporal and spatial distribution of major coal and evaporite deposits during the Jurassic and Cretaceous (Hallam 1982; Parrish *et al.* 1982). Warm saline bottom waters were probably responsible for the major OAEs in the Callovo-Oxfordian and the middle Cretaceous. Poor ventilation appears to have been restricted largely to shallow epicontinental seas and perhaps narrow ocean basins (alpine geocline) during the Kimmeridgian-Tithonian anoxic event. Because of insufficient information on the Bajocian and Liassic pelagic sedimentary record it is impossible to determine the importance of warm saline bottom water formation for the accumulation of black shales during these time periods. Widespread coal deposits of that age in the USSR and in Australia tend to indicate high rainfall at middle and high latitudes. Thus, periodic high organic carbon preservation may have been caused by poor ventilation due to low salinity surface waters during the Liassic and Bajocian. Some saline bottom water could have been derived from the numerous evaporite basins surrounding the Tethys Ocean during the lower and middle Jurassic.

Major regressions and/or better meridional connections at the end of the Jurassic and Cretaceous led to cooling and thus to more vigorous deep ocean circulation and ventilation. Nannofossil diversities were at their lowest levels during periods of major regression.

Conclusions

In summary, pelagic sediment deposition and biologic productivity are ultimately related to tectonic processes through links provided by changes in sea-level, albedo, and the stability of ancient oceans. Pulses in mantle convection (see Sheridan, this volume) were the ultimate cause of first and second order climatic cycles, resulting in biological evolution of marine calcareous plankton and nektobenthos, and major patterns of pelagic sedimentation. Changes in the earth's obliquity, precession and eccentricity caused only minor perturbations (third order cycles) of the earth's climate, ocean circulation, biological evolution and pelagic sedimentation during the Mesozoic Era.

ACKNOWLEDGMENT: The author thanks his colleagues Frank H. Brown, Thure Cerling, A.A. Ekdale, and Kadir Uygur for critical review of this paper.

References

ARTHUR, M.A. 1979. Paleoceanographic events—recognition, resolution, and reconsideration. *Rev. Geophys. Space Phys.* **17,** 1474–94.

—— 1982. The carbon cycle: Controls on atmospheric CO_2 and climate in the geologic past. *In*: BERGER, W.H. & CROWELL, J.C. (eds), *Climate in Earth History*. National Academy Press, Washington, DC. 55–67.

BALTUCK, N. 1982. Provenance and distribution of Tethyan pelagic and hemipelagic siliceous sediments, Pindos Mountains, Greece. *Sedim. Geol.* **31,** 63–88.

—— 1983. Some sedimentary and diagenetic signatures in the formation of bedded radiolarite. *In*: IIJIMA, A., HEIN, J.R. & SIEVER, R. (eds), *Siliceous Deposits in the Pacific Region*. Elsevier, Amsterdam. 299–316.
BERGER, W.H. 1970. Biogenous deep-sediments: Fractionation by deep-sea circulation. *Bull. Geol. Soc. Am.* **81,** 1385–402.
BERNER, R.A., LASAGA, A.C. & GARRELS, R.M. 1983. The carbonate-silicate geochemical cycle and its effect on atmospheric carbon dioxide over the past 100 million years. *Am. J. Sci.* **283,** 641–83.
BRALOWER, T.J. & THIERSTEIN, H.R. 1984. Low productivity and slow deep-water circulation in mid-Cretaceous oceans. *Geology* **12,** 614–18.
—— & THIERSTEIN, H.N. 1986. Organic carbon and metal accumulation in Holocene and mid-Cretaceous marine sediments: Paleoceanographic significance. *In*: BROOKS, J. & FLEET, A. (eds), *Marine Petroleum Source Rocks. Spec. Publ. Geol. Soc. Lond.* Published by Blackwell Scientific Publications, Oxford.
BRASS, G.W., SOUTHAM, J.R. & PETERSON, W.H. 1982. Warm saline bottom water in the ancient ocean. *Nature (Lond)*, **296,** 620–3.
BUDYKO, M.J. & RONOV, A.B. 1979. Chemical evolution of the atmosphere in the Phanerozoic. *Geokhimya*, **5,** 643–53.
COOK, P.J. 1984. Spatial and temporal controls on the formation of phosphate deposits—A review. *In*: NRIAGU, J.O. & MOORE, P.B. (eds), *Phosphate Minerals*. Springer-Verlag, Berlin. 242–74.
DEAN, W.E. & GARDNER, J.V. 1982. Origin and geochemistry of redox cycles of Jurassic to Eocene age, Cape Verde Basin (DSDP site 367), continental margin of north-west Africa. *In*: SCHLANGER, S.O. & CITA, M.B. (eds), *Nature and Origin of Cretaceous Carbon-rich Facies*. Academic Press, London. 55–75.
FORCE, E.R. 1984. A relation among geomagnetic reversals, seafloor spreading rate, paleoclimate, and black shales. *EOS* **65,** 18–19.
GARRELS, R.M. & PERRY, E.A. 1974. Cycling of carbon, sulfur, and oxygen through geologic time. *In*: GOLDGERG, E. (ed.), *The Sea*. John Wiley & Sons, New York. 303–36.
—— & LERMAN, A. 1984. Coupling of the sedimentary sulfur and carbon cycles—an improved model. *Am. J. Sci.* **284,** 989–1007.
GRACIANSKY, P.C. DE., BROSSE, E., DEROO, G., HERBIN, J.P., MONTADERT, L., MULLER, C., SIGAL, J. & SCHAAF, A. 1982. Les formations d'âge Crétacé de l'Atlantique Nord et leur matière organique. Paléogéographie et milieux de dépôt. *Rev. Inst. Franç. Pétr.* **37,** 275–336.
HABIB, D. 1979. Sedimentary origin of North Atlantic Cretaceous palynofacies. *In*: TALWANI, M., HAY, W. & RYAN, W.B.F. (eds), *Deep Drilling Results in the Atlantic Ocean. Continental Margins and Paleoenvironments*. Am. Geophys. Union, Maurice Ewing Series 3. 420–37.
—— 1983. Sedimentation-rate dependent distribution of organic matter in the North Atlantic Jurassic-Cretaceous. *Initial Repts DSDP* **76,** 781–94.
HALLAM, A. 1975. *Jurassic Environments*. Cambridge University Press, Cambridge. p. 269.
—— 1981. A revised sea level curve for the early Jurassic. *J. Geol. Soc. Lond.* **138,** 735–43.
—— 1982. The Jurassic climate. *In*: *Climate in Earth History. Studies in Geophysics*. National Academy Press, Washington, DC. 159–63.
HAMILTON, G.B. 1983. Triassic and Jurassic calcareous nannofossils. *In*: LORD, A.R. (ed.), *A Stratigraphical Index of Calcareous Nannofossils*. Published for the British Micropalaeontological Society by Ellis Horwood Ltd. 17–39.
HOLSER, W.T. 1984. Gradual and abrupt shifts in ocean chemistry during Phanerozoic time. *In*: HOLLAND, H.D. & TRENDALL, A.F. (eds), *Patterns of Change in Earth Evolution. Dahlam Konferenzen 1984*. Springer-Verlag, Berlin. 123–43.
JENKYNS, H.C. 1980. Cretaceous anoxic events: from continents to oceans. *J. geol. Soc. Lond.* **137,** 171–88.
—— & WINTERER, E.L. 1982. Palaeoceanography of Mesozoic ribbon radiolarites. *Earth Planet. Sci. Lett.* **60,** 351–75.
—— 1983. A Jurassic anoxic event? *Abstracts, First International Conference on Paleoceanography, Zürich, 1983*, 32.
KENNEDY, W.J. 1977. Ammonite evolution. *In*: HALLAM, A. (ed.), *Patterns of Evolution*. Elsevier, Amsterdam. 251–304.
KUENEN, PH. H. 1941. Geochemical calculations concerning the total mass of sediments in the earth. *Am. J. Sci.* **239,** 161–90.
MCCAVE, I.N. 1979. Depositional features of organic carbon-rich black and green mudstones at DSDP sites 386 and 387, western North Atlantic. *Init. Repts DSDP* **43,** 411–16.
MEDD, A.W. 1982. Nannofossil zonation of the English middle and upper Jurassic. *Mar. Micropaleont.* **7,** 73–95.
OGG, J.G., ROBERTSON, A.H.F. & JANSA, L.F. 1983. Jurassic sedimentation history of site 534 (western North Atlantic) and of the Atlantic-Tethys Seaway. *Init. Repts DSDP* **76,** 829–84.
PARRISH, J.T. & CURTIS, R.L. 1982. Atmospheric circulation, upwelling and organic-rich rocks in the Mesozoic and Cenozoic eras. *Palaeogeogr. Palaeoclimatol. Palaeoecol.* **40,** 31–66.
——, ZIEGLER, A.M. & SCOTESE, C.R. 1982. Rainfall patterns and the distribution of coals and evaporites in the Mesozoic and Cenozoic. *Palaeogeogr. Palaeoclimatol. Paleoecol.* **40,** 67–101.
ROBERTSON, A.H.F. & BLIEFNICK, D.M. 1983. Sedimentology and origin of lower Cretaceous pelagic carbonates and redeposited clastics, Blake-Bahama Formation. Deep Sea Drilling Project Site 534, western Equatorial Atlantic. *Init. Repts DSDP* **76,** 795–828.
ROTH, P.H. 1973. Calcareous nannofossils—Leg 17, Deep Sea Drilling Project. *Init. Repts DSDP* **17,** 695–795.
—— 1978. Cretaceous nannoplankton biostratigraphy and oceanography of the northwestern Atlantic Ocean. *Init. Repts DSDP* **44,** 731–59.

—— 1981. Mid-Cretaceous calcareous nannoplankton from the central Pacific: Implications for paleoceanography. *Init. Repts DSDP* **62,** 471–89.

—— 1983. Jurassic and lower Cretaceous calcareous nannofossils in the western Atlantic (Site 534): Biostratigraphy, preservation and observations on biogeography and paleoceanography. *Init. Repts DSDP* **76,** 587–621.

—— & BOWDLER, J.L. 1981. Middle Cretaceous calcareous nannoplankton biogeography and oceanography of the Atlantic Ocean. *Spec. Publ. Soc. Econ. Paleont. Mineral.* **32,** 517–46.

——& KRUMBACH, K.R. (in press). Middle Cretaceous calcareous nannofossil biogeography and preservation in the Atlantic and Indian oceans: Implications for paleoceanography. *Mar. Micropaleont.*

RYAN, W.B.F. & CITA, M.B. 1977. Ignorance concerning episodes of ocean-wide stagnation. *Mar. Geol.* **23,** 197–215.

SALTZMAN, E.S. & BARRON, E.J. 1982. Deep circulation in the late Cretaceous: Oxygen isotope paleotemperatures from Inoceramus remains in DSDP cores. *Paleogeogr. Palaeoclimatol. Palaeoecol.* **40,** 167–81.

SAVIN, S.M. 1977. The history of the earth's surface temperature during the past 100 million years. *Ann. Rev. Earth Planet. Sci.* **5,** 319–55.

SCHOLLE, P.A. & ARTHUR, M.A. 1980. Carbon isotopic fluctuations in pelagic limestones: Potential stratigraphic and petroleum exploration tool. *Bull. Am. Ass. Petrol. Geol.* **64,** 67–89.

SIMONEIT, B.R.T. & STUERMER, D.H. 1982. Organic geochemical indicators for sources for organic matter and paleoenvironmental conditions of Cretaceous oceans. *In*: SCHLANGER, S.O. & CITA, M.B. (eds), *Nature and Origin of Cretaceous Carbon-rich Facies.* Academic Press, London. 145–64.

SUMMERHAYES, C.P. 1981. Organic facies of middle Cretaceous black shales in deep North Atlantic. *Bull. Am. Ass. Petrol. Geol.* **65,** 2364–80.

—— 1986. Organic rich Cretaceous sediments from the North Atlantic. *In*: BROOKS, J.R. & FLEET, A.J. (eds), *Marine Petroleum Source Rocks. Spec. Publ. Geol. Soc. Lond.* Published by Blackwell Scientific Publications, Oxford.

—— & MASRAN, Th. C. 1983. Organic facies of Cretaceous and Jurassic sediments from DSDP Site 534 in the Blake-Bahama Basin, western North Atlantic. *Init. Repts DSDP* **76.** 469–80.

THIERSTEIN, H.R. 1981. Late Cretaceous nannoplankton and the change at the Cretaceous-Tertiary boundary. *In*: WARME, J.E. *et al.* (eds), *The Deep Sea Drilling Project: A Decade of Progress.* Spec. Publ. Soc. Econ. Paleont. Miner. **32,** 355–94.

—— 1983. Trends and events in Mesozoic oceans. Proceedings of the Joint Oceanographic Assembly 1982. *General Symposia (Canadian National Committee/Scientific Committee on Ocean Research, Ottawa., Canada).* 127–30.

—— & BERGER, W.H. 1978. Injection events in ocean history. *Nature Lond.* **276,** 461–6.

—— & ROTH, P.H. 1980. Stable isotopes and biogeography of mid-Cretaceous microfossils. *Abstracts 26th Int. geol. Congr. Paris* **293.**

TISSOT, B., DEROO, G. & HERBIN, J.P. 1979. Organic matter in Cretaceous sediments of the North Atlantic: Contribution to sedimentology and paleoceanography. *In*: TALWANI, M., HAY, W. & RYAN, W.B.F. (eds), *Deep Drilling Results in the Atlantic Ocean: Continental Margins and Paleoenvironments.* Am. Geophys. Union, Maurice Ewing Series 3. Washington, DC. 362–74.

——, DEMAISON, G., MASSON, P., DELTEIL, J.R. & COMBAZ, A. 1980. Paleoenvironment and petroleum potential of middle Cretaceous black shales in Atlantic Basin. *Bull. Am. Ass. Petrol. Geol.* **64,** 2051–63.

VAIL, P.R., HARDENBOL, J. & TODD, R.G. 1984. Jurassic unconformities, chronostratigraphy and sea-level changes from seismic stratigraphy and biostratigraphy. *In*: SCHLEE, J.S. (ed.), *Interregional unconformities and Hydrocarbon Accumulation. Am. Ass. Petrol. Geol. Mem.* **36,** 129–44.

WILDE, P. & BERRY, W.B.N. 1982. Progressive ventilation of the oceans-potential for return to anoxic conditions in the post-Paleozoic. *In*: SCHLANGER, S.O. & CITA, M.B. (eds), *Nature and Origin of Cretaceous Carbon-rich Facies.* Academic Press, London. 209–24.

PETER H. ROTH, Department of Geology and Geophysics, University of Utah, Salt Lake City, Utah 84112–1183, USA.

Changes in the organic carbon burial during the Early Cretaceous

P.L. de Boer

SUMMARY: The early and middle Cretaceous exhibit a paradoxical coincidence between low primary production in the oceans and entrapment of large amounts of organic matter in pelagic sediments. The high average temperatures and the decreased intensity of the (haline) circulation led to a strong decrease of the supply of oxygen to deep water, encouraging the accumulation of organic matter in black shales.

The nutritious elements, involved in the excess burial of organic matter, were available because nutrients of the terrestrial cycle of sedimentation and erosion were removed into the marine domain during the global rise of sea-level.

With respect to the warm climate, the sluggish circulation, and the widespread deposition of black shales, it is suggested that an initial rise of CO_2, related to increased spreading and emissions of juvenile carbon, and to increased carbonate production, led to a first 'greenhouse warming'. This was reinforced by a transfer of CO_2 from the warming oceans into the atmosphere. During subsequent cooling, CO_2 re-dissolved within the oceans, and conditions returned to 'normal'.

Model calculations, based on the effect of changes of burial rate of organic matter upon $\delta^{13}C$ of carbon in the ocean, show that the excess storage involves an increase of burial of organic carbon of the order of 20% only.

The deposition of large amounts of organic matter in deep marine sediments during the first half of the Cretaceous (Schlanger & Jenkyns 1976; Arthur 1979) coincided with a global rise of sea-level caused by an increase of volume of oceanic ridges (Hays & Pitman 1973). Attempts to explain the large scale storage have led to suggestions that the oceans were highly productive and/or that large amounts of terrestrial organic matter were transported to the sea (Jenkyns 1980; Habib 1982). Others claim that Cretaceous oceans had a low circulation velocity and low organic production (Berger 1979; De Boer 1983; Bralower & Thierstein 1984).

In different places the factors involved in the deposition of organic C-rich sediment may have interplayed in a different way. The system as a whole, however, was 'poised at relatively low oxygen levels which periodically tipped in favour of organic matter preservation' (Arthur *et al.* 1984).

Estimates by different authors of the present annual primary production on land and in the oceans, and of fossilization of organic matter differ widely. However, the ratio between the amount of fossilized and the amount of primary produced organic matter is large, 1:1000 to 1:10 000 (cf. Pytkowicz 1973; Bolin *et al.* 1979; Hay & Southam 1979). That is, less than 0.1% of the organic matter produced annually is fossilized; 99.9% or more is recycled and remains within the biosphere. To keep the marine chemosphere stable, it is important that, in the long term, the carbon and nutrients removed into the fossil reservoir are replenished. So there must be a link between input and (anoxic and/or aerobic) output of nutrients and carbon.

Organic production in the oceans depends mainly on the amounts of nutrients recycled by oxidation, chiefly in the water column, and by circulation in relation to photosynthesis. In the short term the supply of nutrients from the land is of subordinate importance. The terrestrial supply of, e.g. phosphorus amounts to only 1% of the amount brought up annually from deep ocean water and used in photosynthesis, and it is a factor 10^4 to 10^5 lower than the total dissolved stock in the ocean (Broecker & Peng 1982). Therefore, productivity in the ocean mainly depends on circulation intensity and recycling of nutrients. On the other hand, deposition of slowly accumulating organic-rich sediments depends especially on the ability of the system to provide sufficient oxygen to deep water for the digestion of the sinking and settling organic matter. Demaison & Moore (1980, Figs 1 and 2) indeed demonstrate the absence of a systematic correlation between primary production and the organic carbon content of bottom sediments in the ocean.

In the case of a high rate of sedimentation it is possible that, despite oxidizing conditions in deep water, abundant organic matter is buried (Summerhayes 1983). However, many of the Cretaceous black shales, discussed in literature, represent slowly accumulating sediments. The large scale deposition of these cannot be considered *ipso facto* the result of a high organic production or a large input of terrestrial organic matter. Depletion of the oxygen stock in deep water, and, in the long run, a consistent input of sufficient

From SUMMERHAYES, C.P. & SHACKLETON, N.J. eds), 1986, *North Atlantic Palaeoceanography*, Geological Society Special Publication No. 21, pp. 321–331.

nutrients from the land are the main prerequisites.

Productivity and fossilization of organic matter in the Cretaceous ocean

The Cretaceous rise of sea-level was large, and it occupied tens of millions of years (Hays & Pitman 1973). The average rate of rise was about 0.01 mm/yr. This is two orders of magnitude slower than the rise caused by melting ice caps during the Holocene. Inundation of coastal areas in the Western Interior was suggested to have occurred with rates of up to 500 km/Ma (Hancock & Kauffman 1979). This is rapid in the geological sense, but it does not imply a 'catastrophic flooding of lowlands'. Neither during the Holocene, nor in the Cretaceous was the ocean 'awash with wood'.

At present, deep waters at low and middle latitudes are up to 20°C colder than surface waters at the same latitudes. Cretaceous benthic and planktonic foraminifera mostly show a small difference in $\delta^{18}O$, suggesting much less temperature difference between surface and deep water (Savin 1977). Deep waters were as warm as 15°C (Brass *et al.* 1982).

Polar ice caps were absent, and the North Atlantic and Tethys Ocean had no direct connection with high latitude water masses. The 'coolest' (i.e. minimum estimates) mean annual surface temperature curve in the mid-Cretaceous shown by Barron (1983, Fig. 3) reveals more or less the same values as at present near the Equator, but towards high latitudes the difference gradually increases to over 10°C, the Cretaceous values being the higher ones. The maximum estimates of Barron even show a 5°C higher temperature near the Equator to about 20°C or more at high latitudes.

Models of Wilde & Berry (1982) indicate that the significance of cool high latitude waters driving the deep circulation system vanishes when their temperature exceeds 10°C. Consequently, high density saline shelf waters and atmospheric influences were the main factors for ocean water circulation (Arthur & Natland 1979; Barron & Washington 1982; Brass *et al.* 1982). It was suggested that the Cretaceous haline circulation could have been vigorous (Brass *et al.* 1982), but many arguments plead against such supposition, and deep water circulation apparently was slow as compared to the present. This proposition is sustained by the observation that sea floor erosion and hiatuses became an important aspect only after the Cenomanian (Arthur 1979). Together with the reduced circulation intensity, the reflux of nutrients from the deep ocean must have been low, and the organic production in the ocean must have been less than at present.

Brumsack (1984) compared the content of heavy elements of black shales from highly productive areas to those from the Black Sea, and found samples from organic C rich intervals of Cenomanian–Turonian age to compare best to those from the stagnant and low productive Black Sea. Bralower & Thierstein (1984) compared recent and fossil occurrences of organic C-rich deep marine sediments in the light of productivity of surface waters and preservation. They conclude that productivity and deep water renewal rates during the early Cretaceous were about an order of magnitude lower than today's. Even levels with an extreme high content of organic matter such as the Bonarelli Level at the Cenomanian–Turonian boundary in the Apennines (Italy), containing over 20% of organic C, may have formed below low productive surface waters (De Boer 1982).

Both organic production, defined mainly by the amount of nutrients brought up from the deep, and the downward flux of oxygen, depend on circulation intensity. These relations and their effect upon anoxicity and burial of black shales, however, are not linear (Fig. 1).

In strongly circulating systems large amounts of nutrients are brought to the surface, primary organic production is high, and abundant organic matter sinks down. The solubility of O_2 in water which goes down (elsewhere), however, is limited, putting an upper limit to the replenishment in deep water. Moreover, O_2 is consumed during transport by deep ocean currents. This may result in an oxygen minimum level at mid-water depths in areas of upwelling.

In stagnant systems circulation is ideally nil and the supply of O_2 by diffusion can be neglected. Supply of nutrients and organic matter from outside sources (river input, wind blown dust) continues, allowing some organic production, sinking of organic matter, and the formation of black shales.

At many places in the present deep ocean the supply of oxygen by deep currents from high latitudes is sufficient to ensure the oxidation of the bulk of the settling organic matter.

In the Cretaceous, cool high latitude oxygen supplying waters were largely missing, and the warm and slow, salinity driven waters could transport considerably less oxygen to the deep. Solubility of oxygen decreases 2% for every degree of rise of temperature (cf. Wilde & Berry 1982). Thus the reduced circulation intensity

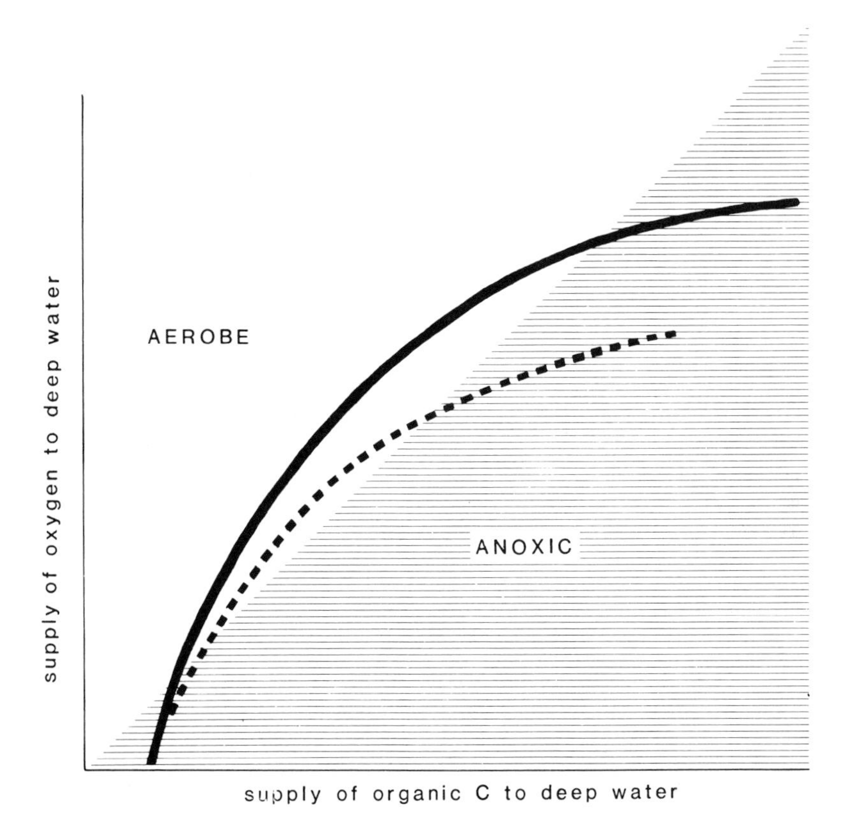

FIG. 1. Inferred relation between ventilation of deep water (vertical axis), and primary organic production and sinking of organic matter (horizontal axis). Both depend on circulation intensity.

In strongly circulating systems (upper right) abundant nutrients are brought to the surface and abundant organic matter is produced and sinks down. Solubility of O_2 in the (elsewhere) downgoing water is limited, putting an upper limit to the replenishment in deep water.

In 'stagnant systems' (lower left) deep circulation is small or absent. Some nutrients and organic matter from outside sources (river input, wind blown dust) are supplied, allowing some production and sinking of organic matter. For the digestion of this organic matter oxygen is insufficiently available in deep water.

Solid line: present day situation. Interrupted line: situation for the Cretaceous; reduced circulation caused a low organic production, and a low oxygen supply to deep water. In addition, the replenishment of O_2 was hampered due to a lower solubility of O_2 in the warmer water.

affected the organic production and in particular the supply of oxygen to deep water. The interrupted line in Fig. 1 schematically shows the inferred relation between the supply of oxygen and organic matter to deep water during the Cretaceous in comparison to the present (continuous line).

High temperatures and slow circulation allowed depletion of oxygen in deep water and burial of organic matter to occur more easily than at present, in stagnant as well as in upwelling type settings. Or, to agree with Fischer (1981), 'the sea suffered from indigestion, and was less able to metabolize the organic matter brought into it or developed within it'.

Supply of carbon and nutritious elements

Carbon

Barron (1984) states that features such as a high sea-level and a different position of the continents can only partly explain high temperatures during the Cretaceous; high atmospheric CO_2 content must be very important (greenhouse effect).

Based on calculations of the amounts of magmatic products, Budyko & Ronov (1979) suggested that atmospheric CO_2 content during the Cretaceous might have been ten times the present value. Other authors give lower estimates, but all strongly exceed the present day amount (cf. Arthur *et al.* 1985). Considering the increased spreading rates and the inferred high atmospheric CO_2 levels during the Early Cretaceous, it is attractive to assume that much of the carbon needed for the excess burial of organic matter was derived ultimately from juvenile sources.

Feedback in the rise of CO_2 level

In view of the presumed high temperature during the Cretaceous, the possibility of positive feedback must be considered. Solubility of CO_2 in sea water is heavily dependent on temperature. Under present conditions, solubility decreases 4% for every degree rise of temperature (Skirrow 1975). So, the effect of a rise of, e.g. 10°C must be considerable.

A primary increase of CO_2 in the atmosphere by, e.g. increased juvenile output thus could have caused an initial greenhouse warming, to be followed and reinforced by an extra addition of CO_2 from the warming ocean. The resulting warm climate favoured slow circulation and widespread formation of black shales.

In addition, the increased shelf area and favourable conditions for the production of shallow marine carbonates could have contributed to the rise of CO_2^- level (cf. 'coral reef hypothesis', Berger 1982). Even, one might argue that a release of CO_2 by increased carbonate production on expanding shelf surfaces during transgressions alone could start up greenhouse warming, reinforced by a positive feedback.

Return to 'normal conditions' is inferred to have started when spreading and juvenile carbon emissions slowed down. Wide-spread anoxia and excess burial of organic matter may have continued initially, thus withdrawing more CO_2 from the system than was added by juvenile input. A first decrease of the greenhouse effect is likely to be reinforced because of an increasing transfer of CO_2 back into the cooling ocean. The inferred succession of events is illustrated in Fig. 2.

It is important to stress that such feedback-related transfer of CO_2 from the ocean to the atmosphere and vice versa, would influence the stable isotopic composition of oceanic dissolved carbon. Fractionation of carbon-isotopes over the air–water interface causes $\delta^{13}C$ of atmospheric CO_2 to be 7–10‰ lower than in dissolved HCO_3^- (Mook *et al.* 1974); e.g. the release of 10% of the oceanic dissolved carbon stock as CO_2 to the atmosphere could cause a rise of $\delta^{13}C$ of ocean dissolved carbon of 0.5–1 permille.

Nutritious elements

Following the discussion above, a change from a fast to a slowly circulating ocean implies either an accumulation of nutrients in deep water, or a loss of nutrients from the system. Assuming a P:C ratio of about 0.01, a 10% increase of fossilization of organic matter would exhaust the oceanic stock of phosphorus in 0.5–1 Ma (cf. Broecker & Peng 1982). Therefore, to maintain an excess storage of organic C for millions of years, as occurred during the Cretaceous, the supply of nutrients to the ocean must be above the average long term flux.

The Cretaceous rise of sea level was large, and it caused a reduction of the land to maybe 20% of the Earth surface (cf. Hallam 1971), that is a reduction of the order of 30% of the present land surface. If part of the land is occupied by the sea, elevated areas, providing the bulk of weathering products, will be largely unaffected. However, low-lying parts of the continents, where net deposition occurs, are flooded, and many nutrients that otherwise would have been locked in terrestrial sediments, are transferred to the marine biosphere. During the lower Cretaceous transgression the surface with net deposition of terrestrial sediments was reduced, as testified by many sedimentary successions of that age. The related nutrient flux was transferred to the ocean (Fig. 3). The early Cretaceous oceanographic conditions ensured that much of these excess nutrients was removed along with organic matter.

The transfer of part of the surficial terrestrial nutrient stock into the marine biosphere, combined with the apparent 'indigestion' of the oceans, must have been an essential feature of the large scale deposition of organic matter within pelagic sediments. When the land surface is reduced by 20%, the extra amount of phosphorus transferred to the marine biosphere can well account for an excess burial of organic matter as during the early Cretaceous.

Amount of organic matter stored

The overall organic production in the oceans was low, but the estimates of organic carbon stored in pelagic sediments during the first half of the Cretaceous are high figures (cf. Irving *et al.* 1974; Tissot 1979). Accurate volumetric calculations of the amount stored in lower and middle Cretaceous pelagic sediments may be a future possibility when more data are available from deep sea drilling. Here another approach is attempted. A model is presented in which the effects of photosynthetic fractionation of stable carbon-isotopes,

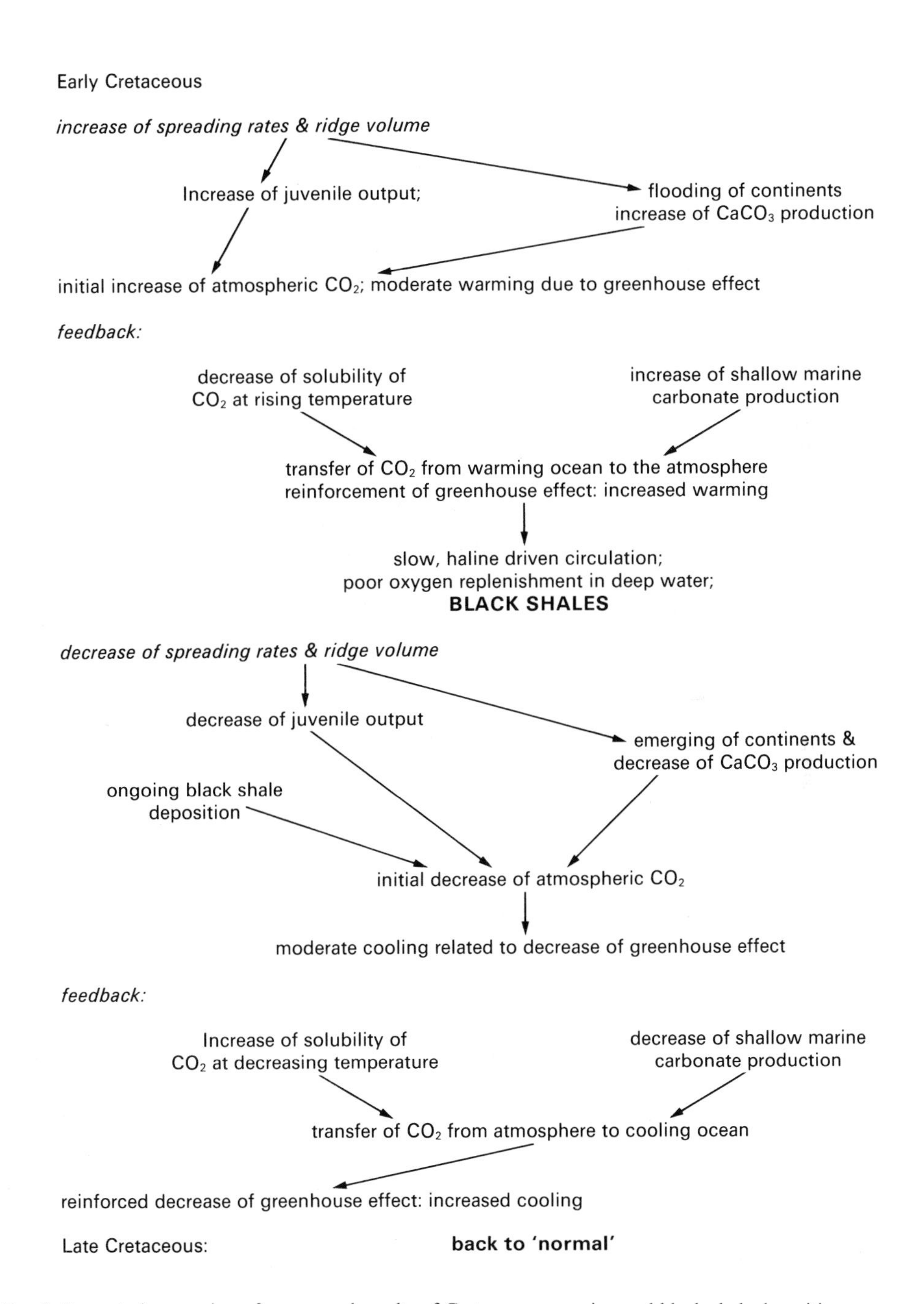

FIG. 2. Suggested succession of causes and results of Cretaceous warming and black shale deposition.

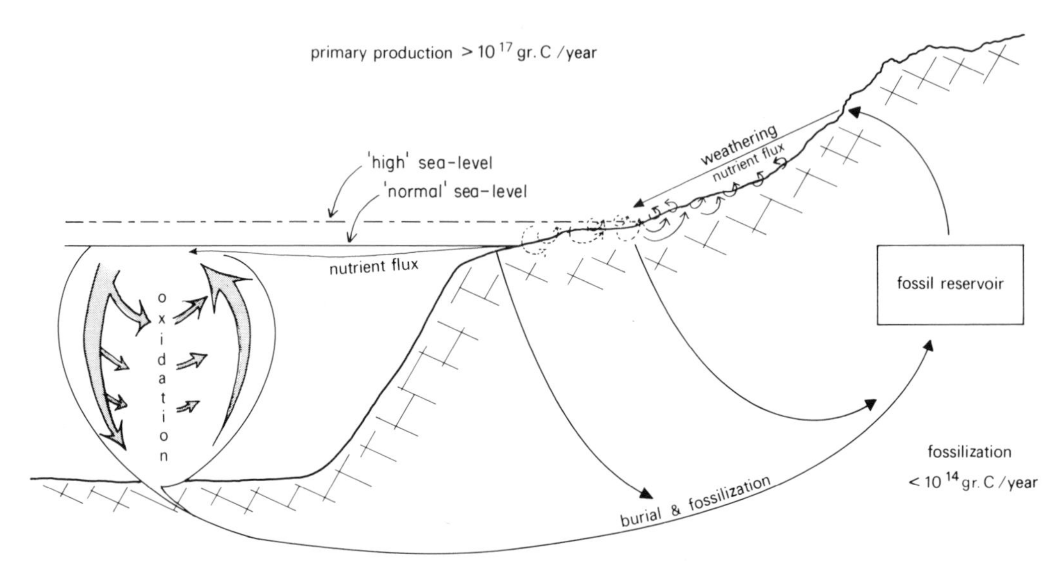

FIG. 3. Terrestrial and marine organic cycles and the transfer of nutrients. On the land and in the sea the organic products and the nutritious elements involved are largely recycled. Only a small part is annually removed into the fossil reservoir. Also the annual flux from the land into the sea is small, being equivalent to less than a permille of the annual organic production in the seas and oceans. During a gradual transgression part of the nutritious elements of the terrestrial sedimentary system are transferred to the marine biosphere. Per surface unit, the production capacity of the marine biosphere is, on the average, lower than on the land. So, a slow transgression leads to an excess of nutrients in the sea, part of which can be removed—if deep water anoxia occurs—as part of organic matter in black shales.

and of the burial of organic matter upon $\delta^{13}C$ of carbon in the ocean are considered. It represents a further elaboration of a previous model (De Boer 1983).

Stable carbon-isotopes

The preference of photosynthetic processes for ^{12}C over the ^{13}C isotope is reflected in ^{13}C depletion of organic matter ($\delta^{13}C = -20$–$-30‰$). Therefore, the carbon remaining in the biosphere becomes more positive and it is enriched in ^{13}C relative to the amount of ^{12}C, when excess organic C is removed. Consequently Scholle & Arthur (1980) showed that the large scale removal of organic matter into the fossil reservoir during the early half of the Cretaceous led to a significant increase in the $\delta^{13}C$ of the carbon in the oceans, reflected in the $\delta^{13}C$ of marine carbonates. Similar positive excursions of the stable carbon-isotope ratio are seen in other periods with a great burial of organic matter (cf. Veizer *et al.* 1980).

Calculations

In the case of a fully steady-state system with constant fluxes and constant climates, the isotope ratios of carbon in the ocean and the atmosphere are constant. The amounts of ^{12}C and ^{13}C added by erosion and juvenile input, and removed by fossilization would be equal. If fluxes change, especially those concerning compounds with strongly deviating carbon-isotope ratios, $\delta^{13}C$ of the reservoir is affected.

The values in Table 1 were used as basic input parameters in calculations on changes of oceanic $\delta^{13}C$. 'Adopted values' were co-ordinated in that $\delta^{13}C$ of juvenile (magmatic) carbon ($-7‰$), of carbonate ($+1‰$) and of organic matter ($-20‰$–$-25‰$) imply a long term ratio of about 2:1 for burial of carbonate and organic matter. The long term fluvial supply of phosphorus is about 10^{12} g P/year (1.7×10^{12}, Garrels *et al.* 1973; 0.7×10^{12}, Delwiche & Likens 1977; 1.08×10^{12}, Froelich *et al.* 1982). Burial of phosphorus in sinks other than as part of organic matter, and the release of P from organic matter during the early burial stage, may lead to the fossilization of P and organic C in a proportion departing from the Redfield ratio of 1:106. However, these features are assumed not to lead to great deviations. Thus an amount of about 0.4×10^{14} g C/year is calculated for the burial in marine sediments. Assuming the burial of organic matter in terrestrial and marine sediments has about the same magnitude

TABLE 1. *Size of different carbon fluxes as used in the model. Figures derived from comparison of data of various authors*

Fluxes	Estimated fluxes in 10^{14} g C/yr		Adopted long term value	$\delta^{13}C$
Juvenile carbon			0.08*	−7‰
	0.082	Pytkowicz 1973		
	0.1–0.5	Freyer 1979		
Cosmic influx	0.0025	Livingstone 1973		
Weathering of fossil carbonates			1.50	+1
	2.85	Pytkowicz 1973		
	1.46	Hay & Southam 1979		
Weathering of fossil organic C			0.70	−25
	0.3	Pytkowicz 1973		
	0.70	Garrels *et al.* 1973		
	0.7	Hay & Southam 1979		
Fossilization of carbonate			1.50	+1**
	2.85	Pytkowicz 1973		
	1.46	Hay & Southam 1979		
	2.1	Javoy *et al.* 1982/83		
	2.2	Berner & Raiswell 1983		
	5.04	Arthur *et al.* 1984b		
Fossilization of organic C			0.7–0.8*	−25**
	0.3	Pytkowicz 1973		
	0.7	Hay & Southam 1979		
near shore:	0.027	Mopper & Degens 1979		
pelagic:	0.92	Mopper & Degens 1979		
marine sediments:	1.1	Mopper & Degens 1979		
marine sediments:	1.26	Berner 1982		
	1.05	Javoy *et al.* 1982/83		
	0.46	Berner & Raiswell 1983		
	1.2	Walker 1983		
	1.26	Arthur *et al.* 1984b		
	1.15	Arthur *et al.* 1985		

* Fluxes of juvenile carbon and of fossilizing organic C are variable in the model.
** In proportion to $\delta^{13}C$ of carbon in the ocean.

(cf. Arthur *et al.* 1985), a total of 0.7–0.8 × 10^{14} g C is believed to be fossilized as organic matter annually.

Arguments for a high CO_2 level during the Cretaceous are fairly convincing (cf. Fischer 1981). Juvenile carbon is released by volcanism associated with spreading and with subduction of oceanic plates. Both types increase with increasing plate motions. Changes of juvenile influx were considered in the model.

The results of measurements of $\delta^{13}C$ of juvenile carbon released from cooling magma differ widely. They generally fall within the range 0 to −30‰. Mattey *et al.* (1984) found oceanic ridge basalt to deliver carbon with a $\delta^{13}C$ typically within the range of −3.5–7.4‰. The figure of −7‰ for the initial mantle, most often cited (see discussion by Javoy *et al.* 1982), was applied in the calculations.

The effect of changes of the amount of juvenile carbon added to the system and the amount of organic matter fossilized (in terrestrial and marine sediments) were measured by calculating the effects of changes of fluxes of ^{12}C and ^{13}C upon the $\delta^{13}C$ of the ocean in annual steps. Fluxes of weathering fossil carbonates and organic matter and of fossilization of carbonate were considered constant. The ocean, containing 3.5 × 10^{18} g C, was assumed to react as a single system.

The response time to sudden and drastic changes of the size of fluxes in the model appears to be 0.5–1 Ma (Fig. 4). The diagram in Fig. 5 shows the effects which changes in the input of juvenile carbon and the amount of organic matter

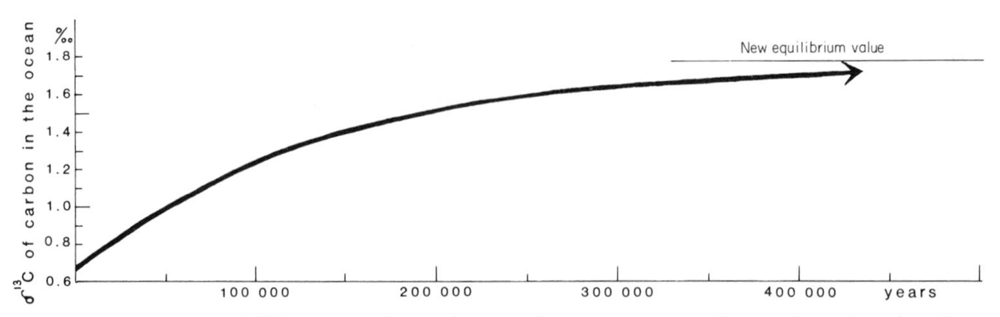

FIG. 4. Tentative change of $\delta^{13}C$ of ocean dissolved carbon due to an increase of input of juvenile carbon from 0.08–0.24 × 10^{14} g C/yr and a parallel increase of the storage of organic matter from 0.78–0.94 × 10^{14} g C/yr. For steady fluxes see 'adopted values' Table 1.

stored annually (in aquatic and terrestrial sediments) have upon $\delta^{13}C$ of the ocean.

The percentages indicate the amount of reduced carbon, expressed relative to the annual juvenile input (horizontal axis), stored in excess (100–200%) of, or less (−200%–+100%) than the amount of carbon brought into the biosphere by weathering and oxidation of fossil organic matter (0.7 × 10^{14} g C/yr). Thus, at the 100% line all of the juvenile C input is stored as organic matter, and the amount of carbon in the biosphere remains stable. Above +100% there is a decrease, and at less than +100% an increase of the sum of atmospheric and ocean dissolved carbon. Additions and removals of other rock components also may have fluctuated, but as these (e.g. $CaCO_3$) show less deviating stable carbon isotope values, this had much less influence upon the oceanic $\delta^{13}C$. Moreover, large scale extraction of carbon from the oxidized carbon stock would have put heavy constraints upon the exogenic chemical cycle.

Results

Marine carbonates show $\delta^{13}C$ values of about −1‰ at the Jurassic–Cretaceous boundary, and a gradual increase to +2–+3‰ occurs in the first half of the Cretaceous (Scholle & Arthur 1980). From the model, it appears (Fig. 5) that this would be the result of variations of storage of organic C of in between 0.6 × 10^{14} to 1.0 × 10^{14} g C/yr. If the value of 0.7–0.8 × 10^{14} g C is considered the long term average, it implies fluctuations up to about 25% at either side of the mean.

It must be noted that a transfer of CO_2 from the ocean into the atmosphere (and vice versa), due to greenhouse-related feedback, could contribute up to a permille to the observed changes of $\delta^{13}C$, so that fluctuations in organic burial could have been less than suggested by the values shown in the diagram below.

Arthur *et al.* (1985) made calculations similar to the ones presented here. Assuming an average Cretaceous $\delta^{13}C$ of 2.05‰, a steady rate of burial of organic C of 1.15 × 10^{14} g C, and about 5 × 10^{14} g C as carbonate, they arrive at a 57% higher burial rate of organic matter (1.8 × 10^{14} g C/year during the Cretaceous). Because of the levelling effect of the steady state fluxes of, e.g. weathering and fossilizing carbonate upon changes of $\delta^{13}C$ in the ocean, the assumption of rather high steady fluxes of weathering and fossilization implies also that relative changes of burial of organic matter must be large to explain an increase of $\delta^{13}C$ to 2‰. Indeed, insertion of steady state fluxes as proposed by Arthur *et al.* (1985) in the author's model gives a result similar to theirs, that is an increase of 60–70%. It would mean an excess burial of about 0.7 × 10^{14} g C/year. This seems rather high, however. For example, the related influx of phosphorus then would have been about 2.5 times the present rate. Therefore, preference is given to the above suggested long term rate of about 0.8 × 10^{14} g C/year for burial of organic matter, with an increase of about 20% during the early and middle Cretaceous.

With respect to the low $\delta^{13}C$ values during the early Cretaceous (Scholle & Arthur 1980), one may think of a first increase of supply of juvenile carbon, excess oxidation of weathering fossil organic matter, or of an overshoot due to the return of carbon dioxide into a cooling ocean following a high sea-level, warm climates and anoxic events during the Jurassic.

The positive values of $\delta^{13}C$ during the lower and middle Cretaceous primarily resulted from the excess storage of organic C of about 20% relative to the long term value. Moreover, a transfer of CO_2 from the sea into the atmosphere, due to greenhouse-related feedback might have contributed to the isotopic effect.

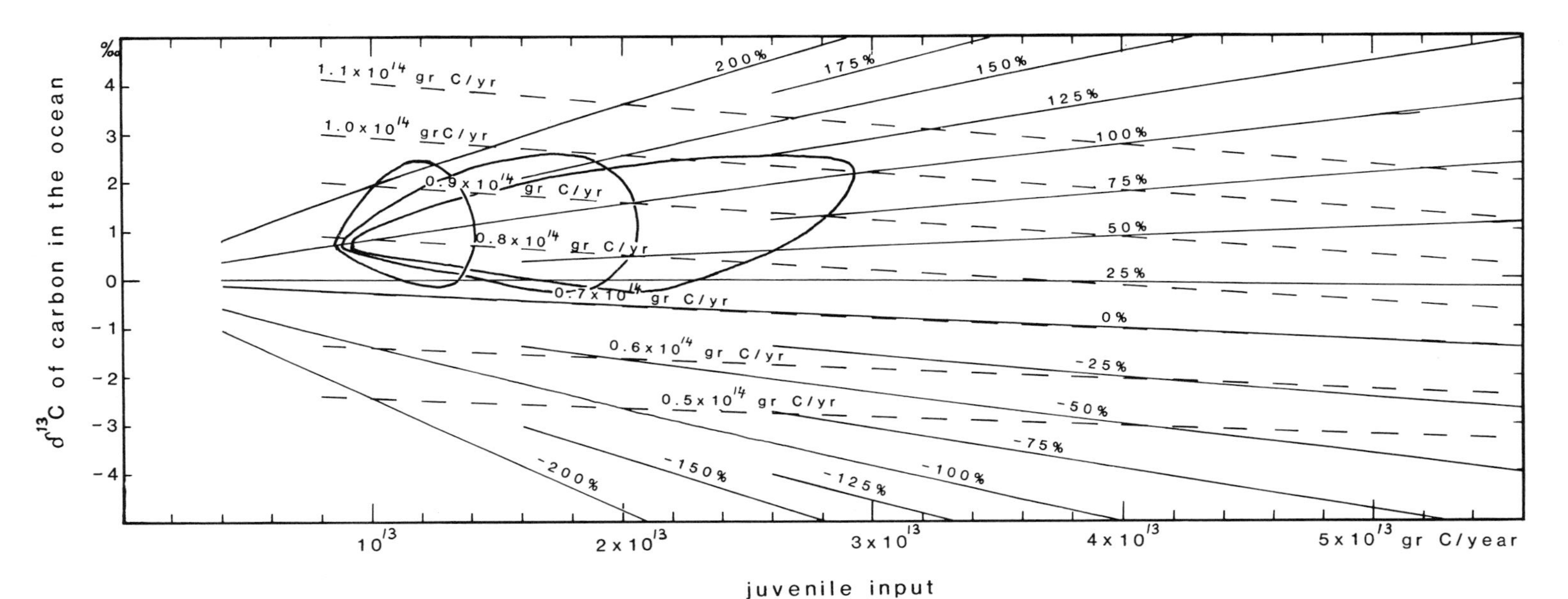

FIG. 5. Effect of size of fluxes of juvenile carbon and of organic matter fossilized (in aquatic and terrestrial environments) upon $\delta^{13}C$ of the ocean. Solid diverging lines: percentage of the amount of carbon, expressed relative to the annual juvenile input, stored in excess of (100–200%) or less than (−200–+100%) the amount of carbon brought into the biosphere by weathering and oxidation of fossil organic matter (0.7×10^{14} gr C/yr). At the 100% line all of the juvenile C input is stored as organic matter, and the amount of C in the biosphere remains stable. Above +100% a decrease, and at less than +100% an increase of the sum of atmospheric and oceanic carbon occurs. Broken lines: amount of organic matter annually fossilized. Curves at left: possible scenarios for the change of $\delta^{13}C$ of ocean dissolved carbon during the early Cretaceous (cf. Scholle & Arthur 1980): increase of juvenile CO_2 output, followed by an increase of storage of organic C, a decrease of juvenile output, and finally a decrease of excess storage of organic matter, and a return to 'normal' conditions. Weathering and burial of $CaCO_3$ are assumed constant in the model (1.5×10^{14} gr C/year).

Causes and results: conclusions

Did excess supply of CO_2 and nutritious elements drive the marine system to anoxia, or did anoxia sequester carbon dioxide and nutrients, which happened to be made available by increase of juvenile production and by the occupation of vast land masses by the sea?

It is suggested (Fig. 2) that excess juvenile production of CO_2 led to an initially moderate greenhouse warming, later reinforced by an escape of CO_2 from the warming ocean. In response, a climate induced decrease in the rate of oceanic circulation led to poor ventilation of deep waters, causing expansion of O_2 depleted waters and an increase of burial or organic matter in pelagic sediments. The transfer into the marine biosystem of nutrients from the terrestrial burial and erosion cycle during the (slow) transgression encouraged the accumulation of organic matter in black shales.

In the long run, a close link between juvenile output and burial of carbon in reduced form seems likely, whereas the absence of such a relation would put heavy constraints on the exogenic chemical cycle.

The increase of burial rate of organic matter of (only) about 20%, and the fact that only a tiny fraction of the primary produced organic matter tends to be fossilized, imply that there is no need to assume increases of primary production in the ocean. On the contrary, in analogy with the observation of Shimkus & Trimonis (1974) that in the present day Black Sea sediments are richest in organic matter below zones of lowest primary production, large parts of the Cretaceous oceans may have had low productivity.

ACKNOWLEDGEMENTS. Critical reviews and comments by H.C. Jenkyns, N.J. Shackleton, and C.P. Summerhayes are gratefully acknowledged. This is Comparative Sedimentology Division publication number 60.

References

ARTHUR, M.A. 1979. North Atlantic black shales: the record at site 398 and a brief comparison with other occurrences. *In:* RYAN, W.B.F., SIBUET, J.C. *et al.* (eds), *Init. Repts DSDP* **47.** US Govt Print. Off., Washington, DC. 719–38.

——, DEAN W.E. & SCHLANGER, S.O. 1985. Variations in the global carbon cycle during the Cretaceous related to climate volcanism, and changes in atmospheric CO_2. *In:* SUNDQUIST, G. & BROECKER, W.S. (eds), *Natural Variations in the Carbon Cycle.* Am. Geophys. Union, in press.

——, DEAN, W.E. & STOW, D.A.V. 1984. Models for the deposition of Mesozoic-Cenozoic fine-grained organic-carbon-rich sediment in the deep sea. *In:* STOW, D.A.V. & PIPER, D.J.W. (eds), *Fine-Grained Sediments: Deep-Water Processes and Facies. Geol. Soc. Lond. Spec. Publ.* **15.** 527–60. Published by Blackwell Scientific Publications, Oxford.

—— & NATLAND, J.H. 1979. Carbonaceous sediments in the North and South Atlantic: the role of salinity in stable stratification of early Cretaceous basins. *In:* TALWANI, M., HAY, W.W. & RYAN, W.B.F. (eds), *Deep Drilling Results in the Atlantic Ocean: Continental Margins and Paleoenvironment.* Am. Geophys. Union. Maurice Ewing Series 3, 375–401.

——, SCHLANGER, S.O. & JENKYNS, H.C. 1986. The Cenomanian-Turonian Oceanic Anoxic Event, II. Paleoceanographic controls on organic matter production and preservation. *In:* BROOKS, J. & FLEET, A. 1984. *Marine Petroleum Source Rocks. Spec. Publ. Geol. Soc. Lond.* Published by Blackwell Scientific Publications, Oxford.

BARRON, E.J. 1983. A warm, equable Cretaceous: the nature of the problem. *Earth Sci. Rev.* **19,** 305–38.

—— 1984. Cause of the Tertiary global cooling trend: results from climate model simulations. *Geol. Soc. Am. Abstr. with Programs 97th Annual Meeting.* 437–8.

—— & WASHINGTON, W.M. 1982. Cretaceous climate: a comparison of atmospheric simulations with the geologic record. *Palaeogeogr., Palaeoclimatol., Palaeoecol.* **40,** 103–33.

BERGER, W.H. 1979. Impact of deep-sea drilling on paleoceanography. *In:* TALWANI, M., HAY, W.W. & RYAN, W.B.F. (eds), *Deep Drilling Results in the Atlantic Ocean: Continental Margins and Paleoenvironment.* Am. Geophys. Union, Maurice Ewing Series 3. 297–314.

—— 1982. Deglacial CO_2 buildup: constraints on the coral-reef model. *Palaeogeogr., Palaeoclimatol., Palaeoecol.* **40,** 235–53.

BERNER, R.A. 1982. Burial of organic carbon and pyrite sulfur in the modern ocean: its geochemical and environmental significance. *Am. J. Sci.* **282,** 451–73.

—— & RAISWELL, R. 1983. Burial of organic carbon and pyrite sulfur in sediments over Phanerozoic time: a new theory. *Geochim. Cosmochim. Acta* **47,** 855–62.

BOLIN, B., DEGENS, E.T., DUVIGNEAUD, P. & KEMPE, S. 1979. The global biochemical carbon cycle. *In:* BOLIN, B., DEGENS, E.T., KEMPE, S. & KETNER, P. (eds), *The Global Carbon Cycle.* John Wiley & Sons, Chichester. 1–56.

BRALOWER, T.J. & THIERSTEIN, H.R. 1984. Low productivity and slow deep-water circulation in mid-Cretaceous oceans. *Geology* **12,** 614–8.

BRASS, G.W., SOUTHAM, J.R. & PETERSON, W.H. 1982. Warm saline bottom water in the ancient ocean. *Nature* **296,** 620–3.

BROECKER, W.S. & PENG, T.-H. 1982. *Tracers in the Sea.*

Lamont-Doherty Geological Observatory, Palisades, New York. 690 p.

BRUMSACK, H.J. 1984. A note on Cretaceous black shales and recent sediments from oxygen deficient environments: paleoceanographic implications. *In:* THIEDE, J. & SUESS, E. (eds), *Coastal Upswelling; Its Sediment Record.* Plenum Press, New York. 471–83.

BUDYKO, M.I. & RONOV, A.B. 1979. Chemical evolution of the atmosphere in the Phanerozoic. *Geokhimiya* **5,** 643–53.

DE BOER, P.L. 1982. Remarks about the stable isotope composition of cyclic pelagic sediments from the Cretaceous in the Apennines (Italy). *In:* SCHLANGER, S.O. & CITA, M.B. (eds), *Nature and Origin of Cretaceous Carbon-Rich Facies.* Academic Press, London. 129–43.

—— 1983. Aspects of Middle Cretaceous pelagic sedimentation in S. Europe. *Geol. Ultraiectina, Utrecht.* **31,** 112 p.

DEMAISON, G.J. & MOORE, G.T. 1980. Anoxic environments and oil source bed genesis. *Bull. Am. Ass. Petrol. Geol.* **64,** 1179–209.

DELWICHE, C.C. & LIKENS, G.E. 1977. Biological response to fossil fuel combustion products. *In:* STUMM, W. (ed.), *Global Chemical Cycles and Their Alterations by Man.* Report Dahlem Workshop, Berlin 1976. 73–88.

FISCHER, A.G. 1981. Climatic oscillations in the biosphere. *In: Biotic Crises in Ecological and Evolutionary Time.* Academic Press, London. 103–31.

FREYER, H.-D. 1979. Variations in the atmospheric CO_2 content. *In:* BOLIN, B., DEGENS, E.T., KEMPE, S. & KETNER, P. (eds), *The Global Carbon Cycle.* John Wiley & Sons, Chichester. 79–99.

FROELICH, P.N., BENDER, M.L., LUEDTKE, N.A., HEATH, G.R. & DEVRIES, P. 1982. The marine phosphorus cycle. *Am. J. Sci.* **282,** 474–511.

GARRELS, R.M., MCKENZIE, F.T. & HUNT, C. 1973. *Chemical Cycles and the Global Environment.* Kaufman Inc., Los Angeles.

HABIB, D. 1982. Sedimentary supply origin of Cretaceous black shales. *In:* SCHLANGER, S.O. & CITA, M.B. (eds), *Nature and Origin of Cretaceous Carbon-Rich Facies.* Academic Press, London. 113–27.

HALLAM, A. 1971. Re-evaluation of the paleographic argument for an expanding Earth. *Nature* **232,** 180–2.

HANCOCK, J.M. & KAUFFMAN, E.G. 1979. The great transgression of the Late Cretaceous. *J. geol. Soc. Lond.* **136,** 175–86.

HAY, W.W. & SOUTHAM, J.R. 1979. Modulation of marine sedimentation by the continental shelves. *In:* ANDERSEN, N.R. & MALAHOFF, A. (eds), *The Fate of Fossil Fuel CO_2 in the Oceans.* Plenum Press, New York. 569–604.

HAYS, J.D. & PITMAN, W.C. 1973. Lithospheric plate motion, sea level changes and ecological consequences. *Nature* **246,** 18–22.

IRVING, E., NORTH, F.K. & COULLARD, R. 1974. Oil, climate and tectonics. *Can. J. Earth Sci.* **11,** 1–15.

JAVOY, M., PINEAU, F. & ALLEGRÉ, C.-J. 1982. Carbon geodynamic cycle. *Nature* **300,** 171–3.

——, PINEAU, F. & ALLEGRÉ, C.-J. 1983. Reply to Walker 1983. *Nature* **303,** 731.

JENKYNS, H.C. 1980. Cretaceous anoxic events: from continents to oceans. *J. geol. Soc. London.* **137,** 171–88.

LIVINGSTONE, D.A. 1973. The biosphere. *In:* WOODWELL, G.M. & PECAN, E.V. (eds), *Carbon and the Biosphere.* Proc. 24th Brookhaven Symp. in Biol. Techn. Inf. Centre. 1–5.

MATTEY, D.P., CARR, R.H., WRIGHT, I.P. & PILLINGER, C.T. 1984. Carbon isotopes in submarine basalts. *Earth Planet. Sci. Lett.* **70,** 196–206.

MOOK, W.G., BOMMERSON, J.C. & STAVERMAN, W.H. 1974. Carbon isotope fractionation between dissolved bicarbonate and gaseous carbon dioxide. *Earth Planet. Sci. Lett.* **22,** 169–76.

MOPPER, K. & DEGENS, E.T. 1979. Organic carbon in the ocean: nature and cycling. *In:* BOLIN, B., DEGENS, E.T., KEMPE, S. & KETNER, P. (eds), *The Global Carbon Cycle.* John Wiley & Sons, Chichester. 293–316.

PYTKOWICZ, R.M. 1973. The carbon dioxide system in the oceans. *Hydrologie* **35,** 8–28.

SAVIN, S.M. 1977. The history of the earth's surface temperature during the past 100 million years. *Ann. Rev. Earth Planet. Sci.* **5,** 319–55.

SCHLANGER, S.O. & JENKYNS, H.C. 1976. Cretaceous oceanic anoxic events—causes and consequences. *Geologie en Mijnbouw* **55,** 179–84.

SCHOLLE, P.A. & ARTHUR, M.A. 1980. Carbon-isotope fluctuations in Cretaceous pelagic limestones: potential stratigraphic and petroleum exploration tool. *Bull. Am. Ass. Petrol. Geol.* **64,** 67–87.

SHIMKUS, K.M. & TRIMONIS, E.S. 1974. Modern sedimentation in Black Sea. *In:* DEGENS, E.T. & ROSS, D.A. (eds), *The Black Sea—Geology, Chemistry and Biology. Mem. Am. Ass. Petrol. Geol.* **20,** 249–78.

SKIRROW, G. 1975. Table of physical and chemical constants relevant to marine chemistry. *In:* RILEY, J.P. & SKIRROW, G. (eds), *Chemical Oceanography.* Vol. 1. 599–631.

SUMMERHAYES, C.P. 1983. Sedimentation of organic matter in upwelling regimes. *In:* THIEDE, J. & SUESS, E. (eds), *Coastal Upwelling; Its Sediment Record.* 29–72.

TISSOT, B. 1979. Effects of prolific petroleum source rocks and major coal deposits caused by sea-level changes. *Nature* **277,** 463–5.

VEIZER, J., HOLSER, W.T. & WILGUS, C.K. 1980. Correlation of $^{13}C/^{12}C$ and $^{34}S/^{32}S$ secular variations. *Geochim. Cosmochim. Acta* **44,** 579–87.

WALKER, J.C.G. 1983. Carbon geodynamic cycle. *Nature* **303,** 730–1. (see Javoy 82/83)

WILDE, P. & BERRY, W.B.N. 1982. Progressive ventilation of the oceans—potential for return to anoxic conditions in the post-Paleozoic. *In:* SCHLANGER, S.O. & CITA, M.B. (eds), *Nature and Origin of Cretaceous Carbon-Rich Facies.* Academic Press, London. 209–24.

P.L. DE BOER, Comparative Sedimentology Division, Institute of Earth Sciences, Budapestlaan 4, P.O. Box 80.021, 3508 TA Utrecht, The Netherlands.

Organic geochemistry of Cretaceous organic-carbon-rich shales and limestones from the western North Atlantic Ocean

Philip A. Meyers, Keith W. Dunham & Pamela L. Dunham

SUMMARY: Organic-carbon-rich black shales are found in Cretaceous sedimentary settings on the North American outer continental rise. Comparison of carbon isotope and molecular biomarker compositions of black shales with adjacent lithologies gives evidence of enhanced preservation of marine organic matter in the darker layers. In middle Cretaceous sequences, the black shales represent short episodes of exceptional preservation of organic materials within an oxygenated, deep-ocean depositional setting. Neocomian black shales show poorer preservation and were deposited within turbiditic carbonates and sandstones, containing large proportions of land-derived organic matter.

Occurrences of dark-coloured layers of Cretaceous rocks having relatively high concentrations of organic matter have been found in numerous locations studied as part of the Deep Sea Drilling Project (DSDP). The distribution of such occurrences in the North Atlantic Ocean has been discussed by Arthur (1979), Tucholke & Vogt (1979), Thierstein (1979), Summerhayes (1981, 1986), Graciansky *et al.* (1981), Weissert (1981), and Waples (1983), among others, with the intention of identifying the palaeoceanographic factors involved in the formation of these unusual strata, commonly called 'black shales'. Improved preservation of organic matter, increased contribution of continental organic matter to oceanic basins, and enhanced production of marine organic matter are some of the factors which have been suggested. Because these three possibilities affect the character of the organic content of black shales, investigations of the type of organic matter in North Atlantic examples have been done and are summarized by Tissot *et al.* (1980), Summerhayes (1981, 1986), Katz & Pheifer (1983), and Graciansky *et al.* (1982). Varying proportions of marine and terrigenous organic constituents are found in sediments deposited at different times and locations in the Cretaceous Atlantic Ocean, and evidence of low levels of dissolved oxygen, which can arise from several processes, is commonly present.

During DSDP Leg 93, black shales were found in rocks ranging in age from Berriasian to Santonian at Site 603 on the outer Hatteras Rise in the North American Basin (Fig. 1). Neither the lithologic settings nor the organic carbon contents of these deposits are uniform over this considerable time span, although the common presence of turbidites and of bioturbation shows certain similarities existed in the depositional environments. The Cretaceous rock formations underlying the North Atlantic as described by Jansa *et al.* (1979) were encountered at Site 603. Late Turonian to Santonian rocks consist of variegated claystones with sparse black shales, corresponding to the Plantagenet Formation. The Aptian to Turonian section contains abundant black shales interspersed among red and green claystones, and it corresponds to the Hatteras Formation. Limestones and sandstones with black claystone turbidites make up the Berriasian–Aptian Blake-Bahama Formation. Cenomanian black shales contain the highest concentrations of organic carbon (Meyers, in press), and all black shales generally exist as thin layers, a few centimetres thick, surrounded by thicker layers of rocks poor in organic carbon.

A major objective of DSDP Leg 93 was to provide new information about the palaeoceanographic conditions leading to the formation of Mesozoic black shales. Characterization of the organic matter contained within black shales and their adjacent organic-carbon-poor lithologies contributes to this information and constitutes the research effort of a number of investigators. In this report, comparisons of analyses done on black shales and adjacent strata from the Hatteras and Blake-Bahama Formations are described, and these new data are compared with the results of earlier investigations in the western North Atlantic Ocean.

Experimental

Samples

Three groups of Hole 603B samples from different ages in the Cretaceous Period were selected on board D/V *Glomar Challenger* for this study and augmented from post-cruise sampling of frozen core sections. Each group contains organic-car-

From SUMMERHAYES, C.P. & SHACKLETON, N.J. (eds), 1986, *North Atlantic Palaeoceanography*, Geological Society Special Publication No. 21, pp. 333–345.

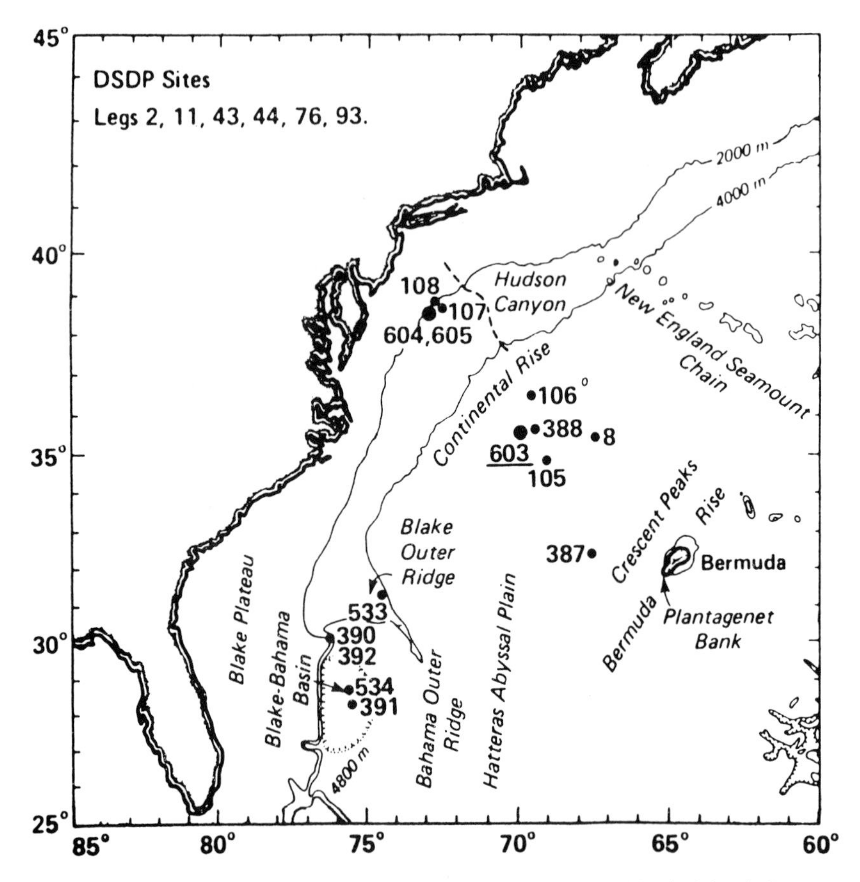

FIG. 1. Location of DSDP Site 603 and some other DSDP sites in the western North Atlantic Ocean.

bon-rich and -lean strata. Hatteras Formation samples consist of three Cenomanian samples from Section 603B-37-4 (sub-bottom depth 1160 m) and five Albian samples from Section 603B-40-2 (1183 m). Three samples of Hauterivian age from Section 603B-66-2 (1425 m) are from the Blake-Bahama Formation. All samples were frozen immediately after collection and remained in this state until analysis began.

Analysis

The samples were freeze-dried for determination of their total carbon contents with a Hewlett-Packard 185B CHN Analyzer. Residual carbon was measured after HCl dissolution of carbonates and was considered to represent the total organic carbon content. Percent calcium carbonate was calculated from the difference between initial and residual carbon contents. Organic matter atomic C/N ratios were determined from residual carbon values. Percent organic carbon contents of the samples were calculated on a dry-weight basis for the original, carbonate-containing sediment.

Stable carbon isotope ratios of the organic carbon content of these samples were determined on carbonate-free samples using a VG Micromass 602 mass spectrometer calibrated with NBS-20 (carbonate) and NBS-21 (graphite) standards. Data are corrected for ^{17}O and are presented in terms of the PDB standard.

Geolipid contents of the freeze-dried samples were obtained by Soxhlet extraction for 24 hours with toluene/methanol. Fatty acids were treated with methanolic boron trifluoride to convert them to their methyl esters, and then the geolipid subfractions were separated by column chromatography on alumina over silica gel. The subfractions so obtained contained alkanes and alkenes, aromatic hydrocarbons, fatty acid methyl esters, and hydroxy lipids, including sterols and alkanols. Hydroxy compounds were silylated with BSTFA prior to gas chromatography.

Splitless injection gas-liquid chromatography was employed to determine the types and amounts of components present in the geolipid subfractions. A Hewlett-Packard 5830 FID gas chromatography equipped with a 20 m SE54 fused silica capillary column was used with hydrogen as carrier gas. Quantification was achieved through the use of known amounts of internal standards added to each sample before column chromatography. Individual compounds are tentatively identified by retention times in this preliminary survey. Reported values have been corrected for the small amounts of laboratory contaminants determined from blanks and for mass discrimination over the wide molecular-weight ranges surveyed.

Results and discussion

Organic carbon, C/N ratios, and carbon isotopic composition

The four examples of black shales listed in Table 1 have significantly higher concentrations of organic carbon than do their adjacent lighter coloured strata. The seven latter samples average 0.3% organic carbon, which is the value compiled by McIver (1975) from data from DSDP Legs 1 through 33 and which can be considered a background level for ancient deep-ocean sediments. The higher concentrations encountered in the black shales are unusual for marine sediments and are not found in modern oceanic environments except where poorly oxygenated conditions prevail (Demaison & Moore 1980).

C/N ratios of Cenomanian-to-Albian black shales from Site 603 have averages over twice as great as their neighbouring green and red claystones, whereas the average ratios of Neocomian black shales and of surrounding lithologies have mutually similar values around 35 (Meyers, in press). The C/N values of the samples listed in Table 1 generally agree with this pattern–black shales are mutually similar, red and green claystones have low values, and the limestone sample has a high C/N ratio. Based upon a survey of marine sediments, Premuzic *et al.* (1982) suggest C/N ratios less than eight indicate mostly marine organic matter and values greater than 15 show a predominance of land-derived material. With increasing time of burial, however, C/N ratios change as a result of diagenesis. Waples & Sloan (1980) report a gradual decrease in C/N values from *c.*10 in Quaternary sediments to *c.*4 in Miocene samples, followed by increases in older sediments. Furthermore, C/N values tend to be high in sediments where marine organic matter is well preserved, such as in Cenomanian black shales from the Angola Basin (Meyers *et al.* 1984). In the case of these Site 603 black shales, the high C/N values may record selective preservation of carbonaceous components of marine organic matter relative to nitrogenous components, combined with varying proportions of continentally derived organic matter.

Organic matter in the organic-carbon-lean green and red claystones of the Hatteras Formation is probably made up of detrital, highly oxidized materials similar to those found in most deep-sea sediments (cf. Degens & Mopper 1976). Such sediments commonly have low C/N ratios in

TABLE 1. *General geochemical characteristics of Hole 603B samples from the Hatteras and Blake-Bahama formations*

Sample	Description	% $CaCO_3$	% C_{org}	C/N	$\delta^{13}C_{org}$
603B-37-4 (Cenomanian)					
119–123 cm	green claystone	4	0.33	6.6	−27.9
123–126 cm	black shale	<1	2.18	19.6	−26.4
126–128 cm	green claystone	<1	0.47	<1	−24.3
603-40-2 (Albian)					
26–29 cm	red claystone	<1	0.19	<1	−24.1
37–41 cm	green claystone	2	0.30	35.0	−23.2
42–43 cm	black shale	<1	0.82	23.9	−27.7
45–48 cm	green claystone	<1	0.20	1.5	−26.7
51–54 cm	red claystone	<1	0.31	<1	−24.8
603B-66-2 (Hauterivian)					
66–70 cm	black shale	17	1.96	6.3	−24.8
123–126 cm	limestone	96	0.04	98.0	−28.7
133–135 cm	black shale	19	0.99	35.6	−25.6

older strata. The high C/N value of the Blake-Bahama limestone suggests this sample may contain substantial amounts of coaly material eroded from continental deposits. The Rock Eval results (Meyers, in press) show the relatively inert character of organic matter in similar limestones, even though some are fairly rich in organic carbon.

Carbon isotope data provide further information about the sources of organic matter in these samples. In general, modern land-derived organic matter is more depleted in ^{13}C than is marine organic matter, although carbon isotope ratios appear to be sensitive to diagenetic modification in black shale deposits and hence should not, by themselves, be considered absolute determinants of source (Dean *et al.* 1984a; Meyers *et al.* 1984). Comparison of carbon isotope compositions of black shales with adjacent organic-carbon-lean samples from Hole 603B shows a positive relationship between higher organic carbon concentration and a larger proportion of marine isotopic character in black shale samples from the Hatteras Formation, but this trend is not present in the Blake-Bahama samples (Dunham *et al.*, in press). The isotope values of rocks surrounding the black shales display a wide range, which suggests varying origins for the organic matter in these layers of turbiditic, organic-carbon-lean rock. Carbon isotope data in Table 1 agree with with general picture. The claystone and limestone samples give a range of $\delta\ ^{13}C$ values from -23.2 to -28.7‰ which is independent of their narrow range in organic carbon contents. Black shale samples richer in organic carbon have isotope values that are isotopically heavier than the less rich black shales. This relationship may imply that the organic matter concentration of these Site 603 black shales is directly related to burial of greater proportions of marine organic matter superimposed upon a background of poorly preserved organic matter from variable sources.

Extractable alkanes, alkanoic acids, and alkanols

The geolipid composition of geological samples provides a wealth of information about organic matter sources and diagenesis, despite constituting only a minor fraction of the total organic content. This information is derived from the distribution of homologous series of similar compounds, such as the *n*-alkanes, and from the presence of individual biomarker molecules, such as the acyclic hydrocarbon pristane or the cyclic alcohol cholesterol. In this comparison of Site 603 samples, the authors concentrate on the distributions of homologous *n*-alkanes, *n*-alkanoic acids, and *n*-alkanols, and also consider their contents of the isoprenoid hydrocarbons pristane and phytane.

The geolipid distributions of the Cenomanian grouping of a black shale and neighbouring green claystones are compared in Fig. 2. Long-chain-length geolipids, such as C_{27} to C_{33} odd-numbered *n*-alkanes and C_{24} to C_{32} even-numbered *n*-alkanoic acids and *n*-alkanols, are present in all samples and are characteristic of land plant waxes (Simoneit 1978). Sample 603B-37-4, 119–123 cm, contains a significant contribution of the C_{17} to C_{21} *n*-alkanes indicative of algal inputs (Simoneit 1978). All of the fatty acid distributions are dominated by *n*-C_{16}, which is a ubiquitous component characteristic of all biota. The ratio of the isoprenoid hydrocarbons, pristane and phytane, is less than one in these Cenomanian samples (pristane is below detection in sample 603B-37-4, 126–128 cm). Although pristane/phytane values below one have been proposed as an indication of anoxic depositional conditions (Didyk *et al.* 1978), methanogenic bacteria generate phytane during anaerobic fermentation of organic matter (Risatti *et al.*, 1984). Because such fermentation can occur under anoxic conditions deeper in sediments as well as at an anoxic water/sediment boundary, low pristane/phytane ratios may record bacterial fermentation rather than a specific depositional condition.

Distributions of geolipids extracted from the Albian and Hauterivian samples (Figs 3 and 4) are generally similar to those of the Cenomanian rocks. Intercomparison of all 11 samples reveals several patterns in geolipid compositions. Foremost among these is the relatively large proportion of terrigenous components in the five black shale samples. The importance of land-derived geolipids is especially conspicuous in the *n*-alkane distributions, which in this regard resemble distributions reported for black shales sampled from the Hatteras and Blake-Bahama Formations at DSDP Sites 391 and 534, farther south of Site 603 in the North American Basin (Deroo *et al.* 1978; Erdman & Schorno 1978; Stuermer & Simoneit 1978; Herbin *et al.* 1983). It is also evident, but to a lesser degree, in the *n*-alkanoic acid distributions of Samples 603B-37-4, 123–126 cm and 603B-40-2, 42–43 cm. These distributions are similar to one reported by Cardoso *et al.* (1978) in a black shale, also from the Hatteras Formation, at Site 391.

Another common pattern is the dominance of the C_{28} and C_{30} *n*-alkanols in all of these Site 603 samples. Long-chain *n*-alkanols such as these have been interpreted to be indicators of terrigenous geolipids in marine sediments (Brassell *et al.* 1982). Because turbidites are common through-

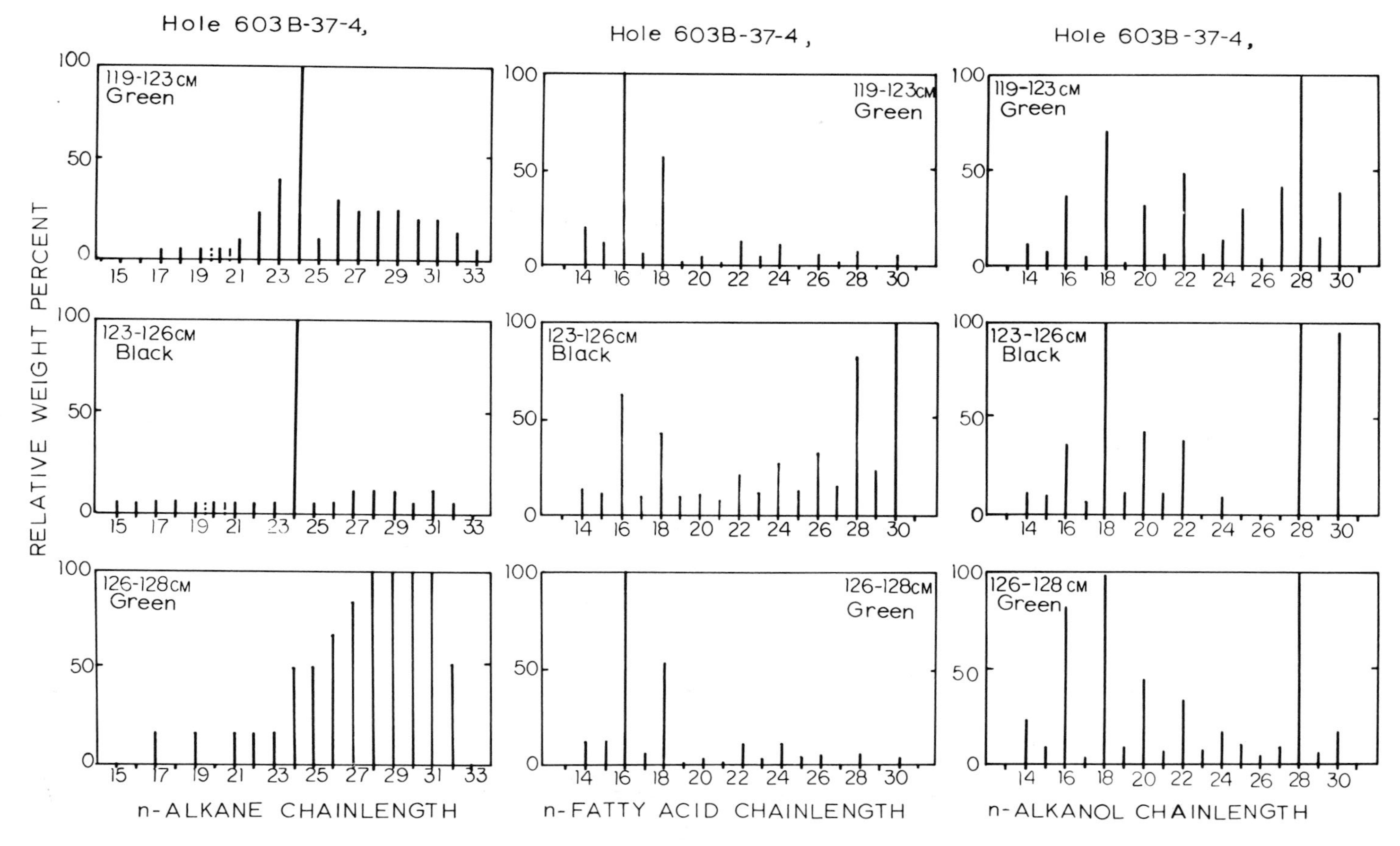

FIG. 2. Geolipid distributions of Cenomanian samples from Section 603B-37-4, Hatteras Formation. Relative abundances are normalized to the major component in each fraction. Isoprenoid hydrocarbons pristane and phytane are represented as dotted and dashed lines, respectively.

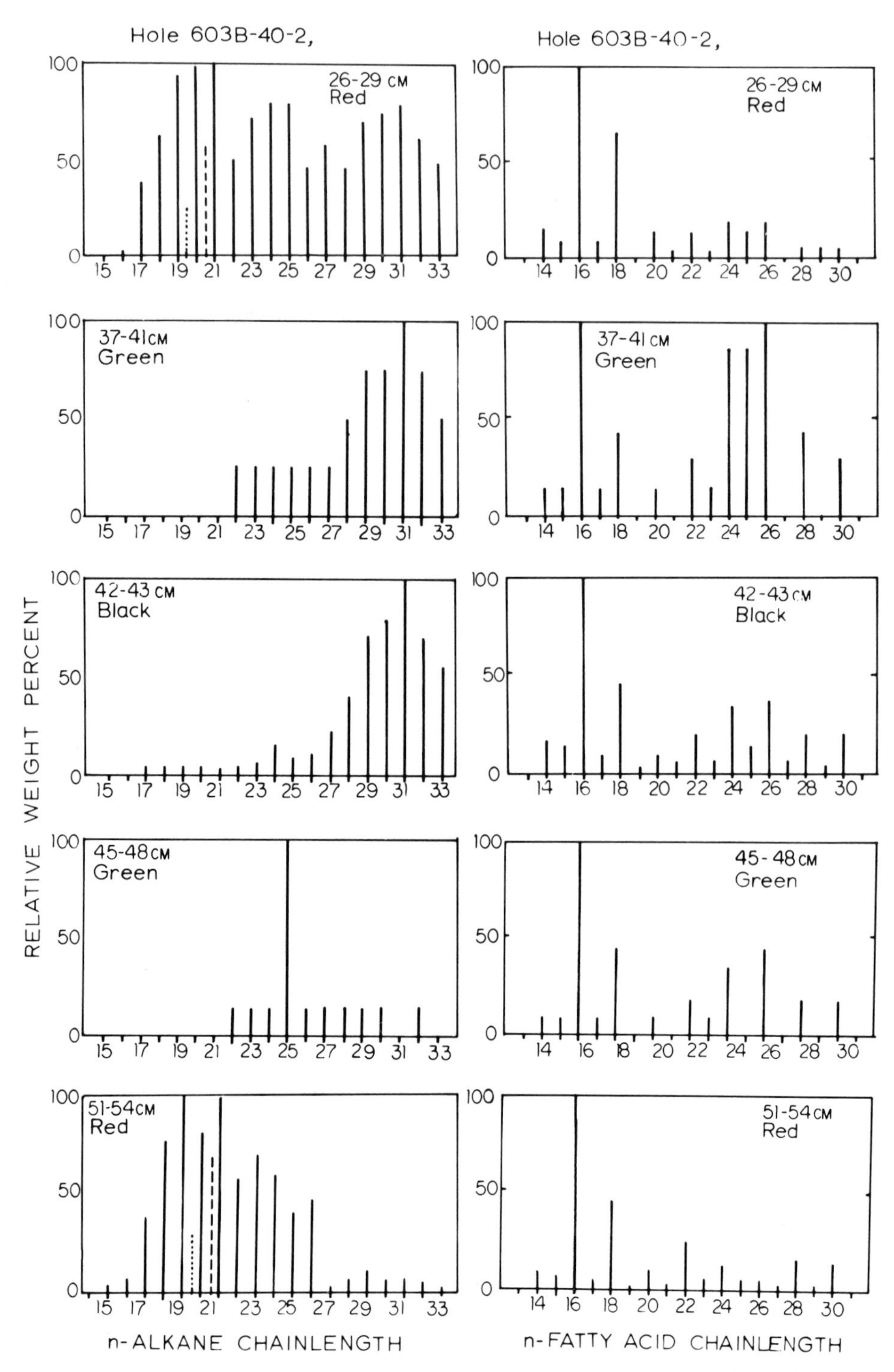

FIG. 3. Geolipid distributions of Albian samples from Section 603B-40-2, Hatteras Formation. Relative abundances are normalized to the major component in each fraction. Isoprenoid hydrocarbons pristane and phytane are represented as dotted and dashed lines, respectively.

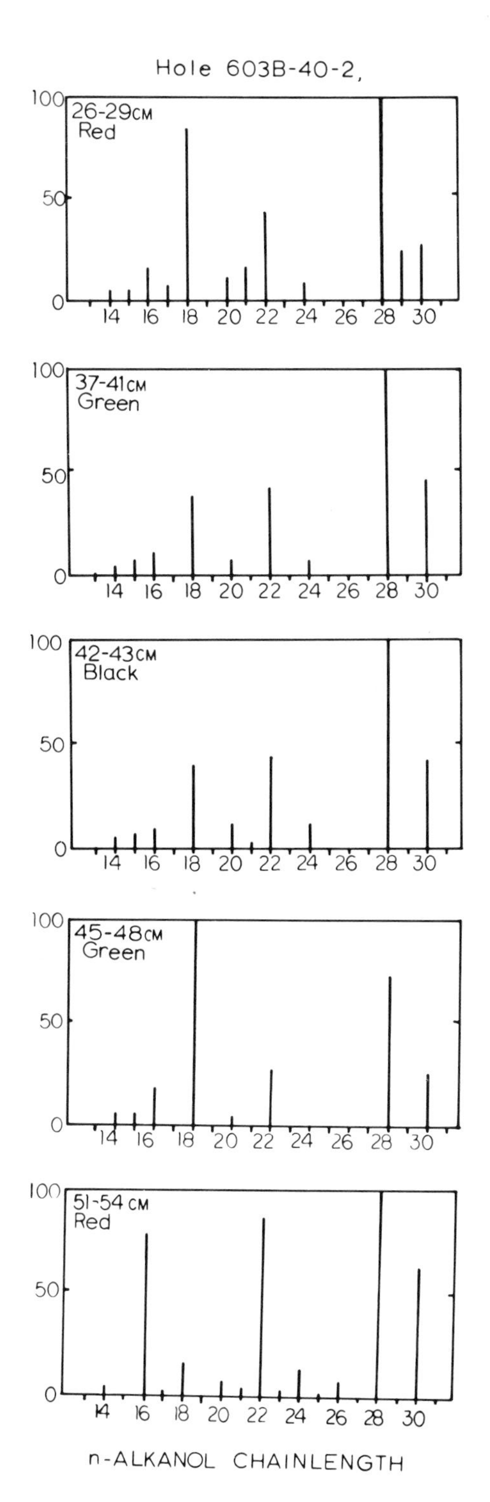

Fig. 3.

out the Hatteras and Blake-Bahama Formations, transport of continental plant components to this deep-sea location is quite feasible. It is surprising, however, that the *n*-alkanol distributions are so similar despite considerable differences in total organic carbon concentrations and in *n*-alkane distributions. Alkanols are considered to be more subject to microbial reworking than are alkanes, with the result that the C_{22} *n*-alkanol postulated to indicate microbial activity is the major alkanol found in open-ocean sediments in which organic carbon concentrations are low (Keswani *et al.* 1984). It seems paradoxical that only the two Hatteras Formation black shales (Figs 2 and 3) display major contributions of the C_{22} *n*-alkanol, unless they are the only samples which originally contained organic matter of the proper type and amount to support microbial activity.

A third, less well-developed pattern in these samples is a crude correspondence between carbon isotope ratios and geolipid distributions. This is best illustrated by the two Hauterivian black shales, which have δ ^{13}C of -24.8 and -25.6‰ and have very similar distributions of *n*-alkanes, *n*-alkanoic acids, and *n*-alkanols (Fig. 4), although they differ by a factor of two in organic carbon content. Within the group of Albian samples, the black shale and underlying green claystone have respective δ ^{13}C values of -27.7 and -26.7‰ and similar *n*-alkane distributions (Fig. 3).The *n*-alkanoic acid and *n*-alkanol distributions bear some mutual similarities, but they are not strong. These samples differ in organic carbon content by a factor of four. Finally, the Cenomanian black shale and its overlying green claystone are somewhat similar in having long-chain *n*-alkanes and *n*-alkanoic acids and important contributions of the C_{22} *n*-alkanol (Fig. 2). Their δ ^{13}C values are -26.4 and -27.9‰, respectively, and the black shale contains seven times the organic carbon content of the green claystone. These correspondences, although weak, suggest that some of the adjacent strata received organic matter from the same sources and differ partly in the proportions from these sources, but mostly in the amount of material preserved in them. In addition, the lack of correspondence in isotope values between the Cenomanian, Albian, and Hauterivian black shales, despite general similarities in geolipid distributions, implies temporal differences in sources of the bulk, non-lipid organic matter or in the isotopic compositions of the carbon reservoirs available to the biosynthetic sources.

Concentrations of extractable geolipids are given in Table 2 in terms of parts per million of dry sample weight and relative to the organic carbon concentrations. Compared with samples

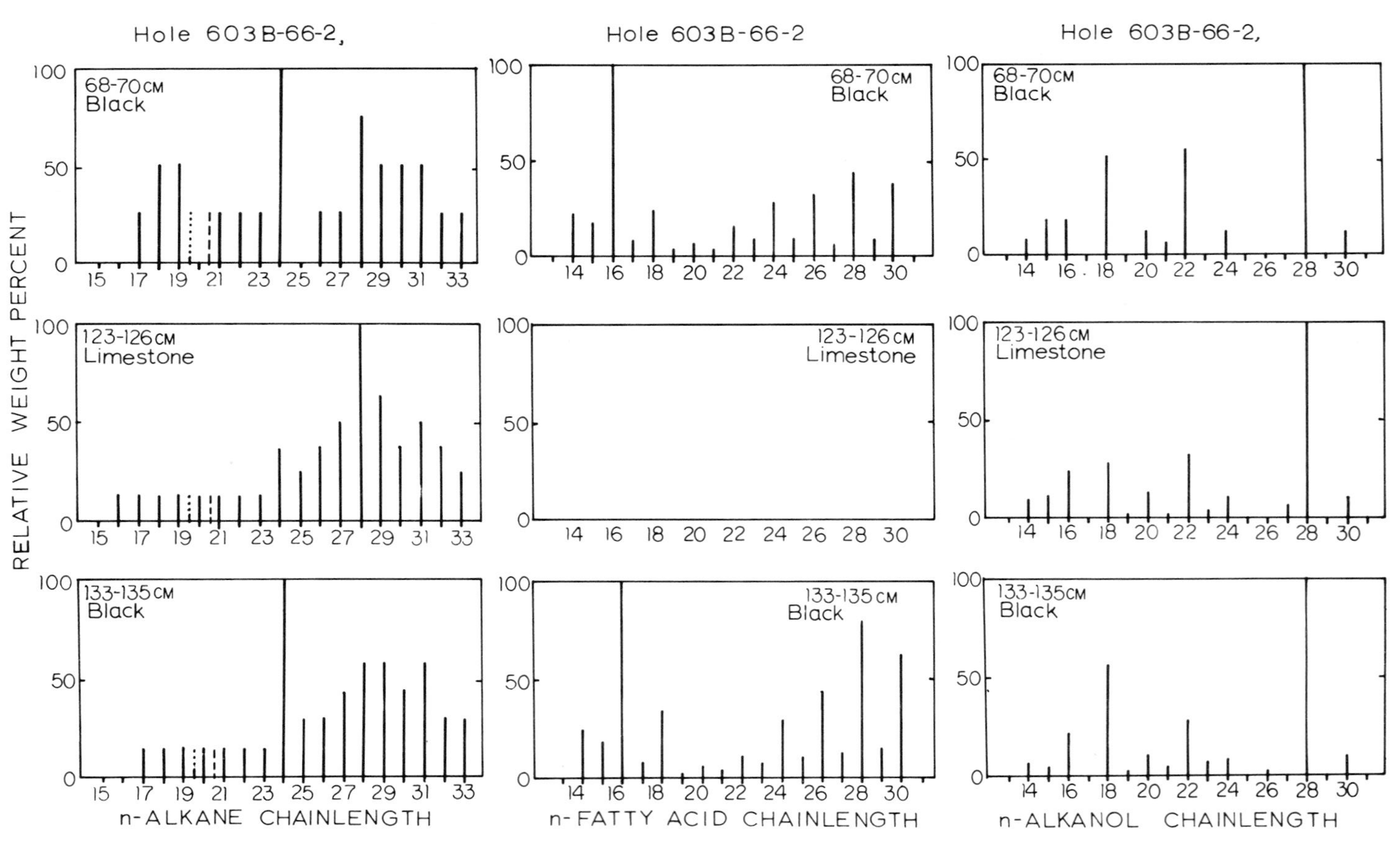

FIG. 4. Geolipid distributions of Hauterivian samples from Section 603B-66-2, Blake-Bahama Formation. Relative abundances are normalized to the major component in each fraction. Isoprenoid hydrocarbons pristane and phytane are represented as dotted and dashed lines, respectively.

TABLE 2. *Concentrations of extractable geolipids in Hole 603B samples from the Hatteras and Blake-Bahama Formations*

Sample	Description	*n*-Alkanes		*n*-Alkanoic Acids		*n*-Alkanols	
		ppm	ppm/C_{org}*	ppm	ppm/C_{org}	ppm	ppm/C_{org}
603B-37-4 (Cenomanian)							
119–123 cm	green claystone	1.6	4.9	10.4	31.5	36.2	109.7
123–126 cm	black shale	3.3	1.5	10.1	4.6	11.2	5.1
126–128 cm	green claystone	0.8	1.7	15.3	32.6	18.5	39.4
603B-40-2 (Albian)							
26–29 cm	red claystone	1.9	10.0	6.5	34.2	15.1	79.5
37–41 cm	green claystone	1.5	5.0	9.2	30.7	13.8	46.0
42–43 cm	black shale	1.9	2.3	1.1	1.3	2.7	3.3
45–48 cm	green claystone	0.5	2.5	4.4	22.0	10.1	50.5
51–54 cm	red claystone	0.8	2.6	8.9	28.7	11.9	38.4
603B-66-2 (Hauterivian)							
66–70 cm	black shale	3.8	1.9	18.0	9.2	17.1	8.7
123–126 cm	limestone	2.8	70.0	9.0	225.0	28.4	710.0
133–135 cm	black shale	4.1	4.1	32.8	33.1	25.5	25.8

* Calculated from ppm extractable geolipid fraction divided by percent organic carbon

from DSDP Site 530 in the Angola Basin, the organic-carbon-lean samples contain about the same amounts of *n*-alkanes but are richer in alkanoic acids and alkanols (Meyers *et al.* 1984). The higher amounts of the latter two geolipid fractions may reflect larger proportions of waxy land plant material in the Site 603 samples, which is consistent with the geolipid distribution (Figs 2, 3 and 4). The black shale samples, in contrast, contain markedly lower concentrations than the black shales from the Angola Basin (Meyers *et al.* 1984) and often less geolipids than adjacent strata. It is evident that these samples of Site 603 black shales are geolipid-lean. Part of the explanation for this characteristic may arise from enhanced preservation of non-lipid components of sediment organic matter in black shales. Such dilution of geolipid concentrations is found in black shales relatively rich in organic carbon (cf. Meyers *et al.* 1984). In black shales with relatively low amounts of organic carbon, such as sample 603B-40-2, 42–43 cm, it appears that even the lipid components have been microbially reworked and nearly all of the remaining organic matter is detrital, inert material.

Palaeoceanographic implications

Both the type and the amount of organic matter contained within black shales and their adjacent strata provide information about the palaeoceanographic processes which participated in the formation of Cretaceous black shales in the western North Atlantic Ocean. Possible explanations of black shale depositional conditions are exemplified by Graciansky *et al.* (1982), Summerhayes & Masran (1983), and Summerhayes (1986) and combine the effects of sea-level changes, changes in continental climates, nutrient availability, and basin morphology to achieve periods of midwater or bottom anoxia in the Mesozoic North Atlantic. It is probable these are the important factors, yet Graciansky *et al.* (1984) caution that a simple model employing one set of conditions appears inadequate to explain the geographically broad extent of Cenomanian black shales. Similar caution may be appropriate in seeking explanations of black shale formation during other times as well.

The proportion of continental and marine organic matter present in black shale samples from the western North Atlantic varies considerably. Although most of the organic matter appears to be terrigenous (Katz & Pheifer 1982), the marine fraction increases with distance from North America (Tissot *et al.* 1980; Summerhayes & Masran 1983). This pattern has been explained by Summerhayes & Masran (1983) to reflect the decrease in turbidite dilution of marine sediments with continental materials as distance from shore increases. An exception to this generality occurs in Cenomanian black shales, where large proportions of marine organic matter are found at Site

105 (Summerhayes 1981) and Site 603 (Dunham *et al.*, in press), which are both located on the continental rise and in turbiditic environments. This exception illustrates the regional and temporal variability that can exist in the mixture of organic matter types in black shales. On the basis of carbon isotope ratio variability noted by Dunham *et al.* (in press), the organic-carbon-lean rocks surrounding black shale samples from Site 603 also appear to demonstrate considerable variation in their proportions of marine and terrigenous organic matter, although Summerhayes (1981), Dean *et al.* (1984a), and Meyers *et al.* (1984) question the accuracy of carbon isotope data alone in identifying source types in Cretaceous samples.

Preservation of organic matter is an important element in forming organic-carbon-rich shales. Anoxic bottom waters have been postulated to have enabled enhanced preservation of organic matter in the Cretaceous North Atlantic (Arthur & Schlanger 1979; Jenkyns 1980; Tissot *et al.* 1980; Summerhayes & Masran 1983; Bralower & Thierstein 1984; as examples). The abundant presence of burrowed, oxidized sediments above and below the black shales in this ocean argue against such bottom water anoxia being extensive in either volume or duration (Katz & Pheifer 1982; Waples 1983). An alternative scenario leading to greater preservation of organic matter calls for the midwater oxygen minimum zone to become intensified and perhaps expanded through sluggish circulation or enhanced influx of organic matter (for example, Demaison & Moore 1980; Waples 1983). Where a midwater anoxic layer intercepts the ocean bottom, sediments rich in organic matter can accumulate. Downslope movement of such sediments can result in formation of black shales within deep-ocean turbiditic sequences, as suggested by Dean *et al.* (1984b) for Site 530 in the Angola Basin, if reburial is sufficiently rapid to preserve the organic matter. This scenario points out another factor important to preservation of organic materials—the sedimentation rate. From a survey of DSDP data, Ibach (1982) shows that organic matter degradation is depressed as a function of sediment accumulation rates in Mesozoic sediments. Quicker burial results in better preservation, even under oxygenated bottom-water conditions, through establishment of anoxic conditions in the sediments whenever organic matter supplies are sufficient to deplete pore-water oxygen levels. This phenomenon has led Habib (1983) and Robertson & Bliefnick (1983) to suggest that Mesozoic black shales at Site 534 result primarily from rapid sedimentation of terrigenous organic matter associated with turbidity currents and that the western Atlantic need not have been anoxic during these episodes.

The high biological productivity associated with upwelling produces elevated concentrations of organic matter in underlying ocean margin sediments (e.g. Demaison & Moore 1980), and Parrish & Curtis (1982) note a correspondence between postulated palaeo-upwelling areas and occurrence of Cretaceous black shales. Nonetheless, the overall productivity of the western North Atlantic was probably low to moderate, based upon sediment compositions (Waples 1983; Bralower & Thierstein 1984; Graciansky *et al.* 1985). Indeed, Parrish & Curtis (1982) discuss conditions under which black shales can form despite the absence of upwelling, and Demaison & Moore (1980) comment on the lack of correlation in modern oceans between productivity and preservation of marine organic matter. In general, conditions in the water column and at the sea bottom seem to be the major elements in black shale formation in the western North Atlantic through preservation, and not production, of organic matter, with the possible exception of Cenomanian times when the influx of marine material was enhanced (Summerhayes 1981; Dunham *et al.* in press).

The palaeoceanographic picture that emerges is that, during the Cretaceous, the western North Atlantic appears to have been a small sea filled with oxygenated water, to have received an abundant supply of terrigenous clastics, and to have had at best moderate levels of marine productivity. How did black shales form under such conditions? While it is possible that bottom-water anoxia existed from time to time in the basins comprising the proto-Atlantic, it is unlikely that enough marine organic matter would survive sinking through the predominantly oxygenated water column to form black shales. It is more likely that the organic material reached the deep-ocean sites of black shale deposition in the company of turbidity flows, and preservation occurred from rapid burial in the oxygenated deep basins.

This scenario of downslope transport and redeposition is essentially the same as earlier proposed by Dean & Gardner (1982) for site 367 in the Cape Verde Basin, by Dean *et al.* (1984b) for Site 530 in the Anglo Basin, and by Robertson & Bliefnick (1983) for Site 534 in the Blake-Bahama Basin, and it helps explain the variability in organic character present in the sediments of the western North Atlantic. This explanation is based upon assuming each occurrence of black shale deposition to be a local or regional event, loosely linked to other such events by palaeoceanographic conditions. Changes in climate, sea-level, or oceanic circulation might destabilize ocean margin sediments and initiate turbidity flows at numerous locations around the North Atlantic. Such flows need not be synchronous and need not contain the same organic matter content. Except for those originating where the oxygen minimum layer intercepts the bottom, most would actually be poor in organic carbon; all would have their proportions of marine and terrigenous materials controlled by local conditions. On the sea floor, these flows would form fan-like deposits which would overlap and interfinger, rarely creating the continuous layers like those formed under shallow epicontinental seas. Although each black shale deposit would be initiated by a common set of global or ocean-wide conditions, its individual characteristics would be determined by regional or local factors.

Summary and conclusions

(1) The content and character of organic matter in Cretaceous black shales from DSDP Site 603 in the western North Atlantic are quite variable. In general, organic carbon contents are not as high as found in the eastern North Atlantic, and the proportion of marine-derived material is lower.

(2) Consistent with their large fraction of terrigenous organic matter, the black shales from Site 603 are relatively poor in extractable alkanes, alkanoic acids, and alkanols.

(3) The black shales represent short episodes of enhanced burial of organic matter within a general setting of poor preservation of carbonaceous materials. A prevailing oxygenated depositional setting is indicated by faunal burrowing, oxidized minerals, and low concentrations of organic carbon in most of the Cretaceous rocks from the western North Atlantic.

(4) A model of downslope transport by turbidity flows from ocean margin locations within oxygen minima and redeposition in deep-ocean settings is proposed to explain black shale formation at Site 603 and other western North Atlantic locations.

ACKNOWLEDGMENTS: The authors thank the Deep Sea Drilling Project for providing the samples used in this study and for giving P.A. Meyers the opportunity to participate in Leg 93. They are especially grateful to K.C. Lohman for the use of the carbon isotope mass spectrometer. Walter Dean and Massimo Sarti kindly reviewed the manuscript. This work was partially supported by the US National Science Foundation.

References

ARTHUR, M.A. 1979. North Atlantic Cretaceous black shales: the record at Site 398 and a brief comparison with other occurrences. *In*: RYAN, W.B.F., SIBUET, J.-C., *et al.* (eds), *Init. Repts DSDP* **47/2.** US Govt. Print. Off., Washington, DC. 719–38.

—— & SCHLANGER, S.O. 1979. Cretaceous 'oceanic anoxic events' as causal factor in development of reef-reservoired giant oil fields. *Am. Ass. Petrol. Geol. Bull.* **63,** 870–85.

BRALOWER, T.J. & THIERSTEIN, H.R. 1984. Low productivity and slow deep-water circulation in mid-Cretaceous oceans. *Geology* **12,** 614–18.

BRASSELL, S.C., EGLINTON, G. & MAXWELL, J.R. 1982. Preliminary lipid analyses of two Quarternary sediments from the Middle America Trench, southern Mexico transect, Deep Sea Drilling Project Leg 66. *In*: WATKINS, J.S., MOORE, J.C. *et al.* (eds), *Init. Repts DSDP* **66.** US Govt. Print. Off., Washington, DC. 557–80.

CARDOSO, J.N., WARDROPER, A.M.K., WATTS, C.D., BARNES, P.J., MAXWELL, J.R., EGLINTON, G., MOUND, D.G. & SPEERS, G.C. 1978. Preliminary organic geochemical analyses; Site 391, Leg 44 of the Deep Sea Drilling Project. *In*: BENSON, W.E., SHERIDAN, R.E. *et al.* (eds), *Init. Repts DSDP* **44.** US Govt. Print. Off., Washington, DC. 617–23.

DEAN, W.E. & GARDNER, J.V. 1982. Origin and geochemistry of redox cycles of Jurassic to Eocene age, Cape Verde Basin (DSDP Site 367), continental margin of northwest Africa. *In*: SCHLANGER, S.O. & CITA, M.B. (eds), *Nature and Origin of Cretaceous Carbon-rich Facies*. Academic Press, London. 55–78.

——, CLAYPOOL, G.E. & THIEDE, J. 1984a. Accumulation of organic matter in Cretaceous oxygen-deficient depositional environments in the central Pacific Ocean. *Org. Geochem.* **7,** 39–51.

——, ARTHUR, M.A. & STOW, D.A.V. 1984b. Origin and geochemistry of Cretaceous deep-sea black shales and multicolored claystones, with emphasis on Deep Sea Drilling Project Site 530, southern Angola Basin. *In*: HAY, W.W., SIBUET, J.C. *et al.* (eds), *Init. Repts DSDP* **75.** US Govt Print. Off., Washington, DC. 819–44.

DEGENS, E.T. & MOPPER, K. 1976. Factors controlling the distribution and early diagenesis of organic material in marine sediments. *In*: RILEY, J.P., CHESTER, R. (eds.), *Chemical Oceanography*, 6. Academic Press, London. 59–113.

DEMAISON, G.J. & MOORE, G.T. 1980. Anoxic environments and oil source bed genesis. *Org. Geochem.* **2,** 9–31.

DEROO, G., HERBIN, J.P., ROUCACHE, J.R., TISSOT, B., ALBRECHT, P. & DASTILLING, M. 1978. Organic geochemistry of some Cretaceous claystones from Site 391, Leg 44, Western North Atlantic. *In*: BENSON, W.E., SHERIDAN, R.E. *et al.* (eds), *Init. Repts DSDP*, **44.** US Govt. Print. Off., Washington, DC. 593–603.

DIDYK, B.M., SIMONEIT, B.R.T., BRASSELL, S.C. & EGLINTON, G. 1978. Geochemical indicators of paleoenvironmental conditions of sedimentation. *Nature* **272,** 216–22.

DUNHAM, K.W., MEYERS, P.A. & DUNHAM, P.L., in press. Organic geochemical comparison of Cretaceous black shales and adjacent strata from Hole 603B, North American Basin. *In*: WISE, S., VAN HINTE, J.E. *et al.* (eds), *Init. Repts DSDP* **93.** US Govt. Print. Off., Washington, DC.

ERDMAN, J.G. & SCHORNO, K.S. 1978. Geochemistry of carbon: Deep Sea Drilling Project Leg 44. *In*: BENSON, W.E., SHERIDAN, R.E. *et al.* (eds), *Init. Repts DSDP* **44.** US Govt. Print. Off., Washington, DC. 605–15.

GRACIANSKY, P.C. DE, BOURBON, M., LEMOINE, M. & SIGAL, J. 1981. The sedimentary record of mid-Cretaceous events in the western Tethys and central Atlantic Oceans and their continental margins. *Eclogae Geol. Helv.* **47,** 353–67.

——, BROSSE, E., DEROO, G., HERBIN, J.-P., MONTADERT, L., MULLER, C., SIGAL, J. & SCHAAF, A. 1982. Les formations d'age Cretace de L'Atlantique Nord et leur matiere organique: paleogeographie et milieux de depot. *Revue de L'Institut Francais du Petrole.* **37/3,** 275–337.

——, DEROO, G., HERBIN, J.P., MONTADERT, L., MULLER, C., SCHAAF, A. & SIGAL, J. 1984. Ocean-wide stagnation episode in the late Cretaceous. *Nature* **308,** 346–9.

HABIB, D. 1983. Sedimentation-rate-dependent distribution of organic matter in the North Atlantic Jurassic-Cretaceous. *In*: Sheridan, R.E., Gradstein, F.M. *et al.* (eds), *Init. Repts DSDP* **76.** US Govt. Print. Off., Washington, DC. 781–94.

HERBIN, J.P., DEROO, G. & ROUCACHE, J. 1983. Organic geochemistry in the Mesozoic and Cenozoic formations of Site 534, Leg 76, Blake-Bahama Basin, and comparison with Site 391, Leg 44. *In*: SHERIDAN, R.E., GRADSTEIN, F.M. *et al.* (eds), *Init. Repts DSDP* **76.** US Govt. Print. Off., Washington, DC. 481–93.

IBACH, L.E.J. 1982. Relationship between sedimentation rate and total organic carbon content in ancient marine sediments. *Am. Ass. Petrol. Geol. Bull.* **66,** 170–88.

JANSA, L.F., ENOS, P., TUCHOLKE, B.E., GRADSTEIN, F.M. & SHERIDAN R.E. 1979. Mesozoic-Cenozoic sedimentary formations of the North American Basin, Western North Atlantic. *In*: TALWANI, M., HAY, W. & RYAN, W.B.F., *Deep Drilling Results in the Atlantic Ocean: Continental Margins and Paleoenvironment.* Am. Geophys. Union, Washington. 1–57.

JENKYNS, H.C. 1980. Cretaceous anoxic events: from continents to oceans. *J. Geol. Soc. London* **137,** 171–88.

KATZ, B.J. & PHEIFER, R.N. 1982. Characteristics of Cretaceous organic matter in the Atlantic. *In*: WATKINS, J. & DRAKE, C., *Geology of Continental Margins, Am. Assoc. Petrol. Geol. Mem.* **34,** 617–28.

KESWANI, S.R., DUNHAM, K.W. & MEYERS, P.A. 1984. Organic geochemistry of late Cenozoic sediments from the subtropical South Atlantic. *Mar. Geol.* **61,** 25–42.

MCIVER, R. 1975. Hydrocarbon occurrences from JOIDES Deep Sea Drilling Project. *Proc. Ninth World Petrol. Congr.* 269–80.

MEYERS, P.A., in press. Organic carbon content of sediments and rocks from Sites 603, 604, and 605, western margin of the North Atlantic. *In*: WISE, S., VAN HINTE, J.E. *et al., Init. Repts DSDP* **75.** US Govt. Print. Off., Washington, DC.

MEYERS, P.A., LEENHEER, M.J., KAWKA, O.E. & TRULL, T.W. 1984. Enhanced preservation of marine-derived organic matter in Cenomanian black shales from the southern Angola Basin. *Nature* **312,** 356–9.

PARRISH, J.T. & CURTIS, R.L. 1982. Atmospheric circulation, upwelling and organic-rich rocks in the Mesozoic and Cenozoic eras. *Paleogeogr., Paleoclimatol., Paleoecol.* **40,** 31–66.

PREMUZIC, E.T., BENKOVITZ, C.M., GAFFNEY, J.S. & WALSH, J.J. 1982. The nature and distribution of organic matter in the surface sediment of world oceans and seas. *Org. Geochem.* **4,** 63–77.

RISATTI, J.B., ROWLAND, S.J., YON, D. & MAXWELL, J.R., 1984. Stereochemical studies of acyclic isoprenoids—XII. Lipids of methanogenic bacteria and possible contributions to sediments. *Org. Geochem.* **6,** 93–104.

ROBERTSON, A.H.F. & BLIEFNICK, D.M. 1983. Sedimentology and origin of Lower Cretaceous pelagic carbonates and redeposited clastics, Blake-Bahama Formation, Deep Sea Drilling Project Site 534, western equatorial Atlantic. *In*: SHERIDAN, R.E., GRADSTEIN, F.M. *et al.* (eds), *Init. Repts DSDP* **76.** US Govt. Print. Off., Washington, DC. 795–828.

SIMONEIT, B.R.T. 1978. The organic chemistry of marine sediments. *In*: RILEY, J.P. & CHESTER, R. (eds), *Chemical Oceanography*, 7. Academic Press, London. 233–311.

STUERMER, D.H. & SIMONEIT, B.R.T. 1978. Varying sources for the lipids and humic substances at Site 391, Blake-Bahama Basin, Deep Sea Drilling Project Leg 44. *In*: BENSON, W.E., SHERIDAN, R.E. *et al.* (eds), *Init. Repts DSDP* **44.** US Govt. Print. Off., Washington, DC. 587–91.

SUMMERHAYES, C.P. 1981. Organic facies of middle Cretaceous black shales in the deep North Atlantic. *Am. Ass. Petrol. Geol. Bull.* **65,** 2364–80.

—— & MASRAN, T.C. 1983. Organic facies of Cretaceous and Jurassic sediments from Deep Sea Drilling Project Site 534 in the Blake-Bahama Basin, western North Atlantic. *In*: SHERIDAN, R.E., GRADSTEIN, F.M. *et al.* (eds), *Init. Repts DSDP* **76.** US Govt. Print. Off., Washington, DC. 469–80.

—— 1986. Organic rich Cretaceous sediments from the North Atlantic. *In*: BROOKS, J. & FLEET, A.J. (eds), *Marine Petroleum Source Rocks. Spec. Publ. Geol. Soc. Lond.* Published by Blackwell Scientific Publications, Oxford.

THIERSTEIN, H.R. 1979. Paleoceanographic implications of organic carbon and carbonate distribution in Mesozoic deep sea sediments. *In*: TALWANI, M., HAY, W., RYAN, W.B.F. (eds), *Deep Drilling Results in the Atlantic Ocean: Continental Margins and Paleoenvironments.* Am. Geophys. Union, Washington. 249–74.

TISSOT, B.P., DEMAISON, G., MASSON, P., DETEIL, J.R. &

COMBAZ, A. 1980. Paleoenvironment and petroleum potential of the mid-Cretaceous black shales in the Atlantic basins. *Am. Ass. Petrol. Geol. Bull.* **64,** 2051–63.
TUCHOLKE, B.E. & VOGT, P.R. 1979. Western North Atlantic: Sedimentary evolution and aspects of tectonic history. *In*: TUCHOLKE, B.E., VOGT, P.R. *et al.* (eds), *Init. Repts DSDP* **43.** US Govt. Print. Off., Washington, DC. 791–825.
WAPLES, D.W. 1983. Reappraisal of anoxia and organic richness, with emphasis on Cretaceous of North Atlantic. *Am. Ass. Petrol. Geol. Bull.* **67,** 963–78.
—— & SLOAN, J.R. 1980. Carbon and nitrogen diagenesis in deep sea sediments. *Geochim. Cosmochim. Acta.* **44,** 1463–70.
WEISSERT, H. 1981. The environment of deposition of black shales in the early Cretaceous: An ongoing controversy. *In*: WARME, J.E., DOUGLAS, R.G., WINTERER, E.L. (eds), *The Deep Sea Drilling Project: A Decade of Progress. Soc. Econ. Paleo. Miner. Spec. Pub.* **32.** 547–60.

PHILIP A. MEYERS, KEITH W. DUNHAM & PAMELA L. DUNHAM, Oceanography Program, Department of Atmospheric and Oceanic Science, The University of Michigan, Ann Arbor, Michigan 48109–2143, USA.

Geolipids of black shales and claystones in Cretaceous and Jurassic sediment sequences from the North American Basin

P. Farrimond, G. Eglinton & S.C. Brassell

SUMMARY: Aliphatic hydrocarbon distributions in DSDP black shales and green claystones from the North American Basin have been examined by gas chromatography—mass spectrometry (GC–MS). The sediment sequences studied were: (i) black shale and claystone horizons (Callovian–Cenomanian) from the Blake-Bahama Basin (Leg 76, Site 534), and (ii) black shales and associated green claystones (Cenomanian–Turonian) from the US East Coast Continental Rise (Leg 93, Site 603). In both series of black shales, specific biological marker compounds provide evidence for contributions of organic matter from terrestrial, marine and bacterial sources. In addition, several of these geolipid distributions (e.g. hopanoids) are markedly similar for both sequences, indicating some common inputs of organic matter, and comparable diagenetic histories. For the Leg 93 samples, the relatively organic-lean green claystones are found to contain mainly terrigenous marker compounds, in constrast to their interbedded black shales which show a larger contribution from marine sources.

All the sediments are immature, with a trend of increasing diagenesis evident from changes in the geolipid distributions down the Leg 76 sequence. One sample, however, is anomalous; its enhanced maturity may be due to an input of reworked, older organic material at the time of deposition.

Molecular organic geochemistry applied to black shales

In recent years, the Deep Sea Drilling Project (DSDP) has revealed the widespread distribution of middle Cretaceous and Late Jurassic black shales over large areas of the world's ocean basins, particularly in the North and South Atlantic Oceans (Arthur & Natland 1979). 'Oceanic Anoxic Events' have been proposed (Schlanger & Jenkyns 1976) to account for the synchronous deposition of such organic-rich sediments over wide areas; in contrast, other authors (e.g. Waples 1983) have argued that localized and sporadic anoxia in restricted marginal basins led to such deposits. Certainly, these, and other, oxygen deficient aquatic environments are conducive to the formation of potential oil source beds (Demaison & Moore 1980).

The high organic contents of black shales make them appropriate for detailed organic geochemical evaluation. In particular, investigations of the molecular distributions of such sediments provide evidence for the sources/inputs of their organic matter (e.g. Brassell & Eglinton 1983), the palaeoenvironment of their deposition (Didyk *et al.* 1978), and the effects of subsequent diagenesis (e.g. Mackenzie *et al.* 1982). Visual kerogen and Rock-Eval pyrolysis studies of Cretaceous black shales from the Atlantic Oceans indicate mixed and variable sources of organic matter (e.g. Tissot *et al.* 1979; Summerhayes 1981). Such differences, together with variations in depositional conditions associated with lithological changes in the sediment sequence, have also been illustrated by geolipid distributions (e.g. Brassell 1984; Meyers *et al.* 1984). The results of organic geochemical investigations, combined with geological observations concerning the regional and temporal distributions of black shale facies, provide an approach towards understanding the abundance of such sediments at distinct time intervals of the Earth's history.

The work presented here is a preliminary survey, being part of an extensive study of the biological marker compounds of black shales, using gas chromatography (GC) and gas chromatography—mass spectrometry (GC-MS) to assess organic geochemical evidence for their mode of formation. The sediments consist of seven black shales and dark grey carbonaceous claystones (Callovian–Cenomanian) from the Blake-Bahama Basin (Leg 76, Site 534), and two black shales and two green claystones (Cenomanian–Turonian) from the US East Coast Continental Rise (Leg 93, Site 603; Fig. 1, Table 1). Stratigraphic sections and lithologic units, together with details of the sedimentary context for the sections examined from Sites 534 and 603 are given in the relevant DSDP Initial Report Volumes (Sheridan *et al.* 1983; van Hinte and Wise *et al.*, in press, respectively). Variations in

From SUMMERHAYES, C.P. & SHACKLETON, N.J. (eds), 1986, *North Atlantic Palaeoceanography*, Geological Society Special Publication No. 21, pp. 347–360.

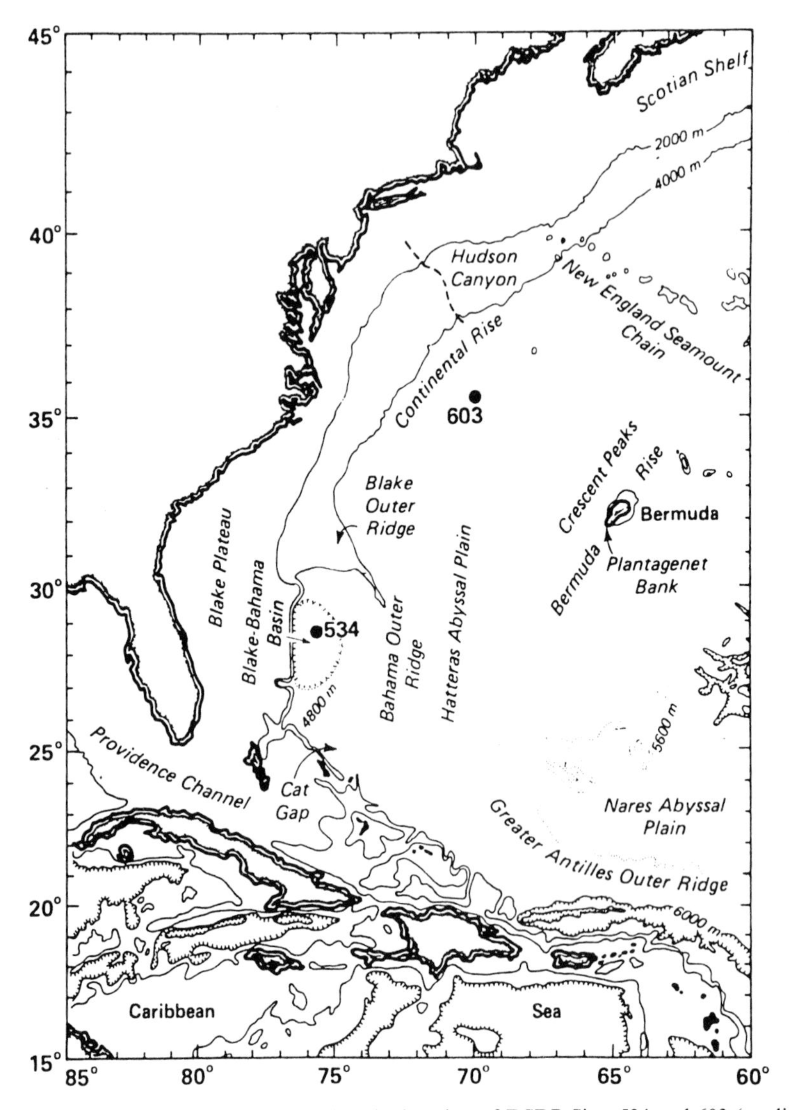

FIG. 1. Map of the North American Basin showing the location of DSDP Sites 534 and 603 (modified from van Hinte, Wise *et al.*, in press).

the character of the organic matter (evident in geolipid distributions) may reflect different depositional conditions, in terms of source, productivity, transport, or subsequent preservation of organic material. The aims of this work were to determine the sources of organic matter, and whether they were (i) similar for different lithologies, and (ii) comparable at both sites.

Experimental procedure

After thawing, each sediment sample was crushed in a Tema disc mill. Total organic carbon measurements were made using a Perkin-Elmer 240 CHN analyser. The extractable lipids of each sediment were obtained by soxhlet extraction (dichloromethane/methanol; 3:1; 24 hours), and subsequently methylated (BF_3/MeOH) to produce methyl esters of any carboxylic acids present. Thin-layer chromatography (TLC) afforded several lipid fractions, of which only the aliphatic hydrocarbons are considered here. Gas chromatography was performed by split/splitless injection on a Carlo Erba 2150 gas chromatograph, fitted with an OV-1 glass capillary column

TABLE 1. *Sample codes, depths, ages and lithologies*

Leg 76, Site 534 (Blake-Bahama Basin)

Sample	Abbreviation	Depth (m)*	Age*	Lithology
28–1 (130–140)	28–1	766	Cenomanian	Black shale
34–1 (120–135)	34–1	832	Mid Albian	Black shale
37–3 (130–140)	37–3	864	Early Albian	Dark grey claystone
41–5 (120–130)	41–5	903	Late Aptian	Dark grey claystone
44–2 (120–127)	44–2	926	Mid Aptian	Dark grey claystone
66–3 (120–124)	66–3	1120	Late Valanginian	Black shale
125–5 (135–138)	125–5	1620	Callovian	Black shale

Leg 93, Site 603 (US East Coast Continental Rise)

Sample	Abbreviation	Depth (m)*	Age	Lithology
33–1 (141–146)	G1	1120	Cenomanian/Turonian	Green claystone
33–2 (16–21)	B1	1120	Cenomanian/Turonian	Black shale
33–2 (29–34)	B2	1120	Cenomanian/Turonian	Black shale
33–2 (47–52)	G2	1120	Cenomanian/Turonian	Green claystone

* Taken from Initial Reports of the DSDP for Leg 76, and Initial Core Descriptions for Leg 93.

(*c.* 13 m × 0.25 mm i.d.). A temperature programme of 50–300°C at 6° min^{-1} was employed, and the helium carrier gas had a flow rate of *c.* 1 ml min^{-1}. Computerized gas chromatography—mass spectrometry (GC-MS) was performed using a Carlo Erba Mega Series 5160 gas chromatograph fitted with an OV-1 fused silica capillary column (25 m × 0.30 mm i.d.), and programmed from 50–300°C at 4°C min^{-1}, with helium as carrier gas, linked to a Finnigan 4000 mass spectrometer (ionizing energy 35eV; ion source temperature 250°C; scan time (m/z 50–550) 1 second). An INCOS 2300 system was used for data acquisition and processing. Compound identifications were made by a combination of their GC retention times, mass fragmentographic responses, and mass spectra, by comparison with literature data.

Chemical structures of the various geolipids recognized in these sediments are not shown herein. They can be found in many of the organic geochemical papers cited in the references.

Blake-Bahama Basin (Leg 76, Site 534)

Results

Only results for aliphatic hydrocarbon analyses are presented here; they are discussed in qualitative rather than quantitative terms, since absolute compound abundances are not determined.

Bulk geochemical data are presented in Table 2. The samples show considerable variability; especially the Callovian black shale (125–5) with its higher total soluble extract (TSE) and carbonate content. Note also, the high TSE relative to organic carbon of 44–2.

The approximate relative abundances of individual compound classes (Table 2) show that the samples are chemically similar. Notable differences are: (i) an increase in diasterene abundance with depth, and (ii) a decrease in fernenes and sterenes with depth.

n-Alkanes

n-Alkane distributions for four of the seven sediment samples are presented in Fig. 2. Specific molecular ratios derived from the *n*-alkane data (Table 2) are largely similar for all the samples, except for variations in their maxima, and their relative contents of long and medium chain length compounds; samples 28–1 and 125–5 are notably different from the others in this latter respect. All the carbon preference indices (CPI) are considerably greater than unity, with those for 37–3, 41–5 and 44–2 being the lowest.

Acyclic isoprenoid alkanes

Figure 2 also shows the distributions of acyclic isoprenoid alkanes. Pristane/phytane ratios (Pr/Ph; Table 2) are similar, and less than unity, for all

TABLE 2. *Bulk sediment compositions, geolipid abundances and alkane parameters for the Leg 76 samples. TSE = Total Soluble Extract; TOC = Total Organic Carbon; '+', '++' and '+++' refer to the approximate relative abundances of the compound classes; Tr = trace amounts found; nd = not detected; CPI = Carbon Preference Index (calculated over the range C_{23} to C_{32}); Pr/Ph = pristane/phytane ratio; 25i + 30i = ratio of acyclic isoprenoid alkanes to the dominant n-alkane*

	TSE (mg/g sed.)	TSE (% of Org. C)	TOC (%)	% Carbonate	*n*-alkanes	Acyclic isoprenoids	Diterpenoids	Triterpenoids	Fernenes	Sterenes	Steranes	Diasterenes	Diasteranes	Maxima	C_{17}/C_{29}	CPI (C_{23}–C_{32})	Pr/Ph	25i + 30i/dominant *n*-alkane
28–1	0.17	2.5	0.7	0.2	+++	+++	++	+++	++	+	+	nd	+	18 31	1.29	2.0	0.76	0.62
34–1	0.09	1.3	0.7	1.6	+++	+++	++	+++	+	+	+	nd	+	18 27	0.34	2.1	0.56	0.28
37–3	0.09	3.6	0.2	0.2	+++	+++	++	+++	+	+	+	nd	+	18 27	0.34	1.7	0.62	0.31
41–5	0.01	1.4	0.1	0.2	+++	+++	++	+++	+	Tr	+	++	+	18 31	0.32	1.5	0.73	0.58
44–2	0.16	11.0	0.1	0.2	+++	+++	++	++	+	+	+	++	+	18 31	0.21	1.7	0.42	0.26
66–3	0.15	1.4	1.1	1.6	+++	+++	++	+++	Tr	nd	+	+++	+	18 27	0.27	2.8	0.55	0.11
125–5	0.55	4.6	1.2	2.4	+++	+++	++	+++	nd	nd	+	+++	+	19 31	1.08	1.8	1.25	0.33

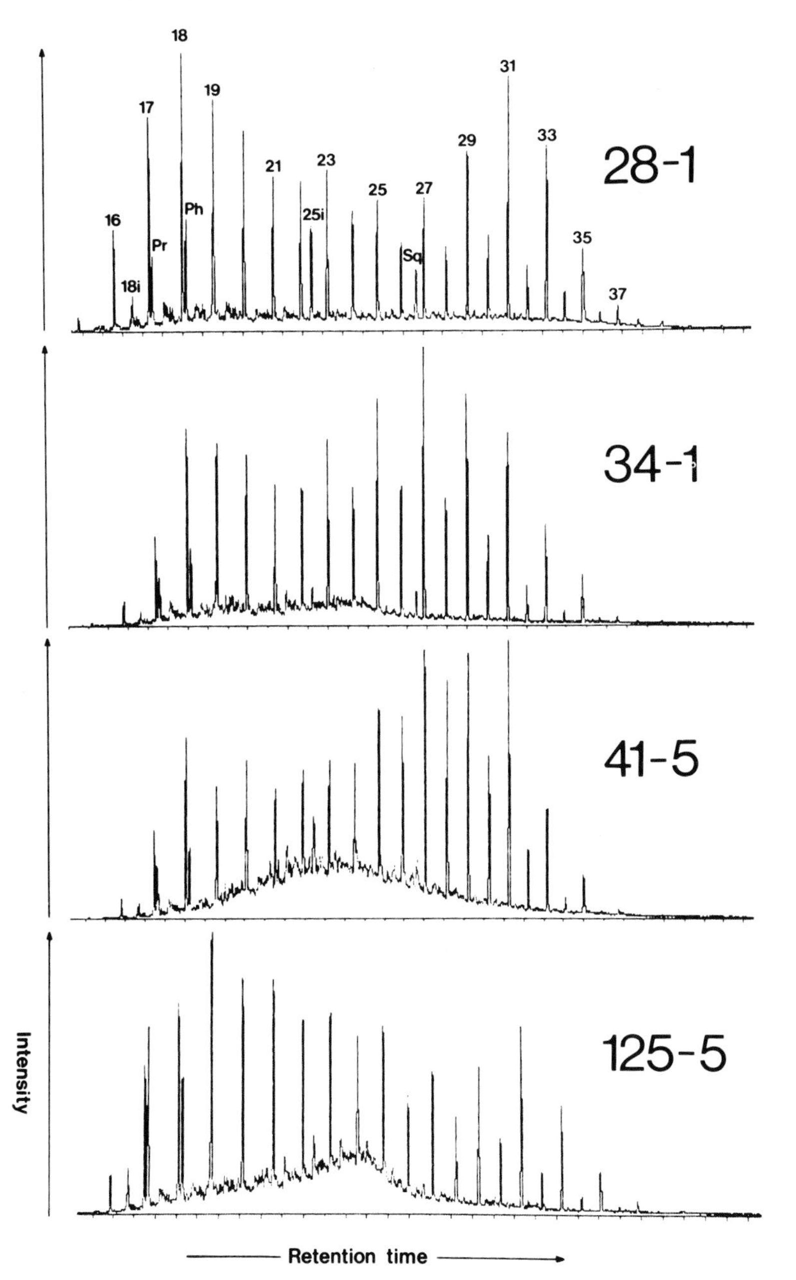

FIG. 2. m/z 85 Mass fragmentograms showing the *n*-alkane and acyclic isoprenoid distributions of four samples from Site 534. *n*-Alkanes are labelled by their carbon numbers. Acyclic isoprenoid alkanes: 18i, Pr = pristane (C_{19}), Ph = phytane (C_{20}), 25i, Sq = squalane (30i).

samples except the Callovian black shale (125–5). The relative abundance of C_{25} and C_{30} (squalane, Sq) acyclic isoprenoids is suggested by the ratio of (25i + 30i)/*n*-alkane (Table 2) and varies markedly.

Diterpenoids

All seven samples show similar distributions of extended tricyclic diterpanes, dominated by the C_{23} component.

Triterpenoids

The triterpenoid distributions (Fig. 3) are largely similar, containing the same series of compounds (17α(H),21β(H)-, 17β(H),21α(H)- and 17β(H), 21β(H)- hopanes, $\Delta^{13(18)}$-neohopenes, $\Delta^{17(21)}$-hopenes and an unknown diene; Table 3), although the relative amounts of specific components do vary. The two deepest horizons (66–3 and 125–5) contain larger amounts of $\Delta^{13(18)}$-neohopenes. In addition, 125–5 contains a series of benzohopanes (Hussler *et al.* 1984). Sediment samples 37–3, 41–5 and 44–2 contain increased abundances of 17α(H), 21β(H)-hopanes, with the triterpane distribution in 41–5 being skewed towards the lower carbon number components (Fig. 3).

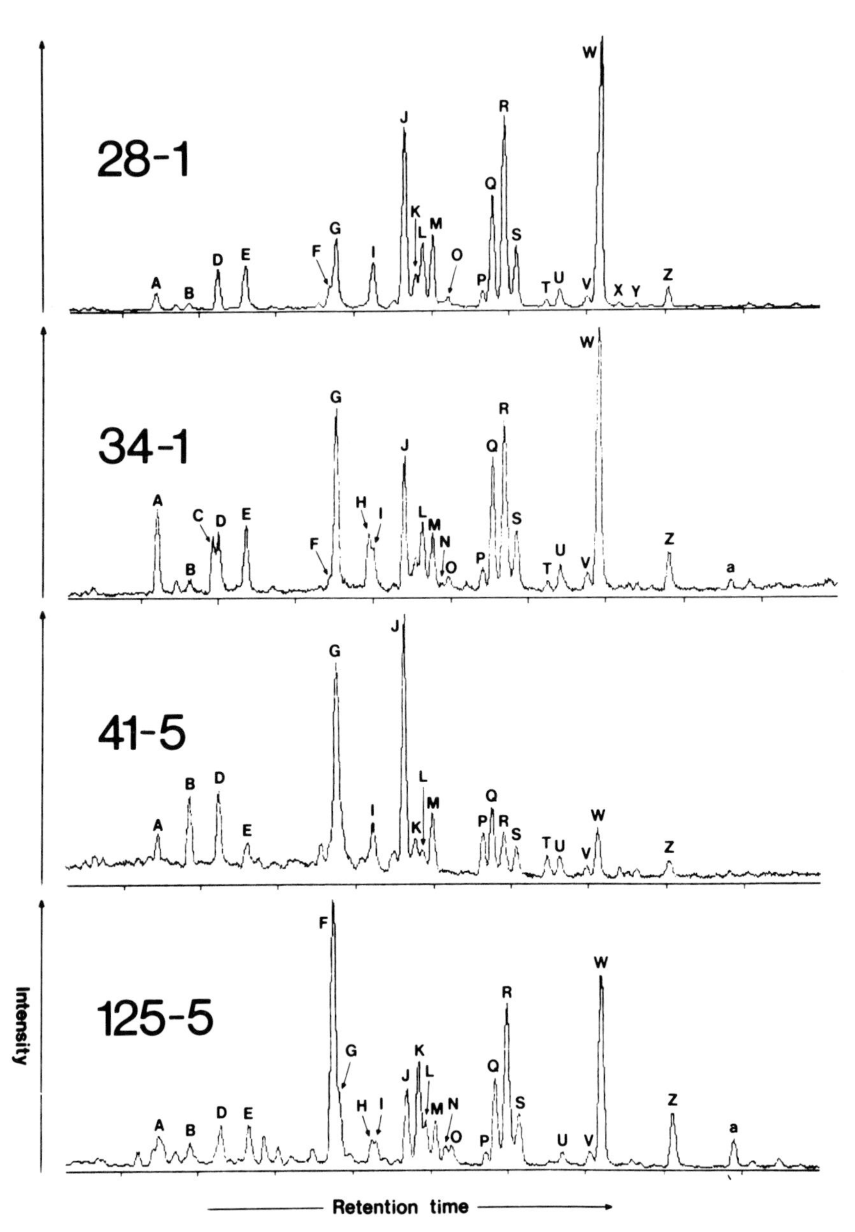

FIG. 3. m/z 191 Mass fragmentograms illustrating the triterpenoid hydrocarbon distributions for four of the seven samples from Site 534 (peak assignments are given in Table 3).

TABLE 3. *Hopane and hopene assignments (*cf. *Figs 3 and 6). a, b=17α(H)- and 17β(H)-22,29,30-trisnorhopane, respectively. c=Compound co-elutes with an unknown C_{30} diene*

Structural type	C_{27}	C_{29}	C_{30}	C_{31} S R	C_{32} S R	C_{33} S R	C_{34} S R
	Carbon number and C–22 stereochemistry						
18α(H)-neohopane	B						
Δ17(21)-hopene	C		H	N O			
Δ13(18)-neohopene	A	F	K				
17α(H),21β(H)-hopane	D^a	G	J^c	P Q	T U	X Y	
17β(H),21α(H)-hopane		I	M	S	V		
17β(H),21β(H)-hopane	E^b	L	R	W	Z	a	b

In addition to the hopanoids, fernenes (Δ^8 and $\Delta^{9(11)}$ isomers) occur in the upper samples, decreasing in abundance with depth (Table 2).

Steroids

Whilst Δ^4- and Δ^5-sterenes are only present in the upper five samples, steranes occur in low concentrations in all the horizons, with C_{27} and C_{29} 5α(H),14α(H),17α(H), 20R isomers dominant (Table 2). Diasterenes are major constituents of the deeper samples (Table 2), but were not detected in the upper horizons. Des-A and A-nor diasterenes (van Grass 1982) occur in 66–3 and 125–5; spirosterenes (Peakman *et al.* 1984) are also present in the latter sample. Like the steranes, diasteranes are found in low concentrations throughout the section (Table 2).

Discussion

The data presented above reflect (i) the extent of sediment diagenesis, and (ii) sources of organic matter, and the palaeoenvironment of deposition.

Diagenesis

Geolipids are sensitive and systematic indicators of the extent of sediment diagenesis (e.g. Mackenzie *et al.* 1982). The general immaturity of these sediments is shown by the presence of 17β(H), 21β(H)-hopanes, hopenes, sterenes, spirosterenes and diasterenes (cf. Brassell *et al.* 1983). As the last two compound types only appear in the deeper horizons, the entire sequence lies within the zone of early diagenesis (i.e. the zone between sediment deposition and the onset of oil generation; cf. Brassell *et al.* 1984).

Systematic diagenetic trends (precursor/product relationships) with increasing burial depth are evident for several hopanoids (Fig. 4(a) & (c)). This trend is complicated by the appearance of a maximum at 41–5, possibly representing an input of reworked (more mature) organic matter (Fig. 4(a) & (b)). Fernene isomerization (Fig. 4(d)) may be at equilibrium although these compounds were not detected in 66–3 and 125–5.

Steroidal compounds also illustrate increasing diagenesis with depth; these diagenetic reactions include: (i) diasterene formation from sterene precursors (cf. Table 2), and (ii) diasterene isomerization at C–20 (Fig. 4(e); note the anomalous maturity of 41–5).

n-Alkanes, acyclic isoprenoids and extended diterpenoids show no maturity-related trends with depth. However, the low CPI values of samples 37–3, 41–5 and 44–2 confirm that this part of the section is anomalously mature (Table 2), as CPI values decrease with increasing maturity (Bray & Evans 1961).

Sources of organic matter

All the samples examined from Leg 76 show evidence for contributions of organic material from algal, bacterial and terrestrial sources in varying proportions (Table 4).

Several parameters linked to the sources of organic matter show no systematic trends with depth (Fig. 5), although this data set is limited since functionalized compounds (e.g. steroidal alcohols and ketones) are not considered. However, some of the parameters serve to illustrate the anomalous nature of 41–5, since deviations from the general norm (Fig. 5(a), (b), (c)) suggest different or additional sources of certain lipids, consistent with an input of reworked organic matter at this stratigraphic level.

The Callovian black shale (125–5) resembles the Cretaceous samples in several respects, especially in its triterpenoid distribution (Fig. 3), which is very similar to that of 66–3 (not shown

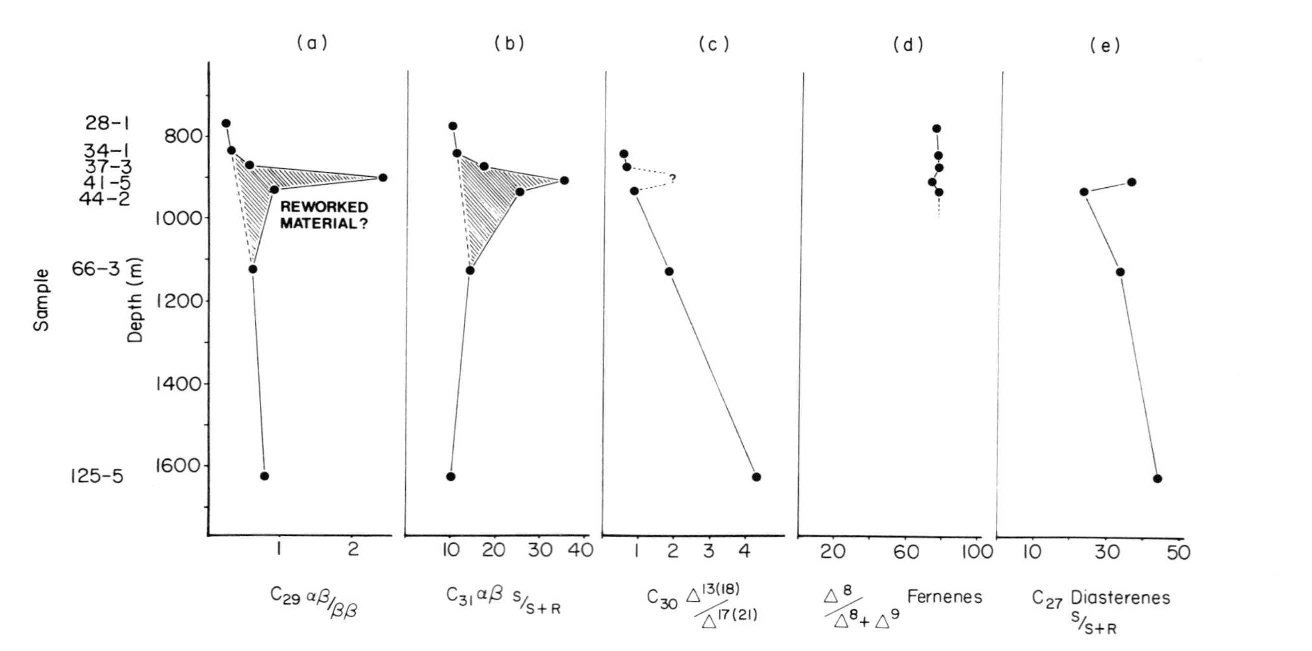

FIG. 4. Downhole plots of selected diagenesis-related parameters (Site 534). (a) Ratio of 17α(H),21β(H)- and 17β(H), 21β(H)-30-norhopanes (Mackenzie *et al.* 1980). (b) Ratio of 22S to 22S+22R 17α(H),21β(H)-homohopanes as a measure of isomerization at C–22 (Mackenzie *et al.* 1980). (c) Ratio of $\Delta^{13(18)}$-neohopene to $\Delta^{17(21)}$-hopene (Brassell *et al.* 1980). (d) Fernene isomerization ratio ($\Delta^8/\Delta^8+\Delta^{9(11)}$) (Brassell & Eglinton 1983). (e) Ratio of diacholestene isomerization at C–20 (Brassell *et al.*, 1984).

TABLE 4. *Geolipid distributions indicative of contributions from terrestrial higher plants, algae and bacteria, and inferred biological sources of specific geolipids (italics) in the Leg 76 samples*

	Terrestrial	Algal	Bacterial	Example references
n-alkanes	High Mwt compounds with high CPI	Low Mwt compounds with little or no odd-over-even predominance		Brassell *et al.* 1978
Acyclic isoprenoids		Pristane Phytane?	$C_{25}C_{30}$ & C_{40} Phytane?	Brassell *et al.* 1981 Brassell *et al.* 1978
Diterpenoids	No evidence of *conifers*		Extended diterpenoids?	Simoneit 1977 Ourisson *et al.* 1982
Steroids	C_{29} compounds? (when dominant)	Sterenes and diasterenes		Mackenzie *et al.* 1982
4-methylsteroids		*Dinoflagellates*		Robinson *et al.* 1984
Triterpenoids	No *higher plant* triterpenoids	Various hopanoids	Fernenes	Brassell & Eglinton 1983

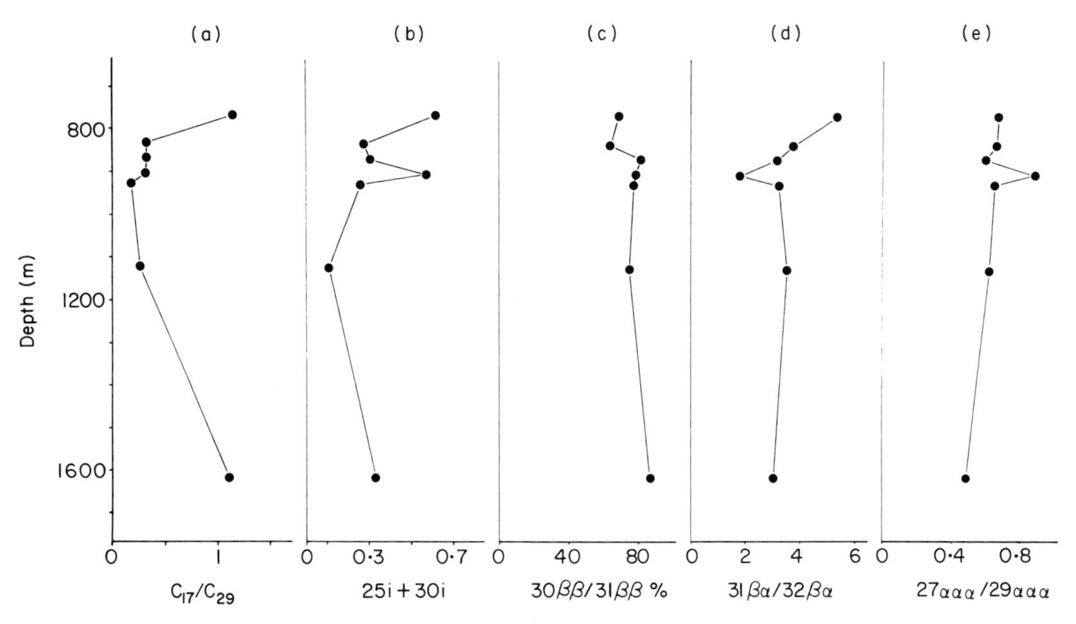

FIG. 5. Downhole plots for source/input-related parameters chosen to minimize effects due to diagenesis. (a) Ratio of C_{17} to C_{29} *n*-alkanes; a rough measure of marine versus terrestrial input (cf. Table 4; e.g. Leenheer & Meyers 1983). (b) C_{25} and C_{30} acyclic isoprenoid alkanes (25i+30i) are markers for methanogenic bacteria. (c, d) Carbon number ratios for 17β(H)-, 21β(H)- and 17β(H)-, 21α(H)-hopanes respectively; these compounds are derived principally from bacterial sources. (e) Ratio of C_{27} to C_{29} 5α(H),14α(H),17α(H) 20R steranes; a rough measure of algal versus terrestrial input (cf. Table 4).

for brevity), suggesting some related inputs of organic matter. Variations in other source-related parameters (Fig. 5) for sample 125-5 may reflect an increase in algal relative to terrigenous organic matter (Table 4). Likewise, 28–1 appears to contain a larger proportion of marine algal material, which distinguishes it from the other Cretaceous sediments which are dominated by terrigenous/bacterial organic matter (cf. Fig. 5 and Tables 2 & 4).

In all samples except 125–5, pristane is dominated by phytane (Table 2), suggesting some

TABLE 5. *Bulk sediment compositions, geolipid abundances and alkane parameters for the Leg 93 samples. n.m. = not measurable. (For details see caption to Table 2.)*

	TSE (mg/g sed.)	TSE (% of Org. C)	TOC (%)	% Carbonate	*n*-alkanes	Acyclic isoprenoids	Diterpenoids	Triterpenoids	Fernenes	Sterenes	Steranes	Diasterenes	Diasteranes	Maxima	C_{17}/C_{29}	CPI (C_{23}–C_{32})	Pr/Ph	25i + 30i/dominant *n*-alkane
G1	0.06	n.m.	0.1	0.2	+++	+++	+	++	Tr	Tr	Tr	Tr	nd	23 31	n.m.	1.8	n.m.	0.20
B1	2.1	4.3	4.7	0.5	+++	+++	+	+++	++	++	+	++	Tr	18 29	1.41	1.1	0.63	0.1
B2	2.9	5.2	5.5	0.4	+++	+++	Tr	+++	++	++	+	++	Tr	29	0.02	1.9	0.20	0.32
G2	0.03	n.m.	0.1	0.2	+++	+++	+	++	nd	Tr	Tr	nd	nd	18 31	0.98	1.3	0.45	0.26

degree of oxygen deficiency during deposition (Didyk *et al.* 1978). The relatively high TOC (1.2%), coupled with evidence suggesting the presence of oxygen (Pr/Ph ratio), indicate that the Callovian sample may have been laid down rapidly, perhaps in the form of a slump or turbidite; certainly, slump structures and graded beds are common in this part of the sequence (Sheridan *et al.* 1983; Tyson 1984). However, it must be noted that the higher Pr/Ph ratio for 125–5 may be at least partly due to source differences rather than increased oxygen concentration.

US East Coast Continental Rise (Leg 93, Site 603)

Results

Bulk geochemical data are presented in Table 5. The black shale horizons (B1 and B2) are the richest in organic carbon, with high TOC and TSE values. In contrast, the carbonate contents of the black shales are comparable to those of the green claystones (G1 and G2).

Table 5 also shows the relative abundances of individual compound classes. The green claystones exhibit similar geolipid class abundances, as do the black shales. The lithologies differ in the higher concentration of polycyclic components (e.g. triterpenoids and steroids; cf. Fig. 6) in the black shales. Abundances of tricyclic diterpanes and acyclic isoprenoid alkanes are comparable in all samples.

n-Alkanes

The parameters in Table 5 suggest two families: G1 & B2, and G2 & B1. The former have low C_{17}/C_{29} ratios and different maxima compared with the latter. All samples show CPI values greater than unity, with G2 and B1 appreciably lower than G1 and B2.

Acyclic isoprenoid alkanes

Pristane/phytane ratios vary (Table 5), but are less than unity for all samples. The abundance of C_{25} plus C_{30} acyclic isoprenoids is comparable (Table 5).

Diterpenoids

These minor components show similar distributions, dominated by the C_{23} tricyclic diterpane in all samples.

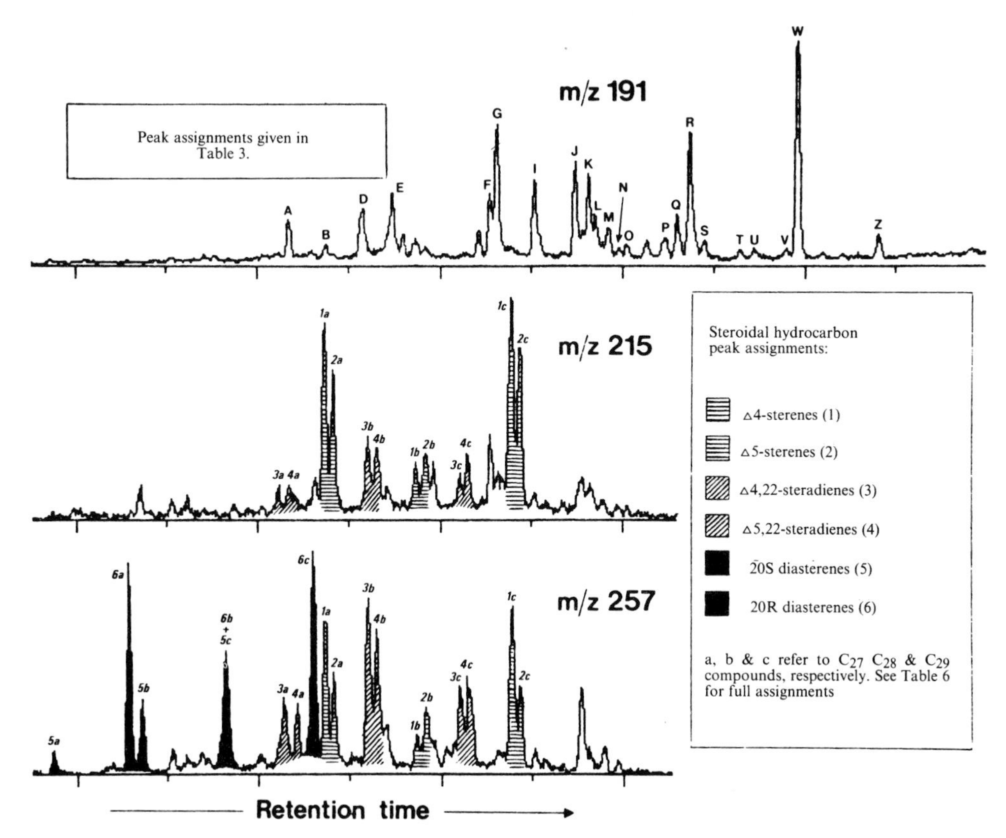

FIG. 6. Distributions of hopanoid hydrocarbons (m/z 191) and unsaturated steroidal hydrocarbons (m/z 215 and 257) of a Leg 93 black shale (B1). For peak assignments see Tables 3 and 6 respectively.

Triterpenoids

Both black shales possess similar hopanoid distributions, dominated by 17β(H),21β(H)-hopanes (e.g. Fig. 6). In contrast, the major components of the green claystone triterpenoids are 17α(H),21β(H)-hopanes; in G1 these dominate a background distribution similar to that seen in the black shales. The major component of both black shales is a C_{30} triterpene of unknown structure (UNK in Fig. 7).

Fernenes are prominent components of the black shales (cf. Table 5), with the Δ^8 isomer dominant over the $\Delta^{9(11)}$.

Steroids

In the two black shale horizons, C_{27}-C_{29} series of Δ^4- and Δ^5-sterenes, $\Delta^{4,22}$- and $\Delta^{5,22}$-steradienes, and 20S- and 20R-diasterenes are observed (Fig. 6). Steranes are present in low concentrations, with C_{27} and C_{29} 5α(H),14α(H),17α(H) 20R isomers dominant. In contrast, the green claystones contain only trace amounts of steroidal hydrocarbons.

Discussion

Diagenesis

The polycyclic components of the black shales indicate that the sediments lie within the zone of early diagenesis (cf. Brassell *et al.* 1984). Their hydrocarbons appear to be indigenous, in contrast to certain triterpanes (17α(H), 21β(H)-hopanes), steranes and diasteranes in the green claystones, whch suggest a partial contribution of organic matter from a mature source (either sampling contaminant such as drilling mud, or reworked organic debris). *n*-Alkanes derived from this anomalous source may be seen in the green claystones, superimposed on a distribution attributable to contributions from higher plants (Fig. 6), and also in one of the black shales (B1; not shown).

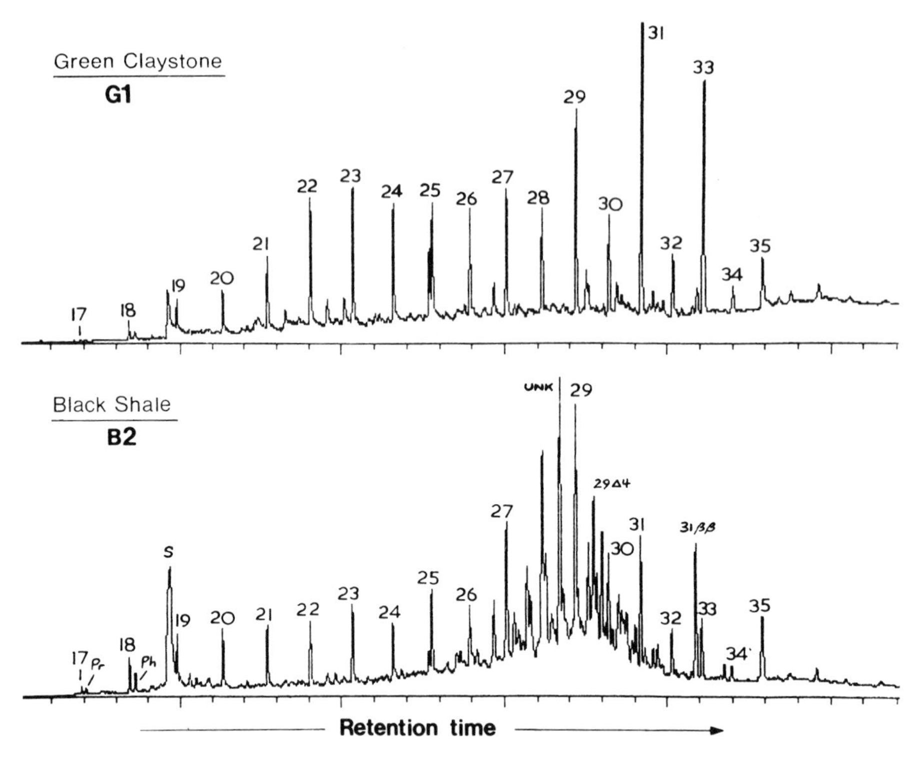

FIG. 7. Reconstructed ion counts (RIC) showing the aliphatic hydrocarbon distributions for a green claystone (G1) and a black shale (B2) from Site 603. *n*-Alkanes are numbered. Also labelled: pristane (Pr), phytane (Ph), sulphur (S), an unknown C_{30} triterpene (UNK), $C_{29}\Delta^4$-sterene ($29\Delta^4$; 1c in Table 6) and C_{31} $\beta\beta$-homohopane ($31\beta\beta$; W in Table 3).

TABLE 6. *Unsaturated steroidal hydrocarbon assignments listed in order of increasing GC elution time* (cf. Fig. 6). *Molecular structures are shown in the Appendix*

Peak	Assignment
5a	20S-diacholest-13(17)-ene
6a	20R-diacholest-13(17)-ene
5b	20S-24-methyldiacholest-13(17)-ene
6b	20R-24-methyldiacholest-13(17)-ene
5c	20S-24-ethyldiacholest-13(17)-ene
3a	cholesta-4,22-diene
4a	cholesta-5,22-diene
6c	20R-24-ethyldiacholest-13(17)-ene
1a	cholest-4-ene
2a	cholest-5-ene
3b	24-methylcholesta-4,22-diene
4b	24-methylcholesta-5,22-diene
1b	24-methylcholest-4-ene
2b	24-methylcholest-5-ene
3c	24-ethylcholesta-4,22-diene
4c	24-ethylcholesta-5,22-diene
1c	24-ethylcholest-4-ene
2c	24-ethylcholest-5-ene

Sources of organic matter

The presence of apparently anomalous hydrocarbons in the samples limits detailed comparison of the sources of organic matter (e.g. Pr/Ph and C_{17}/C_{29} ratios). There is evidence, however, for contributions of organic matter from algal (e.g. steroids), bacterial (e.g. acyclic isoprenoids, hopanoids and fernenes) and terrigenous (e.g. C_{25}-C_{35} *n*-alkanes) sources (cf. Table 4). The key difference between the black shales and green claystones is the much larger amount of algal and bacterial lipids in the former (as seen by the dominance of polycyclic components; Fig. 7). It is unclear whether this is due to (i) an increased supply of algal/bacterial organic material, (ii) enhanced preservation caused by a change in the environment of deposition, or (iii) a combination of both factors.

Conclusions

Comparisons between Leg 76 and Leg 93 hydrocarbons

The features of the hydrocarbon distributions of these two suites of samples can be summarized as follows.

(1) Their mat'ırity levels are similar; both lie within the zone of early diagenesis.
(2) A period of contribution of reworked organic matter is suggested within the Leg 76 sequence.
(3) Black shales from both sites exhibit generally similar relative abundances of the different geolipid classes. The green claystones contain less organic matter, and considerably lower concentrations of polycyclic compounds.
(4) Black shales from both sites contain hydrocarbons predominantly of algal and bacterial origin, diluted to a greater or lesser extent by terrigenous material. The green claystones from Leg 93 are dominated by the latter.
(5) The variability in the *n*-alkane distributions is considerable; indeed, the differences within the Leg 76 sequence are as great as those between the two sites. The distributions of acyclic isoprenoid alkanes, and hopanes and hopenes are largely comparable at both sites, except for the presence of an unknown C_{30} diene in the Leg 76 samples, and an unknown C_{30} triterpene in B1 and B2 (Leg 93). The steroidal hydrocarbons at both sites closely resemble each other, except that C_{27}-C_{29} $\Delta^{4,22}$ and $\Delta^{5,22}$ steradienes are present in the Leg 93 black shales.

Comparisons between the hydrocarbons of Cretaceous and Jurassic black shales from Legs 76/93 and other DSDP Sites

In terms of the extent of diagenesis and the sources of organic matter, the black shales examined herein show generally similar features to those from other locations (e.g. Brassell *et al.* 1983). Specifically:

(1) They all lie within the zone of early diagenesis, except where locally exposed to sills etc. (e.g. Leg 41; Cape Verde Rise; Simoneit *et al.* 1981).
(2) They all contain a prominent signal of compounds derived from marine algal/bacterial sources, which is diluted to a greater or lesser extent by land-derived material.
(3) The black shale and associated green claystones from Site 603 (Leg 93) are similar to those from Site 530 (Leg 75; Angola Basin; Brassell 1984; Meyers *et al.* 1984), with their green claystones being almost entirely composed of organic matter from terrestrial sources. Also, an unknown C_{30} triterpene is the predominant hydrocarbon in black shales from both sites. In contrast, $\Delta^{4,22}$- and $\Delta^{5,22}$-steradienes occur only in the Site 603 samples, whereas pristane and phytane are observed in higher concentrations at Site 530.

ACKNOWLEDGEMENTS: The authors are grateful to SERC for a studentship (PF), and NERC (GR3/2951 & GR3/3758) for GC-MS facilities. We would also like to thank Mrs A.P. Gowar, Miss L. Dyas and Mr C. Saunders for assistance with GC-MS analyses. The samples were supplied with the assistance of the NSF.

References

ARTHUR, M.A. & NATLAND, J.H. 1979. Carbonaceous sediments in the North and South Atlantic: the role of salinity in stable stratification of Early Cretaceous Basins. *In:* TALWANI, M., HAY, W. & RYAN, W.B.F. (eds), *Deep Drilling Results in the Atlantic Ocean: Continental Margins and Paleoenvironment.* Am. Geophys. Union. 375–401.

BRASSELL, S.C. 1984. Aliphatic hydrocarbons of a

Cretaceous black shale and its adjacent green claystone from the Southern Angola Basin, Deep Sea Drilling Project Leg 75. *In:* HAY, W.W., SIBUET, J.C. *et al.* (eds), *Init. Repts DSDP* **75.** US Govt. Print. Off., Washington, DC. 1019–30.

—— & EGLINTON, G. 1983. Steroids and triterpenoids in deep sea sediments as environmental and diagenetic indicators. *In:* BJORØY, M. *et al.* (eds), *Advances in Organic Geochemistry 1981*. John Wiley & Sons, Chichester. 684–97.

——, EGLINTON, G., MAXWELL, J.R. & PHILP, R.P. 1978. Natural background of alkanes in the aquatic environment. *In:* (eds), HUTZINGER, O. *et al. Aquatic pollutants: Transformations and Biological Effects*. 69–86.

——, COMET, P.A., EGLINTON, G., ISAACSON, P.J., MCEVOY, J., MAXWELL, J.R., THOMSON, I.D., TIBBETS, P.J.C. & VOLKMAN, J.K. 1980. The origin and fate of lipids in the Japan Trench. *In:* (eds), DOUGLAS, A.G. & MAXWELL, J.R. *Advances in Organic Geochemistry 1979*. Pergamon Press, Oxford. 375–92.

——, WARDROPER, A.M.K., THOMSON, I.D., MAXWELL, J.R. & EGLINTON, G. 1981. Specific acyclic isoprenoids as biological markers of methanogenic bacteria in marine sediments. *Nature* **290,** 693–6.

——, HOWELL, V.J., GOWAR, A.P. & EGLINTON, G. 1983. Lipid geochemistry of Cretaceous sediments recovered by the Deep Sea Drilling Project. *In* (eds), Bjorøy, M. *et al. Advances in Organic Geochemistry 1981*. John Wiley & Sons, Chichester. 477–84.

——, MCEVOY, J., HOFFMANN, C.F., LAMB, N.A., PEAKMAN, T.M. & MAXWELL, J.R. 1984. Isomerization, rearrangement and aromatization of steroids in distinguishing early stages of diagenesis in sediments. *In:* SCHENCK, P.A., DE LEEUW, J.W. & LIJMBACH, G.W.M. (eds), *Advances in Organic Geochemistry 1983*. Pergamon, Oxford. 11–23.

BRAY, E.E. & EVANS, E.D. 1961. Distribution of *n*-paraffins as a clue to recognition of source beds. *Geochim. Cosmochim. Acta* **22,** 2–22.

DEMAISON, G.J. & MOORE, G.T. 1980. Anoxic environments and oil source bed genesis. *Org. Geochem.* **2,** 9–31.

DIDYK, B.M., SIMONEIT, B.R.T., BRASSELL, S.C. & EGLINTON, G. 1978. Organic geochemical indicators of palaeoenvironmental conditions of sedimentation. *Nature* **272,** 216–22.

VAN GRAAS, G. 1982. *Organic Geochemistry of Cretaceous Black Shale Deposits from Italy and France*. PhD Thesis, Delft University.

VAN HINTE, J.E., WISE, S.W. *et al.*, in press. *Init. Repts DSDP* **93.** US Govt. Print. Off. Washington, DC.

HUSSLER, G., ALBRECHT, P., OURISSON, G., CESARIO, M., GUILHEM. J. & PASCARD, C. 1984. Benzohopanes, a novel family of hexacyclic geomarkers in sediments and petroleums. *Tet. Letts.* **25,** 1179–82.

LEENHEER, M.J. & MEYERS, P.A. 1983. Comparison of lipid compositions in marine and lacustrine sediments. *In:* BJORØY, M. *et al.* (eds), *Advances in Organic Geochemistry 1981*. John Wiley & Sons, Chichester. 309–16.

MACKENZIE, A.S., PATIENCE, R.L., MAXWELL, J.R., VANDENBROUCKE, M. & DURAND, B. 1980. Molecular parameters of maturation in the Toarcian shales, Paris Basin, France. I. Changes in the configurations of acyclic isoprenoid alkanes, steranes and triterpanes. *Geochim. Cosmochim. Acta* **44,** 1709–21.

——, BRASSELL, S.C., EGLINTON, G. & MAXWELL, J.R. 1982. Chemical fossils. The geological fate of steroids. *Science* **217,** 491–504.

MEYERS, P.A., LEENHEER, M.J., KAWKA, O.E. & TRULL, T.W. 1984. Enhanced preservation of marine-derived organic matter in Cenomanian black shales from the southern Angola Basin. *Nature* **312,** 356–9.

OURISSON, G., ALBRECHT, P. & ROHMER, M. 1982. Predictive microbial biochemistry—from molecular fossils to procaryotic membranes. *Trends Biochem. Sci.*, 236–9.

PEAKMAN, T.M., LAMB, N.A. & MAXWELL, J.R. 1984. Naturally occurring spiro steroid hydrocarbons. *Tet. Letts.* **25,** 349–52.

ROBINSON, N., EGLINTON, G., BRASSELL, S.C. & CRANWELL, P.A. 1984. Dinoflagellate origin for sedimentary 4α-methylsteroids and 5α(H)-stanols. *Nature* **308,** 439–42.

SCHLANGER, S.O. & JENKYNS, H.C. 1976. Cretaceous oceanic anoxic events: causes and consequences. *Geol. Mijnbouw* **55,** 179–84.

SHERIDAN, R.E., GRADSTEIN, F.M. *et al.* 1983. Site 534: Blake-Bahama Basin. *In:* SHERIDAN, R.E., GRADSTEIN, F.M. *et al. Init. Repts DSDP* **76.** US Govt. Print. Off., Washington, DC. 141–240.

SIMONEIT, B.R.T. 1977. Diterpenoid compounds and other lipids in deep-sea sediments and their geochemical significance. *Geochim. Cosmochim. Acta* **41,** 463–76.

——, BRENNER, S., PETERS, K.E. & KAPLAN, I.R. 1981. Thermal alteration of Cretaceous black shale by diabase intrusions in the Eastern Atlantic—II. Effects on bitumen and kerogen. *Geochim. Cosmochim. Acta* **41,** 1581–602.

SUMMERHAYES, C.P. 1981. Organic facies of Middle Cretaceous black shales in deep North Atlantic. *Am. Ass. Petrol. Geol. Bull.* **65,** 2364–80.

TISSOT, B., DEROO, G. & HERBIN, J.P. 1979. Organic matter in Cretaceous sediments of the North Atlantic: contribution to sedimentology and paleogeography. *In:* TALWANI, M., HAY, W. & RYAN, W.B.F. (eds), *Deep drilling results in the Atlantic Ocean: Continental Margins and Paleoenvironment*. Am. Geophys. Union. 362–74.

TYSON, R.V. 1984. Palynofacies investigation of Callovian (Middle Jurassic) sediments from DSDP Site 534, Blake-Bahama Basin, Western Central Atlantic. *Mar. Petrol. Geol.* **1,** 3–13.

WAPLES, D.W. 1983. Reappraisal of anoxia and organic richness, with emphasis on Cretaceous of North Atlantic. *Am. Ass. Petrol. Geol. Bull.* **67,** 963–78.

P. FARRIMOND, G. EGLINTON & S.C. BRASSELL. Organic Geochemistry Unit, University of Bristol, School of Chemistry, Cantock's Close, Bristol BS8 1TS.

Turbidites, the principal mechanism yielding black shales in the early deep Atlantic Ocean

E.T. Degens, K.-C. Emeis, B. Mycke & M.G. Wiesner

SUMMARY: A series of organic and inorganic geochemical parameters and facies indicators were studied in Late Jurassic and Cretaceous black shales and organic rich sediment sections from basins offshore Angola, Ghana and North Carolina, and from the western Wealden Basin in NW-Germany. Based on these data the sediments can be grouped into: (i) intracratonic (Bueckeberg Formation, NW Germany), (ii) shelf (Cretaceous basins offshore Ghana), (iii) continental rise turbidite (Hatteras Formation of Leg 93, Hole 603B, Section IV), and (iv) deep basin (Leg 75, Hole 530B, Section VIII) black shales. Trends of geochemical evolution are interpreted as the effect of multiple reworking and equilibration, when organic-rich sediments from shelfbound basins are transferred into oceanic depocentres as turbidites. It is suggested: (i) that during turbidite phases, organic-rich suspensates originating from these primary basins on continental margins are advected along density boundaries within the water column overlying oceanic basins, and (ii) that these substrates are introduced into zones where responding microbial populations create expanded zones of oxygen depletion. Microbial degradation affects predominantly marine organic compounds contained in these recycled sediments, leaving a residual kerogen composed of structureless organic matter with pronounced terrigenous signals.

During slow sinking of these suspensions, chemical equilibration, possibly with the ambient water, leads to an association of certain meals, i.e. vanadium, zinc, and nickel, with the organic fraction. Pyrite placers form as the result of gravitational separation from oxygen depleted suspensates at the bottom of black shale beds. It appears that oceanic anoxia is not a prerequisite for black shale sedimentation in the Cretaceous basins of the Atlantic Ocean.

Following the concept of James Hutton that 'the present is a key to the past', the authors have expanded surveys of recent anoxic environments—the Red Sea (Degens & Ross 1969), the Black Sea (Degens & Ross 1974) and East African Rift lakes (Degens *et al.* 1973)—to palaeoenvironments, for which the sedimentary record indicates conditions favourable for organic carbon preservation. Among the most intriguing questions in this regard is what causes the widespread occurrence of black shales deposited in the Atlantic Ocean during the Cretaceous (Ryan & Cità 1977; Jansa *et al.* 1978; Jenkyns 1980; Schlanger & Cita 1982)? At present, there is no consensus about the mechanisms and environmental settings which cause organic matter to accumulate in the form of black shales. The concept of permanent anoxia in restricted basins (Natland 1978) opposes the concept of an expanded oxygen minimum layer in the open sea (Thiede & van Andel 1977). Recently oscillations about the anoxic/oxic transition at the sediment-water interface have been invoked to explain redox cycles in sediment sequences of the South Atlantic (Hay *et al.* 1982), and black shale beds have been interpreted as distal fan turbidites (Dean *et al.* 1984). Some important questions are:

(1) Does an increased input of terrigenous organic matter trigger the swing from well-oxidized to strongly reduced conditions in the sediments and the water column?

(2) Is this phenomenon controlled by high primary productivity in the euphotic zone (Meyers 1984)?

(3) What are the mechanisms leading to trace metal enrichment in some of the Atlantic black shales (Brumsack 1980)?

Any single geochemical or sedimentological approach is unable to resolve these questions.

In this article, these questions about black shales are addressed by comparing a number of different environmental settings in the realm of the Atlantic where, during the early stage of opening, organic matter accumulated. The depositional environments range from shallow inland seas bordering the young Atlantic, to a series of basins on shelf, slope, rise, and abyssal plain. By comparing their sediment record, the intrinsic relationships between continental and various marine regimes come to light. This article does not intend to give all details on analytical results obtained during the individual studies (Wiesner 1983; Emeis 1985; Emeis *et al.*, submitted), but rather gives a synopsis of the authors' ideas on possible mechanisms leading to these controversial Atlantic deep-basin black shales.

From SUMMERHAYES, C.P. & SHACKLETON, N.J. [eds], 1986, *North Atlantic Palaeoceanography*, Geological Society Special Publication No. 21, pp. 361–376.

Geological setting and geochemical results

The four environments studied are (Fig. 1).

(1) The Bueckeberg Formation of the Wealden Basin, NW-Germany.

(2) Shelf and upper slope basins offshore Ghana.

(3) The Hatteras Formation, cored during Leg 93 of DSDP/IPOD off Cape Hatteras.

(4) Black shale beds cored during Leg 75 of DSDP/IPOD in the deep Angola Basin.

These sequences were chosen because (1) they coincided with specific stages during the evolution of the Atlantic margin, and (2) they included different black shale facies types common to this region.

Figure 2 shows the sample locations in their relative positions on an idealized passive continental margin.

The Bueckeberg Formation in Germany was deposited in a restricted, shallow, intracratonic basin and records limnic, brackish, and brackish/marine events (Fig. 3; Wiesner 1982). The section is of Berriasian age and characterized by cyclic sedimentation of laminated claystones rich in organic matter (oil shales), alternating with bioturbated claystones rich in fossils (bivalves, ostracods) and graded beds of shell debris (lumachelles). Water depth was probably less than 200 m. The organic matter in the oil shales is mainly of *in situ* algal origin. C_{org} averages 5.2%; C/N ratios average 34.4. Maturity of the organic matter is at the onset of catagenesis; traces of microbial

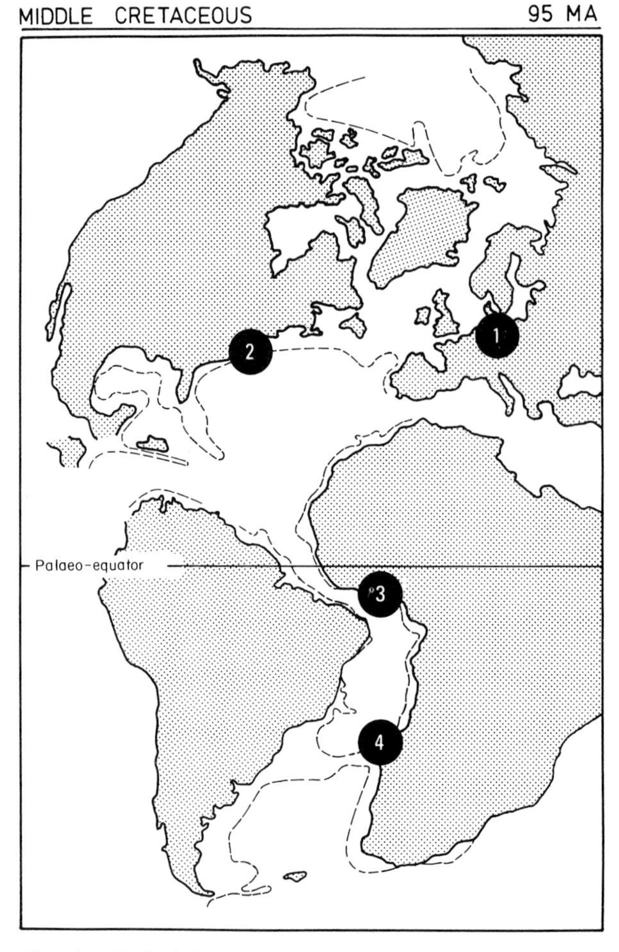

FIG. 1. Position of sampling sites (dark circles) in a palaeogeographic reconstruction of the Atlantic Ocean at 100 Ma before present (modified from Sclater *et al.* 1977). Sampling sites are from North to South: (1) Wealden Basin, (NW Germany), (2) DSDP Site 603 (US east coast continental rise), (3) Cretaceous basins offshore Ghana, and (4) DSDP Site 530 (Angola Basin).

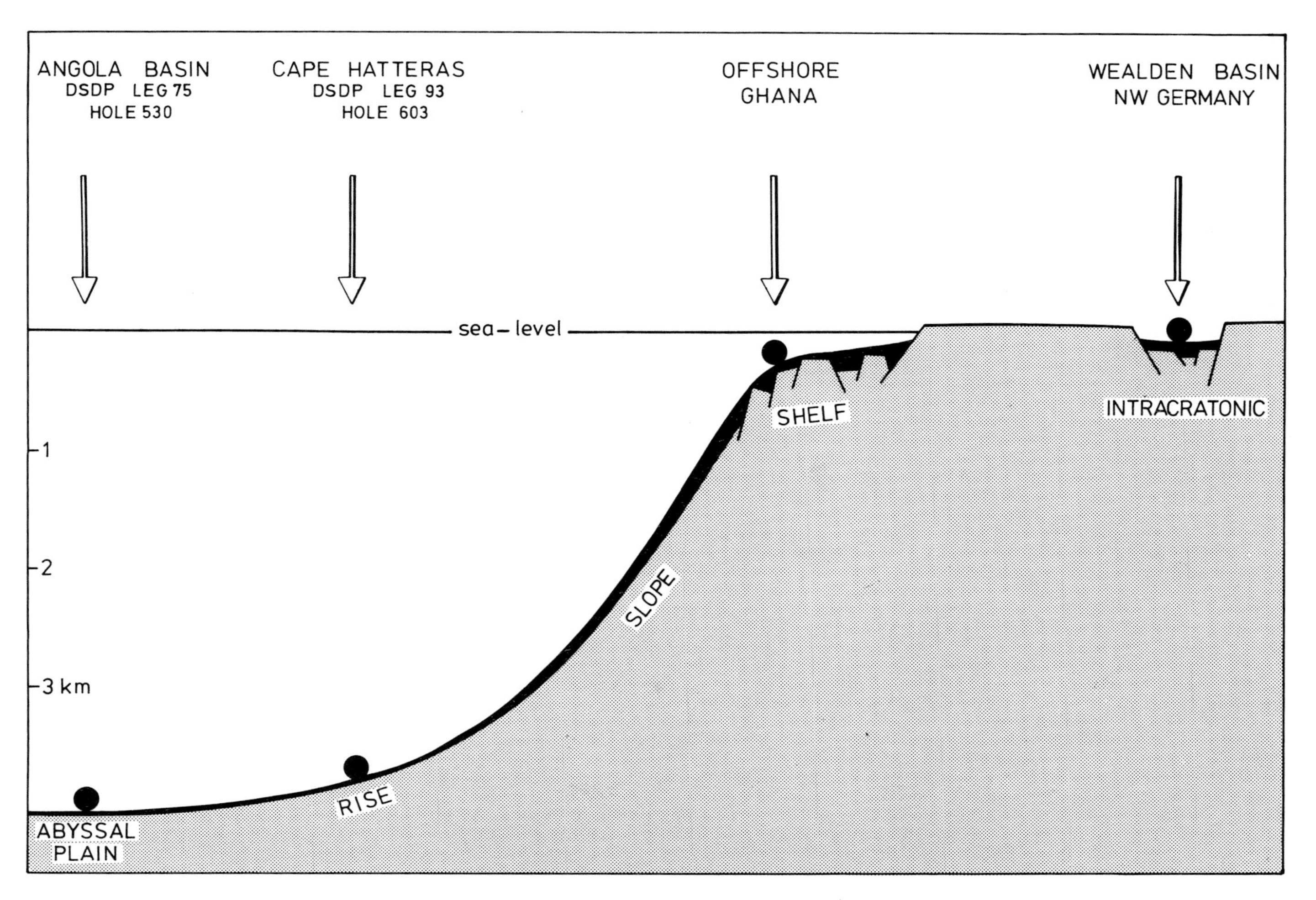

FIG. 2. Relative position of sampling sites on an idealized continental margin. Not to scale!

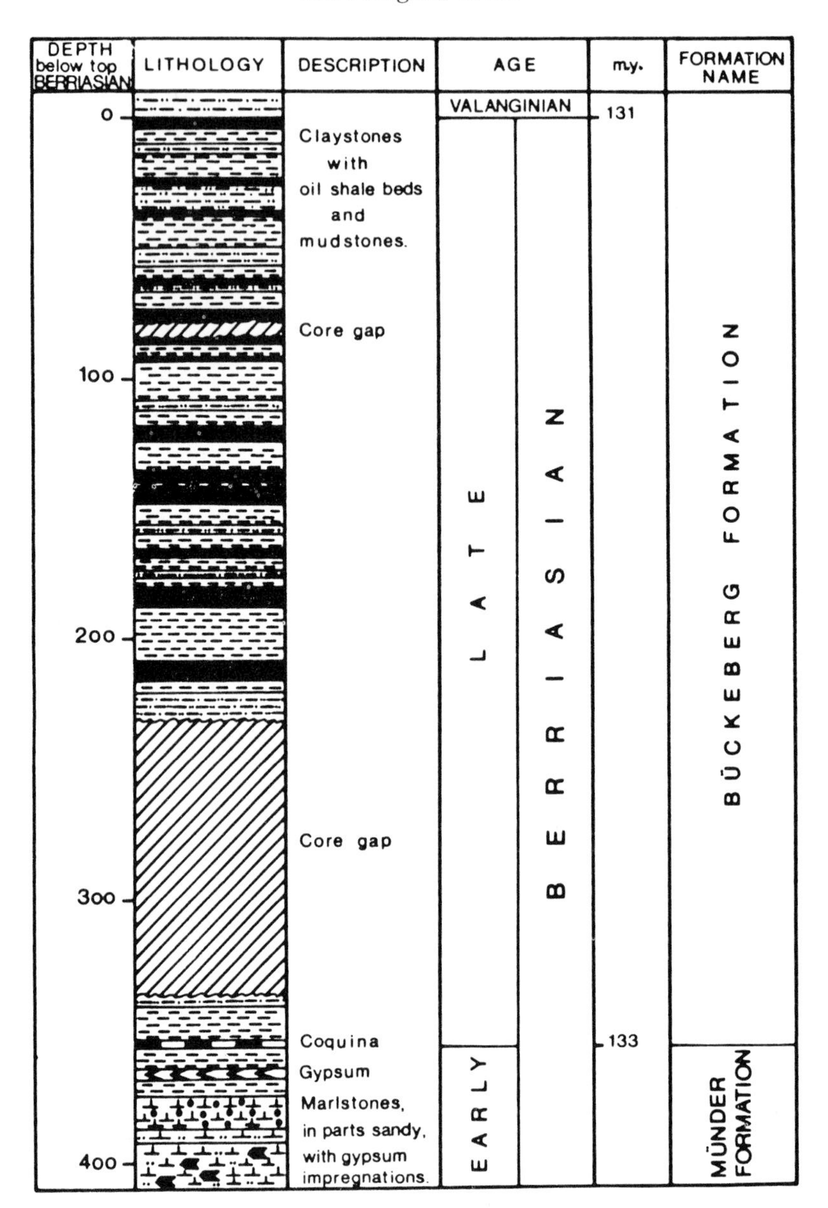

FIG. 3. Summary section of the Berriasian sequence of the western Wealden Basin/NW Germany (generalized after Wiesner 1983).

reworking are apparent from petrographic examinations of kerogen and hydrocarbon analysis. High C/N ratios are attributed to diagenetic loss of N_{org} (Wiesner 1983). No enrichment of metals relative to the claystone standard of Turekian & Wedepohl (1961) is discernible.

The Cretaceous basins offshore Ghana display tectonic features attributable to tensional fracturing and rapid vertical movements which result in small horst and graben structures. Coarse grained clastics intercalated with mudstones, lignite seams, and organic rich (1–2%) claystones of Barrèmian to Albian age, were deposited there under limnic or brackish delta conditions, as well as under open marine conditions on the shelf (Fig. 4). Subsidence led to normal marine conditions becoming established in the Turonian. In the Upper Cretaceous (Turonian–Senonian), dark calcareous and organic-rich claystones and mudstones containing mixed terrestrial and predom-

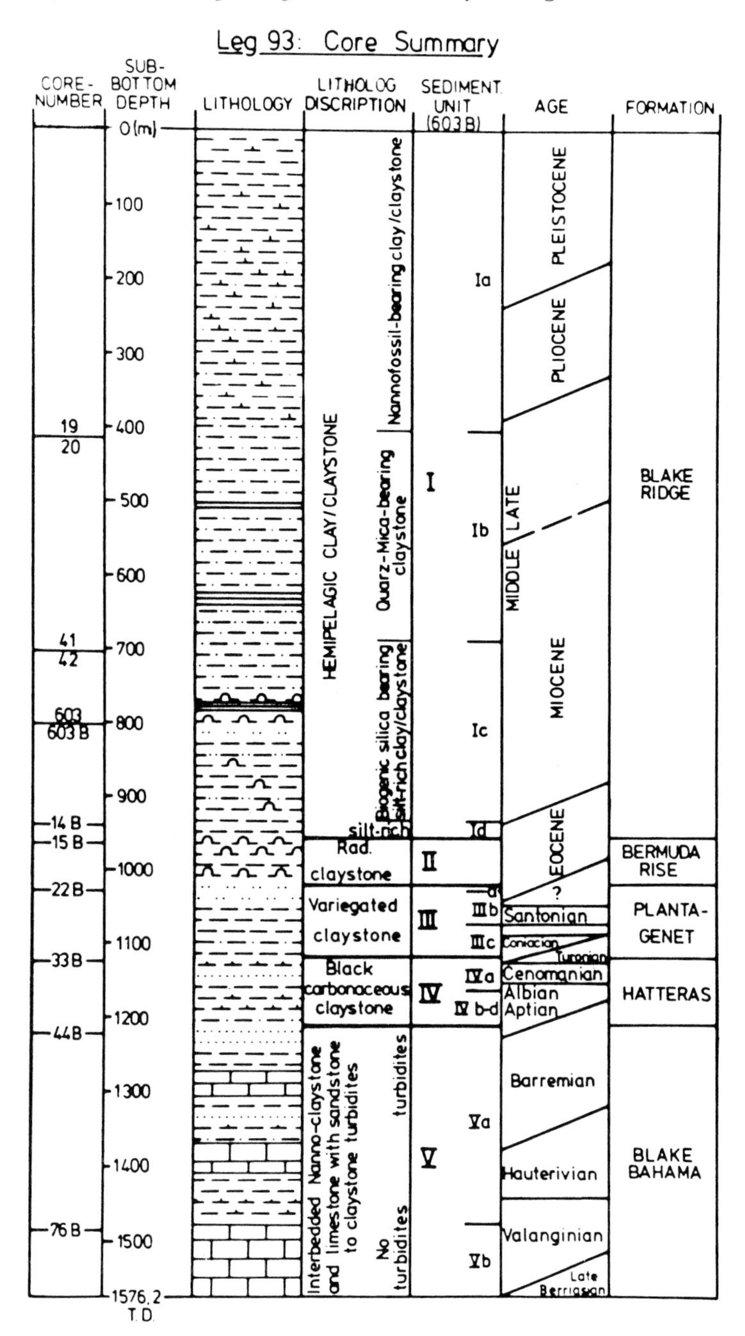

FIG. 4. Summary section of Site 603 sedimentary column, DSDP Leg 93 (US east coast continental rise off Cape Hatteras).

inantly marine organic matter were deposited in water depths of less than 500 m. Initial data indicate no enrichment of metals in the organic-rich members of this sedimentary sequence.

Unit IV of Hole 603B (Leg 93 DSDP), the *Hatteras Formation* of Barrèmian?/Aptian/Albian age (Fig. 5), consists of 106 m of claystones, black carbonaceous claystones, mud turbidites, silt and pyrite laminae, suggesting a depositional environment in which turbidite acti-

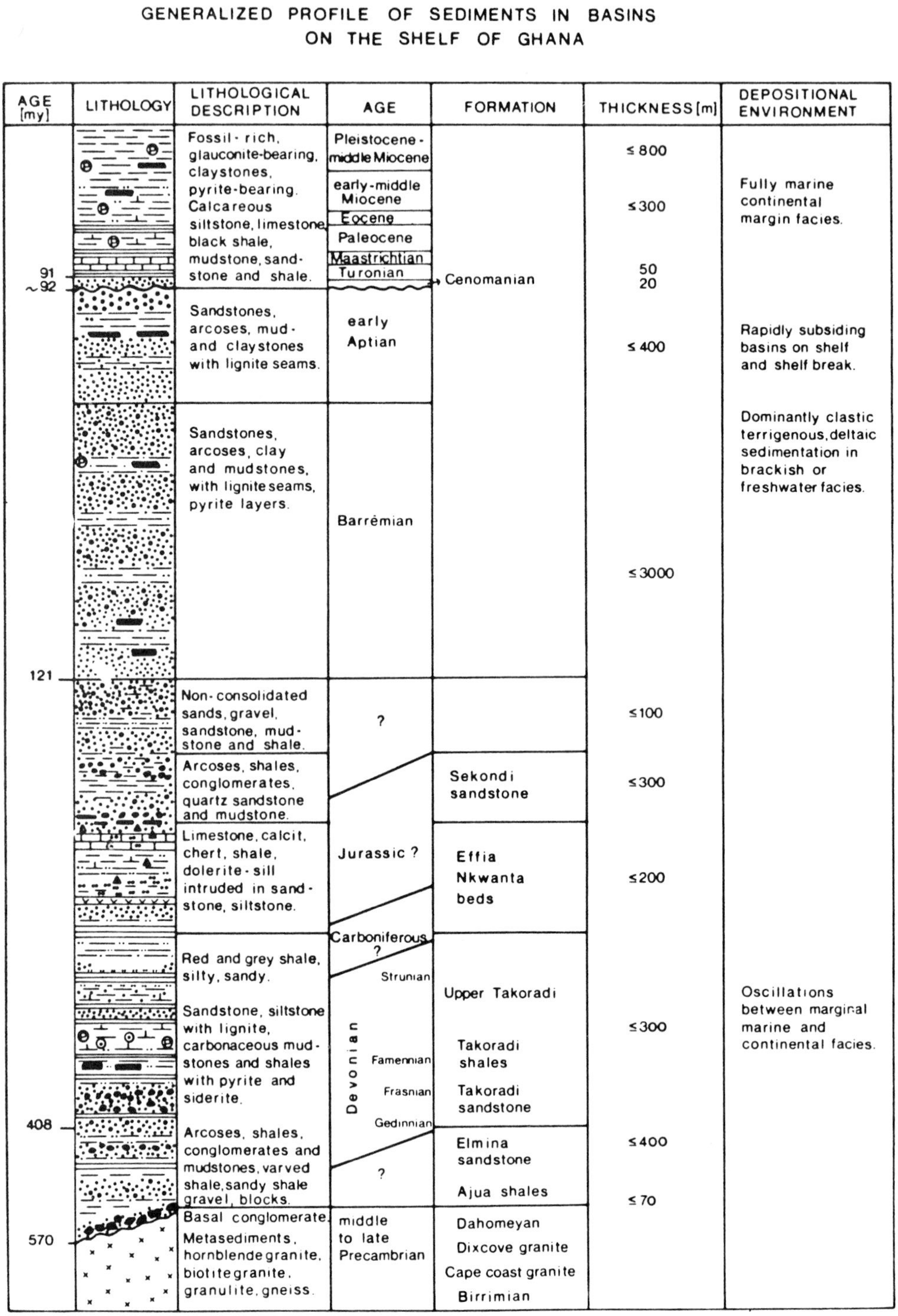

GENERALIZED PROFILE OF SEDIMENTS IN BASINS ON THE SHELF OF GHANA

AGE [my]	LITHOLOGY	LITHOLOGICAL DESCRIPTION	AGE	FORMATION	THICKNESS [m]	DEPOSITIONAL ENVIRONMENT
91 ~92		Fossil-rich, glauconite-bearing, claystones, pyrite-bearing. Calcareous siltstone, limestone, black shale, mudstone, sandstone and shale.	Pleistocene - middle Miocene early-middle Miocene Eocene Paleocene Maastrichtian Turonian	→ Cenomanian	≤ 800 ≤300 50 20	Fully marine continental margin facies.
		Sandstones, arcoses, mud- and claystones with lignite seams.	early Aptian		≤ 400	Rapidly subsiding basins on shelf and shelf break.
121		Sandstones, arcoses, clay and mudstones, with lignite seams, pyrite layers.	Barrêmian		≤ 3000	Dominantly clastic terrigenous, deltaic sedimentation in brackish or freshwater facies.
		Non-consolidated sands, gravel, sandstone, mud-stone and shale.	?		≤100	
		Arcoses, shales, conglomerates, quartz sandstone and mudstone.		Sekondi sandstone	≤300	
		Limestone, calcit, chert, shale, dolerite-sill intruded in sand-stone, siltstone.	Jurassic ?	Effia Nkwanta beds	≤200	
		Red and grey shale, silty, sandy. Sandstone, siltstone with lignite, carbonaceous mud-stones and shales with pyrite and siderite.	Carboniferous ? Devonian: Strunian, Famennian, Frasnian, Gedinnian	Upper Takoradi Takoradi shales Takoradi sandstone	≤300	Oscillations between marginal marine and continental facies.
408		Arcoses, shales, conglomerates and mudstones, varved shale, sandy shale gravel, blocks.	?	Elmina sandstone Ajua shales	≤400 ≤ 70	
570		Basal conglomerate. Metasediments, hornblende granite, biotite granite, granulite, gneiss.	middle to late Precambrian	Dahomeyan Dixcove granite Cape coast granite Birrimian		

FIG. 5. Summary section of sediments in basins offshore Ghana.

vity is superimposed on hemipelagic, clastic sedimentation. In organic-rich turbidite layers organic carbon values average 4.8%, C/N ratios are high (around 25), and no significant enrichment of trace metals can be discerned. Hydrocarbons, sugars, and microscopic inspections of organic matter reveal mixed marine and terrestrial inputs. Marine-derived structureless bituminite dominates kerogen samples of organic-rich turbidites (Emeis *et al.* submitted). Bioturbation is recorded throughout the section. Reddish lithologies and poor conservation of organic matter in the claystones representing 'background' sedimentation indicate oxidizing conditions, when turbidites were not being deposited (Leg 93 Shipboard Party, Site 603 summary).

The 163 m black shale section of Unit VIII, Hole 530A/ Leg 75 DSDP (*Angola Basin*, Fig. 6) ranges from Upper Albian to Santonian, and is characterized by dominantly red and green claystones intercalated with 260 individual black shale beds (Hay *et al.* 1984; Emeis, in prep.). Cyclic sedimentation is indicated. In a typical cycle, colours change from red claystones, which constitute the 'background' sediments, to green claystones, followed by pyrite beds with silt laminae, followed by graded and bioturbated black shale beds. The black shales are on average 43 mm thick and their organic carbon content averages 8.7% (as opposed to an average of 0.5% in the claystones). C/N ratios are around 15 in the red claystones and well above 25 in the black shales. Trace metals, most notably zinc, vanadium, cobalt, and nickel, are significantly enriched in the black shales as compared to the clay standard. Mean values exceed those reported by Brumsack (1980) for North Atlantic black shales (Table 2). Organic matter is dominated by algal material, although visual examinations of kerogen extracts by incident and ultra-violet light microscopy and hydrocarbon analyses reveal a considerable terrigenous input (Meyers *et al.* 1984; Rullkötter *et al.* 1984). The organic material is thermally immature, and severely biodegraded (Emeis, in prep.). There is no clear evidence for significant changes from marine to terrestrial sources of organic matter during the deposition of this interval (Meyers 1984).

Starting with the Angola deep basin sediments (DSDP, Leg 75) relevant details bearing on the black shale problem are briefly enumerated: hemipelagic clays, reddish in colour, represent the bulk of the sediment. They have low to moderate C_{org} values. C/N ratios are low, so are the metal contents. Microscopic studies of the organic matter indicate characteristic residual signals of intertinite. Intercalated between the oxic strata are reduced sediments including greenish clays, pyrite placers, and graded as well as bioturbated black shale beds. The reddish and greenish clays are chemically and mineralogically identical except for a small deficit in iron in the case of the greenish variety (Dean & Parduhn 1984; Emeis, in prep.). The reddish colour is due to hematite coatings, a few molecular layers thick, on top of the smectite surfaces (Konta, pers. comm.). Removal of the Fe_2O_3 micro-layer will bring to light the inherent greenish colour. It is well established that three-valent iron in silicate structures will frequently give a greenish hue to clay-sized sediments. The most conspicuous members of these cycles are pyrite layers (90% of the beds) and bioturbated (60% of the beds) black shale beds (Schallreuther 1984; Stow & Dean 1984). In view of this uniform pattern, *in situ* generation of such placer-like deposits can be excluded. It is unlikely that a pyrite placer is deposited in a distal fan simply due to pyrite's high specific gravity (5 g/cm^3), which would cause sedimentation of dispersed grains already upslope.

The black shales have high contents of refractory organic matter for which mixed marine and terrestrial sources are indicated on the basis of organic geochemical analyses (sugars, hydrocarbon extracts) and examinations of maceral composition. Although structureless organic matter indicative of biodegraded marine tissue (Gutjahr 1983) generally amounts to over 70% relative to terrigenous and figurated marine macerals in kerogen samples, hydrocarbon extracts display distinct terrestrial fingerprints and are relatively depleted in short chain n-alkanes. C/N ratios are high despite the dominance of marine derived organic matter. Carbohydrate fractions are depleted in 'marine' sugars galactose, xylose, rhamnose, and mannose, and are dominated by glucose and fructose. Certain trace metals (V, Ni, Zn) are significantly enriched in the organic fraction. The authors interpret these findings as indications for advanced biodegradation and a particular mode of sedimentation, which altered the original composition to refractory organic matter, and resulted in pronounced metal enrichment.

In contrast to these deep basin-type black shales, autochthonous black shales of the intracratonic Berriasian lack extensive reworking and metal enrichment. Organic matter in mudstones of the Hatteras Formation, representing continental rise turbidite sedimentation, appears to be less refractory as compared to the Leg 75 black-shales, although amorphous marine-type organic material dominates. In some black shale members, vitrinite and inertinite material are abundant. In spite of high C_{org}, metal content in the organic-rich turbidites is low, in contrast to

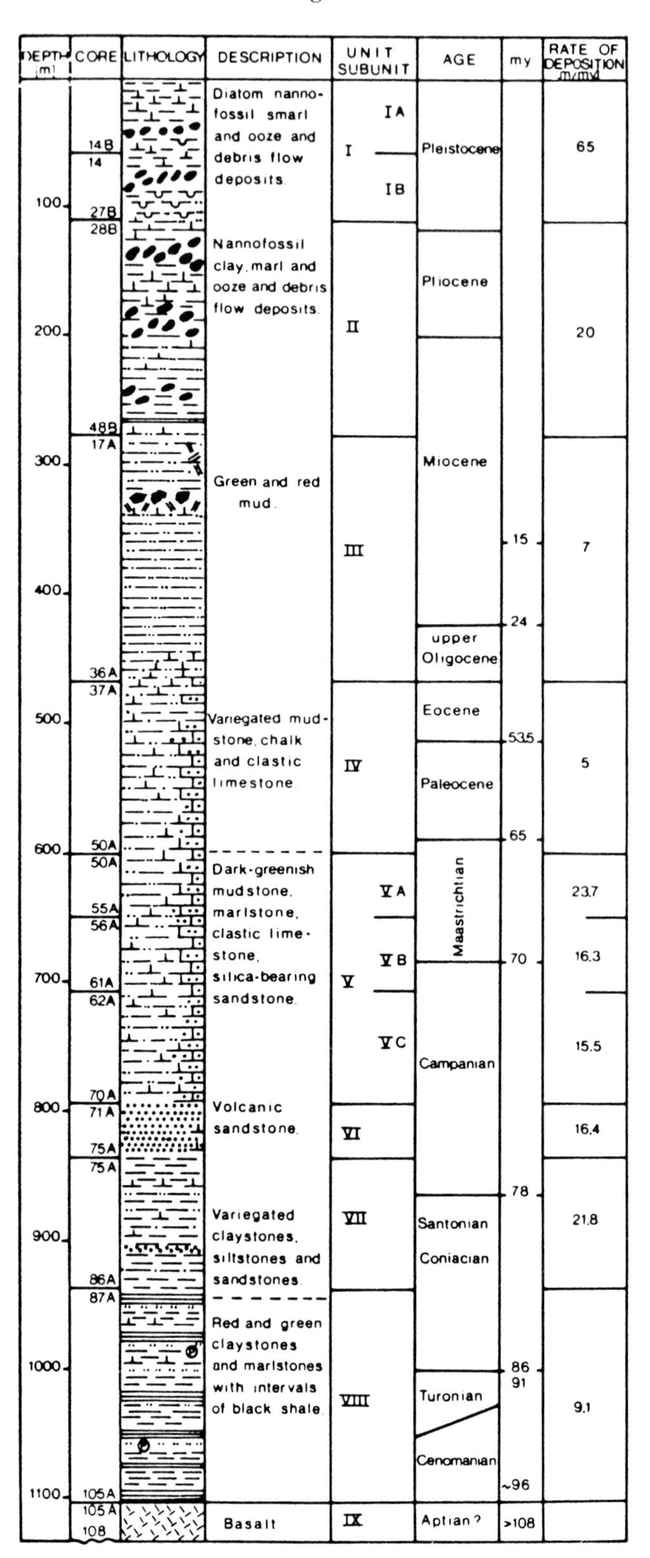

FIG. 6. Summary section of Site 530 sedimentary column, DSDP Leg 75 (Angola Basin).

TABLE 1. *Some properties of the sections studied*

	Bueckeberg Formation NW Germany	Cretaceous of Ghana	Hatteras Formation Leg 93/ Hole 603B	Angola Basin Leg 75/ Hole 530A
Age	Berriasian	Barremian to Maastrichtian	Albian to Coniacian	Albian to Santonian
Environment	Shallow basin on the shelf, alternating oxic and anoxic conditions	Shallow, rapidly subsiding graben structures on the shelf, oxic	Deep water, lower continental rise, oxic	Deep oceanic basin, oxic
Water Depth	Less than 200 m	Less than 200 m	2000–3000 m	3500–4000 m
Main type of organic matter	Algae of freshwater, brackish-marine origin, low terrestrial contribution	Lignite, brackish or lacustrine algae	Marine with low terrestrial contribution	Marine with terrestrial contribution during turbidite activity
C/N ratios	34	15	18	26
Preservation of organic matter and maturity	Good, onset of thermal catagenesis	Good, onset of thermal catagenesis	Moderate, low thermal maturation	Refractory, biodegraded, onset of thermal catagenesis
Metal enrichment	No	No	No	Yes (Ni, V, Zn)
Facies	Lacustrine oil shales	Delta deposits	Hemipelagic turbidite sequences	Hemipelagic/pelagic clastic sediments & sedimentation of of allochthonous black shale beds

TABLE 2. *Geochemistry of black shales*

Element/ oxide	Clay standard[1]	N. Atlantic black shales[2]	Leg 75 black shales[3]	Leg 75 others	Leg 93 black shales	Berriasian[4] oil shales
CaO	2.2%	n.g.	1.5%	1.8%	9.5%	3.6%
Fe_2O_3	4.7%	3.5%	9.0%	6.9%	4.6%	7.0%
C_{org}	0.3%	6.2%	8.7%	0.5%	4.8%	5.2%
S_{total}	n.g.	2.2%	4.0%	1.0%	1.8%	2.0%
Co	19	n.g.	89	39	n.d.	24
Cr	90	202	257	149	106	110
Cu	45	156	355	100	105	33
Ni	68	186	256	80	74	79
V	130	822	918	218	161	170
Zn	95	828	1525	114	305	127

[1] Turekian & Wedepohl 1961. [2] Brumsack 1980. [3] Emeis, in prep. [4] Wiesner, 1983. n.g. = not given in reference. n.d. = not detected.

the deep basin black shales of similar organofacies.

Table 1 summarizes characteristic properties of the different sediment sequences. Table 2 gives mean values for some selected parameters.

By subjecting the data set to a series of statistical analyses, relationships between parameters become apparent: Results are summarized in the form of dendrograms (cluster analysis) (Figs 7, 8 and 9). In the Bueckeberg Formation sediments and the turbidite sections of the Hatteras Formation there is no association between

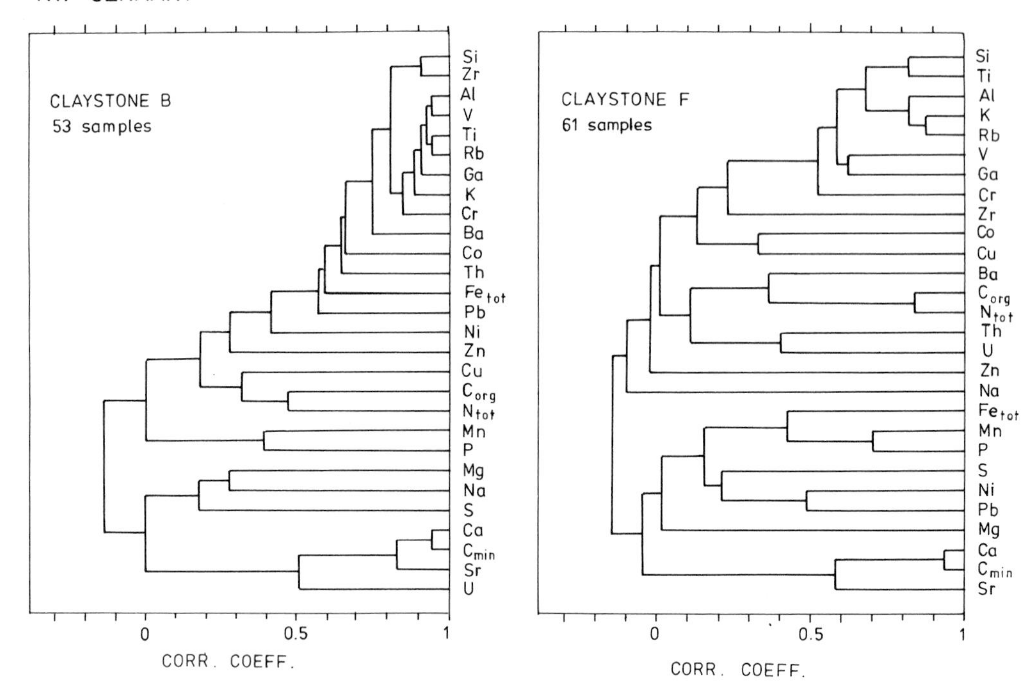

FIG. 7. Dendogram of cluster analysis for parameters, Bueckberg Formation, Wealden Basin, NW Germany (F = Oilshales, B = background claystones).

metals and organic matter (Figs 7 and 8). This contrasts with the Angola Basin results, where the three most abundant trace metals, i.e. V, Ni and Zn, group with the organic fraction (Fig. 9). This association is interpreted to be a characteristic feature of what is called the deep ocean black shale facies.

Discussion

The environment leading to deposition of specific deep-water black shales may be deduced from the sedimentary record, assuming that conditions in the nascent South and North Atlantic basins of the Cretaceous were similar to those found in today's ocean.

Deposition of background lithology

Under normal steady state conditions today (Fig. 10), photosynthetically fixed carbon and detrital minerals are effectively and rapidly removed from the euphotic zone as pellets and aggregates by grazing zooplankton (Honjo *et al.* 1980; Deuser *et al.* 1983). A standing stock of microbes, preferentially settling at pycnoclines in the water column, may remineralize part of the rapidly sinking organic material (Karl *et al.* 1984) and create zones of lower O_2 contents below the thermocline (Codispoti & Richards 1976). Their access to this substrate is limited mainly by high settling rates, and by the small surface areas of the sinking particles. Upon arrival at the bottom (the benthic transition layer of Honjo *et al.* 1982), these particles will serve as the main source of food and energy for benthic organisms (Fig. 10). Carbon will become oxidized to a large extent, and in phase with high surface productivity, anoxic conditions may temporarily be established in the benthic transition layer. Any metabolizable carbon escaping immediate oxidation will be consumed at times when productivity is low, and when bottom waters are normally aerated and sediments are bioturbated (Müller & Suess 1976; Reimers & Suess 1983).

Thus, the sediment finally making up the stratigraphic column exhibits characteristics of a hemipelagic clay, which constitutes the 'background' lithology of the reddish, oxidized claystones encountered in many Cretaceous sections of the Atlantic.

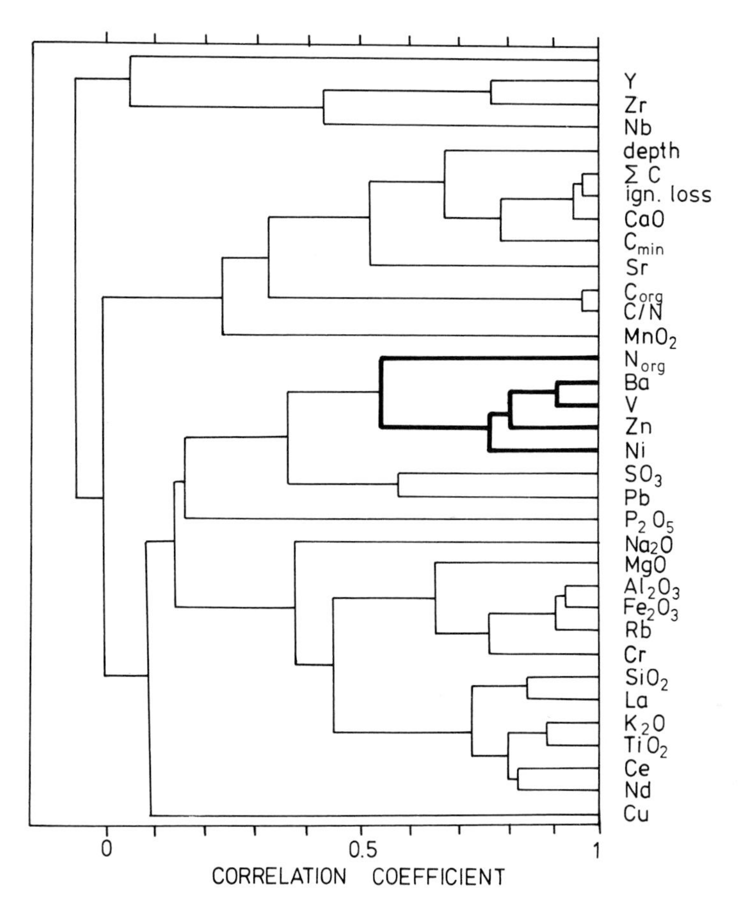

FIG. 8. Dendrogram of cluster analysis for parameters, 14 samples of organic-rich turbidite beds, Hole 603B, DSDP Leg 93.

Basinal black shale deposition

This quasi steady state can become disrupted at times of turbidite events in the aftermath of tectonic activities along continental shelf and slope, or after exogenic processes leading to large-scale resuspension of shelf sediment, e.g. storms (McCave 1972) (Fig. 1). Observations off the American Pacific coast (Honjo *et al.* 1980; Pak *et al.* 1980) show that a specific sediment fraction derived from slope or shelf, and largely composed of certain clay minerals and organic detritus, is advected along density boundaries at midwater depths, and is transported horizontally basinward for hundreds of kilometres. This selective advection of clay sized material rich in organic matter leads to the injection of an ideal substrate for fast growth into the zone of microbial remineralization in mid-water (Karl *et al.* 1984). The residence time of these particles is high compared to that of pellets and aggregates. Eventually, however, these particles are swept to abyssal depths by passing aggregates (Ittekkot *et al.* 1984a,b). The authors believe that microbial reworking, as well as kinetic equilibration of organic matter with dissolved metals, would reach completion by the time these particles reached the sea floor. The original material would have been a mixture of recycled proto-kerogen (marine and terrestrial), and fresh marine matter. This composition is typical for recent shelf and slope material (Premuzik 1980). It is further suggested that, in suspension, the marine diluting

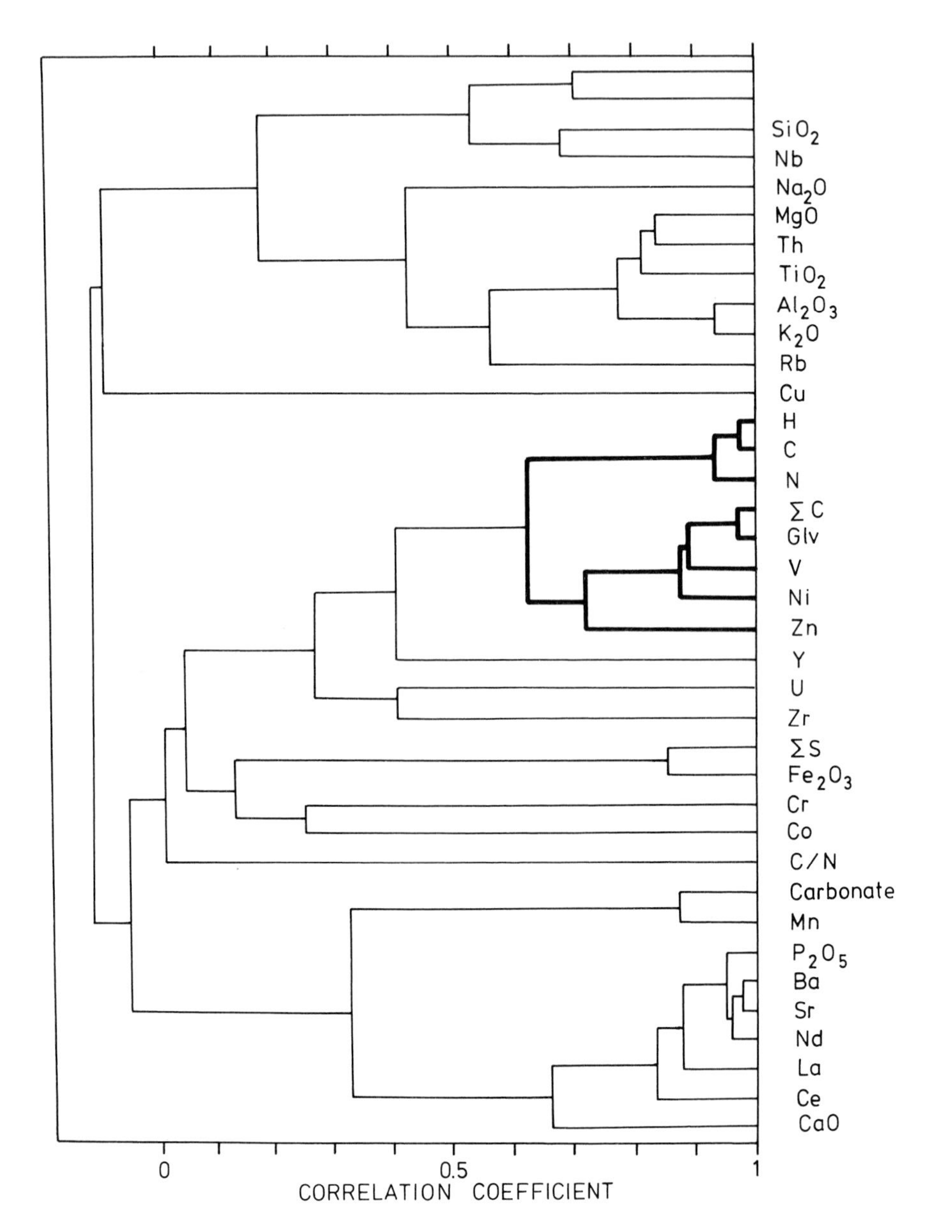

FIG. 9. Dendrogram of cluster analysis for parameters, 44 black shales of Unit 8, Hole 530A, DSDP Leg 75.

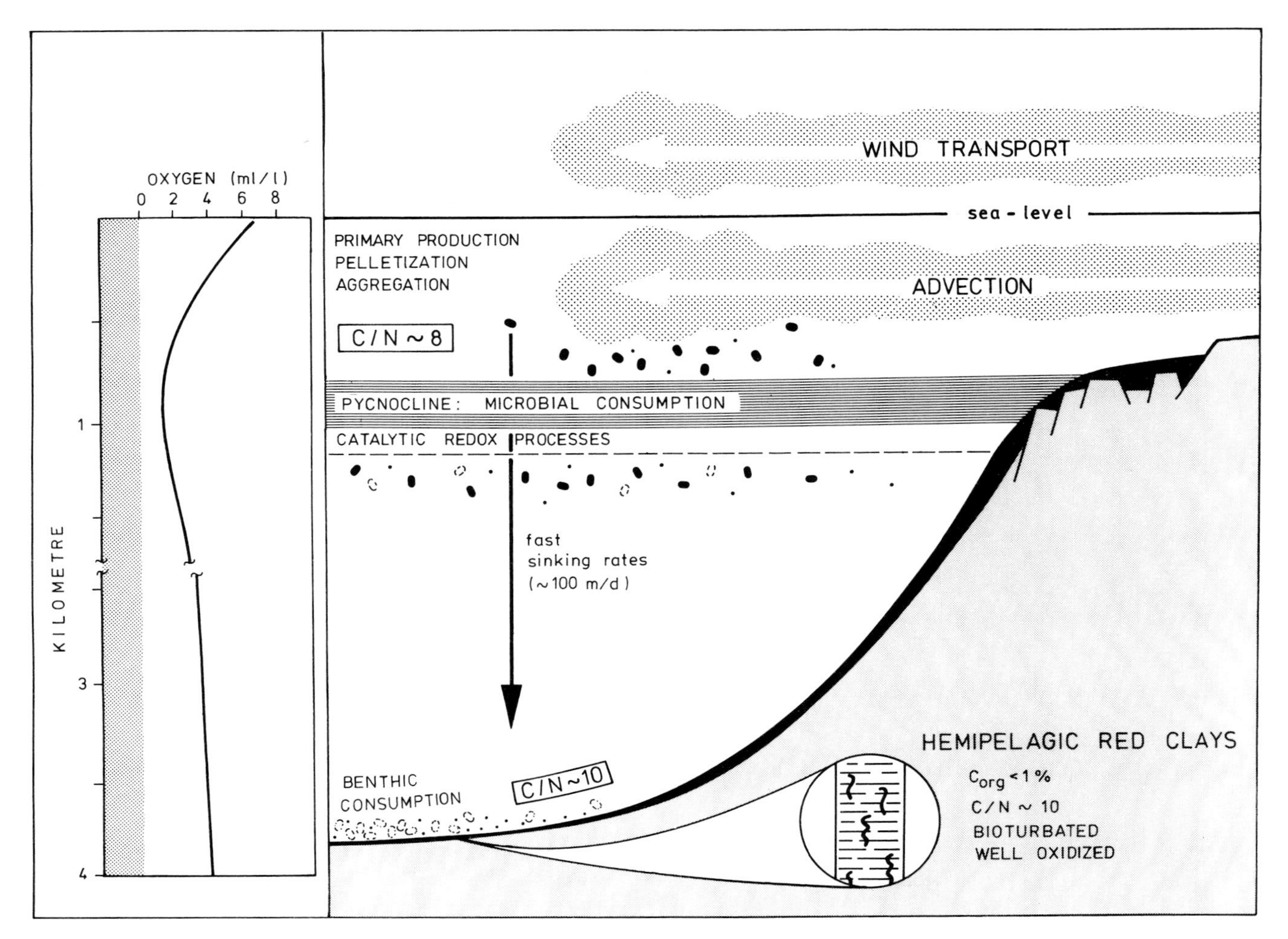

FIG. 10. Sketch of a continental margin during normal, non-turbidite conditions. See text for explanation.

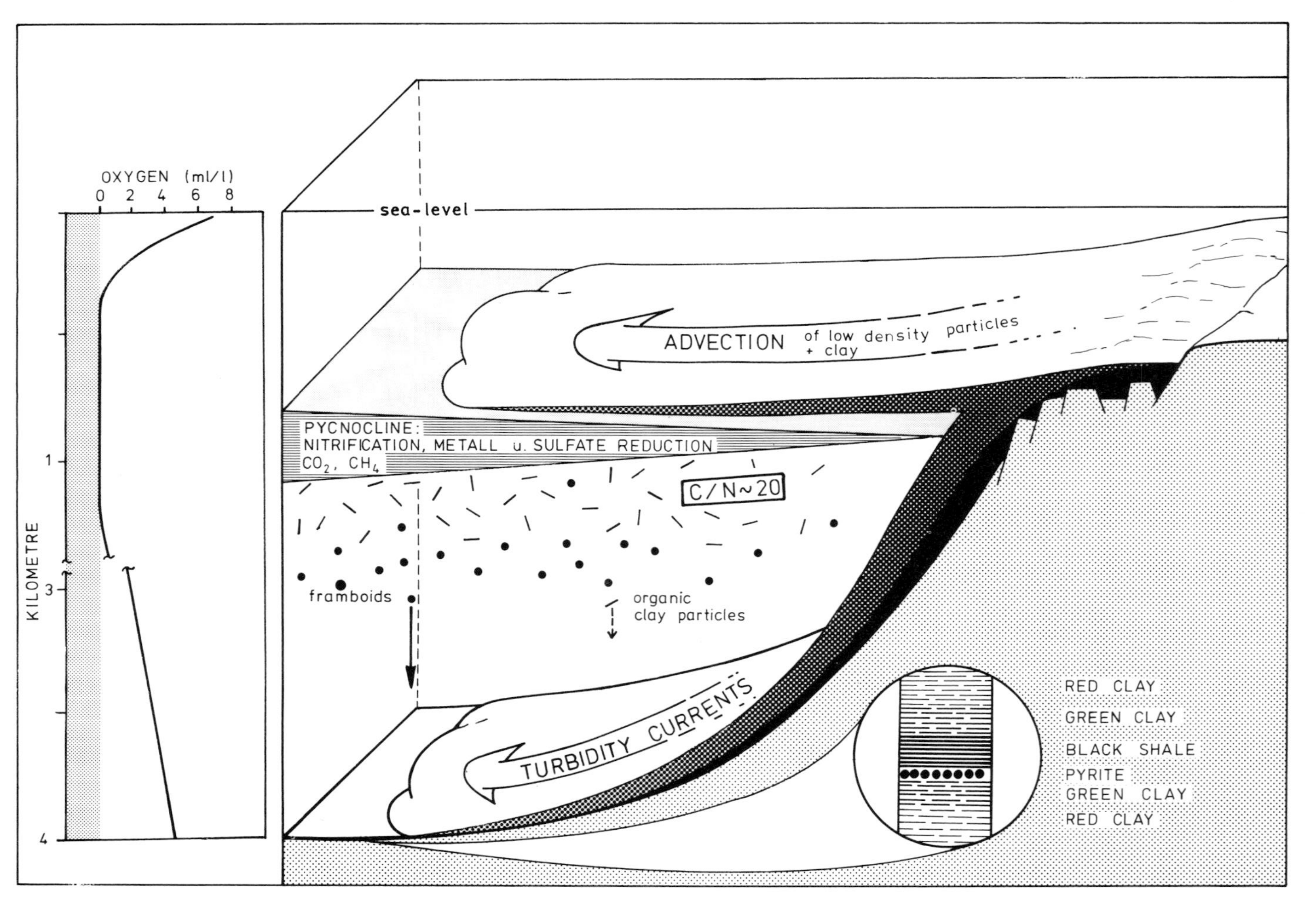

FIG. 11. Sketch of a continental margin during turbidite conditions. See text for explanation.

component would become preferentially metabolized in the course of bacterial reworking, so that formerly low amounts of terrestrial substances would become relatively more prominent in the sediment (Mycke 1981). Furthermore, bacterial activity would cause a prompt and dramatic increase in carbon oxidation at mid-water depths, leading to complete depletion of oxygen in the water. Secondary sources for oxygen, e.g. nitrate, phosphate, oxides of manganese and iron, would become reduced prior to the onset of sulphate reduction (Bender & Heggie 1984). Hematite coatings would be stripped from the clay surface, giving rise to high dissolved iron contents in the surrounding water. This in turn would result in the formation of pyrite framboids, when the sulphide concentration increased within the settling cloud.

Bearing these possibilities in mind it can be seen that the facies pattern in the sediments can be visualized as a reflection of (1) oxygen depletion in the water column associated with turbid suspensions, and of (2) differences in the settling velocities of sediment particles. Pyrite would be the first to reach the sediment-water interface at the front of a settling cloud, followed by less dense organic-rich matter and clay particles stripped of their reddish coatings. These processes are summarized in Fig. 11.

Conclusion

Deep basin-type black shales deposited during the Cretaceous differ from contemporaneous black shales in a number of ways, i.e. trace metal content, type of C_{org}, state of preservation and redox cycles. These differences may be explained by chemical and microbial transformations proceeding in the water column following basinward advection of organic-rich slope and shelf sediments on mid-water density surfaces. The transformations involve the extraction of certain metals from the dissolved phase, the degradation of marine tissue, and the relative enrichment of residue in inert terrestrial organic matter. Oxygen depletion in midwater takes place after advection of organic rich turbidite material and the end result is the deposition of black shales of deep ocean type.

During the slow sinking of resuspended particles from midwater, suspensates of detrital minerals are chemically transformed and new minerals are generated. Differential settling of this advected shelf or slope material, and of authigenic minerals generated in mid-water in slowly sinking detrital cloud complexes, creates a facies pattern characterized by alternating, reddish and greenish layers plus pyrite placers and black shale intercalations. A preliminary review of literature of black shale strata encountered in the Atlantic Ocean during drilling campaigns of the DSDP suggests that many, if not all, organic-rich sediments of the Mesozoic may have formed as the result of turbidites as suggested here.

ACKNOWLEDGEMENTS: Financial support by the Deutsche Forschungsgemeinschaft in the course of this research project is gratefully acknowledged. The authors would like to thank V. Ittekkot for helpful discussion, and the anonymous reviewers for the trouble they took to improve the paper.

References

BENDER, M.L. & HEGGIE, D.T. 1984. The fate of organic carbon reaching the deep sea floor: a status report. *Geochim. Cosmochim. Acta* **48,** 977–86.

BRUMSACK, H.-J. 1980. Geochemistry of Cretaceous black shales from the Atlantic Ocean (DSDP Legs 11, 14, 36 & 41). *Chem. Geol.* **31,** 1–25.

CODISPOTI, L.A. & RICHARDS, F.A. 1976. An analysis of the horizontal regime of denitrification in the eastern tropical North Pacific. *Limnol. Ocean.* **21/3,** 379–88.

DEAN, W.E. & PARDUHN, N.L. 1984. Inorganic geochemistry of sediments and rocks recovered from the southern Angola Basin and adjacent Walvis Ridge, Sites 530 and 532, DSDP Leg 75, *In*: HAY, W.W. & SIBUET, J.C. *et al.* (eds), *Init. Repts DSDP* **75.** US Govt Print. Off., Washington, DC. 923–58.

——, ARTHUR, M.A. & STOW, D.A.V. 1984. Origin and geochemistry of Cretaceous deep sea black shales and multicoloured claystones, with emphasis on DSDP Site 530, southern Angola Basin. *In*: HAY, W.W. & SIBUET, J.-C. *et al.* (eds), *Init. Repts DSDP* **75.** US Govt. Print. Off. Washington, DC. 819–44.

DEGENS, E.T., VON HERZEN, R.P. & WONG, H.K. 1973. Lake Kivu, structure, chemistry and biology of an East African Rift lake. *Geol. Rdsch.* **62/1,** 245–77.

—— & ROSS, D.A. (eds) 1969. *Hot Brines and Recent Heavy Metal Deposits in the Red Sea.* Springer Verlag, New York.

—— & ROSS, D.A. (eds) 1974. *The Black Sea: Geology, Chemistry, and Biology. Am. Ass. Petrol. Geol. Mem.* **20,** Tulsa.

DEUSER, W.G., EMEIS, K.-C., ITTEKKOT, V. & DEGENS, E.T. 1983. Fly-ash particles intercepted in the deep Sargasso Sea. *Nature (Lond.)* **305,** 216–18.

EMEIS, K.-C. 1985. *Geochemie und Fazies von Schwarzschiefern und organisch-reichen Sedimenten des Atlantischen Ozeans.* PhD Thesis, University of Hamburg (unpublished).

——, Mycke, B., Richnow, H.H., Spitzy, A. & Degens, E.T. (submitted). Carbon and metals in Site 603 sediments. *Init. Repts DSDP* **93.**

Gutjahr, C.C.M. 1983. Introduction to incident-light microscopy of oil and gas source rocks. *Geologie en Mijnbouw* **62,** 417–25.

Hay, W.W., Sibuet, J.-C. *et al.* 1982. Sediment and accumulation of organic carbon in the Angola Basin and on Walvis Ridge: Preliminary results of DSDP Leg 75. *Geol. Soc. Am. Bull.* **93,** 1038–50.

Honjo, S., Manganini, S.J. & Cole, J.J. 1980. Sedimentation of biogenic matter in the deep ocean. *Deep Sea Res.* **29/5A,** 609–25.

——, Manganini, S. & Poppe, L.J. 1982. Sedimentation of lithogenic particles in the deep ocean. *Mar. Geol.* **50,** 199–220.

Ittekkot, V., Degens, E.T. & Honjo, S. 1984. Seasonality in the fluxes of sugars, amino acids and amino sugars to deep ocean: Panama Basin. *Deep Sea Res.* **31/9A,** 1057–63.

——, Deuser, W.G. & Degens, E.T. 1984. Seasonality in the fluxes of sugars, amino acids and amino sugars to deep ocean: Sargasso Sea. *Deep-Sea Res.* **31/9A,** 1064–71.

Jansa, L., Gardner, J.V. & Dean, W.E. 1978. Mesozoic sequences of the Central North Atlantic. *In*: Lancelot, Y., Seibold, E. *et al.*, *Init. Repts DSDP* **41.** US Govt. Print. Off., Washington, D.C. 991–1031.

Jenkyns, H.C. 1980. Cretaceous anoxic events: from continents to oceans. *Bull. Geol. Soc. Lond.* **137,** 171–88.

Karl, D.M., Knauer, G.A., Martin, J.H. & Ward, B.B. 1984. Bacterial chemolithotrophy in the ocean is associated with sinking particles. *Nature, Lond.* **309,** 54–6.

McCave, I.N. 1972. Transport and escape of fine-grained sediment from shelf areas. *In*: Swift, D.J.P., Duane, H. & Pilkey, O. (eds), *Shelf Sediment Transport.* Dowden, Hutchinson & Ross, Stroudsburg.

Meyers, P.A. 1984. Organic geochemistry of sediments from Angola Basin and the Walvis Ridge: A synthesis of studies from DSDP Leg 75. *In*: Hay, W.W., Sibuet, J.-C. *et al.* (eds), *Init. Repts. DSDP* **75.** US Govt. Print. Off. Washington, DC. 459–67.

——, Leenheer, M.J., Kawka, O.E. & Trull, T.W. 1984. Enhanced preservation of marine-derived organic matter in Cenomanian black shales from the southern Angola Basin. *Nature (Lond.)*, **312,** 356.

Müller, P.J. & Suess, E. 1979. Productivity, sedimentation rate, and sedimentary organic matter in the oceans—I. Organic carbon preservation. *Deep Sea Res.* **26A,** 1347–62.

Mycke, B. 1980. *Organisch-geochemische Charakterisierung rezenter Wattsedimente.* MSc. Thesis, University of Hamburg (unpublished).

Natland, J.H. 1978. Composition, provenance, and diagenesis of Cretaceous clastic sediments drilled on the Atlantic Continental Rise off Southern Africa, DSDP Site 361-Implications for the early circulation of the South Atlantic. *In*: Bolli, H.M., Ryan, W.B.F. *et al.* (eds), *Init. Repts DSDP.* **40,** US Govt. Print. Off., Washington, DC. 1025–61.

Pak, H., Codispoti, L.A. & Zanefeld, J.R.V. 1980. On the intermediate particle maxima associated with oxygen-poor waters off western South America. *Deep Sea Res.* **27A,** 873–97.

Premuzic, E.T. 1980. Organic carbon and nitrogen in the surface sediments of world oceans and seas: Distribution and relationship to bottom morphology. *Brookhaven National Laboratory Rept.* **51084.**

Reimers, C.E. & Suess, E. 1983. The partitioning of organic carbon fluxes and sedimentary organic matter decomposition rates in the ocean. *Mar. Chem.* **13,** 141–68.

Rullkötter, J., Mukhopadhyay, P.K. & Welte, D.H. 1984. Geochemistry and petrography of organic matter in sediments from Hole 530A, Angola Basin, and 532, Walvis Ridge, Deep Sea Drilling Project. *In*: Hay, W.W. & Sibuet, J.-C. *et al. Init. Repts DSDP*, **75.** US Govt. Print. Off., Washington, DC. 1069–87.

Ryan, W.B.F. & Cita, M.B. 1977. Ignorance concerning episodes of oceanwide stagnation. *Mar. Geol.* **23,** 197–215.

Schallreuter, R. 1984. Framboidal pyrite in deep-sea sediments. *In*: Hay, W.W. & Sibuet, J.-C. *et al.*, *Init. Repts DSDP* **75.** US Govt. Print. Off., Washington, DC. 875–91.

Schlanger, S.O. & Cita, M.B. (eds) 1982. *Nature and Origin of Cretaceous Organic-Rich Facies.* Academic Press, London.

Schlayer, J.G., Hellinger, S. & Tapscott, C. 1977. The palaeobathymetry of The Atlantic Ocean from the Jurassic to the present. *J. Geol.* **85,** 509–52.

Stow, D.A.V. & Dean, W.A. 1984. Middle Cretaceous black shales at Site 530 in the southeastern Angola Basin. *In*: Hay, W.W., Sibuet, J.-C. *et al.* (eds). *Init. Repts DSDP* **75.** US Govt. Print. Off., Washington, DC. 809–17.

Thiede, J. & van Andel, T.H. 1977. The paleoenvironment of anaerobic sediments in the late Mesozoic South Atlantic Ocean. *Earth Planet. Sci. Lett.* **33,** 301–9.

Turekian, K.K. & Wedepohl, K.H. 1961. Distribution of the elements in some major units of the earth's crust. *Bull. Geol. Soc. Am.* **72,** 175–92.

Wiesner, M.G. 1983. *Lithologische und geochemische Faziesuntersuchungen an bituminösen Sedimenten des Berrias im Raum Bentheim-Salzbergen (Emsland)*. PhD Thesis, University of Hamburg (unpublished).

E.T. Degens, K.-C. Emeis, B. Mycke & M.G. Wiesner, Geologisch—Paläontologisches Institut, Bundesstrasse 55, D–2000 Hamburg 13, Federal Republic of Germany.

Comparison of Mesozoic carbonaceous claystones in the western and eastern North Atlantic (DSDP Legs 76, 79 and 93)

Jürgen Rullkötter & Prasanta K. Mukhopadhyay

SUMMARY: Mesozoic carbonaceous claystones were recovered on both margins of the North Atlantic Ocean in a number of holes drilled by the Deep Sea Drilling Project (DSDP). Geochemical investigation revealed that the organic matter in these sediments belongs to different organofacies types, as indicated by varying degrees of preservation and mixing of marine and terrigenous material. The intensity of terrigenous organic matter supply, coastal upwelling, downslope transport events, and local oxygen depletion are major factors controlling the composition and preservation of the organic matter in most of the Mesozoic carbonaceous claystones, whereas an ocean-wide anoxic event, if at all, can only be invoked for the organic-matter-rich sediments deposited at the Cenomanian–Turonian boundary. Examples discussed are from the Mazagan Escarpment (Site 547, DSDP Leg 79), the Blake Bahama Basin (Site 534, DSDP Leg 76), and the continental rise off Cape Hatteras (Site 603, DSDP Leg 93).

Detailed studies of the organofacies types of marine sediments will provide information on the provenance of the organic matter, its modes of transportation to the ultimate place of burial, and the depositional environment. This information may contribute to the reconstruction of palaeoclimate and palaeocirculation patterns in oceanic basins, and help in a better understanding of the geologic history of the world's oceans. Since the recovery of organic-matter-rich deep sea sediments at various locations of the Atlantic Ocean by the efforts of the Deep Sea Drilling Project (DSDP), this approach has often been applied to describe the conditions of sedimentation during different stages of the geologic development of the Atlantic Ocean.

Accumulation of organic matter in deep-water oceanic environments, in a general sense, is not favoured due to the destruction of most of the material during settling through oxygenated water masses and in the oxygenated surface sediments. Where transport is rapid, or in fecal pellets, it is feeding activities at the sediment/water interface that contribute the most to this destruction (e.g. Welte *et al.* 1979). A high surface water bioproductivity alone is not a sufficient criterion for the deposition of organic-matter-rich deep sea sediments. A number of simultaneously active, mutually supporting factors usually control the preservation or (selective) mineralization of organic matter in the deep sea environment. In addition to the marine surface bioproductivity and the intensity of terrigenous organic matter supply, these factors include the settling rate of organic matter through the water column, the rate of sediment accumulation, the oxygen content of the water masses, and downslope sediment transport processes (e.g. Welte *et al.* 1979; Rullkötter *et al.* 1983; Arthur *et al.* in press).

The organofacies of North Atlantic organic-matter-rich deep sea sediments has previously been described by Tissot *et al.* (1977; 1980), Summerhayes (1981; in press), de Graciansky *et al.* (1982), Rullkötter *et al.* (1983), and Rullkötter & Mukhopadhyay (1984) using total organic carbon contents, Rock-Eval pyrolysis yields, and maceral composition by microscopy as the most important parameters. These sediments are often referred to as 'black shales', which in a strict sense implies a high content of marine organic matter (Waples 1983), but is sometimes less specifically used for a wide range of black-coloured sediments which may be organic-carbon-lean or contain mainly residual organic matter.

Oceanic anoxic events have been invoked for the formation of Cretaceous 'black shales' in the Atlantic Ocean (Schlanger & Jenkyns 1976; Arthur & Schlanger 1979; de Graciansky *et al.* 1984). Distinct organic facies differences between the western and eastern North Atlantic (Tissot *et al.* 1980; Summerhayes 1981, in press; Rullkötter & Mukhopadhyay 1984), however, indicate that basin-wide events cannot account for all the occurrences of organic-matter-rich sediments during this time period.

The present study extends the data base for organofacies characterization of deep sea sediments from both sides of the North Atlantic Ocean by a comparison of Mesozoic sediments from the Mazagan Escarpment (Site 547, DSDP Leg 79), the Blake Bahama Basin (Site 534,

From SUMMERHAYES, C.P. & SHACKLETON, N.J. (eds), 1986, *North Atlantic Palaeoceanography*, Geological Society Special Publication No. 22, pp. 377–387.

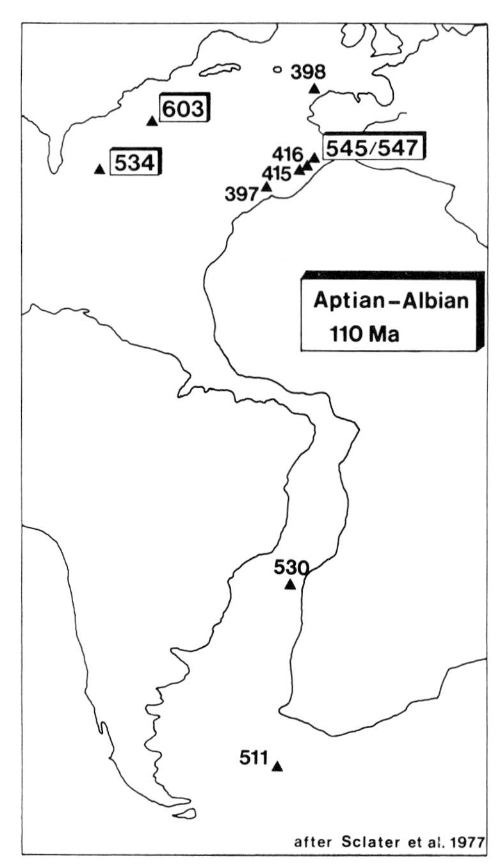

FIG. 1. Location of Deep Sea Drilling Project Sites in the Atlantic Ocean on a middle Cretaceous (Aptian-Albian) palaeogeographic map (after Sclater *et al.* 1977)

DSDP Leg 76) and the continental rise off Cape Hatteras (Site 603, DSDP Leg 93). An attempt is made to delineate the factors influencing the different organofacies types at these sites. Approximate sampling locations are shown on a schematic palaeogeographic map of the Atlantic Ocean for the Aptian-Albian time period (Fig. 1).

Site 547, Mazagan Escarpment

A compilation of organic carbon data for a Triassic through Tertiary sediment section recovered at DSDP Site 547, seaward of the Mazagan Escarpment off Morocco, is given in Fig. 2, together with the hydrogen index (HI) values from Rock-Eval pyrolysis representing yields of hydrocarbon-type compounds upon pyrolysis, normalized to organic carbon (cf. Espitalié *et al.* 1977). The hydrogen index provides a bulk measure of the type of organic matter present; in the relatively immature sediments studied, changes in hydrogen index due to maturity differences may safely be ignored. Combination of these data with kerogen microscopy results has led to the distinction of different organofacies types as shown in Table 1. For a general description of experimental procedures and additional geochemical data of the DSDP Site 547 sediments see Rullkötter *et al.* (1984).

Tertiary

The Tertiary in the North Atlantic is a domain of hemipelagic and pelagic sedimentation under oxygenated water conditions, which is incompatible with the preservation of significant amounts of organic matter in deep sea sediments. This general picture is occasionally disturbed by slump masses, which have transported significant amounts of organic matter of predominantly marine origin down the continental slope to the continental rise. Sediments of this kind were encountered in the Eocene section at Site 547, where organic carbon values exceeding 5% (Fig. 2), and hydrogen index values of more than 600 mg HC/g C_{org}, are indications of deposition under strongly anoxic conditions. Table 1 lists an example with a slightly reduced organic carbon content and hydrogen index, probably due to partial mixing with autochthonous pelagic sediment, but kerogen microscopy still reveals a dominance of maceral components of marine origin.

Palaeontological (Shipboard Party 1984) and organic geochemical data (Rulkötter *et al.* 1984) provide evidence that the Eocene slumps at Site 547 do not consist of synsedimentary material but rather are of Cretaceous (Albian to Cenomanian) age and probably were initially deposited within an oxygen-minimum zone close to the shelf edge at a time when coastal upwelling off Morocco was much more intense than in Recent times. For comparable slump masses recovered within the Miocene section on the continental rise south of the Canary Islands (DSDP Site 397; Fig. 1), a Cretaceous origin (Čepek & Wind 1979) and initial deposition at the outer shelf or upper slope (Arthur *et al.* 1979) was also demonstrated on palaeontological grounds.

Recent examples of high organic matter accumulation in shallow to intermediate water depths in upwelling areas are known for the Senegal and Angola continental margins (Sarnthein *et al.* 1982; Jansen *et al.* 1984), and in the latter case it has been shown that turbidites in the Zaire fan transport major amounts of organic matter into the deep sea (Jansen *et al.* 1984).

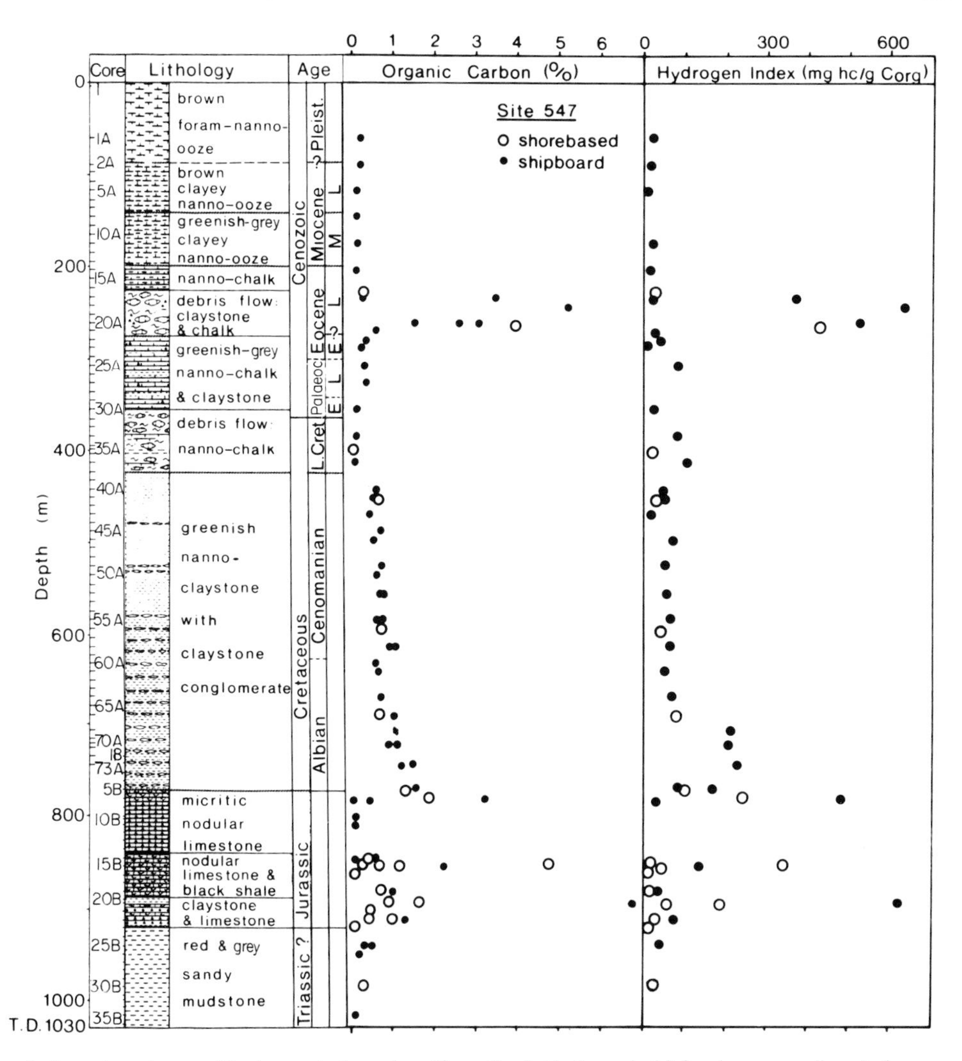

FIG. 2. Organic carbon and hydrogen index values (from Rock-Eval pyrolysis) for deep sea sediments from DSDP Site 547 off the Mazagan Escarpment (after Rullkötter *et al.* 1984).

Cretaceous

The Cretaceous at Site 547 on the Mazagan Escarpment is restricted to mainly Albian through Cenomanian sediments, with a thin organic-carbon-lean Early Cretaceous section on top of the Jurassic, and a Late Cretaceous debris flow section (Fig. 2). The greyish-green nannofossil-bearing claystones of Albian-Cenomanian age commonly contain intraformational flat-pebble mudstone conglomerate layers, slump structures, and repeated sections suggesting slide sheets (Shipboard Party 1984). This indicates that redeposition phenomena were pronounced at this site in the middle Cretaceous.

The organic carbon contents in the Albian sediments start with values around 2%, including an exceptional value of 3% at the bottom, and decrease upwards to slightly below 1%. In the Cenomanian the organic carbon contents are fairly constant around 0.7% (Fig. 2). Roughly parallel to the change in organic carbon content there is a slight decrease in hydrogen index values from Rock-Eval pyrolysis (Fig. 2). Kerogen microscopy reveals a mixture of marine and terrigenous macerals in the lower part of the Albian section and a gradational increase in the proportion of terrigenous organic matter upwards (Table 1). This is confirmed by the

TABLE 1. *DSDP Site 547 organofacies types*

Stratigraphic age	Lithology	C_{org} (%)	HI (mg hc/g C_{org})		Dominant macerals
Eocene	Clayey nannofossil chalk (slump)	4	431		Phyto-, zooclasts; exinite; amorphous marine OM
Cenomanian	Clayey nannofossil chalk (with slumps)	0.7	30–50		Exinite, resinite; amorphous humic matter; vitrinite
Albian	Nannofossil-bearing claystone (frequent slumps)	0.7–3	80–450		Exinite, resinite; amorphous humic matter, vitrinite; amorphous marine OM; phyto- and zooclasts
Jurassic	Claystone/black shale	0.4–10 (gradational)	20–600	Top:	amorphous marine OM; terrigenous liptinite; vitrinite
				Middle:	inertinite; recycled terrigenous liptinite
				Bottom:	inertinite

results of biological marker analysis in the extractable organic matter, which contains abundant steroid hydrocarbons in the lower Albian section and a dominance of higher plant wax alkanes in the Cenomanian sediments (Rullkötter *et al.* 1984). Preservation of both the terrigenous and marine organic matter is excellent, although both types were partially converted to amorphous material by sedimentary bacteria, but without apparent reduction of the hydrogen content. Incorporation of microbial biomass may have also helped to keep the bulk hydrogen content relatively high.

Because recognizable slump clasts have slightly higher organic carbon contents than the surrounding matrix (less than 20% relative difference on the average), and because there is no significant difference in organic matter type between these two sediment types, it has to be assumed that much of the organic matter in the middle Cretaceous sediments on the Mazagan Escarpment has been transported down the continental slope and redeposited on the continental rise. Abundant organic matter from a nearby continental source initially may have been deposited on the outer shelf, leading to anoxic conditions in the sediment and good preservation of a major part of the terrigenous lipid material. Apparently, these conditions also allowed a certain amount of marine organic matter to accumulate and to be preserved; this preservation may have been favoured by coastal upwelling connected with high bioproductivity and oxygen depletion in the water column although the relatively low organic carbon contents do not seem to suggest this (but see the Eocene slump described earlier). These sediments then were transported down the slope and redeposited on the continental rise. Rapid burial allowed much of the organic matter to be preserved even when the bottom water was oxic.

Jurassic

Most of the Jurassic at DSDP Site 547 consists of micritic nodular limestones extremely poor in organic matter, interbedded with several thin intervals of black carbonaceous claystones of Sinemurian to Pliensbachian age (Fig. 2). Gradational increase of organic matter content from the micritic limestone matrix into the black claystones, together with decreasing bioturbation, suggests autochthonous deposition in a progressively more anoxic environment, rather than downslope transport events as in the younger sections at this site. High concentrations of terrigenous organic matter indicate that periodically increasing land-derived supply of lipid-rich organic matter, rather than stagnant anoxic water conditions, caused oxygen depletion in the sediments. This process laid the basis for preservation of significant amounts of (amorphous) marine organic matter in the topmost parts of the black claystone layers, which reach up to 10% organic carbon and 600 mg hydrocarbons/g C_{org} pyrolysis yields (Table 1; Fig. 2). This is an example of the effect of terrigenous organic matter supply on organic matter preservation in the marine environment. This process, however, could only be active in this case because the water depth at the time of deposition was relatively shallow, i.e. less than about 300 m (Winterer & Hinz 1984).

DSDP Site 534, Blake-Bahama Basin

Drilling at DSDP Site 534 in the Blake-Bahama Basin in the western North Atlantic (Fig. 1) has penetrated an 800 m thick, nearly complete Cretaceous section overlying Jurassic sediments of Callovian through Tithonian age (Fig. 3). A great number of organic carbon data have been compiled from the available literature and compared with the authors' own measurements (Fig. 3). Organofacies parameters for these sediments are summarized in Table 2, based on the authors' own investigations.

Cretaceous

A compilation of organic carbon data from various authors (Fig. 3) shows a broad maximum in the middle Cretaceous. Most sediment samples still have less than 2% C_{org}, although the organic carbon content reaches 3% and more in parts of the Barremian to Albian section. The organic matter concentration is highly variable, at least partly because of the common occurrence of turbidites throughout the Cretaceous. Typical values measured for a few samples in our laboratory range between 0.7 and 1.8% C_{org} (Table 2).

Despite the variation of organic carbon content, and in contrast to the middle Cretaceous sediments at the Mazagan Escarpment in the eastern North Atlantic, the Blake-Bahama Basin samples, where they do not exceed 2% organic carbon, consistently have low pyrolysis yields typical of predominantly residual organic matter (Table 2). Only in the most organic-carbon-rich sediments do the hydrogen index values from Rock-Eval pyrolysis exceed 100 mg HC/g C_{org}, and thus indicate the presence of well preserved terrigenous organic matter with the occasional minor admixture of marine material (Herbin *et al.* 1983).

The Valanginian to Albian samples studied under the microscope are similar to each other in maceral composition. About 90% of their kerogen, consisting mainly of vitrinite, inertinite, sporinite, and degraded ligno-cellulose and higher plant liptinites (Table 2), is derived from terrestrial sources. The amorphous organic matter, which exceeds 30% in some of the samples, mostly is of humic nature, but a smaller proportion also is of marine origin and has reached the sediment as fecal pellets (cf. also Habib 1983). The amorphous organic matter is covered with micrinite (small inertinite particles) considered to be oxidative degradation products of liptinite particles (Stach *et al.* 1982). Yellow to brown recycled spores and recycled vitrinites are common in all the Cretaceous samples at Site 534.

The organofacies characteristics of the Cretaceous carbonaceous claystones in the Blake-Bahama Basin indicate that terrigenous organic matter supplied from the neighbouring North American continent reached its place of final deposition only after distal transport into the deep sea. The dominance of terrigenous recycled particulate organic matter and microbially altered amorphous humic matter, is evidence of an oxidative depositional environment with slightly better conditions for organic matter preservation occurring only occasionally during Barremian to Albian times. A similar organofacies type for the Cretaceous sediments at the nearby DSDP Sites 105 and 391 was described by Kendrick *et al.* (1978), Tissot *et al.* (1980) and Summerhayes (1981). Turbidite flows down the continental margin into the Blake-Bahama Basin could not improve the preservation of organic matter, as on the opposite margin of the North Atlantic.

Jurassic

Organic carbon contents in excess of 2% characterize the Callovian black nannofossil claystone at DSDP Site 534 (Fig. 3, Table 2). Rock-Eval pyrolysis indicates a mixed terrigenous/marine kerogen type, and visual examination under the microscope revealed that most of the organic matter consists of unstructured, amorphous liptinite derived from both phyto- and zooclasts and from terrigenous higher plants. The particulate organic matter mainly consists of spores and vitrinite. Similar results for this Callovian section have been obtained by Summerhayes & Masran (1983) and Tyson (1984).

The preservation of the organic matter apparently is better than in the overlying Cretaceous section. Deposition may have occurred in shallower water, and local anoxic conditions near the sediment-water interface in small basin-like structures observed at this site on seismic records cannot be excluded. Still, the hydrogen index values from Rock-Eval pyrolysis are low in view of the visual kerogen data. Strong bottom water currents leading to temporary resuspension of the deposited material into a nepheloid layer (Biscaye & Eittreim 1977) may have caused oxidative reduction of the hydrogen content of the organic matter, without excessive degradation.

DSDP Site 603, off Cape Hatteras

Drilling at DSDP Site 603, on the continental rise

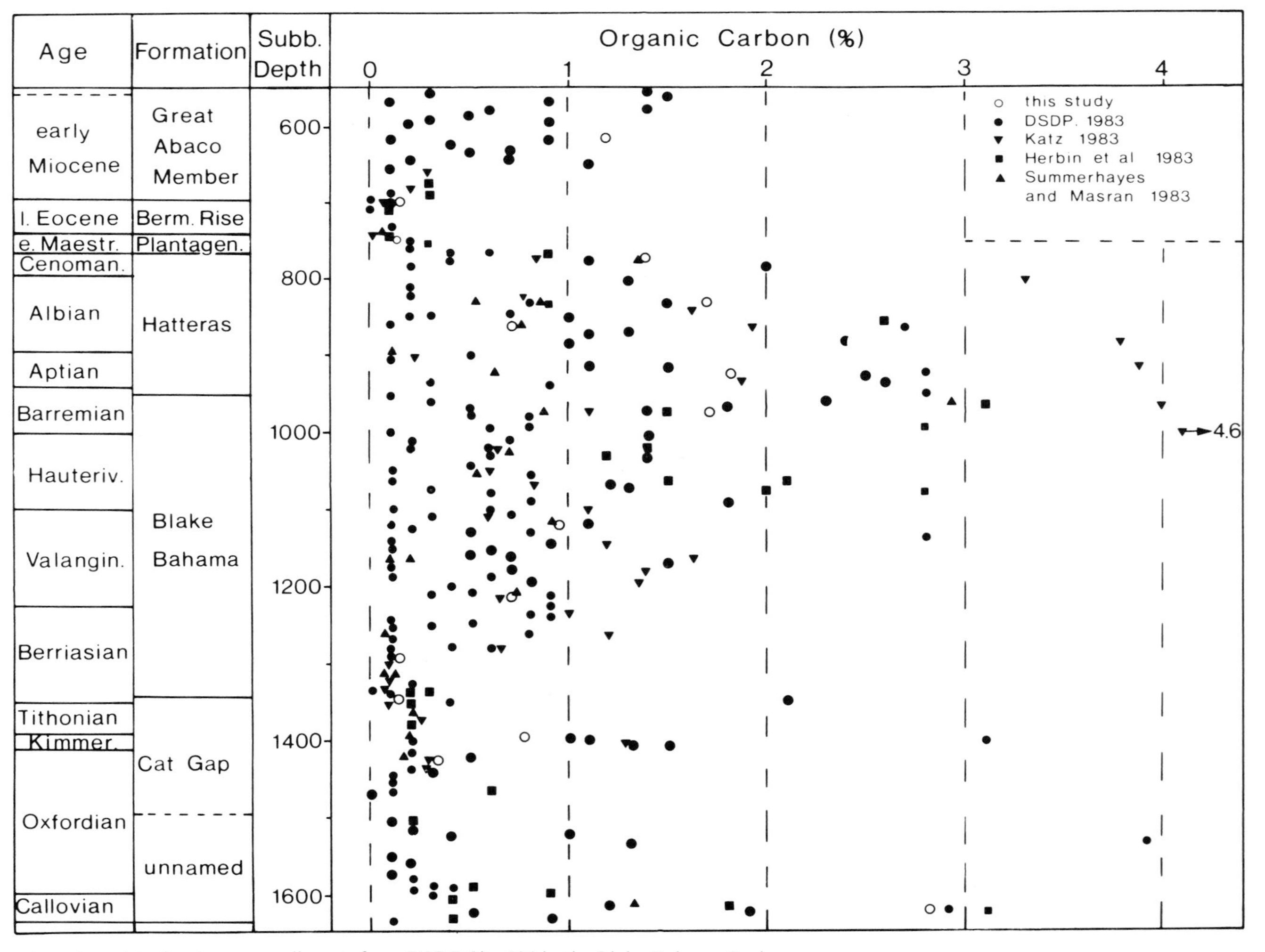

FIG. 3. Organic carbon data for deep sea sediments from DSDP Site 534 in the Blake-Bahama Basin.

TABLE 2. *DSDP Site 534 organofacies types*

Stratigraphic age	Lithology	C_{org} (%)	HI (mg hc/g C_{org})	Dominant macerals
Valanginian-Cenomanian	Grey limestone; calcareous claystone; carbonaceous claystone	0.7–1.8	55	Vitrinite; amorphous marine OM (fecal pellets); inertinite; recycled/oxidized spores
Callovian	Black nannofossil claystone	2.8	238	Amorphous marine OM; amorphous humic OM; vitrinite; spores, pollen

off the northeast American coast near Cape Hatteras, penetrated a nearly complete Cretaceous section about 600 m thick underlying more than 1000 m of Tertiary sediments (Fig. 4).

Cretaceous

Four different organofacies types, listed in five stratigraphic age groups in Table 3, have been recognized in the organic-carbon-rich (> 1% C_{org}) Cretaceous sediments at DSDP Site 603. The Santonian variegated claystone, despite 1.55% organic carbon, has a negligible hydrogen content based on Rock-Eval pyrolysis (Fig. 3). Kerogen microscopy revealed the presence of nearly 80% recycled vitrinite together with about equal amounts of primary vitrinite and inertinite.

The organic matter in the Coniacian to late Turonian variegated claystone (8.5% C_{org}), and in three Turonian to Cenomanian black carbonaceous claystones (6–14.5% C_{org}), is of a nearly uniform type and origin. Hydrogen index values around 400 mg HC/g C_{org} indicate a strong contribution of marine liptinitic material in these samples. Under the microscope this was seen to consist mainly of sapropelinite (structurally degraded liptinite), which reached the sediment in the form of fecal pellets. These are associated with unicellular algae and phytoplankton fragments, but pollen and spores of terrigenous origin were also observed. This assemblage of marine macerals is uncommon for the Cretaceous in the western North Atlantic, where terrigenous material usually dominates in the deep sea sediments (e.g. Tissot *et al.* 1980). The depositional environment at the Cenomanian–Turonian boundary, must have been favourable for the preservation of organic matter. Despite the occurrence of turbidites throughout the Cretaceous at DSDP Site 603, which may have helped in the transport and rapid burial of the delicate marine liptinitic material in the deep sea, it seems plausible that the oxygen content of the water columns was reduced at this location during that period. Similar occurrences of carbonaceous claystones at various places on the North Atlantic continental margins led de Graciansky *et al.* (1984) to suggest an ocean-wide water stagnation at the time of the Cenomanian–Turonian boundary. The organic geochemical data of the Site 603 sediments seem to support this idea, although a more local event cannot be fully excluded.

In the pre-Cenomanian sediments, terrigenous organic matter predominates. Barremian dark grey carbonaceous claystones, with organic carbon contents between 1 and 38% contain coaly material. The very high organic carbon values result from hand-picking small, coaly-looking sediment samples. Hydrogen index values of 50–75 mg HC/g C_{org} are consistent with the microscopic observation of primary huminite/vitrinite. Intimate association of this coaly organic matter with abundant pyrite is unusual for a deep sea sediment deposited in an oxygenated marine environment. It is assumed that formation of the huminite/pyrite mixture occurred in a nearby coastal swamp, and rapid transportation and burial at the continental rise must have prevented oxidation of the framboidal pyrite.

The Berriasian to Valanginian nannofossil limestones contain black stringers of concentrated organic material. A bulk sample with 1.5% organic carbon, and a handpicked sample from one of the stringers with 60% organic carbon, both exhibit the same pyrolysis results when normalized to organic carbon (about 300 mg HC/g C_{org}), and the same maceral composition. The kerogen consists of a mixture of bitumen-impregnated coaly material (saprovitrinite), with well preserved liptinites of both terrigenous (spores, pollen) and aquatic origin (dinoflagellates), which is a rare association in deep sea sediments.

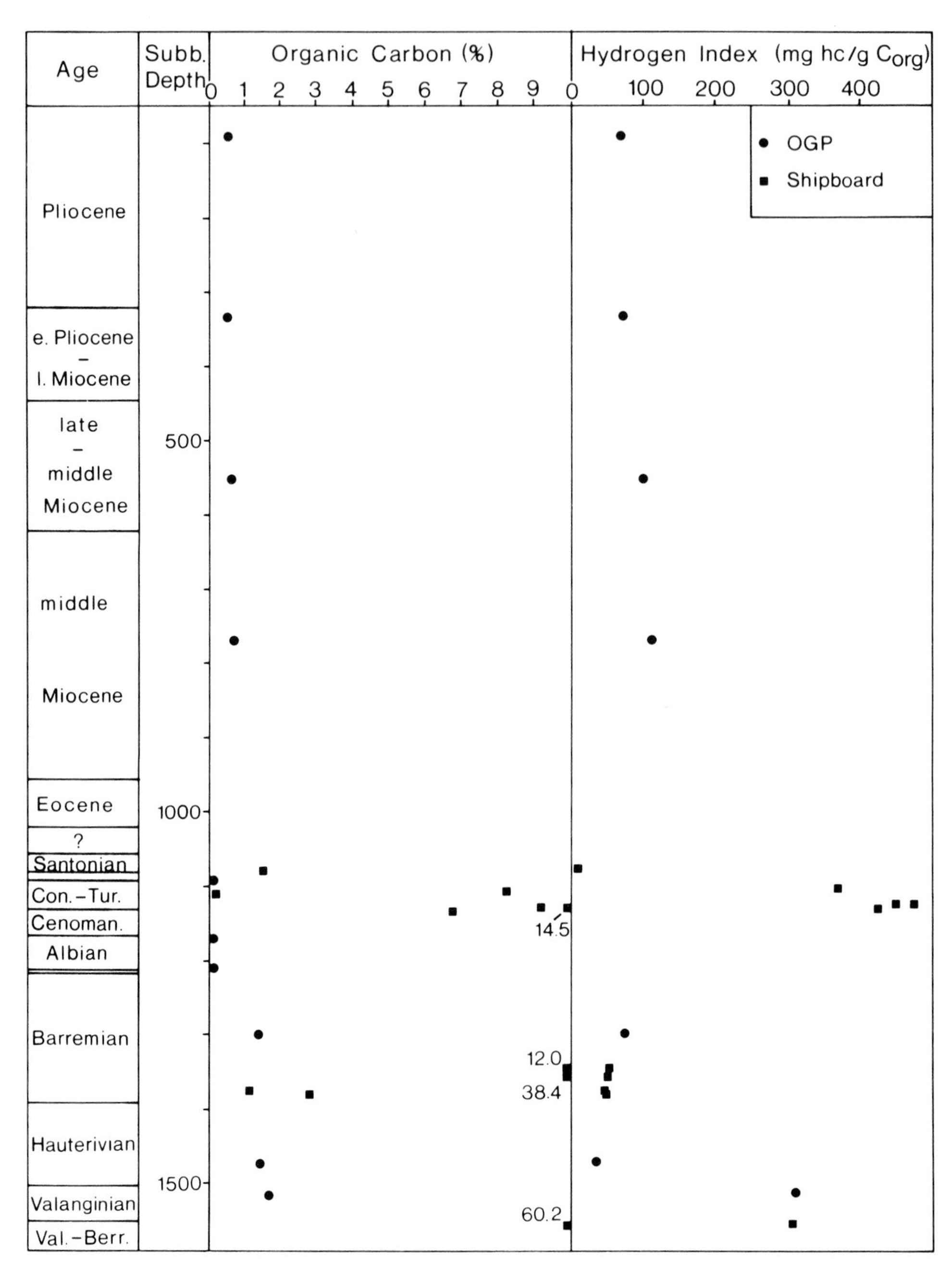

Fig. 4. Organic carbon and hydrogen index values (from Rock-Eval pyrolysis) for deep sea sediments from DSDP Site 603 off Cape Hatteras. Circles indicate frozen organic geochemistry panel samples taken at regular intervals, squares represent small-size samples picked from organic-carbon-rich strata onboard DV *Glomar Challenger*.

TABLE 3. *DSDP Site 603 organofacies types*

Stratigraphic age	Lithology	C_{org} (%)	HI (mc hc/g C_{org})	Dominant macerals
Santonian	Variegated claystone	1.55	18	78% recycled vitrinite
Coniacian –late Turonian	Variegated claystone	8.5	346	> 50% sapropelinite (fecal pellets)
Turonian –Cenomanian	Black carbonaceous claystone	6–14.5	> 400	50% degraded liptinite (fecal pellets?), phytoclasts
Barremian	Dark grey carbonaceous claystone	1–38	50–75	primary vitrinite with pyrite (unusual association)
Berriasian- –Valanginian	Nannofossil limestone	1.5–60	≈ 300	saprovitrinite (coal)

Conclusions

Organic geochemical and organic petrographic investigation of Mesozoic sediments from three selected Deep Sea Drilling Project sites on both sides of the North Atlantic Ocean has demonstrated the variability of organofacies types. The main purpose of this study was to show that there cannot be a simple mechanism by which organic-matter-rich sediments were formed in the deep sea of the North Atlantic during the Mesozoic. Ocean-wide oxygen deficiency will not explain most of the phenomena observed.

The supply of terrigenous organic matter to the ocean was an important control on accumulation of the organic matter in the North Atlantic. Purely marine organic matter was not detected in any of the samples investigated, and sediments containing kerogen macerals of dominantly marine origin were scarce. Marine organic matter (although always associated with terrigenous material) is more commonly found in the sediments of the eastern North Atlantic, probably due to the effect of coastal upwelling on marine organic matter production and preservation on that side of the ocean.

Resedimentation processes on the continental slopes (slumps, turbidites, debris flows) help organic matter preservation by reducing the destructive effect of long-term transport through oxygenated water masses, and by ensuring rapid burial in the deep sea. These processes are particularly effective where conditions favourable for the accumulation of organic matter exist in shallow water, on the shelf or upper slope, e.g. within an oxygen-minimum zone related to coastal upwelling.

Finally, local factors such as a basin-like topography near the continental rise, may have favoured the trapping of organic matter in the deep sea. Indications for this were found in the Callovian sediments in the Blake-Bahama Basin, and in the Mesozoic sediments off the Mazagan Escarpment. Jurassic sediments at this latter location appear to demonstrate the initial effect of terrigenous organic matter supply on the improvement of organic matter preservation conditions.

ACKNOWLEDGEMENTS: Samples for this study were made available through the assistance of the Deep Sea Drilling Project and the Deutsche Forschungsgemeinschaft. Support of this work by Professor D.H. Welte is gratefully acknowledged. The paper benefits from careful reviews of Dr A.S. Mackenzie (British Petroleum Company, Sunbury-on-Thames) and Dr R. Stein (KFA Jüllich). The authors thank the Deutsche Forschungsgemeinschaft, Bonn, for financial support, grant No. We 346/25, P.K.M. acknowledges a grant from the Alexander von Humboldt Foundation, Bonn.

References

ARTHUR, M.A. & SCHLANGER, S.O. 1979. Cretaceous 'oceanic anoxic events' as causal factors in development of reef-reservoired giant oil fields. *Bull. Am. Ass. Petrol. Geol.* **63,** 870–85.

——, DEAN, W.E. & STOW, D.A.V. 1985. Models for the deposition of Mesozoic-Cenozoic fine-grained organic-carbon-rich sediment in the deep sea. *In:* STOW, D.A.V. & PIPER, D.J.W. (eds), *Fine-Grained*

Sediments: Deep-Water Processes and Facies. Spec. Publ. Geol. Soc. **15**. Published for The Geological Society by Blackwell Scientific Publications, Oxford.

BISCAYE, P.E. & EITTREIM, S.L. 1977. Suspended particulate loads and transport in the nepheloid layer of the abyssal Atlantic Ocean. *Mar. Geol.* **23,** 155–72.

ČEPEK, P. & WIND, F.H. 1979. Neogene and Quaternary calcareous nannoplankton from DSDP Site 397 (Northwest African margin). *In:* VON RAD, U., RYAN, W.B.F. *et al.* (eds), *Init. Repts DSDP* **47,** (1). US Govt. Print. Off., Washington, DC. 289–315.

DE GRACIANSKY, P.C., BROSSE, E., DEROO, G., HERBIN, J.P., MONTADERT, L., MÜLLER, C., SIGAL, J. & SCHAAF, A. 1982. Les formations d'âge Crétacé de l'Atlantique du Nord et leur matière organique: palaeogéographie et milieux de dépot. *Rev. Inst. Franç. Pétr.* **37,** 275–336.

——, DEROO, G., HERBIN, J.P., MONTADERT, L., MÜLLER, D., SCHAAF, A. & SIGAL, J. 1984. Ocean-wide stagnation episode in the late Cretaceous. *Nature* **308,** 346–9.

DSDP 1983. Carbon and carbonate analysis, Leg 76. *In:* SHERIDAN, R.E., GRADSTEIN, F.M. *et al.* (eds), *Init. Repts DSDP* **76.** US Govt. Print. Off., Washington, DC. 945–7.

ESPITALIÉ, J., LAPORTE, J.L., MADEC, M., MARQUIS, F., LEPLAT, P., PAULET, F. & BOUTEFEU, A. 1977. Méthode rapide de caractérisation des roches-mères, de leur potentiel pétrolier et de leur degré d'évolution. *Rev. Inst. Franç. Pétr.* **32,** 23–42.

HABIB, D. 1983. Sedimentation-rate-dependent distribution of organic matter in the North Atlantic Jurassic-Cretaceous. *In:* SHERIDAN, R.E., GRADSTEIN, F.M. *et al.* (eds), *Init. Repts DSDP* **76.** US Govt. Print. Off., Washington, DC. 781–94.

HERBIN, J.P., DEROO, G. & ROUCACHE, J. 1983. Organic geochemistry in the Mesozoic and Cenozoic formations of Site 534, Leg 76, Blake-Bahama Basin, and comparison with Site 391, Leg 44. *In:* SHERIDAN, R.E., GRADSTEIN, F.M. *et al.* (eds), *Init. Repts DSDP* **76.** US Govt. Print. Off., Washington, DC. 481–93.

JANSEN, J.H.F., VAN WEERING, T.C.E., GIELES, R. & VAN IPEREN, J. 1984. Middle and late Quaternary oceanography and climatology of the Zaire-Congo Fan and the adjacent eastern Angola Basin. *Netherlands J. Sea Res.* **17,** 201–49.

KATZ, B.J. 1983. Organic geochemical character of some Deep Sea Drilling Project Cores from Legs 76 and 44. *In:* SHERIDAN, R.E., GRADSTEIN, F.M. *et al.* (eds), *Init. Repts DSDP* **76.** US Govt. Print. Off., Washington, DC. 463–8.

KENDRICK, J.W., HOOD, A. & CASTAÑO, J.R. 1978. Petroleum generating potential of sediments from Leg 44, Deep Sea Drilling Project. *In:* BENSON, W.E., SHERIDAN, R.E. *et al.* (eds), *Init. Repts DSDP* **44.** US Govt. Print. Off., Washington, DC. 599–604.

RULLKÖTTER, J. & MUKHOPADHYAY, P.K. 1984. Jurassic and mid-Cretaceous carbonaceous claystones in the western (DSDP Leg 76) and eastern (DSDP Leg 79) North Atlantic. *In:* SCHENCK, P.A., DE LEEUW, J.W. & LIJMBACH, G.W.M. (eds), *Advances in Organic Geochemistry—1983.* Pergamon Press, Oxford. 761–7.

——, VUCHEV, V., HINZ, K., WINTERER, E.L., BAUMGARTNER, P.O., BRADSHAW, M.L., CHANNELL, J.E.T., JAFFREZO, M., JANSA, L.F., LECKIE, R.M., MOORE, J.M., SCHAFTENAAR, C., STEIGER, T.H. & WIEGAND, G.E. 1983. Potential deep sea petroleum source beds related to coastal upwelling. *In:* THIEDE, J. & SUESS, E. (eds), *NATO Advanced Research Institute on 'Coastal Upwelling: Its Sedimentary Record', Part B: Sedimentary Records of Ancient Coastal Upwelling.* Plenum Press, New York. 467–83.

——, MUKHOPADHYAY, P.K., SCHAEFER, R.G. & WELTE, D.H. 1984. Geochemistry and petrography of organic matter in sediments from DSDP Sites 545 and 547, Mazagan Escarpment. *In:* HINZ, K., WINTERER, E.L. *et al.* (eds), *Init. Repts DSDP* **79.** US Govt. Print. Off., Washington, DC. 775–806.

SARNTHEIN, M., THIEDE, J., PFLAUMANN, U., ERLENKEUSER, H., FÜTTERER, D., KOOPMANN, B., LANGE, H. & SEIBOLD, E. 1982. Atmospheric and oceanic circulation patterns off Northwest Africa during the past 25 million years. *In:* VON RAD, U., HINZ, K., SARNTHEIN, M. & SEIBOLD, E. (eds), *Geology of the Northwest African Continental Margin.* Springer-Verlag, Heidelberg. 545–604.

SCHLANGER, S.O. & JENKYNS, H.C. 1976. Cretaceous anoxic events: causes and consequences. *Geologie en mijnbouw* **55,** 179–84.

SCLATER, J.G., HELLINGER, S. & TAPPSCOTT, C. 1977. The paleobathymetry of the Atlantic Ocean from the Jurassic to Present. *J. Geol.* **85,** 509–52.

SHIPBOARD PARTY 1984. Site 547. *In:* HINZ, K., WINTERER, E.L. *et al.* (eds), *Init. Repts DSDP* **79.** US Govt. Print. Off., Washington, DC. 223–361.

STACH, E., MACKOWSKY, M.T., TEICHMÜLLER, M., TAYLOR, G.H., CHANDRA, D. & TEICHMÜLLER, R. 1982. *Coal Petrology.* Gebrüder Bornträger, Stuttgart. 535 p.

SUMMERHAYES, C.P. 1981. Organic facies of middle Cretaceous black shales in deep North Atlantic. *Bull. Am. Ass. Petrol. Geol.* **65,** 2364–80.

——, in press. Organic-rich Cretaceous sediments from the North Atlantic. *In:* BROOKS, J. & FLEET, A.J. (eds), *Marine Petroleum Source Rocks.* Blackwell Scientific Publications, Oxford.

—— & MASRAN, T.C. 1983. Organic facies of Cretaceous and Jurassic sediments from Deep Sea Drilling Project Site 534 in the Blake-Bahama Basin, Western North Atlantic. *In:* SHERIDAN, R.E., GRADSTEIN, F.M. *et al.* (eds), *Init. Repts DSDP* **76.** US Govt. Print. Off., Washington, DC. 469–80.

TISSOT, B., DEROO, G. & HERBIN, J.P. 1979. Organic matter in Cretaceous sediments of the North Atlantic: contribution to sedimentology and paleogeography. *In:* TALWANI, M., HAY, W. & RYAN, W.B.F. (eds), *Deep Drilling Results in the Atlantic Ocean: Continental Margins and paleoenvironment.* Am. Geophys. Union. Maurice Ewing Series 3. 362–74.

——, DEMAISON, G., MASSON, P., DELTEIL, J.R. & COMBAZ, A. 1980. Paleoenvironment and petroleum potential of middle Cretaceous black shales in Atlantic basins. *Bull. Am. Ass. Petrol. Geol.* **64,** 2051–63.

TYSON, R.V. 1984. Palynofacies investigation of Callovian (Middle Jurassic) sediments from DSDP Site

534, Blake Bahama Basin, Western Central Atlantic. *Mar. Petrol. Geol.* **1,** 3–13.

WAPLES, D.W. 1983. Reappraisal of anoxia and organic richness with emphasis on Cretaceous of North Atlantic. *Bull. Am. Ass. Petrol. Geol.* **67,** 963–78.

WELTE, D.H., CORNFORD, C. & RULLKÖTTER, J. 1979. Hydrocarbon source rocks in deep sea sediments. *Proc. 11th Ann. Offshore Techn. Conf. (Houston)* **1,** 457–64.

WINTERER, E.L. & HINZ, K. 1984. The evolution of the Mazagan continental margin: A synthesis of geophysical and geological data with results of the drilling during Deep Sea Drilling Project Leg 79. *In:* HINZ, K., WINTERER, E.L. *et al.* (eds), *Init. Repts DSDP* **79.** US Govt. Print. Off., Washington, DC. 893–919.

JÜRGEN RULLKÖTTER & PRASANTA K. MUKHOPADHYAY, Institute of Petroleum and Organic Geochemistry, Kernforschungsanlage Jülich GmbH, D-5170 Jülich, Federal Republic of Germany.

Organic-rich sedimentation at the Cenomanian-Turonian boundary in oceanic and coastal basins in the North Atlantic and Tethys

J.P. Herbin, L. Montadert, C. Müller, R. Gomez, J. Thurow & J. Wiedmann

SUMMARY: One of the most striking results of the Deep Sea Drilling Project is the proof that organic-rich sediments have a widespread geographical distribution during the period from Upper Cenomanian to Middle Turonian. Such sediments were drilled at North Atlantic DSDP Sites: 105, 135, 137, 138, 367, 398, 551, 603. They are also present (from outcrops or oil wells) on the shelf of the African continental margin (Senegal, Tarfaya and Agadir Basins), in the Tethys (former Alboran Block, Southern Spain, Algeria, Tunisia and Umbrian Apennines/Italy), and in the North Sea. Although these sediments have different lithologies and depositional environments (from shelf areas to the deep sea), their study, mainly based on organic geochemistry with additional data on sedimentology, biostratigraphy and palaeobathymetry, suggests that a unique 'pulse' of organogenic accumulation characterizes the Cenomanian/Turonian Boundary Event (CTBE).

The content and type of organic matter are related to the depositional environment and organic preservation. The organogenic accumulation is distributed according to various trends. Off the African continental margin the organic content increases from onshore areas to the shelf (Casamance area), and, moreover, increases also in deep sea areas, with a gradual transition from terrestrial type III to marine type II (the best preservation of the organic matter being in the deepest areas, i.e. Site 367). Off the American continental margin Site 603 shows the same TOC and type of organic matter as at Site 105. The CTBE is also well recorded in the northern part of the Atlantic (Celtic margin, North Sea) by a drastic lithological change (black shales within chalks), but the type of organic matter is mainly terrestrial. In the Tethyan area the organic matter is of marine origin and well preserved. Results are compared with those of Pratt (1984) from the Western Interior Basin of the USA. Different hypotheses to explain this synchronous widespread accumulation of organic matter are discussed.

Organic-rich sediments are commonly found in the Upper Jurassic to Middle Cretaceous (Bitterli 1963; Schlanger & Jenkyns 1976; Ryan & Cita 1977; Arthur & Natland 1979; Jenkyns 1980). However, for the overall time interval available, data are still too scarce and/or too heterogenous to prove whether the enrichments in organic matter belong to 'oceanic anoxic events' or are related to local conditions.

Previous geochemical studies of Cretaceous black shales from Atlantic DSDP Sites have been carried out to reconstruct the palaeoenvironment, based on the type of organic matter (Herbin & Deroo 1979; Tissot *et al.* 1979; Demaison & Moore 1980; Graciansky *et al.* 1982; Herbin & Deroo 1982; Graciansky *et al.* 1984; Summerhayes 1981, 1985). Nonetheless, a more precise palaeogeographical study can be attempted for a particular period such as the well-defined Cenomanian–Turonian Boundary Event (CTBE), since a large quantity of data is available (DSDP Sites, outcrops, oil wells) from various environments (oceanic or coastal basins) and areas (African, European and American margins).

The present investigation summarizes the geographical distribution of organic matter enrichment in the North Atlantic and Tethys at the CTBE. It focuses on the geochemical similarities, independently from the geological setting, extending the results obtained from the DSDP Sites toward shelf areas. Only few quantitative studies were done to characterize the organogenic content of this event. Data from the literature are mainly descriptive and furnish elusive concepts of TOC content and the nature of the organic matter.

Since the oxicity/anoxity of the sediments and the water column affects the amount and nature of the organic matter incorporated in the sediments during deposition, the depositional environment can be characterized by determining the quantity and nature of the organic matter (Didyk *et al.* 1978). Black shales are thinly laminated carbonaceous, clayey and marly pelagic sediments, without bioturbation and benthonic life and generally rich in organic matter and in various trace elements. As long as the dissolved oxygen in the bottom water exceeds about 0.2 ml/l, benthonic life including burrowers is

From SUMMERHAYES, C.P. & SHACKLETON, N.J. (eds), 1986, *North Atlantic Palaeoceanography*, Geological Society Special Publication No. 21, pp. 389–422.

possible (Rhoads & Morse 1971), whereas at a lower oxygen level most benthonic organisms cannot survive, and sediment laminations are preserved. Modern black shales form in calm oxygen-depleted water, such as stagnant basins (fjords, Black Sea) (Pelet & Debyser 1977), or below zones of high primary production (e.g. upwelling zones off Peru). The level of dissolved oxygen in the bottom waters represents a dynamic equilibrium between depletion (mainly by decay of organic matter), and replenishment (by oxygen-bearing bottom currents) (Waples 1983). The oxygen consumption rate also depends on the supply and type of organic material. Because of different capacities to induce anoxia, the ratio between terrestrial and marine organic matter is important for defining the depositional environment (Tissot *et al.* 1979).

In this study the amount of total organic carbon (TOC) was determined by a new methodology worked out for the Rock Eval apparatus (Espitalié *et al.* 1984). The type of organic matter was determined by pyrolysis (Espitalié *et al.* 1977). For the DSDP Sites, more detailed analyses have been made, including elemental analyses of kerogen, and optical studies. Three types of kerogen can be distinguished from geochemical studies. Types I and II are related to lacustrine or marine reducing environments, and are derived mainly from planktonic organisms, whereas Type III comes from terrestrial plants transported to a marine or non-marine environment with a moderate level of degradation. Intermediate kerogens are common, particularly between Types II and III. They result from a mixture of marine and terrestrially derived organic matter, or from biodegradation of marine organic matter (Tissot & Pelet 1981). A fourth type of organic matter is residual organic matter, either recycled from older sediments by erosion, or deeply altered by subaerial weathering (Tissot *et al.* 1979).

All areas studied in this paper are located on a palaeogeographic map reconstruction of the North Atlantic at 85 Ma (Fig. 1). The increase of sea-floor spreading between the Cenomanian and anomaly 34 suggests intense tectono-magnetic processes on the mid oceanic ridge during the CTBE.

African continental margin: comparison between Casamance area and DSDP Site 367

Casamance area

The Mesozoic basin in Senegal extends from the southern part of Rio de Oro in the north across Mauretania, Senegal and Gambia to Guinea Bissau in the south (Fig. 1). It is the largest of the West African basins, with a total land area of about 340 000 km^2. It extends into a large offshore basin reaching to the Cape Verde Islands, where Jurassic and Cretaceous sediments are exposed on Maio Island (Stahlecker 1934; Robertson & Bernoulli 1982). In the eastern part, the faulted pre-Mesozoic basement dips gradually toward the west (Fig. 2). Near Dakar the structure is complicated by horsts and folds, whereas in the south the Palaeo-Cretaceous embayment of the Casamance area is disturbed only by salt domes of unknown age.

Four Middle Cretaceous formations are described from the Casamance area (Templeton 1972) and are dated by De Klasz (1978). In this paper only the middle/late Cenomanian to early Turonian 'Banc du Large Formation' are dealt with, represented by shales alternating with limestones, clays and sandstones, and the bituminous clays of the 'Brikama Formation' of early Turonian age. Offshore the Casamance Maritime wells are located in a Cretaceous embayment, with CM2 and CM10 being in a more oceanic environment than CM1 and CM4 (Fig. 2). In the latter wells the Cenomanian consists of glauconitic white limestones with fine-grained sandstones and grey to green clays, whereas, during the Turonian, shales with a pelagic fauna predominated. In wells CM2 and CM10 the facies is more shaly and bituminous, with a lower detrital input.

The authors' results show several trends (Fig. 2). Organic matter tends to increase from the continent toward the open ocean basin. In the eastern part of the section (DM1 and NC1), the Cenomanian–Turonian deposits are fine to coarse sands or sandstones rich in dark silty clays with only 0.2–1.9% TOC. On the edge of the embayment the TOC remains low (TOC$\leqslant$2.3% in well CM4), whereas, in the axis of the embayment (wells CM1, CM2, CM10) the TOC reaches 10.6% in bituminous marls (Fig. 2). This enrichment occurs during the CTBE, at the top of the 'Banc du Large Formation' and at the base of the 'Brikama Shales Formation' (Fig. 2). In the underlying Albian and overlying Senonian sediments the TOC is less than 1%.

The type of organic matter follows the same trend. The terrestrial organic matter (Type III) in well DM1 becomes a mixture of Types II and III farther west in wells NC1, CM4 and CM1. In wells CM2 and CM10, Type II is dominant with a hydrogen index of 380 to 570 mg HC/g TOC. There is decreasing dilution by terrigenous material, and an increase in the preservation of organic matter of planktonic origin toward the open

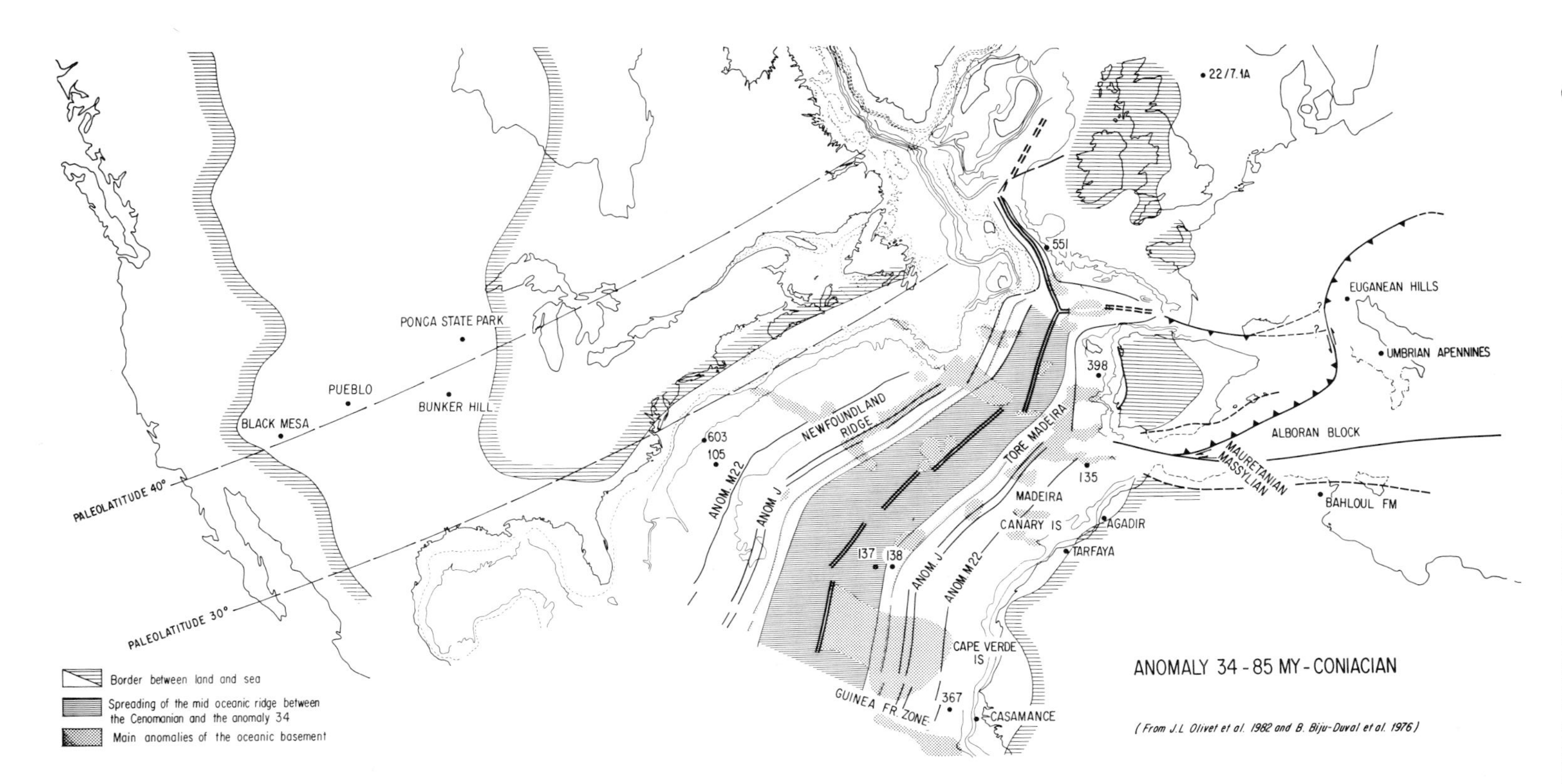

FIG. 1. Location of the areas studied set in a palaeogeographic reconstruction of the North Atlantic at 85 Ma. The increase in sea-floor spreading between the Cenomanian and anomaly 34 suggests intense tectono-magmatic processes on the mid-oceanic ridge during the CTBE.

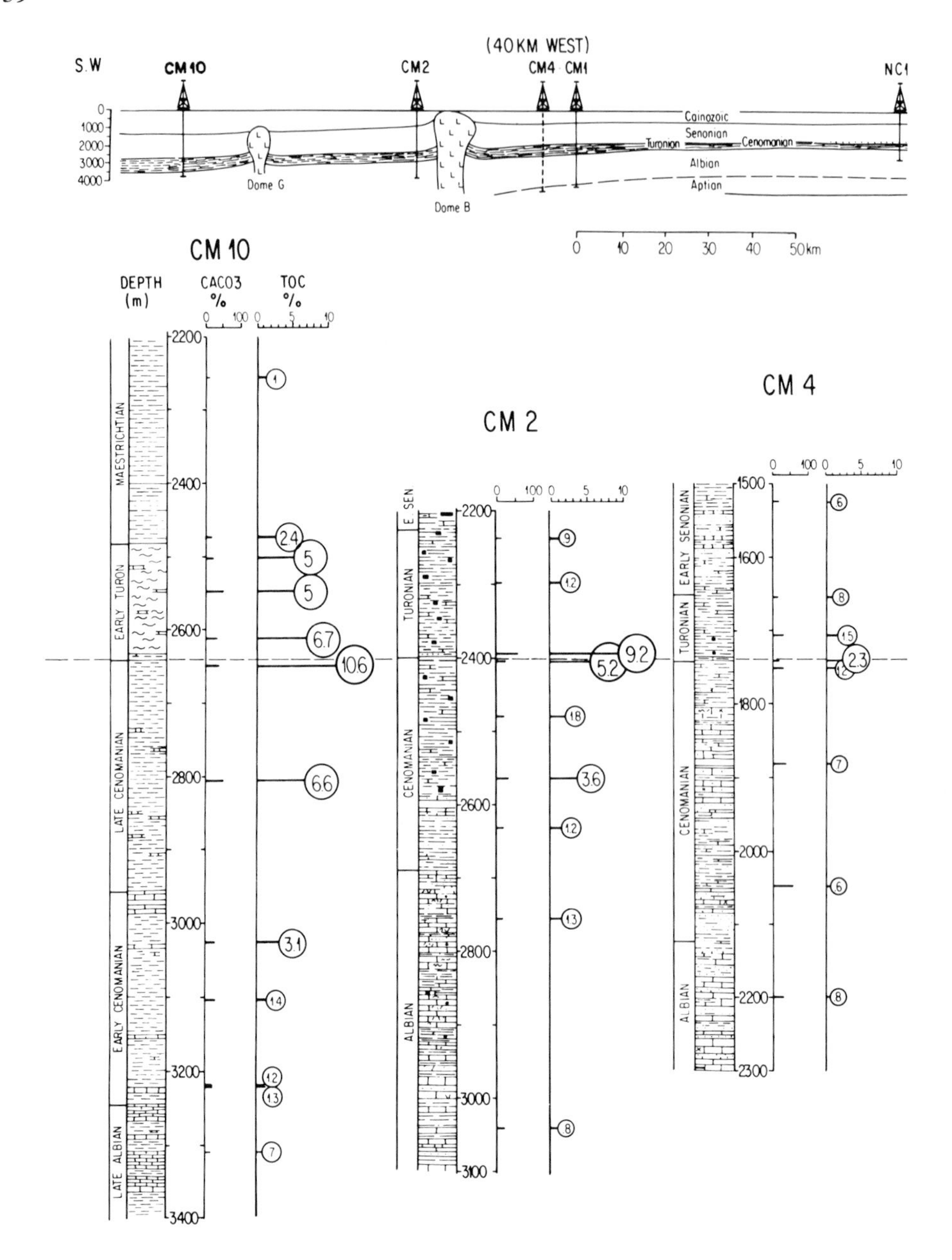
(40 KM WEST)
S.W
CM 10
CM2
CM4
CM1
NC1
Cainozoic
Senonian
Turonian
Cenomanian
Albian
Aptian
Dome G
Dome B
0 10 20 30 40 50 km
CM 10
DEPTH (m)
CACO3 %
TOC %
MAESTRICHTIAN
EARLY TURON
LATE CENOMANIAN
EARLY CENOMANIAN
LATE ALBIAN
CM 2
E. SEN
TURONIAN
CENOMANIAN
ALBIAN
CM 4
EARLY SENONIAN
TURONIAN
CENOMANIAN
ALBIAN

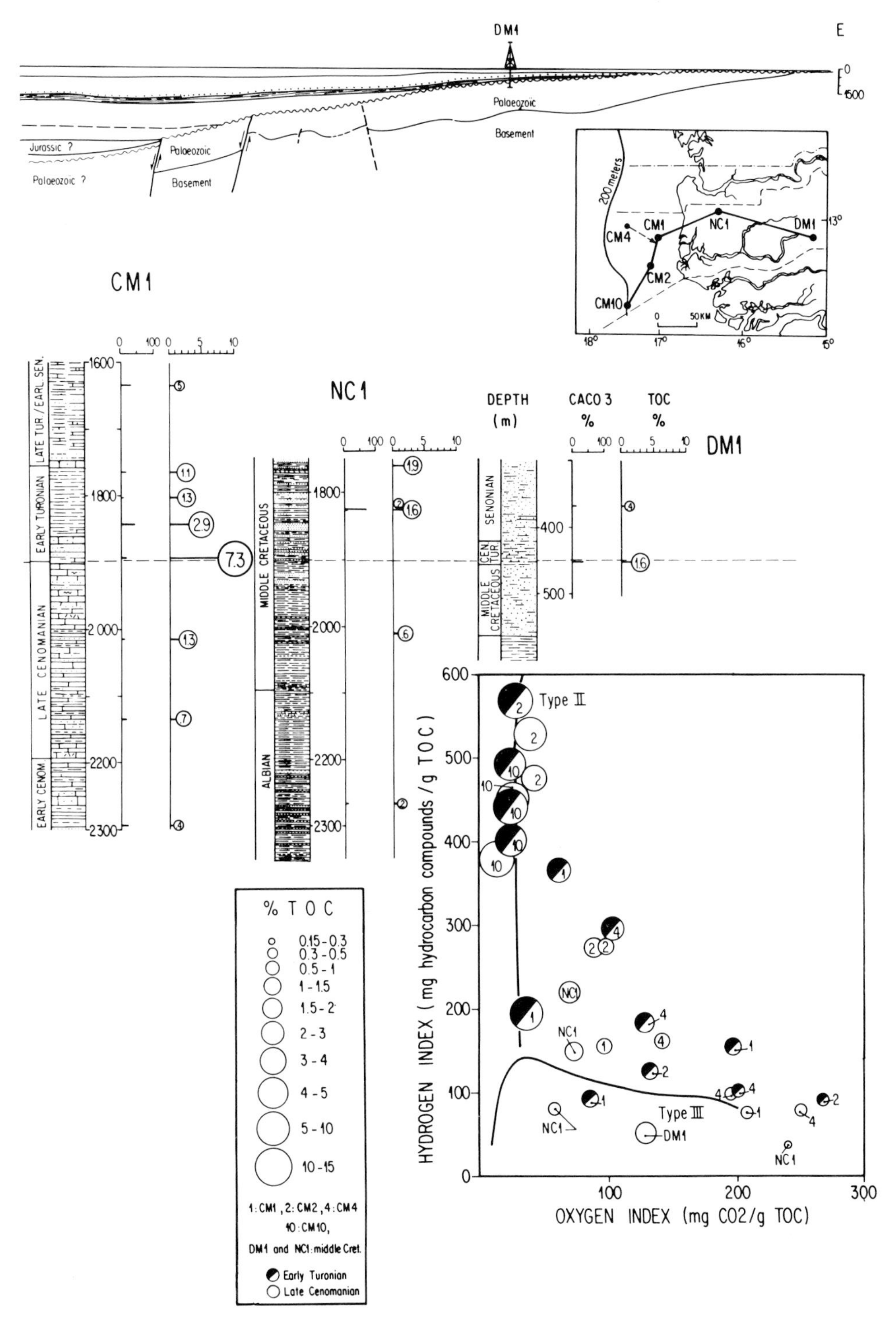

FIG. 2. Cross-sections in the Casamance area between wells DM1 and NC1, CM1, CM4, CM2 and CM10, and characterization of the type of organic matter at the Cenomanian–Turonian boundary.

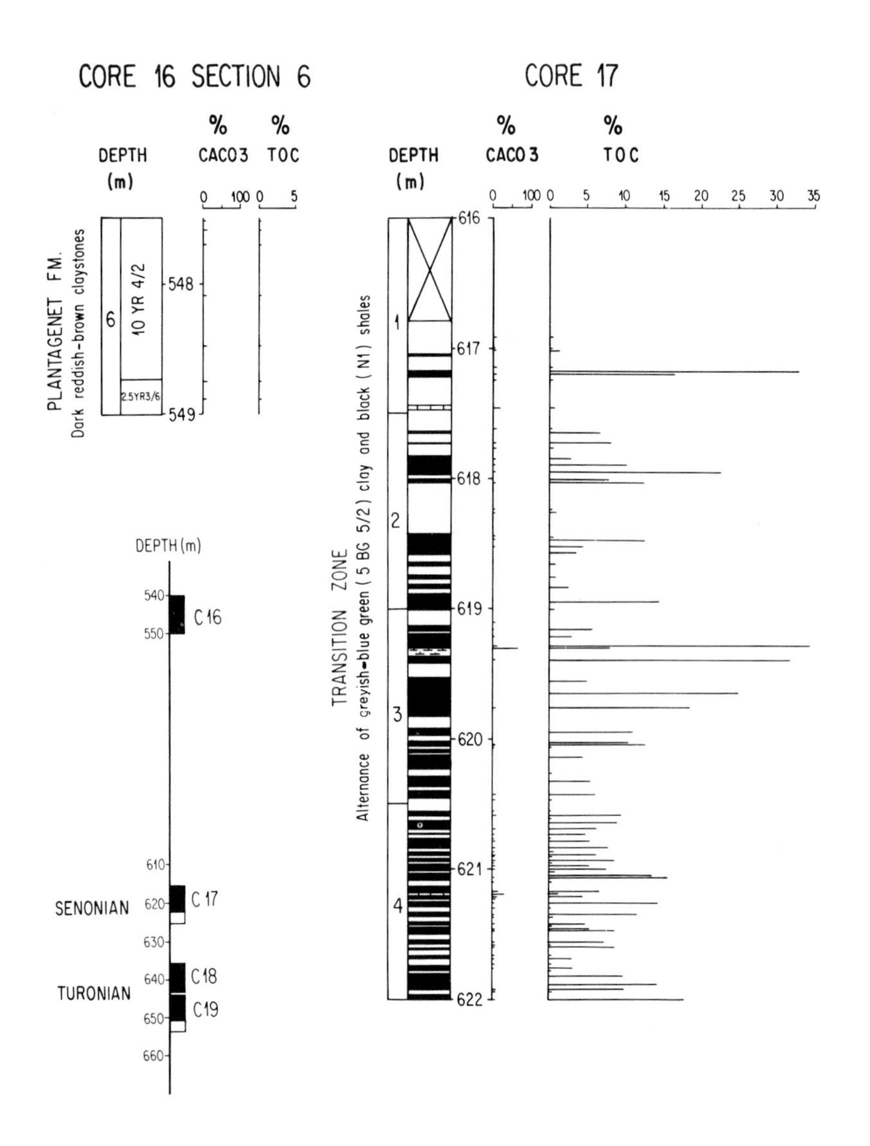

CORE 16 SECTION 6
CORE 17
DEPTH (m)
% CACO3
% TOC
PLANTAGENET FM.
Dark reddish-brown claystones
10 YR 4/2
2.5YR3/6
548
549
TRANSITION ZONE
Alternance of greyish-blue green (5 BG 5/2) clay and black (N1) shales
616
617
618
619
620
621
622
DEPTH (m)
540
550
610
620
630
640
650
660
C 16
C 17
C 18
C 19
SENONIAN
TURONIAN

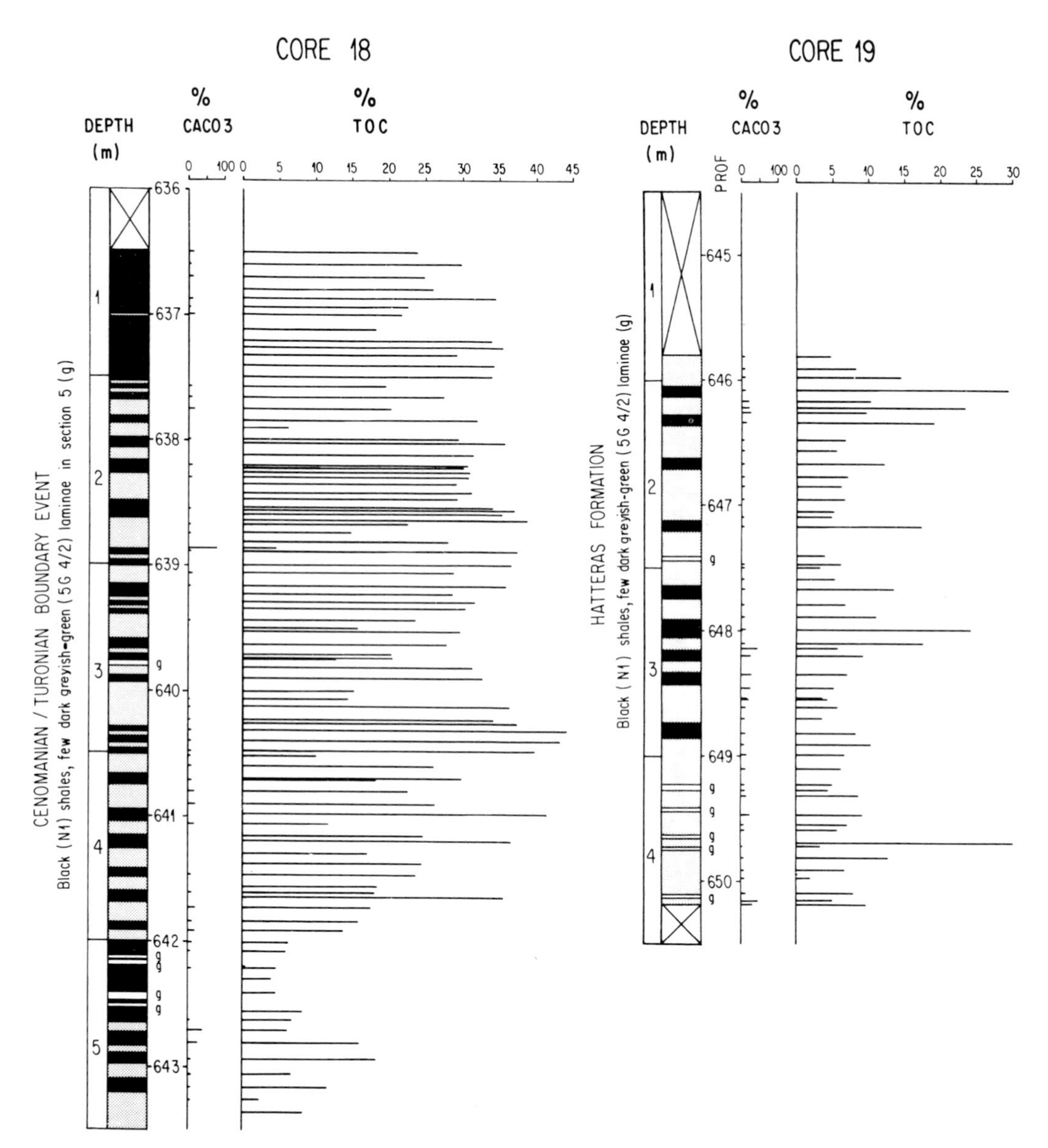

FIG. 3. Cape Verde Basin. Detailed study of the CTBE at Site 367, Core 16 (Section 6) and Cores 17, 18 and 19. Log of total organic carbon and carbonate contents.

ocean (CM2–CM10). The sediment richest in organic matter is also rich in metals (V, Ni, Pb, Sr, Cr, Mo and Zn) and phosphates. These sediments are potential source rocks offshore from Senegal, but their low degree of maturity places them above the oil window.

DSDP Site 367

Site 367 is located at the base of the continental rise in the Cape Verde Basin about 330 km west of wells CM2–CM10 (compare Fig. 1 and 2). Two sites (367 and 368) off the African margin show high enrichment in organic matter in the CTBE (Jansa *et al.* 1978). Unfortunately, at Site 368, local heating by basaltic sills prevents a detailed study of the organic matter.

In cores 23 to 20 from Site 367 the black carbonaceous shales (interbedded with light-green and olive-grey clayey nannofossil chalk) are equivalent in age to the Hatteras Formation (Jansa *et al.* 1979), but richer in organic matter. The regular changes from green to black shale suggest redox cycles. Some silt turbidites, well defined in cores 21 and 22, are superimposed on these cycles (Dean *et al.* 1978). In core 19 some green and olive-grey clayey nannofossil chalks are interbedded in the black carbonaceous shales that form the entire section in core 18. Both cores are of early Turonian age (Muller *et al.* 1983). These cores (19 to 17) were sampled in detail to characterize the CTBE.

In core 19, and in section 5 of core 18, the sediments have an average TOC of 6%, with some beds reaching 10–20% (exceptionally 30%) (Fig. 3). In the upper part of core 18 a layer six metres thick (sections 1 to 4) has a TOC of up to 40%, and an average TOC of 30% (Fig. 3). This enrichment coincides with a change in the type of organic matter. From core 19 to section 5 of core 18 the organic matter is Type II, with an average hydrogen index of 400 mg HC/g TOC (a few beds rich in TOC have an HI of 550–600). Above, in sections 1 to 4 of core 18, the hydrogen indices are higher, about 600, and sometimes reach 850 mg HC/g TOC (Fig. 4). The organic matter in this sequence belongs to uppermost Type II, as is confirmed by the results of the elemental analysis of the kerogen in sample 18–2/80 cm (H/C = 1.41, O/C = 0.13).

A gap of ten metres separates cores 18 and 17. In core 17 the cycles of green to black shales reappear and form a transitional zone below the Plantagenet Formation which is Senonian in age. In the lower part of the core (sections 4 and 3) are numerous black-shale layers, but in sections 2 and 1 most of the sediments are green claystones with low TOC (0.2 to 0.5%) and residual organic matter. The TOC is almost 7% in the black shales, reaching 10 to 20% (exceptionally 30%) in some samples (Fig. 3). In section 3 of core 17 these alternations have been interpreted as typical redox cycles without turbiditic input (Dean *et al.* 1978). The average hydrogen indices of the black shales is about 350 mg HC/g TOC (Fig. 4). However, beds with higher TOC contents (20–30%) have an HI between 460 and 560. Type II organic matter is also confirmed by elemental analysis of the kerogen (H/C = 1.28, O/C = 0.14). An uncored interval of 66.5 metres between cores 17 and 16 prevents detailed study of the transition toward oxidized facies. The TOC of the bioturbated dark reddish brown claystones of the overlying Plantagenet Formation (core 16) is very low (<0.2%).

After Leg 41, two models were proposed to explain this black-shale sedimentation: a sluggish deep-water circulation model (Lancelot *et al.* 1972) and an increased supply of organic matter model (Dean & Gardner 1982; Gardner *et al.* 1978). In the first model, stagnation of bottom water in a restricted environment (euxinic conditions) implies that changes in organic content in the sediments are controlled by the preservation of the organic matter. In the second model the variations in redox conditions in the sediments are seen as the result of varying rates of supply of organic matter.

Reconstruction of the palaeodepths applied to Site 367 (Chénet & Francheteau 1979) places the black-shale sedimentation of the CTBE at a depth of about 3700 metres. At the same time the bituminous 'Banc du Large Formation' and 'Brikama Formation', with up to 10% TOC, were deposited on the shelf about 300 kilometres farther east. If the organogenic sedimentation in cores 17–18 is typical of redox cycles devoid of turbiditic input (Dean *et al.* 1978), the increasing enrichment in Type II organic matter from the shelf toward the deep area, where TOC contents reach up to 40%, would imply that most of the water column was favourable to the preservation of planktonic organic matter during the CTBE.

Moroccan continental margin: comparison between the Aaiun-Tarfaya, Agadir Basins and DSDP Sites 135, 137, 138

Whereas the CTBE can be studied only from cores or cuttings in the DSDP Sites and Casamance oil wells, the outcrops on the Moroccan margin—and all around the Tethys—are of great

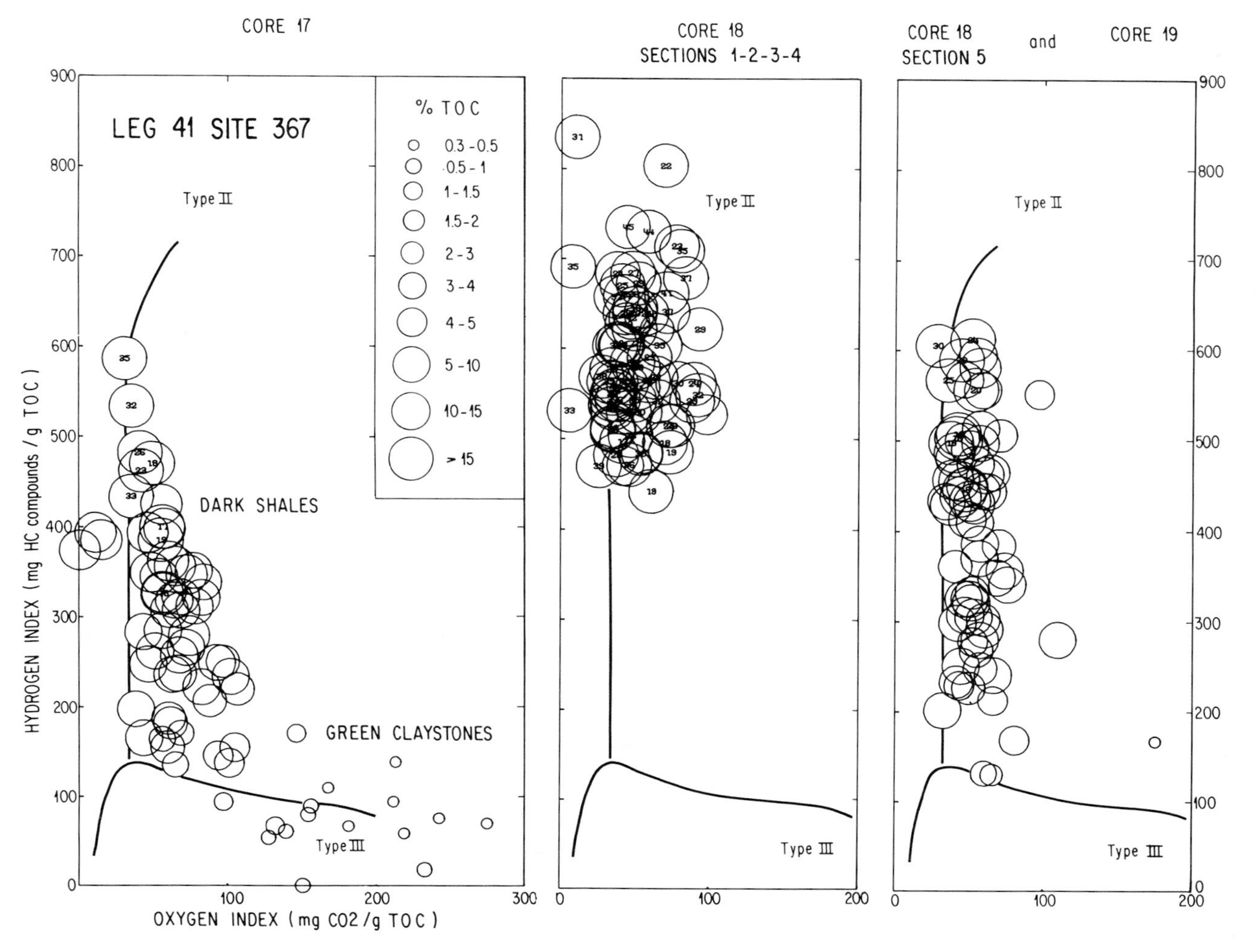

FIG. 4. Cape Verde Basin. Characterization of the type of organic matter at Site 367, Cores 17, 18 and 19. The size of the circle is proportional to the organic carbon content.

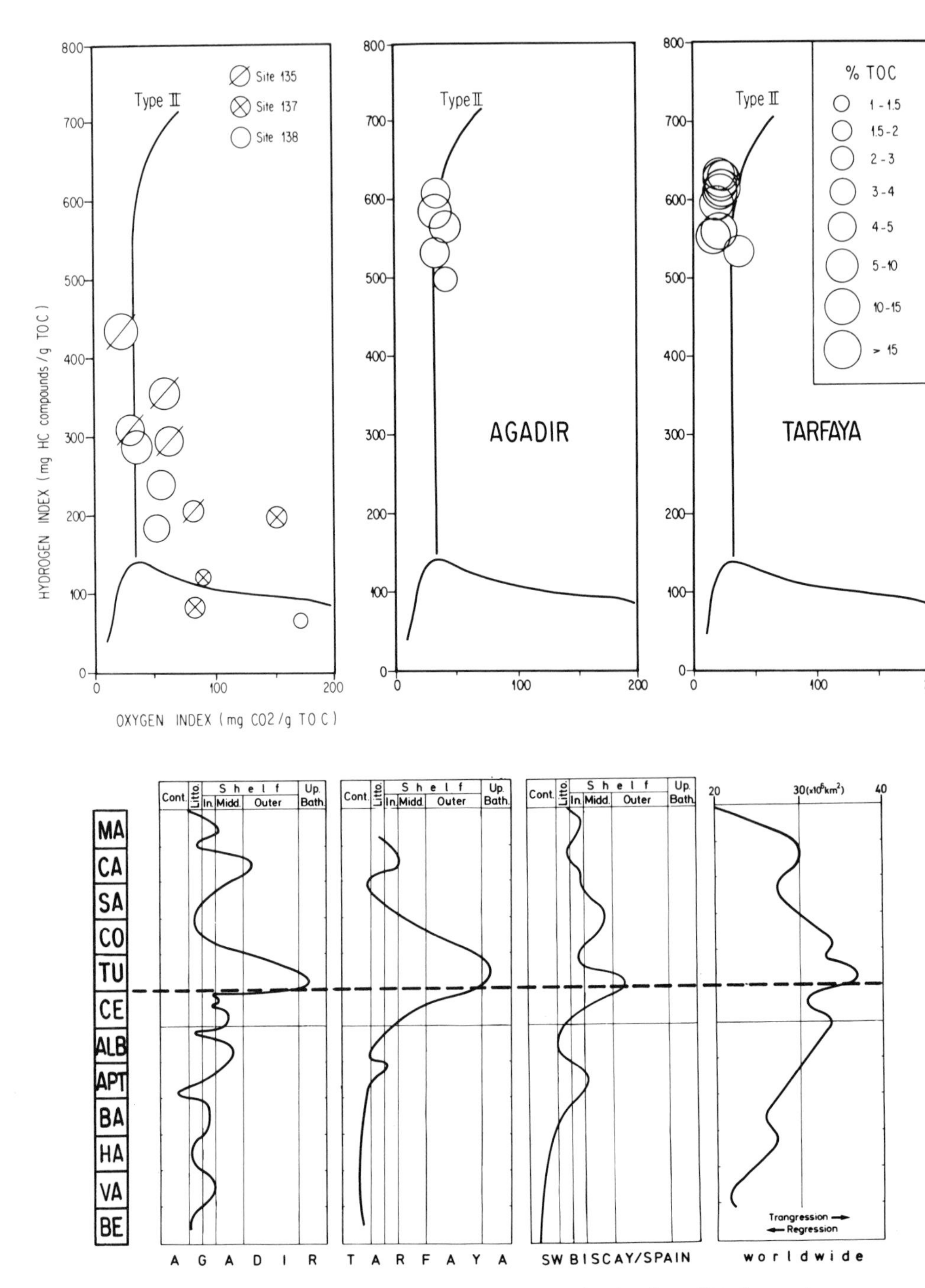

FIG. 5. Type of organic matter in the CTBE at Sites 135, 137 and 138, Agadir and Tarfaya regions. Comparison of the Cretaceous bathymetric curves of the Agadir and Tarfaya basins in Morocco and the Bascocantabric basin in Spain, with Sliter's curve of extension of the world's oceans (from Wiedmann *et al.* 1982).

value because they allow more detailed sedimentological and faunal studies.

Aaiun-Tarfaya Basin

In the Aaiun-Tarfaya Basin (inland from Tarfaya, Fig. 1) most of the Lower Cretaceous sequences are of continental or deltaic facies. A marine transgression began in late middle Albian times, and lasted into the late Turonian–early Coniacian. The facies of the transgressive Cenomanian and Turonian consists of thinly bedded black marls and shales, limestones and siliceous limestones (Wiedmann 1976; Wiedmann *et al.* 1978a, 1978b; Einsele & Wiedmann 1975, 1982, 1983; Thurow *et al.* 1982). The deposits of the Gaada Formation are characterized by enrichment in organic matter (Auxini 1969).

These bituminous marls contain 7–19.6% TOC, whereas the Upper Albian marls have a TOC of 0.1%. The organic matter is Type II, with hydrogen indices reaching 630 mg HC/g TOC (Fig. 5). The sediments also have a high silica content (cherts), originating from diatoms and radiolarians. Phosphate content ranges from 250–4600 ppm, which is less than in recent upwelling sediments off the continental margin of north-western Africa, i.e. 2000 to 10 000 ppm (Summerhayes *et al.* 1976). The enrichment of these sediments in aquatic organic matter correlates with enrichment in heavy metals, such as Zn, Ni and Cr.

The planktonic microfaunas in these black marls indicate an outer shelf environment (water depth of about 200–300 m). The $CaCO_3$ content, varying between 35 and 99%, indicates an environment situated above the calcite compensation depth with low terrestrial input. Faunal evidence sugests lack of dissolved oxygen (small size of macro- and microfaunas, high density and low diversity of the species, preservation of fish debris, scarcity of benthonic macro- and microbiotas restricted to some apparently more oxygenated layers). Moreover, the macrofauna is poor in Tethyan forms and contains northern temperate elements of the North American and northern European Provinces. Exchanges with the neighbouring Algerian and Tunisian Atlas Ranges is rather limited.

The average sedimentation rate for this sequence (carbonaceous laminated marls and interbedded pelagic limestones with cherts) is about 5–20 m per million years.

Agadir Basin

Farther north, the Mid-Cretaceous evolution of the Agadir Basin (near Agadir, Fig. 1) is similar. The faunal assemblages suggest a connection between the High Atlas, the Moroccan Meseta and the Tethys; connection with the Aaiun-Tarfaya Basin to the south is less pronounced (Wiedmann *et al.* 1982). The Mid-Cretaceous facies of black marls, limestones and cherts is restricted to the lower part of the Turonian, with a total thickness of about 140 m. The sediments occasionally contain turbidites and slumps, locally interbedded with thinly laminated bituminous marls. In contrast to the predominantly benthonic macro- and microbiotas, below and above, the Lower Turonian microfaunas and microfloras again are exclusively planktonic (coccoliths, foraminifers, radiolarians). Benthonic associations of possibly cool-water relationships (e.g. astartids) are restricted to the upper part of the Turonian. The nature and widespread distribution of these sediments further inland at the time of maximum transgression (Fig. 5) suggest calm depositional conditions in an outer shelf to upper slope environment.

As in the Aaiun-Tarfaya Basin, the Cenomanian-Turonian black shales of the Agadir Basin contain a considerable amount of organic matter (5.6–11.4% of TOC) of aquatic origin ($520 \leqslant HI \leqslant 620$ mg HC/g TOC, Fig. 5). The maturity of the organic matter is relatively low in the Agadir Basin ($T_{max} = 416°C$) as well as in the Aaiun-Tarfaya Basin ($T_{max} = 412°C$), due to limited burial.

DSDP Sites 135, 137, 138

The organic-rich sediments in the Aaiun-Tarfaya and Agadir Basins can be compared to those of DSDP Sites 135, 137 and 138 (Fig. 1). Unfortunately, in the more proximal Sites off the Moroccan margin, i.e. Sites 369/370 (Lancelot *et al.* 1978) and Site 416 (Lancelot *et al.* 1980), an erosional gap prevents the study of the CTBE (such hiatuses are not restricted to the African margin, they also exist off the American continental margin, e.g. at Site 391, and reflect strong bottom circulation at the Upper Cretaceous/Lower Tertiary boundary). Also the bathyal DSDP Sites of Leg 14 (Hayes *et al.* 1972) can be correlated with the Cenomanian–Turonian onshore facies (Fig. 6).

In Site 135, south of the Tethyan fracture zone, the black shales of late Cenomanian age (Core 8, Section 1) contain 2.6–11.9% TOC of aquatic origin ($200 \leqslant IH \leqslant 430$ mg HC/g TOC). There is a similar high organic content in the lower Turonian sediments overlying the oceanic crust at Site 138 (Core 6, Section 2) (up to 10.7% TOC) and at Site 137 (Core 7, Section 1) (1.2 to 2.6% TOC). The abundance and preservation of the fauna

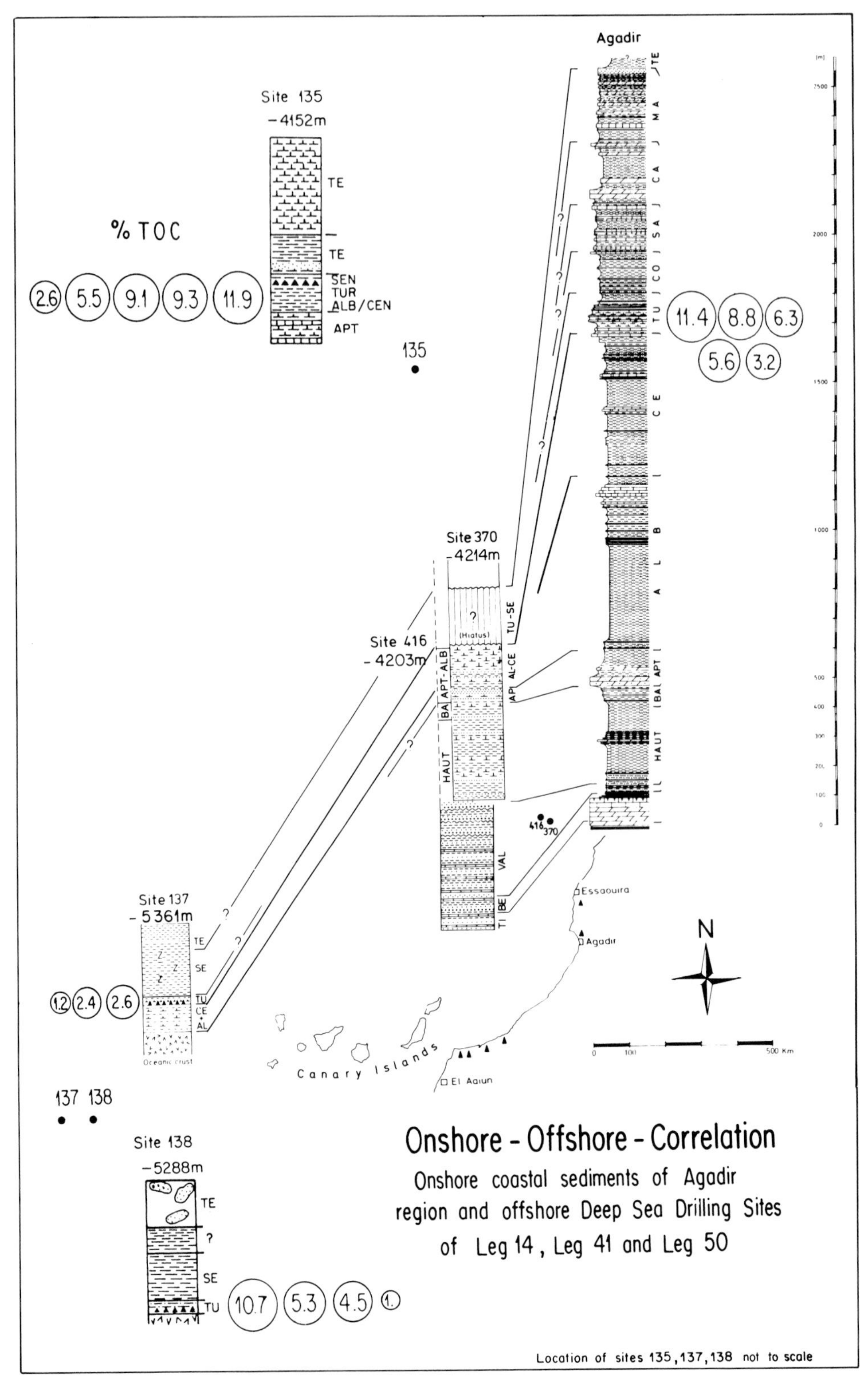

FIG. 6. Correlation between coastal sediments in the Agadir region and DSDP Sites 416/370, 135, 137 and 138 off the Moroccan margin, with total organic carbon content of facies deposited during the CTBE.

locates Site 137 above the calcite compensation depth (CCD). Subsidence below the CCD is indicated by the deposition of deep sea clays of late Turonian to Senonian age. Based on the data obtained from Leg 14 it is possible to extend the CTBE far from the distal area off the Moroccan margin, onto the former Mid-Atlantic Oceanic Ridge (Fig. 1).

Rif Mountains (former 'Alboran Block')

The westernmost part of the Alpine folded Maghrebinian Orogenic Belt—the Rif Mountains in northern Morocco and the Campo de Gibraltar (in its broadcast sense) in southernmost Spain—provides excellent outcrops for studying the CTBE (See Fig. 1 for general location of this area). This fold belt is divided into three major parts, each of which is related to a distinct palaeogeographic domain (Suter 1980). The three major parts are (Fig. 7):

(1) the Internal Domain with the 'Chaînes Calcaires' and the folded Palaeozoic;
(2) the 'flysch domain' with the Numidian, Mauretanian and Massylian nappes (Bouillin *et al.* 1970; Raoult 1972);
(3) the External Domain in the Rif (Intrarif, Mesorif, Prerif) and in Spain (Penibetic, Subbetic, Prebetic). In some areas the whole pile of nappes were affected by younger tectonic movements that led to large-scale gravity sliding, destroying the original setting (e.g. the Campo de Gibraltar and the Grade Kabylie in Algeria).

The Internal Domain is related to the former Alboran Block in the Western Tethys (Wildi 1983). Only a few Cretaceous sediments are present, without any organic-rich deposits. The rare outcrops show pelagic limestones and marls rich in microplankton, condensed sedimentation and stratigraphic gaps. In the Gibraltar Arch area sediments of Middle Cretaceous age are also unknown. Within the Kabylies, Cenomanian to Upper Cretaceous sediments consist of 'Scaglia' facies with chert layers rich in radiolarians at the base of the Turonian. The total thickness of the Middle Cretaceous is a few metres.

In the External Domain, related to the former continental margin of North Africa and Iberia, the CTBE is found in the Intrarif Zone as well as the Mesorif Zone (especially the Loukkos Zone). Calciturbidites alternate with dark shales and marls. Radiolarians and silicification are abundant. Deposition took place on the lower slope above the CCD. The organic rich layers sometimes show features similar to those in the 'flysch domain' (TOC up to 3.6%). Due to the scarcity of outcrops and the small degree of metamorphism in some parts of the External Domain, only a few analyses are available. In Spain the External Domain flanks the Campo de Gibraltar, where the facies of Middle Cretaceous sediments is comparable to that of the Internal Domain. Some sections of the Penibetic Zone include fairly thick sediments of mid-Cretaceous age. Toward the end of the Cenomanian white pelagic limestones, rarely marly, with abundant planktonic forminifers, are interbedded with some thin layers (1–10 cm) of dark grey to black bioturbated marly limestones (TOC up to 0.6%). The base of the Turonian is marked by the first occurrence of a rich radiolarian fauna and black chert nodules and layers. Some of the limestones are of darker colour, but without any increase in the TOC (0.02%).

The best evidence of the CTBE is in the 'flysch domain' in the Mauretanian and Massylian nappes. These nappes comprise the Maghrebinian Fold Belt and extend from Spain in the west over the Rif and Tellian Atlas to Sicily in the east. Biostratigraphic correlation between the different nappes shows that the enrichment of organic matter is synchronous (Thurow *et al.* 1982; Gübeli 1982; Thurow & Kuhnt, this volume).

Mauretanian nappe

In the Campo de Gibraltar area Cretaceous sediments within the Mauretanian series have not been proved. In the Rif, the Mauretanian nappe is divided into the Tisirene nappe (Dogger–Lower Cretaceous) and the Beni Ider nappe (Upper Cretaceous–Tertiary), which is now in a lower tectonic position. During early Cretaceous times (Tisirene nappe) sedimentation was almost entirely turbiditic. In the Beni Ider nappe, sediments of Middle Cretaceous age include carbonate detrital turbidites alternating with multicoloured shales. The Upper Cenomanian and, to a lesser extent, the Turonian are marked by the occurrence of white chert layers and nodules. Radiolarians are abundant, and thin sections show reworked phosphatic material within the turbidites. At the base of the Turonian, coloured shales are replaced by black laminites and shales. The lower Turonian sediments are interpreted as mid to outer-fan deposits below the carbonate compensation depth. Whereas the TOC of the Albian sediments is low, due to dilution by terrigenous material (TOC < 0.5%, exceptionally reaching 4%), a distinct increase in TOC coincides with the CTBE. Deposition of organic-rich sediments continued during late Turonian times in the Gibraltar arch area (Thurow & Kuhnt, this volume). In the

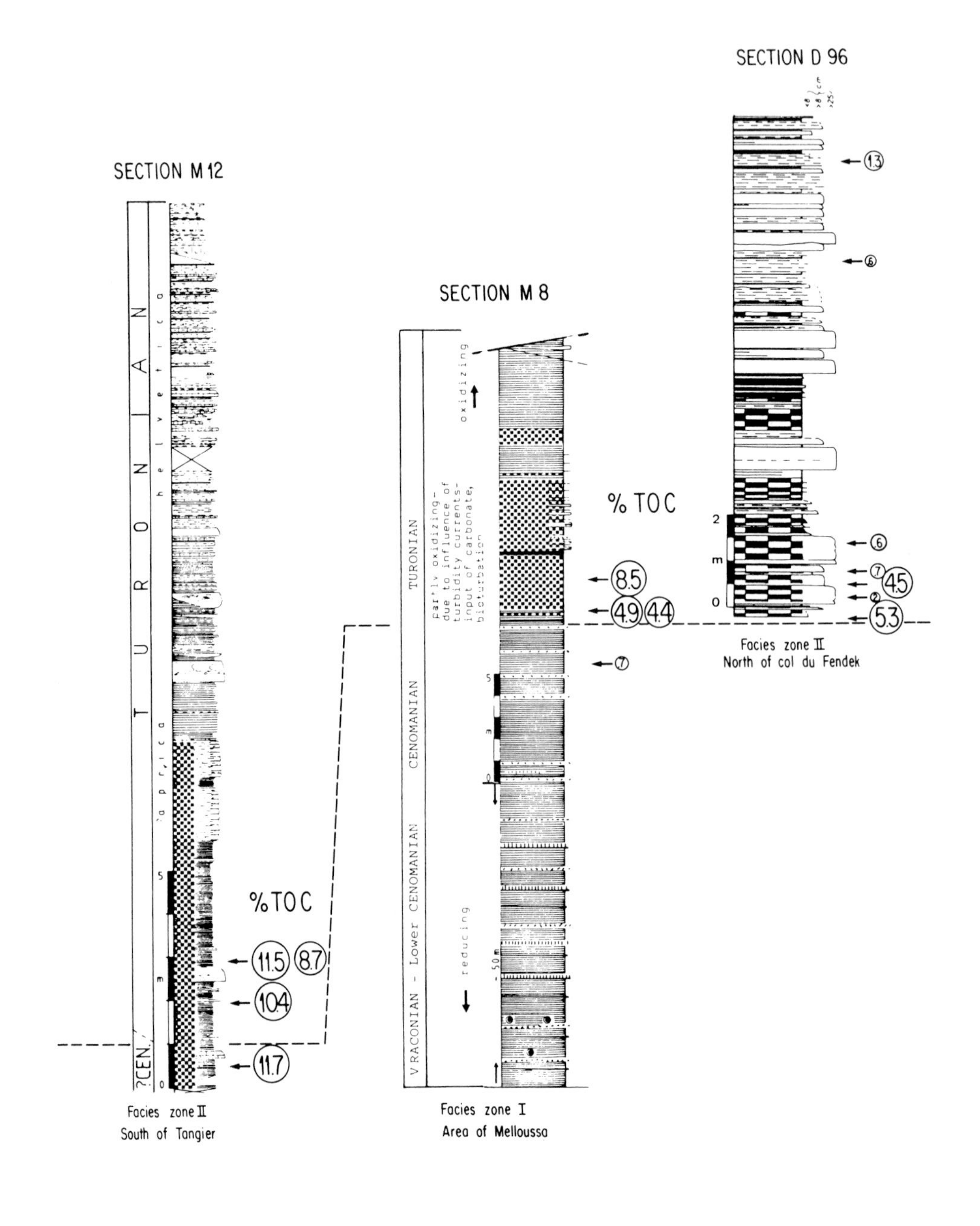
MASSYLIAN
SECTION M 12
TURONIAN
helvetica
?apprica
?CEN.
%TOC
11.5
8.7
10.4
11.7
Facies zone II
South of Tangier
SECTION M 8
TURONIAN
CENOMANIAN
V RACONIAN - Lower CENOMANIAN
oxidizing
partly oxidizing-due to influence of turbidity currents-input of carbonate, bioturbation
reducing
%TOC
8.5
4.9
4.4
7
Facies zone I
Area of Melloussa
SECTION D 96
1.3
6
6
7
4.5
2
5.3
Facies zone II
North of col du Fendek

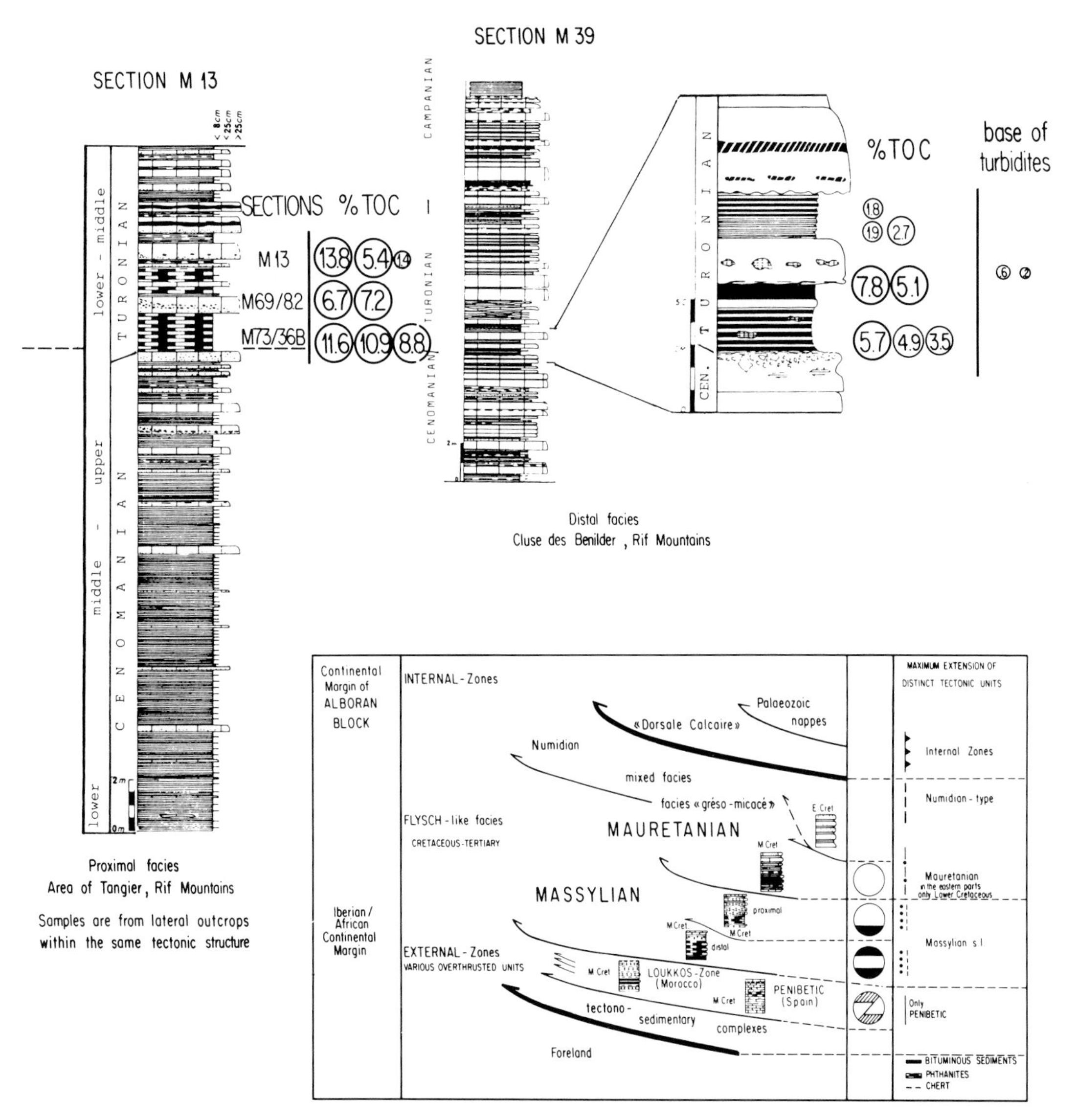

FIG. 7. Former Alboran block: characteristic sections with the TOC content in the Massylian and Mauretanian series. Sketch giving an idea of the sequence of nappes forming the present-day Rif Mountains with the corresponding palaeogeographic domains.

area of Tangiers the Turonian sediments (Sections M73 and M13, proximal facies of the Mauretanian series) have TOCs of 1.4–13.3% (Fig. 7). The hydrogen index fluctuates between 317 and 663 (Fig. 8), exceptionally reaching 797 mg HC/g TOC in a sample analysed by J. Rullkötter (KFA). In this part of the Beni Ider nappe, maturity is relatively low (average $T_{max}=425°C$). In section M69 (20 km SE of M73/13) maturity is higher (average $T_{max}=443°C$). Although the TOC remains high (6.7–7.2%, Fig. 7), the HI decreases due to the evolution of the organic-matter: $410 \leqslant HI \leqslant 464$ mg HC/g TOC (Fig. 8). Farther south, the distal facies of the Mauretanian series (section M39) is quite mature (average $T_{max}=458°C$). In spite of the high TOC in the sediments (1.8–7.8%, Fig. 7), the HI is lower than 73, due to the metamorphism of the series (Fig. 8). Type II organic matter coincides with the CTBE in the unmetamorphized sections of the Mauretanian series (M73, 13 and 69). In the more mature sections (M18, M32, M39, M40) the origin of the organic matter is undetermined. It can be supposed that tectonic compression has been greater in the south than in the north. The Mauretanian series also occurs in Algeria. One sample from section ACM9/13 gives results comparable to those from section M69 in Morocco: TOC=3.7%, HI=420, $T_{max}=441°C$.

Massylian nappe

The Massylian nappes are represented by the Melloussa nappe in the west and the Chouamat nappe in the east. Equivalent formations are found in the Campo de Gibraltar. Albian sedimentation started with siliciclastic turbidites alternating with grey-green to dark shales. Deposits of the Upper Cenomanian are rarely dated. They correspond to condensed sequences. The uppermost Cenomanian to Middle Turonian siliceous dark to black carbonaceous laminites (phthanites of French authors) form a characteristic feature in the landscape. In section M12 (proximal facies, south of Tangier) the TOC ranges from 2–11.7%. The maturity is relatively low (average $T_{max}=433°C$). The hydrogen index reaches 612 and even 718 mg HC/g TOC in a sample analysed by J. Rullkötter (KFA) (Fig. 8). Farther south, in section D96 (proximal facies north of Col du Fendek), as well as in section M8 (distal facies, Melloussa area) maturity is higher (440°C and 445°C respectively) and the hydrogen indices range from 73–213 mg HC/g TOC, although the TOC remains high (up to 5.3% in section D96, up to 8.5% in Section M8) (Fig. 7). On the other side of the Gibraltar Arch, in Spain, the Massylian series (Section A77 proximal facies and section S11 distal facies) contains respectively 3.8 and 1.9% TOC, with organic matter of Type II origin (HI=462 and 513 mg HC/g TOC). The post-Turonian sediments include multicoloured clays of distal facies, and fine-grained carbonate detrital turbidites alternating with multicoloured or grey shales in the more proximal zones.

From a study of the organic matter content no difference appears between the Mauretanian and Massylian sequences and equivalent series in Algeria and Southern Spain. Rich organic matter of Type II origin indicates the anoxicity of the environment. The laminated dark shales of the CTBE are also characterized by low accumulation rates in the turbiditic sequences of the 'flysch domain'.

Tethys

The Tethys provides interesting outcrops for studying the CTBE either on the former continental margin of Africa (Tunisian Atlas), in the present southern Alpine sections (e.g. the Euganean hills), or in the Umbrian Apennines, Italy.

Tunisian Atlas (See Figs 1 and 9 for general location)

Eastern Tunisia and the Pelagian Sea form a vast stable platform which progressively subsided during the Mesozoic and Tertiary (Burollet *et al.* 1978). During late Albian and Cenomanian time the dark grey shales of the Fahdene Formation were deposited. Throughout eastern, central and northern Tunisia no break in sedimentation occurs between the Cenomanian and Turonian, and the boundary is placed within the Bahloul Formation (Burollet 1956). The lower Turonian consists of bituminous blackish marls and well-bedded sublithographic limestones with a thickness of 20–50 m. This facies is widely developed in the Fkirine and Zaghouan regions where bituminous limestones persist up to the Lower Coniacian (Edjehaf facies) (Salaj 1978). The sediments of the Bahloul Formation were deposited in a fairly deep environment. In shallower areas the Bahloul Formation is either not developed or is represented by rare blackish limestone interbeddings. In the area of Maktar the samples studied have relatively high TOCs of 1.1–4.7%, and the hydrogen indices range from 260–670 mg HC/g TOC (Fig. 9) with organic matter of Type II origin. Their maturity is moderate with an average T_{max} of 431°C due to the low post-Turonian accumulation rates. Organic rich sedimentation in the Pelagian Sea during the CTBE extends farther west on the Tethyan margin of Africa in

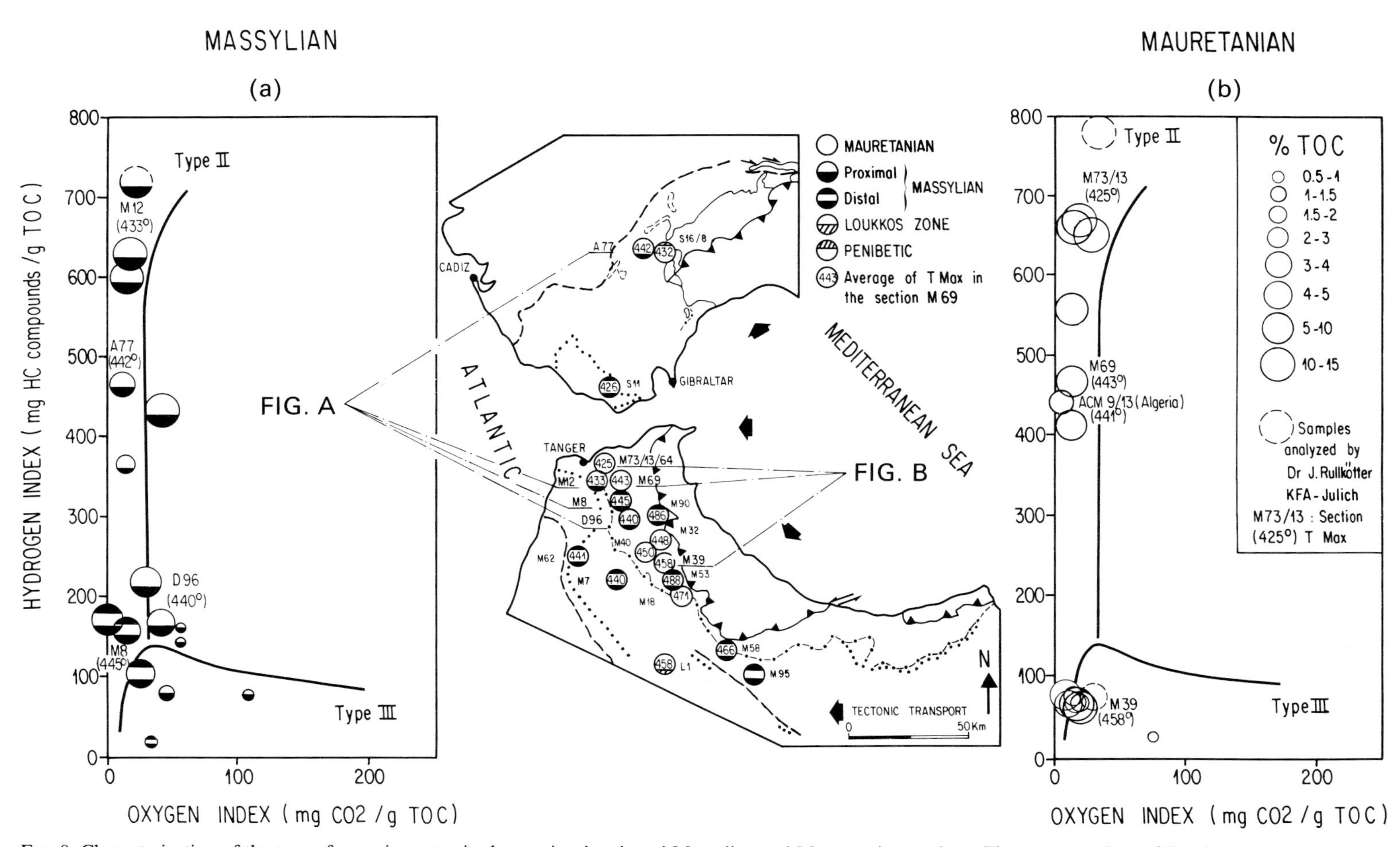

FIG. 8. Characterization of the type of organic matter in the previously selected Massylian and Mauretanian sections. The average values of T_{max} in relation to the different sections are given in a structural sketch of the Gibraltar Arch which gives an idea of the westernmost extension of the different overthrust nappes (caption in Fig. 7).

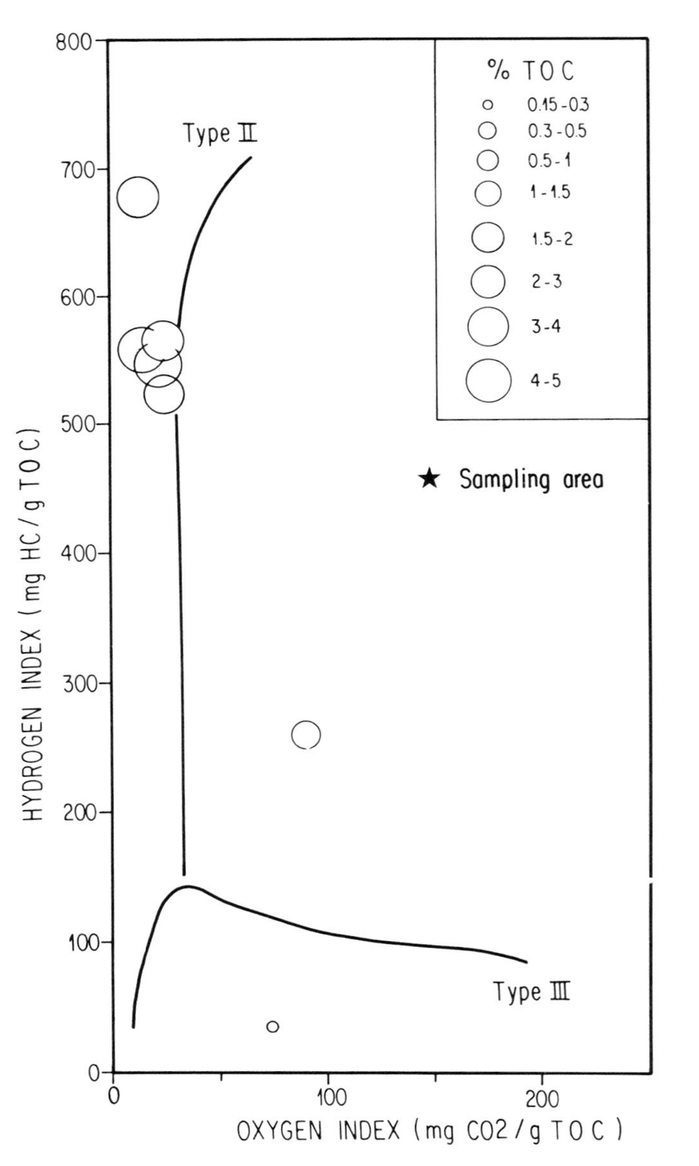

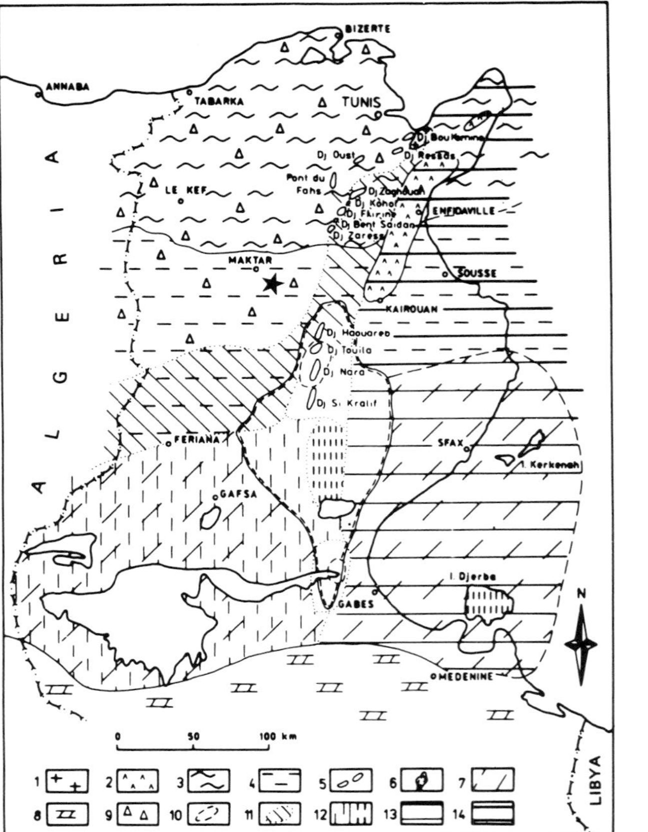

Sketch map of the facies distribution during the Albian to Turonian. In the figure separate symbols are used for the Albian and Cenomanian–Turonian, so that changes of facies may be discerned by the superposition of symbols. *Albian:* (2) zone of condensed sedimentation prior to Upper Albian transgression; (3) pelagic—marls and marly limestones of the Tunisian Trough; (4) pelagic marls; (5) Jurassic massifs of the north-south axis; (7) laguno-neritic zone with dolomites, evaporites, and marls; (8) epineritic carbonates of the Sahara platform. *Cenomanian–Turonian:* (1) reef facies (rudists—Vraconian, Cenomanian, and Lower–Middle Turonian); (9) pelagic, marly facies of the Tunisian Trough; (10) region where Cenomanian is not represented; (11) transition zone between Tunisian Trough and Saharan platform and Pelagian block; (12) epineritic—lagoonal carbonates and evaporites

From SALAJ J. (1978)

Fig. 9. Characterization of the type of organic matter in the Bahloul Formation (Tunisia). The size of the circle is proportional to the organic carbon content

Algeria (Bou Saada area) where Emberger (1960) and Kieken (1962) have identified bituminous marls at the base of the Turonian limestones.

Umbrian Apennines, Italy and Southern Alpine Section (Euganean hills) (See Figs. 1 and 10 for general location)

In the Umbrian region the 'Livello Bonarelli', at the Cenomanian–Turonian boundary (Arthur & Fischer 1977), divides the Scaglia Bianca Formation (Cenomanian age) from the Scaglia Rossa Formation (Coniacian–Santonian age). The green-grey limestones with cherts in the Scaglia Bianca and Scaglia Rossa Formation contain almost no organic matter ($<0.2\%$). In contrast the very thin 'Livello Bonarelli' (50–100 cm) consists of radiolarian sands with interbedded black laminated mudstones with a high TOC (up to 20%). This Type II origin, with hydrogen indices ranging from 450–260 mg HC/g TOC (Fig. 10), is confirmed by the amorphous nature of the organic matter (Arthur & Premoli-Silva 1982).

Their enrichment in marine organic matter can be followed farther north in the region of the Euganean Hills near Padova (Leo, Fig. 10), where a sample of earliest Turonian age has a TOC of 2% and an HI of 370 mg HC/g TOC, whereas in the Tuscan Nappe (Go) the samples are poor in organic carbon.

North-east Atlantic: Iberian and Celtic margins

Two DSDP Sites (398 and 551) provide valuable data concerning the CTBE off the European continental margin.

Iberian margin

Site 398 (Leg 47B) is located at the base of the slope south of Vigo Seamount (Fig. 1). At this site a sedimentological break resulting from condensed sedimentation (Rehault & Mauffret 1979) occurs in sample 56–2/19 cm between two lithologic units equivalent to the Hatteras and Plantagenet Formations. This condensed sequence in section 2 of core 56 is interpreted as being a consequence of the increasing rate of sea floor spreading in the Atlantic Ocean (Hays & Pitman 1973; Hart & Tarling 1974). The decrease in terrigenous input is assumed to be related to the removal of source areas when the Portuguese shelf became submerged by the Cenomanian–Turonian transgression (Muselec 1974; Mougenot 1976). A Turonian age has been determined by dinoflagellates in sample 56–2/122–124 cm (Taugourdeau-Lantz *et al.* 1982; Masure 1984), whereas sample 56–3/45 cm is of Cenomanian age (Sigal 1979). Above the break, sediments with an assemblage of agglutinated foraminifers were deposited below the carbonate compensation depth. Sigal has proposed a Senonian age, probably Santonian to Campanian.

The deposits of the Hatteras Formation consist of dark, relatively organic-rich and carbonate-poor shales and mudstones. During early and middle Albian time the sedimentation rate was very high (greater than 100 m/MA), and the organic matter of terrestrial origin does not exceed 1% TOC. This high sedimentation rate is interpreted as the consequence of regional subsidence. During the late Albian and early Cenomanian the sedimentation rate decreased (15 m/Ma) as the transgression began, and became lower than 1 m/Ma during the CTBE.

Detailed sampling of sections 2 to 5 of core 56 has been undertaken to characterize the CTBE. In the thin black levels of sections 3 to 5, the TOC is less than 3.5%, and often about 1.5% (Fig. 11). This organic matter, which is common in the black shales of the Hatteras Formation, is Type III, with an HI ranging between 50 and 150 mg HC/g TOC. All the interbedded greenish claystones, representing the thicker facies, contain 0.1–0.3% TOC. In section 2 the laminated dark grey to greyish black (N2 to N3) claystones are thicker and relatively rich in TOC (2.5–13%). The organic matter is a mixture of Types II and III, and the richer levels have hydrogen indices ranging from 200–380 mg HC/g TOC (Fig. 11). The amorphous material is composed partly of sapropelic types, as shown by fluorescence (Doerenkamp & Robert 1979). This mixture of Types II and III is confirmed by the elemental analysis of the kerogen (when $TOC=5.9\%$, $H/C=1.2$ and $O/C=0.24$; when $TOC=8.7\%$; $H/C=1.18$ and $O/C=0.23$). The amount of Type II is slightly greater in section 2 (Deroo *et al.* 1979), indicating anoxic conditions in the bottom water (Arthur 1979). The upper contact with the Plantagenet Formation is clearly marked in section 2/19 cm, with the beginning of the yellowish-brown and dark reddish-brown zeolitic unfossiliferous mudstones and claystones being almost devoid of organic carbon (0.1%). The oxidation of the sediments is linked to the opening of the ocean, between the North and South Atlantic, and the initiation of deep oceanic circulation.

Celtic margin

Site 551 (Leg 80) is located on top of a volcanic basement high, 250 km south-west of Ireland, in the northern part of the Bay of Biscay, on the edge

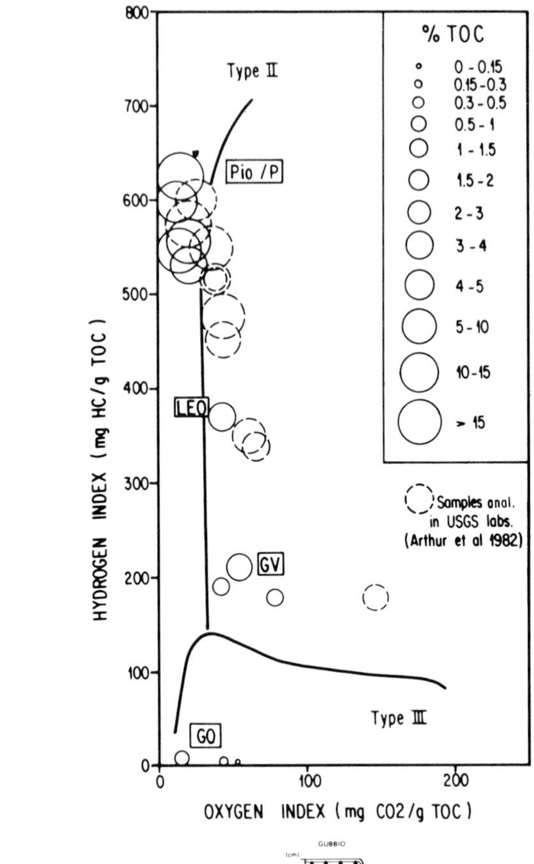

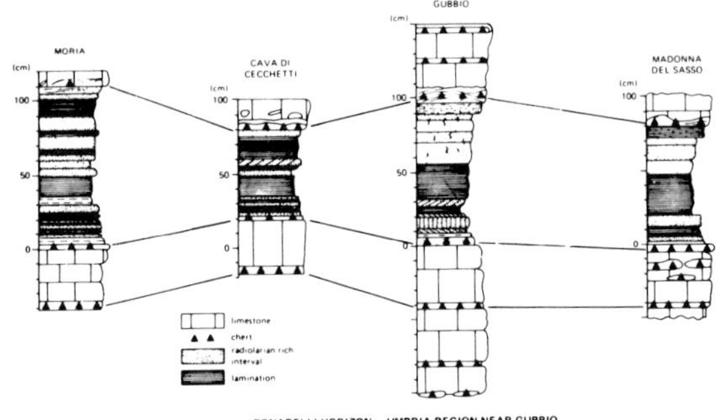

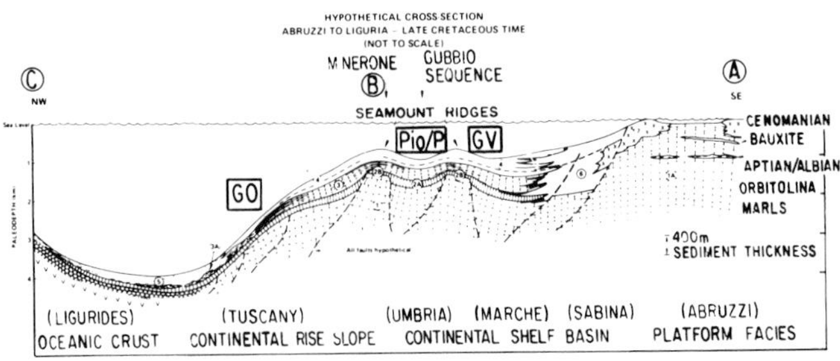

From ARTHUR M.A. & PREMOLI SILVA I. (1982)

Fig. 10.

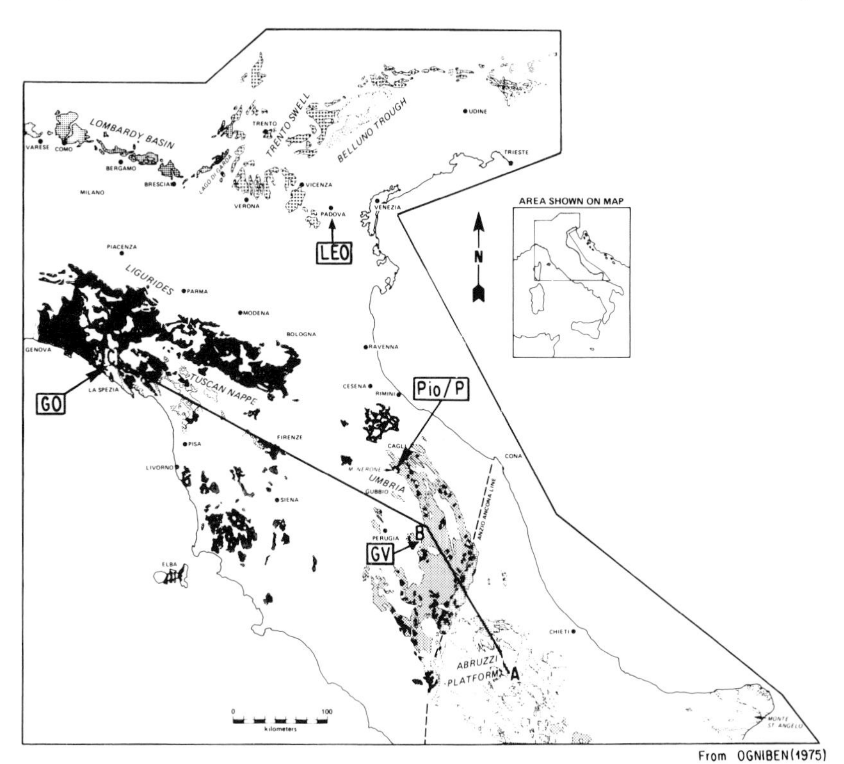

FIG. 10. Characterization of the type of organic matter in the Southern Alpine sections (Euganean hills) and Umbrian Apennines (Italy). The size of the circle is proportional to the organic carbon content.

of the Goban Spur (Fig. 1). An organic-rich black shale level of early Turonian age was found in Core 5, Section 2, 60–120 cm, and in the core catcher (Graciansky *et al.* 1985). The underlying Upper Cenomanian sediments consist of white to pale yellow chalks. Above (Core 5, Sections 1 and 2/0–60 cm) the facies consists of white to pale green nannofossil chalk and siliceous mudstone of Turonian age. The black shales were probably deposited in deep water (1500–2000 m). In these sediments the TOC contents range between 4 and 10%, and the organic matter is mainly a mixture of Types II and III, with hydrogen indices ranging from 200–300 mg HC/g TOC (Fig. 12).

Other black shales of Turonian age were recovered in Core 27 from Site 549, located 20 km north-eastward. Two black layers (Core 27, Section 1, 20–41 cm and 49–48 cm) are separated by an ash layer several centimetres thick. Organic carbon contents are relatively high (3.4% in the upper level, 3.5% in the lower one), and the organic matter is a mixture of Types II and III as at Site 551 (Waples & Cunningham 1985).

These accumulations of organic matter are equivalent to the 'Black Band' of Yorkshire and Humberside (Hart & Bigg 1981) or to the 'Plenus Marl Formation' in the North Sea (Burnhill & Ramsay 1981). In the 'Black Band' as well as in the DSDP Sites the fauna indicates a change in palaeoceanographic conditions associated with the transgression. The areal distribution of these facies in northern Europe is shown by Hart & Bigg (1981, Fig. 14–3A). As in Site 549 the volcanic origin of the montmorillonite of the 'Black Band' indicates contemporaneous volcanic activity. A sample of the Plenus Marl Formation from well 22/7–1A (North Sea) shows a very high TOC of 30% (Fig. 12). The organic matter is of terrestrial origin (HI = 160, OI = 111). However, this single analysis cannot be generalized to the whole formation. The problem of the origin of the organic matter in the dark carbonaceous mudstones having a time equivalence with the CTBE still has to be determined on the North-western Platform of Europe (England, North Sea, Germany, Denmark). The present data (Sites 398 and 551) suggest that the environment was not so favourable to the preservation of aquatic organic matter in the northern part of the North Atlantic as in the southern one (African continental margin).

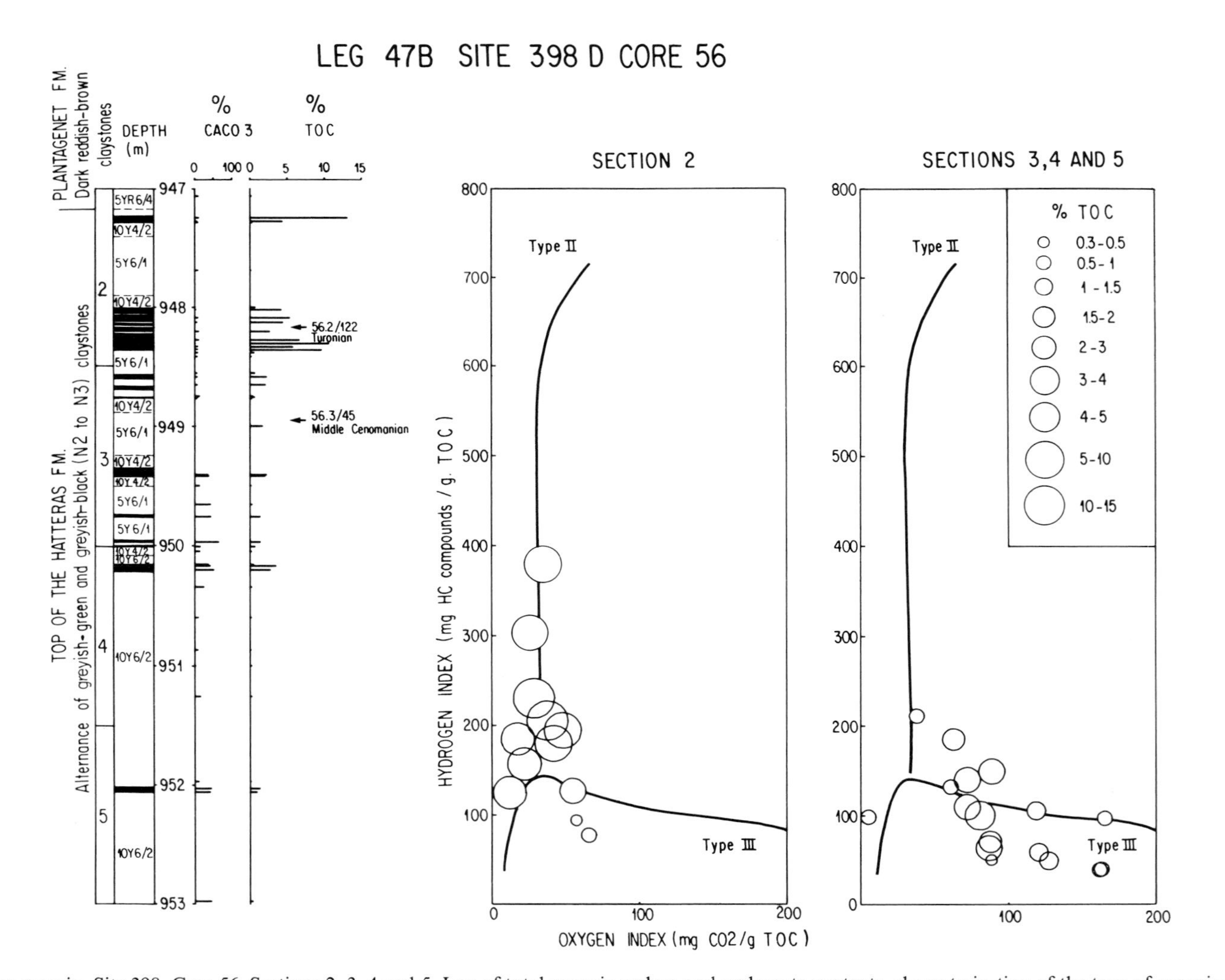

FIG. 11. Iberian margin. Site 398, Core 56, Sections 2, 3, 4 and 5. Log of total organic carbon and carbonate contents, characterization of the type of organic matter. The size of the circle is proportional to the organic carbon content.

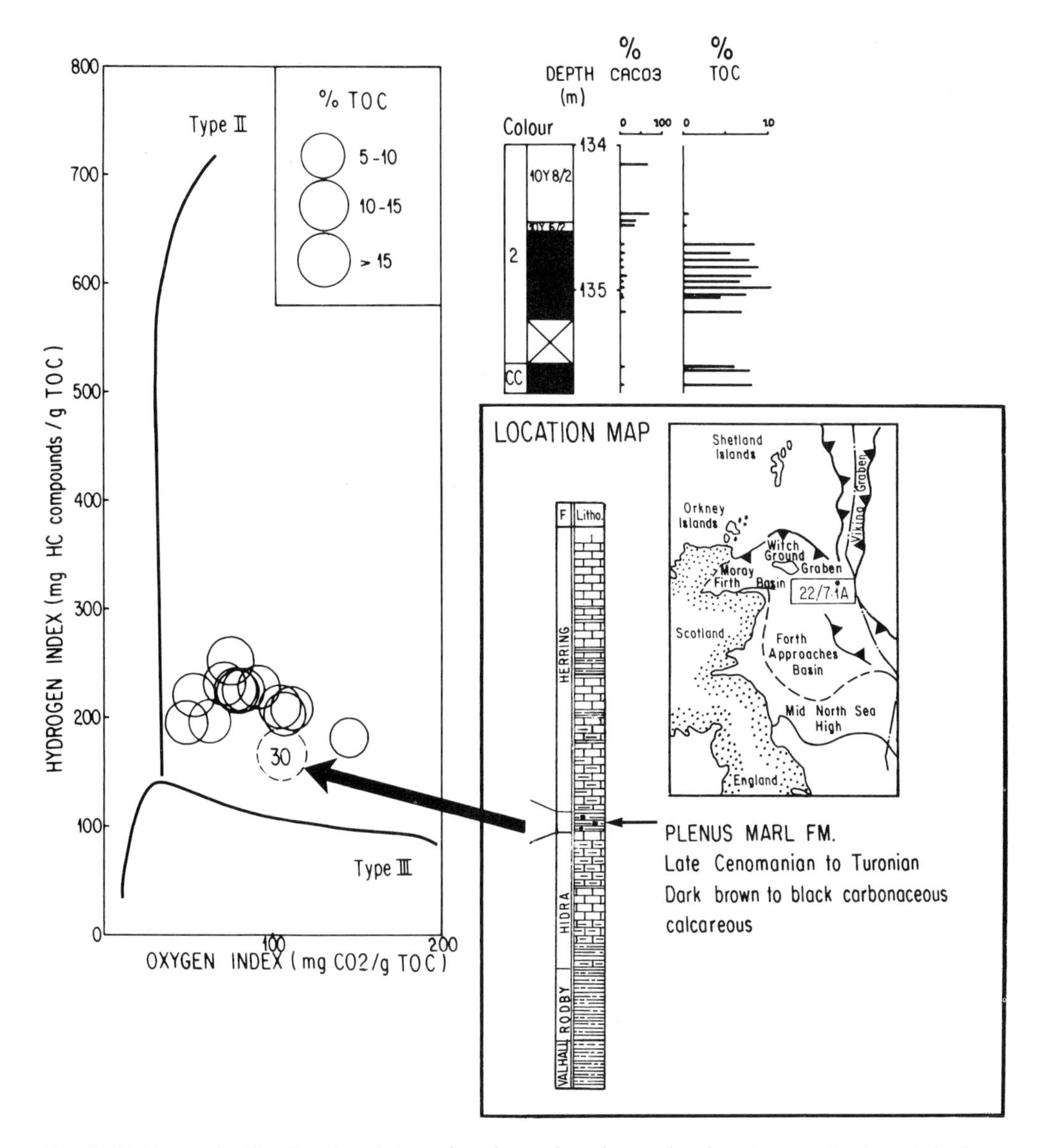

FIG. 12. Celtic margin. Site 551, Core 5. Log of total organic carbon and carbonate contents; characterization of the type of organic matter. Comparison with the Plenus Marl Formation in well 22/7–1A in the North Sea.

American continental margin and Western Interior Basin

To demonstrate the widespread nature of the organic-rich sedimentation in the North Atlantic, it is important also to consider the CTBE on the American margin.

Hills of the lower continental rise off Cape Hatteras

In Site 105 (Leg 11—Ewing & Hollister 1972) (Fig. 1) a major lithological break marks the end of the 'Black shale' sedimentation at the top of the Hatteras Formation, in Sections 3 and 4 of Core 9 (Fig. 13). A stratigraphic study indicates a late

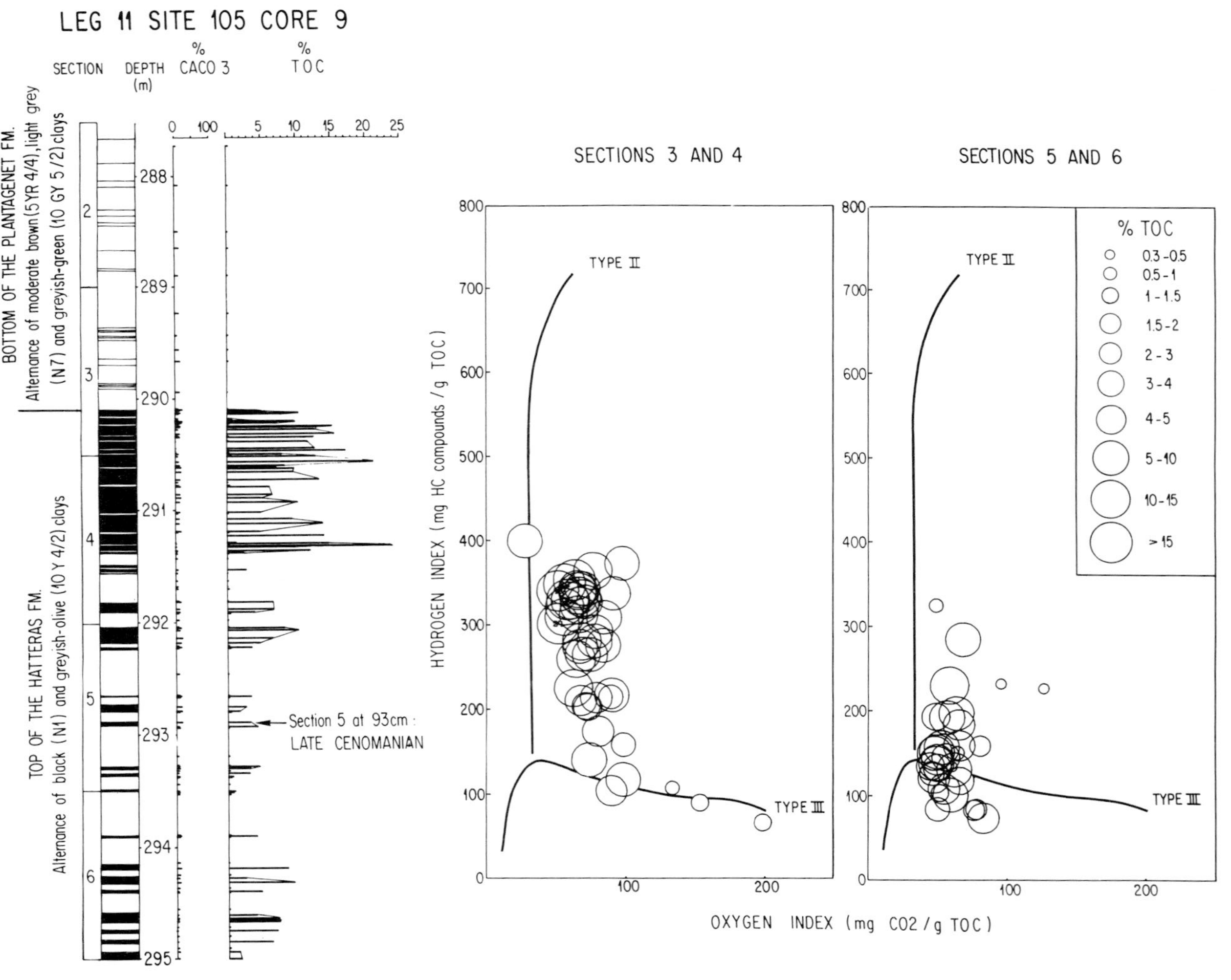

FIG. 13. Hills of the lower continental rise off Cape Hatteras. Site 105, Core 9, Section 3, 4, 5 and 6. Log of total organic carbon and carbonate contents; characterization of the type of organic carbon content.

Cenomanian age for sample 9–5/93 cm and a Senonian age for Core 9, Section 1 (Muller *et al.* 1983), placing the sediments of Sections 3 and 4 within the CTBE. Deeper in the 'black shales' of the Hatteras Formation (Cores 17 to 10), the TOC ranges from 1–5%. TOC values are richest in Core 9 which contains a condensed sequence, where the accumulation rate is less than 1 m/Ma.

Black and green claystones occur in Section 6 to Section 4–87 cm of Core 9. As in the deeper Hatteras Formation, the green layers are thicker than the black ones and the organic matter is mainly of Type III origin (Fig. 13). In contrast black claystones with only rare centimetre or millimetre-scale green layers predominate from Section 4–86 cm to Section 3–110 cm. These black sediments are rich in TOC (10–25%), and their organic matter is a mixture of Types II and III

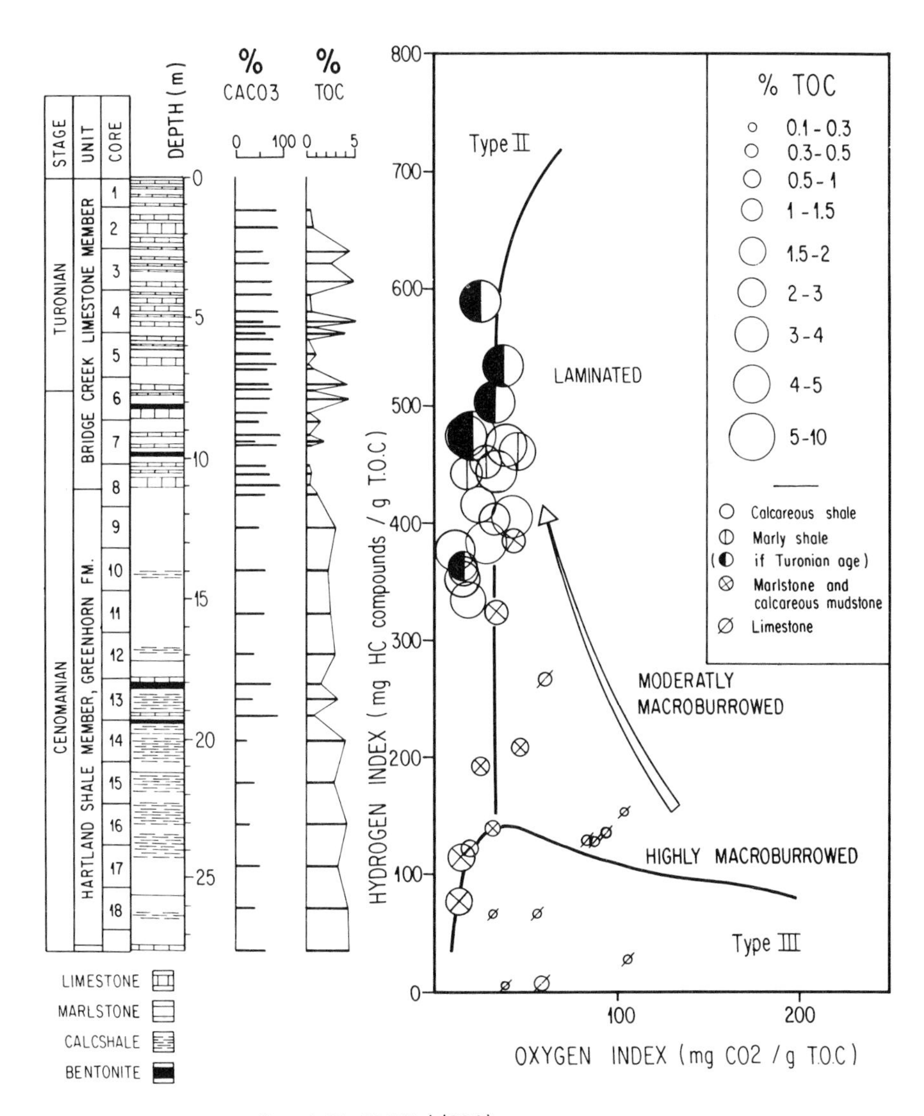

FIG. 14. Western Interior Basin. Pueblo Core. Log of total organic carbon and carbonate contents; characterization of the type of organic matter. The size of the circle is proportional to the organic carbon content (from Pratt 1984).

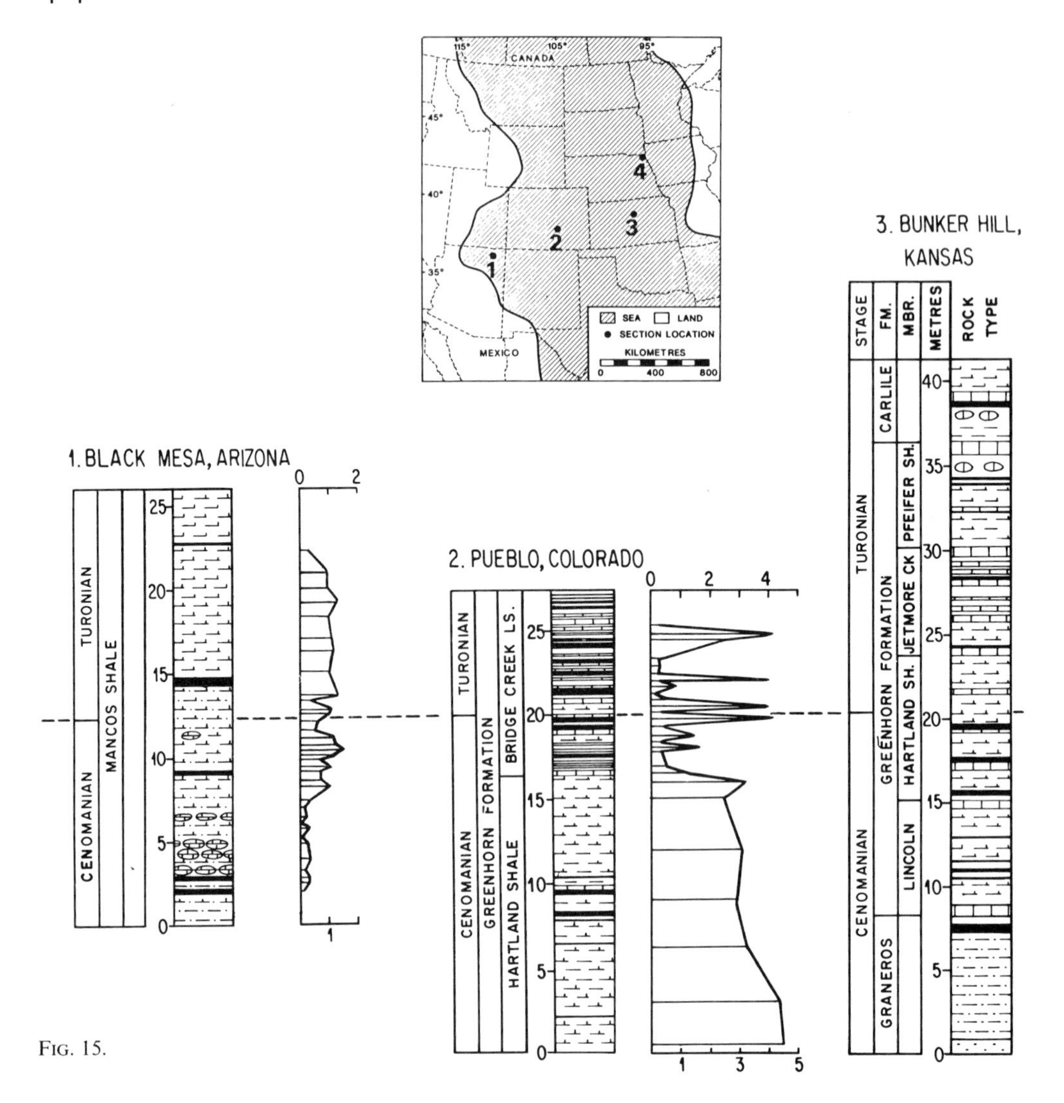

FIG. 15.

(Fig. 13), as confirmed by elemental analysis of the kerogen in sample 9–4/13 cm (H/C = 1.10 and O/C = 0.19). The interbedded green claystones have a low TOC (0.35%), like the moderate brown claystones of the Plantagenet Formation at the top of the core (Fig. 13). All facies are poor in carbonate (0–17% $CaCO_3$).

The same enrichment in organic matter occurs at the top of the Hatteras Formation at Site 603 (Leg 93) (Fig. 1). Initial results show that in cores 33 and 34 the TOC reaches 20%, with an organic matter of aquatic origin (HI = 483 mg HC/g TOC, OI = 26 mg CO_2/g TOC—H/C = 1.22, O/C = 0.18). These organic rich sediments are dated as Cenomanian–Turonian and represent the CTBE (Herbin *et al.*, in press). Detailed study in core 34 shows that the CTBE is not a single homogeneous black bed but a condensation of several thin beds (nineteen in core 34) rich in organic matter of Type II, interbedded with thinner layers of green claystones. This alternation is identical to that in the Hatteras Formation, but at the CTBE preservation of aquatic organic matter is better, indicating the presence of an anoxic environment.

During the CTBE these two DSDP Sites were located in deep water (about 4100 m for Site 105) (Chénet & Francheteau 1979).

Western Interior Basin

The CTBE in the Western Interior Basin of America was demonstrated by Pratt (1981), who

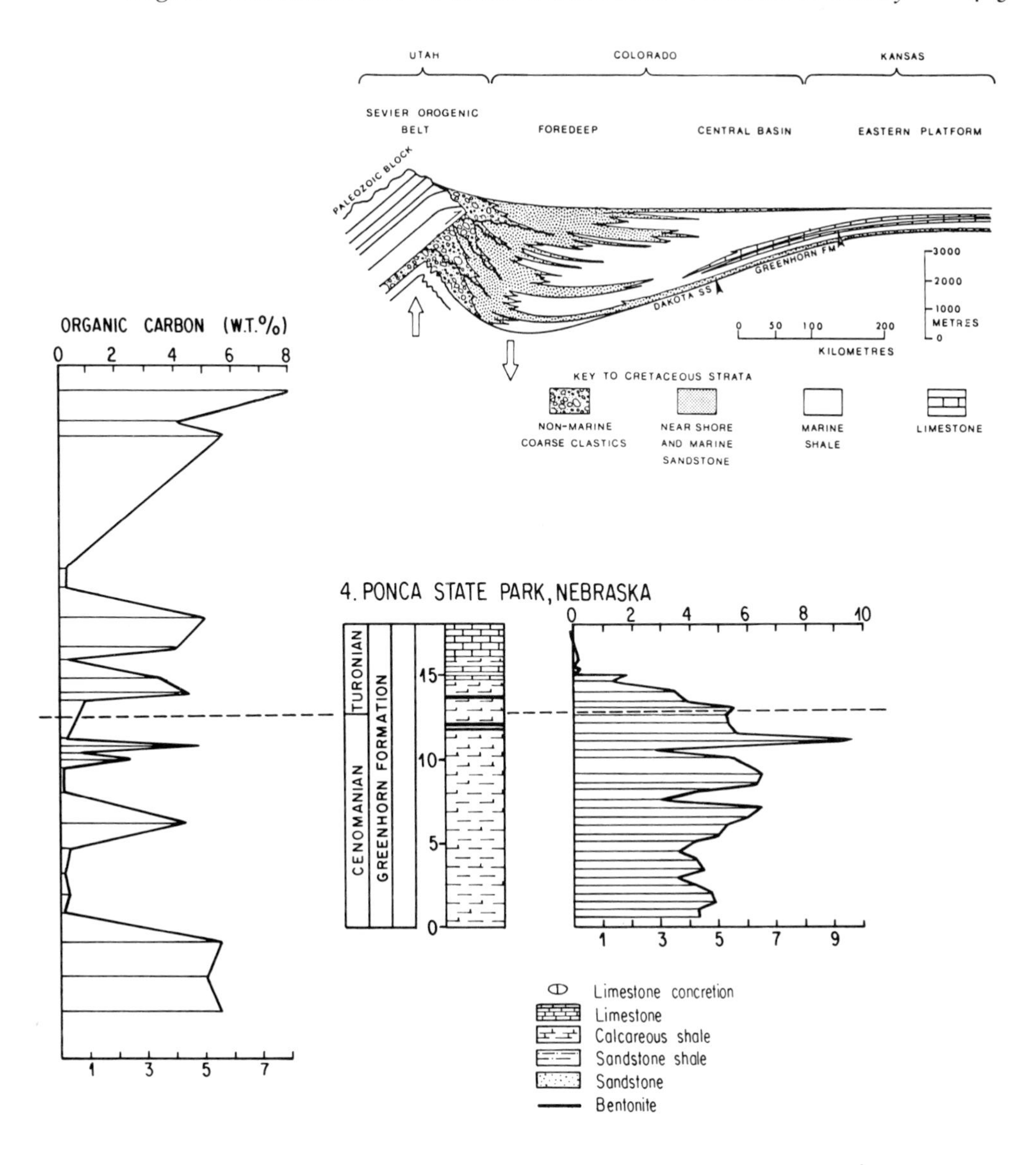

FIG. 15. Western Interior Basin. Correlation between sections of Black Mesa (Arizona), Pueblo (Colorado), Bunker Hill (Kansas) and Ponca State Park (Nebraska) (from Pratt & Threlkel 1984).

proposed a palaeoceanographic model for the middle Cretaceous sediments of the Greenhorn Formation (Upper Cenomanian–Lower Turonian). During the mid-Cretaceous the Western Interior Basin was a broad seaway across the Great Plains of the USA, extending to Arctic Canada and Alaska in the north and to the Gulf of Mexico in the south (Fig. 1). In the central part of the basin, the Hartland Shale Member (Upper Cenomanian) and the Bridge Creek Limestone Member (upper part of the Upper Cenomanian to Lower Turonian) were deposited during the maximum transgression of the Greenhorn cycle (Kauffman 1977). The Bridge Creek Limestone Member consists of highly bioturbated limestones and microburrowed marlstones. The underlying Hartland Shale Member is represented by moderately bioturbated calcareous mudstones and laminated calcareous shales.

In the limestones (73–92% $CaCO_3$) the TOC ranges from 0.1–0.6%. Limestones are interbedded with marlstones ($CaCO_3$=46–72%), or calcareous mudstones ($CaCO_3$=39–45%), that are richer in TOC (0.9–5.1%). The calcareous shales, which are relatively poor in $CaCO_3$ (20–39%), have the same range of TOC content, i.e. 2.9–4.3% (Fig. 14). In the highly bioturbated limestones or marlstones the organic matter is terrestrial (Type III), with small fragments of structured and opaque residual organic matter, while most of the richer levels of marlstones, calcareous mudstones and calcareous shales contain dominantly amorphous organic matter of aquatic origin (Type II). The highest hydrogen indices ($605 \leqslant HI \leqslant 646$ mg HC/g TOC) belong to the marlstones of the Bridge Creek Formation (early Turonian) (Fig. 14).

Faunal assemblages show seasonal or longer-term variations in surface-water salinity, and the water column seems to have been controlled by two distinct water masses (Pratt 1984). A weakly to moderately stratified water column without any bottom currents allowed the deposition of marlstones and calcareous shales rich in organic matter of Type II origin. Based on sedimentological and faunal evidence, the palaeodepth of the Central Western Interior Seaway varied between 50 and 500 m. A water depth of about 100–150 m is estimated for the Bridge Creek Formation.

Other sites in Arizona, Kansas and Nebraska show the same enrichment in organic matter during the CTBE (Pratt & Threlkeld 1984). The calcareous shales from the Pueblo, Bunker Hill and Ponca State Park sections have a TOC content ranging from 2–10%, while the Black Mesa section, near the Sevier Orogenic belt, is more detrital (sandstone shale) and has a lower TOC < 2% (Fig. 15).

Conclusions

The Cenomanian–Turonian Boundary Event (CTBE) which coincides with the Cenomanian–Turonian sea-level rise, is most probably a global synchronous phenomenon representing a short period of time. Results obtained by this detailed study have shown that the sediments deposited during this time interval have certain features in common:

(1) very high total organic carbon contents, especially at the base of the Turonian;
(2) organic matter mainly of aquatic (Type II) origin in pelagic and hemipelagic sediments (whereas mainly terrestrial Type III and residual organic matter occurred in the formations of Albian and lower Cenomanian age);
(3) high accumulation of organic matter deposition independent of given palaeogeographic settings (e.g. east margin of America, west margin of Africa, and even on the mid oceanic ridge) and palaeobathymetry (from shallow water—less than 500 m—to deep water—up to 4000 m);
(4) occurrence even in depositional environments with high detrital input and strong dilution (e.g. deep sea fans);
(5) co-occurrence with biogenic silica (phthanites) and, often, volcanic ashes;
(6) condensed sedimentary sequence with very low sediment accumulation rate in DSDP Sites (1 m/Ma), reaching 5–20 m/Ma on the shelf.

Such deposition of aquatic organic matter induces anoxia in the depositional environment because renewal of oxygen is insufficient to oxidize sinking organic matter. Furthermore most of the finely laminated black shales are devoid of bioturbation, which demonstrates the absence of benthonic life.

This facies has been identified on the African, European and American continental margins, but it exists also on the Venezuelan margin in the La Luna and Querecual Formations (Hedberg 1937, 1950) and on the Demerara Rise (Site 144 Core 8) (Herbin & Deroo 1982). The CTBE spreads beyond the North Atlantic to the South Atlantic (Site 356, Site 364 and Site 530; Deroo *et al.* 1984) and the Pacific (Site 585; Moberly & Schlanger 1985).

Modelling the mechanism that allows widespread simultaneous formation of organic rich sediments is a controversial enterprise. Some authors dealing with the more local aspects of the CTBE have explained such enrichment by invoking an increase in biogenic surface productivity. Others have argued for restricted circulation, with widespread stagnant conditions caused by stratification of the water column. Hypotheses based on resedimentation of organic matter from shallow environments into deeper areas of deep-sea fan systems have also been suggested (Thurow & Kuhnt, this volume). Each of these hypotheses can explain a local geological record, but to explain the widespread synchronous distribution of the CTBE something with wider dimensions must be envisaged. It is hard to imagine how such vast areas could have been stagnant during the CTBE, but it is also difficult to see how an upwelling system could wind around the different ocean margins—east and west—and even occur

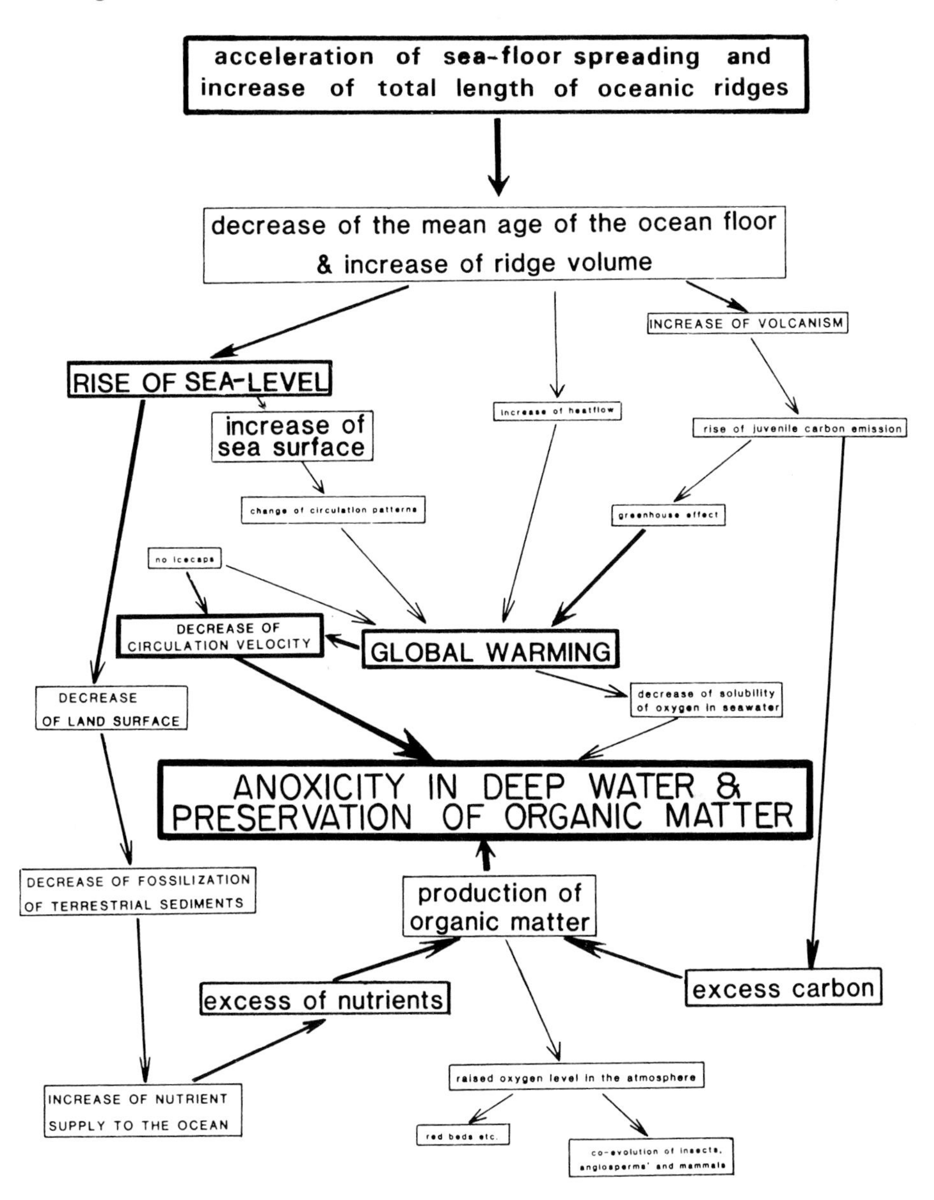

FIG. 16. Interaction of the processes connected to acceleration of sea floor spreading during middle Cretaceous times, causing anoxia in deep water and the preservation of the organic matter (from De Boer 1983).

simultaneously on the mid oceanic ridge. Furthermore, results of organic geochemistry, and isotope studies (Brumsack, this volume; De Boer, this volume) suggest a period of oxygen depletion of Black Sea type without high primary productivity.

Schlanger *et al.* (1985) and Arthur *et al.* (1985) suggest that warm saline bottom water formed in areas where evaporation was higher than precipitation plus inflow (Brass *et al.* 1982; Busson 1984). Rates of upwelling would have increased with rates of deep warm saline bottom water formation, thereby increasing primary productivity which caused expansion of a midwater oxygen minimum zone. But the model of predicted areas of upwelling for the mid-Cretaceous does not fit with the data obtained off the American continental margin (Sites 105–603) and in the South Atlantic on the Rio Grande Rise (up to 15% of type II organic matter in Core 41 Section 3 of Site 356). The symmetry of organogenic sedimentation between the east and west margins gives a global character to the anoxicity during the CTBE.

In fact, phenomena which characterize the middle Cretaceous are: increased rate of sea floor spreading and volcanism, high sea-levels, and low water circulation. Due to low latitudinal differentiation in climate and the presence of land masses separating the North Atlantic and Tethyan Domains from polar areas, the deep currents of cold and oxygenated water of the present day were absent from the deep ocean basins (Berger 1979; Frakes 1979). Thus poor circulation may be one of the main reasons for the lack of oxygen replenishment and the formation of black shales. However, all these palaeoenvironmental features, also existing in the Lower Cretaceous, did not permit deposition of highly organically enriched black shales in the Blake Bahama and Hatteras Formations, during Barremian, Aptian and Albian ages.

On the other hand the temperatures were warmer than those of modern deep waters, reaching a peak in the middle Cretaceous (Savin 1984), which also encouraged a decrease of dissolved oxygen content during the CTBE (Fig. 16).

The extreme widespread anoxia typical of the CTBE coincided with an acceleration of the sea-level rise and transgression onto the shelves (Vail *et al.* 1977). However the main consequence of transgressions was a decreasing input of detrital material to the basins, resulting in lower sedimentation rates, and there is no evidence that they induce global anoxic events. This acceleration of the sea-level rise at the Cenomanian–Turonian boundary, linked to a sudden change in the mid oceanic ridge volume, is attributed to an increase in global tectono-magmatic processes (Rona 1973; Matsumoto 1977; Pitman 1978). The effects of intense volcanic activity on the mid oceanic ridge and on the fracture zones in a basin experiencing a low rate of oxygen renewal in deep waters are unknown. Large amounts of silica released into the sea water through sustained volcanic activity and resultant alteration processes, might induce high production of siliceous fauna and thereby cause accumulation of organic material, e.g. on the Mid Pacific Mountains in the Aptian/Albian sediments on Site 463 (Mélières *et al.* 1981). The quantity of volcanic rocks formed per unit time during the Cretaceous was at least two times larger than at the present time. This activity would have increased the CO_2 level more than five times (Budyko & Ronov 1978). Such an increased addition of juvenile carbon to the exogenic cycle could have furnished an important part of the extra quantity of fossilized organic matter during this period (De Boer 1983). But the question, whether the anoxic deep ocean waters acted as an active trap or the anoxicity was induced by a relative increase of organic production (too high for the low ocean water circulation) remains undetermined. In De Boer's hypotheses the anoxicity allowing the preservation of marine organic matter would be a consequence of the acceleration of sea floor spreading and volcanism in a basin without intense oceanic circulation. The hypothesis of increasing CO_2 level could furthermore explain the lithological change from carbonated toward uncarbonated sedimentation, e.g. the Celtic margin (Site 551) where black shales are within chalks. Moreover the exceptional preservation and accumulation of organic matter in the southern part of the North Atlantic (Site 367) could be related to the vicinity of the Equatorial Fracture Zones, an area of intense tectonic and magmatic processes during the middle Cretaceous. In such a model the preservation of black shales would be a consequence of internal phenomenon, in relation with the theory of 'pulsation tectonics' (Sheridan, this volume).

The detailed analytic results given in this paper are the base for a better knowledge of the CTBE. They might help to find a convincing model for this phenomenon from a 'cartography' of the organic matter (age, TOC, type of organic matter, thickness of the organogenic sequences), which remains one of the most important stages in a geological approach.

ACKNOWLEDGEMENTS: The authors thank the National Science Foundation for providing many of the samples. Laboratory studies were performed at the Institut Français du Pétrole under a contract with the Comité d'Etudes Pétrolières Marines (CEPM). Descriptions of cores and appropriate sampling were done at the Lamont Doherty Geological Observatory, with the approval of the Chief Scientist of the Deep Sea Drilling Project, and with the help of the curatorial staff of the East Coast Repository. The authors also thank Mrs Bourdon, P.F. Burollet, R.S.M. Templeton and A. Verdier from CFP for their advice on the study of the Casamance area and Tunisia. The authors thank also Dr. H. Jenkyns and Prof. B.M. Funnell for their comments and review. The assistance of C.P. Summerhayes is acknowledged in the final editing of the manuscript.

References

ARTHUR, M.A. & FISCHER, A.G. 1977. Upper Cretaceous–Paleocene magnetic stratigraphy at Gubbio, Italy. I. Lithostratigraphy and Sedimentology. *Geol. Soc. Am. Bull.* **88,** 367–89.

—— 1979. North Atlantic Cretaceous black shales: The record at Site 398 and a brief comparison with other occurrences: *In*: SIBUET J.C & RYAN W.B.F. (eds), *Init. Repts DSDP* **47.** Part 2. US Govt. Print. Off., Washington, DC. 719–52.

—— & NATLAND, J.H. 1979. Carbonaceous sediments in the North and South Atlantic: the role of salinity in stable stratification of Early Cretaceous Basins. *In*: TALWANI, M., HAY W. & RYAN W.B.F. (eds), *Deep Drilling Results in the Atlantic Ocean: Continental Margins and Paleoenvironments.* Am. Geophys. Union, Maurice Ewing Series **3.** 297–344.

—— & PREMOLI SILVA, I. 1982. Development of widespread organic carbon rich strata in the Mediterranean Tethys. *In*: SCHLANGER S.O. & CITA M.C (eds), *Nature and Origin of Cretaceous Carbon Rich Facies.* Academic Press, London. 7–54.

——, SCHLANGER, S.O. & JENKYNS, H.C. 1985. The Cenomanian–Turonian oceanic anoxic event, II. Paleoceanographic controls on organic matter production and preservation. *In*: BROOKS, J. & FLEET, A. (eds), *Marine Petroleum Source Rocks. Spec. Publ. Geol. Soc. Lond.* Published by Blackwell Scientific Publications, Oxford.

AUXINI, 1969. Correlacion estratigrafica de los sondeos perforados en el Sahara Espanol. *Boletin Geologico y Minero* **83,** 235–51.

BERGER, W.H. 1979. Impact of deep-sea drilling on paleoceanography. *In*: TALWANI, M., HAY, W. & RYAN, W.B.F. (eds), *Deep Drilling Results in the Atlantic Ocean: Continental Margins and Paleoenvironment.* Am. Geophys. Union, Washington, DC. Maurice Ewing Series **3.** 297–314.

BIJU-DUVAL, B., DERCOURT, J. & LE PICHON, X. 1977. From the Tethys Ocean to the Mediterranean seas: A plate tectonic model of the evolution of the western Alpine System. *In*: BIJU-DUVAL, B. & MONTADERT, L. (eds), *International Symposium on the Structural History of the Mediterranean Basins.* Ed. Technip, Paris. 143–64.

BITTERLI, P. 1963. Aspects of the Genesis of Bituminous rock Sequences. *Geologie en Mijnbouw* **42,** (6), 183–201.

BOUILLIN, J.P., DURAND DELGA, M., GELARD, J.P., LEIKINE, M., RAOULT, J.F., RAYMOND, D., TEFIANI, M. & VILA, J.M. 1970. Definition d'un flysch Massylien et d'un flysch Maurétanien au sein des flyschs allochtones de l'Algérie. *C.R. Acad. Sc. Fr.* **270,** 2249–52.

BRASS, G.W., SOUTHAM, J.R. & PETERSON, W.H. 1982. Warm saline bottom water in the ancient ocean. *Nature* **296,** 620–28.

BUDYKO, M.I. & RONOV, A.B. 1979. Chemical evolution of the atmosphere in the Phanerozoic. *Geokhimiya* **5,** 643–53 (Translated in *Geochem. Intern.* **1979,** 1–9.)

BURNHILL, T.J. & RAMSAY, W.V. 1981. Mid Cretaceous Palaeontology and Stratigraphy, Central North Sea. *In*: ILLING, L.V. & HOBSON, G.D. (eds), *Petroleum Geology of the Continental Shelf of Northwest Europe.* Institute of Petroleum, London. 245–54.

BUROLLET, P.F. 1956. Contribution à l'étude stratigraphique de la Tunisie centrale. *Ann. Mines Géol., Tunis* **18,** 1–350.

——, MUGNIOT, J.M. & SAVEENEY, P. 1978. The geology of the Pelagian Block, the margins and basins off Southern Tunisia and Tripolitania. *In*: NAIRN, A.E.M., KANES, W.H. & STEHLI, F.G. (eds), *The Ocean Basins and Margins, Volume 4B, The Western Mediterranean.* Plenum Press, New York. 331–59.

BUSSON, G. 1984. Relations entre la sédimentation du Crétacé moyen et supérieur de la plateforme du Nord-Ouest africain et les dépôts contemporains de l'Atlantique centre et nord. *Eclogae Geol. Helv.* **77,** 2, 221–35.

CHÉNET, P.Y. & FRANCHETEAU, J. 1979. Bathymetric reconstruction method: application to the Central Atlantic Basin between 10°N and 40°N. *In*: DONELLY, T., FRANCHETEAU, J., BRYAN, W., ROBINSON, P., FLOWER, M. & SALISBURY, M. *Init. Repts DSDP* **51, 52, 53,** part 2. US Govt. Print. Off., Washington, DC. 1501–14.

DEAN, W.E., GARDNER, J.V., JANSA, L.F., CEPEK, P. & SEIBOLD, E. 1978. Cyclic sedimentation along the continental margin of Northwest Africa. *In*: LANCELOT, Y. & SEIBOLD, E. (eds), *Init. Repts DSDP* **41.** US Govt. Print. Off., Washington, DC. 965–90.

—— & GARDNER, J.V. 1982. Origin and geochemistry of redox of Jurassic to Eocene Age, Cape Verde Basin (DSDP Site 367), Continental Margin of North West Africa. *In*: SCHLANGER, S.O. & CITA, M.B. (eds), *Nature and Origin of Cretaceous Carbon Rich Facies.* Academic Press, London. 57–78.

DE BOER, P.L. 1983. Aspects of Middle Cretaceous pelagic sedimentation in Southern Europe; production and storage of organic matter, stable isotopes, and astronomical influences. Thesis. *Instituut voor Aardwetens-chappen der Rijksuniversiteit. Utrecht* **31,** 112 p.

DEMAISON, G.J. & MOORE, G.T. 1980. Anoxic environments and oil source bed benesis. *Bull. Am. Ass. Petrol. Geol.* **64,** 1179–209.

DEROO, G., HERBIN, J.P., ROUCACHÉ, J. & TISSOT, B. 1979. Organic geochemistry of Cretaceous shales from DSDP Site 398, Leg 47B, Eastern North Atlantic. *In*: SIBUET, J.C. & RYAN, W.B.F. (eds), *Init. Repts DSDP* **47,** Part 2. US Govt. Print. Off., Washington, DC. 513–22.

DEROO, G., HERBIN, J.P. & HUC, A.Y. 1984. Organic geochemistry of Cretaceous black shales from Deep Sea Drilling Project Site 530, Leg 75. In: HAY, W.W. & SIBUET, J.C. (eds). *Init. Repts DSDP.* US Govt. Print. Off., Washington, DC. 983–99.

DIDYCK, B.M., SIMONEIT, B.R.T., BRASSELL, S.C. & EGLINTON, G. 1978. Organic geochemical indicators of palaenvironmental conditions of sedimentation. *Nature (Lond).* **272,** 216–22.

DOERENKAMP, A. & ROBERT, P. 1979. Optical study of Organic Matter from some samples of Cretaceous age, Leg 47B, Hole 398D. *In*: SIBUET, J.C. & RYAN,

W.B.F. (eds), *Init. Repts DSDP* **47,** Part 2. US Govt. Print. Off., Washington, DC. 529–32.

EINSELE, G. & WIEDMANN, J. 1975. Faunal and sedimentological evidence for upwelling in the Upper Cretaceous coastal basins of Tarfaya, Morocco. *9th Congr. Intern. Ass. Sedimentologists, Nice. Theme* **1,** 67–74.

——, WIEDMANN, J. 1982. Turonian black shales in the Moroccan coastal basins: First upwelling in the Atlantic Ocean? *In*: RAD, U.V. *et al.* (eds), *Geology of the Northwest African Continental Margin*. Springer-Verlag, Berlin. 396–414.

——, WIEDMANN, J. 1983. Cretaceous upwelling off Northwest Africa: a summary. *In*: THIEDE, J. & SUESS, W. (eds), *Coastal Upwelling*. Part B. Plenum Press, New York. 485–99.

EMBERGER, J. 1960. Esquisse géologique de la partie orientale des monts de Ouled Nails. *Bull. Serv. Carte géol. Algérie Serie* **27.** Thése Algerie 1959. 398 pp.

ESPITALIÉ, J., MADEC, M., TISSOT, B., MENNIG, J.H. & LEPLAT, P. 1977. Source rock characterization method for petroleum exploration. *Offshore Technology Conference, Houston, paper* **2935,** 6 pp.

——, MARQUIS, F. & BARSONY, I. 1984. Geochemical logging. *In*: VOORHEES, K.J. (ed), *Analytical Pyrolysis, Techniques and Applications*. Butterworths, London. 276–304.

EWING, J.I. & HOLLISTER, C.D. 1972. Regional Aspects of Deep Sea Drilling in the Western North Atlantic. *In*: HOLLISTER, C.D. & EWING, J.I. (eds), *Init. Repts DSDP* **11.** US Govt. Print. Off., Washington, DC. 951–73.

FRAKES, L.A. 1979. *Climates throughout Geologic Time*. Elsevier, Amsterdam. 310 p.

GARDNER, J.V., DEAN, W.E. & JANSA, L. 1978. Sediments recovered from the Northwest African Continental Margin. *In*: LANCELOT, Y. & SEIBOLD, E. (eds), *Init. Repts DSDP* **41.** US Govt. Print. Off., Washington, DC. 1121–34.

GRACIANSKY, P.C. DE, BROSSE, E., DEROO, G., HERBIN, J.P., MONTADERT, L., MULLER, C., SCHAAF, A. & SIGAL, J. 1982. Les formations d'âge Crétacé de l'Atlantique Nord et leur matière organique, paléogéographie et milieux de dépôt. *Rev. Int. Fr. du Pétrole* **37** (3), 275–337.

——, DEROO, G., HERBIN, J.P., MONTADERT, L., MULLER, C., SCHAAF, A. & SIGAL, J. 1984. Ocean wide stagnation episode in the late Cretaceous. *Nature, (Lond)*. **308** (5957), 346–9.

——, POAG, W. *et al.* 1985. Site 551. *In*: GRACIANSKY, P.C. DE & POAG, W. (eds), *Init. Repts DSDP* **80.** US Govt. Print. Off., Washington, DC.

GUBELI, A. 1983. *Stratigraphische und Sedimentologische Untersuchung der detritischen Unterkreide-Serien des zentralen Rif (Marokko)*. Thesis, University Zürich. 192 pp.

HART, H.B. & TARLING, D.H. 1974. Cenomanian paleogeography in the North Atlantic and possible mid-Cenomanian eustatic movements and their implication. *Paleogeogr., Paleoclimatal, Paleoecol.* **15,** 95–108.

—— & BIGG, P.J. 1981. Anoxic events in the late Cretaceous chalk seas of North-West Europe. *In*: NEALE, J.W. & BRASIER, M.D. (eds), *Microfossils from Recent and Fossil Seas*. The British Micropaleo. Society. 177–85.

HAYES, D.E., PIMM, A.C. *et al.* 1972. *Init. Repts DSDP.* **14,** US Govt. Print. Off., Washington, DC. 975 pp.

HAYS, J.D. & PITMAN, W.L., III. 1973. Lithospheric plate motion, sea level changes and climatic and ecological consequence. *Nature, Lond.* **246** 18–22.

HEDBERG, H.D. 1937. Stratigraphy of the Rio Querecual section of northeastern Venezuela. *Bull. Geol. Soc. Am.* **8,** 1971–2204.

—— 1950. Geology of eastern Venezuela. *Bull. Geol. Soc. Am.* **61,** 1173–215.

HERBIN, J.P. & DEROO, G. 1979. Etude sédimentologique de la matière organique dans les argilites noires crétacées de l'Atlantique Sud. *Doc. Lab. Géol. Fac. Sc. Lyon* **75,** 71–87.

—— & DEROO G. 1982. Sédimentologie de la matière organique dans les formations du Mésozoique de l'Atlantique Nord. *Bull. Soc. Géol. Fr.*, (7), **24,** (3), 497–510.

——, MASURE, E. & ROUCACHÉ, J., in press. Organic geochemistry and dinoflagellate cysts of Cretaceous formations from Site 603, Leg 93, Lower Continental Rise of Cape Hatteras. Comparison of the Cenomanian Turonian Boundary Event (CTBE) with Site 105, Leg 11. *In*: VAN HINTE, J.E. & WISE, S.W. (eds), *Init. Repts DSDP* **93.** US Govt. Print. Off., Washington, DC.

JANSA, L.F., ENOS, P., TUCHOLKE, B.E., GRADSTEIN, F.H. & SHERIDAN, R.E. 1979. Mesozoic-Cenozoic sedimentary formations of the North American Basin; western North Atlantic. *In*: TALWANI, M., HAY, W. & RYAN, W.B.F. (eds), *Deep Drilling Results in the Atlantic Ocean: Continental Margins and Paleoenvironment*. Am. Geophys. Union, Washington, DC. Maurice Ewing Series **3.** 1–57.

——, GARDNER, J.V. & DEAN, W.E. 1978. Mesozoic sequences of the central North Atlantic. *In*: LANCELOT, Y. & SEIBOLD, E. (eds), *Init. Repts DSDP* **41.** US Govt. Print. Off., Washington, DC. 991–1032.

JENKYNS, H.C. 1980. Cretaceous anoxic events: from continents to oceans. *J. Geol. Soc. Lond.* **137,** 171–88.

KAUFFMAN, E.G. 1977. Geological and biological overview: Western Interior Cretaceous Basin. *The Mountain Geologist* **14,** 75–99.

KIEKEN, M. 1962. Les traits essentiels de la Géologie Algérienne. *In*: *Livre à la Mémoire du Pr. Fallot, P.* **1.** *Soc. Géol. Fr.*, 545–614.

KLASZ, I. DE. 1978. The West African sedimentary Basins. *In*: MOULLADE M. & NAIRN A.E.M. (eds), *The Phanerozoic Geology of the World, Volume II, The Mesozoic*. Elsevier, New York. 371–99.

LANCELOT, Y., HATHAWAY, J.C. & HOLLISTER, C.D. 1972. Lithology of sediments from the Western North Atlantic. *In*: HOLLISTER, C.D. & EWING, J.I. (eds), *Init. Rept. DSDP* **11.** US Govt. Print. Off., Washington, DC. 901–50.

——, SEIBOLD, E. *et al.*1978. The evolution of the central north-eastern Atlantic. Summary of results of DSDP Leg 41. *In*: LANCELOT, Y. & SEIBOLD, E. (eds), *Init. Repts DSDP* **41.** US Govt. Print. Off., Washington, DC. 1215–45.

——, WINTERER, E.L. *et al.* 1980. *Init. Repts* **50.** US Govt. Print. Off., Washington. 868 pp.

MASURE, E. 1984. L'indice de diversité et les dominances des 'communautés' de kystes de Dinoflagellés: marqueurs bathymétriques; forage 398 D, croisière 47B. *Bull. Soc. Geol. Fr.* (7), **26,** (1), 93–111.

MATSUMOTO, T. 1977. On the so-called Cretaceous Transgressions. *Palaeont. Soc. Japan. Spec. Pap.* **21,** 75–84.

MELIÈRES, F., DEROO, G. & HERBIN, J.P. 1981. Organic matter rich and hypersiliceous Aptian sediments from Western Mid-Pacific Mountains, Deep Sea Drilling Project, Leg 62. *In*: THIEDE J., VALLIER, T.L. *et al.* (eds), *Init. Repts DSDP* **62,** US Govt. Print. Off., Washington, DC. 903–921.

MOBERLY, R. & SCHLANGER, S.O. 1985. Site chapter 585. *In*: MOBERLY R., SCHLANGER, S.O. *et al* (eds), *Init. Repts DSDP* **89,** US Govt. Print. Off., Washington, DC. in press.

MOUGENOT, D. 1976. *Géologie du Plateau Continental Portugais (entre le Cap Carvoeio et le Cap de Sines).* Thèse 3ème cycle Rennes, 2 volumes, 81 pp + 16 pp.

MULLER, C., SCHAAF, A. & SIGAL, J. 1983. Biochronostratigraphie des formations d'âge Crétacé dans les forages du DSDP dans l'Océan Atlantique Nord. *In*: *Rev. Inst. Fr. Pétrol.* **38** (6), 683–708 and **39,** (1), 3–23.

MUSELEC, P. 1974. *Géologie du plateau Continental Portugais au Nord du Cap Carvoeio.* Thèse 3ème cycle Rennes, Geol. Sous-marine VI, 150 pp.

OLIVET, J.L., BONNIN, J., BEUZART, P. & AUZENDE, J.M. 1982. *Cinématique de l'Atlantique Nord et Centrale, CNEXO, Cartes éditées par le Beicip.*

PELET, R. & DEBYSER, Y. 1977. Organic geochemistry of Black Sea Cores. *Geochim Coschim. Acta* **41,** 1575–86.

PITMANN, W.C. 1978. Relationship between eustacy and stratigraphic sequences of passive margins. *Geol. Soc. Am. Bull.* **89,** 1389–1403.

PRATT, L.M. 1981. *A Paleo-Oceanographic Interpretation of the Sedimentary Structures, Clay Minerals, and Organic Matter in a Core of the Middle Cretaceous Greenhorn Formation drilled near Pueblo, Colorado.* PhD Thesis, Princeton University. 186 pp.

—— 1984. Influence of paleoenvironmental factors on preservation of organic matter in the Mid-Cretaceous Greenhorn Formation, Pueblo, Colorado. *Am. & Ass. Petrol. Geol. Bull.* **68,** (9), 1146–59.

—— & THRELKELD, C.N. 1984. Stratigraphic significance of 13C/12C ratios in mid-Cretaceous rocks of the Western Interior, USA. *In*: The Mesozoic of Middle North America. *Can. Soc. Petrol. Geol.*, in press.

RAOULT, J.F. 1972. Précisions sur le flysch Massylien: série stratigraphique, variations de faciès, nature du matériel remanié (Nord du Constantinois, Algérie). *Bull. Soc. Hist. Nat. Afr. Nord* **63,** 73–92.

REHAULT, J.P. & MAUFFRET, A. 1979. Relationships between tectonics and sedimentation around the Northwestern Iberian Margin. *In*: SIBUET, J.C. & RYAN, W.B.F., *Init. Repts DSDP* **47,** Part 2. US Govt. Print. Off., Washington, DC. 663–82.

RHOADS, D.C. & MORSE, J.W. 1971. Evolutionary and ecologic significance of oxygen deficient marine basins. *Lethaia* **4,** 413–28.

ROBERTSON, A.M.F. & BERNOULLI, D. 1982. Stratigraphy, facies and significance of Late Mesozoic and Early Tertiary sedimentary rocks of Fuerteventura (Canary Islands) and Maio (Cape Verde Islands). *In*: VON RAD U., HINZ K., SARNTHEIN, M. & SEIBOLD E. (eds), *Geology of the Northwest African Continental margin.* Springer-Verlag, Berlin. 498–525.

RONA, P.E. 1973. Relations between rates of sediment accumulation on continental shelves, sea floor spreading, and eustacy inferred from the central North Atlantic. *Geol. Soc. Am. Bull.* **84,** 2851–82.

RYAN, W.B.F. & CITA, M.B. 1977. Ignorance concerning episodes of ocean wide stagnation. *Mar. Geol.* **23,** 197–215.

SALAJ, J. 1978. The geology of the Pelagian Block: The eastern Tunisian Platform. *In*: NAIRN, A.E.M., KANES, W.H. & STEHLI, F.G. (eds), *The Oceans Basins and Margins. Volume 4B, The Western Mediterranean.* Plenum Press, New York. 361–416.

SAVIN, 1984. The history of the earth's surface temperature. *In*: DURAND (ed), *Thermal Phenomena in Sedimentary Basins.* Ed. Technip, Paris. 11–20.

SIGAL, J. 1979. Chronostratigraphy and ecostratigraphy of Cretaceous Formations recovered on DSDP Leg 47B, Site 398. *In*: SIBUET, J.C. & RYAN, W.B.F., *Init. Repts DSDP* **47.** US Govt. Print. Off., Washington, DC. 287–326.

SCHLANGER, S.O. & JENKYNS, H.C. 1976. Cretaceous oceanic anoxic events, causes and consequences. *Geologie en Mijnbouw* **55,** (3–4), 179–84.

——, ARTHUR, M.A., JENKYNS, H.C. & SCHOLLE, P.A. 1986. The Cenomanian Turonian Oceanic Anoxic Events. I. Stratigraphy and Distribution of Organic carbon Rich Beds and the Marine δ 13C Excursion. *In*: BROOKS, J. & FLEET, A. (eds), *Marine Petroleum Source Rocks. Spec. Publ. Geol. Soc. Lond.* Published by Blackwell Scientific Publications, Oxford.

STAHLECKER, R. 1934. Neocom aur der Kapverden-Insel Maio. *Neues Jahrb. Mineral. Geol. Palaeont., Beil. Bd. Abt.* **B73,** 265–301.

SUMMERHAYES, C.P., MILLIMAN, J.D., BRIGGS, S.R., BEE, A.G. & HOGAN, C. 1976. Northwest African Shelf Sediments: influence of climate and sedimentary Processes. *J. Geol.* **84,** 277–300.

—— 1981. Organic facies of Middle Cretaceous black shales in deep North Atlantic. *Bull. Am. Ass. Petrol. Geol.* **65,** 2364–80.

—— 1986. Organic rich Cretaceous Sediments from the North Atlantic. *In*: BROOKS, J. & FLEET, A.J. (eds), *Marine Petroleum Source Rocks. Spec. Publ. Geol. Soc. Lond.* Published by Blackwell Scientific Publications, Oxford.

SUTER, G. 1980. Carte structurale de la chaîne Rifaine, 1: 500 000—Notes. *Mém. Géol. Maroc* **245b.**

TAUGOURDEAU-LANTZ, J., AZEMA, C., HASENBOEHLER, B., MASURE, E. & MORON, J.M. 1982. Evolution des domaines continentaux et marins de la marge Portugaise (Leg 47B, Site 398D) au cours du Crétacé: essai d'interprétation par l'analyse palynologique comparée. *Bull. Soc. Geol. Fr.* (7), **24,** (3), 447–59.

TEMPLETON, R.S.M. 1971. Geology of the Continental Margin between Dakar and Cape Palmas. *In*: DELAMY, F.M. (ed.), *The Geology of the East Atlantic Continental margin. (ICSU/SCOR Working Party 31 Symp. Cambridge 1970), 4, Africa Nat. Environ. Res. Counc. Inst. Geol. Sci. Rep.* **70/16,** 47–60.

THUROW, J. KUHNT, W. & WIEDMANN, J. 1982. Zeitlicher und paläogeographischer Rahmen der Phthanit und Black Shale-Sedimentation in Marokko. *Neuers Jb. Geol. Paläont. Abh.* **165,** 1, 147–176.

TISSOT, B., DEROO, G. & HERBIN, J.P. 1979. Organic matter in Cretaceous sediments of the North Atlantic: contribution to sedimentology and paleogeography. *In*: TALWANI, M., HAY, W., RYAN, W.B.F. (eds), *Deep Drilling Results in the Atlantic Ocean: Continental Margins and Paleoenvironments.* Am. Geophys. Union, Maurice Ewing Series **3.** 362–74.

—— & PELET, R. 1981. Source and fate of organic matter in ocean sediments. *Oceanol Acta, Actes 26 Int. Congr. Geol., Paris*, 97–103.

VAIL, P.R., MITCHUM, R.H. & THOMPSON, S. 1977. Global cycles of relative changes of sea level, *Am. Ass. Petrol. Geol. Mem.* **26,** 83–103.

WAPLES, D.W. 1983. A reappraisal of anoxic and organic richness with emphasis on the Cretaceous North Atlantic. *Am. Ass. Petrol. Geol. Bull.* **67,** 963–78.

—— & CUNNINGHAM, R. 1985. Leg 80 Shipboard organic geochemistry. Deep Sea Drilling Project. *Init. Repts DSDP* **80.** US Govt. Print. Off., Washington, DC.

WIEDMANN, J. 1976. Geo-und hydrodynamische Prozesse im Schelfbereich in ihrer Auswirkung auf mesozoische Fossil-Vergesellschaftungen. *Zbl. Geol. Paläont. Teil II* **1976,** 424–39.

——, BUTT, A. & EINSELE, G. 1978a. Vergleich von Marokkanischen Kreide-Küstenaufschlussen und Tiefseebohrungen (DSDP), Stratigraphie, Paläoenvironment und Subsidenz an einem passiven Kontinentalrand. *Geol. Rdsch.* **67,** 454–508.

——, EINSELE, G. & IMMEL, H. 1978b. Evidence faunistique et sedimentologique pour un upwelling dans le Bassin Côtier de Tarfaya/Maroc dans le Crétacé supérieur. *Actes 6ème Coll. Afr. Micropaléontol., Tunis 1974, Ann. Mines Géol.* **28/II,** 415–41.

——, BUTT, A. & EINSELE, G. 1982. Cretaceous stratigraphy environment, and subsidence history at the Moroccan continental margin. *In*: VON RAD, U., HINZ, K., SARNTHEIN, M. & SEIBOLD, E. (eds), *Geology of the North-West African Continental Margin.* Springer-Verlag, Berlin, 366–95.

WILDI, W. 1983. La chaîne tello-rifaine (Algérie, Maroc, Tunisie): structure, stratigraphie et évolution du Trias au Miocène. *Rev. Géol. Dyn. Géogr. Phys.* **24** (3), 201–97.

J.P. HERBIN, L. MONTADERT & C. MULLER, Institut Français du Pétrole; Boite Postale 311, 95206 Rueil Malmaison CEDEX, France.

R. GOMEZ, Department de Géologie, Faculté des Sciences, Université de Dakar, Dakar-Fann, Senegal.

J. THUROW & J. WIEDMANN, Universität Tübingen, Institut und Museum für Geologie und Paläontologie, 7400 Tübingen 1, Sigwartstrasse 10, Federal Republic of Germany.

Mid-Cretaceous of the Gibraltar Arch Area

J. Thurow & W. Kuhnt

SUMMARY: Mid-Cretaceous depositional environments of the Gibraltar Arch and adjacent areas exhibit at the Cenomanian–Turonian boundary a special type of organic carbon-rich siliceous sedimentation which is intercalated in all the different environments—from the shelf down to the deep—of this area. This will be called the Cenomanian–Turonian Boundary Event (CTBE). Summarizing the most important features of the CTBE it can be pointed out: (1) the initiation of the CTBE is nearly coeval; (2) very high TOCs, especially at the base of Turonian; (3) kerogen is exclusively of type (I-) II in the undiluted strata; (4) independent of given palaeogeographic setting (e.g. E. and W. margin of the Atlantic ocean); (5) bituminous sedimentation is also very pronounced in environments with high detrital input and strong dilution (e.g. deep sea fans). Here too, most strata comprise kerogen of type II; (6) as far as we know, there is a striking co-occurrence of these sediments with strong enrichment in biogenic silica from diagenetically altered radiolaria, but there is no link between good preservation of radiolaria and bituminous sediments (e.g. Scaglia-facies); (7) rich radiolarian faunas are common in the carbon-rich and carbon-free sediments of shallow marine and deep sea environments; (8) the important change in the evolution of planktonic organisms (foraminifera/radiolaria) which marks the Cenomanian–Turonian boundary coincides with the peak of anoxic conditions and biogenic silica-rich sedimentation.

All these features of the CTBE observed in the Gibraltar Arch area are comparable with coeval DSDP-facies in the North Atlantic and allow on-shore–off-shore tracing of the CTBE.

In this paper the results of palaeoenvironmental studies of the mid-Cretaceous sediments of the Gibraltar Arch at the junction between the North Atlantic and the Tethys are summarized (Fig. 1 (a) shows location). The authors focus their attention on sedimentation at the Cenomanian–Turonian boundary, which can be studied in distinct and different settings, depths and environments that occur close together there (Fig. 1 (b)). Among the different settings and their deposits are:

(1) Slope associations (North African Continental Margin)
(2) Pelagic limestones (Iberian Continental Margin)
(3) Turbiditic associations (Alboran Margin and oceanic basin)

Usually the facies of these different environments are quite dissimilar, but this is not so at the Cenomanian–Turonian boundary, where the various environments show evidence of high productivity of biogenic siliceous material and organic rich sedimentation. These organic-rich sediments at the Cenomanian–Turonian boundary are coeval with similar facies from the North Atlantic, western Morocco, and the Umbrian Appennines. This phenomenon is referred to as the Cenomanian–Turonian boundary event (CTBE), as also in Herbin *et al.* (this volume).

Certain features are peculiar to the CTBE sediments in the Gibraltar Arch area:

(1) re-sedimentation of organic matter, together with biogenic debris derived from oxic environments;
(2) alternation of fine-grained, flaser laminated and kerogen-rich turbidites (silt-mud turbidites), radiolarites and thin layers of true black shales. Erosional contacts between single layers are detectable;
(3) beds revealing short-term oxic conditions (pale beds with bioturbation).

The authors' detailed studies of these deposits in the Gibraltar Arch area allow them to discriminate between competing models that attempt to explain this palaeoceanographic event.

Geology and palaeogeography

The Gibraltar Arch is the westerly termination of the Alpine Maghrebinian foldbelt (Fig. 1 (a)). Its tectonic structure is complex (Fig. 1 (b)), the major features are nappes comprising sediments deposited in three different palaeogeographic domains—the External Continental Margins of Africa and Iberia, the basinal 'flysch'-series and, in uppermost tectonic position, the remains of the former Alboran 'microplate'. The Internal Zones (remains of the Alboran Block) are continuous throughout the Rif and the Betic Chains. The 'flysch nappes' are generally thrust southward or westward onto the External Zones representing the former African Margin (Fig. 1 (b)).

The palaeogeography of this area during the Cretaceous is still unclear, although the positions of the Iberian- and African plates, as well as their motions with respect to the Mid-Atlantic ridge, can be reconstructed. The position of the two

From SUMMERHAYES, C.P. & SHACKLETON, N.J. (eds), 1986, *North Atlantic Palaeoceanography*, Geological Society Special Publication No. 22, pp. 423–445.

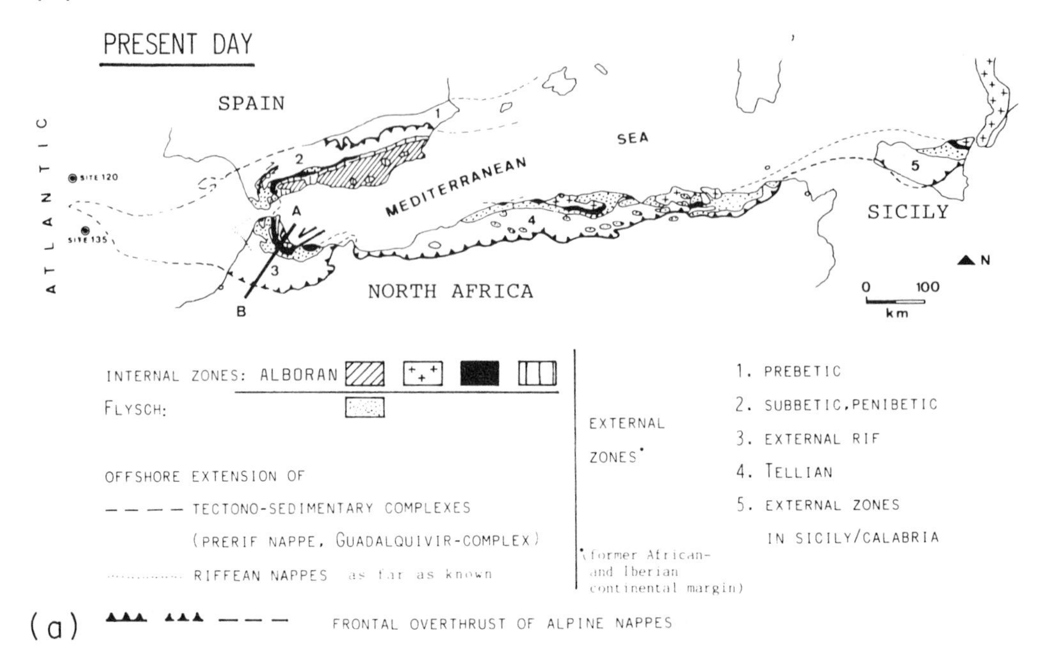

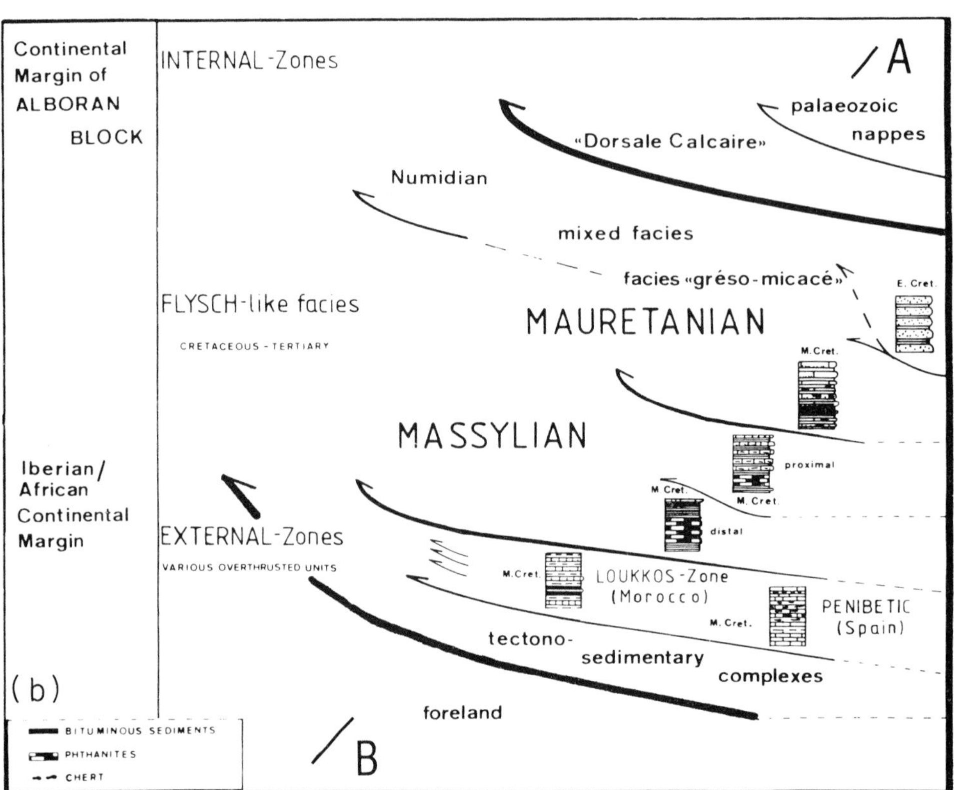

FIG. 1. (a) Present day geological setting of the Betic-Rif-Tell orogenic belt. (b) Idealized section through the pile of nappes of the Rif Mountains (A–B in (a)). Tectonic units with outcropping Upper Cenomanian–Middle Turonian deposits are indicated by small lithological columns. Internal nappes are in the uppermost tectonic position, overthrusting the 'flysch-nappes', which themselves overthrust the external nappes.

plates relative to one another suggests that there was a broad mid-Cretaceous 'Palaeostrait of Gibraltar', which allowed an exchange of Atlantic and Tethyan waters (Fig. 2). The third plate in the area was the Alboran Block (? microplate), which was moving westward along strike-slip faults, folding the basinal 'flysch'-series, and colliding gradually with the external realms to form a mountain range now appearing as the Gibraltar Arch (Fig. 2).

Palaeoenvironment

Mid-Cretaceous organic-rich sediments were deposited at the CTBE in nearly all the sedimentary environments of the Gibraltar Arch area, and may be reworked into younger deposits, especially in slope areas. Four main features characterize the CTBE:

(1) Deposition of organic carbon-rich sediments, partly true black shales, sometimes with extremely high values of TOC (kerogen type II), indicating anoxic conditions in the depositional environment. These sediments lack bioturbation and microbenthos, attesting to anoxic conditions.
(2) Organic-rich sedimentation is not confined to a particular palaeogeographic setting.
(3) It is associated with a major change in the planktonic foraminiferal assemblages.
(4) It shows a sharp rise in the radiolaria/planktonic forminifera ratio in open marine environments. This has led to limestone chertification and the formation of siliceous sediments like radiolarites and phthanites.

Sedimentary environments and facies

Shelf and slope of the External Rif (former North African Continental Margin)

Cretaceous facies ranging from inner shelf in the south (Rides Prérifaines) to lower slope in the north are exposed within the different structural units of the External Rif (Fig. 1).

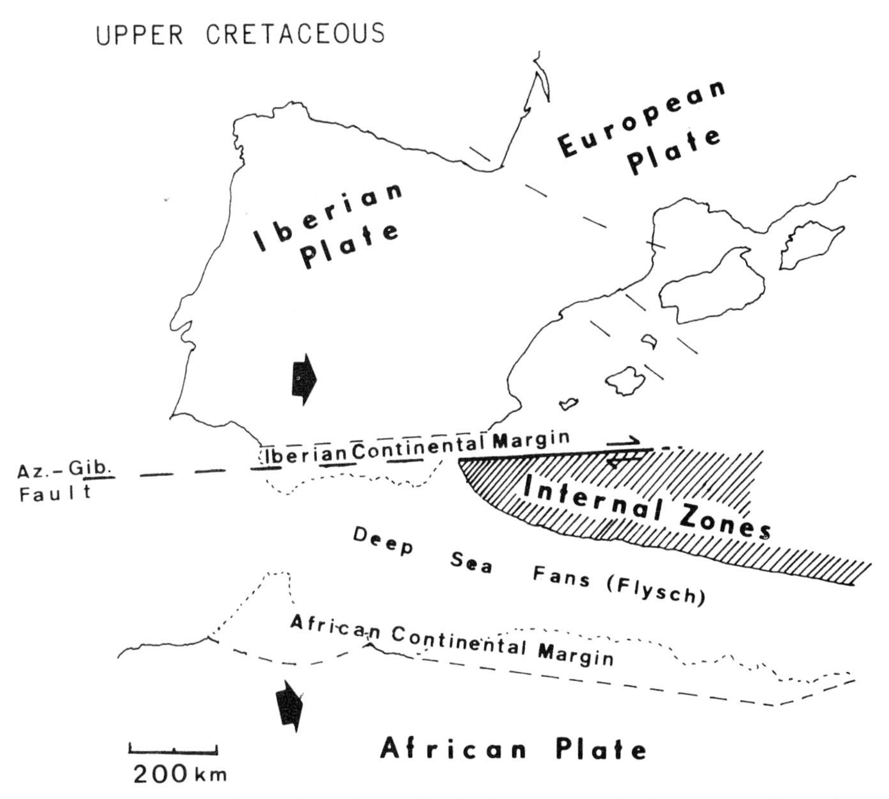

FIG. 2. The mid-Cretaceous position of the Alboran Block with respect to the Iberian and African plates (modified from Leblanc & Olivier 1984).

Despite different palaeobathymetric settings, there are certain obvious patterns common to Cretaceous sedimentation here, for instance:

(1) There were pelagic limestones, with gradually increasing siliciclastic inputs in slope areas, during the Berriasian–Barremian.

(2) The Aptian–Albian had pelitic and siliciclastic sediments, with reduced sedimentation rates and hiatuses on the shelf, and small deep sea fans in the deeper parts of the basin.

(3) The Upper Albian is represented by transgressive sediments, with bivalve and oyster facies in the shallow shelf areas of the Rides Prérifaines, and with pelagic limestones and marls on the outer shelf and slope. The limestones and marls continue into the Cenomanian. Intercalated dark, laminated, carbonaceous marls, with up to 1.3% TOC and with kerogen of terrestrial or mixed marine and terrestrial origin occur too, but are restricted to the Uppermost Albian.

(4) The Turonian of the Rides Prérifaines is represented by dolomitic limestones with pelagic macrofaunas (*Inoceramus*, ammonites). It is quite comparable with the Turonian of the Moroccan Meseta (Wiedmann *et al.* 1978). Outer shelf and slope environments of the External Rif generally show a hiatus in the Turonian. There are, as far as is known, two sections of Turonian sediments within the Loukkos-Zone (sections F and G of Fig. 3). Limestones, marls and cherts, showing intense slump features and bearing abundant and well preserved planktonic microfaunas occur in the southern section (G). Silicified calciturbidites, some black shales, and hemipelagic sediments occur in the more northerly section (F). Calcareous microfossils are partly dissolved in section F, thus the depositional environment was probably on the lower slope close to the CCD (see Fig. 4).

Although the basal Turonian sediments of section F show features of the CTBE (laminated, black, silicified layers bearing radiolaria and pyrite) the TOC is very low (0.7%). In the Middle to Upper Turonian (helvetica-schneegansi zone) dark turbiditic strata occur with increased TOC (up to 3.6%) of terrestrial or mixed marine/terrestrial origin.

(5) Pelitic sedimentation characterized all deposits of the External Rif during the Coniacian to Lower Campanian. Foraminiferal assemblages from the deeper part of the slope are typical of sub-CCD environment.

(6) Beginning in the Upper Campanian, and culminating in the Lower Maastrichtian, slumping and sliding (olistostromes, pebbly mudstones, debris flows, and calciturbidites) were frequent. Pelagic cherty limestones with Turonian microfossils form boulders in the resedimented materials, indicating reworking of Turonian sediments in the late Cretaceous, and partly explaining the stratigraphic hiatuses of the outer shelf and slope environments of the External Rif.

Continental Margin Basin (Iberian Continental Margin)

Within the external part of the Western Betic Cordillera, two palaeogeographic domains can be distinguished. Both are related to the Cretaceous Iberian Continental Margin: the Subbetic Zone in the north, and the Penibetic Zone in the south (Fig. 1 (a)).

The Cretaceous of the Subbetic Zone is characterized by marly sediments deposited in slope environments. Ammonite-bearing, rhythmic, marl-limestone couplets of Neocomian age; thin greenish-dark shales of Aptian–Albian age; and pelagic marls and limestones of the Upper Albian to Lower Cenomanian, characterize the Lower Cretaceous. Calciturbidites, laminites, and ripple-laminated beds in the Middle to Upper Cenomanian indicate increased current activity and/or slope instability. Generally, in the section studied, Lower Turonian sediments are missing, most probably due to erosion. The Coniacian to Lower Campanian is marked by red marls, with planktonic forminiferal assemblages. Campanian and Maastrichtian sediments are similar, but with increasing calciturbiditic intercalations.

The Upper Cretaceous marls and limestones of the Penibetic are quite comparable to coeval Subbetic facies. Additional Zoophycus ichnofacies, and comparable foraminiferal assemblages indicate an analogous palaeobathymetric position for the two zones. However, there are no calciturbidites in the Penibetic Zone. The Lower Cretaceous sediments have sedimentary features typical of deposition on a pelagic swell (Gonzalez-Donoso *et al.* 1983), being a condensed sequence with several hiatuses. During the mid-Cretaceous, evidence for increasing subsidence is given by a complete sedimentary record beginning in the Upper Albian, with low, but increasing, sedimentation rates.

The most obvious influence of the CTBE on Lower Turonian sediments in the Penibetic zone is the occurrence of nodular chert, which sometimes forms irregular cherty layers within a white micritic limestone background (the Scaglia facies). This chertification, which usually begins at the base of the *archaeocretacea*-zone and reaches up to the lower part of the *helvetica*-zone, correlates with an enormous increase in the radiolarian/planktonic forminiferal ratio, and is presumably caused by early diagenetic migration of amorphous silica.

Fine lamination, or raised organic carbon content, have not been observed within this interval (0.02% TOC). Some darker, laminated layers, occur in the Upper Cenomanian, where there is a slight increase in organic carbon (0.6% TOC) of terrestrial or mixed marine/terrestrial origin.

The lack of detrital input during the Upper Cretaceous indicates the persistence of the Penibetic Zone as a deep swell with pelagic facies in the southern part of the Iberian Continental Margin (section A in Figs 3 and 4); this interpretation is supported by a very high planktonic/benthic ratio and the dominance of large sized, keeled foraminifera, which indicate deep-water environments above the carbonate lysocline.

Deep sea fans and basin plain associations (Mauretanian- and Massylian 'Flysch' Basin)

The Cretaceous 'flysch'-deposits can be divided into the Mauretanian- and Massylian series, within their corresponding Mauretanian- and Massylian nappes (Bouillin *et al.* 1970; Raoult 1972; see Fig 1 (b)). The mainly flyschoid sediments can be ascribed to parts of deep sea fan to basin plain associations. They comprise sediments deposited between the external realms and the former Alboran Block (Figs 2 and 4), on the margins of a rather narrow elongated basin with detrital input from various borderlands. Deep sea fans along the Alboran Block are well documented, and can be detected in the Mauretanian nappes (Fig. 1 (b)—Gübeli 1983; Thurow 1984). The Massylian nappes and their respective series (Fig. 1b), represent more distal regions with reduced sediment thickness.

The source areas of the detrital components, and the structural relationships, confirm the existence of a single flysch basin, which, at the most, may have had a basin and swell morphology in the mid-Cretaceous, as suggested in Fig. 3. Inner fan and slope deposits are unknown in the Gibraltar Arch, but do occur in the Mauretanian series in Eastern Algeria.

In the Rif, the Mauretanian is divided into the Tisirene Nappe (?Dogger–Lower Cretaceous) and the Beni Ider Nappe (Upper Cretaceous–Tertiary) now found in the tectonically lower position. In the Campo de Gibraltar area (Spain), Cretaceous sediments of the Mauretanian series are not clearly proved. Sedimentation within the Tisirene Nappe is almost entirely turbiditic. Early Cretaceous pre-flysch with calciturbidites alternates with marls and shales. Late–early Cretaceous sediments consist of siliciclastic-terrigenous turbidites and shales. At the base of the Upper Albian there is a marked compositional change in the detritus. The sedimentation of siliciclastics decreases rather abruptly, whilst carbonate-detrital influx increases rapidly. Two early Cretaceous sediments consist of siliciclastic-terrigenous turbidites and shales. At the base of the Upper Albian there is a marked compositional change in the detritus. The sedimentation of siliciclastics decreases rather abruptly, whilst carbonate-detrital influx increased rapidly. Two source areas can be distinguished, one provides shallow water debris (Urgonian facies), the other shows reworked Mesozoic sediments of the Austro-Alpine facies type (see Bernoulli & Jenkyns 1974), which can be clearly related to the Alboran Block. These differences in composition, together with variations in total thickness, and analyses of palaeocurrents, allow us to differentiate between a more proximal mid-Cretaceous depositional environment (section B, Fig. 3 and 4), now located in the area of Tangier, and a more distal environment (section C, Fig. 3 and 4), now located farther to the south-east.

As stated above, the mid-Cretaceous comprises carbonate-detrital turbidites alternating with red, olive-green or multicoloured shales. Upper Cenomanian, and, to a lesser extent, Turonian, are marked by the occurrence of white chert layers/nodules, and finely dispersed pyrite. Radiolarians are common in the Turonian, and thin sections show a certain amount of reworked phosphatic material within the Cenomanian–Turonian. At the base of the Turonian coloured shales disappear. They are replaced by black laminites and black shales; additionally, the upper (fine-grained) parts of turbidites are darkened, and layers or organic-rich stringers are intercalated in the coarse-grained parts of turbidites.

The Upper Cretaceous again comprises carbonate-detrital turbidites alternating with red or multicoloured clays, but grain sizes increase, and in the Campanian–Maastrichtian coarse-grained debris flows and channel-fill deposits (boulders up to 1 m) were probably initiated by tectonic movements. Faunal patterns, ichnofacies, and mineralogy or autochthonous layers indicate a deep depositional environment below the CCD.

Massylian series are represented by the Melloussa-type Nappes in the west and the Choua-

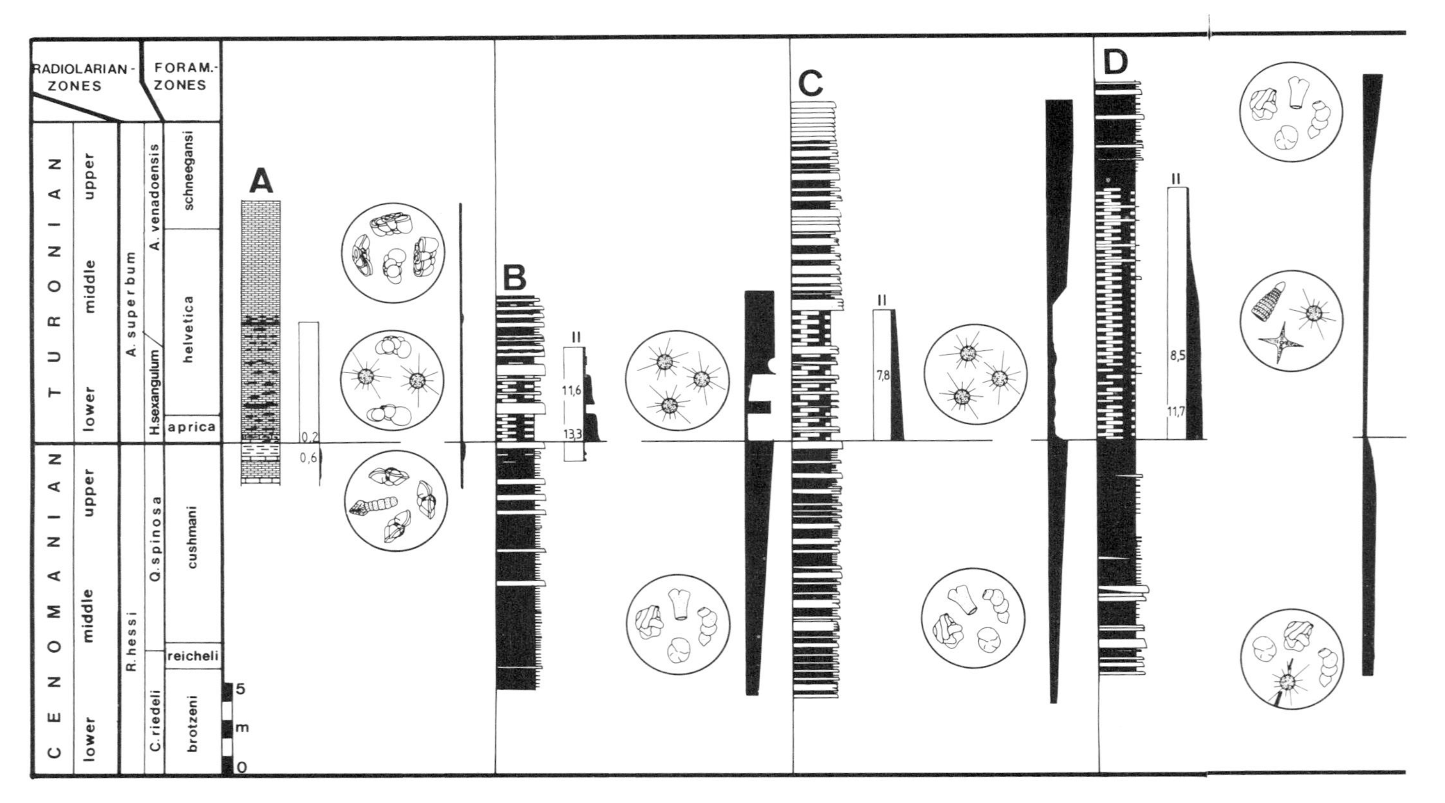
RADIOLARIAN-ZONES
FORAM.-ZONES
TURONIAN
CENOMANIAN
lower
middle
upper
A. superbum
R. hessi
H. sexangulum
A. venadoensis
Q. spinosa
C. riedeli
aprica
helvetica
schneegansi
brotzeni
reicheli
cushmani
5
m
0
A
B
C
D
0,2
0,6
11,6
13,3
7,8
8,5
11,7

Fig. 3.

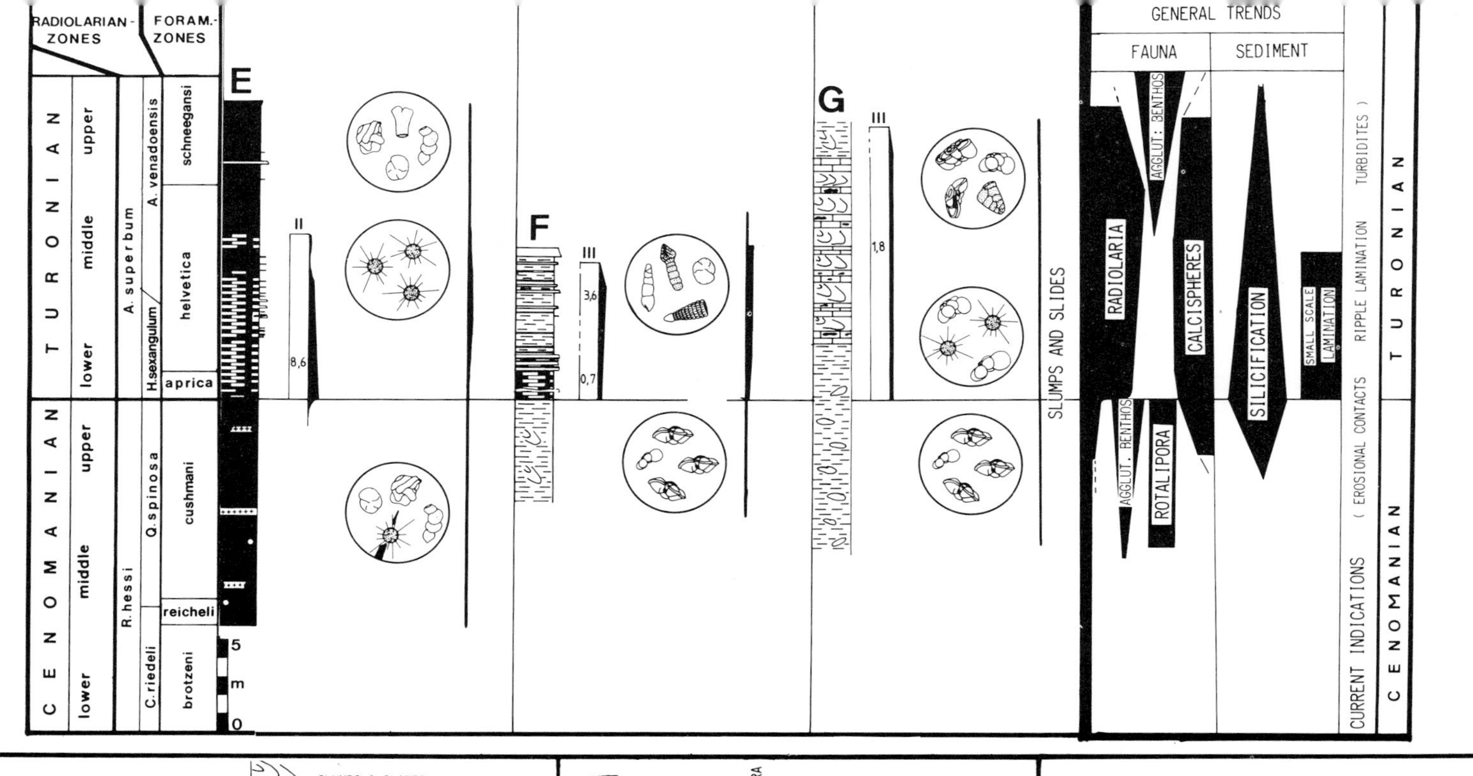

FIG. 3. Stratigraphic correlation and CTBE effects in selected sections in the Gibraltar Arch area. A is Penibetic, B and C Mauretanian, D and E Massylian, F and G Loukkos Zone. The important *autochthonous* microfaunal assemblages are plotted alongside the lithological columns (uppermost fauna in section G is mixed). At least three and up to several tens of washing residues are represented by each circle. Additionally, thin sections were used in columns A and G. The overall biostratigraphy is also based on redistributed planktonic foraminifera (which are not figured).

mat Nappe in the east, which exhibit similar facies evolution. Corresponding formations (Unités de Facinas/Almarchal) crop out in the Campo de Gibraltar. Parts of the External Zones—in the Tanger Intern Unit—also belong to the Massylian series. Most of them range from Aptian–Albian to Maastrichtian. They can be arranged in three facies zones—two of them comprising mid-Cretaceous deposits. The corresponding nappes can directly be related to their former series. The deeper one reveals deposits of very distal character (section E/Figs 3 and 4), the higher one is more proximal and shows sedimentary relationships to Mauretanian sedimentation (section D/Figs 3 and 4). Sedimentation within both facies is turbiditic, with major shale sections, starting in the Albian with siliciclastic turbidites alternating with grey-green to dark shales. The change from siliciclastic to carbonate-detrital sedimentation (seen in the Mauretanian) also occurs here in the uppermost Albian, but is not as abrupt and rapid as in the Mauretanian series. In the more proximal environment (section D) siliciclastic and carbonate-detrital layers alternate, whilst in the distal environment (section E) carbonate detritus is concentrated in scour fills and small channels. Sedimentary components are of shallow marine origin ('Urgonian' facies). Shaly sections are characterized by various diagenetic concretions. The Upper Cenomanian is thin, poor in detritus and rarely outcropping.

In the uppermost Cenomanian sedimentation changed: about 10 m of siliceous, dark to black carbonaceous shales were deposited (phthanites). Post-Turonian sedimentation includes multicoloured clays in the distal environment, and fine-grained calciturbidites alternating with multicoloured shales or grey shales in the more proximal environment. During the late Cretaceous coarse-grained resedimentation took place as in the Mauretanian series.

The depositional environment of the Massylian facies zones was clearly below the CCD, and analogous to the Mauretanian, which can be ascribed to an outer fan to basin plain association. However, deposition within deep sea fans is not proved.

In the 'flysch' basin, O_2-depleted to anoxic conditions occurred during the CTBE. Yet, there are some indications that comparable but less pronouned conditions had been established earlier (Fig. 5).

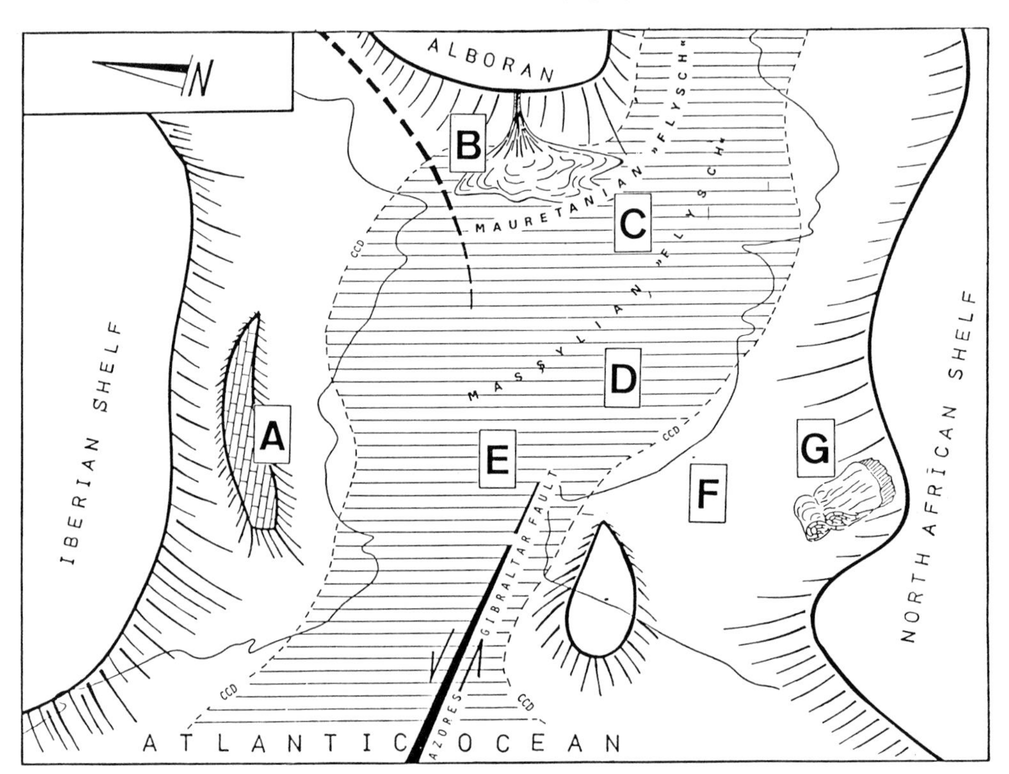

FIG. 4. Palinspastic reconstruction of the Gibraltar Arch area during mid-Cretaceous. A view is presented from the Atlantic passing the 'Palaeostrait of Gibraltar' into the Western Tethys up to the internal realm (Alboran Block). Letters A–G correspond with Fig. 3.

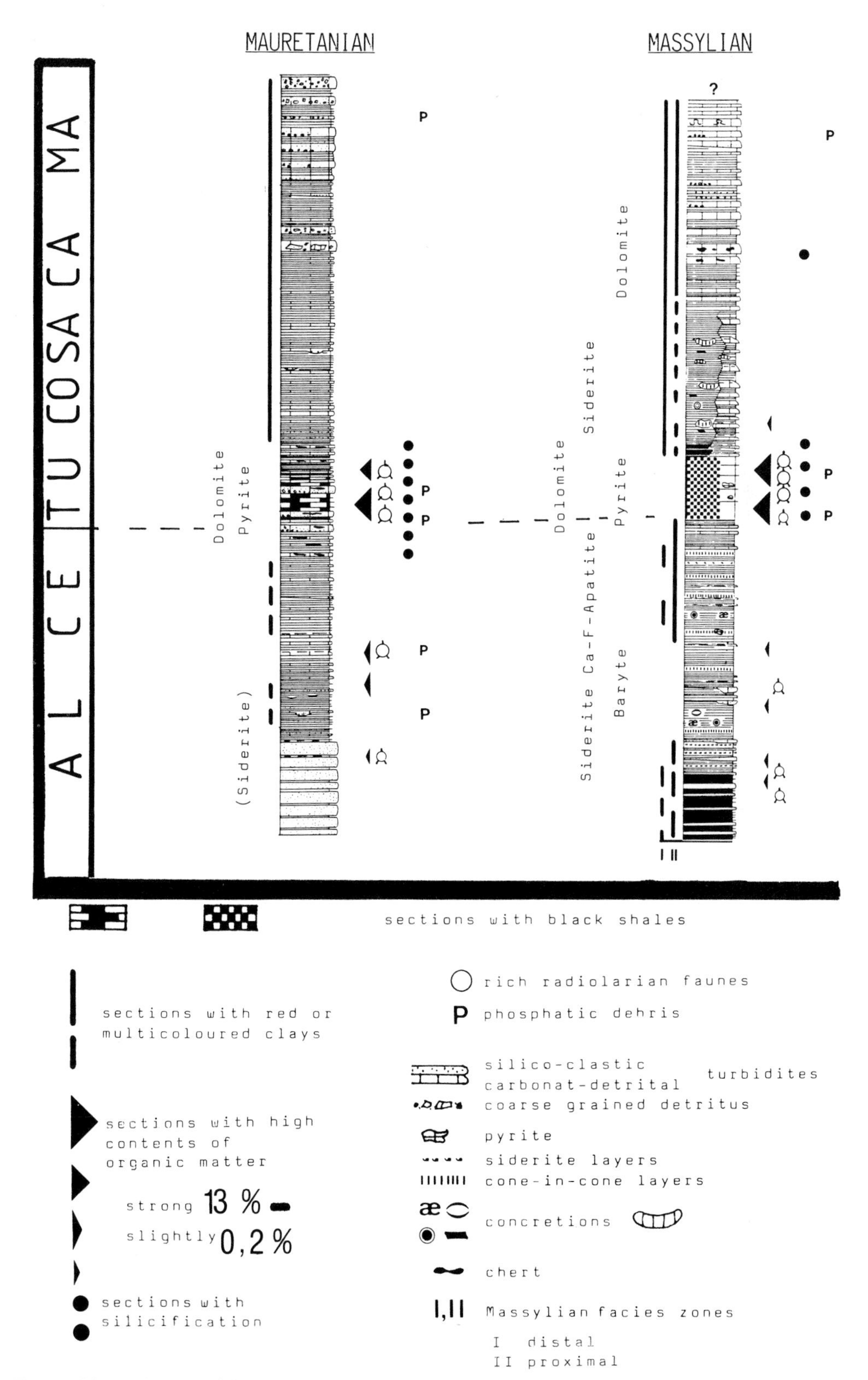

FIG. 5. Lithostratigraphy of the Mauretanian and Massylian series, summarized in two synthetic sections showing all the characteristic features related to the deposition of bituminous sediments in the mid-Cretaceous. There is a striking co-occurrence of anomalous high TOC, rich radiolarian faunas, silicification, and phosphatic debris, as well as the absence of multicoloured shales.

Mauretanian

Analyses are rare due to high dilution with terrigenous material during the Lower Cretaceous. Some of the Upper Albian shaly layers between turbidites contain up to 0.35% TOC; algae (*Leiosphaera*—common in black shales) are common, as well as large amounts of woody debris. Furthermore some radiolarians (preserved in pyrite), and mixed floras of Neocomian–Upper Albian age may be found. Some of the terrigenous turbidites are charged with plant fragments that form lignite layers in places. In the transitional, shaly, detritus-poor deposits of Vraconian to Cenomanian age, organic matter reaches 0.9% TOC. Pyrite-preserved radiolarians are common. In more distal environments some layers of dark limestones are enriched in organic carbon (up to 4% TOC). These strata are redeposited sediments (distal turbidites Tde), with mixed floras of Neocomian–Upper Albian age. The high TOC in this re-sedimented material indicates that organic matter was redistributed from shallower areas.

Massylian

Albian shales, some with minute layers of tiny plant fragments, contain in places up to 0.5% TOC. Floras do not indicate reworking of older material. Pyritic radiolarians are common, as well as pyrite crystals, and many diagenetic concretions—baryte-sphaerolites, siderites, cone in cone of various mineralogies—and replacement of primary minerals/sediments by siderite and pyrite. Such minerals may indicate reducing environments in the Upper Albian–Lower Cenomanian.

CTBE in Mauretanian and Massylian

The Uppermost Cenomanian to Middle Turonian coincides with the most striking increase in organic enrichment. The kerogen is type II, but sometimes degraded by slight catagenesis (see Herbin *et al.*, this volume).

Within the Mauretanian the CTBE is characterized by turbidite/laminite alternations. Coarse-grained beds (> 1 cm) are organic carbon-free. Fine-grained ones are somewhat enriched in organic carbon, especially in the upper BOUMA-sequences (Tde) (up to 2.7% TOC). The laminites consist of alternating silty and kerogen-rich layers on a millimetre scale, with some coarser intercalations (? thin turbidites). Pyrite nodules occur, and the TOC reaches 13%.

The most substantial accumulations of organic carbon in the CTBE occur in the dark siliceous carbonaceous 'phthanites' of the Massylian series. These 'phthanites' comprise thin, very fine-grained turbidites (up to 0.9% TOC), laminites (up to 3% TOC), true black shales (12% TOC), and radiolarites (up to 2% TOC). Pyrite nodules with diameters up to 10 cm are common.

Sediments of the CTBE

Sedimentation during the CTBE is generally distinct from strata deposited before and after the event. Besides the remarkably regular bedding visible in outcrops (Fig. 6 (a)), there is alternation between background sediments and organic carbon-rich sediments. Background sedimentation (mostly turbidites and grey-green shales) is obviously controlled by the depositional environment, whereas the factors affecting organic carbon-rich sedimentation are not always clear. The non-bituminous turbidites comprise components of the Austro-Alpine facies type, together with abundant peloids, benthic and planktonic foraminifera, radiolaria and phosphatic debris. There are no indications for anoxic conditions in the source area. The grey-green autochthonous shales (illite, chlorite), which are always strongly silicified, may contain scattered radiolaria.

Although intense diagenesis (and metamorphism) has obscured the sedimentary record of depositional conditions, different types of organic carbon-rich sediments can be distinguished:

(a) Black shales
(b) Laminites and turbidites
(c) Radiolarian-rich laminites, radiolarites
(d) Turbidites with mixed shallow water detritus and marine organic matter.

True black shales are common in the Massylian and occur also in the Mauretanian series. They are very thin bedded and often interrupted by one

FIG. 6. Selected sedimentary features. (a) Typical Massylian 'Phthanite' outcrop. Type Massylian—Oued bou Chouk, Petite Kabylie, Algeria. (b) Cyclicity in phthanites. Pale, bioturbated layers alternate with dark organic-rich laminated layers. Type Massylian, see (a). (c) Erosional contact between two mud turbidites with elongated mud chips. Base of eroding turbidite consists of a single layer of quartz. Massylian-Rif Mountains. (d) Alternation of radiolarites—(base and top), black shales and laminated, radiolarian-bearing carbonaceous shale. Massylian-Rif Mountains. (e) Mixed turbidite. Note amalgamation with black shale (5% TOC) at base. Turbidite reworked amorphous organic matter (black, flasered rip-up clasts at the base of the turbidite). The upper part (BOUMA b-sequence, 1.5% TOC) shows alternation of shallow water detritus with flasers rich in organic matter. Mauretanian-Rif Mountains.

(a)
(c)
(b)
2 mm
(d)
2 mm
(e)

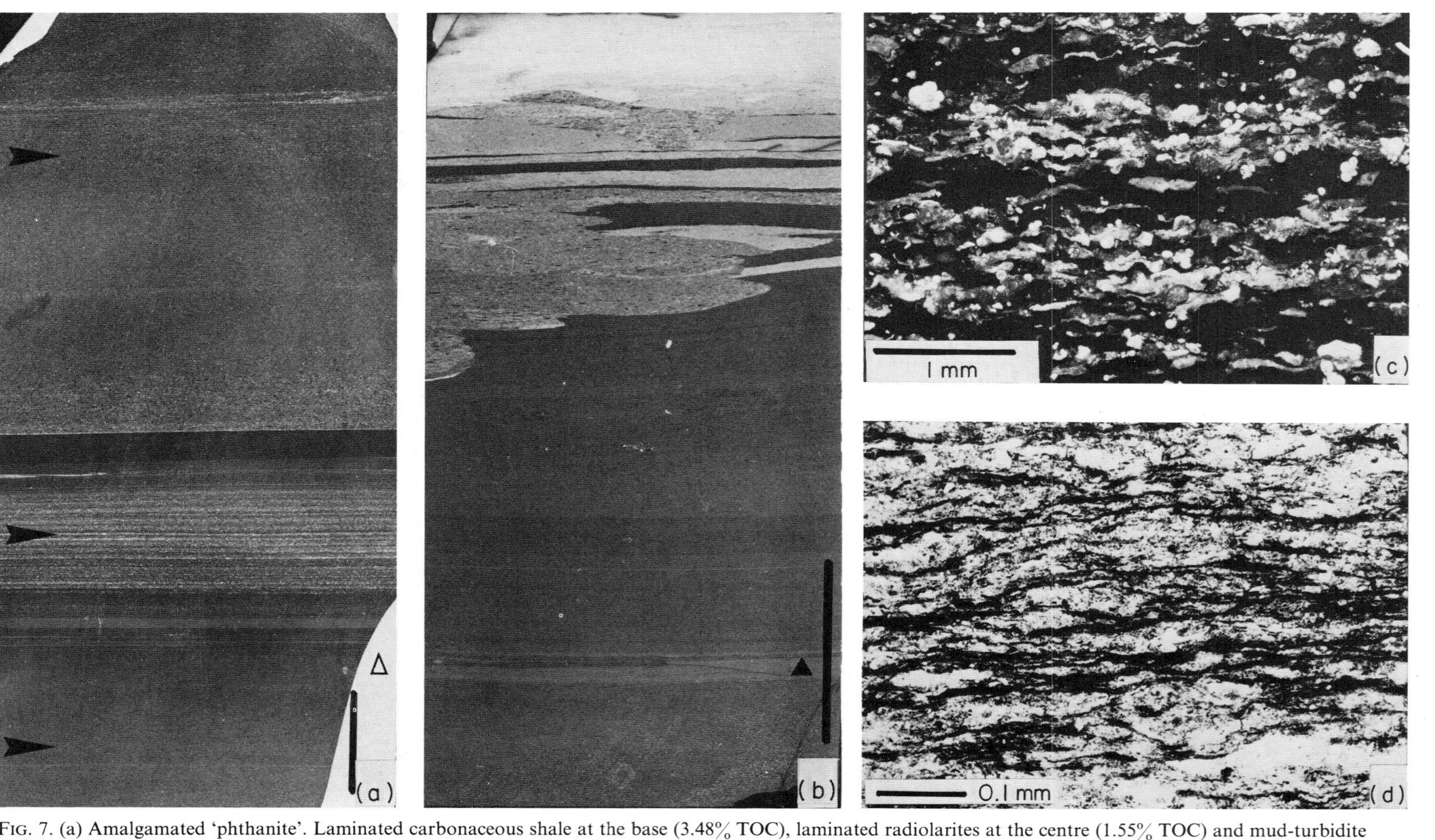

FIG. 7. (a) Amalgamated 'phthanite'. Laminated carbonaceous shale at the base (3.48% TOC), laminated radiolarites at the centre (1.55% TOC) and mud-turbidite (3.39% TOC) at the top. Note basal grading in the mud-turbidite and transition to laminite at the top (BOUMA b-sequence?, amalgamation?). Massylian-Rif Mountains. (b) Laminated, carbonaceous shale ('phthanite') at the base and oxic layer at the top. Note burrowing from oxic into anoxic part. Massylian—Rif Mountains. (c) Mixed turbidite—alternating of flasers rich in amorphous organic matter with layers comprising quartz-silt, foraminifera, carbonate-silt, pellets and abundant calcispheres. Mauretanian—Rif Mountains. (d) Microfacies of a true black shale (12% TOC). Note stringers of organic material. Massylian—Rif Mountains. Scale bar of polished slabs is 2 cm.

of the sediment-types mentioned above (Fig. 6 (d); Fig. 7 (a), (c)). The typical, minute, flaser-like bedding of the kerogen is distinct (Fig. 7 (d)), and the TOC of immature samples is always higher than 10%. Layers or aggregates of bright pyrite can be seen. Sedimentary textures suggest autochthonous sedimentation, but relicts of dinoflagellates and tasmanites in unaltered samples indicate more shallow environments than could be inferred from the depositional environment.

The terms 'laminites' and 'mud-turbidites' are the collective names for the bulk of organic carbon-rich sediments, as it is often very difficult to distinguish individual beds within a section. Furthermore, it may not be clear whether laminites are true cyclic layers of rhythmically deposited silt and organic matter, or the upper BOUMA-sequences of turbidites. The scale of a single lamina ranges from a one-grain silt layer up to a few millimetres. Quartz and clay minerals form the detrital part, whereas the other part is formed by amorphous, often degraded kerogen. Seasonal sedimentary cycles can be excluded. It is more likely that contour currents provided the

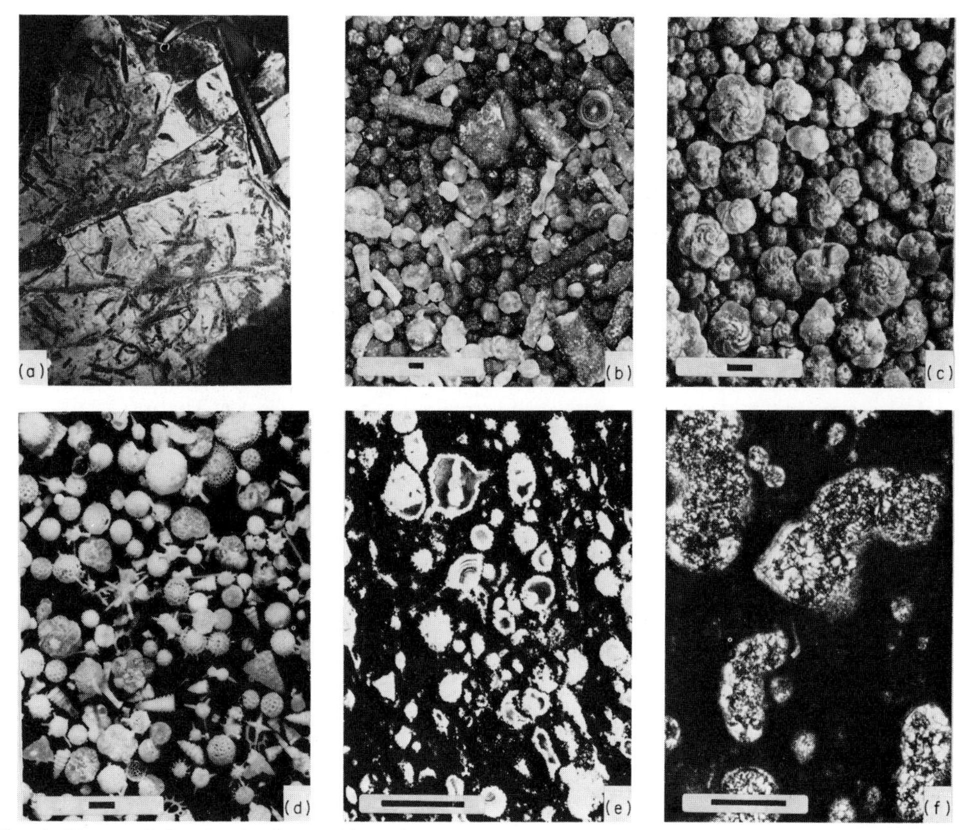

FIG. 8. Characteristic microfossil assemblages from the three main mid-Cretaceous palaeoenvironments within the Gibraltar Arch. Iberian and North African (c) slope and Massylian/Mauretanian deep-sea 'flysch'-sediments (b). Scale bar is 250 μ. (a) Bioturbated, white, oxic layer sandwiched between two laminated organic-rich layers. Type–Massylian. Oued bou Chouk, Petite-Kabylie, Algeria.

Faunal patterns of the background sedimentation ((b), (c)). (b) Upper Turonian of autochthonous shales (Massylian, Rif Mountains). Primitive agglutinating foraminiferal assemblages. Faunal composition is very similar to Alpine and Carpathian Flysch-assemblages which are deposited below CCD. (c) Cenomanian of the Southern Loukkos Zone (Morocco). 100% planktonic foraminiferal assemblages. Upper slope environment.

Faunal patterns of the Lower Turonian CTBE deposits ((d), (e), (f)). (d) HCl-residue of cherty pelagic limestone (Scaglia facies). Radiolarian tests and chalcedony-replaced planktonic foraminifera. Lower Turonian, Penibetic Zone, Spain. (e) Radiolarite with chalcedony-replaced radiolarians. Lower Turonian, Massylian, Rif Mountains, Morocco (thin section). (f) Chalcedony replacement of planktonic foraminifera. Micrite limestones with abundant radiolarians and planktonic foraminifera—basal Turonian, Penibetic Zone, Spain (thin section, crossed Nicols). Within the rims of chert nodules, even the foraminifera are filled by silica—they can be extracted and allow a direct calibration of the radiolarian-zonation during the CTBE.

detrital components, and formed the typical sedimentary texture. A significant framboidal pyrite content is common, and there are some remains of fish-skeletons. True mud-turbidites, rich in organic carbon (Fig. 6 (c); Fig. 7 (a)) are probably the most common sediment-type, but difficult to recognize due to small particle size. If this is true, one can assume that much of the organic matter is re-sedimented.

Radiolarians are the major planktonic components during the CTBE (see Fig. 6 (d); Fig. 8 (d), (e); Fig. 9). They are common in all sediment-types except in true black shales, but are preferentially enriched in layered radiolarites (Fig. 6 (d); Fig. 7 (a)). Lamination of the radiolarites and occasional bioturbation may indicate that they were formed during short-termed oxic conditions, perhaps in oxygenated bottom currents. Radiolarians were deposited by a continuous organic 'rain' from the surface, and were subsequently accumulated by currents (Fig. 7 (a)). Sea-floor conditions must often have been oxic, as shown by intense bioturbation (Fig. 7 (b); Fig. 8 (a)).

Mixed turbidites are the most puzzling sediment-type of the CTBE. They are enriched in amorphous organic matter in two ways. Either the kerogen is concentrated in the upper BOUMA-sequences, or it alternates with shallow-water detritus of the background-type (Fig. 6 (e); Fig. 7 (c)) (with algal-kerogen?). However, it is not clear whether the organic material is re-sedimented *together* with shallow-water detritus, or whether it is 'whirled up' by gravity flows and subsequently re-deposited. The overall sedimentary texture suggests that clearly graded turbidites are of the second type, while turbidites with organic material/detritus couplets are of the first type. Figure 6 (b) shows a typical sequence of turbidite-free 'phthanites' deposited somewhere in the outer fan position. Note that there is a 'cyclicity' with dark, organic carbon-rich and light, bioturbated layers. This is a common feature in all outcrops. Whatever explains the CTBE, the cyclicity in environmental O_2-shows that long-term stagnation is unlikely.

Biostratigraphy and faunal patterns

A detailed biostratigraphical analysis of mid-Cretaceous, dark, siliceous carbonaceous sediments from the different structural units within the Gibraltar Arch and adjacent areas has led to a good correlation of the base of this facies. This level occurs at the top of the *Rotalipora cushmani*-zone and the first appearance of certain radiolarians—*Crucella cachensis*, *Alievum superbum* (Fig. 9). Comparable studies in North and North-West African coastal basins (S. Morocco, Tunisia), and in parts of the former southern margin of the Tethys in Italy (Gubbio-area, Umbrian Appennines; Euganean Hills near Padova) corroborate this correlation of radiolarian and planktonic foraminiferal biostratigraphies (e.g. Thurow *et al.* 1982).

Table 1 and Fig. 8 (b)–(d) give a brief summary of the major characteristics and differences of

FIG. 9. Selected Radiolaria—typical of the CTBE. 1 and 2. *Acanthocircus* sp. nov. Lower to Middle Turonian (as far as known)—Morocco, Spain, Italy. 3. *Cavaspongia antelopensis* PES. Turonian of California, Morocco and Italy. 4. *Pyramispongia glascockensis* PES. Cenomanian–Turonian of California, Morocco, Spain and Italy. 5 and 6. *Crucella cachensis* PES. The first appearance of this form coincides with the CTBE peak. In sections where planktonic forminifera are also present, its first appearance is after the extinction of Rotalipora. Uppermost Cenomanian—Upper Turonian in Morocco, Spain and Italy; Turonian of California. 7. *Alievum superbum* PES. Same remarks as *C. cachensis* Index fossil for *superbum*-zone (Lower Turonian) of Pessagno (1976). Turonian of Morocco, Spain and Italy (Greece?), Turonian/Lower Coniacian of California, Cenomanian? of Costa Rica. 8. *Pseudoaulophacus puthahensis* PES. Turonian of Morocco, Spain, Italy and Califotnia. 9. *Parvicingulidae*. Common in Turonian—Morocco, Spain, Italy. 10. *Dictyomitra cf. multicostata* ZITTEL. Upper Cretaceous, as far as known not in Cenomanian. 11. *Pseudodictyomitra pseudomacrocephala* (SQUIN.) Common in mid Cretaceous—worldwide. 12. *Rhopalosyringium sp. A*. Turonian, as far as known—Agadir coastal basin/Morocco. 13. *Parvicingulidae*. Turonian, as far as known—Agadir coastal basin/Morocco. 14. *Archaeospongoprunum sp. A*. Turonian, as far as known—Agadir coastal basin/Morocco. 15. *Halesium sexangulum* PES. Common in Upper Cenomanian and Lower Turonian Morocco, Spain, Italy, California. 16. *Halesium quadratum* PES. See *H. sexangulum* 17. *Patulibracchium davisi* PES. Cenomanian of California, Turonian of Morocco. 18. *Paronella salonensis* PES. Coniacian of California, Turonian of Morocco and Italy. 19. *Triactoma echoides* FORE. Cretaceous. 20. *Neosciadiocapsidae*. Such forms with narrow meshwork and without apical horn(s) are common in Turonian of Morocco, Spain and Italy. 21. *Hemicryptocapsa polyhedra*. DUM. Common in Cenomanian/Turonian. 22. *Praeconocaryomma universa* PES. Recorded from the Cenomanian to Coniacian-California, Costa Rica, Morocco, Spain, Italy, Greece. 23. *Dumitricaia maxwellensis* PES. Common in Upper Cenomanian and Turonian Morocco, Spain, Italy, California.

All figured species are from Lower Turonian CTBE deposits—Rif Mountains (N. Morocco), Agadir Coastal Basin (SW Morocco), Penibetic Zone (S. Spain) and Euganean Hills (N. Italy). Scale bar gives standard magnification, different magnifications are marked with own scale bars, each are 100 μ.

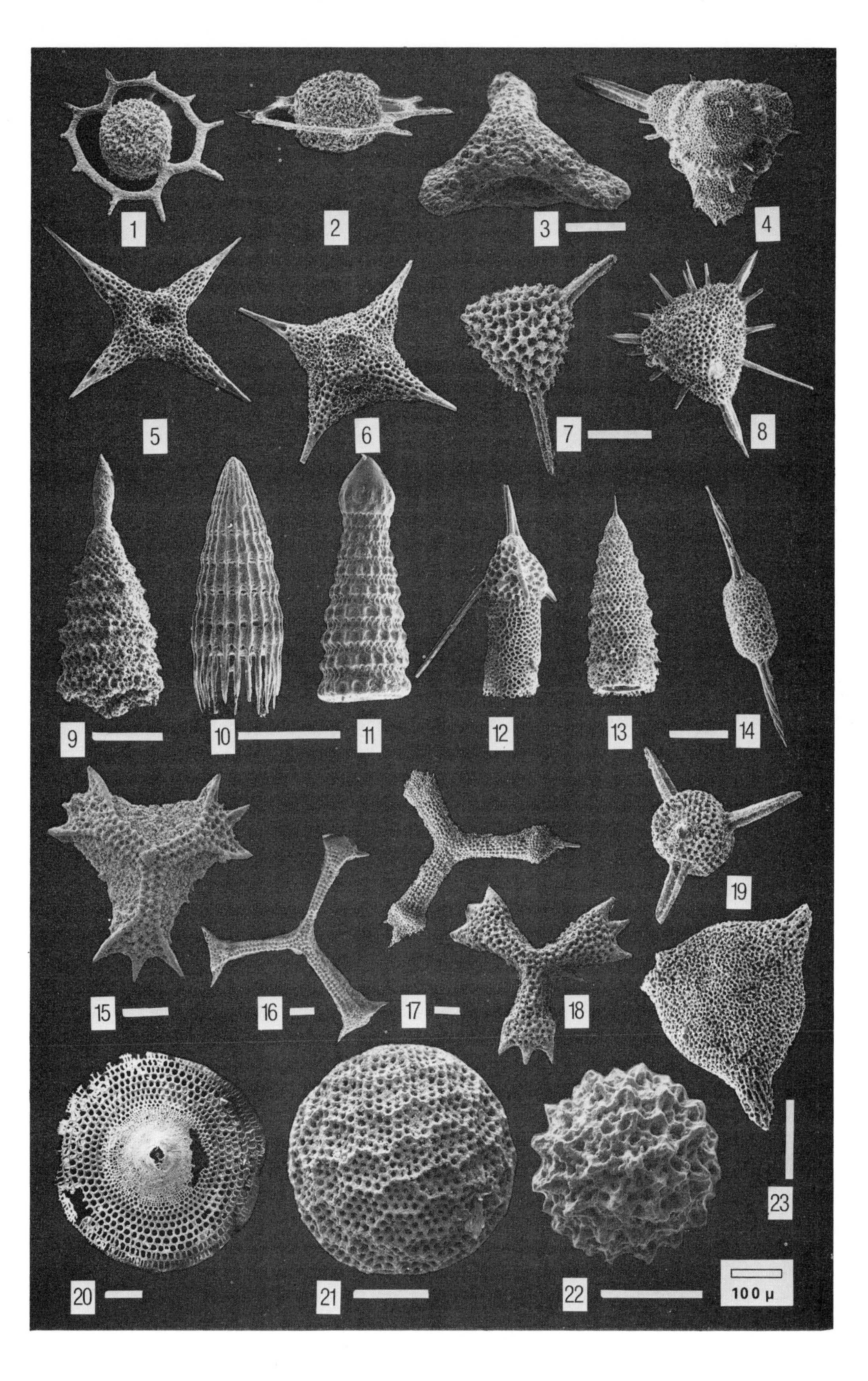

1
2
3
4
5
6
7
8
9
10
11
12
13
14
15
16
17
18
19
20
21
22
23
100 μ

TABLE 1. *General characteristics of the microfaunal assemblages in the major palaeogeographic domains before, during, and after the CTBE. Numbers and letters correspond to Fig. 8*

	Penibetic Zone (Iberian margin)	'Flysch' Mauretanian and Massylian basin	Loukkos Zone (North African Margin)
Upper Turonian	100% planktonic foraminiferal assemblages	Primitive agglutinating forms of high diversity (typical abyssal assemblages of sub CCD-environment) reworked foraminifera in turbidites (Fig. 8 (b))	Redeposited (?) benthonic forms dominating, a small amount of planktonic and primitive agglutinating species seem to represent the autochthonous fauna
Lower Turonian (CTBE)	Abundant radiolaria and planktonic foraminiferal assemblages of low diversity (Fig. 8 (d), (f))	Rich radiolarian assemblages of high diversity (foraminifera only in turbidites) (Fig. 8 (e))	Planktonic foraminifera and radiolaria, benthonic forms occur rarely
Cenomanian	Assemblages consisting of 100% planktonic foraminifera or with a small content of textulariid agglutinating forms	Primitive agglutinating foraminiferal assemblages and pyritized radiolaria. Planktonic foraminiferal assemblages reworked in turbidites	100% planktonic foraminifera; in some samples containing radiolarians or benthonic foraminifera (Fig. 8 (c))

important microfaunal assemblages in relation to the different depositional environments in the mid-Cretaceous of the Gibraltar Arch area. Although these characteristics are generalized, they give an overview of the faunal response to the bathymetry of the environment. Note that important faunal changes occurred during the CTBE in all environments.

Some remarks on faunal patterns

Radiolaria

Apart from the organic carbon-rich sedimentation, the most obvious effect of the CTBE is a 'bloom' in radiolaria. Modern radiolaria are marine planktonic protozoans. They are common in all oceanic waters of normal salinity. Most live in the uppermost few hundred metres of the water column, but a few inhabit deeper levels. Distinct faunal-associations can be related to latitude and to the occurrence of upwelling. In spite of this, little is known about the palaeobiogeography and palaeoecology of mid-Cretaceous radiolaria. They are scarce and badly preserved in the North Atlantic (Foreman 1977). But this statement is not true of the CTBE. The rich and sometimes well-preserved radiolarian faunas there provide an excellent basis for biostratigraphic studies, and allow the CTBE to be traced into areas where other indications for this event are scarce.

Californian mid-Cretaceous radiolarian assemblages (Pessagno 1976, 1977) are quite comparable with those of the Gibraltar Arch area (Thurow *et al.* 1982). The first appearance and extinction of many marker species are coeval, thus enabling the use of the California zonation in the Gibraltar Arch. In contrast to planktonic foraminifers, the diversity of radiolarian assemblages is high during the CTBE (Fig. 8 (d)), and there are no faunal breaks on a generic or higher level. Some of the species most characteristic of the CTBE are shown in Fig. 9.

The abundance of preserved radiolarian remains in all the different lithologies (Fig. 8 (d), (e)), and the common silicification of the sediment (Fig. 8 (f)) suggest that production of biogenic silica was high at this time.

Planktonic foraminifera

The CTBE in the Gibraltar Arch and adjacent areas coincides with important changes in the evolution of planktonic forminifera (Table 2). One major and two minor evolutionary changes can be pointed out:

(1) First signs of CTBE are coeval with the first appearance of the genus *Whiteinella* and new species of *Praeglobotruncana* in the Upper Cenomanian cushmani-zone.
(2) The peak of the CTBE is coeval with the extinction of the single-keeled genus *Rotalipora* and the first appearance of important marker species of the genera *Whiteinella* and *Praeglobotruncana*. This distinct datum level, the base of the *archaeocretacea*-zone, is called the Cenomanian–Turonian boundary in this paper.
(3) The end of the CTBE is accompanied by the appearance and radiation of the double-keeled genera *Marginotruncana* and *Dicarinella*.

This coincidence of the CTBE and these important events in the evolution of planktonic foraminifera cannot be accidental, and probably reflects some palaeoceanographic control.

Two interpretations are given:

(1) Extinction of the bathypelagic genus *Rotalipora* by development of an ocean-wide O_2-minimum layer unusually depleted in oxygen (Wonders 1980).
(2) The unkeeled and low diversity planktonic foraminiferal fauna of the Lower Turonian 'zone a grandes globigérines' in the shelf basins of the north-east Atlantic was interpreted as a boreal faunal assemblage (Butt 1982). This idea is based on the distribution of recent planktonic foraminifera, where keeled forms are restricted to the tropical warm water areas, while small-sized, low diversity faunas of unkeeled forms are characteristic of boreal regions. If this idea is correct, then faunal change at the Cenomanian–Turonian boundary may have been caused by the cooling of the surface waters over the outer shelf and upper slope areas. Such cooling may have been caused by upwelling currents, which may explain also the high radiolarian/planktonic foraminiferal ratios in the Lower Turonian of the Gibraltar Arch. Similar high ratios occur in recent sediments of the north-eastern Atlantic continental margin in the region of intense upwelling (Diester-Haas 1977).

Correlation of the CTBE in the different units and comparison with adjacent areas

Figure 10 shows the authors' attempt to compare and correlate biostratigraphy, lithology, organic geochemistry, and faunal patterns within seven typical simplified sections of the different palaeogeographic domains.

Both the sedimentary and faunal patterns of the section and their position in the structural

TABLE 2. *Ranges of important species of planktonic foraminifera in the Upper Cenomanian—Turonian in the Gibraltar Arch and adjacent areas. Time scale according to Harland et al. (1982) in Ma.*

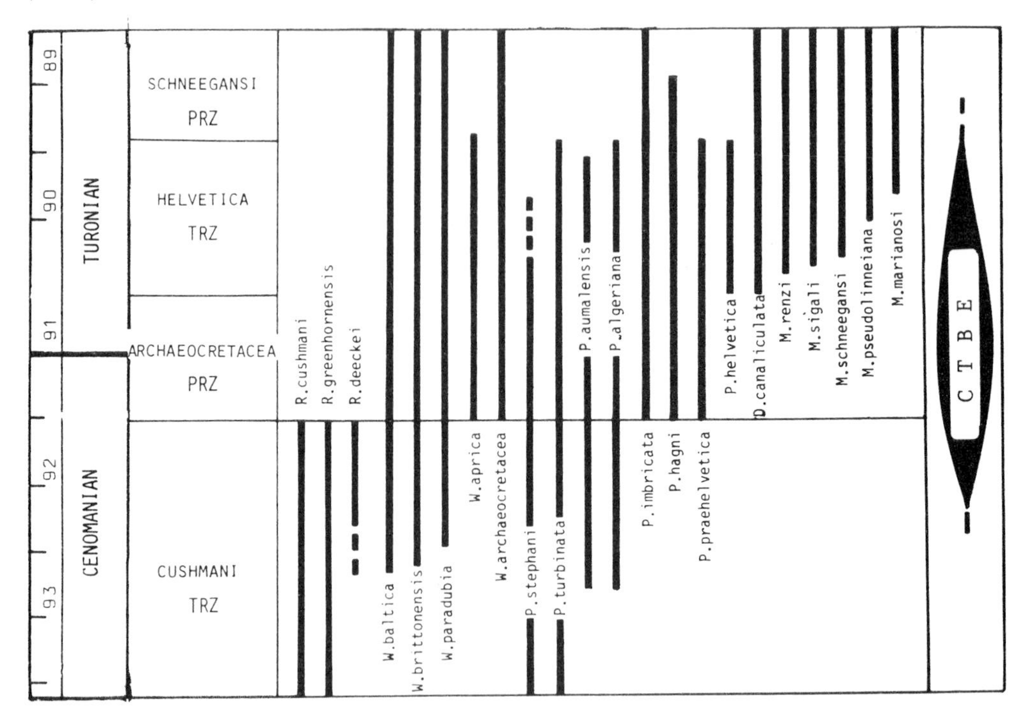

frame of the area (Fig. 1) allow us to reconstruct the former setting of the related series, and to establish (together with many sections not presented here) an idea of the basin and margin topography of the Gibraltar Arch area during the mid-Cretaceous (Fig. 4).

One of the most striking results of the Deep Sea Drilling Project is the recognition that organic carbon-rich sediments were widespread in mid-Cretaceous times. These sediments have been drilled in the Pacific as well as in the Atlantic Ocean, and can be traced on-shore onto the known palaeo-continental margins of alpine fold belts, especially the Maghrebinian Fold Belt of Southern Spain and North Africa, and also on to unfolded stable areas. Within the mid-Cretaceous, organic carbon-rich, sedimentation event, the Upper Cenomanian to Middle Turonian interval is characterized by an increase in organic enrichment (Summerhayes 1986). Despite highly variable lithologies and depositional environments (from shelf areas down to the deep sea) the organic rich sediments of the CTBE are surprisingly similar (Table 3).

Discussion and conclusion

The conditions that led to the deposition of sediments rich in organic carbon and in biogenic silica at the Cenomanian–Turonian boundary are the subject of much discussion (Schlanger *et al.* 1986). Some authors call upon high biogenic surface productivity caused by seasonal, coastal, upwelling of cold waters to the surface, and subsequent development of anoxic conditions in subsurface waters (e.g. Einsele & Wiedmann 1982; Wiedmann *et al.* 1982); others argue for restricted anoxic environments, stagnant basins (de Graciansky *et al.* 1982), or stratification of the water column leading to intensification of the O_2-minimum zone at intermediate depths (e.g. Schlanger & Jenkyns 1976; Ryan & Cita 1977; Demaison & Moore 1980; Jenkyns 1980; Busson 1984). Jenkyns (1980) suggested that the oxygen minimum zone expanded as a result of the global Turonian transgression coupled with climatic effects that led to sluggish bottom water circulation and depletion of oxygen in basal water masses. Tucholke & Vogt (1979) suggest the

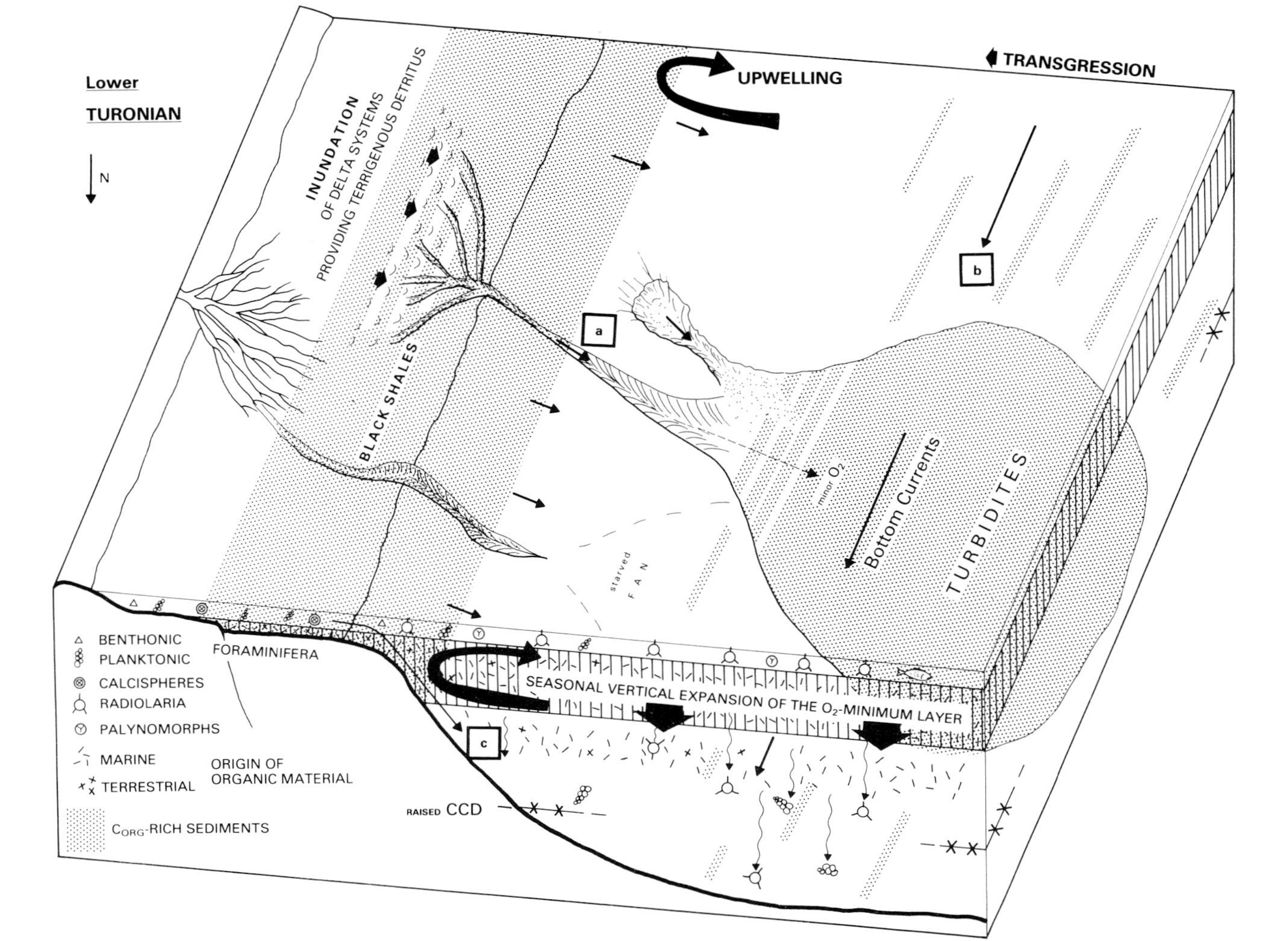

FIG. 10. Continental margin model representing the mid-Cretaceous of the Alboran Block. The depositional environment is characterized by seasonal upwelling and resulting high biogenic productivity, and development of a de-oxygenated O_2-minimum zone that is vertically pulsing allowing preservation of organic matter in greater depth. Additional organic matter was transported to deep water environments by processes of re-deposition. The crystalline hinterland (source of detritus) was flooded over large areas preventing formation of the sedimentary facies characteristic of the Lower Cretaceous; shallow water material and carbonate-detritus were common in tectonically active areas within the margin (due to strike-slip faulting). Bottom currents were common on the basin floor, allowing the formation of layered radiolarites. (a) detrital input. (b) bottom currents. (c) down-slope transport.

TABLE 3. *Compilation of the important sedimentological/faunistical characteristics of some important on-shore deposits of the CTBE in the Gibraltar Arch and adjacent areas (data for Tarfaya basin are from Wiedmann et al. (1982)). For an additional comparison with off-shore data see De Graciansky et al. (1982) and Herbin et al. (this volume)*

	Coastal basins Agadir/Tarfaya Upper Cenomanian–Lower Coniacian	Deep sea fans Gibraltar Arch (Mauretanian/Massylian) Upper Cenomanian–Middle Turonian	Slope associations Gibraltar Arch (Loukkos-Zone) Lower Turonian–Middle Turonian	Oceanic basins Appennines Upper Cenomanian–Lower-Middle Turonian
Resediments (due to gravity flows)		× ×	×	
Reduction of overall grain size		× ×	?	
Enrichment of C_{org} in resediments		×	×	
Black shales/C_{org}-rich sediments	× ×	× ×	×	× ×
Fine lamination	× ×	× ×	×	× ×
Bioturbation (within C_{org}-rich layers)	–	–	–	–
Strong silicification	× ×	× ×	×	0
Formation of chert	×	–	×	× ×
in turbidites		× ×	—	
Pyrite	0 (Radiolarians preserved in pyrite)	× ×	0	0
Radiolarians	× ×	× ×	× (Radiolarians preserved in calcite)	× ×
Calcispheres	× ×	× × (in resediments)	?	?
Calcareous plankton	× ×	–	×	
in turbidites		× ×	—	
		× ×	0	
Microbenthos	–	– (common in resediments)	–	–
'Boreal faunas'	×	?	?	?
Phosphorite	×	×	?	?
Fish debris	× ×	× ×	0	×
Rate of sedimentation (m/my)	5–20	2–5	5	<1
CCD	above CCD	below CCD	above CCD	below CCD

× × = Abundant. × = Common. 0 = Rare.

establishment of a modern current system led to ocean-wide upwelling. Thurow *et al.* (1982) claim that organic matter was redeposited from areas of coastal upwelling into deeper waters. Some authors assume that Cenomanian–Turonian surface waters were highly fertile; others suggest decreased surface fertility and high preservation rates (DeBoer 1982; Bralower & Thierstein 1984). Detailed studies of the CTBE-Sedimentation in the Gibraltar Arch produced information to help to discriminate between competing models that attempt to explain this palaeoceanographic event:

(1) Organic-rich facies occurs in very different palaeobathymetric settings, from upper slope-environments to basin plain sediments with sub-CCD-character.
(2) Organic-rich facies in greater water depths seems to be more the result of turbidity current-transport than of an expanded O_2-minimum layer (alternation of oxic environments).
(3) Current indications (e.g. erosional contacts, reworking, redistribution and bioturbation) are common within the CTBE-sediments of the Gibraltar Arch. This does not fit with a stagnant basin model.
(4) In all cases the organic-rich facies is accompanied by a very distinct 'bloom' of planktonic organisms (e.g. radiolarians, calcispheres).
(5) Rich radiolarian faunas are common in the carbon-rich and carbon-free sediments of shallow marine and deep sea environments, which most probably is due to high productivity in surface waters and not to favourable conditions for preservation.
(6) Maximum sea-level highstand during CTBE can be proven by the transition from neritic to pelagic facies in the shelf environments (e.g. Rides Prérifaines).

The model which the authors prefer (Fig. 7) involves seasonal upwelling and the development of a de-oxygenated O_2-minimum zone. Most of the organic matter was transported to the deep water environments by processes of re-deposition. This model suggests that some of the black shales are autochthonous, while others are allochthonous organic-rich upper BOUMA-sequences of turbidites formed by suspension and redeposition of organic matter from shallow areas. Rapid transport of the organic matter into deep water facies inhibits oxygenation or bacterial degradation and decomposition en route. Once deposited the decomposition begins, but rapidly strips oxygen from pore waters and from the few centimetres above the bottom, so preservation of organic material is ensured (e.g. Meyers *et al.* 1984). The sedimentary features that the authors observed can be explained by erosion of autochthonous layers, or of the upper parts of turbidites, by later turbidity currents charged with shallow water debris.

Autochthonous organic-rich layers even occur below CCD in the flysch basin, suggesting that the base of the O_2-minimum layer was sometimes below CCD. This enlarges the model proposed by Arthur *et al.* (1986—their Fig. 5 (b)), indicating a vertical expansion of the minimum layer at least to 3 km depth at times. But such conditions and subsequent autochthonous 'true' black shales are rare compared with the total of CTBE-sedimentation.

The faunal characteristics of the CTBE (radiolarian-'bloom' and low diversity of unkeeled planktonic foraminiferal assemblages) can be explained by upwelling conditions, which cause a strong increase in surface productivity, resulting in an intensified O_2-minimum layer followed by deterioration of the life habitat of bathypelagic keeled planktonic foraminifera (*Rotalipora*).

The considerable transgression at the Cenomanian–Turonian boundary results in the deep ocean becoming a sediment starved basin. Arthur *et al.* (1986) present a model in which they point out that the maximum sea-level high stand during this time is the ultimate driving force for organic-rich deposition in globally widespread basins under different climatic and ocean circulation regimes. Increased C_{org}-preservation in epicontinental seas, continental slopes and deep sea fans suggests formation under an expanded and intensified oxygen-minimum layer.

The most probable reason for the nearly coeval inception of CTBE-sedimentation is changes in the configuration of oceanic basins:

(1) a pre-existing oceanic connection (since Middle Albian—Wiedmann & Neugebauer 1978; Moullae & Guérin 1982) was enlarged during Cenomanian–Turonian.
(2) a maximum transgression flooded large areas, expanding shallow marine environments. Both changes may have been caused by strongly increased spreading rates during the mid-Cretaceous (see Sheridan, this volume), or by other volcano-tectonic processes—as yet unknown—attaining their maximum during the Upper Cenomanian–Lower Turonian.

ACKNOWLEDGMENTS: The authors thank the German National Science Foundation (DFG) for providing financial support, and Dr J.P. Herbin (Institute Français de Pétrole) for providing Kerogen analysis.

The authors also thank H.C. Jenkyns and B.M.

Funnell for their comments and review. The assistance of C.P. Summerhayes is acknowledged in the final editing of the manuscript.

Note added in proof:

In the Penibetic (Iberian continental margin) we recently found a thick CTBE section in Sierra de los Canutos, north of Gibraltar in southern Spain. The sediments are black shales and radiolarites with up to 33% TOC, intercalated in white pelagic limestones.

References

ARTHUR, M.A., SCHLANGER, S.O. & JENKYNS, H.C. 1986. The Cenomanian–Turonian oceanic anoxic event, II. Paleoceanographic controls on organic matter production and preservation. *In*: BROOKS, J. & FLEET, A. (eds), *Marine Petroleum Source Rocks. Spec. Publ. Geol. Soc. Lond.* Published by Blackwell Scientific Publications, Oxford.

BERNOULLI, D. & JENKYNS, H.C. 1974. Alpine, Mediterranean and central Atlantic Mesozoic facies in relation to the early evolution of the Tethys. *In*: DOTT, R.H. & SHAVER, R.H. (eds), *Modern and Ancient Geosynclinal Sedimentation. Spec. Publ. Soc. Econ. Paleont. Miner.* **19,** 129–60.

BOUILLIN, J.P., DURAND DELGA, M., GELARD, J.P., LEIKINE, M., RAOULT, J.F., RAYMOND, D., TEFIANI, M. & VILLA, J.M. 1970. Définition d'un flysch Massylian et d'un flysch Maurétanian au sein des flyschs allochtones de l'Algérie. *C.R. Acad. Sc. Fr.* **270,** 2249–52.

BRALOWER, T.J. & THIERSTEIN, H.R. 1984. Low productivity and slow deep-water circulation in Mid-Cretaceous oceans. *Geology* **12,** 614–18.

BUSSON, G. 1984. Relations entre la sédimentation de Crétacé moyen et supérieur de la plate-forme du nord ouest africain et les dépôts contemporain de l'Atlantique centre et nord. *Eclogae geol. Helv.* **77/2,** 221–35.

BUTT, A. 1982. Micropaleontological Bathymetrie of the Cretaceous of Western Morocco. *Palaeogr. Paleoclimatol. Paleoecol.* **37,** 235–75.

DE BOER, P.L. 1982. Aspects of Middle Cretaceous pelagic sedimentation in Southern Europe. *Geologica Ultraiectina, Utrecht* **31,** 112 pp.

DEMAISON, G.L. & MOORE, G.T. 1980. Anoxic environments and oil source bed genesis. *Am. Ass. Petrol. Geol. Bull.* **64,** 1179–209.

DIESTER-HAAS, L. 1977. Radiolarian/planktonic foraminferal ratios in a coastal upwelling region. *J. Foram. Res.* **7,** 26–33.

EINSELE, G. & WIEDMANN, J. 1982. Turonian black shales in the Moroccan coastal basins: First upwelling in the Atlantic Ocean. *In*: VON RAD, U. *et al.* (eds), *Geology of the North-west African Continental Margin.* Springer-Verlag, Berlin. 396–414.

FOREMAN, H.P. 1977. Mesozoic radiolaria from the Atlantic Basin and its borderlands. *In*: SWAIN, F.M. (ed), *Stratigraphic Micropalaeontology of Atlantic Basin and Borderlands.* Elsevier, Amsterdam. 305–20.

GONZALEZ-DONOSO, J.M., LINARES, D., MARTIN-ALGARRA, A., REBOLLO, M., SERRANO, F. & VERA, J.A. 1983. Discontinuidades estratigráphicas durante el Cretácico en el Penibético (Cordillera Bética). *Estudios Geologicos* **39** 275–36.

DE GRACIANSKY, P.C., BROSSE, M., DEROO, G. *et al.* 1982. Les formations d'âge Crétacé de l'Atlantique nord et leur matière organique: Paléogéographie et milieux de dèpôt. *Rev. de l'Inst. Fr. Pétrole* **39,** 275–336.

GÜBELI, A. 1983. *Stratigraphische und sedimentologische Untersuchungen der detritischen Unterkreide-Serien des Zentralen Rif (Marokko).* Thesis, University of Zürich. 192 pp.

HARLAND, W.B., COX, A., LLEWELLYN, P.G., PICKTON, C.A.S., SMITH, A.G. & WALTERS, R. 1983. *A Geologic Time Scale.* Cambridge University Press, New York. 112 pp.

JENKYNS, H.C. 1980. Cretaceous anoxic events: from continents to oceans. *J. Geol. Soc. Lond.* **137,** 171–88.

LEBLANC, D. & OLIVIER, PH. 1984. Role of strike-slip faults in the Betic-Rifian orogeny. *Tectonophysics* **101,** 345–55.

MEYERS, P.A., BRASSELL, S.C. & HUC, A.Y. 1984. Geochemistry of organic carbon in South Atlantic sediments from Deep Sea Drilling Project Leg 75. *In*: HAY, W.W., SIBUET, J.-C. *et al.* (eds), *Init. Repts DSDP* **75,** 967–81.

MOULLADE, M. & GUÉRIN, S. 1982. Le problème des relations de l'Atlantique Sud et de l'Atlantique Central au Crétacé moyen: nouvelles données microfauniques d'après les forages D.S.D.P. *Bull. Soc. géol. France,* (7), **24,** (3), 511–17.

PESSAGNO, E.A., JR. 1976. Radiolarian zonation and stratigraphy of the Upper Cretaceous portion of the Great Valley Sequence, California Coast Ranges. *Micropaleontology, Spec. Publ.* **2,** 1–95.

—— 1977. Lower Cretaceous radiolarian biostratigraphy of the Great Valley Sequence and Franciscan Complex, California Coast Ranges. *Cushman Found. Foram. Res., Spec. Paper* **15,** 170 pp.

RAOULT, J.F. 1972. Précisions sur le flysch Massylien: série stratigraphique, variations de faciès, nature du matériel remanié (Nord du Constantinois, Algérie). *Bull. Soc. Hist. Nat. Afr. Nord, Alger.* **63,** 73–92.

RYAN, W.B.F. & CITA, M.B. 1977. Ignorance concerning episodes of ocean-wide stagnation. *Mar. Geol.* **23,** 197–215.

SCHLANGER, S.O., ARTHUR, M.A., JENKYNS, H.C. & SCHOLLE, P.A. 1986. The Cenomanian–Turonian Anoxic Event. I. Stratigraphy and distribution of Organic Carbon-rich beds and the marine ^{13}C Excursion. *In*: BROOKS, J. & FLEET, A. (eds), *Marine Petroleum Source Rocks. Spec. Publ. Geol. Soc. Lond.* Published by Blackwell Scientific Publications, Oxford.

—— & JENKYNS, H.C. 1976. Cretaceous oceanic anoxic events: causes and consequences. *Geologie en Mijnbouw* **55,** 179–84.

SUMMERHAYES, C.P. 1986. Organic-rich Cretaceous sediments from the North Atlantic. *In*: BROOKS, J. & FLEET, A. (eds), *Marine Petroleum Source Rocks. Spec. Publ. Geol. Soc. Lond.* Published by Blackwell Scientific Publications, Oxford.

THUROW, J. 1984. Unterkretazische Turbiditserien im Raum des Gibraltarbogens (N-Marokko, S-Spanien). *Int. Ass. Sedimentol., 5th European Meeting (Marseille), Abstract*, 433–4.

——, KUHNT, W. & WIEDMANN, J. 1982. Zeitlicher und paläogeographischer Rahmen der Phthanit- und Black Shale-Sedimentation in Marokko. *N. Jb. Geol. Palaeont. Abh.* **165,** 147–76.

TUCHOLKE, B.E. & VOGT, P.R. 1979. Western North Atlantic: Sedimentary evolution and aspects of tectonic history. *In*: TUCHOLKE, B.E. & VOGT, P.R. *et al., Init. Repts DSDP* **43,** 791–825.

WIEDMANN, J., BUTT, A. & EINSELE, G. 1978. Vergleich von marokkanischen Küstenaufschlüssen und Tiefseebohrungen (DSDP): Stratigraphie, Paläontologie und Subsidenz an einem passiven Kontinentalrand. *Geol. Rdsch.* **67,** 454–508.

—— & NEUGEBAUER, J. 1978. Lower Cretaceous Ammonites from the South Atlantic Leg 40 (DSDP), their stratigraphic value and sedimentologic properties. *In*: BOLLI, H.M. & RYAN, W.B.F. *et al.* (eds), *Init. Repts DSDP* Suppl. **38–41,** 709–21.

——, BUTT, A. & EINSELE G. 1982. Cretaceous stratigraphy, environment, and subsidence history at the Moroccan continental margin. *In*: VON RAD, U. *et al.* (eds), *Geology of the North-west African Continental Margin.* Springer-Verlag, Berlin. 366–95.

WONDERS, A.A.H. 1980. Middle and Late Cretaceous planktonic foraminifera of the Western Mediterranean Area. *Utrecht Micropaleontol. Bull.* **24,** 157 pp.

J. THUROW & W. KUHNT, Institut und Museum für Geologie und Paläontologie der Universität Tübingen, Sigwartstrasse 10, 7400 Tübingen 1, Federal Republic of Germany.

The inorganic geochemistry of Cretaceous black shales (DSDP Leg 41) in comparison to modern upwelling sediments from the Gulf of California

Hans J. Brumsack

SUMMARY: Recent trace element data for sea water and plankton reflect the importance of biological activity for the distribution of these elements in the water column and for their accumulation in sediments underlying areas of high productivity. Many trace metals in sea water are highly correlated to nutrients like phosphate, nitrate, or silica. By considering the ability of an element to be involved in nutrient cycling, as demonstrated by its deep/surface sea water concentration ratio, and its availability in sea water, a relationship between plankton and sea water chemistry is evident.

Fifty sediment samples from the Gulf of California, an area of exceptionally high primary productivity, have been analysed for all major and 15 minor elements, including Cd, Zn, Cu, Ni, Ba, Se, Cr, As, V, Mo, Co, Pb, Tl, and Bi. The 'excess' of nutrient and trace elements in upwelling sediments from the Gulf of California reflects the plankton chemistry. Among those elements found enriched in the diatomaceous oozes from the Guaymas Basin Slope relative to average shale are P, S, Mo, Se, Cd, and possibly Bi. For Mo and V an additional accumulation process besides bioconcentration seems to be active during early diagenesis. For these two elements their concentration ratios to aluminum increase with burial depth in the upper 30 cm of the sedimentary column, indicating that sea water seems to be a major source, besides plankton.

The chemistry of Ba in pore-waters and solids is controlled by barite solubility. A front of barite precipitation and dissolution (the 'barite front') is moving through the sedimentary column in the depth range of near zero pore-water sulphate, which at a drastic reduction of the sediment accumulation rate would lead to the formation of larger barite crystals or even nodules in this depth range.

Cretaceous black shales from the Cape Verde Basin (Site 367, Cores 17 to 21) are characterized by extremely high (see maximum values in brackets) concentrations of Mo (280 p.p.m.), V (3600 p.p.m.), Zn (7000 p.p.m.), and a number of other trace elements, besides S (7.4%) and organic carbon (35.6%) which cannot be explained by upwelling. If these enrichments are of primary origin, then complete stagnation of the Cape Verde Basin during intervals in Cretaceous time seems likely. The marine environment is metal-poor, therefore stagnation and corresponding low accumulation rates of detrital material would favour the accumulation of organic carbon and trace metals.

Sediments rich in organic carbon often are characterized by very unusual trace element abundances, in particular the so called black shales may have extremely high metal contents (Vine & Tourtelot 1970). The origin of these metal enrichments is still under debate. Some authors (e.g. Brongersma-Sanders 1965) have proposed biogenous metal accumulation, as presently occurs in areas of upwelling, to explain the metal content of some black shales, like the Kupferschiefer. But Calvert & Price (1970) demonstrated that sediments underlying areas of upwelling generally do not have extreme metal contents.

In order to understand bioconcentration of elements under high productivity conditions and corresponding processes within the water column and during early diagenesis, and whether these processes might be responsible for the chemistry of certain Cretaceous black shale layers, a detailed study of upwelling sediments from the Gulf of California was undertaken. This study included both interstitial water and sediment chemistry and followed the pathways of some elements from sea water to their incorporation in plankton and final burial in sediments.

The Gulf of California, a long marginal sea of the Pacific Ocean, is characterized by an exceptionally high primary productivity (Zeitschel 1969). Laminated, diatomaceous sediments are found mainly on the continental slopes of the central and southern Gulf (van Andel & Shor 1964; Schrader *et al.* 1980), where the upwelling-induced oxygen-minimum zone impinges on the sediment–sea water interface at water depths ranging from 400–800 m (Calvert 1966). The laminae result from annual changes in the proportion of the biogenous and terrigenous input. A light lamina with higher abundances of oceanic diatoms forms during the dry winter season of NW winds and corresponding upwelling at the

From SUMMERHAYES, C.P. & SHACKLETON, N.J. (eds), 1986, *North Atlantic Palaeoceanography*, Geological Society Special Publication No. 21, pp. 447–462.

mainland side. A dark lamina is deposited during the wet summer season with SE winds accompanied by higher terrigenous input (Donegan & Schrader 1982).

During the past 5–10 years reliable sea water and plankton data have been generated by avoiding contamination during sampling and applying new analytical techniques. These 'oceanographically consistent' data are very important for the understanding of the cycling of metals through the oceans and have to be considered when trying to interpret upwelling sediments or metal enrichments in organic carbon rich sediments in general.

In the following sections a very brief overview about the more recent findings of the behaviour of trace elements in sea water is given. It is then demonstrated that there is a relation between the behaviour of certain nutrient and trace elements in sea water and their abundance in marine plankton, and that the chemistry of upwelling sediments reflects processes taking place in the water column. Finally the unusual trace element chemistry of certain black shale layers from Site 367, DSDP Leg 41, is discussed in the light of the findings about the origin of trace metals in upwelling sediments.

Sea water and plankton chemistry

In Fig. 1 three typical trace element depth profiles for Pacific waters are shown. One has to distinguish elements which show surface depletion and deep water enrichment, like Zn, from those which show depletion in deeper waters and an enrichment in the surface layer, like Co.

Surface depletion in general is explained by the uptake of an element during plankton growth. The element is released in deeper waters from decaying plankton remains when they settle through the water column. The behaviour of these elements is similar to that of nutrients like nitrate, phosphate or silica. The concentration ratio of an element in deep water versus surface water gives an idea of the magnitude of the involvement of an element in the biological cycle. This ratio (so far known) is highest for Cd and Zn (Bruland 1980; see also Table 1).

Another group of elements, including Co, Mn, Pb, and Sn shows an opposite trend. The surface enrichment is the result of fluvial and/or aeolian input. The elements are rather reactive and scavenged by particles. Because the scavenging rate is higher than the release rate, sea water concentrations decrease with depth.

A third group of elements, including V, As, Cr, and probably Mo shows relatively little concentration change with depth, except for minor surface depletion. These elements are involved in biocycling to a lesser extent and do not show pronounced deep or intermediate water regeneration, unlike Cd or Zn.

Elements of the first group (Zn, Cd, etc.) are also known as the biolimiting elements, whereas elements of the third group belong to the biointermediate group (Broecker & Peng 1982). The shape of trace element profiles in sea water may not be that easy to explain, when mid-water scavenging, supply of elements from the sediment–sea water interface, remineralization in the oxygen-minimum, or different behaviour of element species is evident.

Relation between metal concentrations in sea water and plankton

As for sea water, reliable trace element data for marine plankton have only been produced during the past ten years (Table 2). The variability of metal concentrations in plankton is higher than in sea water. A simple correlation between the mean sea water concentration of trace metals and the average metal content of marine plankton is not observable. But if one considers the ability of an element to be involved in nutrient cycling, as demonstrated by the deep/surface sea water ratio of an element, a relationship between plankton and sea water chemistry is more likely.

Fig. 2 shows the correlation (log scale) of metal concentrations in plankton (Table 2) with a factor A (Table 3). This factor is the product of the deep/surface enrichment factor of an element and its mean sea water concentration, and combines the availability of an element with its ability to be involved in biological processes.

All biolimiting or biointermediate elements fall on a regression line, the only exceptions being Mo, whose sea water concentration is rather high compared to its abundance in plankton, and Pb and Co which do not show nutrient-like sea water profiles.

Chemistry of upwelling sediments from the Gulf of California

A total of 50 sediment samples, comprising 39 laminated samples from 3 cores from within the oxygen-minimum zone (620–650 m water depth), and 11 homogenous samples from one core slightly below the oxygen minimum (800 m water depth) have been analysed for all major and 15

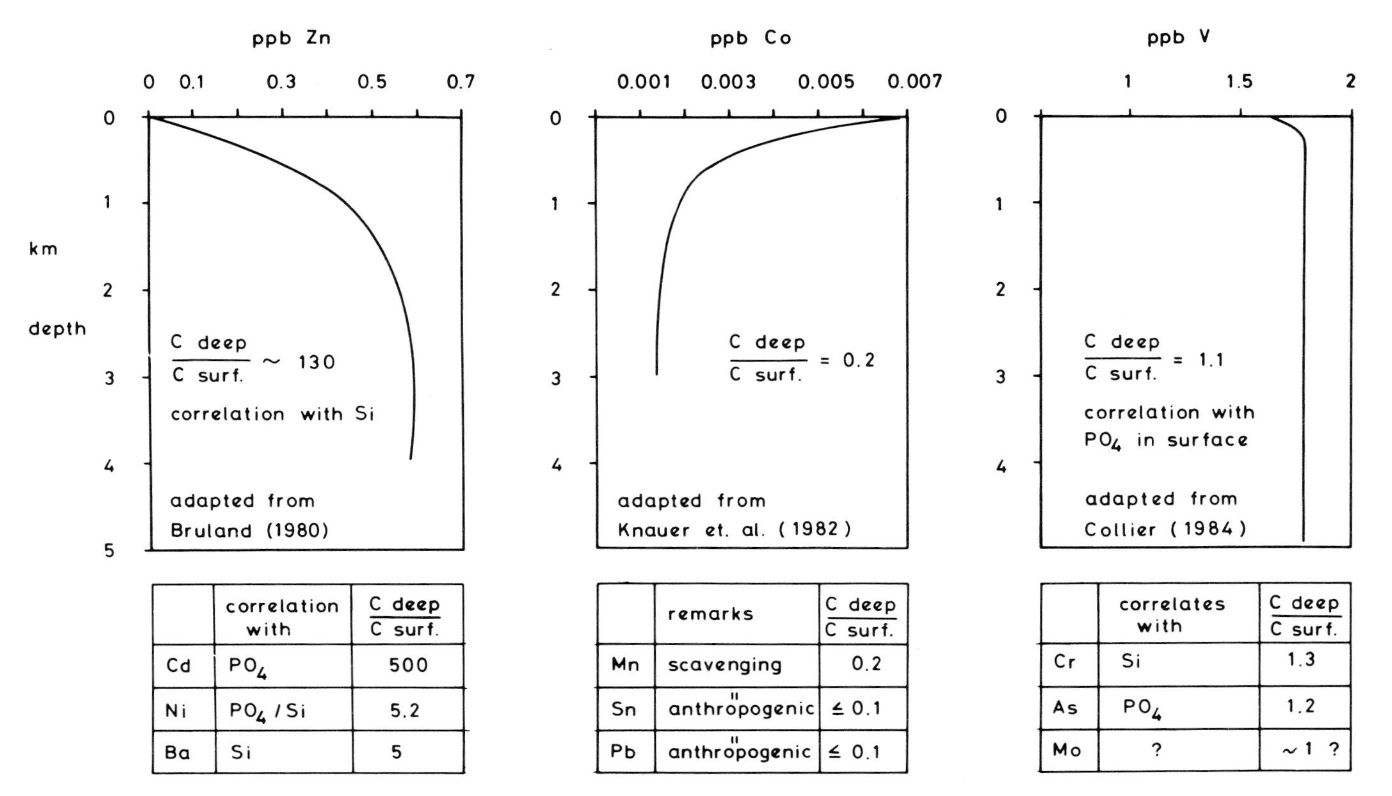

	correlation with	C deep / C surf.
Cd	PO_4	500
Ni	PO_4 / Si	5.2
Ba	Si	5

	remarks	C deep / C surf.
Mn	scavenging	0.2
Sn	" anthropogenic	≤ 0.1
Pb	" anthropogenic	≤ 0.1

	correlates with	C deep / C surf.
Cr	Si	1.3
As	PO_4	1.2
Mo	?	~1 ?

FIG. 1. Idealized concentration–depth profiles for trace metals in sea water.

TABLE 1. *Deep/surface trace metal and nutrient concentration ratios for sea water*

Element	C_{deep}/C_{surf}	Correlation with	References
Cd	150.0 (20.0–500.0)	phosphate, nitrate	Boyle *et al.* (1976) Bruland (1980)
Zn	130.0	silica	Bruland (1980)
Cu	12.0	nutrients, scavenging, bottom source	Boyle *et al.* (1977) Bruland (1980)
Ni	5.2	phosphate, silica	Bruland (1980)
Ba	5.0	silica	Chan *et al.* (1976)
Se	4.3	phosphate, silica	Measures *et al.* (1980)
Cr	1.4	silica	Cranston & Murray (1978) Campbell & Yeats (1981)
As	1.2	phosphate	Andreae (1978)
V	1.1	phosphate	Collier (1984)
Mn	0.2	scavenged, mobilized in oxygen-minimum	Klinkhammer & Bender (1980) Landing & Bruland (1980)
Co	0.2	scavenged	Knauer *et al.* (1982)
Sn	0.1	scavenged, input anthropogenic	Byrd & Andreae (1982)
Pb	0.1	scavenged, input anthropogenic	Schaule & Patterson (1981)
Si	35.0		Broecker & Peng (1982)
P	20.0		Broecker & Peng (1982)

TABLE 2. *Compilation of reliable trace metal and nutrient data for marine plankton from the literature*

Element	Reported values (in p.p.m. dry weight)	Used value
Cd	22.0 (0.2–54.0)[a] 12.0 ± 6.0[b] 3.2 (0.4–6.4)[d]	12.0
Zn	131.0 (21.0–400.0)[a] 47.0 ± 24.0[b] 44.0 ± 27.0[c]	110.0
Cu	13.5 (3.0–18.0)[a] 7.0 ± 3.5[b] 6.0 ± 3.0[c] 16.6 (6.5–57.5)[d]	11.0
Ni	12.0 (1.0–25.0)[a] 6.4 ± 4.3[b] 6.6 ± 5.6[c] 5.2 (2.0–11.6)[d] 8.1[e]	7.5
Ba	55.0 (28.0–71.0)[a] 147.0 ± 200.0[b] 38.0 (17.0–323.0)[d]	80.0
Se	2.7[c] 0.2–0.8[f]	2.0
Cr	1.0[c] 4.9[e]	2.0
As	4.0[c] 10.0[f]	6.0
V	3.0[d] 3.0[f]	3.0
Mo	2.0[d] 2.0[f]	2.0
Co	1.8 ± 1.3[c] 1.0[d] 0.87[e]	1.0
Pb	5.4 ± 4.5[b] 6.0 (2.1–31.0)[d]	6.0
Tl	0.1[g]	0.1
Bi	2.0–7.7?[f] 0.04[g]	0.1
P	7600.0 ± 2400.0[b]	7600.0
Si	36000.0 ± 50000[b]	36000.0

[a] = Collier & Edmond (1984). [b] = Martin *et al.* (1976). [c] = Trefry & Presley (1976), stations 10–18, 2 values used for As. [d] = Martin & Knauer (1973), weighted mean from their Table 6. [e] = Fowler (1977), microplankton. [f] = Eisler (1981). [g] = Bowen 1979), values for fish.

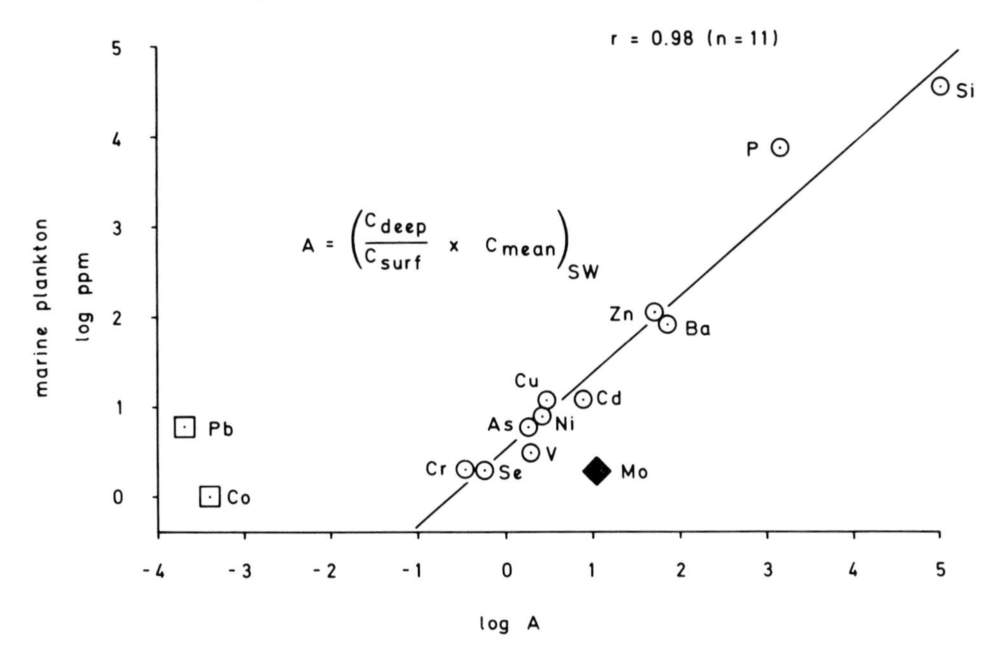

FIG. 2. Correlation (log scale) of trace metal and nutrient concentrations in marine plankton and the factor A. Factor A is the product of the deep/surface ratio of an element and its mean sea water concentration.

TABLE 3. *Data base for the calculation of the factor A, which is the product of the deep/surface concentration ratio of an element and its average concentration in sea water*

Element	Mean sea water[a] concentration	C_{deep}/C_{surf}[c]	A
Cd	0.08	150.0	12.0
Zn	0.39	130.0	51.0
Cu	0.25	12.0	3.0
Ni	0.5	5.2	2.6
Ba	14.0	5.0	70.0
Se	0.13	4.3	0.56
Cr	0.21	1.4	0.29
As	1.7	1.2	2.04
V	1.8[b]	1.1	1.98
Mo	10.6	1.0	10.6
Co	0.002	0.2	0.0004
Pb	0.002	0.1	0.0002
P	71.0	20.0	1420.0
Si	2800.0	35.0	98000.0

[a] = Broecker & Peng (1982). [b] = Collier (1984). [c] = see references in Table 1, this study. Sea water concentrations in p.p.b.

minor elements. These samples represent the squeezed cakes from a previous pore-water study of the same cores (Brumsack & Gieskes 1983). The analytical data presented in this study have been produced by XRF, AA and ICP-AES methods which were checked with international rock standards. All data are corrected for pore-water salts.

In order to compare the chemistry of these sediments with average shale, the element/Al ratios have been plotted on a log scale in Fig. 3, bar-length represents the (one sigma) standard

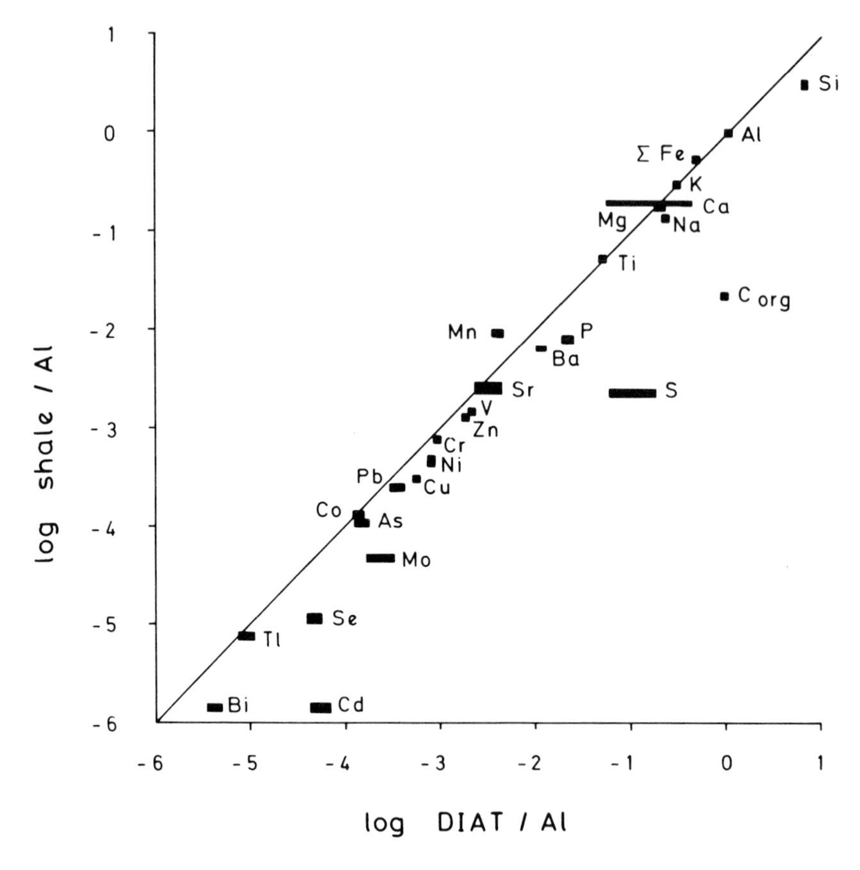

FIG. 3. Comparison of the chemistry (element/Al ratios) of diatomaceous sediments from the Gulf of California (DIAT) with average shale, low in organic carbon.

deviation. The elements Fe, K, Mg, Ti, and ±Co are of detrital origin and can be completely accounted for by the 'average shale' component of the sediments. The low Mn/Al ratio reflects the reducing character of the environment of deposition.

A large number of elements is slightly enriched in this sediment type, due to a biogenous input. Higher enrichments of Si, organic carbon, P, S, Mo, Se, Cd, and possibly Bi are characteristic for the Gulf of California sediments. The high Si/Al ratio reflects the presence of around 35% opaline silica. The high Cd/Al ratio may be used as an indicator for high productivity, because Cd is a rather rare element in average shale (mean value 0.13 p.p.m.; Heinrichs *et al.* 1980), but is concentrated to high levels in marine biota (Martin *et al.* 1976).

Plankton as the source for the excess of trace elements in upwelling sediments

Figure 4 shows the plot of some major and minor element concentrations in marine plankton (see Table 2) compared to the excess of these elements in the diatomaceous sediments from the Gulf of California (DIAT). The excess concentrations have been calculated by subtracting the average shale contribution from the bulk concentration values, by assuming that all Al is of terrigenous origin (see Table 4).

There is a strong relationship between plankton, the element source, and the non-detrital fraction of diatomaceous sediments, the element sink. Those elements, which are regenerated at

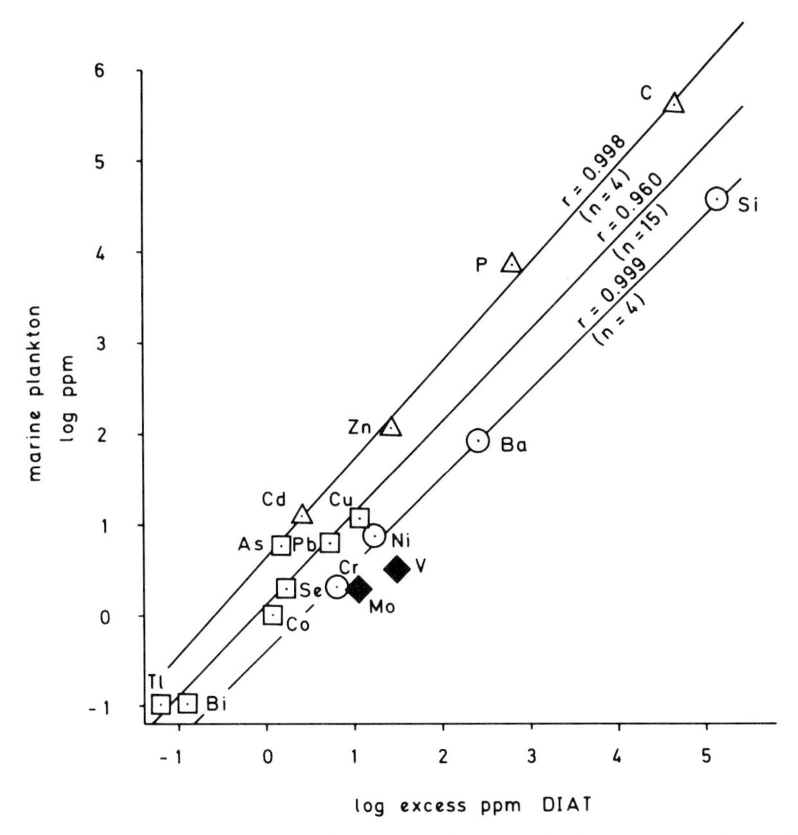

FIG. 4. Relation between the 'excess' concentrations of several elements in diatomaceous sediments from the Gulf of California (DIAT) and marine plankton.

shallow depth in the water column (C, P, Cd, partly Zn), have relatively higher concentrations in plankton than in the biogenous fraction of the underlying sediments. Si and Ba, by contrast, are involved in deeper regeneration cycle and therefore are concentrated to comparably higher levels in this sediment type.

It seems very likely indeed, that the excess of most trace metals and nutrient elements in upwelling sediments originates from plankton. The only elements which fall slightly off the regression lines are Mo and V. For these two elements bio-concentration may not be the only mechanism responsible for their accumulation in organic carbon rich sediments.

Behaviour of Mo and V during early diagenesis

In the previous sections it was mentioned that Mo especially shows a behaviour different from most other trace elements. Its sea water concentration is high (11 ppb) compared to other trace elements, but plankton does not seem to concentrate Mo to very high levels. Nevertheless most organic carbon rich sediments are enriched in this element, compared to average shale. Therefore processes within the water column or during early diagenesis have to be responsible for the additional incorporation of this element in those sediments.

Figure 5 shows the depth profile of the Mo/Al ratio in the Gulf of California sediments. A steady increase of Mo with depth is evident in the upper 30 cm of the sediments. The Mo/Al ratio (by weight) changes from $1.4 * 10^{-4}$ in the upper 12 cm to around $3 * 10^{-4}$ below 30 cm depth. It seems reasonable to assume that Mo is diffusing from the water column into the sediments, where it is reduced from the +6 to the +5 or +4 state and bound to either organic matter or sulphide. The major source for Mo in organic carbon rich sediments therefore seems to be sea water and not plankton. Evidence for this mechanism is given by the high Mo content of marine humic and

TABLE 4. *Major and minor element concentrations in 'average shale', diatomaceous sediments from the Gulf of California (DIAT) calculated 'excess' element concentrations in DIAT*

Element	ppm average shale[a]	ppm DIAT[b]	ppm DIAT excess[c]
Cd	0.13[d]	2.53	2.45
Zn	115.0[d]	88.0	26.0
Cu	28.0[f]	27.0	12.0
Ni	40.0[f]	38.0	16.5
Ba	580.0	566.0	255.0
Se	1.0[e]	2.2	1.7
Cr	70.0[f]	44.0	6.5
As	10.0	6.9	1.5
V	130.0	101.0	31.0
Mo	2.6	11.9	10.5
Co	10.0[f]	6.6	1.2
Pb	22.0[d]	17.0	5.0
Tl	0.68[d]	0.43	0.07
Bi	0.13[d]	0.20	0.13
P	700.0	1040.0	665.0
Si	275000.0	293000.0	145500.0
C_{org}	2000.0	43500.0	42400.0
Al	88000.0	47200.0	

[a]=Wedepohl (1970). [b]=mean value of 50 samples, corrected for pore water salt. [c]=excess values calculated from DIAT, assuming Al to represent the 'average shale' contribution. [d]=Heinrichs *et al.* 1980. [e]=Keltsch (1983). [f]=values from core G–32 off Mazatlan, corrected for the same Al content like 'average shale'.

fulvic acids extracted from sediments from the Namibian shelf (Calvert & Morris 1977), and the correlation of Mo and 'yellow substance' (DOC) in the porewaters from the samples used for this study (Brumsack & Giskes 1983).

But the Mo/Al ratio is also correlated to the S/Al ratio in the surface samples from the Gulf of California (Fig. 6). This does not necessarily mean that Mo is bound to sulphide, because this correlation simply might reflect that molybdate reduction correlates with sulphate reduction.

The other element which shows slight evidence of an enrichment with increasing burial depth is V (Fig. 7). Like Mo, V also exhibits a rather pronounced redox chemistry. The relatively mobile metavanadate anion (VO_3^-) may be reduced to the vanadyl cation ($(VO)^{2+}$), which is easily fixed by organic matter (Szalay & Szilagyi 1967). V, like Mo, is quite well correlated to 'yellow substance' in the pore-waters.

These findings are quite important in view of interpreting the genesis of 'old' organic carbon rich sediments. When the major amount of the excess Mo and V in black shales originates from sea water and enters the sediment by a diffusion-controlled process, then the sediment accumulation rate must be an important factor limiting the total amount of Mo and V in this sediment type. Diffusion takes time and therefore black shales with rather high concentrations of both metals have most likely been deposited slowly.

The behaviour of Ba during early diagenesis

In sea water, Ba, like silica (Chan *et al.* 1977), is involved in a deep regeneration cycle. High Ba concentrations are reported in the literature from deep-sea diatomaceous oozes (Puchelt 1972). The diatomaceous sediments from the Gulf of California are slightly enriched in Ba; on average about 45% of the Ba is of biogenous origin. Figure 8 shows the concentration–depth profile of Ba in the solids and pore-waters, and the pore-water sulphate concentration and isotopic composition.

A steady increase in pore-water Ba is evident at

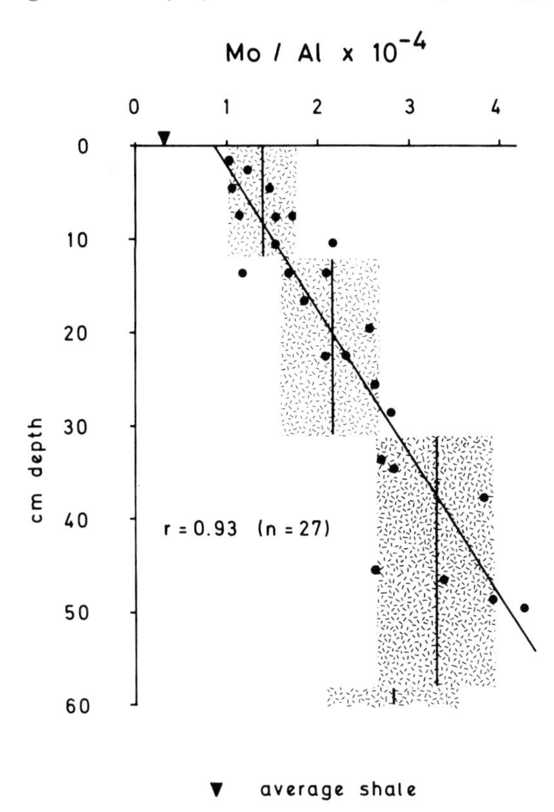

FIG. 5. Depth profile of the Mo/Al ratio for diatomaceous sediments from the Gulf of California. The stippled area represents the one sigma standard deviation.

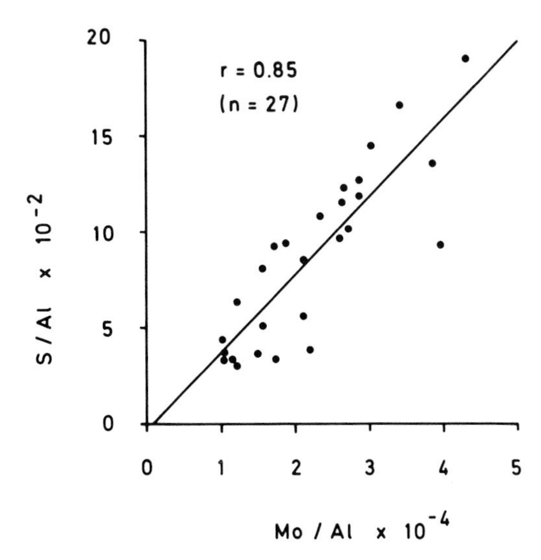

FIG. 6. Correlation of the Mo/Al and S/Al ratios for diatomaceous sediments from the Gulf of California.

depths greater than 2 m; a maximum value of 7.8 ppm Ba is reached at the base of Core E-17. These data indicate that the Ba concentration in pore-waters may be controlled by barite solubility (Church & Wolgemuth 1972). The pore-water sulphate values decrease from sea water values at the core top to near zero values at depths around 3 m. The sulphur isotopic composition of the pore-water sulphate becomes heavier with increasing burial depth due to the action of sulphate reducing bacteria. The 'residual' sulphate reservoir reaches a sulphur isotopic composition around +60‰ (rel. CDT) at near zero sulphate concentrations around 3 m depth.

From the analysis of the corresponding solids (squeezed cakes) a marked increase in the Ba/Al ratio (respectively excess Ba concentration) can be seen in the depth range from 2.5–4 m. This increase does not seem to be related to changes in productivity, because the opaline silica concentration does not show any increase in this depth interval.

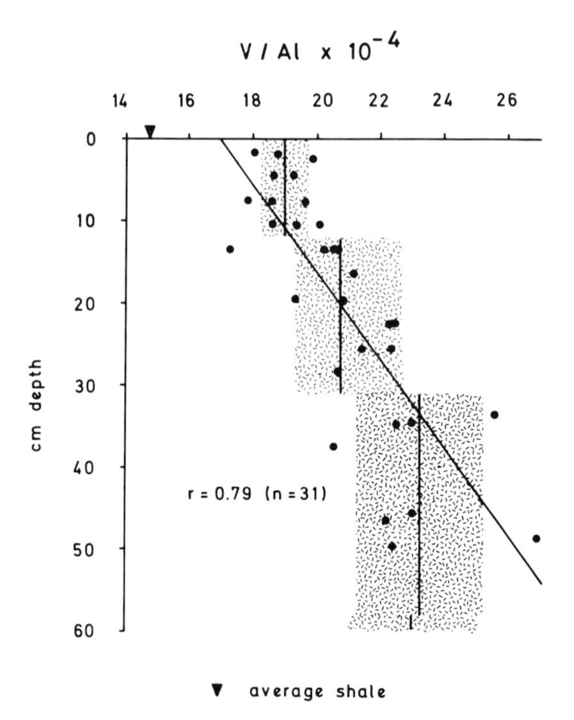

FIG. 7. Depth profile of the V/Al ratio for diatomaceous sediments from the Gulf of California. The stippled area represents the one sigma standard deviation.

It therefore seems appropriate to propose that a front of barite precipitation and dissolution is moving through the sedimentary column in the depth range of near zero pore-water sulphate. Barite crystals around 10 micron diameter could be detected by microprobe scanning in a sample from 3.58 m depth. These barites are approximately 10 times larger than those found in the suspended matter (around 1 micron) in the Atlantic and Pacific oceans by Dehairs *et al.* (1980). Therefore the precipitation of barites seems likely to happen in this depth range. In deeper layers dissolution due to the complete absence of sulphate is evident.

Under special conditions barite nodule formation may be possible in this sediment type. In any case the 'barite front' has to stay at a fixed depth layer relative to the sediment surface. This implies that a drastic reduction of the sediment accumulation rate could provide enough time necessary for nodule formation. Pore-water sulphate would be supplied by diffusion from the sediment–sea water interface and Ba by diffusion and the compactional flow from deeper layers of the sedimentary column. At the same time sulphate reduction would keep the sulphate reservoir enriched in the heavy isotope ^{34}S, and the nodules formed in this depth range therefore should be isotopically heavy with respect to the contemporaneous sea water.

Sakai (1971) analysed barite concretions from banks in the Japan Sea and other locations. He proposes a rather complicated model to explain the extremely heavy sulphur isotopic composition of the barites. The model presented above might give an alternative explanation for the process of barite nodule formation in these sediments. The problem of the nodules being found at the sediment surface might be overcome by assuming that the initiation of strong bottom currents may have resuspended the sedimentary material which was originally present above the 'barite front'.

The model presented suggests that barite formation may take place within the sediments during early diagenesis, and that the sulphur isotopic composition of such nodules has to be heavier than the corresponding sea water.

Geochemistry of Cretaceous black shales from DSDP Site 367

During DSDP Leg 41 black shales of Cretaceous age were recovered at Site 367 in the Cape Verde Basin. Eleven black shale samples from Cores 17 to 21, with organic carbon contents ranging from

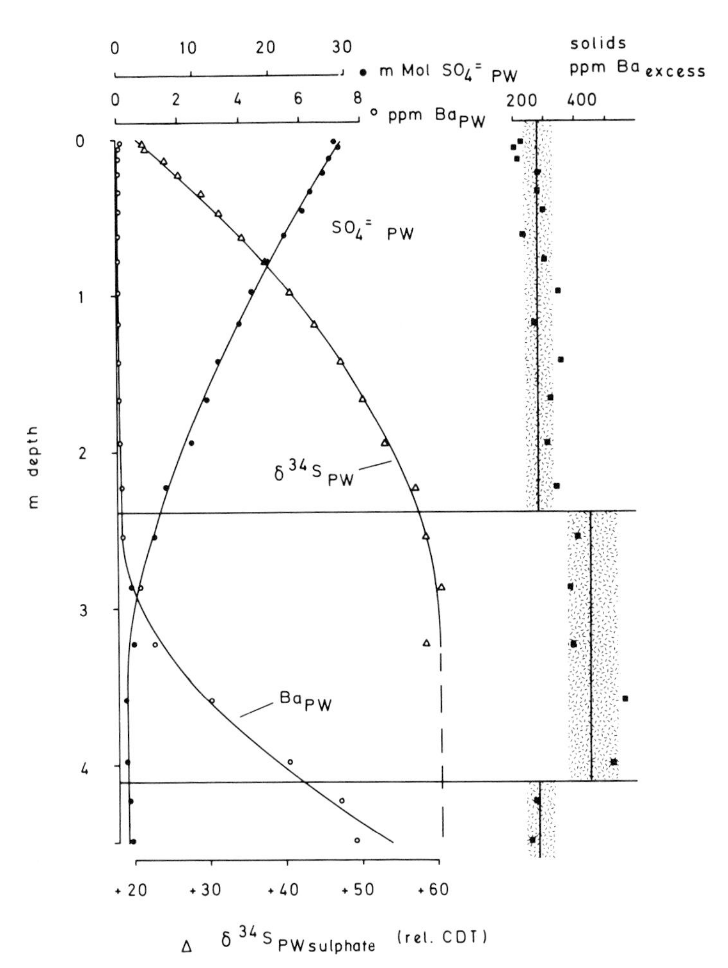

FIG. 8. Concentration–depth profiles for pore-water sulphate, sulphur isotopic composition, and Ba as well as 'excess' Ba concentrations in the solids from Core E-17 from the Gulf of California.

0.5–35‰, have been analysed for some major and minor elements. These samples show a very special trace element chemistry, in particular very high concentrations of Zn, V, Mo, Ni, and Cu compared to average shale (Table 5). In Fig. 9 the element/K ratios are plotted, because Al data are not available for these black shale samples. Compared with average shale these black shales are enriched by 1 to 2 orders of magnitude in several trace elements. Even comparing the black shales with upwelling sediments from the Gulf of California (Fig. 10), we find that several elements are concentrated to levels an order of magnitude higher in the black shales; among those elements are V and Mo. The only elements which seem to show normal concentrations in the black shales are the majors and Pb, besides the redox sensitive element Mn.

The Ba concentration seems to be only slightly enriched, but Dean & Schreiber (1976) report that barites are present in organic-rich Lower Cretaceous sediments at Sites 369 and 370, immediately below major unconformities. Pseudomorphs of calcite after barite are also reported from Section 3 in Core 17, Site 367. Unfortunately, this section was not sampled for this study. Nevertheless these findings might be important in view of the barite formation model presented in the previous section. If the overlying black shale layer at Section1, 120 cm, results from a stagnation event, the barites (resp. pseudomorphs after barite) might represent the former 'barite front'. This black

TABLE 5. *Major and minor element concentrations in 11 Cretaceous black shale samples from Site 367, Cores 17–21, Leg 41 DSDP, in comparison to diatomaceous sediments from the Gulf of California (DIAT)*

Element	11 Black shales	DIAT (n = 50)
% K	1.53 ± 0.51	1.38 ± 0.14
% Na	0.82 ± 0.16	1.06 ± 0.07
% Fe	4.20 ± 1.24	2.15 ± 0.59
% C_{org}	10.82 ± 10.72	4.35 ± 0.41
% S	3.47 ± 1.56	0.57 ± 0.28
p.p.m. Ba	682.0 ± 155.0	566.0 ± 71.0
p.p.m. Pb	19.5 ± 7.7	17.0 ± 3.1
p.p.m. Mo	88.0 ± 86.0	11.9 ± 3.9
p.p.m. Mn	348.0 ± 237.0	193.0 ± 27.0
p.p.m. Co	40.0 ± 27.0	6.6 ± 0.9
p.p.m. Cu	291.0 ± 169.0	27.0 ± 3.0
p.p.m. Ni	250.0 ± 210.0	38.0 ± 4.0
p.p.m. Cr	284.0 ± 92.0	44.0 ± 4.0
p.p.m. V	1434.0 ± 1038.0	101.0 ± 11.0
p.p.m. Zn	1540.0 ± 2161.0	88.0 ± 9.0

shale layer is characterized by some of the highest concentrations of organic carbon (35.6%), sulphur (7.4%), Zn (0.7%), V (0.36%), Ni (870 ppm), Cr (300 ppm), etc. reported so far in the literature for black shales from the Cretaceous Atlantic Ocean. Horizons with similar enrichments have been detected at Site 105, Core 9, Section 4 (DSDP Leg 11), Site 135, Core 8, Section 1 (DSDP Leg 14), Site 144 A, Cores A5 and A6, Section 1 and Core Catcher (DSDP Leg 14), and Site 368, Core 60, Section 5 (DSDP Leg 41), as reported by Brumsack (1979).

The origin of metal enrichments in Cretaceous black shales from DSDP Site 367, Leg 41

Several processes may lead to metal enrichments in black shales and an answer has to be given to the question whether these enrichments result from processes which took place during sedimentation or early diagenesis, or whether later diagenetic mobilization- and fixation-processes are responsible for the high concentrations of some metals in certain black shale layers.

Dean & Gardner (1982) quote Vine & Tourtelot (1970) who mention that 'if thin beds of black shale are interstratified with other lithologies, these (black shale) beds may become sinks for elements mobilized and transported in pore-waters during compaction'. But the interbedded

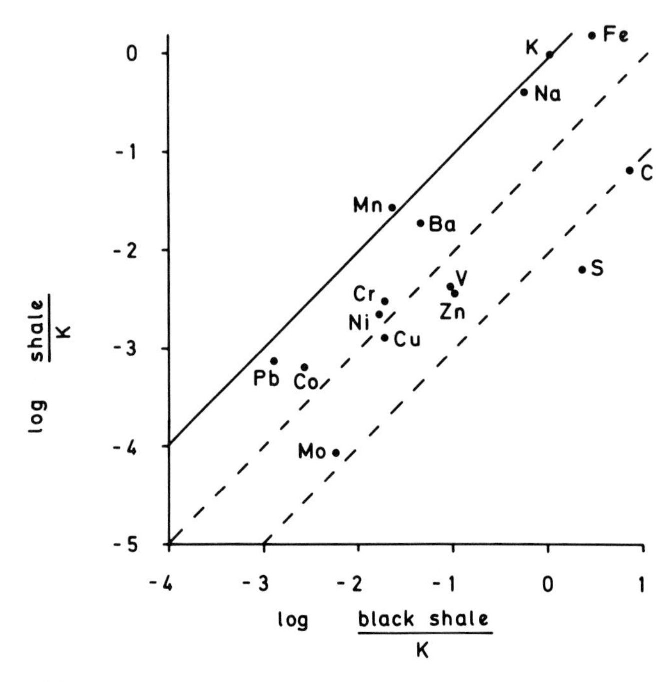

FIG. 9. Comparison of the element/K ratios of Cretaceous black shales from Site 367, Leg 41 DSDP, with average shale, low in organic carbon.

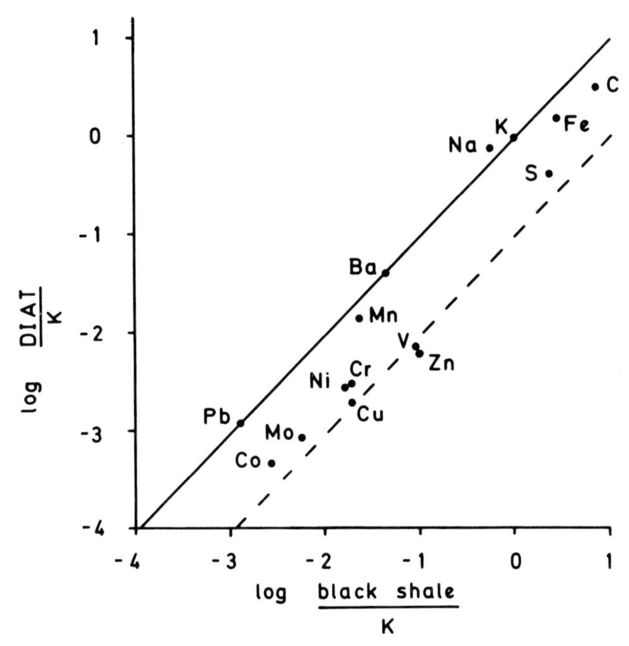

FIG. 10. Comparison of the element/K ratios of Cretaceous black shales from Site 367, Leg 41 DSDP, with diatomaceous sediments from the Gulf of California (DIAT).

organic carbon poorer sequences of the Cretaceous black shale facies at Site 367 are still enriched in metals rather than depleted (Brumsack 1980; Dean *et al.* 1984) and therefore do not seem to represent the source material for the metals found in the overlying black shale layers. Those elements showing the relatively highest enrichments in the black shales (Mo, V, Zn, Cu, Ni, etc.; see Table 5) should be fixed rather than mobilized under reducing conditions. It therefore seems justified to assume that the same process which leads to the accumulation of extremely high amounts of organic carbon also favours the accumulation of trace metals.

The marine environment is metal-poor: metal concentrations in normal sea water are very low (see Table 3). The accumulation of metals during sedimentation or early diagenesis therefore needs long periods of time. On the other hand higher concentrations of organic carbon in sediments deposited in oxygenated waters are only possible where accumulation rates are high enough to prevent oxidation. Therefore the combination of high organic carbon and high trace element concentrations in black shales seems impossible under normal oxygenated conditions in deep waters. Only an anoxic water column would provide the environment necessary to prevent complete oxidation of organic matter and at the same time to favour the accumulation of several redox sensitive trace metals.

The strong correlations of V, Mo, and Zn with organic carbon in black shales from Site 367 (see Fig. 11) indicate that the accumulation of metals may be primary. The correlations of the same metals with sulphur are much less pronounced, even though it could be demonstrated that, for instance, Zn is present as sphalerite (Brumsack 1980). Sulphur enters the sediment predominantly during early diagenesis. This process is diffusion-controlled. The very high sulphur concentrations in most of the metal-rich black shale sections may indicate low bulk accumulation rates.

In summary, the combination of high organic carbon, sulphur, and trace metal concentrations might indicate periods of complete stagnation of the Cape Verde Basin during intervals in Cretaceous time. Similar conclusions were drawn by de Graciansky *et al.* (1984) who proposed a completely anoxic North Atlantic during the deposition of the Cenomanian–Turonian black shale horizon, based on biostratigraphy. Redeposited sediments from the upper slope, which might have been deposited in an oxygen minimum zone under high productivity conditions similar to the Gulf of California sediments, are a rather unlikely source for the metals found enriched in the black shale sequences from Site 367. The extremely high concentrations of elements, like V, Mo, Zn, Ni, Cu, Cr, etc. in black shales cannot be explained by upwelling.

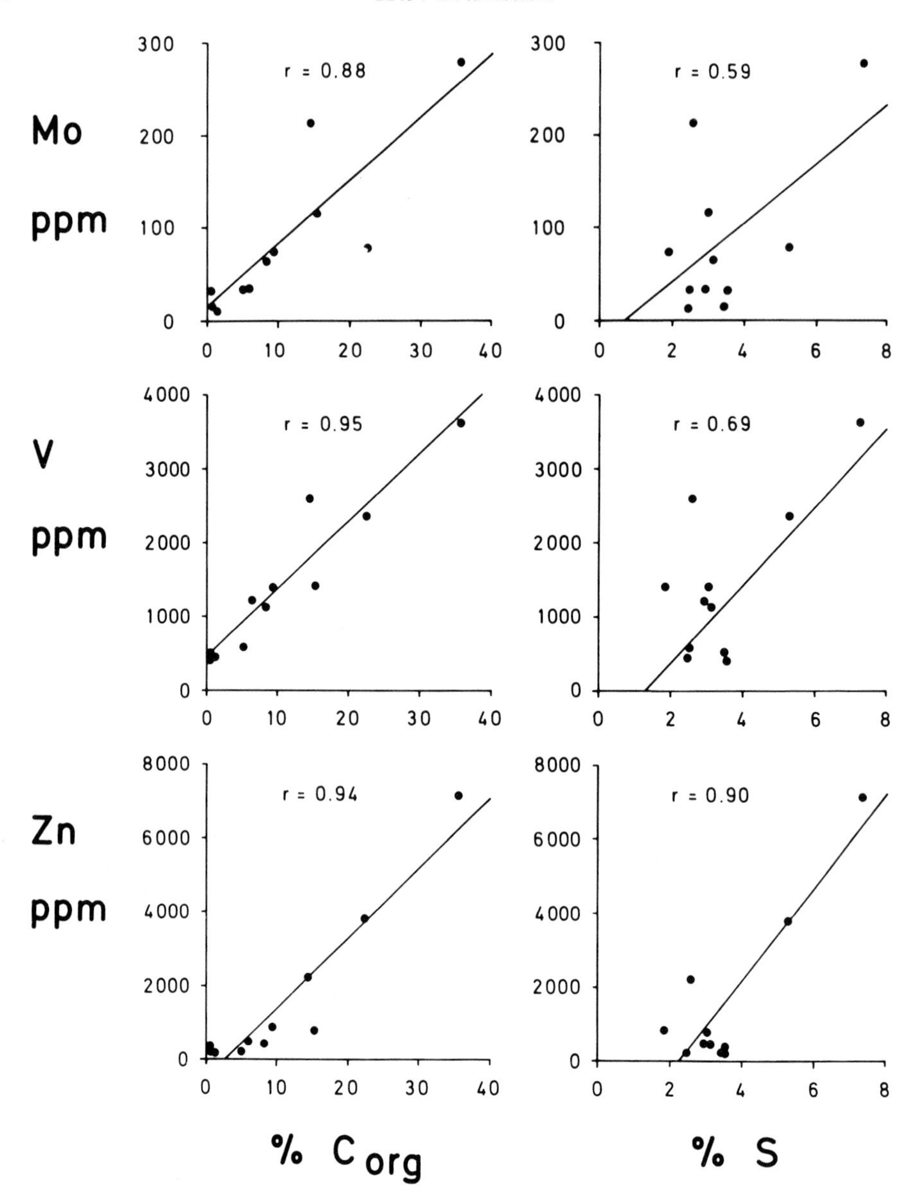

FIG. 11. Correlations of Mo, V, and Zn with organic carbon and sulphur for Cretaceous black shales from Site 367, core 17–21 DSDP, Leg 41 Cape Verde Basin.

We need to understand much more about the behaviour of metals in the oceans and during diagenesis, especially in Black-Sea-type anoxic basins, before a definite answer may be given to the black shale problem.

ACKNOWLEDGEMENTS: The author would like to thank K.H. Wedepohl and H. Nielsen for helpful discussions. Thanks also to H. Schrader and Captain and crew of the Mexican R/V *Mariano Matamoros* for their help and co-operation in sampling the Gulf of California. The paper benefited from two anonymous reviewers. Financial assistance was provided by the German Science Foundation (DFG).

References

ANDREAE, M.O. 1978. Distribution and speciation of arsenic in natural waters and some marine algae. *Deep-Sea Res.* **25,** 391–402.

BOWEN, H.J.M. 1979. *Environmental Chemistry of the Elements*. Academic Press, London. 333 pp.

BOYLE, E.A., SCLATER, F. & EDMOND, J.M. 1976. On the marine geochemistry of cadmium. *Nature* **263,** 42–4.

——, SCLATER, F. & EDMOND, J.M. 1977. The distribution of dissolved copper in the Pacific. *Earth Planet. Sci. Lett.* **37,** 38–54.

BROECKER, W.S. & PENG, T.-H. 1982. *Tracers in the Sea*. Lamont-Doherty Geological Observatory Publication, Columbia University, Palisades, New York. 690 pp.

BRONGERSMA-SANDERS, M. 1965. Metals of the Kupferschiefer supplied by normal seawater. *Geol. Rundschau* **55,** 365–75.

BRULAND, K.W. 1980. Oceanographic distributions of cadmium, zinc, nickel, and copper in the North Pacific. *Earth Planet. Sci. Lett.* **47,** 176–98.

BRUMSACK, H.J. 1979. *Geochemische Untersuchungen an Kretazischen Atlantik-Schwarzschiefern der Legs 11, 14, 36 und 41 (DSDP)*. Ph.D. Thesis, University of Göttingen, FRG. 71 pp.

—— 1980. Geochemistry of Cretaceous black shales from the Atlantic Ocean (DSDP Legs 11, 14, 36, and 41). *Chem. Geol.* **31,** 1–25.

—— & GIESKES, J.M. 1983. Interstitial water trace-element chemistry of laminated sediments from the Gulf of California, Mexico. *Mar. Chem.* **14,** 89–106.

BYRD, J.T. & ANDREAE, M.O. 1982. Tin and methyltin species in seawater: concentrations and fluxes. *Science* **218,** 565–9.

CALVERT, S.E. 1966. Accumulation of diatomaceous silica in the sediments of the Gulf of California. *Geol. Soc. Am. Bull.* **77,** 569–96.

—— & PRICE, N.B. 1970. Minor metal contents of recent organic-rich sediments off South West Africa. *Nature* **227,** 593–5.

—— & MORRIS, R.J. 1977. Geochemical studies of organic-rich sediments from the Namibian shelf. II. Metal-organic associations. *In:* ANGEL, M.V. (ed.), *A Voyage of Discovery*. Pergamon Press, Oxford. 667–80.

CAMPBELL, J.A. & YEATS, P.A. 1981. Dissolved chromium in the northwest Atlantic Ocean. *Earth Planet. Sci. Lett.* **53,** 427–33.

CHAN, L.H., EDMOND, J.M., STALLARD, R.F., BROECKER, W.S., CHUNG, Y.C., WEISS, R.F. & KU, T.L. 1976. Radium and barium at GEOSECS stations in the Atlantic and Pacific. *Earth Planet. Sci. Lett.* **32,** 258–67.

——, DRUMMOND, D., EDMOND, J.M. & GRANT, B. 1977. On the barium data from the Atlantic GEOSECS expedition. *Deep-Sea Res.* **24,** 613–49.

CHURCH, T.M. & WOLGEMUTH, K. 1972. Marine barite saturation. *Earth Planet. Sci. Lett.* **15,** 35–44.

COLLIER, R.W. 1984. Particulate and dissolved vanadium in the North Pacific Ocean. *Nature* **309,** 441–4.

—— & EDMOND, J. 1984. The trace element geochemistry of marine particulate matter. *Prog. Oceanogr.* **13,** 113–99.

CRANSTON, R.E. & MURRAY, J.W. 1978. The determination of chromium species in natural waters. *Anal. Chim. Acta* **99,** 275–82.

DEAN, W.E. & SCHREIBER, B.C. 1976. Authigenic barite, Leg 41 DSDP. *Init. Repts DSDP* **41.** US Govt. Print. Off., Washington, DC. 915–31.

—— & GARDNER, J.V. 1982. Origin and geochemistry of redox cycles of Jurassic to Eocene age, Cape Verde Basin (DSDP site 367), continental margin of Northwest Africa. *In:* SCHLANGER, S.O. & CITA, M.B. (eds), *Nature and Origin of Cretaceous Carbon-Rich Facies*. Academic Press, London. 55–78.

——, ARTHUR, M.A. & STOW, D.A.V. 1984. Origin and geochemistry of Cretaceous deep-sea black shales and multicolored claystones, with emphasis on DSDP site 530, southern Angola Basin. *Init. Repts DSDP* **75.** US Govt Print. Off., Washington, DC. 819–44.

DE GRACIANSKY, P.C., DEROO, G., HERBIN, J.P., MONTADERT, L., MÜLLER, C., SCHAAF, A. & SIGAL, J. 1984. Ocean-wide stagnation episode in the late Cretaceous. *Nature* **308,** 346–9.

DEHAIRS, F., CHESSELET, R. & JEDWAB, J. 1980. Discrete suspended particles of barite and the barium cycle in the open ocean. *Earth Planet. Sci. Lett.* **49,** 528–50.

DONEGAN, D. & SCHRADER, H. 1982. Biogenic and abiogenic components of laminated hemipelagic sediments in the central Gulf of California. *Mar. Geol.* **48,** 215–37.

EISLER, R. 1981. *Trace Metal Concentrations in Marine Organisms*. Pergamon Press, Oxford. 687 pp.

FOWLER, S.W. 1977. Trace elements in zooplankton particulate products. *Nature* **269,** 51–553.

HEINRICHS, H., SCHULZ-DOBRICK, B. & WEDEPOHL, K.H. 1980. Terrestrial geochemistry of Cd, Bi, Tl, Zn and Rb. *Geochim. Cosmochim. Acta* **44,** 1519–33.

KELTSCH, H. 1983. *Die Verteilung des Selens und das Schwefel-Selen Verhältnis in den Gesteinen der kontinentalen Erdkruste*. Ph.D. Thesis, University of Göttingen, FRG. 93 pp.

KLINKHAMMER, G.P. & BENDER, M.L. 1980. The distribution of manganese in the Pacific Ocean. *Earth Planet. Sci. Lett.* **46,** 361–84.

KNAUER, G.A., MARTIN, J.H. & GORDON, R.M. 1982. Cobalt in north-east Pacific waters. *Nature* **297,** 49–51.

LANDING, W.M. & BRULAND, K.W. 1980. Manganese in the North Pacific. *Earth Planet Sci. Lett.* **49,** 45–56.

MARTIN, J.H. & KNAUER, G.A. 1973. The elemental composition of plankton. *Geochim. Cosmochim. Acta* **37,** 1639–53.

——, BRULAND, K.W. & BROENKOW, W.W. 1976. Cadmium transport in the California current. *In:* WINDOM, H.L. & DUCE, R.A. (eds), *Marine Pollutant Transfer*. Lexington Books. 159–84.

MEASURES, C.I., MC DUFF, R.E. & EDMOND, J.M. 1980. Selenium redox chemistry at GEOSECS I re-occupation. *Earth Planet Sci. Lett.* **49,** 102–8.

PUCHELT, H. 1972. Barium. *In:* WEDEPOHL, K.H. (ed.), *Handbook of Geochemistry*. Springer-Verlag, Berlin. Volume II-3, section 56 K.

SAKAI, H. 1971. Sulfur and oxygen isotopic study of barite concretions from banks in the Japan Sea off the Northeast Honshu. *Japan Geochem. J.* **5,** 79–93.

SCHAULE, B.K. & PATTERSON, C.C. 1981. Lead concentrations in the north-east Pacific: evidence for global anthropogenic perturbations. *Earth Planet. Sci. Lett.* **54,** 97–116.

SCHRADER, H., KELTS, K., CURRAY, J., MOORE, D., AGUAYO, E., AUBRY, M.-P., EINSELE, G., FORNARI, D., GIESKES, J., GUERRERO, J., KASTNER, M., LYLE, M., MATOBA, Y., MOLINA-CRUZ, A., NIEMITZ, J., RUEDA, J., SAUNDERS, A., SIMONEIT, B. & VAQUIER, V. 1980. Laminated diatomaceous sediments from the Guaymas Basin slope (central Gulf of California): 250 000-year climate record. *Science* **207,** 1207–9.

SZALAY, A. & SZILAGYI, M. 1967. The association of vanadium with humic acids. *Geochim. Cosmochim. Acta* **31,** 1–6.

TREFRY, J.H. & PRESLEY, B.J. 1976. Heavy metal transport from the Mississippi river to the Gulf of Mexico. *In:* WINDOM, H.L. & DUCE, R.A. (eds), *Marine Pollutant Transfer*. Lexington Books. 159–84.

VAN ANDEL, T.H. & SHOR, G.G. 1964. Marine geology of the Gulf of California. *AAPG Mem.* **3,** 408 pp.

VINE, J.D. & TOURTELOT, E.B. 1970. Geochemistry of black shale deposits—a summary report. *Econ. Geol.* **65,** 253–72.

WEDEPOHL, K.H. 1970. Environmental influences on the chemical composition of shales and clays. *In:* AHRENS, L.H., PRESS, F., RUNCORN, S.K. & UREY, H.C. (eds), *Physics and Chemistry of the Earth*. Pergamon Press, Oxford. 307–33.

ZEITSCHEL, B. 1969. Primary productivity in the Gulf of California. *Mar. Biol.* **3,** 201–7.

HANS J. BRUMSACK, Geochemisches Institut, Goldschmidtstrasse 1, D–3400 Göttingen, Federal Republic of Germany.

Index